*Recursive Macroeconomic Theory*

# *Recursive Macroeconomic Theory*



LARS LJUNGQVIST
*Stockholm School of Economics*

THOMAS J. SARGENT
*Stanford University*
*and*
*Hoover Institution*

The MIT Press
Cambridge, Massachusetts
London, England

Printed and bound in the United States of America.

Library of Congress Cataloging-in-Publication Data

Ljungqvist, Lars.
Recursive macroeconomic theory / Lars Ljungqvist, Thomas J. Sargent.
p. cm.
Includes bibliographical references and index.
ISBN 0-262-19451-1
1. Macroeconomics. 2. Recursive functions. 3. Statics and dynamics (Social sciences)
I. Sargent, Thomas J. II. Title.
HB172.5 .L59 2000
339'.01'51135--dc21 00-056067

# Contents

# *Acknowledgments*

We wrote this book during the 1990's while teaching graduate courses in macro and monetary economics. We owe a substantial debt to the students in these classes for learning with us. We would especially like to thank Marco Bassetto, Victor Chernozhukov, William Dupor, George Hall, Cristobal Huneeus, Hanno Lustig, Sergie Morozov, Monika Piazzesi, Navin Kartik, Martin Schneider, Yongseok Shin, Christopher Sleet, Stijn Van Nieuwerburgh, Laura Veldkamp, Neng E. Wang, Chao Wei, Mark Wright, Sevin Yeltekin, and Lei Zhang. Each of these people made substantial suggestions for improving this book. We expect much from members of this group, as we did from an earlier group of students that Sargent (1987b) thanked.

We received useful comments and criticisms from Jesus Fernandez-Villaverde, Gary Hansen, Jonathan Heathcote, Mark Huggett, Charles Jones, Dirk Krueger, Per Krusell, Rodolfo Manuelli, Beatrix Paal, and Jonathan Thomas.

Rodolfo Manuelli kindly allowed us to reproduce a number of his exercises. We indicate the exercises that he donated. Some of the exercises in chapters 5, 8, and 18 are versions of ones in Sargent (1987b). François Velde provided substantial help with the TeX and Unix macros that produced this book. Angelita Dehe and Maria Bharwada helped typeset it. We thank P.M. Gordon Associates for copy editing.

For providing good environments to work on this book, Ljungqvist thanks the Stockholm School of Economics and Sargent thanks the Hoover Institution and the departments of economics at the University of Chicago and Stanford University.

*To our parents, Zabrina, and Carolyn*

# *Preface*

## *Recursive Methods*

Much of this book is about how to use recursive methods to study macroeconomics. Recursive methods are very important in the analysis of dynamic systems in economics and other sciences. They originated after World War II in diverse literatures promoted by Wald (sequential analysis), Bellman (dynamic programming), and Kalman (Kalman filtering).

### *Dynamics*

Dynamics studies sequences of vectors of random variables indexed by time, called *time series*. Time series are immense objects, with as many components as the number of variables times the number of time periods. A dynamic economic model characterizes and interprets the mutual covariation of all of these components in terms of the purposes and opportunities of economic agents. Agents *choose* components of the time series in light of their opinions about other components.

Recursive methods break a dynamic problem into pieces by forming a sequence of problems, each one posing a constrained choice between utility today and utility tomorrow. The idea is to find a way to describe the position of the system now, where it might be tomorrow, and how agents care now about where it is tomorrow. Thus, recursive methods study dynamics indirectly by characterizing a pair of *functions*: a transition function mapping the *state* of the model today into the state tomorrow, and another function mapping the state into the other endogenous variables of the model. The *state* is a vector of variables that characterizes the system's current position. Time series are generated from these objects by iterating the transition law.

*Recursive approach*

Recursive methods constitute a powerful approach to dynamic economics due to their described focus on a tradeoff between the current period's utility and a continuation value for utility in all future periods. As mentioned, the simplification arises from dealing with the evolution of state variables that capture the consequences of today's actions and events for all future periods, and in the case of uncertainty, for all possible realizations in those future periods. This is not only a powerful approach to characterizing and solving complicated problems, but it also helps us to develop intuition, conceptualize and think about dynamic economics. Students often find that half of the job in understanding how a complex economic model works is done once they understand what the set of state variables is. Thereafter, the students are soon on their way formulating optimization problems and transition equations. Only experience from solving practical problems fully conveys the power of the recursive approach. This book provides many applications.

Still another reason for learning about the recursive approach is the increased importance of numerical simulations in macroeconomics, and most computational algorithms rely on recursive methods. When such numerical simulations are called for in this book, we give some suggestions for how to proceed but without saying too much on numerical methods.[1]

## Philosophy

This book mixes tools and sample applications. Our philosophy is to present the tools with enough technical sophistication for our applications, but little more. We aim to give readers a taste of the power of the methods and to direct them to sources where they can learn more.

Macroeconomic dynamics has become an immense field with diverse applications. We do not pretend to survey the field, only to sample it. We intend our sample to equip the reader to approach much of the field with confidence. Fortunately for us, there are several good recent books covering parts of the field that we neglect, for example, Aghion and Howitt (1998), Barro and Sala-i-Martin (1995), Blanchard and Fischer (1989), Cooley (1995), Farmer (1993), Azariadis (1993), Romer (1996), Altug and Labadie (1994), Walsh (1998), Cooper (1999), Pissarides (1990), and Woodford (2000). Stokey, Lucas, and Prescott (1989) and

---

[1] Judd (1998) provides a good treatment of numerical methods in economics.

Bertsekas (1976) remain standard references for recursive methods in macroeconomics. Chapters 5 and 20 in this book revise material appearing in Chapter 2 of Sargent (1987).

## Ideas

Beyond emphasizing recursive methods, the economics of this book revolves around several main ideas.

1. The competitive equilibrium model of a dynamic stochastic economy: This model contains complete markets, meaning that all commodities at different dates that are contingent on alternative random events can be traded in a market with a centralized clearing arrangement. In one version of the model, all trades occur at the beginning of time. In another, trading in one-period claims occurs sequentially. The model is a foundation for asset pricing theory, growth theory, real business cycle theory, and normative public finance. There is no room for fiat money in the standard competitive equilibrium model, so we shall have to alter the model to let fiat money in.

2. A class of incomplete markets models with heterogeneous agents: The models arbitrarily restrict the types of assets that can be traded, thereby possibly igniting a precautionary motive for agents to hold those assets. Such models have been used to study the distribution of wealth and the evolution of an individual or family's wealth over time. One model in this class lets money in.

3. Several models of fiat money: We add a shopping time specification to a competitive equilibrium model to get a simple vehicle for explaining ten doctrines of monetary economics. These doctrines depend on the government's intertemporal budget constraint and the demand for fiat money, aspects that transcend many models. We also use Samuelson's overlapping generations model, Bewley's incomplete markets model, and Townsend's turnpike model to perform a variety of policy experiments.

4. Restrictions on government policy implied by the arithmetic of budget sets: Most of the ten monetary doctrines reflect properties of the government's budget constraint. Other important doctrines do too. These doctrines, known as Modigliani-Miller and Ricardian equivalence theorems, have a common structure. They embody an equivalence class of government policies that produce the same allocations. We display the structure of such theorems with an eye

to finding the features whose absence causes them to fail, letting particular policies matter.

5. Ramsey taxation problem: What is the optimal tax structure when only distorting taxes are available? The primal approach to taxation recasts this question as a problem in which the choice variables are allocations rather than tax rates. Permissible allocations are those that satisfy resource constraints and implementability constraints, where the latter are budget constraints in which the consumer and firm first-order conditions are used to substitute out for prices and tax rates. We study labor and capital taxation, and examine the optimality of the inflation tax prescribed by the Friedman rule.

6. Social insurance with private information and enforcement problems: We use the recursive contracts approach to study a variety of problems in which a benevolent social insurer must balance providing insurance against providing proper incentives. Applications include the provision of unemployment insurance and the design of loan contracts when the lender has an imperfect capacity to monitor the borrower.

7. Time consistency and reputational models of macroeconomics: We study how reputation can substitute for a government's ability to commit to a policy. The theory describes multiple systems of expectations about its behavior to which a government wants to conform. The theory has many applications, including implementing optimal taxation policies and making monetary policy in the presence of a temptation to inflate offered by a Phillips curve.

8. Search theory: Search theory makes some assumptions opposite to ones in the complete markets competitive equilibrium model. It imagines that there is no centralized place where exchanges can be made, or that there are not standardized commodities. Buyers and/or sellers have to devote effort to search for commodities or work opportunities, which arrive randomly. We describe the basic McCall search model and various applications. We also describe some equilibrium versions of the McCall model and compare them with search models of another type that postulates the existence of a matching function. A matching function takes job seekers and vacancies as inputs, and maps them into a number of successful matches.

## Theory and evidence

Though this book aims to give the reader the tools to read about applications, we spend little time on empirical applications. However, the empirical failures of one model have been a main force prompting development of another model. Thus, the perceived empirical failures of the standard complete markets general equilibrium model stimulated the development of the incomplete markets and recursive contracts models. For example, the complete markets model forms a standard benchmark model or point of departure for theories and empirical work on consumption and asset pricing. The complete markets model has these empirical problems: (1) there is too much correlation between individual income and consumption growth in micro data (e.g., Cochrane, 1991 and Attanasio and Davis, 1995); (2) the equity premium is larger in the data than is implied by an representative agent asset pricing model with reasonable risk-aversion parameter (e.g., Mehra and Prescott, 1985); and (3) the risk-free interest rate is too low relative to the observed aggregate rate of consumption growth (Weil, 1989). While there have been numerous attempts to explain these puzzles by altering the preferences in the standard complete markets model, there has also been work that abandons the complete markets assumption and replaces it with some version of either exogenously or endogenously incomplete markets. The Bewley models of chapters 13 and 14 are examples of exogenously incomplete markets. By ruling out complete markets, this model structure helps with empirical problems 1 and 3 above (e.g., see Huggett, 1993), but not much with problem 2. In chapter 15, we study some models that can be thought of as having endogenously incomplete markets. They can also explain puzzle 1 mentioned earlier in this paragraph; at this time it is not really known how far they take us toward solving problem 2, though Alvarez and Jermann (1999) report promise.

## Micro foundations

This book is about micro foundations for macroeconomics. Browning, Hansen and Heckman (2000) identify two possible justifications for putting microfoundations underneath macroeconomic models. The first is aesthetic and preempirical: models with micro foundations are by construction coherent and explicit. And because they contain descriptions of agents' purposes, they allow us to analyze policy interventions using standard methods of welfare economics. Lucas (1987) gives a distinct second reason: a model with micro foundations broadens the sources of empirical evidence that can be used to assign numerical values to the model's parameters. Lucas endorses Kydland and Prescott's (1982) procedure of borrowing parameter values from micro studies. Browning, Hansen, and Heckman (2000) describe some challenges to Lucas's recommendation for an empirical strategy. Most seriously, they point out that in many contexts the specifications underlying the microeconomic studies cited by a calibrator conflict with those of the macroeconomic model being "calibrated." It is typically not obvious how to transfer parameters from one data set and model specification to another data set, especially if the theoretical and econometric specification differs.

Although we take seriously the doubts about Lucas's justification for microeconomic foundations that Browning, Hansen and Heckman raise, we remain strongly attached to micro foundations. For us, it remains enough to appeal to the first justification mentioned, the coherence provided by micro foundations and the virtues that come from having the ability to "see the agents" in the artificial economy. We see Browning, Hansen, and Heckman as raising many legitimate questions about empirical strategies for implementing macro models with micro foundations. We don't think that the clock will soon be turned back to a time when macroeconomics was done without micro foundations.

## *Road map*

An economic agent is a pair of objects: a utility function (to be maximized) and a set of available choices. Chapter 1 has no economic agents, while chapters 2 through 5 and chapter 13 each contain a single agent. The remaining chapters all have multiple agents, together with an equilibrium concept rendering their choices coherent.

Chapter 1 describes two basic models of a time series: a Markov chain and a linear first-order difference equation. In different ways, these models use the algebra of first-order difference equations to form tractable models of time series. Each model has its own notion of the state of a system. These time series models define essential objects in terms of which the choice problems of later chapters are formed and their solutions are represented.

Chapters 2, 3, and 4 introduce aspects of dynamic programming, including numerical dynamic programming. Chapter 2 describes the basic functional equation of dynamic programming, the Bellman equation, and several of its properties. Chapter 3 describes some numerical ways for solving dynamic programs, based on Markov chains. Chapter 4 describes linear quadratic dynamic programming and some uses and extensions of it, including how to use it to approximate solutions of problems that are not linear quadratic. This chapter also describes the Kalman filter, a useful recursive estimation technique that is mathematically equivalent to the linear quadratic dynamic programming problem.[2] Chapter 5 describes a classic two-action dynamic programming problem, the McCall search model, as well as Jovanovic's extension of it, a good exercise in using the Kalman filter.

While single agents appear in chapters 2 through 5, systems with multiple agents, whose environments and choices must be reconciled through markets, appear for the first time in chapters 6 and 7. Chapter 6 uses linear quadratic dynamic programming to introduce two important and related equilibrium concepts: rational expectations equilibrium and Nash Markov equilibrium. Each of these equilibrium concepts can be viewed as a fixed point in a space of beliefs about what other agents intend to do; and each is formulated using recursive methods. Chapter 7 introduces two notions of competitive equilibrium in dynamic stochastic pure exchange economies, then applies them to pricing various consumption streams.

Chapter 8 first introduces the overlapping generations model as a version of the general competitive model with a peculiar preference pattern. Then it goes on to use a sequential formulation of equilibria to display how the overlapping generations model can be used to study issues in monetary and fiscal economics, including social security.

Chapter 9 compares an important aspect of an overlapping generations model with an infinitely lived agent model with a particular kind of incomplete market structure. This chapter is thus our first encounter with an incomplete markets model. The chapter analyzes the Ricardian equivalence theorem in two distinct but isomorphic settings: one a model with infinitely lived agents who face borrowing constraints, another with overlapping generations of two-period-lived agents with a bequest motive. We describe situations in which the timing of taxes does or does not matter, and explain how binding borrowing constraints in the infinite-lived model correspond to nonoperational bequest motives in the overlapping generations model.

Chapter 10 studies asset pricing and a host of practical doctrines associated with asset pricing, including Ricardian equivalence again and Modigliani-Miller theorems for private and government finance. Chapter 11 is about economic growth. It describes the basic growth model, and analyzes the key features of the specification of the technology that allows the model to exhibit balanced growth.

---

[2] The equivalence is through duality, in the sense of mathematical programming.

Chapter 12 studies competitive equilibria distorted by taxes and our first mechanism design problems, namely, ones that seek to find the optimal temporal pattern of distorting taxes. In a nonstochastic economy, the most startling finding is that the optimal tax rate on capital is zero in the long run.

Chapter 13 is about self-insurance. We study a single agent whose limited menu of assets gives him an incentive to self-insure by accumulating assets. We study a special case of what has sometimes been called the "savings problem," and analyze in detail the motive for self-insurance and the surprising implications it has for the agent's ultimate consumption and asset holdings. The type of agent studied in this chapter will be a component of the incomplete markets models to be studied in chapter 14.

Chapter 14 studies incomplete markets economies with heterogeneous agents and imperfect markets for sharing risks. The models of market incompleteness in this chapter come from simply ruling out markets in many assets, without motivating the absence of those asset markets from the physical structure of the economy. We must wait until chapter 15 for a study of some of the reasons that such markets may not exist.

Chapter 15 describes models in the mechanism design tradition, work that starts to provide a foundation for incomplete assets markets, and that recovers specifications bearing an incomplete resemblance to the models of Chapter 14. Chapter 15 is about the optimal provision of social insurance in the presence of information and enforcement problems. Relative to earlier chapters, Chapter 15 escalates the sophistication with which recursive methods are applied, by utilizing promised values as state variables.

Chapter 16 applies some of the same ideas to problems in "reputational macroeconomics," using promised values to formulate the notion of credibility. We study how a reputational mechanism can make policies sustainable even when the government lacks the commitment technology that was assumed to exist in the policy analysis of chapter 12. This reputational approach is later used in chapter 17 to assess whether or not the Friedman rule is a sustainable policy.

Chapter 17 switches gears by adding money to a very simple competitive equilibrium model, in a most superficial way; the excuse for that superficial device is that it permits us to present and unify ten more or less well known monetary doctrines. Chapter 18 presents a less superficial model of money, the turnpike model of Townsend, which is basically a special nonstochastic version of one of the models of Chapter 14. The specialization allows us to focus on a variety of monetary doctrines.

Chapter 19 describes multiple agent models of search and matching. Except for a section on money in a search model, the focus is on labor markets as a central application of these theories. To bring out the economic forces at work in different frameworks, we examine the general equilibrium effects of layoff taxes.

Two appendixes (chapters 20 and 21) collect various technical results on functional analysis and linear control and filtering.

*Alternative uses of the book*

We have used parts of this book to teach both first- and second-year courses in macroeconomics and monetary economics at the University of Chicago, Stanford University, and the Stockholm School of Economics. Here are some alternative plans for courses:

1. A one-semester first-year course: chapters 1-5, 7, 8, 9, and either chapter 10, 11, or 12.

2. A second-semester first-year course: add chapters 6, 10, 11, 12, parts of 13 and 14, and all of 15.

3. A first course in monetary economics: chapters 8, 16, 17, 18, and last section of 19.

4. A second-year macroeconomics course: select from chapters 10-19.

As an example, Sargent used the following structure for a one-quarter first-year course at the University of Chicago: For the first and last weeks of the quarter, students were asked to read the monograph by Lucas (1987). Students were "prohibited" from reading the monograph in the intervening weeks. During the middle eight weeks of the quarter, students read material from chapters 5 (about search theory), chapter 7 (about complete markets), chapters 8, 17, and 18 (about models of money), and a little bit of chapter 15 (on social insurance with incentive constraints). The substantive theme of the course was the issues set out in a non-technical way by Lucas (1987). However, to understand Lucas's arguments, it helps to know the tools and models studied in the middle weeks of the course. Those weeks also exposed students to a range of alternative models that could be used to measure Lucas's arguments against some of the criticisms made, for example, by Manuelli and Sargent (1988).

Another one-quarter course would assign Lucas's (1992) article on efficiency and distribution in the first and last weeks. In the intervening weeks of the course, assign chapters 13, 14, 15.

As another example, Ljungqvist used the following material in a four-week segment on employment/unemployment in first-year macroeconomics at the Stockholm School of Economics. Labor market issues command a strong interest especially in Europe. Those issues help motivate studying the tools in chapters 5 and 19 (about search and matching models), and parts of 15 (on the optimal provision of unemployment compensation). On one level, both chapters 5 and 19 focus on labor markets as a central application of the theories presented, but on another level, the skills and understanding acquired in these chapters transcend the specific topic of labor market dynamics. For example, the thorough practice on formulating and solving dynamic programming problems in chapter 5 is generally useful to any student of economics, and the models of chapter 19 are an entry-pass to other heterogeneous-agent models like those in chapter 14. Further, an excellent way to motivate the study of recursive contracts in chapter 15 is to ask how unemployment compensation should optimally be provided in the presence of incentive problems.

## *Matlab programs*

Various exercises and examples use Matlab programs. These programs are referred to in a special index at the end of the book. They can be downloaded via anonymous ftp from the web site for the book:
ftp://zia.stanford.edu/pub/sargent/webdocs/matlab.

## *Answers to exercises*

We have created a web site with additional exercises and answers to the exercises in the text. It is at http://www.stanford.edu/˜sargent.

## *Notation*

We use the symbol ▮ to denote the conclusion of a proof. The editors of this book requested that where possible, brackets and braces be used in place of multiple parentheses to denote composite functions. Thus the reader will often encounter $f[u(c)]$ to express the composite function $f \circ u$.

## *Brief history of the notion of the state*

This book reflects progress economists have made in refining the notion of state so that more and more problems can be formulated recursively. The art in applying recursive methods is to find a convenient definition of the state. It is often not obvious what the state is, or even whether a finite-dimensional state *exists* (e.g., maybe the entire infinite history of the system is needed to characterize its current position). Extending the range of problems susceptible to recursive methods has been one of the major accomplishments of macroeconomic theory since 1970. In diverse contexts, this enterprise has been about discovering a convenient state and constructing a first-order difference equation to describe its motion. In models equivalent to single-agent control problems, state variables are either capital stocks or information variables that help predict the future.[3] In single-agent models of optimization in the presence of measurement errors, the true state vector is latent or "hidden" from the optimizer and the economist, and needs to be estimated. Here *beliefs* come to serve as the patent state. For example, in a Gaussian setting, the mathematical expectation and covariance matrix of the latent state vector, conditioned on the available history of observations, serves as the state. In authoring his celebrated filter, Kalman (1960) showed how an estimator of the hidden state could be constructed recursively by means of a difference equation that uses the current observables to update the estimator of last period's hidden state.[4] Muth (1960), Lucas (1972), Kareken, Muench, and Wallace (1973), Jovanovic (1979) and Jovanovic and Nyarko (1996) all used versions of the Kalman filter to study systems in which agents make decisions with imperfect observations about the state.

For a while, it seemed that some very important problems in macroeconomics could not be formulated recursively. Kydland and Prescott (1977) argued that it would be difficult to apply recursive methods to macroeconomic policy design problems, including two examples about taxation and a Phillips curve. As Kydland and Prescott formulated them, the problems were not recursive: the fact that the public's forecasts of the government's future decisions influence the public's

---

[3] Any available variables that *Granger cause* variables impinging on the optimizer's objective function or constraints enter the state as information variables. See C.W.J. Granger (1969).

[4] In competitive multiple-agent models in the presence of measurement errors, the dimension of the hidden state threatens to explode because beliefs about beliefs about ... naturally enter, a problem studied by Townsend (1983). This threat has been overcome through thoughtful and economical definitions of the state. For example, one way is to give up on seeking a purely "autoregressive" recursive structure and to include a moving average piece in the descriptor of beliefs. See Sargent (1991). Townsend's equilibria have the property that prices fully reveal the private information of diversely informed agents.

current decisions made the government's problem simultaneous, not sequential. But soon Kydland and Prescott (1980) and Hansen, Epple, and Roberds (1985) proposed a recursive formulation of such problems by expanding the state of the economy to include a Lagrange multiplier or *costate* variable associated with the government's budget constraint. The co state variable acts as the marginal cost of keeping a promise made earlier by the government. Recently Marcet and Marimon (1999) have extended and formalized a recursive version of such problems.

A significant breakthrough in the application of recursive methods was achieved by several researchers including Spear and Srivastava (1987), Thomas and Worrall (1988), and Abreu, Pearce, and Stacchetti (1990). They discovered a state variable for recursively formulating an infinitely repeated moral hazard problem. That problem requires the principal to track a history of outcomes and to use it to construct statistics for drawing inferences about the agent's actions. Problems involving self-enforcement of contracts and a government's reputation share this feature. A *continuation value* promised by the principal to the agent can summarize the history. Making the promised valued a state variable allows a recursive solution in terms of a function mapping the inherited promised value and random variables realized today into an action or allocation today and a promised value for tomorrow. The sequential nature of the solution allows us to recover history-dependent strategies just as we use a stochastic difference equation to find a 'moving average' representation.[5]

It is now standard to use a continuation value as a state variable in models of credibility and dynamic incentives. We shall study several such models in this book, including ones for optimal unemployment insurance and for designing loan contracts that must overcome information and enforcement problems.

[5] Related ideas are used by Shavell and Weiss (1979), Abreu, Pearce, and Stacchetti (1986, 1990) in repeated games and Green (1987) and Phelan and Townsend (1991) in dynamic mechanism design. Andrew Atkeson (1991) extended these ideas to study loans made by borrowers who cannot tell whether they are making consumption loans or investment loans.

# 1
# Time Series

## Two workhorses

This chapter describes two tractable models of time series: Markov chains and first-order stochastic linear difference equations. These models are organizing devices that put particular restrictions on a sequence of random vectors. They are useful because they describe a time series with parsimony. In later chapters, we shall make two uses each of Markov chains and stochastic linear difference equations: (1) to represent the exogenous information flows impinging on an agent or an economy, and (2) to represent an optimum or equilibrium outcome of agents' decision making. Both the Markov chain and the first-order stochastic linear difference use a sharp notion of a state vector. A state vector summarizes all of the information about the current position that is relevant for determining the future statistics of the system. The Markov chain and the stochastic linear difference equation will be useful tools for studying dynamic optimization problems.

## Markov chains

A stochastic process is a sequence of random vectors. For us, the sequence will be ordered by a time index, taken to be the integers in this book. So we study discrete time models. We study a discrete state stochastic process with the following property:

MARKOV PROPERTY: A stochastic process $\{x_t\}$ is said to have the *Markov property* if for all $k \geq 2$ and all $t$,

$$\text{Prob}(x_{t+1}|x_t, x_{t-1}, \ldots, x_{t-k}) = \text{Prob}(x_{t+1}|x_t).$$

We assume the Markov property and characterize the process by a *Markov chain.* A time-invariant Markov chain is defined by a triple of objects, namely, an $n$-dimensional vector $\bar{x} \in R^n$ that records the possible values of the *state* of the system; an $n \times n$ *transition matrix* $P$, which records the probabilities of moving from one value of the state to another in one period; and an $(n \times 1)$ vector $\pi_0$ recording the probabilities of being in each state $i$ at time 0. The vector $\pi_0$ has the interpretation

$$\pi_{0i} = \text{Prob}(x_0 = \bar{x}_i).$$

The matrix $P$ has the interpretation

$$P_{ij} = \text{Prob}(x_{t+1} = \bar{x}_j | x_t = \bar{x}_i).$$

For these interpretations to be valid, the matrix $P$ and the vector $\pi$ must satisfy the following assumption:

Assumption M:
a. For $i = 1, \ldots, n$, the matrix $P$ satisfies

$$\sum_{j=1}^{n} P_{ij} = 1. \tag{1.1}$$

b. The vector $\pi_0$ satisfies

$$\sum_{i=1}^{n} \pi_{0i} = 1.$$

A matrix $P$ that satisfies property (1.1) is called a *stochastic matrix.* A stochastic matrix defines the probabilities of moving from any value of the state to any other in one period. The probability of moving from any value of the state to any other in *two* periods is determined by $P^2$ because

$$\begin{aligned}
&\text{Prob}(x_{t+2} = \bar{x}_j | x_t = \bar{x}_i) \\
&= \sum_{h=1}^{n} \text{Prob}(x_{t+2} = \bar{x}_j | x_{t+1} = \bar{x}_h) \text{Prob}(x_{t+1} = \bar{x}_h | x_t = \bar{x}_i) \\
&= \sum_{h=1}^{n} P_{ih} P_{hj} = P_{ij}^2,
\end{aligned}$$

where $P_{ij}^2$ is the $i,j$ element of $P^2$. Let $P_{i,j}^k$ denote the $i,j$ element of $P^k$. By iterating on the preceding equation, we discover that

$$\text{Prob}(x_{t+k} = \bar{x}_j | x_t = \bar{x}_i) = P_{ij}^k.$$

The unconditional probability distributions of $x_t$ are determined from

$$\begin{aligned}
\pi_1' &= \text{Prob}(x_1) = \pi_0' P \\
\pi_2' &= \text{Prob}(x_2) = \pi_0' P^2 \\
&\vdots \\
\pi_k' &= \text{Prob}(x_k) = \pi_0' P^k,
\end{aligned}$$

where $\text{Prob}(x_t)$ is the $(1 \times n)$ vector whose $i$th element is $\text{Prob}(x_t = \bar{x}_i)$.

*Stationary distributions*

Evidently, unconditional probability distributions evolve according to

$$\pi_{t+1}' = \pi_t' P. \tag{1.2}$$

A distribution is called *stationary* if it satisfies

$$\pi_{t+1} = \pi_t,$$

that is, if the distribution remains unaltered with the passage of time. Evidently from the law of motion (1.2) for unconditional distributions, a stationary distribution must satisfy

$$\pi' = \pi' P$$

or

$$\pi'(I - P) = 0.$$

Transposing both sides of this equation gives

$$(I - P')\pi = 0, \tag{1.3}$$

which determines $\pi$ as an eigenvector (normalized to satisfy $\sum \pi_i = 1$) associated with a unit eigenvalue of $P'$.

The fact that $P$ is a stochastic matrix (i.e., it has nonnegative elements and satisfies $\sum_j P_{ij} = 1$) guarantees that $P$ has at least one unit eigenvalue, and that there is some $\pi$ that satisfies equation (1.3). Depending on $P$, this stationary distribution may or may not be unique ($P$ may have repeated eigenvalues of unity).

*Example 1.* A Markov chain

$$P = \begin{bmatrix} 1 & 0 & 0 \\ .2 & .5 & .3 \\ 0 & 0 & 1 \end{bmatrix}$$

has two unit eigenvalues with associated stationary distributions $\pi' = [1 \quad 0 \quad 0]$ $\pi' = [0 \quad 0 \quad 1]$.

*Example 2.* A Markov chain

$$P = \begin{bmatrix} .7 & .3 & 0 \\ 0 & .5 & .5 \\ 0 & .9 & .1 \end{bmatrix}$$

has one unit eigenvalue with associated stationary distribution $\pi' = [0 \quad .6429 \quad .3571]$

*Asymptotic stationarity*

We often ask the following question about a Markov process: for an arbitrary initial distribution $\pi_0$, do the unconditional distributions $\pi_t$ approach a stationary distribution

$$\lim_{t\to\infty} \pi_t = \pi_\infty,$$

where $\pi_\infty$ solves equation (1.3)? If the answer is yes, then does the limit distribution $\pi_\infty$ depend on the initial distribution $\pi_0$? If the limit $\pi_\infty$ is independent of the initial distribution $\pi_0$, we say that the process is *asymptotically stationary with a unique invariant distribution.* We call a solution $\pi_\infty$ a *stationary distribution* or an *invariant distribution* of $P$.

We state these concepts formally in the following definition:

DEFINITION: Let $\pi_\infty$ be the unique vector that satisfies $(I - P')\pi_\infty = 0$. If for all initial distributions $\pi_0$ it is true that $P^{t\prime}\pi_0$ converges to the same $\pi_\infty$, we say that the Markov chain is asymptotically stationary with a unique invariant distribution.

The following theorems can be used to show that a Markov chain is asymptotically stationary.

THEOREM 1: Let $P$ be a stochastic matrix with $P_{ij} > 0 \ \forall (i,j)$. Then $P$ has a unique stationary distribution, and the process is asymptotically stationary.

THEOREM 2: Let $P$ be a stochastic matrix for which $P^n_{ij} > 0 \ \forall (i,j)$ for some value of $n \geq 1$. Then $P$ has a unique stationary distribution, and the process is asymptotically stationary.

*Expectations*

From the conditional and unconditional probability distributions that we have listed, it follows that the unconditional expectations of $x_t$ for $t \geq 0$ are determined by $Ex_t = (\pi_0' P^t)\bar{x}$, or

$$\begin{aligned} Ex_0 &= \pi_0' \bar{x} \\ Ex_1 &= \pi_0' P \bar{x} \\ &\vdots \\ Ex_k &= \pi_0' P^k \bar{x}. \end{aligned}$$

Conditional expectations are determined by s

$$E(x_{t+k} | x_t = \bar{x}) = P^k \bar{x}.$$

Notice that

$$\begin{aligned} E(x_t) &= \pi_t' \bar{x} = (\pi_0' P^t) \bar{x} \\ &= \pi_0' (P^t \bar{x}) \\ &= E[E(x_t | x_0 = \bar{x})]. \end{aligned}$$

The statement that $E(x_t) = E(Ex_t | x_0 = \bar{x})$ is an example of the *law of iterated expectations.*

*Forecasting functions*

There are powerful formulas for forecasting functions of a Markov process. Let $h(\bar{x})$ be a function of the state represented by an $(n \times 1)$ vector $h$. Then we have the following two useful formulas:

$$E[h(x_{t+k}) | x_t = \bar{x})] = P^k h$$

$$E\left[\sum_{k=0}^{\infty}\beta^k h(x_{t+k})|x_t=\bar{x})\right]=(I-\beta P)^{-1}h(\bar{x}),$$

where $\beta\in(0,1)$ guarantees existence of $(I-\beta P)^{-1}=(I+\beta P+\beta^2P^2+\ldots)$.

There is a sense in which one-step-ahead forecasts of a sufficiently rich set of functions characterize a Markov chain. In particular, one-step-ahead conditional expectations of $n$ independent functions (i.e., $n$ linearly independent vectors $h_1,\ldots,h_n$) uniquely determine the transition matrix $P$. Thus, let $E[h_i(x_{t+1})|x_t=\bar{x}]=Ph_i=y_i$. Then $Ph=y$ or $P=yh^{-1}$, where $e=[y_1\ \ldots\ y_n]$ and $h=[h_1\ \ldots\ h_n]$.

*Simulating a Markov chain*

It is easy to simulate a Markov chain using a random number generator. The Matlab program `markov.m` does the job. We'll use this program in some later chapters.[1]

*The likelihood function*

Let $P$ be an $n\times n$ stochastic matrix. For convenience, let the states be indexed $1,2,\ldots,n$. Let $\pi_0$ be an $n\times 1$ vector with nonnegative elements summing to $1$, with $\pi_0,i$ being the probability that the state is $i$ at time $0$. Let $i_t$ index the state at time $t$. The Markov property implies that the probability of drawing the path $(x_0,x_1,\ldots,x_{T-1},x_T)=(x_{i_0},x_{i_1},\ldots,x_{i_{T-1}},x_{i_T})$ is

$$\begin{aligned}L&\equiv \text{Prob}(x_{i_T},x_{i_{T-1}},\ldots,x_{i_1},x_{i_0})\\&=P_{i_{T-1},i_T}P_{i_{T-2},i_{T-1}}\cdots P_{i_0,i_1}\pi_{0,i_0}.\end{aligned}\tag{1.4}$$

The probability $L$ is called the *likelihood.* It is a function of both the sample realization $x_0,\ldots,x_T$ and the parameters of the stochastic matrix $P$. For a sample $x_0,x_1,\ldots,x_T$, let $n_{ij}$ be the number of times that there occurs a one-period transition from state $i$ to state $j$. Then the likelihood function can be written

$$L=\pi_{0,i_0}\prod_i\prod_j P_{i,j}^{n_{ij}},$$

a *multinomial* distribution.

[1] An index in the back of the book lists Matlab programs that can downloaded from the textbook web site `ftp://zia.stanford.edu/sargent/webdocs/matlab`.

Formula (1.4) has two uses. A first, which we shall encounter often, is to describe the probability of alternative histories of a Markov chain. We shall use this formula to study prices and allocations in competitive equilibria.

A second use is for estimating the parameters of a model whose solution is a Markov chain. Maximum likelihood estimation for free parameters $\theta$ of a Markov process works as follows. Let the transition matrix $P$ be a function and the initial distribution $\pi_0$ be functions $P(\theta), \pi_0(\theta)$ of a vector of free parameters $\theta$. Given a sample $\{x_t\}_{t=0}^T$, regard the likelihood function as a function of the parameters $\theta$. As the estimator of $\theta$, choose the value that maximizes the likelihood function $L$.

## *Stochastic linear difference equations*

We now turn our attention to a stochastic process whose state $x_t \in I\!R^n$. Thus, we consider the first-order stochastic linear difference equation

$$x_{t+1} = A_o x_t + C w_{t+1} \tag{1.5}$$

for $t = 0, 1, \ldots$, where $x_t$ is an $n \times 1$ state vector, $x_0$ is a given initial condition, $A_o$ is an $n \times n$ matrix, $C$ is an $n \times m$ matrix, and $w_{t+1}$ is an $m \times 1$ random vector satisfying:

$$E w_{t+1} | J_t = 0 \tag{1.6a}$$

$$E w_{t+1} w_{t+1}' | J_t = I, \tag{1.6b}$$

where $J_t = [\, w_t \quad \ldots \quad w_1 \quad x_0 \,]$ is the information set at $t$, and $E[\,\cdot\,|J_t]$ denotes the conditional expectation. A sequence $\{w_{t+1}\}$ satisfying equation (1.6) is said to be a martingale difference sequence adapted to $J_t$.

We shall often append an observation equation $y_t = G x_t$ to equation (1.5) and deal with the augmented system

$$x_{t+1} = A_o x_t + C w_{t+1} \tag{1.7a}$$

$$y_t = G x_t. \tag{1.7b}$$

Here $y_t$ is a vector of variables observed at $t$, which may include only some linear combinations of $x_t$. The system (1.7) is often called a *state-space system.*

*Example 1.* Scalar second-order autoregression: Assume that $y_t$ and $w_t$ are scalar processes and that

$$y_{t+1} = \alpha + \rho_1 y_t + \rho_2 y_{t-1} + w_{t+1}.$$

Represent this relationship as the system

$$\begin{bmatrix} y_{t+1} \\ y_t \\ 1 \end{bmatrix} = \begin{bmatrix} \rho_1 & \rho_2 & \alpha \\ 1 & 0 & 0 \\ 0 & 0 & 1 \end{bmatrix} \begin{bmatrix} y_t \\ y_{t-1} \\ 1 \end{bmatrix} + \begin{bmatrix} 1 \\ 0 \\ 0 \end{bmatrix} w_{t+1}$$

$$y_t = \begin{bmatrix} 1 & 0 & 0 \end{bmatrix} \begin{bmatrix} y_t \\ y_{t-1} \\ 1 \end{bmatrix}$$

which has form (1.7).

*Example 2.* First-order scalar mixed moving average and autoregression: Let

$$y_{t+1} = \rho y_t + w_{t+1} + \gamma w_t.$$

Express this relationship as

$$\begin{bmatrix} y_{t+1} \\ w_{t+1} \end{bmatrix} = \begin{bmatrix} \rho & \gamma \\ 0 & 0 \end{bmatrix} \begin{bmatrix} y_t \\ w_t \end{bmatrix} + \begin{bmatrix} 1 \\ 1 \end{bmatrix} w_{t+1}$$

$$y_t = \begin{bmatrix} 1 & 0 \end{bmatrix} \begin{bmatrix} y_t \\ w_t \end{bmatrix}.$$

*Example 3.* Vector autoregression: Let $y_t$ be an $n \times 1$ vector of random variables. We define a vector autoregression by a stochastic difference equation

$$y_{t+1} = \sum_{j=1}^{4} A_j y_{t+1-j} + C_y w_{t+1}, \tag{1.8}$$

where $w_{t+1}$ is an $n \times 1$ martingale difference sequence satisfying equation (1.6) with $x_0' = \begin{bmatrix} y_0 & y_{-1} & y_{-2} & y_{-3} \end{bmatrix}$ and $A_j$ is an $n \times n$ matrix for each $j$. We can map equation (1.8) into equation (1.5) as follows:

$$\begin{bmatrix} y_{t+1} \\ y_t \\ y_{t-1} \\ y_{t-2} \end{bmatrix} = \begin{bmatrix} A_1 & A_2 & A_3 & A_4 \\ I & 0 & 0 & 0 \\ 0 & I & 0 & 0 \\ 0 & 0 & I & 0 \end{bmatrix} \begin{bmatrix} y_t \\ y_{t-1} \\ y_{t-2} \\ y_{t-3} \end{bmatrix} + \begin{bmatrix} C_y \\ 0 \\ 0 \\ 0 \end{bmatrix} w_{t+1}. \tag{1.9}$$

Define $A_o$ as the state transition matrix in equation (1.9). Assume that $A_o$ has all of its eigenvalues bounded in modulus below unity. Then equation (1.8) can be initialized so that $y_t$ is "covariance stationary," a term we now define.

*Moments*

We can use equation (1.5) to deduce the first and second moments of the sequence of random vectors $\{x_t\}_{t=0}^{\infty}$. A random sequence of vectors is called a stochastic process.

DEFINITION: A stochastic process $\{x_t\}$ is said to be *covariance stationary* if it satisfies the following two properties: (a) the mean is independent of time, $Ex_t = Ex_0$ for all $t$, and (b) the sequence of autocovariance matrices $E(x_{t+j} - Ex_{t+j})(x_t - Ex_t)'$ depends on the separation between dates $j = 0, \pm 1, \pm 2, \ldots$ but not on $t$.

We will make assumptions that guarantee that the stochastic process $x_t$ has a unique stationary distribution, in particular, that all eigenvalues of $A_o$ not affiliated with the constant are bounded in modulus strictly from above by unity. To discover the statistics of the stationary distribution, it is useful to regard the initial condition $x_0$ as being drawn from a distribution with mean $\mu_0 = Ex_0$ and covariance $E(x - Ex_0)(x - Ex_0)'$. We shall deduce starting values for the mean and covariance that make the process covariance stationary.

Taking mathematical expectations on both sides of equation (1.5) gives

$$\mu_{t+1} = A_o \mu_t \tag{1.10}$$

where $\mu_t = Ex_t$. We will assume that all of the eigenvalues of $A_o$ are strictly less than unity in modulus, except possibly for one that is affiliated with the constant terms in the various equations. Then $x_t$ possesses a stationary mean defined to satisfy $\mu_{t+1} = \mu_t$, which from equation (1.10) evidently satisfies

$$(I - A_o)\mu = 0, \tag{1.11}$$

which characterizes the mean $\mu$ as the eigenvector associated with the single unit eigenvalue of $A_o$. Notice that

$$x_{t+1} - \mu = A_o(x_t - \mu) + Cw_{t+1}. \tag{1.12}$$

Also, the fact that the remaining eigenvalues of $A_o$ are less than unity in modulus implies that starting from any $\mu_0$, $\mu_t \to \mu$.[2]

---

[2] To see this point, assume that the eigenvalues of $A_o$ are distinct, and use the representation $A_o = P \Lambda P^{-1}$ where $\Lambda$ is a diagonal matrix of the eigenvalues of $A_o$, arranged in descending

From equation (1.12) we can compute that the stationary variance matrix satisfies

$$C_x(0) \equiv E(x_t - \mu)(x_t - \mu)' = A_o C_x(0) A_o' + CC'. \tag{1.13}$$

Note that

$$(x_{t+j} - \mu) = A_o^j (x_t - \mu) + C w_{t+j} + \ldots + A_o^{j-1} C w_{t+1}.$$

Postmultiplying both sides by $(x_t - \mu)'$ and taking expectations shows that the autocovariance sequence satisfies

$$C_x(j) \equiv E(x_{t+j} - \mu)(x_t - \mu)' = A_o^j C_x(0). \tag{1.14}$$

The autocovariance sequence is also called the *autocovariogram.* Equation (1.13) is a *discrete Lyapunov* equation in the $n \times n$ matrix $C_x(0)$. It can be solved with the Matlab program `doublej.m`. Once it is solved, the remaining moments $C_x(j)$ can be deduced from equation (1.14).[3]

Suppose that $y_t = G x_t$. Then $E y_t = G\mu$ and

$$E(y_{t+j} - \mu_y)(y_t - \mu_y)' = G C_x(j) G', \tag{1.15}$$

for $j = 0, 1, \ldots$. Equations (1.15) are matrix versions of the Yule-Walker equations, according to which autocovariograms for a process governed by a stochastic linear difference equation obey the nonstochastic version of the same difference equation.

### *Impulse response function*

Suppose that the eigenvalues of $A_o$ not associated with the constant are bounded above in modulus by unity. Using the lag operator $L$ defined by $L x_{t+1} \equiv x_t$, express equation (1.5) as

$$(I - A_o L) x_{t+1} = C w_{t+1}. \tag{1.16}$$

order in magnitude, and $P$ is a matrix composed of the corresponding eigenvectors. Then equation (1.10) can be represented as $\mu_{t+1}^* = \Lambda \mu_t^*$, where $\mu_t^* \equiv P^{-1} \mu_t$, which implies that $\mu_t^* = \Lambda^t \mu_0^*$. When all eigenvalues but the first are less than unity, $\Lambda^t$ converges to a matrix of zeros except for the $(1,1)$ element, and $\mu_t^*$ converges to a vector of zeros except for the first element, which stays at $\mu_{0,1}^*$, its initial value, which equals $1$, to capture the constant. Then $\mu_t = P \mu_t^*$ converges to $P_1 \mu_{0,1}^* = P_1$, where $P_1$ is the eigenvector corresponding to the unit eigenvalue.

[3] Notice that $C_x(-j) = C_x(j)'$.

Recall the Neumann expansion $(I - A_oL)^{-1} = (I + A_oL + A_oL^2 + \ldots)$ and apply it to both sides of equation (1.16) to get

$$x_{t+1} = \sum_{j=0}^{\infty} A_o^j C w_{t+1-j}, \tag{1.17}$$

which is the solution of equation (1.5) assuming that equation (1.5) has been operating for the infinite past before $t = 0$. Alternatively, iterate equation (1.5) forward from $t = 0$ to get

$$x_t = A_o^t x_0 + \sum_{j=0}^{t-1} A_o^j C w_{t-j} \tag{1.18}$$

Equations (1.17) and (1.18) are alternative versions of a moving average representation or impulse response function. Each shows how $x_{t+1}$ is affected by lagged values of the shocks, the $w_{t+1}$'s. Thus, the effect of an innovation $w_{t-j}$ on $x_t$ is $A^j C$.[4]

*Prediction and discounting*

From equation (1.5) we can compute the useful prediction formulas

$$E_t x_{t+j} = A_o^j x_t \tag{1.19}$$

for $j \geq 1$, where $E_t(\cdot)$ denotes the mathematical expectation conditioned on $x^t = (x_t, x_{t-1}, \ldots, x_0)$. Let $y_t = Gx_t$, and suppose that we want to compute $E_t \sum_{j=0}^{\infty} \beta^j y_{t+j}$. Evidently,

$$E_t \sum_{j=0}^{\infty} \beta^j y_{t+j} = G(I - \beta A_o)^{-1} x_t, \tag{1.20}$$

provided that the eigenvalues of $\beta A_o$ are less than unity in modulus. Equation (1.20) tells us how to compute an expected discounted sum, where the discount factor $\beta$ is constant. We now turn to a setting where the discount factor itself is random.

[4] The Matlab programs `dimpulse.m` and `impulse.m` compute impulse response functions.

*Stochastic discount factor*

Let $y_t$ be governed by the state-space system (1.7). In addition, assume that there is a scalar random process $z_t$ given by

$$z_t = Hx_t.$$

Suppose that we want to calculate

$$\alpha_t = E_t \sum_{j=0}^{\infty} (\beta^j z_{t+j}) y_{t+j}. \tag{1.21}$$

Here the process $\beta^t z_t$ is serving as a stochastic discount factor. We shall have cause to evaluate expressions like equation (1.21) when we study asset prices. The right side of equation (1.21) is a conditional expectation of a geometric distributed lead of a quadratic form in the state, namely,

$$\alpha_t = E_t \sum_{j=0}^{\infty} \beta^j x'_{t+j} H' G x_{t+j}$$

To get a formula for $\alpha_t$, we guess that it can be written in the form

$$\alpha_t = x_t' \nu x_t + \sigma, \tag{1.22}$$

where $\nu$ is an $(n \times n)$ matrix, and $\sigma$ is a scalar. The definition of $\alpha_t$ and the guess (1.22) imply

$$\begin{aligned}
\alpha_t &= x_t' H' G x_t + \beta E_t (x'_{t+1} \nu x_{t+1} + \sigma) \\
&= x_t' H' G x_t + \beta E_t \Big[ (A_o x_t + C w_{t+1})' \, \nu \, (A_o x_t + C w_{t+1}) + \sigma \Big] \\
&= x_t' (H' G + \beta A_o' \nu A_o) \, x_t + \beta \operatorname{trace}(\nu C C') + \beta \sigma.
\end{aligned}$$

It follows that $\nu$ and $\sigma$ satisfy

$$\begin{aligned}
\nu &= H' G + \beta A_o' \nu A_o \\
\sigma &= \beta \sigma + \beta \operatorname{trace} \nu C C'.
\end{aligned} \tag{1.23}$$

The first equation of (1.23) is a *discrete Lyapunov equation* in the square matrix $\nu$, and can be solved by using one of several algorithms.[5] After $\nu$ has been

[5] The Matlab control toolkit has a program called `dlyap.m`; also see the program called `doublej.m`.

computed, the second equation can be solved for the scalar $\sigma$. The term $\sigma$ has aspects of a risk premium and it is zero when $C = 0$.

*Population regression*

This section explains the notion of a regression equation. Suppose that we have a state-space system (1.7) with initial conditions that make it covariance stationary. We can use the preceding formulas to compute the second moments of any pair of random variables. These moments let us compute a linear regression. Thus, let $X$ be a $1 \times N$ vector of random variables somehow selected from the stochastic process $\{y_t\}$ governed by the system (1.7). For example, let $N = 2 \times m$, where $y_t$ is an $m \times 1$ vector, and take $X = [\, y_t' \quad y_{t-1}' \,]$ for any $t \geq 1$. Let $Y$ be any scalar random variable selected from the $m \times 1$ stochastic process $\{y_t\}$. For example, take $Y = y_{t+1,1}$ for the same $t$ used to define $X$, where $y_{t+1,1}$ is the first component of $y_{t+1}$.

We consider the following least squares approximation problem: find an $N \times 1$ vector of real numbers $\beta$ that attain

$$\min_{\beta} E(Y - X\beta)^2 \tag{1.24}$$

Here $X\beta$ is being used to estimate $Y$, and we want the value of $\beta$ that minimizes the expected squared error. The first-order necessary condition for minimizing $E(Y - X\beta)^2$ with respect to $\beta$ is

$$EX'(Y - X\beta) = 0, \tag{1.25}$$

which can be rearranged as $EX'Y = EX'X\beta$ or[6]

$$\beta = [E(X'X)]^{-1}(EX'Y). \tag{1.26}$$

By using the formulas (1.11), (1.13), (1.14), and (1.15), we can compute $EX'X$ and $EX'Y$ for whatever selection of $X$ and $Y$ we choose. The condition (1.25) is called the least squares normal equation. It states that the projection error $Y - X\beta$ is orthogonal to $X$. Therefore, we can represent $Y$ as

$$Y = X\beta + \epsilon \tag{1.27}$$

[6] That $EX'X$ is nonnegative semidefinite implies that the second-order conditions for a minimum of condition (1.24) are satisfied.

where $EX'\epsilon = 0$. Equation (1.27) is called a regression equation, and $X\beta$ is called the least squares projection of $Y$ on $X$ or the least squares regression of $Y$ on $X$. The vector $\beta$ is called the population least squares regression vector. Econometrics describes conditions under which sample moments converge to population moments, that is, $\frac{1}{S}\sum_{s=1}^{S} X_s' X_s \to EX'X$ and $\frac{1}{S}\sum_{s=1}^{S} X_s' Y_s \to EX'Y$. Under those conditions, sample least squares estimates converge to $\beta$.

There are as many such regressions as there are ways of selecting $Y, X$. We have shown how a model (e.g., a triple $A_o, C, G$, together with an initial distribution for $x_0$) restricts a regression. Going backward, that is, telling what a given regression tells about a model, is more difficult. Often the regression tells little about the model. The likelihood function encodes what a given data set says about the model.

### *The spectrum*

For a covariance stationary stochastic process, all second moments can be encoded in a complex-valued matrix called the *spectral density* matrix. The autocovariance sequence for the process determines the spectral density. Conversely, the spectral density can be used to determine the autocovariance sequence.

Under the assumption that the only eigenvalue of $A_o$ not strictly less than unity in modulus is that associated with the constant, the state $x_t$ converges to a unique stationary probability distribution as $t$ approaches infinity. The spectral density matrix of this stationary distribution $S_x(\omega)$ is defined to be the Fourier transform of the covariogram of $x_t$:

$$S_x(\omega) \equiv \sum_{\tau=-\infty}^{\infty} C_x(\tau) e^{-i\omega\tau}. \tag{1.28}$$

For the system (1.5), the spectral density of the stationary distribution is given by the formula

$$S_x(\omega) = [I - A_o e^{-i\omega}]^{-1} C C' [I - A_o' e^{+i\omega}]^{-1}, \quad \forall \omega \epsilon [-\pi, \pi].$$

The spectral density contains all of the information about the covariances. They can be recovered from $S_x(\omega)$ by the Fourier inversion formula[7]

[7] Spectral densities for continuous-time systems are discussed by Kwakernaak and Sivan (1972). For an elementary discussion of discrete-time systems, see Sargent (1987a). Also see Sargent (1987a, chap. 11) for definitions of the spectral density function and methods of evaluating this integral.

$$C_x(\tau) = (1/2\pi) \int_{-\pi}^{\pi} S_x(\omega) e^{+i\omega\tau} d\omega.$$

Setting $\tau = 0$ in the inversion formula gives

$$C_x(0) = (1/2\pi) \int_{-\pi}^{\pi} S_x(\omega) d\omega,$$

which shows that the spectral density decomposes covariance across frequencies.[8] A formula used in the process of generalized method of moments (GMM) estimation emerges by setting $\omega = 0$ in equation (1.28), which gives

$$S_x(0) \equiv \sum_{\tau=-\infty}^{\infty} C_x(\tau).$$

*Examples*

To give some practice in reading spectral densities, we used the Matlab program `bigshow.m` to generate Figures 1.1, 1.2, 1.3, and 1.4. The program takes as an input a univariate process of the form

$$a(L) y_t = b(L) w_t,$$

where $w_t$ is a univariate martingale difference sequence with unit variance, where $a(L) = 1 - a_2 L - a_3 L^2 - \ldots - a_n L^{n-1}$ and $b(L) = b_1 + b_2 L + \ldots + b_n L^{n-1}$, and where we require that $a(z) = 0$ imply that $|z| > 1$. The program computes and displays a realization of the process, the impulse response function from $w$ to $y$, and the spectrum of $y$. By using this program, a reader can teach himself to read spectra and impulse response functions. Figure 1.1 is for the pure autoregressive process with $a(L) = 1 - .9L, b = 1$. The spectrum sweeps downward in what C.W.J. Granger (1966) called the "typical spectral shape" for an economic time series. Figure 1.2 sets $a = 1 - .8L^4, b = 1$. This is a process with a strong seasonal component. Note that the spectrum peaks at $\pi$ and $\pi/2$, telltale signs of a strong seasonal component. Figure 1.3 sets a $= 1 - .98L, b = 1 - .7L$. This is a version of a process studied by Muth (1960). After the first lag, the

[8] More interestingly, the spectral density achieves a decomposition of covariance into components that are orthogonal across frequencies.

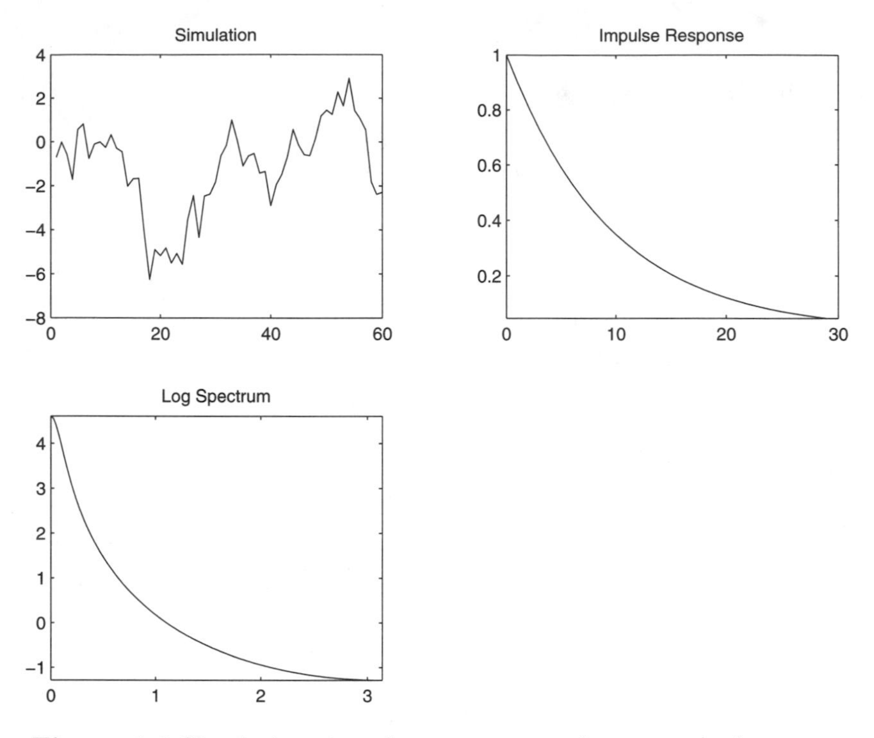


**Figure 1.1** Simulation, impulse response, and spectrum of process $(1 - .9L)y_t = w_t$.

impulse response declines as $.99^j$, where $j$ is the lag length. Figure 1.4 sets $a = 1 - 1.3 + .7L^2, b = 1$. This is a process that has a spectral peak and cycles in its covariogram. See Sargent (1987a) for a more extended discussion.

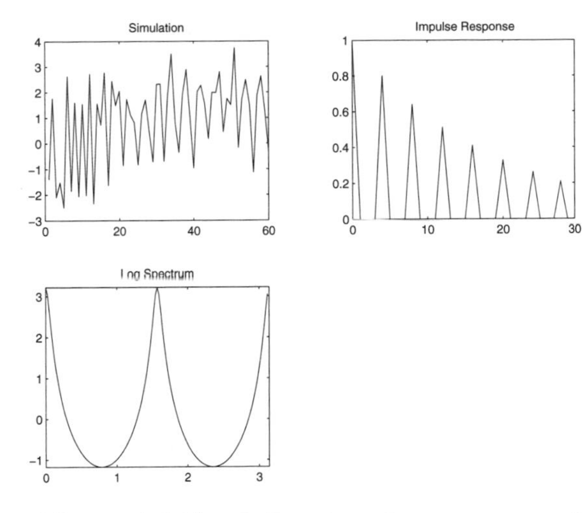


**Figure 1.2** Simulation, impulse response, and spectrum of process $(1 - .8L^4)y_t = w_t$.

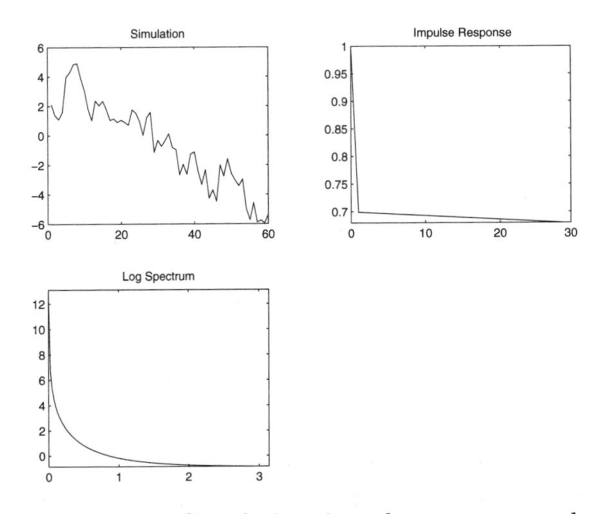


**Figure 1.3** Simulation, impulse response, and spectrum of process $(1 - .98L)y_t = (1 - .7L)w_t$.

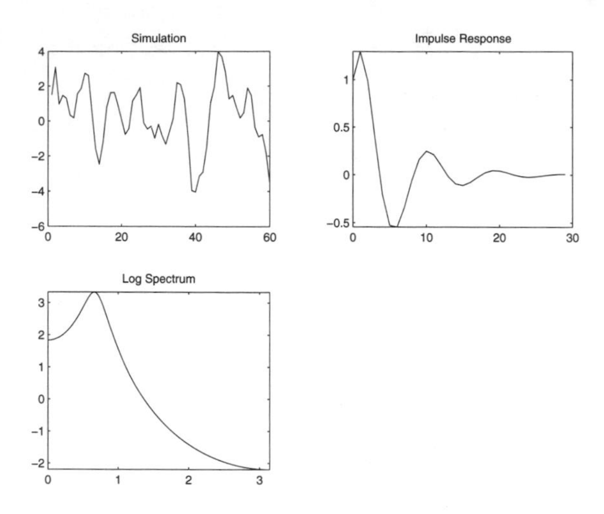


**Figure 1.4** Simulation, impulse response, and spectrum of process $(1-1.3L+.7L^2)y_t = w_t$.

*Estimation*

We have shown how to map the matrices $A_o, C$ into all of the second moments of the stationary distribution of the stochastic process $\{x_t\}$. Linear economic models typically give $A_o, C$ as functions of a set of deeper parameters $\theta$. We shall give examples of some such models in chapters 3 and 4. Those theories and the formulas of this chapter give us a mapping from $\theta$ to these theoretical moments of the $\{x_t\}$ process. That mapping is an important ingredient of econometric methods designed to estimate a wide class of linear rational expectations models (see Hansen and Sargent, 1980, 1981). Briefly, these methods use the following procedures for matching observations with theory. To simplify, we shall assume that in any period $t$ that an observation is available, observations are available on the entire state $x_t$. As discussed in the following paragraphs, the details are more complicated if only a subset or a noisy signal of the state is observed, though the basic principles remain the same.

Given a sample of observations for $\{x_t\}_{t=0}^T \equiv x_t, t = 0, \ldots, T$, the likelihood function is defined as the joint probability distribution $f(x_T, x_{T-1}, \ldots, x_0)$. The likelihood function can be *factored* using

$$f(x_T,\ldots,x_0) = f(x_T|x_{T-1},\ldots,x_0)f(x_{T-1}|x_{T-2},\ldots,x_0)\cdots \\ f(x_1|x_0)f(x_0), \tag{1.29}$$

where in each case $f$ denotes an appropriate probability distribution. For system (1.5), $f(x_{t+1}|x_t,\ldots,x_0) = f(x_{t+1}|x_t)$, which follows from the Markov property possessed by equation (1.5). Then the likelihood function has the recursive form

$$f(x_T,\ldots,x_0) = f(x_T|x_{T-1})f(x_{T-1}|x_{T-2})\cdots f(x_1|x_0)f(x_0). \tag{1.30}$$

If we assume that the $w_t$'s are Gaussian, then the conditional distribution $f(x_{t+1}|x_t)$ is Gaussian with mean $A_o x_t$ and covariance matrix $CC'$. Thus, under the Gaussian distribution, the log of the conditional density of $x_{t+1}$ becomes

$$\log f(x_{t+1}|x_t) = -.5\log(2\pi) - .5\ \det\ (CC') \\ - .5(x_{t+1} - A_o x_t)'(CC')^{-1}(x_{t+1} - A_o x_t) \tag{1.31}$$

Given an assumption about the distribution of the initial condition $x_0$, equations (1.30) and (1.31) can be used to form the likelihood function of a sample of observations on $\{x_t\}_{t=0}^T$. One computes maximum likelihood estimates by using a hill-climbing algorithm to maximize the likelihood function with respect to the free parameters $A_o, C$.

When observations of only a subset of the components of $x_t$ are available, we need to go beyond the likelihood function for $\{x_t\}$. One approach uses filtering methods to build up the likelihood function for the subset of observed variables.[9] We describe the Kalman filter in chapter 4 and the appendix on filtering an control, chapter 21.[10]

## Concluding remarks

In addition to giving us tools for thinking about time series, the Markov chain and the stochastic linear difference equation have each introduced us to the notion

[9] See Hamilton (1994) or Hansen and Sargent (in press).

[10] See Hansen (1982), Eichenbaum (1991) Christiano and Eichenbaum (1992), Burnside, Eichenbaum, and Rebelo (1993), and Burnside and Eichenbaum (1996a, 1996b)for alternative estimation strategies.

of the state vector as a description of the present position of a system.[11] Subsequent chapters use both Markov chains and stochastic linear difference equations. In the next chapter we study decision problems in which the goal is optimally to manage the evolution of a state vector that can be partially controlled.

## *Appendix: A linear difference equation*

This appendix describes the solution of a linear first-order scalar difference equation. First, let $|\lambda| < 1$, and let $\{u_t\}_{t=-\infty}^{\infty}$ be a bounded sequence of scalar real numbers. Then

$$(1 - \lambda L) y_t = u_t, \forall t \tag{1.32}$$

has the solution

$$y_t = (1 - \lambda L)^{-1} u_t + k \lambda^t \tag{1.33}$$

for any real number $k$. You can verify this fact by applying $(1 - \lambda L)$ to both sides of equation (1.33) and noting that $(1 - \lambda L)\lambda^t = 0$. To pin down $k$ we need one condition imposed from outside (e.g., an initial or terminal condition) on the path of $y$.

Now let $|\lambda| > 1$. Rewrite equation (1.32) as

$$y_{t-1} = \lambda^{-1} y_t - \lambda^{-1} u_t, \forall t \tag{1.34}$$

or

$$(1 - \lambda^{-1} L^{-1}) y_t = -\lambda^{-1} u_{t+1}. \tag{1.35}$$

A solution is

$$y_t = -\lambda^{-1} \left( \frac{1}{1 - \lambda^{-1} L^{-1}} \right) u_{t+1} + k \lambda^t \tag{1.36}$$

for any $k$. To verify that this is a solution, check what operating on both sides of equation (1.36) by $(1 - \lambda L)$ does.

Solution (1.33) exists for $|\lambda| < 1$ because the distributed lag in $u$ converges. Solution (1.36) exists when $|\lambda| > 1$ because the distributed lead in $u$ converges. When $|\lambda| > 1$, the distributed lag in $u$ in (1.33) may diverge, so that a solution of this form does not exist. The distributed lead in $u$ in (1.36) need not converge when $|\lambda| < 1$.

[11] See Quah (1990) for an application of some of the tools of this chapter and of chapter 4 to studying some puzzles associated with a permanent income model.

## *Exercises*

*Exercise 1.1* Consider the Markov chain $(P, \pi_0) = \left( \begin{bmatrix} .9 & .1 \\ .3 & .7 \end{bmatrix}, \begin{bmatrix} .5 \\ .5 \end{bmatrix} \right)$, where the state space is $\bar{x} = \begin{bmatrix} 1 \\ 5 \end{bmatrix}$. Compute the likelihood of the following three histories for $t = 0, 1, \ldots, 4$:

**a.** $1, 5, 1, 5, 1$.

**b.** $1, 1, 1, 1, 1$.

**c.** $5, 5, 5, 5, 5$.

*Exercise 1.2* A Markov chain has state space $\bar{x} = \begin{bmatrix} 1 \\ 5 \end{bmatrix}$. It is known that $E(x_{t+1}|x_t = \bar{x}) = \begin{bmatrix} 1.8 \\ 3.4 \end{bmatrix}$ and that $E(x_{t+1}^2|x_t = \bar{x}) = \begin{bmatrix} 5.8 \\ 15.4 \end{bmatrix}$. Find a transition matrix consistent with these conditional expectations. Is this transition matrix unique (i.e., can you find another one that is consistent with these conditional expectations)?

*Exercise 1.3* Consumption is governed by an $n$ state Markov chain $P, \pi_0$ where $P$ is a stochastic matrix and $\pi_0$ is an initial probability distribution. Consumption takes one of the values in the $n \times 1$ vector $\bar{c}$. A consumer ranks stochastic processes of consumption $t = 0, 1 \ldots$ according to

$$E \sum_{t=0}^{\infty} \beta^t u(c_t)$$

where $E$ is the mathematical expectation and $u(c) = \frac{c^{1-\gamma}}{1-\gamma}$ for some parameter $\gamma \geq 1$. Let $u_i = u(\bar{c}_i)$. Let $v_i = E[\sum_{t=0}^{\infty} \beta^t u(c_t) | c_0 = \bar{c}_i]$ and $V = Ev$, where $\beta \in (0, 1)$ is a discount factor.

**a.** Let $u$ and $v$ be the $n \times 1$ vectors whose $i$th components are $u_i$ and $v_i$, respectively. Verify the following formulas for $v$ and $V$: $v = (I - \beta P)^{-1} u$, and $V = \sum_i \pi_{0,i} v_i$.

**b.** Consider the following two Markov processes:

Process 1: $\pi_0 = \begin{bmatrix} .5 \\ .5 \end{bmatrix}$, $P = \begin{bmatrix} 1 & 0 \\ 0 & 1 \end{bmatrix}$.

Process 2: $\pi_0 = \begin{bmatrix} .5 \\ .5 \end{bmatrix}$, $P = \begin{bmatrix} .5 & .5 \\ .5 & .5 \end{bmatrix}$.

For both Markov processes, $\bar{c} = \begin{bmatrix} 1 \\ 5 \end{bmatrix}$.

Assume that $\gamma = 2.5, \beta = .95$. Compute unconditional discounted expected utility $V$ for each of these processes. Which of the two processes does the consumer prefer? Redo the calculations for $\gamma = 4$. Now which process does the consumer prefer?

**c.** An econometrician observes a sample of 10 observations of consumption rates for our consumer. He knows that one of the two preceding Markov processes generates the data, but not which one. He assigns equal "prior probability" to the two chains. Suppose that the 10 successive observations on consumption are as follows: $1, 1, 1, 1, 1, 1, 1, 1, 1, 1$. Compute the likehood of this sample under process 1 and under process 2. Denote the likelihood function $\text{Prob}(\text{data}|\text{Model}_i), i = 1, 2$.

**d.** Suppose that the econometrician uses Bayes' law to revise his initial probability estimates for the two models, where in this context Bayes' law states:

$$\text{Prob}(\text{Model}_i|\text{data}) = \frac{\text{Prob}(\text{data}|\text{Model}_i) \cdot \text{Prob}(\text{Model}_i)}{\sum_j \text{Prob}(\text{data}|\text{Model}_j) \cdot \text{Prob}(\text{Model}_j)}.$$

The denominator of this expression is the unconditional probability of the data. After observing the data sample, what probabilities does the econometrician place on the two possible models?

**e.** Repeat the calculation in part d, but now assume that the data sample is $1, 5, 5, 1, 5, 5, 1, 5, 1, 5$.

*Exercise 1.4* Consider the univariate stochastic process

$$y_{t+1} = \alpha + \sum_{j=1}^{4} \rho_j y_{t+1-j} + c w_{t+1}$$

where $w_{t+1}$ is a scalar martingale difference sequence adapted to $J_t = [w_t, \ldots, w_1, y_0, y_{-1}, y_{-2}, y_{-3}]$, $\alpha = \mu(1 - \sum_j \rho_j)$ and the $\rho_j$'s are such that the matrix

$$A = \begin{bmatrix} \rho_1 & \rho_2 & \rho_3 & \rho_4 & \alpha \\ 1 & 0 & 0 & 0 & 0 \\ 0 & 1 & 0 & 0 & 0 \\ 0 & 0 & 1 & 0 & 0 \\ 0 & 0 & 0 & 0 & 1 \end{bmatrix}$$

has all of its eigenvalues in modulus bounded below unity.

**a.** Show how to map this process into a first-order linear stochastic difference equation.

**b.** For each of the following examples, if possible, assume that the initial conditions are such that $y_t$ is covariance stationary. For each case, state the appropriate initial conditions. Then compute the covariance stationary mean and variance of $y_t$ assuming the following parameter sets of parameter values:

*i.* $\rho = [\,1.2 \quad -.3 \quad 0 \quad 0\,]$, $\mu = 10, c = 1$.

*ii.* $\rho = [\,1.2 \quad -.3 \quad 0 \quad 0\,]$, $\mu = 10, c = 2$.

*iii.* $\rho = [\,.9 \quad 0 \quad 0 \quad 0\,]$, $\mu = 5, c = 1$.

*iv.* $\rho = [\,.2 \quad 0 \quad 0 \quad .5\,]$, $\mu = 5, c = 1$.

*v.* $\rho = [\,.8 \quad .3 \quad 0 \quad 0\,]$, $\mu = 5, c = 1$.

*Hint 1:* The Matlab program `doublej.m`, in particular, the command `X=doublej(A,C*C')` computes the solution of the matrix equation $A'XA + C'C = X$. This program can be downloaded from `ftp://zia.stanford.edu/pub/sargent/webdocs/matlab`.

*Hint 2:* The mean vector is the eigenvector of $A$ associated with a unit eigenvalue, scaled so that the mean of unity in the state vector is unity.

**c.** For each case in part b, compute the $h_j$'s in $E_t y_{t+5} = \gamma_0 + \sum_{j=0}^{3} h_j y_{t-j}$.

**d.** For each case in part b, compute the $\tilde{h}_j$'s in $E_t \sum_{k=0}^{\infty} .95^k y_{t+k} = \sum_{j=0}^{3} \tilde{h}_j y_{t-j}$.

**d.** For each case in part b, compute the autocovariance $E(y_t - \mu_y)(y_{t-k} - \mu_y)$ for the three values $k = 1, 5, 10$.

*Exercise 1.5* A consumer's rate of consumption follows the stochastic process

$$
(1) \quad
\begin{aligned}
c_{t+1} &= \alpha_c + \sum_{j=1}^{2} \rho_j c_{t-j+1} + w_{t+1} + \sum_{j=1}^{2} \delta_j z_{t+1-j} + \psi_1 w_{1,t+1} \\
z_{t+j} &= \sum_{j=1}^{2} \gamma_j c_{t-j+1} + \sum_{j=1}^{2} \phi_j z_{t-j+1} + \psi_2 w_{2,t+1}
\end{aligned}
$$

where $w_{t+1}$ is a $2 \times 1$ martingale difference sequence, adapted to $J_t = [\,w_t \quad \ldots w_1 \quad c_0 \quad c_{-1} \quad z_0 \quad z_{-1}\,]$, with contemporaneous covariance matrix $E w_{t+1} w'_{t+1} | J_t = I$, and the coefficients $\rho_j, \delta_j, \gamma_j, \phi_j$ are such that the matrix

$$
A = \begin{bmatrix} \rho_1 & \rho_2 & \delta_1 & \delta_2 & \alpha_c \\ 1 & 0 & 0 & 0 & 0 \\ \gamma_1 & \gamma_2 & \phi_1 & \phi_2 & 0 \\ 0 & 0 & 1 & 0 & 0 \\ 0 & 0 & 0 & 0 & 1 \end{bmatrix}
$$

has eigenvalues bounded strictly below unity in modulus.

The consumer evaluates consumption streams according to

$$
(2) \qquad V_0 = E_0 \sum_{t=0}^{\infty} .95^t u(c_t),
$$

where the one-period utility function is

$$
(3) \qquad u(c_t) = -.5(c_t - 60)^2.
$$

**a.** Find a formula for $V_0$ in terms of the parameters of the one-period utility function (3) and the stochastic process for consumption.

**b.** Compute $V$ for the following two sets of parameter values:

*i.* $\rho = [\,.8 \quad -.3\,], \alpha_c = 1, \delta = [\,.2 \quad 0\,], \gamma = [\,0 \quad 0\,], \phi = [\,.7 \quad -.2\,],\ \psi_1 = \psi_2 = 1\,.$

*ii.* Same as for part i except now $\psi_1 = 2, \psi_2 = 1\,.$

*Hint:* Remember `doublej.m`.

*Exercise 1.6* Consider the stochastic process $\{c_t, z_t\}$ defined by equations (1) in exercise 1.5. Assume the parameter values described in part b, item i. If possible, assume the initial conditions are such that $\{c_t, z_t\}$ is covariance stationary.

**a.** Compute the initial mean and covariance matrix that make the process covariance stationary.

**b.** For the initial conditions in part a, compute numerical values of the following population linear regression:

$$c_{t+2} = \alpha_0 + \alpha_1 z_t + \alpha_2 z_{t-4} + \epsilon_t$$

where $E\epsilon_t \begin{bmatrix} 1 & z_t & z_{t-4} \end{bmatrix} = \begin{bmatrix} 0 & 0 & 0 \end{bmatrix}$.

*Exercise 1.7* Get the Matlab programs `bigshow.m` and `freq.m` from `ftp://zia.stanford.edu/pub/sargent/webdocs/matlab`.
Use `bigshow` to compute and display a simulation of length 80, an impulse response function, and a spectrum for each of the following scalar stochastic processes $y_t$. In each of the following, $w_t$ is a scalar martingale difference sequence adapted to its own history and the initial values of lagged $y$'s.

a. $y_t = w_t$.

b. $y_t = (1 + .5L) w_t$.

c. $y_t = (1 + .5L + .4L^2) w_t$.

d. $(1 - .999L) y_t = (1 - .4L) w_t$.

e. $(1 - .8L) y_t = (1 + .5L + .4L^2) w_t$.

f. $(1 + .8L) y_t = w_t$.

g. $y_t = (1 - .6L) w_t$.

Study the output and look for patterns. When you are done, you will be well on your way to knowing how to read spectral densities.

*Exercise 1.8* This exercise deals with Cagan's money demand under rational expectations. A version of Cagan's (1956) demand function for money is

$$m_t - p_t = -\alpha(p_{t+1} - p_t), \alpha > 0, \ t \geq 0, \tag{1}$$

where $m_t$ is the log of the nominal money supply and $p_t$ is the price level at $t$. Equation (1) states that the demand for real balances varies inversely with the expected rate of inflation, $(p_{t+1} - p_t)$. There is no uncertainty, so the expected inflation rate equals the actual one. The money supply obeys the difference equation

$$(1 - L)(1 - \rho L)m_t^s = 0 \tag{2}$$

subject to initial condition for $m_{-1}^s, m_{-2}^s$. In equilibrium,

$$m_t \equiv m_t^s \quad \forall t \geq 0 \tag{3}$$

(i.e., the demand for money equals the supply). For now assume that

$$|\rho\alpha/(1 + \alpha)| < 1. \tag{4}$$

An *equilibrium* is a $\{p_t\}_{t=0}^{\infty}$ that satisfies equations (1), (2), and (3) for all $t$.

**a.** Find an expression an equilibrium $p_t$ of the form

$$p_t = \sum_{j=0}^{n} w_j m_{t-j} + f_t. \tag{5}$$

Please tell how to get formulas for the $w_j$ for all $j$ and the $f_t$ for all $t$.

**b.** How many equilibria are there?

**c.** Is there an equilibrium with $f_t = 0$ for all $t$?

**d.** Briefly tell where, if anywhere, condition (4) plays a role in your answer to part a.

**e.** For the parameter values $\alpha = 1, \rho = 1$, compute and display all the equilibria.

*Exercise 1.9* The $n \times 1$ state vector of an economy is governed by the linear stochastic difference equation

$$x_{t+1} = Ax_t + C_t w_{t+1} \tag{1}$$

where $C_t$ is a possibly time varying matrix (known at $t$) and $w_{t+1}$ is an $m \times 1$ martingale difference sequence adapted to its own history with $E w_{t+1} w_{t+1}' | J_t = I$, where $J_t = [\, w_t \quad \ldots \quad w_1 \quad x_0 \,]$. A scalar one-period payoff $p_{t+1}$ is given by

$$p_{t+1} = P x_{t+1} \tag{2}$$

The stochastic discount factor for this economy is a scalar $m_{t+1}$ that obeys

$$m_{t+1} = \frac{M x_{t+1}}{M x_t}. \tag{3}$$

Finally, the price at time $t$ of the one-period payoff is given by $q_t = f_t(x_t)$, where $f_t$ is some possibly time-varying function of the state. That $m_{t+1}$ is a stochastic discount factor means that

$$E(m_{t+1} p_{t+1} | J_t) = q_t. \tag{4}$$

**a.** Compute $f_t(x_t)$, describing in detail how it depends on $A$ and $C_t$.

**b.** Suppose that an econometrician has a time series data set $X_t = [\, z_t \quad m_{t+1} \quad p_{t+1} \quad q_t \,]$, for $t = 1, \ldots, T$, where $z_t$ is a strict subset of the variables in the state $x_t$. Assume that investors in the economy see $x_t$ even though the econometrician only sees a subset $z_t$ of $x_t$. Briefly describe a way to use these data to test implication (4). (Possibly but perhaps not useful hint: recall the law of iterated expectations.)

# 2
# *Dynamic Programming*

This chapter introduces basic ideas and methods of dynamic programming.[1] It sets out the basic elements of a recursive optimization problem, describes the functional equation (the Bellman equation), presents three methods for solving the Bellman equation, and gives the Benveniste-Scheinkman formula for the derivative of the optimal value function. Let's dive in.

## *Sequential problems*

Let $\beta \in (0,1)$ be a discount factor. ¡sequential: recursive¿ We want to choose an infinite sequence of "controls" $\{u_t\}_{t=0}^{\infty}$ to maximize

$$\sum_{t=0}^{\infty} \beta^t r(x_t, u_t), \tag{2.1}$$

subject to $x_{t+1} = g(x_t, u_t)$, with $x_0$ given. We assume that $r_t(x_t, u_t)$ is a concave function and that the set $\{(x_{t+1}, x_t) : x_{t+1} \leq g_t(x_t, u_t), u_t \in R^k\}$ is convex and compact. Dynamic programming seeks a time-invariant *policy function* $h$ mapping the *state* $x_t$ into the control $u_t$, such that the sequence $\{u_s\}_{s=0}^{\infty}$ generated by iterating the two functions

$$\begin{aligned} u_t &= h(x_t) \\ x_{t+1} &= g(x_t, u_t), \end{aligned} \tag{2.2}$$

starting from initial condition $x_0$ at $t = 0$ solves the original problem. A solution in the form of equations (2.2) is said to be *recursive.* To find the policy function

[1] This chapter is written in the hope of getting the reader to start using the methods quickly. We hope to promote demand for further and more rigorous study of the subject. In particular see Bertsekas (1976), Bertsekas and Shreve (1978), Stokey and Lucas (with Prescott) (1989), Bellman (1957), and Chow (1981). This chapter covers much of the same material as Sargent (1987b, chapter 1).

$h$ we need to know another function $V(x)$ that expresses the optimal value of the original problem, starting from an arbitrary initial condition $x \in X$. This is called the *value function.* In particular, define

$$V(x_0) = \max_{\{u_s\}_{s=0}^{\infty}} \sum_{t=0}^{\infty} \beta^t r(x_t, u_t), \tag{2.3}$$

where again the maximization is subject to $x_{t+1} = g(x_t, u_t)$, with $x_0$ given. Of course, we cannot possibly expect to know $V(x_0)$ until after we have solved the problem, but let's proceed on faith. If we knew $V(x_0)$, then the policy function $h$ could be computed by solving for each $x \in X$ the problem

$$\max_u \{r(x, u) + \beta V(\tilde{x})\}, \tag{2.4}$$

where the maximization is subject to $\tilde{x} = g(x, u)$, with $x$ given. Thus, we have exchanged the original problem of finding an infinite *sequence* of controls that maximizes expression (2.1) for the problem of finding the optimal value function $V(x)$ and a function $h$ that solves the continuum of maximum problems (2.4)—one maximum problem for each value of $x$. This exchange doesn't look like progress, but we shall see that it often is.

Our task has become jointly to solve for $V(x), h(x)$, which are linked by the *Bellman equation*

$$V(x) = \max_u \{r(x, u) + \beta V[g(x, u)]\}. \tag{2.5}$$

The maximizer of the right side of equation (2.5) is a *policy function* $h(x)$ that satisfies

$$V(x) = r[x, h(x)] + \beta V\{g[x, h(x)]\}. \tag{2.6}$$

Equation (2.5) or (2.6) is a *functional equation* to be solved for the pair of unknown functions $V(x), h(x)$.

Methods for solving the Bellman equation are based on mathematical structures that vary in their details depending on the precise nature of the functions

$r$ and $g$.[2] All of these structures contain versions of the following four findings. Under various particular assumptions about $r$ and $g$, it turns out that

1. The functional equation (2.5) has a unique strictly concave solution.

2. This solution is approached in the limit as $j \to \infty$ by iterations on

$$V_{j+1}(x) = \max_u \{r(x,u) + \beta V_j(\tilde{x})\}, \tag{2.7}$$

subject to $\tilde{x} = g(x,u), x$ given, starting from any bounded and continuous initial $V_0$.

3. There is a unique and time invariant optimal policy of the form $u_t = h(x_t)$, where $h$ is chosen to maximize the right side of (2.5).[3]

4. Off corners, the limiting value function $V$ is differentiable with

$$V'(x) = \frac{\partial r}{\partial x}[x, h(x)] + \beta \frac{\partial g}{\partial x}[x, h(x)] V'\{g[x, h(x)]\}. \tag{2.8}$$

This is a version of a formula of Benveniste and Scheinkman (1979). We often encounter settings in which the transition law can be formulated so that the state $x$ does not appear in it, so that $\frac{\partial g}{\partial x} = 0$, which makes equation (2.8) become

$$V'(x) = \frac{\partial r}{\partial x}[x, h(x)]. \tag{2.9}$$

At this point, we describe three broad computational strategies that apply in various contexts.

---

[2] There are alternative sets of conditions that make the maximization (2.4) well behaved. One set of conditions is as follows: (1) $r$ is concave and bounded, and (2) the constraint set generated by $g$ is convex and compact, that is, the set of $\{(x_{t+1}, x_t) : x_{t+1} \leq g(x_t, u_t)\}$ for admissible $u_t$ is convex and compact. See Stokey, Lucas, and Prescott (1989), and Bertsekas (1976) for further details of convergence results. See Benveniste and Scheinkman (1979) and Stokey, Lucas, and Prescott (1989) for the results on differentiability of the value function. In an appendix on functional analysis, chapter 20, we describe the mathematics for one standard set of assumptions about $(r, g)$. In chapter 4, we describe it for another set of assumptions about $(r, g)$.

[3] The time invariance of the policy function $u_t = h(x_t)$ is very convenient econometrically, because we can impose a single decision rule for all periods. This lets us pool data across period to estimate the free parameters of the return and transition functions that underlie the decision rule.

*Three computational methods*

There are three main types of computational methods for solving dynamic programs. All aim to solve the functional equation (2.4).

**Value function iteration.** The first method proceeds by constructing a sequence of value functions and associated policy functions. The sequence is created by iterating on the following equation, starting from $V_0 = 0$, and continuing until $V_j$ has converged:[4]

$$V_{j+1}(x) = \max_u \{r(x,u) + \beta V_j(\tilde{x})\}, \tag{2.10}$$

subject to $\tilde{x} = g(x,u), x$ given.[5] This method is called *value function iteration* or *iterating on the Bellman equation.*

**Guess and verify.** A second method involves guessing and verifying a solution $V$ to equation (2.5). This method relies on the uniqueness of the solution to the equation, but because it relies on luck in making a good guess, it is not generally available.

**Howard's improvement algorithm.** A third method, known as *policy function iteration* or *Howard's improvement algorithm*, consists of the following steps:

1. Pick a feasible policy, $u = h_0(x)$, and compute the value associated with operating forever with that policy:

$$V_{h_j}(x) = \sum_{t=0}^{\infty} \beta^t r[x_t, h_j(x_t)],$$

   where $x_{t+1} = g[x_t, h_j(x_t)]$, with $j = 0$.
2. Generate a new policy $u = h_{j+1}(x)$ that solves the two-period problem

$$\max_u \{r(x,u) + \beta V_{h_j}[g(x,u)]\},$$

   for each $x$.
3. Iterate over $j$ to convergence on steps 1 and 2.

---

[4] See the appendix on functional analysis for what it means for a sequence of functions to converge.

[5] A proof of the uniform convergence of iterations on equation (2.10) is contained in the appendix on functional analysis, chapter 20.

In the appendix on functional analysis, chapter 20, we describe some conditions under which the improvement algorithm converges to the solution of Bellman's equation. The method often converges faster than does value function iteration (e.g., see exercise 2.1 at the end of this chapter).[6] The policy improvement algorithm is also a building block for the methods for studying government policy to be described in chapter 16.

Each of these methods has its uses. Each is "easier said than done," because it is typically impossible analytically to compute even *one* iteration on equation (2.10). This fact thrusts us into the domain of computational methods for approximating solutions: pencil and paper are insufficient. The following chapter describes some computational methods that can be used for problems that cannot be solved by hand. Here we shall describe the first of two special types of problems for which analytical solutions *can* be obtained. It involves Cobb-Douglas constraints and logarithmic preferences. Later in chapter 4, we shall describe a specification with linear constraints and quadratic preferences. For that special case, many analytic results are available. These two classes have been important in economics as sources of examples and as inspirations for approximations.

### *Cobb-Douglas transition, logarithmic preferences*

Brock and Mirman (1972) used the following optimal growth example.[7] A planner chooses sequences $\{c_t, k_{t+1}\}_{t=0}^{\infty}$ to maximize

$$\sum_{t=0}^{\infty} \beta^t \ln(c_t)$$

subject to a given value for $k_0$ and a transition law

$$k_{t+1} + c_t = A k_t^{\alpha}, \tag{2.11}$$

where $A > 0, \alpha \in (0,1), \beta \in (0,1)$.

This problem can be solved "by hand," using any of our three methods. We begin with iteration on the Bellman equation. Start with $v_0(k) = 0$, and solve the one-period problem: choose $c$ to maximize $\ln(c)$ subject to $c + \tilde{k} = Ak^{\alpha}$. The solution is evidently to set $c = Ak^{\alpha}, \tilde{k} = 0$, which produces an optimized

[6] The quickness of the policy improvement algorithm is linked to its being an implementation of Newton's method, which converges quadratically while iteration on the Bellman equation converges at a linear rate. See chapter 3 and the appendix on functional analysis, chapter 20.

[7] See also Levhari and Srinivasan (1969).

value $v_1(k) = \ln A + \alpha \ln k$. At the second step, we find $c = \frac{1}{1+\beta\alpha} A k^\alpha, \tilde{k} = \frac{\beta\alpha}{1+\beta\alpha} A k^\alpha, v_2(k) = \ln \frac{A}{1+\alpha\beta} + \beta \ln A + \alpha\beta \ln \frac{\alpha\beta A}{1+\alpha\beta} + \alpha(1+\alpha\beta) \ln k$. Continuing, and using the algebra of geometric series, gives the limiting policy functions $c = (1-\beta\alpha) A k^\alpha, \tilde{k} = \beta\alpha A k^\alpha$, and the value function $v(k) = (1-\beta)^{-1}\{\ln[A(1-\beta\alpha)] + \frac{\beta\alpha}{1-\beta\alpha} \ln(A\beta\alpha)\} + \frac{\alpha}{1-\beta\alpha} \ln k$.

Here is how the guess-and-verify method applies to this problem. Since we already know the answer, we'll guess a function of the correct form, but leave its coefficients undetermined.[8] Thus, we make the guess

$$v(k) = E + F \ln k, \tag{2.12}$$

where $E$ and $F$ are undetermined constants. The left and right sides of equation (2.12) must agree for all values of $k$. For this guess, the first-order necessary condition for the maximum problem on the right side of equation (2.10) implies the following formula for the optimal policy $\tilde{k} = h(k)$, where $\tilde{k}$ is next period's value and $k$ is this period's value of the capital stock:

$$\tilde{k} = \frac{\beta F}{1+\beta F} A k^\alpha. \tag{2.13}$$

Substitute equation (2.13) into the Bellman equation and equate the result to the right side of equation (2.12). Solving the resulting equation for $E$ and $F$ gives $F = \alpha/(1-\alpha\beta)$ and $E = (1-\beta)^{-1}[\ln A(1-\alpha\beta) + \frac{\beta\alpha}{1-\alpha\beta} \ln A\beta\alpha]$. It follows that

$$\tilde{k} = \beta\alpha A k^\alpha. \tag{2.14}$$

Note that the term $F = \alpha/(1-\alpha\beta)$ can be interpreted as a geometric sum $\alpha[1 + \alpha\beta + (\alpha\beta)^2 + \ldots]$.

Equation (2.14) shows that the optimal policy is to have capital move according to the difference equation $k_{t+1} = A\beta\alpha k_t^\alpha$, or $\ln k_{t+1} = \ln A\beta\alpha + \alpha \ln k_t$. That $\alpha$ is less than 1 implies that $k_t$ converges as $t$ approaches infinity for any positive initial value $k_0$. The stationary point is given by the solution of $k_\infty = A\beta\alpha k_\infty^\alpha$, or $k_\infty^{\alpha-1} = (A\beta\alpha)^{-1}$.

---

[8] This is called the method of undetermined coefficients.

*Euler equations*

In many problems, there is no unique way of defining states and controls, and several alternative definitions lead to the same solution of the problem. Sometimes the states and controls can be defined in such a way that $x_t$ does not appear in the transition equation, so that $\partial g_t / \partial x_t \equiv 0$. In this case, the first-order condition for the problem on the right side of the Bellman equation in conjunction with the Benveniste-Scheinkman formula implies

$$\frac{\partial r_t}{\partial u_t}(x_t, u_t) + \frac{\partial g_t}{\partial u_t}(u_t) \cdot \frac{\partial r_{t+1}(x_{t+1}, u_{t+1})}{\partial x_{t+1}} = 0, \qquad x_{t+1} = g_t(u_t).$$

The first equation is called an *Euler equation.* Under circumstances in which the second equation can be inverted to yield $u_t$ as a function of $x_{t+1}$, using the second equation to eliminate $u_t$ from the first equation produces a second-order difference equation in $x_t$, since eliminating $u_{t+1}$ brings in $x_{t+2}$.

*A sample Euler equation*

As an example of an Euler equation, consider the Ramsey problem of choosing $\{c_t, k_{t+1}\}_{t=0}^{\infty}$ to maximize $\sum_{t=0}^{\infty} \beta^t u(c_t)$ subject to $c_t + k_{t+1} = f(k_t)$, where $k_0$ is given and the one-period utility function satisfies $u'(c) > 0, u''(c) < 0, \lim_{c_t \searrow 0} u'(c_t) = \infty$; and where $f'(k) > 0, f''(k) < 0$. Let the state be $k$ and the control be $k'$, where $k'$ denotes next period's value of $k$. Substitute $c = f(k) - k'$ into the utility function and express the Bellman equation as

$$v(k) = \max_{k'} \{u[f(k) - k'] + \beta v(k')\}. \tag{2.15}$$

Application of the Benveniste-Scheinkman formula gives

$$v'(k) = u'[f(k) - k'] f'(k). \tag{2.16}$$

Notice that the first-order condition for the maximum problem on the right side of equation (2.15) is $-u'[f(k) - k'] + \beta v'(k') = 0$, which, using equation (2.16), gives

$$u'[f(k) - k'] = \beta u'[f(k') - k''] f'(k'), \tag{2.17}$$

where $k''$ denotes the "two-period-ahead" value of $k$. Equation (2.17) can be expressed as

$$1 = \beta \frac{u'(c_{t+1})}{u'(c_t)} f'(k_{t+1}),$$

an Euler equation that is exploited extensively in the theories of finance, growth, and real business cycles.

## Stochastic control problems

We now consider a modification of problem (2.1) to permit uncertainty. Essentially, we add some well-placed shocks to the previous non-stochastic problem. So long as the shocks are either independently and identically distributed or Markov, straightforward modifications of the method for handling the nonstochastic problem will work.

Thus, we modify the transition equation and consider the problem of maximizing

$$E_0 \sum_{t=0}^{\infty} \beta^t r(x_t, u_t), \qquad 0 < \beta < 1, \tag{2.18}$$

subject to

$$x_{t+1} = g(x_t, u_t, \epsilon_{t+1}), \tag{2.19}$$

with $x_0$ known and given at $t = 0$, where $\epsilon_t$ is a sequence of independently and identically distributed random variables with cumulative probability distribution function $\text{prob}\{\epsilon_t \leq e\} = F(e)$ for all $t$; $E_t(y)$ denotes the mathematical expectation of a random variable $y$, given information known at $t$. At time $t$, $x_t$ is assumed to be known, but $x_{t+j}, j \geq 1$ is not known at $t$. That is, $\epsilon_{t+1}$ is realized at $(t+1)$, after $u_t$ has been chosen at $t$. In problem (2.18)–(2.19), uncertainty is injected by assuming that $x_t$ follows a random difference equation.

Problem (2.18)–(2.19) continues to have a recursive structure, stemming jointly from the additive separability of the objective function (2.18) in pairs $(x_t, u_t)$ and from the difference equation characterization of the transition law (2.19). In particular, controls dated $t$ affect returns $r(x_s, u_s)$ for $s \geq t$ but not earlier. This feature implies that dynamic programming methods remain appropriate.

The problem is to maximize expression (2.18) subject to equation (2.19) by choice of a "policy" or "contingency plan" $u_t = h(x_t)$. The Bellman equation (2.5) becomes

$$V(x) = \max_u \{r(x, u) + \beta E[V[g(x, u, \epsilon)] | x]\}, \tag{2.20}$$

where $E\{V[g(x, u, \epsilon)] | x\} = \int V[g(x, u, \epsilon)] dF(\epsilon)$ and where $V(x)$ is the optimal value of the problem starting from $x$ at $t = 0$. The solution $V(x)$ of equation (2.20) can be computed by iterating on

$$V_{j+1}(x) = \max_u \{r(x, u) + \beta E[V_j[g(x, u, \epsilon)] | x]\}, \tag{2.21}$$

starting from any bounded continuous initial $V_0$. Under various particular regularity conditions, there obtain versions of the same four properties listed earlier.[9]

The first-order necessary condition for the problem on the right side of equation (2.20) is

$$\frac{\partial r(x,u)}{\partial u} + \beta E \left[ \frac{\partial g}{\partial u}(x,u,\epsilon) V'[g(x,u,\epsilon)] | x \right] = 0,$$

which we obtained simply by differentiating the right side of equation (2.20), passing the differentiation operation under the $E$ (an integration) operator. Off corners, the value function satisfies

$$V'(x) = \frac{\partial r}{\partial x}[x, h(x)] + \beta E \left\{ \frac{\partial g}{\partial x}[x, h(x), \epsilon] V'(g[x, h(x), \epsilon]) | x \right\}.$$

In the special case in which $\partial g / \partial x \equiv 0$, the formula for $V'(x)$ becomes

$$V'(x) = \frac{\partial r}{\partial x}[x, h(x)].$$

Substituting this formula into the first-order necessary condition for the problem gives the stochastic Euler equation

$$\frac{\partial r}{\partial u}(x,u) + \beta E \left[ \frac{\partial g}{\partial u}(x,u,\epsilon) \frac{\partial r}{\partial x}(\tilde{x}, \tilde{u}) | x \right] = 0,$$

where tildes over $x$ and $u$ denote next-period values.

## Concluding remarks

This chapter has put forward basic tools and findings: the Bellman equation and several approaches to solving it; the Euler equation; and the Beneveniste-Scheinkman formula. To appreciate and believe in the power of these tools requires more words and more practice than we have yet supplied. In the next several chapters, we put the basic tools to work in different contexts with particular specification of return and transition equations designed to render the Bellman equation susceptible to further analysis and computation.

---

[9] See Stokey and Lucas (with Prescott) (1989), or the framework presented in the appendix on functional analysis, chapter 20.

## Exercise

*Exercise 2.1* **Howard's policy iteration algorithm**

Consider the Brock-Mirman problem: to maximize

$$E_0 \sum_{t=0}^{\infty} \beta^t \ln c_t,$$

subject to $c_t + k_{t+1} \leq A k_t^\alpha \theta_t$, $k_0$ given, $A > 0$, $1 > \alpha > 0$, where $\{\theta_t\}$ is an i.i.d. sequence with $\ln \theta_t$ distributed according to a normal distribution with mean zero and variance $\sigma^2$.

Consider the following algorithm. Guess at a policy of the form $k_{t+1} = h_0(A k_t^\alpha \theta_t)$ for any constant $h_0 \in (0,1)$. Then form

$$J_0(k_0, \theta_0) = E_0 \sum_{t=0}^{\infty} \beta^t \ln(A k_t^\alpha \theta_t - h_0 A k_t^\alpha \theta_t).$$

Next choose a new policy $h_1$ by maximizing

$$\ln(A k^\alpha \theta - k') + \beta E J_0(k', \theta'),$$

where $k' = h_1 A k^\alpha \theta$. Then form

$$J_1(k_0, \theta_0) = E_0 \sum_{t=0}^{\infty} \beta^t \ln(A k_t^\alpha \theta_t - h_1 A k_t^\alpha \theta_t).$$

Continue iterating on this scheme until successive $h_j$ have converged.

Show that, for the present example, this algorithm converges to the optimal policy function in one step.

# 3
# *Practical Dynamic Programming*

## *The curse of dimensionality*

We often encounter problems where it is impossible to attain closed forms for iterating on the Bellman equation. Then we have to adopt some numerical approximations. This chapter describes two popular methods for obtaining numerical approximations. The first method replaces the original problem with another problem by forcing the state vector to live on a finite and discrete grid of points, then applies discrete-state dynamic programming to this problem. The "curse of dimensionality" impels us to keep the number of points in the discrete state space small. The second approach uses polynomials to approximate the value function. Judd (1998) is a comprehensive reference about numerical analysis of dynamic economic models and contains many insights about ways to compute dynamic models.

## *Discretization of state space*

We introduce the method of discretization of the state space in the context of a particular discrete-state version of an optimal saving problem. An infinitely lived household likes to consume one good, which it can acquire by using labor income or accumulated savings. The household has an endowment of labor at time $t$, $s_t$, that evolves according to an $m$-state Markov chain with transition matrix $\mathcal{P}$. If the realization of the process at $t$ is $\bar{s}_i$, then at time $t$ the household receives labor income of amount $w\bar{s}_i$. The wage $w$ is fixed over time. We shall sometimes assume that $m$ is $2$, and that $s_t$ takes on value 0 in an unemployed state and 1 in an employed state. In this case, $w$ has the interpretation of being the wage of employed workers.

The household can choose to hold a single asset in discrete amount $a_t \in \mathcal{A}$ where $\mathcal{A}$ is a grid $[a_1 < a_2 < \ldots < a_n]$. How the model builder chooses the end points of the grid $\mathcal{A}$ is important, as we describe in detail in chapter 14 on incomplete market models. The asset bears a gross rate of return $r$ that is fixed over time.

The household's maximum problem, for given values of $(w, r)$ and given initial values $(a_0, s_0)$, is to choose a policy for $\{a_{t+1}\}_{t=0}^{\infty}$ to maximize

$$E\sum_{t=0}^{\infty}\beta^t u(c_t), \tag{3.1}$$

subject to

$$\begin{aligned} c_t + a_{t+1} &= (r+1)a_t + ws_t \\ c_t &\geq 0 \\ a_{t+1} &\in \mathcal{A} \end{aligned} \tag{3.2}$$

where $\beta \in (0,1)$ is a discount factor and $r$ is fixed rate of return on the assets. We assume that $\beta(1+r) < 1$. Here $u(c)$ is a strictly increasing, concave one-period utility function. Associated with this problem is the Bellman equation

$$v(a,s) = \max_{a' \in \mathcal{A}}\{u[(r+1)a + ws - a'] + \beta E v(a', s')|s\},$$

or for each $i \in [1, \ldots, m]$ and each $h \in [1, \ldots, n]$,

$$v(a_h, \bar{s}_i) = \max_{a' \in \mathcal{A}}\{u[(r+1)a_h + w\bar{s}_i - a'] + \beta\sum_{j=1}^{m}\mathcal{P}_{ij}v(a', \bar{s}_j)\}, \tag{3.3}$$

where $a'$ is next period's value of asset holdings, and $s'$ is next period's value of the shock; here $v(a,s)$ is the optimal value of the objective function, starting from asset, employment state $(a,s)$. A solution of this problem is a value function $v(a,s)$ that satisfies equation (3.3) and an associated policy function $a' = g(a,s)$ mapping this period's $(a,s)$ pair into an optimal choice of assets to carry into next period.

## *Discrete-state dynamic programming*

For discrete-state space of small size, it is easy to solve the Bellman equation numerically by manipulating matrices. Here is how to write a computer program to iterate on the Bellman equation in the context of the preceding model of asset accumulation.[1] Let there be $n$ states $[a_1, a_2, \ldots, a_n]$ for assets and two states $[s_1, s_2]$ for employment status. Define two $n \times 1$ vectors $v_j, j = 1, 2$, whose $i$th rows are determined by $v_j(i) = v(a_i, s_j), i = 1, \ldots, n$. Let $\mathbf{1}$ be the $n \times 1$ vector consisting entirely of ones. Define two $n \times n$ matrices $R_j$ whose $(i, h)$ element is

$$R_j(i, h) = u[(r+1)a_i + ws_j - a_h], \quad i = 1, \ldots, n, h = 1, \ldots, n.$$

Define an operator $T([v_1, v_2])$ that maps a pair of vectors $[v_1, v_2]$ into a pair of vectors $[tv_1, tv_2]$:[2]

$$\begin{aligned} tv_1 &= \max\{R_1 + \beta\mathcal{P}_{11}\mathbf{1}v_1' + \beta\mathcal{P}_{12}\mathbf{1}v_2'\} \\ tv_2 &= \max\{R_2 + \beta\mathcal{P}_{21}\mathbf{1}v_1' + \beta\mathcal{P}_{22}\mathbf{1}v_2'\}. \end{aligned} \tag{3.4}$$

Here it is understood that the "max" operator applied to an $(n \times m)$ matrix $M$ returns an $(n \times 1)$ vector whose $i$th element is the maximum of the $i$th row of the matrix $M$. These two equations can be written compactly as

$$\begin{bmatrix} tv_1 \\ tv_2 \end{bmatrix} = \max\left\{ \begin{bmatrix} R_1 \\ R_2 \end{bmatrix} + \beta(\mathcal{P} \otimes \mathbf{1}) \begin{bmatrix} v_1' \\ v_2' \end{bmatrix} \right\}, \tag{3.5}$$

where $\otimes$ is the Kronecker product.

The Bellman equation can be represented

$$[v_1 v_2] = T([v_1, v_2]),$$

and can be solved by iterating to convergence on

$$[v_1, v_2]_{m+1} = T([v_1, v_2]_m).$$

[1] Matlab versions of the program have been written by Gary Hansen, Selahattin İmrohoroğlu, George Hall, and Chao Wei.

[2] Programming languages like Gauss and Matlab execute maximum operations over vectors very efficiently. For example, for an $n \times m$ matrix $A$, the Matlab command `[r,index] =max(A)` returns the two $(1 \times m)$ row vectors `r,index`, where $r_j = \max_i A(i,j)$ and $\text{index}_j$ is the row $i$ that attains $\max_i A(i,j)$ for column $j$ [i.e., $\text{index}_j = \text{argmax}_i A(i,j)$]. This command performs $m$ maximizations simultaneously.

## Application of Howard improvement algorithm

Often computation speed is important. We saw in an exercise in chapter 1 that the policy improvement algorithm can be much faster than iterating on the Bellman equation. It is also easy to implement the Howard improvement algorithm in the present setting. At time $t$, the system resides in one of $N$ predetermined positions, denoted $x_i$ for $i = 1, 2, \ldots, N$. There exists a predetermined class $\mathcal{M}$ of $(N \times N)$ stochastic matrices $P$, which are the objects of choice. Here $P_{ij} = \text{Prob}\,[x_{t+1} = x_j \mid x_t = x_i]$, $i = 1, \ldots, N$; $j = 1, \ldots, N$.

The matrices $P$ satisfy $P_{ij} \geq 0$, $\sum_{j=1}^N P_{ij} = 1$, and additional restrictions dictated by the problem at hand that determine the class $\mathcal{M}$. The one-period return function is represented as $c_P$, a vector of length $N$, and is a function of $P$. The $i$th entry of $c_P$ denotes the one-period return when the state of the system is $x_i$ and the transition matrix is $P$. The Bellman equation is

$$v_P(x_i) = \max_{P \in \mathcal{M}} \{c_P(x_i) + \beta \sum_{j=1}^N P_{ij}\, v_P(x_j)\}$$

or

$$v_P = \max_{P \in \mathcal{M}} \{c_P + \beta P v_P\}\ . \tag{3.6}$$

We can express this as

$$v_P = T v_P\ ,$$

where $T$ is the operator defined by the right side of (3.6). Following Putterman and Brumelle (1979) and Putterman and Shin (1978), define the operator

$$B = T - I,$$

so that

$$Bv = \max_{P \in \mathcal{M}} \{c_P + \beta P v\} - v.$$

In terms of the operator $B$, the Bellman equation is

$$Bv = 0. \tag{3.7}$$

The policy improvement algorithm consists of iterations on the following two steps.

1. For fixed $P_n$, solve
$$(I - \beta P_n)\, v_{P_n} = c_{P_n} \tag{3.8}$$
for $v_{P_n}$.

2. Find $P_{n+1}$ such that
$$c_{P_{n+1}} + (\beta P_{n+1} - I)\, v_{P_n} = B v_{P_n} \tag{3.9}$$

Step 1 is accomplished by setting
$$v_{P_n} = (I - \beta P_n)^{-1} c_{P_n}. \tag{3.10}$$

Step 2 amounts to finding a policy function (i.e., a stochastic matrix $P_{n+1} \in \mathcal{M}$) that solves a two-period problem with $v_{P_n}$ as the terminal value function.

Following Putterman and Brumelle, the policy improvement algorithm can be interpreted as a version of Newton's method for finding the zero of $Bv = v$. Using equation (3.8) for $n+1$ to eliminate $c_{P_{n+1}}$ from equation (3.9) gives
$$(I - \beta P_{n+1})\, v_{P_{n+1}} + (\beta P_{n+1} - I)\, v_{P_n} = B v_{P_n}$$
which implies
$$v_{P_{n+1}} = v_{P_n} + (I - \beta P_{n+1})^{-1} B v_{P_n}. \tag{3.11}$$
From equation (3.9), $(\beta P_{n+1} - I)$ can be regarded as the gradient of $B v_{P_n}$, which supports the interpretation of equation (3.11) as implementing Newton's method.[3]

## *Numerical implementation*

We shall illustrate Howard's policy improvement algorithm by applying it to our savings example. Consider a given feasible policy function $k' = f(k, s)$. For each $h$, define the $n \times n$ matrices $J_h$ by
$$J_h(a, a') = \begin{cases} 1 & \text{if } g(a, s_h) = a' \\ 0 & \text{otherwise}. \end{cases}$$

---

[3] Newton's method for finding the solution of $G(z) = 0$ is to iterate on $z_{n+1} = z_n - G'(z_n)^{-1} G(z_n)$.

Here $h = 1,2,\ldots,m$ where $m$ is the number of possible values for $s_t$, and $J_h(a,a')$ is the element of $J_h$ with rows corresponding to initial assets $a$ and columns to terminal assets $a'$. For a given policy function $a' = g(a,s)$ define the $n \times 1$ vectors $r_h$ with rows corresponding to

$$r_h(a) = u[(r+1)a + ws_h - g(a,s_h)], \tag{3.12}$$

for $h = 1,\ldots,m$.

Suppose the policy function $a' = g(a,s)$ is used forever. Let the value associated with using $g(a,s)$ forever be represented by the $m$ $(n \times 1)$ vectors $[v_1,\ldots,v_m]$, where $v_h(a_i)$ is the value starting from state $(a_i, s_h)$. Suppose that $m = 2$. The vectors $[v_1, v_2]$ obey

$$\begin{bmatrix} v_1 \\ v_2 \end{bmatrix} = \begin{bmatrix} r_1 \\ r_2 \end{bmatrix} + \begin{bmatrix} \beta\mathcal{P}_{11}J_1 & \beta\mathcal{P}_{12}J_1 \\ \beta\mathcal{P}_{21}J_2 & \beta\mathcal{P}_{22}J_2 \end{bmatrix} \begin{bmatrix} v_1 \\ v_2 \end{bmatrix}.$$

Then

$$\begin{bmatrix} v_1 \\ v_2 \end{bmatrix} = \left[I - \beta \begin{pmatrix} \mathcal{P}_{11}J_1 & \mathcal{P}_{12}J_1 \\ \mathcal{P}_{21}J_2 & \mathcal{P}_{22}J_2 \end{pmatrix}\right]^{-1} \begin{bmatrix} r_1 \\ r_2 \end{bmatrix}. \tag{3.13}$$

Here is how to implement the Howard policy improvement algorithm.

Step 1. For an initial feasible policy function $g_j(k,j)$ for $j = 1$, form the $r_h$ matrices using equation (3.12), then use equation (3.13) to evaluate the vectors of values $[v_1^j, v_2^j]$ implied by using that policy forever.

Step 2. Use $[v_1^j, v_2^j]$ as the terminal value vectors in equation (3.5), and perform one step on the Bellman equation to find a new policy function $g_{j+1}(k,s)$ for $j+1 = 2$. Use this policy function, update $j$, and repeat step 1.

Step 3. Iterate to convergence on steps 1 and 2.

### *Modified policy iteration*

Researchers have had success using the following modification of policy iteration: for $k \geq 2$, iterate $k$ times on Bellman's equation. Take the resulting policy function and use equation (3.13) to produce a new candidate value function. Then starting from this terminal value function, perform another $k$ iterations on the Bellman equation. Continue in this fashion until the decision rule converges.

## *Sample Bellman equations*

This section presents some examples. The first two examples involve no optimization, just computing discounted expected utility. The appendix to chapter 5 describes some related examples based on search theory.

### *Example 1: Calculating expected utility*

Suppose that the one-period utility function is the constant relative risk aversion form $u(c) = c^{1-\gamma}/(1-\gamma)$. Suppose that $c_{t+1} = \lambda_{t+1} c_t$ and that $\{\lambda_t\}$ is an $n$-state Markov process with transition matrix $P_{ij} = \text{Prob}(\lambda_{t+1} = \bar{\lambda}_j | \lambda_t = \bar{\lambda}_i)$. Suppose that we want to evaluate discounted expected utility

$$V(c_t, \lambda_t) = E_t \sum_{t=0}^{\infty} \beta^t u(c_t), \tag{3.14}$$

where $\beta \in (0,1)$. We can express this equation recursively:

$$V(c_t, \lambda_t) = u(c_t) + \beta E_t V(c_{t+1}, \lambda_{t+1}) \tag{3.15}$$

We use a guess-and-verify technique to solve equation (3.15) for $V(c_t, \lambda_t)$. Guess that $V(c_t, \lambda_t) = u(c_t) w(\lambda_t)$ for some function $w(\lambda_t)$. Substitute the guess into equation (3.15), divide both sides by $u(c_t)$, and rearrange to get

$$w(\lambda_t) = 1 + \beta E_t \left(\frac{c_{t+1}}{c_t}\right)^{1-\gamma} w(\lambda_{t+1})$$

or

$$w_i = 1 + \beta \sum_j P_{ij} (\lambda_j)^{1-\gamma} w_j. \tag{3.16}$$

Equation (3.16) is a system of linear equations in $w_i, i = 1, \ldots, n$. It can be solved by linear algebra.

### *Example 2: Risk-sensitive preferences*

Suppose we modify the preferences of the previous example to be of the recursive form

$$V(c_t, \lambda_t) = u(c_t) + \beta \mathcal{R}_t V(c_{t+1}, \lambda_{t+1}), \tag{3.17}$$

where $\mathcal{R}_t(V) = \left(\frac{2}{\sigma}\right) \log E_t \left[\exp\left(\frac{\sigma V_{t+1}}{2}\right)\right]$ is an operator used by Jacobson (1973), Whittle (1990), and Hansen and Sargent (1995) to induce a preference for robustness to model misspecification.[4] Here $\sigma \leq 0$; when $\sigma < 0$, it represents a concern for model misspecification, or an extra sensitivity to risk.

Let's apply our guess-and-verify method again. If we make a guess of the same form as before, we now find

$$w(\lambda_t) = 1 + \beta \left(\frac{2}{\sigma}\right) \log E_t \left\{ \exp\left[\frac{\sigma}{2} \left(\frac{c_{t+1}}{c_t}\right)^{1-\gamma} w(\lambda_t)\right]\right\}$$

or

$$w_i = 1 + \beta \frac{2}{\sigma} \log \sum_j P_{ij} \exp\left(\frac{\sigma}{2} \lambda_j^{1-\gamma} w_j\right). \tag{3.18}$$

Equation (3.18) is a nonlinear system of equations in the $n \times 1$ vector of $w$'s. It can be solved by an iterative method: guess at an $n \times 1$ vector $w^0$, use it on the right side of equation (3.18) to compute a new guess $w_i^1, i = 1, \ldots, n$, and iterate.

*Example 3: Costs of business cycles*

Robert E. Lucas, Jr., (1987) proposed that the cost of business cycles be measured in terms of a proportional upward shift in the consumption process that would be required to make a representative consumer indifferent between its random consumption allocation and a nonrandom consumption allocation with the same mean. This measure of business cycles is the fraction $\Omega$ that satisfies

$$E_0 \sum_{t=0}^{\infty} \beta^t u[(1+\Omega)c_t] = \sum_{t=0}^{\infty} \beta^t u[E_0(c_t)]. \tag{3.19}$$

Suppose that the utility function and the consumption process are as in example 1. Then for given $\Omega$, the calculations in example 1 can be used to calculate the left side of equation (3.19). In particular, the left side just equals $u[(1+\Omega)c_0]w(\lambda)$, where $w(\lambda)$ is calculated from equation (3.16). To calculate the right side, we have to evaluate

$$E_0 c_t = c_0 \sum_{\lambda_t, \ldots, \lambda_1} \lambda_t \lambda_{t-1} \cdots \lambda_1 \pi(\lambda_t | \lambda_{t-1}) \pi(\lambda_{t-1} | \lambda_{t-2}) \cdots \pi(\lambda_1 | \lambda_0), \tag{3.20}$$

[4] Also see Epstein and Zin (1989) and Weil (1989) for a version of the $\mathcal{R}_t$ operator.

where the summation is meant to be over all possible *paths* of growth rates between 0 and $t$. In the case of i.i.d. $\lambda_t$, this expression simplifies to

$$E_0 c_t = c_0 (E\lambda)^t, \tag{3.21}$$

where $E\lambda_t$ is the unconditional mean of $\lambda$. Under equation (3.21), the right side of equation (3.19) is easy to evaluate.

Given $\gamma, \pi$, a procedure for constructing the cost of cycles—more precisely the costs of deviations from mean trend—to the representative consumer is first to compute the right side of equation (3.19). Then we solve the following equation for $\Omega$:

$$u[(1+\Omega)c_0]w(\lambda_0) = \sum_{t=0}^{\infty} \beta^t u[E_0(c_t)].$$

Using a closely related but somewhat different stochastic specification, Lucas (1987) calculated $\Omega$. He assumed that the endowment is a geometric trend with growth rate $\mu$ plus an i.i.d. shock with mean zero and variance $\sigma_z^2$. Starting from a base $\mu = \mu_0$, he found $\mu, \sigma_z$ pairs to which the household is indifferent, assuming various values of $\gamma$ that he judged to be within a reasonable range.[5] Lucas found that for reasonable values of $\gamma$, it takes a very small adjustment in the trend rate of growth $\mu$ to compensate for even a substantial increase in the "cyclical noise" $\sigma_z$, which meant to him that the costs of business cycle fluctuations are small.

Subsequent researchers have studied how other preference specifications would affect the calculated costs. Tallarini (1996, 2000) used a version of the preferences described in example 2, and found larger costs of business cycles. Alvarez and Jermann (1999) considered other measures of the cost of business cycles, and provided ways to link them to the equity premium puzzle, to be studied in chapter 10.

## Polynomial approximations

Judd (1998) describes a method for iterating on the Bellman equation using a polynomial to approximate the value function and a numerical optimizer to perform the optimization at each iteration. We describe this method in the

[5] See chapter 10 for a discussion of reasonable values of $\gamma$. See Table 1 of Manuelli and Sargent (1988) for a correction to Lucas's calculations.

context of the Bellman equation for a particular problem that we shall encounter later.

In chapter 15, we shall study Hopenhayn and Nicolini's (1997) model of optimal unemployment insurance, which leads to a Bellman equation of the form of $V$ and the continuation value $V^u$. Using these functions allows us to write the Bellman equation in $C(V)$ as

$$C(V) = \min_{V^u} \{c + \beta[1 - p(a)]C(V^u)\} \tag{3.22}$$

where $c, a$ are given by

$$c = u^{-1} [\max(0, V + a - \beta\{p(a)V^e + [1 - p(a)]V^u\})]. \tag{3.23}$$

and

$$a = \max \left\{0, \frac{\log[r\beta(V^e - V^u)]}{r}\right\}. \tag{3.24}$$

Here $V$ is a discounted present value that an insurer has promised to an unemployed worker, $V_u$ is a value for next period that the insurer promises the worker if he remains unemployed, $1 - p(a)$ is the probability of remaining unemployed if the worker exerts search effort $a$, and $c$ is the worker's consumption level. Hopenhayn and Nicolini assume that $p(a) = 1 - \exp(ra)$, $r > 0$. Equation (3.22) is the Bellman equation associated with a cost minimization problem for the insurer.

*Recommended computational strategy*

To approximate the solution of the Bellman equation (3.22), we apply a computational procedure described by Judd (1996, 1998). The method uses a polynomial to approximate the $i$th iterate $C_i(V)$ of $C(V)$. This polynomial is stored on the computer in terms of $n+1$ coefficients. Then at each iteration, the Bellman equation is to be solved at a small number $m \geq n+1$ values of $V$. This procedure gives values of the $i$th iterate of the value function $C_i(V)$ at those particular $V$'s. Then we interpolate (or "connect the dots") to fill in the continuous function $C_i(V)$. Substituting this approximation $C_i(V)$ for $C(V)$ in equation (3.22), we pass the minimum problem on the right side of equation (3.22) to a numerical minimizer. Programming languages like Matlab and Gauss have easy-to-use algorithms for minimizing continuous functions of several variables. We solve one such numerical problem minimization for each node value for $V$. Doing so yields optimized value $C_{i+1}(V)$ at those node points. We then interpolate to build up

$C_{i+1}(V)$. We iterate on this scheme to convergence. Before summarizing the algorithm, we provide a brief description of Chebyshev polynomials.

*Chebyshev polynomials*

Where $n$ is a nonnegative integer and $x \in I\!R$, the $n$th Chebyshev polynomial, is

$$T_n(x) = \cos(n \cos^{-1} x). \tag{3.25}$$

Given coefficients $c_j, j = 0, \ldots, n$, the $n$th-order Chebyshev polynomial approximator is

$$C_n(x) = c_0 + \sum_{j=1}^{n} c_j T_j(x). \tag{3.26}$$

We are given a real valued function $f$ of a single variable $x \in [-1, 1]$. For computational purposes, we want to form an approximator to $f$ of the form (3.26). Note that we can store this approximator simply as the $n+1$ coefficients $c_j, j = 0, \ldots, n$. To form the approximator, we evaluate $f(x)$ at $n+1$ carefully chosen points, then use a least squares formula to form the $c_j$'s in equation (3.26). Thus, to interpolate a function of a single variable $x$ with domain $x \in [-1, 1]$, Judd (1996, 1998) recommends evaluating the function at the $m \geq n+1$ points $x_k, k = 1, \ldots, m$, where

$$x_k = cos\left(\frac{2k-1}{2m}\pi\right), k = 1, \ldots, m. \tag{3.27}$$

Here $x_k$ is the zero of the $k$th Chebyshev polynomial on $[-1, 1]$. Given the $m \geq n+1$ values of $f(x_k)$ for $k = 1, \ldots, m$, choose the "least squares" values of $c_j$

$$c_j = \frac{\sum_{k=1}^{m} f(x_k) T_j(x_k)}{\sum_{k=1}^{m} T_j(x_k)^2}, \quad j = 0, \ldots, n \tag{3.28}$$

*Algorithm: summary*

In summary, applied to the Hopenhayn-Nicolini model, the numerical procedure consists of the following steps:

1. Choose upper and lower bounds for $V^u$, so that $V$ and $V^u$ will be understood to reside in the interval $[\underline{V}^u, \overline{V}^u]$. In particular, set $\overline{V}^u = V^e - \frac{1}{\beta p'(0)}$, the

bound required to assure positive search effort, computed in chapter 15. Set $\underline{V}^u = V_{rmaut}$.

2. Choose a degree $n$ for the approximator, a Chebyshev polynomial, and a number $m \geq n+1$ of nodes or grid points.
3. Generate the $m$ zeros of the Chebyshev polynomial on the set $[1,-1]$, given by (3.27).
4. By a change of scale, transform the $z_i$'s to corresponding points $V_\ell^u$ in $[\underline{V}^u, \overline{V}^u]$.
5. Choose initial values of the $n+1$ coefficients in the Chebyshev polynomial, for example, $c_j = 0, \ldots, n$. Use these coefficients to define the function $C_i(V^u)$ for iteration number $i = 0$.
6. Compute the function $\tilde{C}_i(V) \equiv c + \beta[1-p(a)]C_i(V^u)$, where $c, a$ are determined as functions of $(V, V^u)$ from equations (3.23) and (3.24). This computation builds in the functional forms and parameters of $u(c)$ and $p(a)$, as well as $\beta$.
7. For each point $V_\ell^u$, use a numerical minimization program to find $C_{i+1}(V_\ell^u) = \min_{V^u} \tilde{C}_i(V_u)$.
8. Using these $m$ values of $C_{j+1}(V_\ell^u)$, compute new values of the coefficients in the Chebyshev polynomials by using "least squares" [formula (3.28)]. Return to step 5 and iterate to convergence.

*Shape preserving splines*

Judd (1998) points out that because they do not preserve concavity, using Chebyshev polynomials to approximate value functions can cause problems. He recommends the Schumaker quadratic shape-preserving spline. It ensures that the objective in the maximization step of iterating on a Bellman equation will be concave and differentiable (Judd, 1998, p. 441). Using Schumaker splines avoids the type of internodal oscillations associated with other polynomial approximation methods. The exact interpolation procedure is described in Judd (1998) on p. 233. A relatively small number of evaluation nodes usually is sufficient. Judd and Solnick (1994) find that this approach outperforms linear interpolation and discrete state approximation methods in a deterministic optimal growth problem.[6]

[6] The Matlab program `schumaker.m` (written by Leonardo Rezende of Stanford University) can be used to compute the spline. Use the Matlab command `ppval` to evaluate the spline.

## Concluding remarks

This chapter has described two of three standard methods for approximating solutions of dynamic programs numerically: discretizing the state space and using polynomials to approximate the value function. The next chapter describes the third method: making the problem have a quadratic return function and linear transition law. A benefit of making the restrictive linear-quadratic assumptions is that they make solving a dynamic program easy by exploiting the ease with which stochastic linear difference equations can be manipulated.

# 4
# *Linear Quadratic Dynamic Programming*

## *Introduction*

This chapter describes the class of dynamic programming problems in which the return function is quadratic and the transition function is linear. This specification leads to the widely used optimal linear regulator problem, for which the Bellman equation can be solved quickly using linear algebra. We consider the special case in which the return function and transition function are both time invariant, though the mathematics is almost identical when they are permitted to be deterministic functions of time.

Linear quadratic dynamic programming has two uses for us. A first is to study optimum and equilibrium problems arising for linear rational expectations models. Here the dynamic decision problems naturally take the form of an optimal linear regulator. A second is to use a linear quadratic dynamic program to approximate one that is not linear quadratic.

Later in the chapter, we also describe a filtering problem of great interest to macroeconomists. Its mathematical structure is identical to that of the optimal linear regulator, and its solution is the Kalman filter, a recursive way of solving linear filtering and estimation problems. Suitably reinterpreted, formulas that solve the optimal linear regulator also describe the Kalman filter.

## *The optimal linear regulator problem*

The undiscounted optimal linear regulator problem is to maximize over choice of $\{u_t\}_{t=0}^{\infty}$ the criterion

$$\sum_{t=0}^{\infty} \{x_t' R x_t + u_t' Q u_t\}, \tag{4.1}$$

subject to $x_{t+1} = Ax_t + Bu_t$, $x_0$ given. Here $x_t$ is an $(n \times 1)$ vector of state variables, $u_t$ is a $(k \times 1)$ vector of controls, $R$ is a negative semidefinite symmetric matrix, $Q$ is a negative definite symmetric matrix, $A$ is an $(n \times n)$ matrix, and $B$ is an $(n \times k)$ matrix. We guess that the value function is quadratic, $V(x) = x'Px$, where $P$ is a negative semidefinite symmetric matrix.

Using the transition law to eliminate next period's state, the Bellman equation becomes

$$x'Px = \max_u \{x'Rx + u'Qu + (Ax + Bu)'P(Ax + Bu)\}. \tag{4.2}$$

The first-order necessary condition for the maximum problem on the right side of equation (4.2) is[1]

$$(Q + B'PB)u = -B'PAx, \tag{4.3}$$

which implies the feedback rule for $u$:

$$u = -(Q + B'PB)^{-1}B'PAx \tag{4.4}$$

or $u = -Fx$, where

$$F = (Q + B'PB)^{-1}B'PA. \tag{4.5}$$

Substituting the optimizer (4.4) into the right side of equation (4.2) and rearranging gives

$$P = R + A'PA - A'PB(Q + B'PB)^{-1}B'PA. \tag{4.6}$$

Equation (4.6) is called the *algebraic matrix Riccati* equation. It expresses the matrix $P$ as an implicit function of the matrices $R, Q, A, B$. Solving this equation for $P$ requires a computer whenever $P$ is larger than a $2 \times 2$ matrix.

In exercise 4.1, you are asked to derive the Riccati equation for the case where the return function is modified to

$$x_t'Rx_t + u_t'Qu_t + 2u_t'Wx_t.$$

---

[1] We use the following rules for differentiating quadratic and bilinear matrix forms: $\frac{\partial x'Ax}{\partial x} = (A + A')x$; $\frac{\partial y'Bz}{\partial y} = Bz$, $\frac{\partial y'Bz}{\partial z} = B'y$.

*Value function iteration*

Under particular conditions to be discussed in the section on stability, equation (4.6) has a unique negative semidefinite solution, which is approached in the limit as $j \to \infty$ by iterations on the matrix Riccati difference equation:[2]

$$P_{j+1} = R + A'P_jA - A'P_jB(Q + B'P_jB)^{-1}B'P_jA, \tag{4.7a}$$

starting from $P_0 = 0$. The policy function associated with $P_j$ is

$$F_{j+1} = (Q + B'P_jB)^{-1}B'P_jA. \tag{4.7b}$$

Equation (4.7) is derived much like equation (4.6) except that one starts from the iterative version of the Bellman equation rather than from the asymptotic version.

*Discounted linear regulator problem*

The discounted optimal linear regulator problem is to maximize

$$\sum_{t=0}^{\infty} \beta^t \{x_t'Rx_t + u_t'Qu_t\}, \qquad 0 < \beta < 1, \tag{4.8}$$

subject to $x_{t+1} = Ax_t + Bu_t, x_0$ given. This problem leads to the following matrix Riccati difference equation modified for discounting:

$$P_{j+1} = R + \beta A'P_jA - \beta^2 A'P_jB(Q + \beta B'P_jB)^{-1}B'P_jA. \tag{4.9}$$

The algebraic matrix Riccati equation is modified correspondingly. The value function for the infinite horizon problem is simply $V(x_0) = x_0'Px_0$, where $P$ is the limiting value of $P_j$ resulting from iterations on equation (4.9) starting from $P_0 = 0$. The optimal policy is $u_t = -Fx_t$, where $F = \beta(Q + \beta B'PB)^{-1}B'PA$.

The Matlab program `olrp.m` can be used to solve the discounted optimal linear regulator problem. Matlab has a variety of other programs that solve both discrete and continuous time versions of undiscounted optimal linear regulator problems. The program `policyi.m` solves the undiscounted optimal linear regulator problem using policy iteration.

[2] If the eigenvalues of $A$ are bounded in modulus below unity, this result obtains, but much weaker conditions suffice. See Bertsekas (1976, chap. 4) and Sargent (1980).

*Policy improvement algorithm*

The policy improvement algorithm can be applied to solve the discounted optimal linear regulator problem. Starting from an initial $F_0$ for which the eigenvalues of $A - BF_0$ are less than $1/\sqrt{\beta}$ in modulus, the algorithm iterates on the two equations

$$P_j = R + F_j'QF_j + \beta(A - BF_j)'P_j(A - BF_j) \tag{4.10}$$

$$F_{j+1} = \beta(Q + \beta B'P_jB)^{-1}B'P_jA. \tag{4.11}$$

The first equation is an example of a *discrete Lyapunov* or *Sylvester* equation, which is to be solved for the matrix $P_j$ that determines the value $x_t'P_jx_t$ that is associated with following policy $F_j$ forever. The solution of this equation can be represented in the form

$$P_j = \sum_{k=0}^{\infty} \beta^k (A - BF_j)'^k (R + F_j'QF_j)(A - BF_j)^k.$$

If the eigenvalues of the matrix $A - BF_j$ are bounded in modulus by $1/\sqrt{\beta}$, then a solution of this equation exists. There are several methods available for solving this equation.[3] The Matlab program `policyi.m` solves the undiscounted optimal linear regulator problem using policy iteration. This algorithm is typically much faster than the algorithm that iterates on the matrix Riccati equation. Later we shall present a third method for solving for $P$ that rests on the link between $P$ and shadow prices for the state vector.

## The stochastic optimal linear regulator problem

The stochastic discounted linear optimal regulator problem is to choose a decision rule for $u_t$ to maximize

$$E_0 \sum_{t=0}^{\infty} \beta^t \{x_t'Rx_t + u_t'Qu_t\}, \qquad 0 < \beta < 1, \tag{4.12}$$

subject to $x_0$ given, and the law of motion

$$x_{t+1} = Ax_t + Bu_t + \epsilon_{t+1}, \qquad t \geq 0, \tag{4.13}$$

[3] The Matlab programs `dlyap.m` and `doublej.m` solve discrete Lyapunov equations. See Anderson, Hansen, McGrattan, and Sargent (1996).

where $\epsilon_{t+1}$ is an $(n \times 1)$ vector of random variables that is independently and identically distributed through time and obeys the normal distribution with mean vector zero and covariance matrix

$$E\epsilon_t \epsilon_t' = \Sigma. \tag{4.14}$$

(See Kwakernaak and Sivan, 1972, for an extensive study of the continuous-time version of this problem; also see Chow, 1981.) The matrices $R, Q, A$, and $B$ obey the assumption that we have described.

The value function for this problem is

$$v(x) = x'Px + d, \tag{4.15}$$

where $P$ is the unique negative semidefinite solution of the discounted algebraic matrix Riccati equation corresponding to equation (4.9). As before, it is the limit of iterations on equation (4.9) starting from $P_0 = 0$. The scalar $d$ is given by

$$d = \beta(1-\beta)^{-1} \text{tr } P\Sigma \tag{4.16}$$

where "tr" denotes the trace of a matrix. Furthermore, the optimal policy continues to be given by $u_t = -Fx_t$, where

$$F = \beta(Q + \beta B'P'B)^{-1} B'PA. \tag{4.17}$$

A notable feature of this solution is that the feedback rule (4.17) is identical with the rule for the corresponding nonstochastic linear optimal regulator problem. This outcome is the *certainty equivalence* principle.

Certainty Equivalence Principle: The feedback rule that solves the stochastic optimal linear regulator problem is identical with the rule for the corresponding nonstochastic linear optimal regulator problem.

Proof: Substitute guess (4.15) into the Bellman equation to obtain

$$v(x) = \max_u \Big\{ x'Rx + u'Qu + \beta E \Big[ (Ax + Bu + \epsilon)' P (Ax + Bu + \epsilon) \Big] + \beta d \Big\},$$

where $\epsilon$ is the realization of $\epsilon_{t+1}$ when $x_t = x$ and where $E\epsilon|x = 0$. The preceding equation implies

$$\begin{aligned} v(x) = & \max_u \{x'Rx + u'Qu + \beta E\{x'A'PAx + x'A'PBu \\ & + x'A'P\epsilon + u'B'PAx + u'B'PBu + u'B'P\epsilon \\ & + \epsilon'PAx + \epsilon'PBu + \epsilon'P\epsilon\} + \beta d\}. \end{aligned}$$

Evaluating the expectations inside the braces and using $E\epsilon|x = 0$ gives

$$\begin{aligned} v(x) = & \max \{x'Rx + u'Qu + \beta x'A'PAx + \beta 2x'A'PBu \\ & + \beta u'B'PBu + \beta E\epsilon'P\epsilon\} + \beta d. \end{aligned}$$

The first-order condition for $u$ is

$$(Q + \beta B'PB)u = -\beta B'PAx,$$

which implies equation (4.17). Using $E\epsilon'P\epsilon = \operatorname{tr} E\epsilon'P\epsilon = \operatorname{tr} PE\epsilon\epsilon' = \operatorname{tr} P\Sigma$, substituting equation (4.17) into the preceding expression for $v(x)$, and using equation (4.15) gives

$$P = R + \beta A'PA - \beta^2 A'PB(Q + \beta B'PB)^{-1}B'PA,$$

and

$$d = \beta(1-\beta)^{-1}\operatorname{tr} P\Sigma. \quad \blacksquare$$

*Discussion of certainty equivalence*

The remarkable thing about this solution is that, although through $d$ the objective function (4.14) depends on $\Sigma$, the optimal decision rule $u_t = -Fx_t$ is independent of $\Sigma$. This is the message of equation (4.17) and the discounted algebraic Riccati equation for $P$, which are identical with the formulas derived earlier under certainty. In other words, the optimal decision rule $u_t = h(x_t)$ is independent of the problem's noise statistics.[4] The certainty equivalence principle is a special property of the optimal linear regulator problem and comes from the quadratic objective function, the linear transition equation, and the property

[4] Therefore, in linear quadratic versions of the optimum savings problem, there are no precautionary savings. See chapters 13 and 14.

$E(\epsilon_{t+1}|x_t) = 0$. Certainty equivalence does not characterize stochastic control problems generally.

For the remainder of this chapter, we return to the nonstochastic optimal linear regulator, remembering the stochastic counterpart.

## *Shadow prices in the linear regulator*

For several purposes,[5] it is helpful to interpret the gradient $2Px_t$ of the value function $x_t'Px_t$ as a shadow price or Lagrange multiplier. Thus, associate with the Bellman equation the Lagrangian

$$x_t'Px_t = V(x_t) = \max_{u_t}\Big\{ x_t'Rx_t + u_t'Qu_t + x_{t+1}'Px_{t+1} + 2\mu_{t+1}'[Ax_t + Bu_t - x_{t+1}]\Big\},$$

where $\mu_{t+1}$ is a vector of Lagrange multipliers. The first-order necessary conditions for an optimum with respect to $u_t$ and $x_t$ are

$$\begin{aligned} 2Qu_t + 2B'\mu_{t+1} &= 0 \\ 2Px_{t+1} - 2\mu_{t+1} &= 0. \end{aligned} \tag{4.18}$$

Using the transition law and rearranging gives the usual formula for the optimal decision rule, namely, $u_t = -(Q + B'PB)^{-1}B'PAx_t$. Notice that the shadow price vector satisfies $\mu_{t+1} = Px_{t+1}$, where $P$ is the value function.

Later in this chapter, we shall describe a computational strategy that solves for $P$ by directly finding the optimal multiplier process $\{\mu_t\}$ and representing it as $\mu_t = Px_t$. This strategy exploits the *stability* properties of optimal solutions of the linear regulator problem, which we now briefly take up.

### *Stability*

Upon substituting the optimal control $u_t = -Fx_t$ into the law of motion $x_{t+1} = Ax_t + Bu_t$, we obtain the optimal "closed-loop system" $x_{t+1} = (A - BF)x_t$. This difference equation governs the evolution of $x_t$ under the optimal

[5] The gradient of the value function has information from which prices can be coaxed where the value function is for a planner in a linear quadratic economy. See Hansen and Sargent (2000).

control. The system is said to be stable if $\lim_{t\to\infty} x_t = 0$ starting from any initial $x_0 \in R^n$. Assume that the eigenvalues of $(A - BF)$ are distinct, and use the eigenvalue decomposition $(A - BF) = C\Lambda C^{-1}$ where the columns of $C$ are the eigenvectors of $(A - BF)$ and $\Lambda$ is a diagonal matrix of eigenvalues of $(A - BF)$. Write the "closed-loop" equation as $x_{t+1} = C\Lambda C^{-1} x_t$. The solution of this difference equation for $t > 0$ is readily verified by repeated substitution to be $x_t = C\Lambda^t C^{-1} x_0$. Evidently, the system is stable for all $x_0 \in R^n$ if and only if the eigenvalues of $(A - BF)$ are all strictly less than unity in absolute value. When this condition is met, $(A - BF)$ is said to be a "stable matrix."[6]

A vast literature is devoted to characterizing the conditions on $A, B, R$, and $Q$ under which the optimal closed-loop system matrix $(A - BF)$ is stable. These results are surveyed by Anderson, Hansen, McGrattan, and Sargent (1996) and can be briefly described here for the undiscounted case $\beta = 1$. Roughly speaking, the conditions on $A, B, R$, and $Q$ that are required for stability are as follows: First, $A$ and $B$ must be such that it is *possible* to pick a control law $u_t = -Fx_t$ that drives $x_t$ to zero eventually, starting from any $x_0 \in R^n$ ["the pair $(A, B)$ must be stabilizable"]. Second, the matrix $R$ must be such that the controller *wants* to drive $x_t$ to zero as $t \to \infty$.

It would take us too far afield to go deeply into this body of theory, but we can give a flavor for the results by considering some very special cases. The following assumptions and propositions are too strict for most economic applications, but similar results can obtain under weaker conditions relevant for economic problems.[7]

ASSUMPTION A.1: The matrix $R$ is negative definite.

There immediately follows:

*Proposition 1:* Under Assumption A.1, if a solution to the undiscounted regulator exists, it satisfies $\lim_{t\to\infty} x_t = 0$.

*Proof:* If $x_t \not\to 0$, then $\sum_{t=0}^{\infty} x_t' R x_t \to -\infty$. ∎

ASSUMPTION A.2: The matrix $R$ is negative semidefinite.

---

6 It is possible to amend the statements about stability in this section to permit $A - BF$ to have a single unit eigenvalue associated with a constant in the state vector. See chapter 1 for examples.

7 See Kwakernaak and Sivan (1972) and Anderson, Hansen, McGrattan, and Sargent (1996).

Under Assumption A.2, $R$ is similar to a triangular matrix $R^*$:

$$R = T' \begin{pmatrix} R_{11}^* & 0 \\ 0 & 0 \end{pmatrix} T$$

where $R_{11}^*$ is negative definite and $T$ is nonsingular. Notice that $x_t' R x_t = x_{1t}^* R_{11}^* x_{1t}^*$ where $x_t^* = T x_t = \begin{pmatrix} T_1 \\ T_2 \end{pmatrix} x_t = \begin{pmatrix} x_{1t}^* \\ x_{2t}^* \end{pmatrix}$. Let $x_{1t}^* \equiv T_1 x_t$. These calculations support the proposition:

*Proposition 2:* Suppose that a solution to the optimal linear regulator exists under Assumption A.2. Then $\lim_{t \to \infty} x_{1t}^* = 0$.

The following definition is used in control theory:

DEFINITION: The pair $(A, B)$ is said to be *stabilizable* if there exists a matrix $F$ for which $(A - BF)$ is a stable matrix.

The following is illustrative of a variety of stability theorems from control theory:[8, 9]

THEOREM: If $(A, B)$ is stabilizable and $R$ is negative definite, then under the optimal rule $F$, $(A - BF)$ is a stable matrix.

In the next section, we assume that $A, B, Q, R$ satisfy conditions sufficient to invoke such a stability propositions, and we use that assumption to justify a solution method that solves the undiscounted linear regulator by searching among the many solutions of the *Euler equations* for a stable solution.

---

[8] These conditions are discussed under the subjects of controllability, stabilizability, reconstructability, and detectability in the literature on linear optimal control. (For continuous-time linear system, these concepts are described by Kwakernaak and Sivan, 1972; for discrete-time systems, see Sargent, 1980). These conditions subsume and generalize the transversality conditions used in the discrete-time calculus of variations (see Sargent, 1987a). That is, the case when $(A - BF)$ is stable corresponds to the situation in which it is optimal to solve "stable roots backward and unstable roots forward." See Sargent (1987a, chap. 9). Hansen and Sargent (1981) describe the relationship between Euler equation methods and dynamic programming for a class of linear optimal control systems. Also see Chow (1981).

[9] The conditions under which $(A - BF)$ is stable are also the conditions under which $x_t$ converges to a unique stationary distribution in the stochastic version of the linear regulator problem.

## A Lagrangian formulation

This section describes a Lagrangian formulation of the optimal linear regulator.[10] Besides being useful computationally, this formulation carries insights about the connections between stability and optimality and also opens the way to constructing solutions of dynamic systems not coming directly from an intertemporal optimization problem.[11]

For the undiscounted optimal linear regulator problem, form the Lagrangian

$$\begin{aligned} J = \sum_{t=0}^{\infty} \Bigl\{ & x_t' R x_t + u_t' Q u_t \\ & + 2\mu_{t+1}' [A x_t + B u_t - x_{t+1}] \Bigr\}. \end{aligned}$$

First-order conditions for maximization with respect to $\{u_t, x_{t+1}\}$ are

$$\begin{aligned} 2Qu_t + 2B'\mu_{t+1} &= 0 \\ \mu_t &= Rx_t + A'\mu_{t+1} \ , \ t \geq 0. \end{aligned} \tag{4.19}$$

The Lagrange multiplier vector $\mu_{t+1}$ is often called the *costate* vector. Solve the first equation for $u_t$ in terms of $\mu_{t+1}$; substitute into the law of motion $x_{t+1} = Ax_t + Bu_t$; arrange the resulting equation and the second equation of (4.19) into the form

$$L \begin{pmatrix} x_{t+1} \\ \mu_{t+1} \end{pmatrix} = N \begin{pmatrix} x_t \\ \mu_t \end{pmatrix} \ , \ t \geq 0,$$

where

$$L = \begin{pmatrix} I & BQ^{-1}B' \\ 0 & A' \end{pmatrix} , N = \begin{pmatrix} A & 0 \\ -R & I \end{pmatrix} .$$

When $L$ is of full rank (i.e., when $A$ is of full rank), we can write this system as

$$\begin{pmatrix} x_{t+1} \\ \mu_{t+1} \end{pmatrix} = M \begin{pmatrix} x_t \\ \mu_t \end{pmatrix} \tag{4.20}$$

---

[10] Such formulations are recommended by Chow (1997) and Anderson, Hansen, McGrattan, and Sargent (1996).

[11] Blanchard and Kahn (1980), Whiteman (1983), Hansen, Epple, and Roberds (1985), and Anderson, Hansen, McGrattan and Sargent (1996) use and extend such methods.

where

$$M \equiv L^{-1}N = \begin{pmatrix} A + BQ^{-1}B'A'^{-1}R & -BQ^{-1}B'A'^{-1} \\ -A^{-1}R & A'^{-1} \end{pmatrix} \tag{4.21}$$

To exhibit the properties of the $(2n \times 2n)$ matrix $M$, we introduce a $(2n \times 2n)$ matrix

$$J = \begin{pmatrix} 0 & -I_n \\ I_n & 0 \end{pmatrix}.$$

The rank of $J$ is $2n$.

DEFINITION: A matrix $M$ is called *symplectic* if

$$MJM' = J. \tag{4.22}$$

It can be verified directly that $M$ in equation (4.21) is symplectic. It follows from equation (4.22) and $J^{-1} = J' = -J$ that for any symplectic matrix $M$,

$$M' = J^{-1}M^{-1}J. \tag{4.23}$$

Equation (4.23) states that $M'$ is related to the inverse of $M$ by a similarity transformation. For square matrices, recall that (a) similar matrices share eigenvalues; (b) the eigenvalues of the inverse of a matrix are the inverses of the eigenvalues of the matrix; and (c) a matrix and its transpose have the same eigenvalues. It then follows from equation (4.23) that the eigenvalues of $M$ occur in reciprocal pairs: if $\lambda$ is an eigenvalue of $M$, so is $\lambda^{-1}$.

Write equation (4.20) as

$$y_{t+1} = My_t \tag{4.24}$$

where $y_t = \begin{pmatrix} x_t \\ \mu_t \end{pmatrix}$. Consider the following triangularization of $M$

$$V^{-1}MV = \begin{pmatrix} W_{11} & W_{12} \\ 0 & W_{22} \end{pmatrix}$$

where each block on the right side is $(n \times n)$, where $V$ is nonsingular, and where $W_{22}$ has all its eigenvalues exceeding 1 and $W_{11}$ has all of its eigenvalues less than 1. The *Schur decomposition* and the *eigenvalue decomposition* are two possible such decompositions.[12] Write equation (4.24) as

$$y_{t+1} = VWV^{-1} y_t. \tag{4.25}$$

The solution of equation (4.25) for arbitrary initial condition $y_0$ is evidently

$$y_{t+1} = V \begin{bmatrix} W_{11}^t & W_{12,t} \\ 0 & W_{22}^t \end{bmatrix} V^{-1} y_0 \tag{4.26}$$

where $W_{12,t}$ obeys the recursion

$$W_{12,t+1} = W_{11}^t W_{12,t} + W_{12} W_{22}^t$$

and where $W_{ii}^t$ is $W_{ii}$ raised to the $t$th power.

Write equation (4.26) as

$$\begin{pmatrix} y_{1t+1}^* \\ y_{2t+1}^* \end{pmatrix} = \begin{bmatrix} W_{11}^t & W_{12,t}^t \\ 0 & W_{22}^t \end{bmatrix} \begin{pmatrix} y_{10}^* \\ y_{20}^* \end{pmatrix}$$

where $y_t^* = V^{-1} y_t$, and in particular where

$$y_{2t}^* = V^{21} x_t + V^{22} \mu_t, \tag{4.27}$$

and where $V^{ij}$ denotes the $(i,j)$ piece of the partitioned $V^{-1}$ matrix.

Because $W_{22}$ is an unstable matrix, unless $y_{20}^* = 0$, $y_t^*$ will diverge. Let $V^{ij}$ denote the $(i,j)$ piece of the partitioned $V^{-1}$ matrix. To attain stability, we must impose $y_{20}^* = 0$, which from equation (4.27) implies

$$V^{21} x_0 + V^{22} \mu_0 = 0$$

or

$$\mu_0 = -(V^{22})^{-1} V^{21} x_0.$$

But notice that because $(V^{21}\ V^{22})$ is the second row block of the inverse of $V$,

$$(V^{21}\ V^{22}) \begin{pmatrix} V_{11} \\ V_{21} \end{pmatrix} = 0$$

[12] Evan Anderson's Matlab program `schurg.m` attains a convenient Schur decomposition and is very useful for solving linear models with distortions. See McGrattan (1994) for some examples of distorted economies that could be solved with the Schur decomposition.

which implies

$$V^{21}V_{11} + V^{22}V_{21} = 0.$$

Therefore

$$-(V^{22})^{-1}V^{21} = V_{21}V_{11}^{-1}.$$

So we can write

$$\mu_0 = V_{21}V_{11}^{-1}x_0 \tag{4.28}$$

and

$$\mu_t = V_{21}V_{11}^{-1}x_t.$$

However, we know from equations (4.18) that $\mu_t = Px_t$, where $P$ occurs in the matrix that solves the Riccati equation (4.6). Thus, the preceding argument establishes that

$$P = V_{21}V_{11}^{-1}. \tag{4.29}$$

This formula provides us with an alternative, and typically very efficient, way of computing the matrix $P$.

This same method can be applied to compute the solution of any system of the form (4.20), if a solution exists, even if the eigenvalues of $M$ fail to occur in reciprocal pairs. The method will typically work so long as the eigenvalues of $M$ split half inside and half outside the unit circle.[13] Systems in which the eigenvalues (adjusted for discounting) fail to occur in reciprocal pairs arise when the system being solved is an equilibrium of a model in which there are distortions that prevent there being any optimum problem that the equilibrium solves. See Woodford (1999) for an application of such methods to solve for linear approximations of equilibria of a monetary model with distortions.

## The Kalman filter

Suitably reinterpreted, the same recursion (4.7) that solves the optimal linear regulator also determines the celebrated *Kalman filter*. The Kalman filter is a recursive algorithm for computing the mathematical expectation $E[x_t|y_t, \ldots, y_0]$ of a hidden state vector $x_t$, conditional on observing a history $y_t, \ldots, y_0$ of a vector of noisy signals on the hidden state. The Kalman filter can be used to formulate or simplify a variety of signal-extraction and prediction problems in

[13] See Whiteman (1983), Blanchard and Kahn (1980), and Anderson, Hansen, McGrattan, and Sargent (1996) for applications and developments of these methods.

economics. After giving the formulas for the Kalman filter, we shall describe two examples.[14]

The setting for the Kalman filter is the following linear state space system. Given $x_0$, let

$$x_{t+1} = Ax_t + Cw_{t+1} \tag{4.30a}$$
$$y_t = Gx_t + v_t \tag{4.30b}$$

where $x_t$ is an $(n \times 1)$ state vector, $w_t$ is an i.i.d. sequence Gaussian vector with $Ew_t w_t' = I$, and $v_t$ is an i.i.d. Gaussian vector orthogonal to $w_s$ for all $t, s$ with $Ev_t v_t' = R$; and $A, C$, and $G$ are matrices conformable to the vectors they multiply. Assume that the initial condition $x_0$ is unobserved, but is known to have a Gaussian distribution with mean $\hat{x}_0$ and covariance matrix $\Sigma_0$. At time $t$, the history of observations $y^t \equiv [y_t, \ldots, y_0]$ is available to estimate the location of $x_t$ and the location of $x_{t+1}$. The Kalman filter is a recursive algorithm for computing $\hat{x}_{t+1} = E[x_{t+1}|y^t]$. The algorithm is

$$\hat{x}_{t+1} = (A - K_t G)\hat{x}_t + K_t y_t \tag{4.31}$$

where

$$K_t = A\Sigma_t G'(G\Sigma_t G' + R)^{-1} \tag{4.32a}$$
$$\Sigma_{t+1} = A\Sigma_t A' + CC' - A\Sigma_t G'(G\Sigma_t G' + R)^{-1} G\Sigma_t A. \tag{4.32b}$$

Here $\Sigma_t = E(x_t - \hat{x}_t)(x_t - \hat{x}_t)'$, and $K_t$ is called the Kalman gain. Sometimes the Kalman filter is written in terms of the "observer system"

$$\hat{x}_{t+1} = A\hat{x}_t + K_t a_t \tag{4.33a}$$
$$y_t = G\hat{x}_t + a_t \tag{4.33b}$$

where $a_t \equiv y_t - G\hat{x}_t \equiv y_t - E[y_t|y^{t-1}]$. The random vector $a_t$ is called the *innovation* in $y_t$, being the part of $y_t$ that cannot be forecast linearly from its own past. Subtracting equation (4.33b) from (4.30b) gives $a_t = G(x_t - \hat{x}_t) + v_t$; multiplying each side by its own transpose and taking expectations gives the following formula for the innovation covariance matrix:

$$Ea_t a_t' = G\Sigma_t G' + R. \tag{4.34}$$

---

[14] See Hamilton (1994) and Kim and Nelson (1999) for diverse applications of the Kalman filter. The appendix of this book on dual filtering and control (chapter 21) briefly describes a discrete-state nonlinear filtering problem.

Equations (4.32) display extensive similarities to equations (4.7), the recursions for the optimal linear regulator. Note that equation (4.32b) is a Riccati equation. Indeed, with the judicious use of matrix transposition and reversal of time, the two systems of equations (4.32) and (4.7) can be made to match. In chapter 21 on dual filtering and control, we compare versions of these equations and describe the concept of duality that links them. Chapter 21 also contains a formal derivation of the Kalman filter. We now put the Kalman filter to work, leaving its derivation until chapter 21.[15]

### *Muth's example*

Phillip Cagan (1956) and Milton Friedman (1956) posited that when people wanted to form expectations of future values of a scalar $y_t$ they would use the following "adaptive expectations" scheme:

$$y^*_{t+1} = K \sum_{j=0}^{\infty} (1-K)^j y_{t-j} \tag{4.35a}$$

or

$$y^*_{t+1} = (1-K) y^*_t + K y_t, \tag{4.35b}$$

where $y^*_{t+1}$ is people's expectation. Friedman used this scheme to describe people's forecasts of future income. Cagan used it to model their forecasts of inflation during hyperinflations. Cagan and Friedman did not assert that the scheme is an optimal one, and so did not fully defend it. Muth (1960) wanted to understand the circumstances under which this forecasting scheme would be optimal. Therefore, he sought a stochastic process for $y_t$ such that equation (4.35) would be optimal. In effect, he posed and solved an "inverse optimal prediction" problem of the form "You give me the forecasting scheme; I have to find the stochastic process that makes the scheme optimal." Muth solved the problem using classical (non-recursive) methods. The Kalman filter was first described in print in the same year as Muth's solution of this problem (Kalman, 1960). The Kalman filter lets us present the solution to Muth's problem quickly.

Muth studied the model

$$x_{t+1} = x_t + w_{t+1} \tag{4.36a}$$

$$y_t = x_t + v_t, \tag{4.36b}$$

---

[15] The Matlab program **kfilter.m** computes the Kalman filter. Matlab has several other programs that compute the Kalman filter for discrete and continuous time models.

where $y_t, x_t$ are scalar random processes, and $w_{t+1}, v_t$ are mutually independent i.i.d. Gaussian random process with means of zero and variances $Ew_{t+1}^2 = Q, Ev_t^2 = R$, and $Ev_s w_{t+1} = 0$ for all $t, s$. The initial condition is that $x_0$ is Gaussian with mean $\hat{x}_0$ and variance $\Sigma_0$. Muth sought formulas for $\hat{x}_{t+1} = E[x_{t+1}|y^t]$, where $y^t = [y_t, \ldots, y_0]$.

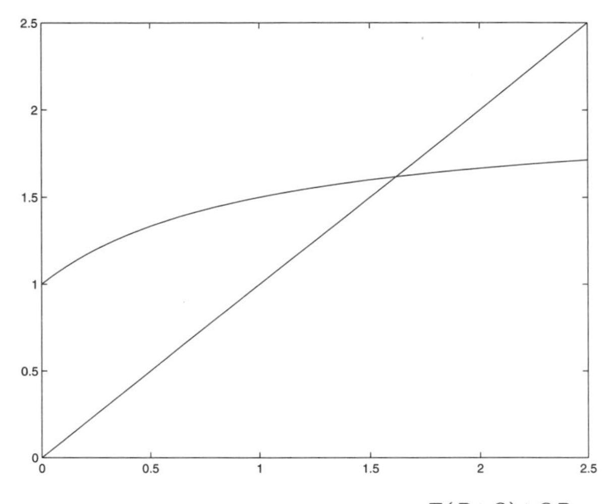


**Figure 4.1** Graph of $f(\Sigma) = \frac{\Sigma(R+Q)+QR}{\Sigma+R}$, $Q = R = 1$, against the 45-degree line. Iterations on the Riccati equation for $\Sigma_t$ converge to the fixed point.

For this problem, $A = 1, CC' = Q, G = 1$, causing the Kalman filtering equations to become

$$K_t = \frac{\Sigma_t}{\Sigma_t + R} \tag{4.37a}$$

$$\Sigma_{t+1} = \Sigma_t + Q - \frac{\Sigma_t^2}{\Sigma_t + R}. \tag{4.37b}$$

The second equation can be rewritten

$$\Sigma_{t+1} = \frac{\Sigma_t(R+Q) + QR}{\Sigma_t + R}. \tag{4.38}$$

For $Q = R = 1$, Figure 4.1 plots the function $f(\Sigma) = \frac{\Sigma(R+Q)+QR}{\Sigma+R}$ appearing on the right side of equation (4.38) for values $\Sigma \geq 0$ against the 45-degree line.

Note that $f(0) = Q$. This graph identifies the fixed point of iterations on $f(\Sigma)$ as the intersection of $f(\cdot)$ and the 45-degree line. That the slope of $f(\cdot)$ is less than unity at the intersection assures us that the iterations on $f$ will converge as $t \to +\infty$ starting from any $\Sigma_0 \geq 0$.

Muth studied the solution of this problem as $t \to \infty$. Evidently, $\Sigma_t \to \Sigma_\infty \equiv \Sigma$ is the fixed point of a graph like Figure 4.1. Then $K_t \to K$ and the formula for $\hat{x}_{t+1}$ becomes

$$\hat{x}_{t+1} = (1 - K)\hat{x}_t + K y_t \tag{4.39}$$

where $K = \frac{\Sigma}{\Sigma+R} \in (0,1)$. This is a version of Cagan's adaptive expectations formula. Iterating backward on equation (4.39) gives $\hat{x}_{t+1} = K \sum_{j=0}^{t} (1-K)^j y_{t-j} + K(1-K)^{t+1}\hat{x}_0$, which is a version of Cagan and Friedman's geometric distributed lag formula. Using equations (4.36), we find that $E[y_{t+j}|y^t] = E[x_{t+j}|y^t] = \hat{x}_{t+1}$ for all $j \geq 1$. This result in conjunction with equation (4.39) establishes that the adaptive expectation formula (4.39) gives the optimal forecast of $y_{t+j}$ for all horizons $j \geq 1$. This finding itself is remarkable and special because for most processes the optimal forecast will depend on the horizon. That there is a single optimal forecast for all horizons in one sense justifies the term "permanent income" that Milton Friedman (1955) chose to describe the forecast.

The dependence of the forecast on horizon can be studied using the formulas

$$E\left[x_{t+j}|y^{t-1}\right] = A^j \hat{x}_t \tag{4.40a}$$

$$E\left[y_{t+j}|y^{t-1}\right] = G A^j \hat{x}_t \tag{4.40b}$$

In the case of Muth's example,

$$E\left[y_{t+j}|y^{t-1}\right] = \hat{y}_t = \hat{x}_t \ \ \forall j \geq 0.$$

### *Jovanovic's example*

In chapter 5, we will describe a version of Jovanovic's (1979) matching model, at the core of which is a "signal-extraction" problem that simplifies Muth's problem. Let $x_t, y_t$ be scalars with $A = 1, C = 0, G = 1, R > 0$. Let $x_0$ be Gaussian with mean $\mu$ and variance $\Sigma_0$. Interpret $x_t$ (which is evidently constant with this specification) as the hidden value of $\theta$, a "match parameter." Let $y^t$ denote the history of $y_s$ from $s = 0$ to $s = t$. Define $m_t \equiv \hat{x}_{t+1} \equiv E[\theta|y^t]$. Then in this particular case the Kalman filter becomes

$$m_t = (1 - K_t)m_{t-1} + K_t y_t \tag{4.41a}$$

$$K_t = \frac{\Sigma_t}{\Sigma_t + R} \tag{4.41b}$$

$$\Sigma_{t+1} = \frac{\Sigma_t R}{\Sigma_t + R}. \tag{4.41c}$$

The recursions are to be initiated from $(m_{-1}, \Sigma_0)$, a pair that embodies all "prior" knowledge about the position of the system. It is easy to see from Figure 4.1 that when $Q = 0$, $\Sigma = 0$ is the limit point of iterations on equation (4.41c) starting from any $\Sigma_0 \geq 0$. Thus, the value of the match parameter is eventually learned.

It is instructive to write equation (4.41c) as

$$\frac{1}{\Sigma_{t+1}} = \frac{1}{\Sigma_t} + \frac{1}{R}. \tag{4.42}$$

The reciprocal of the variance is often called the precision of the estimate. According to equation (4.42) the precision increases without bound as $t$ grows, and $\Sigma_{t+1} \to 0$.[16]

We can represent the Kalman filter in the form (4.33) as

$$m_{t+1} = m_t + K_{t+1} a_{t+1}$$

which implies that

$$E(m_{t+1} - m_t)^2 = K_{t+1}^2 \sigma_{a,t+1}^2$$

where $a_{t+1} = y_{t+1} - m_t$ and the variance of $a_t$ is equal to $\sigma_{a,t+1}^2 = (\Sigma_{t+1} + R)$ from equation (4.34). This implies

$$E(m_{t+1} - m_t)^2 = \frac{\Sigma_{t+1}^2}{\Sigma_{t+1} + R}.$$

For the purposes of our discrete time counterpart of the Jovanovic model in chapter 5, it will be convenient to represent the motion of $m_{t+1}$ by means of the equation

$$m_{t+1} = m_t + g_{t+1} u_{t+1}$$

[16] As a further special case, consider when there is zero precision initially ($\Sigma_0 = +\infty$). Then solving the difference equation (4.42) gives $\frac{1}{\Sigma_t} = t/R$. Substituting this into equations (4.41) gives $K_t = (t+1)^{-1}$, so that the Kalman filter becomes $m_0 = y_0$ and $m_t = [1 - (t+1)^{-1}] m_{t-1} + (t+1)^{-1} y_t$, which implies that $m_t = (t+1)^{-1} \sum_{s=0}^{t} y_t$, the sample mean, and $\Sigma_t = R/t$.

where $g_{t+1} \equiv \left(\frac{\Sigma_{t+1}^2}{\Sigma_{t+1}+R}\right)^{.5}$ and $u_{t+1}$ is a standardized i.i.d. normalized and standardized with mean zero and variance 1 constructed to obey $g_{t+1}u_{t+1} \equiv K_{t+1}a_{t+1}$.

## Concluding remarks

In exchange for the restrictions that they impose, the linear quadratic dynamic optimization models of this chapter acquire tractability. The Bellman equation leads to Riccati difference equations that are so easy to solve numerically that the curse of dimensionality loses most of its force. It is easy to solve linear quadratic control or filtering with many state variables. That it is difficult to solve those problems otherwise is why linear quadratic approximations are used so widely. We describe those approximations in appendix B to this chapter.

In the next chapter, we go beyond the single-agent optimization problems of this chapter and the previous one to study systems with multiple agents simultaneously solving such problems. We introduce two equilibrium concepts for restricting how different agents' decisions are reconciled. To facilitate the analysis, we describe and illustrate those equilibrium concepts in contexts where each agent solves an optimal linear regulator problem.

## Appendix A: Matrix formulas

Let $(z, x, a)$ each be $n \times 1$ vectors, $A, C, D$, and $V$ each be $(n \times n)$ matrices, $B$ an $(n \times m)$ matrix, and $y$ an $(m \times 1)$ vector. Then $\frac{\partial a'x}{\partial x} = a$, $\frac{\partial x'Ax}{\partial x} = (A + A')x$, $\frac{\partial^2 (x'Ax)}{\partial x \partial x'} = (A + A')$, $\frac{\partial x'Ax}{\partial A} = xx'$, $\frac{\partial y'Bz}{\partial y} = Bz$, $\frac{\partial y'Bz}{\partial z} = B'y$, $\frac{\partial y'Bz}{\partial B} = yz'$.

The equation

$$A'VA + C = V$$

to be solved for $V$, is called a *discrete Lyapunov equation*; and its generalization

$$A'VD + C = V$$

is called the discrete *Sylvester equation.* The discrete Sylvester equation has a unique solution if and only if the eigenvalues $\{\lambda_i\}$ of $A$ and $\{\delta_j\}$ of $D$ satisfy the condition $\lambda_i \delta_j \neq 1 \ \forall \ i, \ j$.

## Appendix B: Linear-quadratic approximations

This appendix describes an important use of the optimal linear regulator: to approximate the solution of more complicated dynamic programs.[17] Optimal linear regulator problems are often used to approximate problems of the following form: maximize over $\{u_t\}_{t=0}^{\infty}$

$$E_0 \sum_{t=0}^{\infty} \beta^t r(z_t) \tag{4.43}$$

$$x_{t+1} = Ax_t + Bu_t + Cw_{t+1} \tag{4.44}$$

where $\{w_{t+1}\}$ is a vector of i.i.d. random disturbances with mean zero and finite variance, and $r(z_t)$ is a concave and twice continuously differentiable function of $z_t \equiv \begin{pmatrix} x_t \\ u_t \end{pmatrix}$. All nonlinearities in the original problem are absorbed into the composite function $r(z_t)$.

### *An example: the stochastic growth model*

Take a parametric version of Brock and Mirman's stochastic growth model, whose social planner chooses a policy for $\{c_t, a_{t+1}\}_{t=0}^{\infty}$ to maximize

$$E_0 \sum_{t=0}^{\infty} \beta^t \ln c_t$$

where

$$\begin{aligned} c_t + i_t &= A a_t^{\alpha} \theta_t \\ a_{t+1} &= (1-\delta) a_t + i_t \\ \ln \theta_{t+1} &= \rho \ln \theta_t + w_{t+1} \end{aligned}$$

where $\{w_{t+1}\}$ is an i.i.d. stochastic process with mean zero and finite variance, $\theta_t$ is a technology shock, and $\tilde{\theta}_t \equiv \ln \theta_t$. To get this problem into the form (4.43)–(4.44), take $x_t = \begin{pmatrix} a_t \\ \tilde{\theta}_t \end{pmatrix}$, $u_t = i_t$, and $r(z_t) = \ln(A a_t^{\alpha} \exp \tilde{\theta}_t - i_t)$, and we

[17] Kydland and Prescott (1982) used such a method, and so do many of their followers in the real business cycle literature. See King, Plosser, and Rebelo (1988) for related methods of real business cycle models.

write the laws of motion as

$$\begin{pmatrix} 1 \\ a_{t+1} \\ \tilde{\theta}_{t+1} \end{pmatrix} = \begin{pmatrix} 1 & 0 & 0 \\ 0 & (1-\delta) & 0 \\ 0 & 0 & \rho \end{pmatrix} \begin{pmatrix} 1 \\ a_t \\ \tilde{\theta}_t \end{pmatrix} + \begin{pmatrix} 0 \\ 1 \\ 0 \end{pmatrix} i_t + \begin{pmatrix} 0 \\ 0 \\ 1 \end{pmatrix} w_{t+1}$$

where it is convenient to add the constant 1 as the first component of the state vector.

*Kydland and Prescott's method*

We want to replace $r(z_t)$ by a quadratic $z_t' M z_t$. We choose a point $\bar{z}$ and approximate with the first two terms of a Taylor series:[18]

$$\begin{aligned} \hat{r}(z) = r(\bar{z}) &+ (z-\bar{z})' \frac{\partial r}{\partial z} \\ &+ \frac{1}{2}(z-\bar{z})' \frac{\partial^2 r}{\partial z \partial z'} (z-\bar{z}). \end{aligned} \tag{4.45}$$

If the state $x_t$ is $n \times 1$ and the control $u_t$ is $k \times 1$, then the vector $z_t$ is $(n+k) \times 1$. Let $e$ be the $(n+k) \times 1$ vector with 0's everywhere except for a 1 in the row corresponding to the location of the constant unity in the state vector, so that $1 \equiv e' z_t$ for all $t$.

Repeatedly using $z'e = e'z = 1$, we can express equation (4.45) as

$$\hat{r}(z) = z' M z,$$

where

$$\begin{aligned} M =& e\left[r(\bar{z}) - \left(\frac{\partial r}{\partial z}\right)' \bar{z} + \frac{1}{2}\bar{z}' \frac{\partial^2 r}{\partial z \partial z'} \bar{z}\right] e' \\ &+ \frac{1}{2}\left(\frac{\partial r}{\partial z} e' - e \bar{z}' \frac{\partial^2 r}{\partial z \partial z'} - \frac{\partial^2 r}{\partial z \partial z'} \bar{z} e' + e \frac{\partial r}{\partial z}'\right) \\ &+ \frac{1}{2}\left(\frac{\partial^2 r}{\partial z \partial z'}\right) \end{aligned}$$

[18] This setup is taken from McGrattan (1994) and Anderson, Hansen, McGrattan, and Sargent (1996).

where the partial derivatives are evaluated at $\bar{z}$. Partition $M$, so that

$$z'Mz \equiv \begin{pmatrix} x \\ u \end{pmatrix}' \begin{pmatrix} M_{11} & M_{12} \\ M_{21} & M_{22} \end{pmatrix} \begin{pmatrix} x \\ u \end{pmatrix}$$
$$= \begin{pmatrix} x \\ u \end{pmatrix}' \begin{pmatrix} R & W \\ W' & Q \end{pmatrix} \begin{pmatrix} x \\ u \end{pmatrix}.$$

*Determination of $\bar{z}$*

Usually, the point $\bar{z}$ is chosen as the (optimal) stationary state of the *non-stochastic* version of the original nonlinear model:

$$\sum_{t=0}^{\infty} \beta^t r(z_t)$$
$$x_{t+1} = Ax_t + Bu_t.$$

This stationary point is obtained in these steps:

1. Find the Euler equations.
2. Substitute $z_{t+1} = z_t \equiv \bar{z}$ into the Euler equations and transition laws, and solve the resulting system of nonlinear equations for $\bar{z}$. This purpose can be accomplished, for example, by using the nonlinear equation solver `fsolve.m` in Matlab.

*Log linear approximation*

For some problems Christiano (1990) has advocated a quadratic approximation in logarithms. We illustrate his idea with the stochastic growth example. Define

$$\tilde{a}_t = \log a_t \ , \ \tilde{\theta}_t = \log \theta_t.$$

Christiano's strategy is to take $\tilde{a}_t, \tilde{\theta}_t$ as the components of the state and write the law of motion as

$$\begin{pmatrix} 1 \\ \tilde{a}_{t+1} \\ \tilde{\theta}_{t+1} \end{pmatrix} = \begin{pmatrix} 1 & 0 & 0 \\ 0 & 0 & 0 \\ 0 & 0 & \rho \end{pmatrix} \begin{pmatrix} 1 \\ \tilde{a}_t \\ \tilde{\theta}_t \end{pmatrix}$$
$$+ \begin{pmatrix} 0 \\ 1 \\ 0 \end{pmatrix} u_t \quad + \begin{pmatrix} 0 \\ 0 \\ 1 \end{pmatrix} w_{t+1}$$

where the control $u_t$ is $\tilde{a}_{t+1}$.

Express consumption as

$$c_t = A(\exp \tilde{a}_t)^{\alpha}(\exp \tilde{\theta}_t) + (1-\delta)\exp \tilde{a}_t - \exp \tilde{a}_{t+1}.$$

Substitute this expression into $\ln c_t$ , $\equiv r(z_t)$, and proceed as before to obtain the second-order Taylor series approximation about $\bar{z}$.

*Trend removal*

It is conventional in the real business cycle literature to specify the law of motion for the technology shock $\theta_t$ by

$$\tilde{\theta}_t = \log\left(\frac{\theta_t}{\gamma^t}\right), \quad \gamma > 1$$

$$\tilde{\theta}_{t+1} = \rho\tilde{\theta}_t + w_{t+1}, \qquad |\rho| < 1. \tag{4.46}$$

This inspires us to write the law of motion for capital as

$$\gamma \frac{a_{t+1}}{\gamma^{t+1}} = (1-\delta)\frac{a_t}{\gamma^t} + \frac{i_t}{\gamma^t}$$

or

$$\gamma \exp \tilde{a}_{t+1} = (1-\delta)\exp \tilde{a}_t + \exp(\tilde{i}_t) \tag{4.47}$$

where $\tilde{a}_t \equiv \log\left(\frac{a_t}{\gamma^t}\right), \tilde{i}_t = \log\left(\frac{i_t}{\gamma_t}\right)$. By studying the Euler equations for a model with a growing technology shock $(\gamma > 1)$, we can show that there exists a steady state for $\tilde{a}_t$, but not for $a_t$. Researchers often construct linear-quadratic approximations around the nonstochastic steady state of $\tilde{a}$.

## Exercises

*Exercise 4.1* Consider the modified version of the optimal linear regulator problem where the objective is to maximize

$$\sum_{t=0}^{\infty} \beta^t \{x_t' R x_t + u_t' Q u_t + 2u_t' H x_t\}$$

subject to the law of motion:

$$x_{t+1} = Ax_t + Bu_t.$$

Here $x_t$ is an $n \times 1$ state vector, $u_t$ is a $k \times 1$ vector of controls, and $x_0$ is a given initial condition. The matrices $R, Q$ are negative definite and symmetric. The maximization is with respect to sequences $\{u_t, x_t\}_{t=0}^{\infty}$.

**a.** Show that the optimal policy has the form

$$u_t = -(Q + \beta B'PB)^{-1}(\beta B'PA + H)x_t,$$

where $P$ solves the algebraic matrix Riccati equation

$$P = R + \beta A'PA - (\beta A'PB + H')(Q + \beta B'PB)^{-1}(\beta B'PA + H). \tag{4.48}$$

**b.** Write a Matlab program to solve equation (4.48) by iterating on $P$ starting from $P$ being a matrix of zeros.

*Exercise 4.2* Verify that equations (4.10) and (4.11) implement the policy improvement algorithm for the discounted linear regulator problem.

*Exercise 4.3* A household seeks to maximize

$$-\sum_{t=1}^{\infty} \beta^t \left\{(c_t - b)^2 + \gamma i_t^2\right\}$$

subject to

$$c_t + i_t = ra_t + y_t \tag{4.49a}$$

$$a_{t+1} = a_t + i_t \tag{4.49b}$$

$$y_{t+1} = \rho_1 y_t + \rho_2 y_{t-1}. \tag{4.49c}$$

Here $c_t, i_t, a_t, y_t$ are the household's consumption, investment, asset holdings, and exogenous labor income at $t$; while $b > 0, \gamma > 0, r > 0, \beta \in (0,1)$, and $\rho_1, \rho_2$ are parameters, and $y_0, y_{-1}$ are initial conditions. Assume that $\rho_1, \rho_2$ are such that $(1 - \rho_1 z - \rho_2 z^2) = 0$ implies $|z| > 1$.

**a.** Map this problem into an optimal linear regulator problem.

**b.** For parameter values $[\beta, (1+r), b, \gamma, \rho_1, \rho_2] = (.95, .95^{-1}, 30, 1, 1.2, -.3)$, compute the household's optimal policy function using your Matlab program from exercise 4.1.

*Exercise 4.4* Modify exercise 4.3 by assuming that the household seeks to maximize

$$-\sum_{t=1}^{\infty} \beta^t \left\{ (s_t - b)^2 + \gamma i_t^2 \right\}$$

Here $s_t$ measures consumption services that are produced by durables or habits according to

$$s_t = \lambda h_t + \pi c_t \tag{4.50a}$$

$$h_{t+1} = \delta h_t + \theta c_t \tag{4.50b}$$

where $h_t$ is the stock of the durable good or habit, $(\lambda, \pi, \delta, \theta)$ are parameters, and $h_0$ is an initial condition.

**a.** Map this problem into a linear regulator problem.

**b.** For the same parameter values as in exercise 4.3 and $(\lambda, \pi, \delta, \theta) = (1, .05, .95, 1)$, compute the optimal policy for the household.

**c.** For the same parameter values as in exercise 4.3 and $(\lambda, \pi, \delta, \theta) = (-1, 1, .95, 1)$, compute the optimal policy.

**d.** Interpret the parameter settings in part b as capturing a model of durable consumption goods, and the settings in part c as giving a model of habit persistence.

*Exercise 4.5* A household's labor income follows the stochastic process

$$y_{t+1} = \rho_1 y_t + \rho_2 y_{t-1} + w_{t+1} + \gamma w_t,$$

where $w_{t+1}$ is a Gaussian martingale difference sequence with unit variance. Calculate

$$E \sum_{j=0}^{\infty} \beta^j \left[ y_{t+j} | y^t, w^t \right], \tag{4.51}$$

where $y^t, w^t$ denotes the history of $y, w$ up to $t$.

**a.** Write a Matlab program to compute expression (4.51).

**b.** Use your program to evaluate expression (4.51) for the parameter values $(\beta, \rho_1, \rho_2, \gamma) = (.95, 1.2, -.4, .5)$.

*Exercise 4.6* **Dynamic Laffer curves**

The demand for currency in a small country is described by

$$M_t/p_t = \gamma_1 - \gamma_2 p_{t+1}/p_t, \tag{1}$$

where $\gamma_1 > \gamma_2 > 0$, $M_t$ is the stock of currency held by the public at the end of period $t$, and $p_t$ is the price level at time $t$. There is no randomness in the country, so that there is perfect foresight. Equation (1) is a Cagan-like demand function for currency, expressing real balances as an inverse function of the expected gross rate of inflation.

Speaking of Cagan, the government is running a permanent real deficit of $g$ per period, measured in goods, all of which it finances by currency creation. The government's budget constraint at $t$ is

$$(M_t - M_{t-1})/p_t = g, \tag{2}$$

where the left side is the real value of the new currency printed at time $t$. The economy starts at time $t = 0$, with the initial level of nominal currency stock $M_{-1} = 100$ being given.

For this model, define an *equilibrium* as a pair of *positive* sequences $\{p_t > 0, M_t > 0\}_{t=0}^{\infty}$ that satisfy equations (1) and (2) (portfolio balance and the government budget constraint, respectively) for $t \geq 0$, and the initial condition assigned for $M_{-1}$.

**a.** Let $\gamma_1 = 100, \gamma_2 = 50, g = .05$. Write a computer program to compute equilibria for this economy. Describe your approach and display the program.

**b.** Argue that there exists a continuum of equilibria. Find the *lowest* value of the initial price level $p_0$ for which there exists an equilibrium. (*Hint Number 1:* Notice the positivity condition that is part of the definition of equilibrium. *Hint Number 2:* Try using the general approach to solving difference equations described in the section "A Lagrangian formulation."

**c.** Show that for all of these equilibria except the one that is associated with the minimal $p_0$ that you calculated in part b, the gross inflation rate and the gross money creation rate both eventually converge to the *same* value. Compute this value.

**d.** Show that there is a unique equilibrium with a lower inflation rate than the one that you computed in part b. Compute this inflation rate.

**e.** Increase the level of $g$ to .075. Compare the (eventual or asymptotic) inflation rate that you computed in part b and the inflation rate that you computed in part c. Are your results consistent with the view that "larger permanent deficits cause larger inflation rates"?

**f.** Discuss your results from the standpoint of the "Laffer curve."

*Hint:* A Matlab program `dlqrmon.m` performs the calculations. It is available from the web site for the book.

# 5
# Search, Matching, and Unemployment

## Introduction

This chapter applies dynamic programming to a choice between only two actions, to accept or reject a take-it-or-leave-it job offer. An unemployed worker faces a probability distribution of wage offers or job characteristics, from which a limited number of offers are drawn each period. Given his perception of the probability distribution of offers, the worker must devise a strategy for deciding when to accept an offer.

The theory of search is a tool for studying unemployment. Search theory puts unemployed workers in a setting where they *choose* to reject available offers and to remain unemployed now because they prefer to wait for better offers later. We want to use the theory to study how workers' choices would respond to variations in the rate of unemployment compensation, the perceived riskiness of wage distributions, the quality of information about jobs, and the frequency with which the wage distribution can be sampled.

This chapter provides an introduction to the techniques used in the search literature and a sampling of search models. The chapter studies ideas introduced in two important papers on search by McCall (1970) and Jovanovic (1979a). These papers differ in the search technology with which they confront an unemployed worker.[1]

## Preliminaries

This section describes elementary properties of probabilty distributions that are used extensively in search theory.

[1] Stigler's (1961) important early paper studied a search technology different from both McCall's and Jovanovic's. In Stigler's model, an unemployed worker has to choose in advance a number $n$ of offers to draw, from which he takes the highest wage offer. Stigler's formulation of the search problem was not sequential.

*Nonnegative random variables*

We begin with some characteristics of nonnegative random variables that possess first moments. Consider a random variable $p$ with a cumulative probability distribution function $F(P)$ defined by $\text{prob}\{p \leq P\} = F(P)$. We assume that $F(0) = 0$, that is, that $p$ is nonnegative. We assume that $F(\infty) = 1$ and that $F$, a nondecreasing function, is continuous from the right. We also assume that there is an upper bound $B < \infty$ such that $F(B) = 1$, so that $p$ is bounded with probability 1.

The mean of $p$, $Ep$, is defined by

$$Ep = \int_0^B p\, dF(p). \tag{5.1}$$

Letting $u = 1 - F(p)$ and $v = p$, and using the integration-by-parts formula

$$\int_a^b u\, dv = uv\Big|_a^b - \int_a^b v\, du,$$

we verify that

$$\int_0^B [1 - F(p)]dp = \int_0^B p\, dF(p).$$

Thus we have the following formula for the mean of a nonnegative random variable:

$$Ep = \int_0^B [1 - F(p)]dp = B - \int_0^B F(p)dp. \tag{5.2}$$

Now consider two independent random variables $p_1$ and $p_2$ drawn from the distribution $F$. Consider the event $\{(p_1 < p) \cap (p_2 < p)\}$, which by the independence assumption has probability $F(p)^2$. The event $\{(p_1 < p) \cap (p_2 < p)\}$ is equivalent to the event $\{\max(p_1, p_2) < p\}$, where "max" denotes the maximum. Therefore, if we use formula (5.2), the random variable $\max(p_1, p_2)$ has mean

$$E \max(p_1, p_2) = B - \int_0^B F(p)^2 dp. \tag{5.3}$$

Similarly, if $p_1, p_2, \ldots, p_n$ are $n$ independent random variables drawn from $F$, we have $\text{prob}\{\max(p_1, p_2, \ldots, p_n) < p\} = F(p)^n$ and

$$M_n \equiv E \max(p_1, p_2, \ldots, p_n) = B - \int_0^B F(p)^n dp, \tag{5.4}$$

where $M_n$ is defined as the expected value of the maximum of $p_1, \ldots, p_n$.

*Mean-preserving spreads*

Rothschild and Stiglitz have introduced the notion of mean-preserving spreads as a convenient way of characterizing the riskiness of two distributions with the same mean. Consider a class of distributions with the same mean. We index this class by a parameter $r$ belonging to some set $R$. For the $r$th distribution we denote $\text{prob}\{p \leq P\} = F(P, r)$. We assume that there is a single finite $B$ such that $F(B, r) = 1$ for all $r$ in $R$ and continue to assume as before that $F(0, r) = 0$ for all $r$ in $R$, so that we are considering a class of distributions $R$ for nonnegative, bounded random variables.

From equation (5.2), we now have

$$Ep = B - \int_0^B F(p, r)dp. \tag{5.5}$$

Therefore, two distributions with the same value of $\int_0^B F(\theta, r)d\theta$ have identical means. We write this as the identical means condition:

(i) $$\int_0^B [F(\theta, r_1) - F(\theta, r_2)]d\theta = 0.$$

Two distributions $r_1, r_2$ are said to satisfy the single-crossing property if there exists a $\hat{\theta}$ with $0 < \hat{\theta} < B$ such that

(ii) $$F(\theta, r_2) - F(\theta, r_1) \leq 0 (\geq 0) \qquad \text{when} \quad \theta \geq (\leq) \hat{\theta}.$$

Figure 5.1 illustrates the single-crossing property. If two distributions $r_1$ and $r_2$ satisfy properties (i) and (ii), we can regard distribution $r_2$ as having been obtained from $r_1$ by a process that shifts probability toward the tails of the distribution while keeping the mean constant.

Properties (i) and (ii) imply (iii), the following property:

(iii) $$\int_0^y [F(\theta, r_2) - F(\theta, r_1)]d\theta \geq 0, \qquad 0 \leq y \leq B.$$

Rothschild and Stiglitz regard properties (i) and (iii) as defining the concept of a "mean-preserving increase in spread." In particular, a distribution indexed

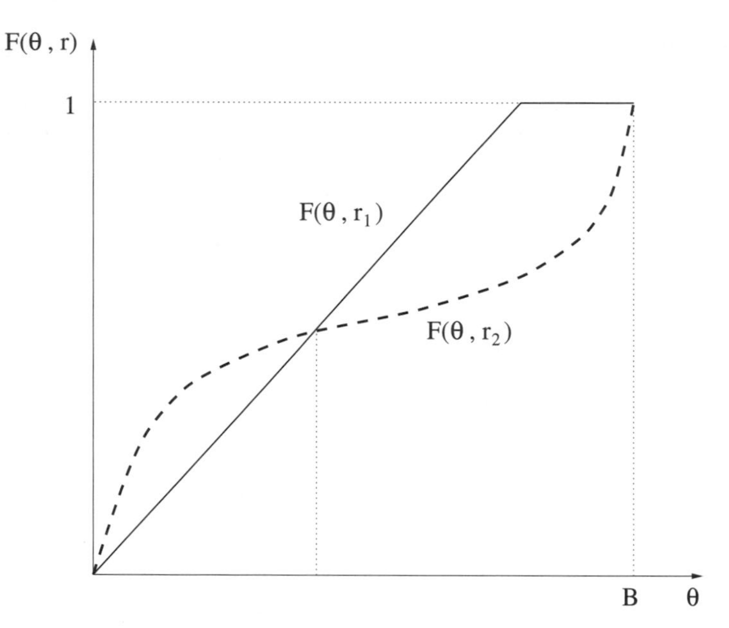


**Figure 5.1** Two distributions, $r_1$ and $r_2$, that satisfy the single-crossing property.

by $r_2$ is said to have been obtained from a distribution indexed by $r_1$ by a mean-preserving increase in spread if the two distributions satisfy (i) and (iii).[2]

For infinitesimal changes in $r$, Diamond and Stiglitz use the differential versions of properties (i) and (iii) to rank distributions with the same mean in order of riskiness. An increase in $r$ is said to represent a mean-preserving increase in risk if

(iv) $$\int_0^B F_r(\theta, r) d\theta = 0$$

[2] Rothschild and Stiglitz (1970, 1971) use properties (i) and (iii) to characterize mean-preserving spreads rather than (i) and (ii) because (i) and (ii) fail to possess transitivity. That is, if $F(\theta, r_2)$ is obtained from $F(\theta, r_1)$ via a mean-preserving spread in the sense that the term has in (i) and (ii), and $F(\theta, r_3)$ is obtained from $F(\theta, r_2)$ via a mean-preserving spread in the sense of (i) and (ii), it does not follow that $F(\theta, r_3)$ satisfies the single crossing property (ii) vis-à-vis distribution $F(\theta, r_1)$. A definition based on (i) and (iii), however, does provide a transitive ordering, which is a desirable feature for a definition designed to order distributions according to their riskiness.

$$\text{(v)} \qquad \int_0^y F_r(\theta, r)d\theta \geq 0, \qquad 0 \leq y \leq B \,,$$

where $F_r(\theta, r) = \partial F(\theta, r)/\partial r$.

## McCall's model of intertemporal job search

We now consider an unemployed worker who is searching for a job under the following circumstances: Each period the worker draws one offer $w$ from the same wage distribution $F(W) = \text{prob}\{w \leq W\}$, with $F(0) = 0$, $F(B) = 1$ for $B < \infty$. The worker has the option of rejecting the offer, in which case he or she receives $c$ this period in unemployment compensation and waits until next period to draw another offer from $F$; alternatively, the worker can accept the offer to work at $w$, in which case he or she receives a wage of $w$ per period forever. Neither quitting nor firing is permitted.

Let $y_t$ be the worker's income in period $t$. We have $y_t = c$ if the worker is unemployed and $y_t = w$ if the worker has accepted an offer to work at wage $w$. The unemployed worker devises a strategy to maximize $E \sum_{t=0}^{\infty} \beta^t y_t$ where $0 < \beta < 1$ is a discount factor.

Let $v(w)$ be the expected value of $\sum_{t=0}^{\infty} \beta^t y_t$ for a worker who has offer $w$ in hand, who is deciding whether to accept or to reject it, and who behaves optimally. We assume no recall. The Bellman equation for the worker's problem is

$$v(w) = \max \left\{ \frac{w}{1-\beta}, c + \beta \int v(w') dF(w') \right\}, \tag{5.6}$$

where the maximization is over the two actions: (1) *accept* the wage offer $w$ and work forever at wage $w$, or (2) *reject* the offer, receive $c$ this period, and draw a new offer $w'$ from distribution $F$ next period. Figure 5.2 graphs the functional equation (5.6) and reveals that its solution will be of the form

$$v(w) = \begin{cases} \dfrac{\bar{w}}{1-\beta} = c + \beta \displaystyle\int_0^B v(w') dF(w') & \text{if} \quad w \leq \bar{w} \\ \dfrac{w}{1-\beta} & \text{if} \quad w \geq \bar{w}. \end{cases} \tag{5.7}$$

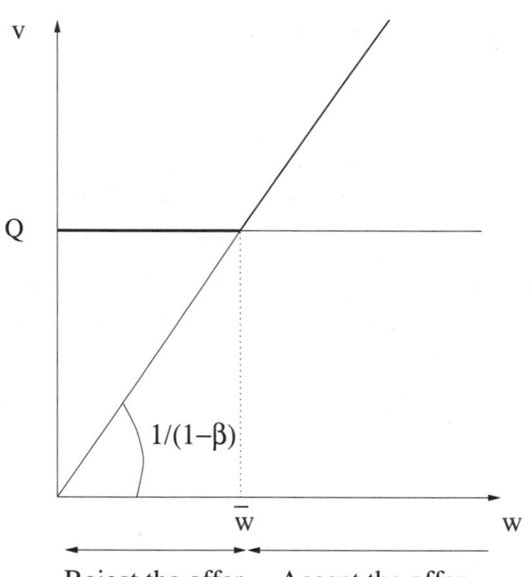


**Figure 5.2** The function $v(w) = \max\{w/(1-\beta), c + \beta \int_0^B v(w')dF(w')\}$. The reservation wage $\bar{w} = (1-\beta)[c + \beta \int_0^B v(w')dF(w')]$.

Using equation (5.7), we can convert the functional equation (5.6) into an ordinary equation in the reservation wage $\bar{w}$. Evaluating $v(\bar{w})$ and using equation (5.7), we have

$$\frac{\bar{w}}{1-\beta} = c + \beta \int_0^{\bar{w}} \frac{\bar{w}}{1-\beta} dF(w') + \beta \int_{\bar{w}}^B \frac{w'}{1-\beta} dF(w')$$

or

$$\frac{\bar{w}}{1-\beta} \int_0^{\bar{w}} dF(w') + \frac{\bar{w}}{1-\beta} \int_{\bar{w}}^B dF(w')$$
$$= c + \beta \int_0^{\bar{w}} \frac{\bar{w}}{1-\beta} dF(w') + \beta \int_{\bar{w}}^B \frac{w'}{1-\beta} dF(w')$$

or

$$\bar{w} \int_0^{\bar{w}} dF(w') - c = \frac{1}{1-\beta} \int_{\bar{w}}^B (\beta w' - \bar{w}) dF(w').$$

Adding $\bar{w} \int_{\bar{w}}^{B} dF(w')$ to both sides gives

$$(\bar{w} - c) = \frac{\beta}{1-\beta} \int_{\bar{w}}^{B} (w' - \bar{w}) dF(w'). \tag{5.8}$$

Equation (5.8) is often used to characterize the determination of the reservation wage $\bar{w}$. The left side is the cost of searching one more time when an offer $\bar{w}$ is in hand. The right side is the expected benefit of searching one more time in terms of the expected present value associated with drawing $w' > \bar{w}$. Equation (5.8) instructs the agent to set $\bar{w}$ so that the cost of searching one more time equals the benefit.

Let us define the function on the right side of equation (5.8) as

$$h(w) = \frac{\beta}{1-\beta} \int_{w}^{B} (w' - w) dF(w'). \tag{5.9}$$

Notice that $h(0) = \beta/(1-\beta)Ew$, that $h(B) = 0$, and that $h(w)$ is differentiable, with derivative given by [3]

$$h'(w) = -\frac{\beta}{1-\beta}[1 - F(w)] < 0.$$

We also have

$$h''(w) = +\frac{\beta}{1-\beta} F'(w) > 0,$$

so that $h(w)$ is convex to the origin. Figure 5.3 graphs $h(w)$ against $(w-c)$ and indicates how $\bar{w}$ is determined. From Figure 5.3 it is apparent that an increase

[3] To compute $h'(w)$, we apply Leibniz' rule to equation (5.9). Let $\phi(t) = \int_{\alpha(t)}^{\beta(t)} f(x,t)dx$ for $t \in [c,d]$. Assume that $f$ and $f_t$ are continuous and that $\alpha, \beta$ are differentiable on $[c,d]$. Then Leibniz' rule asserts that $\phi(t)$ is differentiable on $[c,d]$ and

$$\phi'(t) = f[\beta(t),t]\beta'(t) - f[\alpha(t),t]\alpha'(t) + \int_{\alpha(t)}^{\beta(t)} f_t(x,t)dx.$$

To apply this formula to the equation in the text, let $w$ play the role of $t$.

in $c$ leads to an increase in $\bar{w}$.

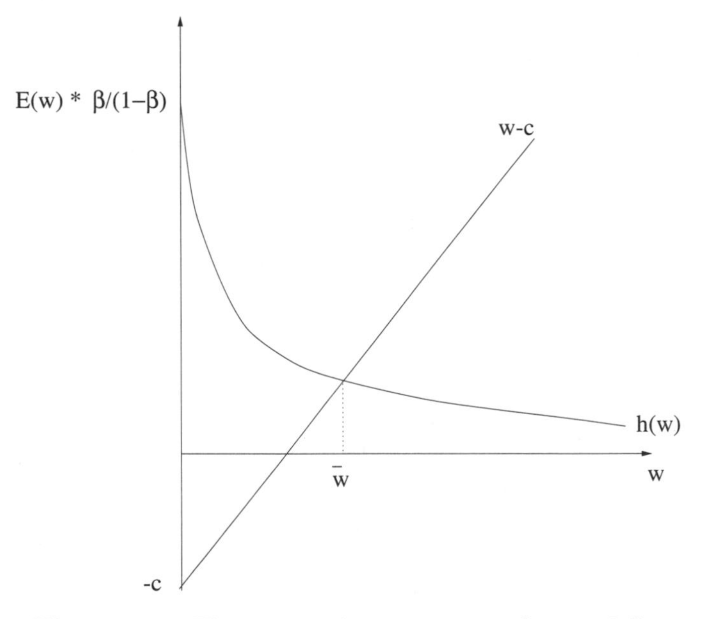


**Figure 5.3** The reservation wage, $\bar{w}$, that satisfies $\bar{w} - c = [\beta/(1-\beta)] \int_{\bar{w}}^{B} (w' - \bar{w}) dF(w') \equiv h(\bar{w})$.

To get an alternative characterization of the condition determining $\bar{w}$, we return to equation (5.8) and express it as

$$\begin{aligned}\bar{w} - c =& \frac{\beta}{1-\beta} \int_{\bar{w}}^{B} (w' - \bar{w}) dF(w') + \frac{\beta}{1-\beta} \int_{0}^{\bar{w}} (w' - \bar{w}) dF(w') \\ & - \frac{\beta}{1-\beta} \int_{0}^{\bar{w}} (w' - \bar{w}) dF(w') \\ =& \frac{\beta}{1-\beta} Ew - \frac{\beta}{1-\beta} \bar{w} - \frac{\beta}{1-\beta} \int_{0}^{\bar{w}} (w' - \bar{w}) dF(w')\end{aligned}$$

or

$$\bar{w} - (1-\beta) c = \beta E w - \beta \int_{0}^{\bar{w}} (w' - \bar{w}) dF(w').$$

Applying integration by parts to the last integral on the right side and rearranging, we have

$$\bar{w} - c = \beta (Ew - c) + \beta \int_{0}^{\bar{w}} F(w') dw'. \tag{5.10}$$

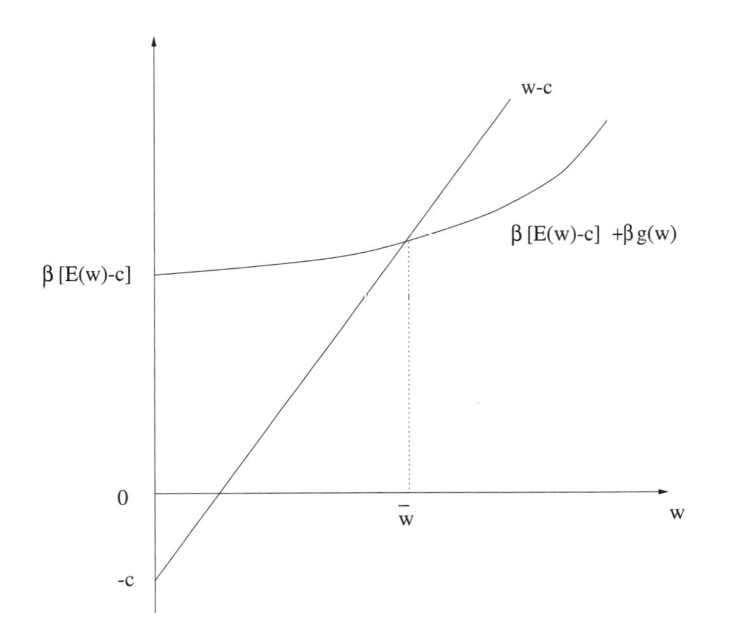


**Figure 5.4** The reservation wage, $\bar{w}$, that satisfies $\bar{w} - c = \beta(Ew - c) + \beta \int_0^{\bar{w}} F(w')dw' \equiv \beta(Ew - c) + \beta g(\bar{w})$.

At this point it is useful to define the function

$$g(s) = \int_0^s F(p)dp. \tag{5.11}$$

This function has the characteristics that $g(0) = 0$, $g(s) \geq 0$, $g'(s) = F(s) > 0$, and $g''(s) = F'(s) > 0$ for $s > 0$. Then equation (5.10) can be expressed alternatively as $\bar{w} - c = \beta(Ew - c) + \beta g(\bar{w})$, where $g(s)$ is the function defined by equation (5.11). In Figure 5.4 we graph the determination of $\bar{w}$, using equation (5.10).

*Effects of unemployment compensation and mean preserving spreads*

Figure 5.4 can be used to establish two propositions about $\bar{w}$. First, given $F$, $\bar{w}$ increases when the rate of unemployment compensation $c$ increases. Second, given $c$, a mean-preserving increase in risk causes $\bar{w}$ to increase. This second proposition follows directly from Figure 5.4 and the characterization (iii) or (v) of a mean-preserving increase in risk. From the definition of $g$ in equation (5.11) and the characterization (iii) or (v), a mean-preserving spread causes an upward shift in $\beta(Ew - c) + \beta g(w)$.

Since either an increase in unemployment compensation or a mean-preserving increase in risk raises the reservation wage, it follows from the expression for the value function in equation (5.7) that unemployed workers are also better off in those situations. It is obvious that an increase in unemployment compensation raises the welfare of unemployed workers but it might seem surprising in the case of a mean-preserving increase in risk. Intuition for this latter finding can be gleaned from the result in option pricing theory that the value of an option is an increasing function of the variance in the price of the underlying asset. This is so because the option holder receives payoffs only from the tail of the distribution. In our context, the unemployed worker has the option to accept a job and the asset value of a job offering wage rate $w$ is equal to $w/(1-\beta)$. Under a mean-preserving increase in risk, the higher incidence of very good wage offers increases the value of searching for a job while the higher incidence of very bad wage offers is less detrimental because the option to work will in any case not be exercised at such low wages.

*Bathtub model*

In the interests of getting a simple equilibrium model of unemployment, assume now that each period the worker faces a fixed probability $\alpha \in (0,1)$ of surviving into the next period. The worker cares about $y_t$ only so long as he is alive. All other features of his problem remain the same. The worker's Bellman equation becomes

$$v(w) = \max \left\{ \frac{w}{1-\beta\alpha}, c + \beta\alpha \int_0^B v(w')dF(w') \right\}, \tag{5.12}$$

which is equivalent to equation (5.6) with an adjustment to the discount factor to account for the survival hazard. Let $\bar{w}$ be the reservation wage associated with an optimal policy.

Now assume that there is a continuum of constant measure of ex ante identical workers. Assume that each period a fraction $1-\alpha$ new workers are born. They replace an equal number of newly departed workers, so that the measure of workers remains constant. Assume that all newborn workers start out being unemployed. Let $U_t$ denote the unemployment rate at the beginning of time $t$. Evidently, the unemployment rate obeys the law of motion

$$U_t = (1 - \alpha) + U_{t-1}\alpha F(\bar{w}).$$

The right side is the sum of the fraction of newborn workers and the fraction of surviving workers who were unemployed at the end of last period (i.e., those who rejected offers because they were less than $\bar{w}$). A constant stationary level of unemployment is then

$$U = \frac{1 - \alpha}{1 - \alpha F(\bar{w})} \tag{5.13}$$

Equation (5.13) gives the constant unemployment rate in which flows into unemployment from birth match flows out of unemployment due to successful search. Such a model is sometimes called a *bathtub model*. It provides a simple way of studying how alterations in $F$ and $c$ affect $U$ through their effects on $\bar{w}$.

*Waiting times*

It is straightforward to derive the probability distribution of the waiting time until a job offer is accepted. Let $N$ be the random variable "length of time until a successful offer is encountered," with the understanding that $N = 1$ if the first job offer is accepted. Let $\lambda = \int_0^{\bar{w}} dF(w')$ be the probability that a job offer is rejected. Then we have $\text{prob}\{N = 1\} = (1 - \lambda)$. The event that $N = 2$ is the event that the first draw is less than $\bar{w}$, which occurs with probability $\lambda$, and that the second draw is greater than $\bar{w}$, which occurs with probability $(1 - \lambda)$. By virtue of the independence of successive draws, we have $\text{prob}\{N = 2\} = (1-\lambda)\lambda$. More generally, $\text{prob}\{N = j\} = (1-\lambda)\lambda^{j-1}$, so the waiting time is geometrically distributed. The mean waiting time is given by

$$\sum_{j=1}^{\infty} j \cdot \text{prob}\{N = j\} = \sum_{j=1}^{\infty} j(1-\lambda)\lambda^{j-1} = (1-\lambda)\sum_{j=1}^{\infty}\sum_{k=1}^{j} \lambda^{j-1}$$
$$= (1-\lambda)\sum_{k=0}^{\infty}\sum_{j=1}^{\infty} \lambda^{j-1+k} = (1-\lambda)\sum_{k=0}^{\infty} \lambda^k (1-\lambda)^{-1} = (1-\lambda)^{-1}.$$

That is, the mean waiting time to a successful job offer equals the reciprocal of the probability of an accepted offer on a single trial.[4]

We invite the reader to prove that, given $F$, the mean waiting time increases with increases in the rate of unemployment compensation, $c$.

*Quitting*

Thus far, we have supposed that the worker cannot quit. It happens that had we given the worker the option to quit and search again, after being unemployed one period, he would never exercise that option. To see this point, recall that the reservation wage $\bar{w}$ satisfies

$$v(\bar{w}) = \frac{\bar{w}}{1-\beta} = c + \beta \int v(w')dF(w').$$

Suppose the agent has some wage $w$ in hand. We compute the lifetime utility associated with three mutually exclusive alternative ways of responding to that offer (assuming that the agent behaves optimally after any rejection of this wage $w$).

A1. Accept the wage and keep the job forever:

$$\frac{w}{1-\beta}.$$

A2. Accept the wage but quit after $t$ periods:

$$\frac{w - \beta^t w}{1-\beta} + \beta^t \left(c + \beta \int v(w')dF(w')\right) = \frac{w}{1-\beta} - \beta^t \frac{w - \bar{w}}{1-\beta}.$$

A3. Reject the wage:

$$c + \beta \int v(w')dF(w') = \frac{\bar{w}}{1-\beta}.$$

We conclude that if $w < \bar{w}$,

$$A1 \prec A2 \prec A3,$$

[4] An alternative way of deriving the mean waiting time is to use the algebra of $z$ transforms, we say that $h(z) = \sum_{j=0}^{\infty} h_j z^j$ and note that $h'(z) = \sum_{j=1}^{\infty} j h_j z^{j-1}$ and $h'(1) = \sum_{j=1}^{\infty} j h_j$. (For an introduction to $z$ transforms, see Gabel and Roberts 1973.) The $z$ transform of the sequence $(1-\lambda)\lambda^{j-1}$ is given by $\sum_{j=1}^{\infty}(1-\lambda)\lambda^{j-1}z^j = (1-\lambda)z/(1-\lambda z)$. Evaluating $h'(z)$ at $z=1$ gives, after some simplification, $h'(1) = 1/(1-\lambda)$. Therefore we have that the mean waiting time is given by $(1-\lambda)\sum_{j=1}^{\infty} j\lambda^{j-1} = 1/(1-\lambda)$.

and if $w > \bar{w}$,

$$A1 \succ A2 \succ A3.$$

The three alternatives yield the same lifetime utility when $w = \bar{w}$.

*Firing*

We now briefly consider a modification of the job search model in which each period after the first period on the job the worker faces probability $1 > \alpha > 0$ of being fired. The probability $\alpha$ of being fired next period is assumed to be independent of tenure. The worker continues to sample wage offers from a time-invariant and known probability distribution $F$ and to receive unemployment compensation in the amount $c$. The worker receives a time-invariant wage $w$ on a job until she is fired. A worker who is fired becomes unemployed for one period before drawing a new wage.

We let $v(w)$ be the expected present value of income of a previously unemployed worker who has offer $w$ in hand and who behaves optimally. If she rejects the offer, she receives $c$ in unemployment compensation this period and next period draws a new offer $w'$, whose value to her now is $\beta \int v(w')dF(w')$. If she rejects the offer, $v(w) = c + \beta \int v(w')dF(w')$. If she accepts the offer, she receives $w$ this period, with probability $1-\alpha$ that she is not fired next period, in which case she receives $\beta v(w)$ and with probability $\alpha$ that she is fired, and after one period of unemployment draws a new wage, receiving $\beta[c + \beta \int v(w')dF(w')]$. Therefore, if she accepts the offer, $v(w) = w + \beta(1-\alpha)v(w) + \beta\alpha[c + \beta \int v(w')dF(w')]$. Thus the Bellman equation becomes

$$v(w) = \max\{w + \beta(1-\alpha)v(w) + \beta\alpha[c + \beta Ev], c + \beta Ev\},$$

where $Ev = \int v(w')dF(w')$. This equation has a solution of the form[5]

$$v(w) = \begin{cases} \dfrac{w + \beta\alpha[c + \beta Ev]}{1 - \beta(1-\alpha)}, & w \geq \bar{w} \\ c + \beta Ev, & w \leq \bar{w}, \end{cases}$$

where $\bar{w}$ solves

$$\frac{\bar{w} + \beta\alpha[c + \beta Ev]}{1 - \beta(1-\alpha)} = c + \beta Ev.$$

[5] That it takes this form can be established by guessing that $v(w)$ is nondecreasing in $w$. This guess implies the equation in the text for $v(w)$, which is nondecreasing in $w$. This argument verifies that $v(w)$ is nondecreasing, given the uniqueness of the solution of the Bellman equation.

The optimal policy is of the reservation wage form. The reservation wage $\bar{w}$ will not be characterized here as a function of $c$, $F$, and $\alpha$; the reader is invited to do so by pursuing the implications of the preceding formula.

## A model of career choice

This section describes Derek Neal's (1999) model of career choice, which he used to study the employment histories of recent high school graduates. Neal wanted to explain why young men switch jobs and "careers" often early in their work history, then later focus their job choice on a single career. Neal's model can be regarded as a simplified version of Miller's (1984) model.

A worker chooses career-job $(\theta, \epsilon)$ pairs subject to the following conditions: There is no unemployment. The worker's earnings at time $t$ are $\theta_t + \epsilon_t$. The worker maximizes $E\sum_{t=0}^{\infty} \beta^t(\theta_t + \epsilon_t)$. A *career* is a draw of $\theta$ from c.d.f. $\sim F$; a *job* is a draw of $\epsilon$ from c.d.f. $\sim G$. Successive draws are independent, and $G(0) = F(0) = 0$, $G(B_\epsilon) = F(B_\theta) = 1$. The worker can draw a new career only if he also draws a new job. However, the worker is free to retain his existing career ($\theta$), and to draw a new job ($\epsilon'$). The worker decides at the beginning of a period whether to stay in the current career-job pair, stay in his current career but draw a new job, or to draw a new career-job pair. There is no recalling past jobs or careers.

At the beginning of a period, let $v(\theta, \epsilon)$ be the optimal value of the problem for a worker with career-job pair $(\theta, \epsilon)$ and who is about to decide whether to draw a new career or job. The Bellman equation is

$$\begin{aligned} v(\theta, \epsilon) = \max\Bigg\{&\theta + \epsilon + \beta v(\theta, \epsilon),\ \theta + \int [\epsilon' + \beta v(\theta, \epsilon')]\, dG(\epsilon'), \\ &\int\int [\theta' + \epsilon' + \beta v(\theta', \epsilon')]\, dF(\theta') dG(\epsilon')\Bigg\}, \end{aligned} \tag{5.14}$$

where the maximization is over the three actions: (1) retain the present job-career pair; (2) retain the present career but draw a new job; and (3) draw both a new job and a new career. The value function is increasing in both $\theta$ and $\epsilon$.

Figures 5.5 and 5.6 display the optimal value function and the optimal decision rule for an example of Neal's model where $F$ and $G$ are each distributed according to uniform distributions with 15 nodes and $\beta = .95$. The value function was computed by iterating to convergence on the Bellman equation. The optimal policy is characterized by two curves in the $\theta, \epsilon$ space. For high values of both, the worker stays put. For high $\theta$ but low $\epsilon$, the worker retains his career but searches for a better job.

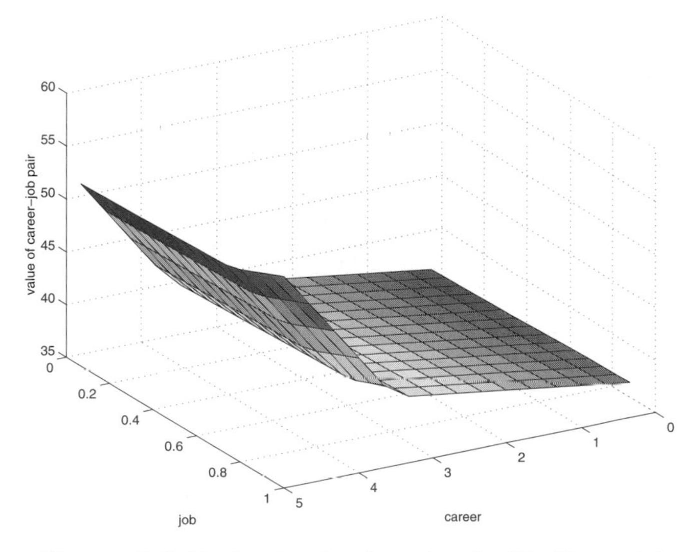


**Figure 5.5** Optimal value function for Neal's model with $\beta = .95$.

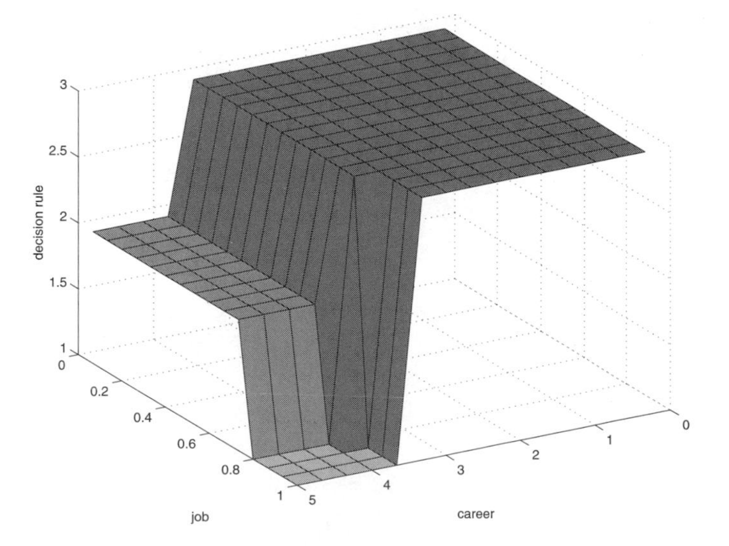


**Figure 5.6** Optimal decision rule for Neal's model. The ordinates $1, 2, 3$ denote keep job and career, keep career but draw a new job, and draw a new job-career pair, respectively.

When the career-job pair $(\theta, \epsilon)$ is such that the worker chooses to stay put, the value function in (5.14) attains the value $(\theta + \epsilon)/(1 - \beta)$. Of course, this happens when the decision to stay put weakly dominates the other two actions;

$$\frac{\theta + \epsilon}{1 - \beta} \geq \max\left\{\theta + \int [\epsilon' + \beta v(\theta, \epsilon')]\, dG(\epsilon'),\ Q\right\}, \tag{5.15}$$

where Q is the value of drawing both a new job and a new career,

$$Q \equiv \int\int [\theta' + \epsilon' + \beta v(\theta', \epsilon')]\, dF(\theta') dG(\epsilon').$$

For a given career $\theta$, we can solve for the job $\epsilon$ that makes equation (5.15) hold with equality, let $\bar{\epsilon}(\theta)$ denote this critical value. It follows that the decision to stay put is optimal for any career, job pair $(\theta, \epsilon)$ that satisfies $\epsilon \geq \bar{\epsilon}(\theta)$. When this condition is not satisfied, the worker will either draw a new career-job pair $(\theta', \epsilon')$ or only a new job $\epsilon'$. The latter decision to retain the current career $\theta$ is optimal when

$$\theta + \int [\epsilon' + \beta v(\theta, \epsilon')]\, dG(\epsilon') \geq Q. \tag{5.16}$$

Here we can solve for the critical career value $\bar{\theta}$ such that equation (5.16) holds with equality. That is, independent of $\epsilon$, the worker will never abandon any career $\theta \geq \bar{\theta}$. The decision rule for accepting the current career can thus be expressed as follows: accept the current career $\theta$ if $\theta \geq \bar{\theta}$ or if the current career-job pair $(\theta, \epsilon)$ satisfies $\epsilon \geq \bar{\epsilon}(\theta)$.

Probably the most interesting feature of the model is that it is possible to draw a $(\theta, \epsilon)$ pair such that the value of keeping the career $(\theta)$ and drawing a new job match $(\epsilon')$ exceeds both the value of stopping search, and the value of starting again to search from the beginning by drawing a new $(\theta', \epsilon')$ pair. This outcome occurs when a large $\theta$ is drawn with a small $\epsilon$. In this case, it can occur that $\theta \geq \bar{\theta}$ and $\epsilon < \bar{\epsilon}(\theta)$.

Viewed as a normative model for young workers, Neal's model tells them: don't shop for a firm until you have found a career you like. As a positive model, it predicts that workers will not switch careers after they have settled on one. Neal presents data indicating that while this prediction is too stark, it is a tolerable first approximation. He suggests that extending the model to include learning, along the lines of Jovanovic's model to be described next, could help explain the later career switches that his model misses.[6]

[6] Neal's model can be used to deduce waiting times to the event $(\theta \geq \bar{\theta}) \cup [\epsilon \geq \bar{\epsilon}(\theta)]$ and $\epsilon \geq \bar{\epsilon}(\theta)$. The first event is choosing a career that is never abandoned. The second is choosing

## A simple version of Jovanovic's matching model

The preceding models invite questions about how we envision the determination of the wage distribution $F$. Given $F$, we have seen that the worker sets a reservation wage $\bar{w}$ and refuses all offers less than $\bar{w}$. If homogeneous firms were facing a homogeneous population of workers all of whom used such a decision rule, no wages less than $\bar{w}$ would ever be recorded. Furthermore, it would seem to be in the interest of each firm simply to offer the reservation wage $\bar{w}$ and never to make an offer exceeding it. These considerations reveal a force that would tend to make the wage distribution collapse to a trivial one concentrated at $\bar{w}$. This situation, however, would invalidate the assumptions under which the reservation wage policy was derived. It is thus a serious challenge to imagine an equilibrium context in which there survive both a distribution of wage or price offers and optimal search activity by individual agents in the face of that distribution. A number of attempts have been made to meet this challenge.

One interesting effort stems from matching models, in which the main idea is to reinterpret $w$ not as a wage but instead, more broadly, as a parameter characterizing the entire quality of a match occurring between a pair of agents. The parameter $w$ is regarded as a summary measure of the productivities or utilities jointly generated by the activities of the match. We can consider pairs consisting of a firm and a worker, a man and a woman, a house and an owner, or a person and a hobby. The idea is to analyze the way in which matches form and maybe also dissolve by viewing both parties to the match as being drawn from populations that are statistically homogeneous to an outside observer, even though the match is idiosyncratic from the perspective of the parties to the match.

Jovanovic (1979a) has used a model of this kind supplemented by a hypothesis that both sides of the match behave optimally but only gradually learn about the quality of the match. Jovanovic was motivated by a desire to explain three features of labor market data: (1) on average, wages rise with tenure on the job, (2) quits are negatively correlated with tenure (that is, a quit has a higher probability of occurring earlier in tenure than later), and (3) the probability of a subsequent quit is negatively correlated with the current wage rate. Jovanovic's insight was that each of these empirical regularities could be interpreted as reflecting the operation of a matching process with gradual learning about match quality. We consider a simplified version of Jovanovic's model of matching. (Prescott and Townsend, 1980, describe a discrete-time version of Jovanovic's model, which has

---

a permanent job. Neal used the model to approximate and interpret observed career and job switches of young workers.

been simplified here.) A market has two sides that could be variously interpreted as consisting of firms and workers, or men and women, or owners and renters, or lakes and fishermen. Following Jovanovic, we shall adopt the firm-worker interpretation here. An unmatched worker and a firm form a pair and jointly draw a random match parameter $\theta$ from a probability distribution with cumulative distribution function $\text{prob}\{\theta \leq s\} = F(s)$. Here the match parameter reflects the marginal productivity of the worker in the match. In the first period, before the worker decides whether to work at this match or to wait and to draw a new match next period from the same distribution $F$, the worker and the firm both observe only $y = \theta + u$, where $u$ is a random noise that is uncorrelated with $\theta$. Thus in the first period, the worker-firm pair receives only a noisy observation on $\theta$. This situation corresponds to that when both sides of the market form only an error-ridden impression of the quality of the match at first. On the basis of this noisy observation, the firm, which is imagined to operate competitively under constant returns to scale, offers to pay the worker the conditional expectation of $\theta$, given $(\theta + u)$, for the first period, with the understanding that in subsequent periods it will pay the worker the expected value of $\theta$, depending on whatever additional information both sides of the match receive. Given this policy of the firm, the worker decides whether to accept the match and to work this period for $E[\theta|(\theta + u)]$ or to refuse the offer and draw a new match parameter $\theta'$ and noisy observation on it, $(\theta' + u')$, next period. If the worker decides to accept the offer in the first period, then in the second period both the firm and the worker are assumed to observe the true value of $\theta$. This situation corresponds to that in which both sides learn about each other and about the quality of the match. In the second period the firm offers to pay the worker $\theta$ then and forever more. The worker next decides whether to accept this offer or to quit, be unemployed this period, and draw a new match parameter and a noisy observation on it next period.

We can conveniently think of this process as having three stages. Stage 1 is the "predraw" stage, in which a previously unemployed worker has yet to draw the one match parameter and the noisy observation on it that he is entitled to draw after being unemployed the previous period. We let $Q$ denote the expected present value of wages, before drawing, of a worker who was unemployed last period and who behaves optimally. The second stage of the process occurs after the worker has drawn a match parameter $\theta$, has received the noisy observation of $(\theta + u)$ on it, and has received the firm's wage offer of $E[\theta|(\theta + u)]$ for this period. At this stage, the worker decides whether to accept this wage for this period and the prospect of receiving $\theta$ in all subsequent periods. The third stage occurs in the next period, when the worker and firm discover the true value of $\theta$ and the worker must decide whether to work at $\theta$ this period and in all subsequent periods that he remains at this job (match).

We now add some more specific assumptions about the probability distribution of $\theta$ and $u$. We assume that $\theta$ and $u$ are independently distributed random variables. Both are normally distributed, $\theta$ being normal with mean $\mu$ and variance $\sigma_0^2$, and $u$ being normal with mean 0 and variance $\sigma_u^2$. Thus we write

$$\theta \sim N(\mu, \sigma_0^2), \qquad u \sim N(0, \sigma_u^2) . \tag{5.17}$$

In the first period, after drawing a $\theta$, the worker and firm both observe the noise-ridden version of $\theta$, $y = \theta + u$. Both worker and firm are interested in making inferences about $\theta$, given the observation $(\theta + u)$. They are assumed to use Bayes' law and to calculate the "posterior" probability distribution of $\theta$, that is, the probability distribution of $\theta$ conditional on $(\theta + u)$. The probability distribution of $\theta$, given $\theta + u = y$, is known to be normal, with mean $m_0$ and variance $\sigma_1^2$. Using the Kalman filtering formula in chapter 1 and the appendix on filtering, chapter 21, we have[7]

$$\begin{aligned} m_0 = E(\theta|y) &= E(\theta) + \frac{\operatorname{cov}(\theta, y)}{\operatorname{var}(y)}[y - E(y)] \\ &= \mu + \frac{\sigma_0^2}{\sigma_0^2 + \sigma_u^2}(y - \mu) \equiv \mu + K_0(y - \mu) , \\ \sigma_1^2 = E[(\theta - m_0)^2|y] &= \frac{\sigma_0^2}{\sigma_0^2 + \sigma_u^2}\sigma_u^2 = K_0 \sigma_u^2 . \end{aligned} \tag{5.18}$$

After drawing $\theta$ and observing $y = \theta + u$ the first period, the firm is assumed to offer the worker a wage of $m_0 = E[\theta|(\theta + u)]$ the first period and a promise to pay $\theta$ for the second period and thereafter. (Jovanovic assumed firms to be risk neutral and to maximize the expected present value of profits. They compete for workers by offering wage contracts. In a long-run equilibrium the payments practices of each firm would be well understood, and this fact would support the described implicit contract as a competitive equilibrium.) The worker has the choice of accepting or rejecting the offer.

From equation (5.18) and the property that the random variable $y - \mu = \theta + u - \mu$ is normal, with mean zero and variance $(\sigma_0^2 + \sigma_u^2)$, it follows that $m_0$

[7] In the special case in which random variables are jointly normally distributed, linear least squares projections equal conditional expectations.

is itself normally distributed, with mean $\mu$ and variance $\sigma_0^4/(\sigma_0^2+\sigma_u^2)=K_0\sigma_0^2$:

$$m_0 \sim N\left(\mu, K_0\sigma_0^2\right). \tag{5.19}$$

Note that $K_0\sigma_0^2 < \sigma_0^2$, so that $m_0$ has the same mean but a smaller variance than $\theta$.

The worker seeks to maximize the expected present value of wages. We now proceed to solve the worker's problem by working backward. At stage 3, the worker knows $\theta$ and is confronted by the firm with an offer to work this period and forever more at a wage of $\theta$. We let $J(\theta)$ be the expected present value of wages of a worker at stage 3 who has a known match $\theta$ in hand and who behaves optimally. The worker who accepts the match this period receives $\theta$ this period and faces the same choice at the same $\theta$ next period. (The worker can quit next period, though it will turn out that the worker who does not quit this period never will.) Therefore, if the worker accepts the match, the value of match $\theta$ is given by $\theta+\beta J(\theta)$, where $\beta$ is the discount factor. The worker who rejects the match must be unemployed this period and must draw a new match next period. The expected present value of wages of a worker who was unemployed last period and who behaves optimally is $Q$. Therefore, the Bellman equation is $J(\theta)=\max\{\theta+\beta J(\theta),\beta Q\}$. This equation is graphed in Figure 5.7 and evidently has the solution

$$J(\theta)=\begin{cases}\theta+\beta J(\theta)=\frac{\theta}{1-\beta} & \text{for } \theta\geq\bar{\theta}\\ \beta Q & \text{for } \theta\leq\bar{\theta}.\end{cases} \tag{5.20}$$

The optimal policy is a reservation wage policy: accept offers $\theta\geq\bar{\theta}$, and reject offers $\theta\leq\bar{\theta}$, where $\theta$ satisfies

$$\frac{\bar{\theta}}{1-\beta}=\beta Q. \tag{5.21}$$

We now turn to the worker's decision in stage 2, given the decision rule in stage 3. In stage 2, the worker is confronted with a current wage offer $m_0 = E[\theta|(\theta+u)]$ and a conditional probability distribution function that we write as $\text{prob}\{\theta\leq s|\theta+u\}=F(s|m_0,\sigma_1^2)$. (Because the distribution is normal, it can be characterized by the two parameters $m_0,\sigma_1^2$.) We let $V(m_0)$ be the expected present value of wages of a worker at the second stage who has offer $m_0$ in hand and who behaves optimally. The worker who rejects the offer is unemployed this period and draws a new match parameter next period. The expected present

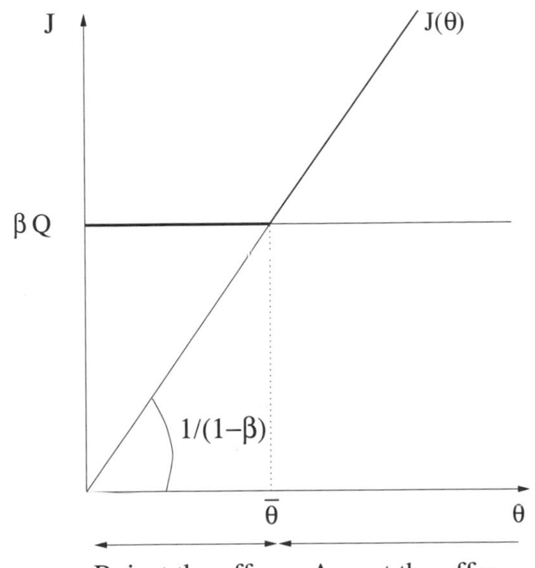


**Figure 5.7** The function $J(\theta) = \max\{\theta + \beta J(\theta), \beta Q\}$. The reservation wage in stage 3, $\bar{\theta}$, satisfies $\bar{\theta}/(1-\beta) = \beta Q$.

value of this option is $\beta Q$. The worker who accepts the offer receives a wage of $m_0$ this period and a probability distribution of wages of $F(\theta'|m_0, \sigma_1^2)$ for next period. The expected present value of this option is $m_0 + \beta \int J(\theta') dF(\theta'|m_0, \sigma_1^2)$. The Bellman equation for the second stage therefore becomes

$$V(m_0) = \max\left\{ m_0 + \beta \int J(\theta') dF(\theta'|m_0, \sigma_1^2), \beta Q \right\}. \tag{5.22}$$

Note that both $m_0$ and $\beta \int J(\theta') dF(\theta'|m_0, \sigma_1^2)$ are increasing in $m_0$, whereas $\beta Q$ is a constant. For this reason a reservation wage policy will be an optimal one. The functional equation evidently has the solution

$$V(m_0) = \begin{cases} m_0 + \beta \int J(\theta') dF(\theta'|m_0, \sigma_1^2) & \text{for} \quad m_0 \geq \bar{m}_0 \\ \beta Q & \text{for} \quad m_0 \leq \bar{m}_0. \end{cases} \tag{5.23}$$

If we use equation (5.23), an implicit equation for the reservation wage $\bar{m}_0$ is then

$$V(\bar{m}_0) = \bar{m}_0 + \beta \int J(\theta') dF(\theta'|\bar{m}_0, \sigma_1^2) = \beta Q. \tag{5.24}$$

Using equations (5.24) and (5.20), we shall show that $\bar{m}_0 < \bar{\theta}$, so that the worker becomes choosier over time with the firm. This force makes wages rise with tenure.

Using equations (5.20) and (5.21) repeatedly in equation (5.24), we obtain

$$\begin{aligned} &\bar{m}_0 + \beta \frac{\bar{\theta}}{1-\beta} \int_{-\infty}^{\bar{\theta}} dF(\theta'|\bar{m}_0, \sigma_1^2) + \frac{\beta}{1-\beta} \int_{\bar{\theta}}^{\infty} \theta' dF(\theta'|\bar{m}_0, \sigma_1^2) \\ &= \frac{\bar{\theta}}{1-\beta} = \frac{\bar{\theta}}{1-\beta} \int_{-\infty}^{\bar{\theta}} dF(\theta'|\bar{m}_0, \sigma_1^2) \\ &+ \frac{\bar{\theta}}{1-\beta} \int_{\bar{\theta}}^{\infty} dF(\theta'|\bar{m}_0, \sigma_1^2). \end{aligned}$$

Rearranging this equation, we get

$$\bar{\theta} \int_{-\infty}^{\bar{\theta}} dF(\theta'|\bar{m}_0, \sigma_1^2) - \bar{m}_0 = \frac{1}{1-\beta} \int_{\bar{\theta}}^{\infty} (\beta\theta' - \bar{\theta}) dF(\theta'|\bar{m}_0, \sigma_1^2). \tag{5.25}$$

Now note the identity

$$\bar{\theta} = \int_{-\infty}^{\bar{\theta}} \bar{\theta} dF(\theta'|\bar{m}_0, \sigma_1^2) + \left(\frac{1}{1-\beta} - \frac{\beta}{1-\beta}\right) \int_{\bar{\theta}}^{\infty} \bar{\theta} dF(\theta'|\bar{m}_0, \sigma_1^2). \tag{5.26}$$

Adding equation (5.26) to (5.25) gives

$$\bar{\theta} - \bar{m}_0 = \frac{\beta}{1-\beta} \int_{\bar{\theta}}^{\infty} (\theta' - \bar{\theta}) dF(\theta'|\bar{m}_0, \sigma_1^2). \tag{5.27}$$

The right side of equation (5.27) is positive. The left side is therefore also positive, so that we have established that

$$\bar{\theta} > \bar{m}_0. \tag{5.28}$$

Equation (5.27) resembles equation (5.8) and has a related interpretation. Given $\bar{\theta}$ and $\bar{m}_0$, the right side is the expected benefit of a match $\bar{m}_0$, namely, the expected present value of the match in the event that the match parameter eventually turns out to exceed the reservation match $\bar{\theta}$ so that the match endures. The left side is the one-period cost of temporarily staying in a match paying less than the eventual reservation match value $\bar{\theta}$: having remained unemployed for a period in order to have the privilege of drawing the match parameter $\theta$, the worker has made an investment to acquire this opportunity and must make a similar investment to acquire a new one. Having only the noisy observation of $(\theta + u)$ on $\theta$, the worker is willing to stay in matches $m_0$ with $\bar{m}_0 < m_0 < \bar{\theta}$ because it is worthwhile to speculate that the match is really better than it seems now and will seem next period.

Now turning briefly to stage 1, we have defined $Q$ as the predraw expected present value of wages of a worker who was unemployed last period and who is about to draw a match parameter and a noisy observation on it. Evidently $Q$ is given by

$$Q = \int V(m_0) dG\left(m_0 | \mu, K_0 \sigma_0^2\right). \tag{5.29}$$

where $G(m_0|\mu, K_0\sigma_0^2)$ is the normal distribution with mean $\mu$ and variance $K_0\sigma_0^2$, which, as we saw before, is the distribution of $m_0$.

Collecting some of the equations, we see that the worker's optimal policy is determined by

$$J(\theta) = \begin{cases} \theta + \beta J(\theta) = \frac{\theta}{1-\beta} & \text{for } \theta \geq \bar{\theta} \\ \beta Q & \text{for } \theta \leq \bar{\theta} \end{cases} \tag{5.30}$$

$$V(m_0) = \begin{cases} m_0 + \beta \int J(\theta') dF(\theta' | m_0, \sigma_1^2) & \text{for} \quad m_0 \geq \bar{m}_0 \\ \beta Q & \text{for} \quad m_0 \leq \bar{m}_0. \end{cases} \tag{5.31}$$

$$\bar{\theta} - \bar{m}_0 = \frac{\beta}{1-\beta} \int_{\bar{\theta}}^{\infty} (\theta' - \bar{\theta}) dF(\theta' | \bar{m}_0, \sigma_1^2). \tag{5.32}$$

$$Q = \int V(m_0) dG\left(m_0 | \mu, K_0 \sigma_0^2\right). \tag{5.33}$$

To analyze formally the existence and uniqueness of a solution to these equations, one would proceed as follows. Use equations (5.30), (5.31), and (5.32) to write a single functional equation in $V$,

$$V(m_0) = \max\left\{ m_0 + \beta \int \max\left[\frac{\theta}{1-\beta}, \beta \int V(m_1') \, dG\left(m_1' | \mu, K_0\sigma_0^2\right)\right] dF(\theta | m_0, \sigma_1^2), \; \beta \int V(m_1') dG\left(m_1' | \mu, K_0\sigma_0^2\right) \right\}.$$

The expression on the right defines an operator, $T$, mapping continuous functions $V$ into continuous functions $TV$. This functional equation can be expressed $V = TV$. The operator $T$ can be directly verified to satisfy the following two properties: (1) it is monotone, that is, $v(m) \geq z(m)$ for all $m$ implies $(Tv)(m) \geq (Tz)(m)$ for all $m$; (2) for all positive constants $c$, $T(v+c) \leq Tv + \beta c$. These are Blackwell's sufficient conditions for the functional equation $Tv = v$ to have a unique continuous solution. See the appendix on functional analysis, chapter 20.

We now proceed to calculate probabilities and expectations of some interesting events and variables. The probability that a previously unemployed worker accepts an offer is given by

$$\text{prob}\{m_0 \geq \bar{m}_0\} = \int_{\bar{m}_0}^{\infty} dG(m_0 | \mu, K_0\sigma_0^2).$$

The probability that a previously unemployed worker accepts an offer and then quits the second period is given by

$$\text{prob}\{(\theta \leq \bar{\theta}) \cap (m_0 \geq \bar{m}_0)\} = \int_{\bar{m}_0}^{\infty} \int_{-\infty}^{\bar{\theta}} dF(\theta | m_0, \sigma_1^2) \cdot dG(m_0 | \mu, K_0\sigma_0^2).$$

The probability that a previously unemployed worker accepts an offer the first period and also elects not to quit the second period is given by

$$\text{prob}\{(\theta \geq \bar{\theta}) \cap (m_0 \geq \bar{m})\} = \int_{\bar{m}_0}^{\infty} \int_{\bar{\theta}}^{\infty} dF(\theta | m_0, \sigma_1^2) dG(m_0 | \mu, K_0\sigma_0^2).$$

The mean wage of those employed the first period is given by

$$\bar{w}_1 = \frac{\displaystyle\int_{\bar{m}_0}^{\infty} m_0 \, dG(m_0|\mu, K_0\sigma_0^2)}{\displaystyle\int_{\bar{m}_0}^{\infty} dG(m_0|\mu, K_0\sigma_0^2)}, \tag{5.34}$$

whereas the mean wage of those workers who are in the second period of tenure is given by

$$\bar{w}_2 = \frac{\displaystyle\int_{\bar{m}_0}^{\infty}\int_{\bar{\theta}}^{\infty} \theta \, dF(\theta|m_0, \sigma_1^2)\, dG(m_0|\mu, K_0\sigma_0^2)}{\displaystyle\int_{\bar{m}_0}^{\infty}\int_{\bar{\theta}}^{\infty} dF(\theta|m_0, \sigma_1^2)\, dG(m_0|\mu, K_0\sigma_0^2)}. \tag{5.35}$$

We shall now prove that $\bar{w}_2 > \bar{w}_1$, so that wages rise with tenure. After substituting $m_0 \equiv \int \theta dF(\theta|m_0, \sigma_1^2)$ into equation (5.34),

$$\begin{aligned}
\bar{w}_1 &= \frac{\displaystyle\int_{\bar{m}_0}^{\infty}\int_{-\infty}^{\infty} \theta \, dF(\theta|m_0, \sigma_1^2)\, dG(m_0|\mu, K_0\sigma_0^2)}{\displaystyle\int_{\bar{m}_0}^{\infty} dG(m_0|\mu, K_0\sigma_0^2)} \\
&= \frac{1}{\displaystyle\int_{\bar{m}_0}^{\infty} dG(m_0|\mu, K_0\sigma_0^2)} \left\{ \int_{\bar{m}_0}^{\infty}\int_{-\infty}^{\bar{\theta}} \theta \, dF(\theta|m_0, \sigma_1^2)\, dG(m_0|\mu, K_0\sigma_0^2) \right. \\
&\qquad \left. + \; \bar{w}_2 \int_{\bar{m}_0}^{\infty}\int_{\bar{\theta}}^{\infty} dF(\theta|m_0, \sigma_1^2)\, dG(m_0|\mu, K_0\sigma_0^2) \right\} \\
&< \frac{\displaystyle\int_{\bar{m}_0}^{\infty} \left\{ \bar{\theta}\, F(\bar{\theta}|m_0, \sigma_1^2) + \bar{w}_2[1 - F(\bar{\theta}|m_0, \sigma_1^2)] \right\} dG(m_0|\mu, K_0\sigma_0^2)}{\displaystyle\int_{\bar{m}_0}^{\infty} dG(m_0|\mu, K_0\sigma_0^2)} \\
&< \bar{w}_2.
\end{aligned}$$

It is quite intuitive that the mean wage of those workers who are in the second period of tenure must exceed the mean wage of all employed in the first period. The former group is a subset of the latter group where workers with low productivities, $\theta < \bar{\theta}$, have left. Since the mean wages are equal to the true average productivity in each group, it follows that $\bar{w}_2 > \bar{w}_1$.

The model thus implies that "wages rise with tenure," both in the sense that mean wages rise with tenure and in the sense that $\bar{\theta} > \bar{m}_0$, which asserts that the lower bound on second-period wages exceeds the lower bound on first-period wages. That wages rise with tenure was observation 1 that Jovanovic sought to explain.

Jovanovic's model also explains observation 2, that quits are negatively correlated with tenure. The model implies that quits occur between the first and second periods of tenure. Having decided to stay for two periods, the worker never quits.

The model also accounts for observation 3, namely, that the probability of a subsequent quit is negatively correlated with the current wage rate. The probability of a subsequent quit is given by

$$\text{prob}\{\theta' < \bar{\theta} | m_0\} = F(\bar{\theta} | m_0, \sigma_1^2),$$

which is evidently negatively correlated with $m_0$, the first-period wage. Thus the model explains each observation that Jovanovic sought to interpret. In the version of the model that we have studied, a worker eventually becomes permanently matched with probability 1. If we were studying a population of such workers of fixed size, all workers would eventually be absorbed into the state of being permanently matched. To provide a mechanism for replenishing the stock of unmatched workers, one could combine Jovanovic's model with the "firing" model of an earlier section. By letting matches $\theta$ "go bad" with probability $\lambda$ each period, one could presumably modify Jovanovic's model to get the implication that, with a fixed population of workers, a fraction would remain unmatched each period because of the dissolution of previously acceptable matches.

## *A longer horizon version of Jovanovic's model*

Here we consider a $T+1$ period version of Jovanovic's model, in which learning about the quality of the match continues for $T$ periods before the quality of the match is revealed by "nature." (Jovanovic assumed that $T = \infty$.) We use the recursive projection technique (the Kalman filter) of chapter 4 to handle the firm's and worker's sequential learning. The prediction of the true match quality can then easily be updated with each additional noisy observation.

A firm-worker pair jointly draws a match parameter $\theta$ at the start of the match, which we call the beginning of period 0. The value $\theta$ is revealed to the pair only at the beginning of the $(T+1)$th period of the match. After $\theta$

is drawn but before the match is consummated, the firm-worker pair observes $y_0 = \theta + u_0$, where $u_0$ is random noise. At the beginning of each period of the match, the worker-firm pair draws another noisy observation $y_t = \theta + u_t$ on the match parameter $\theta$. The worker then decides whether or not to continue the match for the additional period. Let $y^t = \{y_0, \dots, y_t\}$ be the firm's and worker's information set at time $t$. We assume that $\theta$ and $u_t$ are independently distributed random variables with $\theta \sim \mathcal{N}(\mu, \Sigma_0)$ and $u_t \sim \mathcal{N}(0, \sigma_u^2)$. For $t \geq 0$ define $m_t = E[\theta | y^t]$ and $m_{-1} = \mu$. The conditional means $m_t$ and variances $E(\theta - m_t)^2 = \Sigma_{t+1}$ can be computed with the Kalman filter via the formulas from chapter 1:

$$m_t = (1 - K_t) m_{t-1} + K_t y_t \tag{5.36a}$$

$$K_t = \frac{\Sigma_t}{\Sigma_t + R} \tag{5.36b}$$

$$\Sigma_{t+1} = \frac{\Sigma_t R}{\Sigma_t + R}, \tag{5.36c}$$

where $R = \sigma_u^2$. The recursions are to be initiated from $m_{-1} = \mu$, and given $\Sigma_0$. Using the formulas from chapter 1, we have that $m_{t+1} \sim \mathcal{N}(m_t, K_{t+1}\Sigma_{t+1})$ and $\theta \sim \mathcal{N}(m_t, \Sigma_{t+1})$.

*The Bellman equations*

For $t \geq 0$, let $v_t(m_t)$ be the value of the worker's problem at the beginning of period $t$ for a worker who optimally estimates that the match value is $m_t$ after having observed $y^t$. At the start of period $T+1$, we suppose that the value of the match is revealed without error. Thus, at time $T$, $\theta \sim \mathcal{N}(m_T, \Sigma_{T+1})$. The firm-worker pair estimates $\theta$ by $m_t$ for $t = 0, \dots, T$, and by $\theta$ for $t \geq T+1$. Then the following functional equations characterize the solution of the problem:

$$v_{T+1}(\theta) = \max\left\{\frac{\theta}{1-\beta}, \beta Q\right\}, \tag{5.37}$$

$$v_T(m) = \max\{m + \beta \int v_{T+1}(\theta) dF(\theta \mid m, \Sigma_{T+1}), \beta Q\}, \tag{5.38}$$

$$v_t(m) = \max\{m + \beta \int v_{t+1}(m') dF(m' | m, K_{t+1}\Sigma_{t+1}), \beta Q\} \; t = 0, \dots, T-1, \tag{5.39}$$

$$Q = \int v_1(m) dF(m|\mu, K_1\Sigma_1), \tag{5.40}$$

with $K_t$ and $\Sigma_t$ from the Kalman filter. Starting from $v_{T+1}$ and reasoning backward, it is evident that the worker's optimal policy is to set reservation wages $\bar{m}_t, t = 0, \ldots, T$ that satisfy

$$\begin{aligned}
&\bar{m}_{T+1} = \bar{\theta} = \beta(1-\beta)Q, \\
&\bar{m}_T + \beta \int v_{T+1}(\theta) dF(\theta|\bar{m}_T, \Sigma_{T+1}) = \beta Q, \\
&\bar{m}_t + \beta \int v_{t+1}(m') dF(m' \mid \bar{m}_t, K_{t+1}\Sigma_{t+1}) = \beta Q, \qquad t = 1, \ldots, T-1.
\end{aligned} \tag{5.41}$$

To compute a solution to the worker's problem, we can define a mapping from $Q$ into itself, with the property that a fixed point of the mapping is the optimal value of $Q$. Here is an algorithm:

a. Guess a value of $Q$, say $Q^i$ with $i = 1$.
b. Given $Q^i$, compute sequentially the value functions in equations (5.37) through (5.39). Let the solutions be denoted $v^i_{T+1}(\theta)$ and $v^i_t(m)$ for $i = 1, \ldots, T$.
c. Given $v^i_1(m)$, evaluate equation (5.40) and call the solution $\tilde{Q}^i$.
d. For a fixed "relaxation parameter" $g \in (0,1)$, compute a new guess of $Q$ from
$$Q^{i+1} = gQ^i + (1-g)\tilde{Q}^i.$$
e. Iterate on this scheme to convergence.

We now turn to the case where the true $\theta$ is never revealed by nature, that is, $T = \infty$. Note that $(\Sigma_{t+1})^{-1} = (\sigma_u^2)^{-1} + (\Sigma_t)^{-1}$, so $\Sigma_{t+1} < \Sigma_t$ and $\Sigma_{t+1} \to 0$ as $t \to \infty$. In other words, the accuracy of the prediction of $\theta$ becomes arbitrarily good as the information set $y^t$ becomes large. Consequently, the firm and worker eventually learn the true $\theta$, and the value function "at infinity" becomes

$$v_\infty(\theta) = \max\left\{\frac{\theta}{1-\beta}, \beta Q\right\},$$

and the Bellman equation for any finite tenure $t$ is given by equation (5.39), and $Q$ in equation (5.40) is the value of an unemployed worker. The optimal policy is a reservation wage $\bar{m}_t$, one for each tenure $t$. In fact, in the absence of a final date $T+1$ when $\theta$ is revealed by nature, the solution is actually a time-invariant policy function $\bar{m}(\sigma_t^2)$ with an acceptance and a rejection region in the space of $(m, \sigma^2)$.

To compute a numerical solution when $T = \infty$, we would still have to rely on the procedure that we have outlined based on the assumption of some finite date when the true $\theta$ is revealed, say in period $\hat{T}+1$. The idea is to choose a sufficiently large $\hat{T}$ so that the conditional variance of $\theta$ at time $\hat{T}$, $\sigma_{\hat{T}}^2$, is close to zero. We then examine the approximation that $\sigma_{\hat{T}+1}^2$ is equal to zero. That is, equations (5.37) and (5.38) are used to truncate an otherwise infinite series of value functions.

## Concluding remarks

The situations analyzed in this chapter are ones in which a currently unemployed worker rationally chooses to refuse an offer to work, preferring to remain unemployed today in exchange for better prospects tomorrow. The worker is voluntarily unemployed in one sense, having chosen to reject the current draw from the distribution of offers. In this model, the activity of unemployment is an investment incurred to improve the situation faced in the future. A theory in which unemployment is voluntary permits an analysis of the forces impinging on the choice to remain unemployed. Thus we can study the response of the worker's decision rule to changes in the distribution of offers, the rate of unemployment compensation, the number of offers per period, and so on.

Chapter 15 studies the optimal design of unemployment compensation. That issue is a trivial one in the present chapter with risk neutral agents and no externalities. Here the government should avoid any policy that affects the workers' decision rules since it would harm efficiency, and the first-best way of pursuing distributional goals is through lump-sum transfers. In contrast, chapter 15 assumes risk-averse agents and incomplete insurance markets which together with information asymmetries make for an intricate contract design problem in the provision of unemployment insurance.

Chapter 19 presents various equilibrium models of search and matching. We study workers searching for jobs in an island model, workers and firms forming matches in a model with a "matching function," and how a medium of exchange can overcome the problem of "double coincidence of wants" in a search model of money.

## Appendix: More practice with numerical dynamic programming

This appendix describes two more examples using the numerical methods of chapter 3.

### Example 4: Search

An unemployed worker wants to maximize $E_0 \sum_{t=0}^{\infty} \beta^t y_t$ where $y_t = w$ if the worker is employed at wage $w$, $y_t = 0$ if the worker is unemployed, and $\beta \in (0,1)$. Each period an unemployed worker draws a positive wage from a discrete state Markov chain with transition matrix $P$. Thus, wage offers evolve according to a Markov process with transition probabilities given by

$$P(i,j) = \text{Prob}(w_{t+1} = \tilde{w}_j | w_t = \tilde{w}_i).$$

Once he accepts an offer, the worker works forever at the accepted wage. There is no firing or quitting. Let $v$ be an $(n \times 1)$ vector of values $v_i$ representing the optimal value of the problem for a worker who has offer $w_i, i = 1, \ldots, n$ in hand and who behaves optimally. The Bellman equation is

$$v_i = \max_{\text{accept,reject}} \left\{ \frac{w_i}{1-\beta}, \beta \sum_{j=1}^{n} P_{ij} v_j \right\}$$

or

$$v = \max\{\tilde{w}/(1-\beta), \beta P v\}.$$

Here $\tilde{w}$ is an $(n \times 1)$ vector of possible wage values. This matrix equation can be solved using the numerical procedures described earlier. The optimal policy depends on the structure of the Markov chain $P$. Under restrictions on $P$ making $w$ positively serially correlated, the optimal policy has the following reservation wage form: there is a $\bar{w}$ such that the worker should accept an offer $w$ if $w \geq \bar{w}$.

### Example 5: A Jovanovic model

Here is a simplified version of the search model of Jovanovic (1979a). A newly unemployed worker draws a job offer from a distribution given by $\mu_i = \text{Prob}(w_1 = \tilde{w}_i)$, where $w_1$ is the first-period wage. Let $\mu$ be the $(n \times 1)$ vector with $i$th component $\mu_i$. After an offer is drawn, subsequent wages associated with the job evolve according to a Markov chain with time-varying transition matrices

$$P_t(i,j) = \text{Prob}(w_{t+1} = \tilde{w}_j | w_t = \tilde{w}_i),$$

for $t = 1, \ldots, T$. We assume that for times $t > T$, the transition matrices $P_t = I$, so that after $T$ a job's wage does not change anymore with the passage of time. We specify the $P_t$ matrices to capture the idea that the worker-firm pair is learning more about the quality of the match with the passage of time. For example, we might set

$$P_t = \begin{bmatrix} 1-q^t & q^t & 0 & 0 & \ldots & 0 & 0 \\ q^t & 1-2q^t & q^t & 0 & \ldots & 0 & 0 \\ 0 & q^t & 1-2q^t & q^t & \ldots & 0 & 0 \\ \vdots & \vdots & \vdots & \vdots & \vdots & \vdots & \vdots \\ 0 & 0 & 0 & 0 & \ldots & 1-2q^t & q^t \\ 0 & 0 & 0 & 0 & \ldots & q^t & 1-q^t \end{bmatrix},$$

where $q \in (0,1)$. In the following numerical examples, we use a slightly more general form of transition matrix in which (except at end-points of the distribution),

$$\begin{aligned} \text{Prob}(w_{t+1} = \tilde{w}_{k \pm m} | w_t = \tilde{w}_k) = P_t(k, k \pm m) = q^t \\ P_t(k,k) = 1 - 2q^t. \end{aligned} \tag{5.42}$$

Here $m \geq 1$ is a parameter that indexes the spread of the distribution.

At the beginning of each period, a previously matched worker is exposed with probability $\lambda \in (0,1)$ to the event that the match dissolves. We then have a set of Bellman equations

$$v_t = \max\{\tilde{w} + \beta(1-\lambda)P_t v_{t+1} + \beta\lambda Q, \beta Q + \bar{c}\}, \tag{5.43a}$$

for $t = 1, \ldots, T$, and

$$v_{T+1} = \max\{\tilde{w} + \beta(1-\lambda)v_{T+1} + \beta\lambda Q, \beta Q + \bar{c}\}, \tag{5.43b}$$

$$\begin{aligned} Q = \mu' v_1 \otimes \mathbf{1} \\ \bar{c} = c \otimes \mathbf{1} \end{aligned}$$

where $\otimes$ is the Kronecker product, and $\mathbf{1}$ is an $(n \times 1)$ vector of ones. These equations can be solved by using calculations of the kind described previously. The optimal policy is to set a sequence of reservation wages $\{\bar{w}_j\}_{j=1}^n$.

*Wage distributions*

We can use recursions to compute probability distributions of wages at tenures $1, 2, \ldots, n$. Let the reservation wage for tenure $j$ be $\bar{w}_j \equiv \tilde{w}_{\rho(j)}$, where $\rho(j)$ is the index associated with the cutoff wage. For $i \geq \rho(1)$, define

$$\delta_1(i) = \text{Prob}\{w_1 = \tilde{w}_i \mid w_1 \geq \bar{w}_1\} = \frac{\mu_i}{\sum_{h=\rho(1)}^{n} \mu_h}.$$

Then

$$\gamma_2(j) = \text{Prob}\{w_2 = \tilde{w}_j \mid w_1 \geq \bar{w}_1\} = \sum_{i=\rho(1)}^{n} P_1(i,j)\delta_1(i).$$

For $i \geq \rho(2)$, define

$$\delta_2(i) = \text{Prob}\{w_2 = \tilde{w}_i \mid w_2 \geq \bar{w}_2 \cap w_1 \geq \bar{w}_1\}$$

or

$$\delta_2(i) = \frac{\gamma_2(i)}{\sum_{h=\rho(2)}^{n} \gamma_2(h)}.$$

Then

$$\begin{aligned}\gamma_3(j) &= \text{Prob}\{w_3 = \tilde{w}_j \mid w_2 \geq \bar{w}_2 \cap w_1 \geq \bar{w}_1\}\\ &= \sum_{i=\rho(2)}^{n} P_2(i,j)\delta_2(i).\end{aligned}$$

Next, for $i \geq \rho(3)$, define $\delta_3(i) = \text{Prob}\{w_3 = \tilde{w}_i \mid (w_3 \geq \bar{w}_3) \cap (w_2 \geq \bar{w}_2) \cap (w_1 \geq \bar{w}_1)\}$. Then

$$\delta_3(i) = \frac{\gamma_3(i)}{\sum_{h=\rho(3)}^{n} \gamma_3(h)}$$

Continuing in this way, we can define the wage distributions $\delta_1(i), \delta_2(i), \delta_3(i), \ldots$. The mean wage at tenure $k$ is given by

$$\sum_{i \geq \rho(k)} \tilde{w}_i \delta_k(i).$$

*Separation probabilities*

The probability of rejecting a first period offer is $Q(1) = \sum_{h<\rho(1)} \mu_h$. The probability of separating at the beginning of period $j \geq 2$ is $Q(j) = \sum_{h<\rho(j)} \gamma_j(h)$.

*Numerical examples*

Figures 5.8, 5.9, and 5.10 report some numerical results for three versions of this model. For all versions, we set $\beta = .95, c = 0, q = .5$ and $T+1 = 21$. For all three examples, we used a wage grid with sixty equispaced points on the interval $[0, 10]$. For the initial distribution $\mu$ we used the uniform distribution. We used a sequence of transition matrices of the form (5.42), with a "gap" parameter of $m$. For the first example, we set $m = 6$ and $\lambda = 0$, while the second sets $m = 10$ and $\lambda = 0$ and third sets $m = 10$ and $\lambda = .1$.

Figure 5.8 shows the reservation wage falls as $m$ increases from 6 to 10, and that it falls further when the probability of being fired $\lambda$ rises from zero to .1. Figure 5.9 shows the same pattern for average wages. Figure 5.10 displays quit probabilities for the first two models. They fall with tenure, with shapes and heights that depend to some degree on $m, \lambda$.

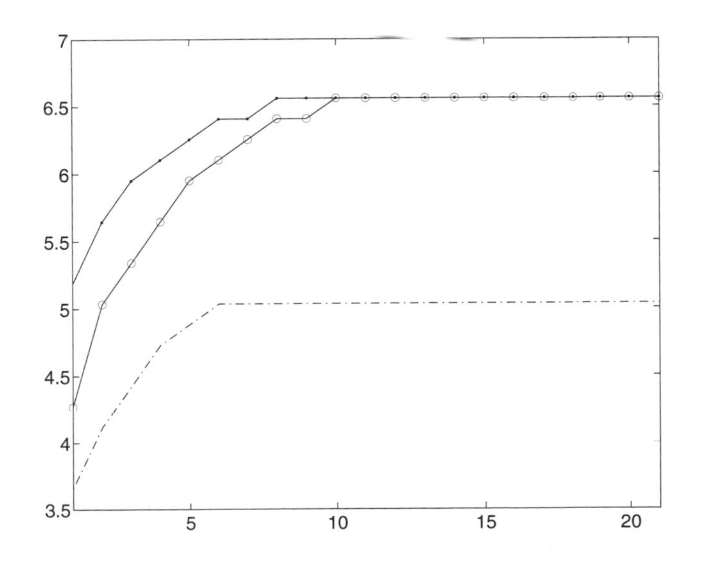


**Figure 5.8** Reservation wages as function of tenure for model with three different parameter settings $[m = 6, \lambda = 0]$ (the dots), $[m = 10, \lambda = 0]$ (the line with circles), and $[m = 10, \lambda = .1]$ (the dashed line).

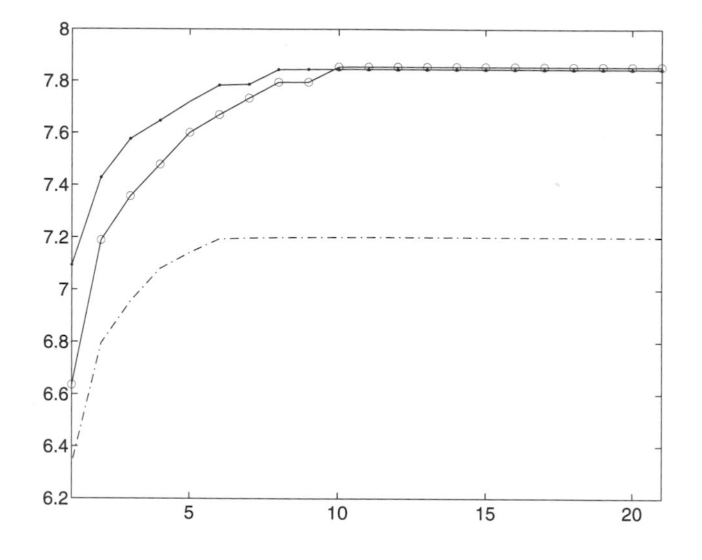


**Figure 5.9** Mean wages as function of tenure for model with three different parameter settings $[m = 6, \lambda = 0]$ (the dots), $[m = 10, \lambda = 0]$ (the line with circles), and $[m = 10, \lambda = .1]$ (the dashed line).

## *Exercises*

### *Exercise 5.1* **Being unemployed with a chance of an offer**

An unemployed worker samples wage offers on the following terms: Each period, with probability $\phi$, $1 > \phi > 0$, she receives no offer (we may regard this as a wage offer of zero forever). With probability $(1 - \phi)$ she receives an offer to work for $w$ forever, where $w$ is drawn from a cumulative distribution function $F(w)$. Successive draws across periods are independently and identically distributed. The worker chooses a strategy to maximize

$$E \sum_{t=0}^{\infty} \beta^t y_t, \qquad \text{where} \quad 0 < \beta < 1,$$

$y_t = w$ if the worker is employed, and $y_t = c$ if the worker is unemployed. Here $c$ is unemployment compensation, and $w$ is the wage at which the worker is

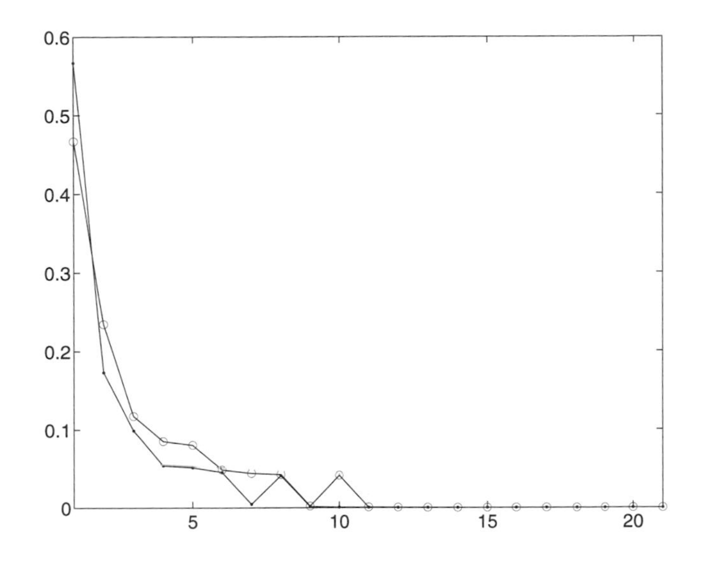


**Figure 5.10** Quit probabilities as a function of tenure for Jovanovic model with $[m = 6, \lambda = 0]$ (line with dots) and $[m = 10, \lambda = .1]$ (the line with circles).

employed. Assume that, having once accepted a job offer at wage $w$, the worker stays in the job forever.

Let $v(w)$ be the expected value of $\sum_{t=0}^{\infty} \beta^t y_t$ for an unemployed worker who has offer $w$ in hand and who behaves optimally. Write the Bellman equation for the worker's problem.

*Exercise 5.2* **Two offers per period**

Consider an unemployed worker who each period can draw *two* independently and identically distributed wage offers from the cumulative probability distribution function $F(w)$. The worker will work forever at the same wage after having once accepted an offer. In the event of unemployment during a period, the worker receives unemployment compensation $c$. The worker derives a decision rule to maximize $E \sum_{t=0}^{\infty} \beta^t y_t$, where $y_t = w$ or $y_t = c$, depending on whether she is employed or unemployed. Let $v(w)$ be the value of $E \sum_{t=0}^{\infty} \beta^t y_t$ for a currently unemployed worker who has best offer $w$ in hand.

**a.** Formulate the Bellman equation for the worker's problem.

**b.** Prove that the worker's reservation wage is *higher* than it would be had the worker faced the same $c$ and been drawing only *one* offer from the same distribution $F(w)$ each period.

### *Exercise 5.3* A random number of offers per period

An unemployed worker is confronted with a random number, $n$, of job offers each period. With probability $\pi_n$, the worker receives $n$ offers in a given period, where $\pi_n \geq 0$ for $n \geq 1$, and $\sum_{n=1}^{N} \pi_n = 1$ for $N < +\infty$. Each offer is drawn independently from the same distribution $F(w)$. Assume that the number of offers $n$ is independently distributed across time. The worker works forever at wage $w$ after having accepted a job and receives unemployment compensation of $c$ during each period of unemployment. He chooses a strategy to maximize $E\sum_{t=0}^{\infty} \beta^t y_t$ where $y_t = c$ if he is unemployed, $y_t = w$ if he is employed.

Let $v(w)$ be the value of the objective function of an unemployed worker who has best offer $w$ in hand and who proceeds optimally. Formulate the Bellman equation for this worker.

### *Exercise 5.4* Cyclical fluctuations in number of job offers

Modify exercise 5.3 as follows: Let the number of job offers $n$ follow a Markov process, with

$$\begin{aligned} &\text{prob}\{\text{Number of offers next period} = m | \text{Number of offers this} \\ &\quad \text{period} = n\} = \pi_{mn}, \qquad m = 1, \ldots, N, \quad n = 1, \ldots, N \\ &\sum_{m=1}^{N} \pi_{mn} = 1 \qquad \text{for} \quad n = 1, \ldots, N. \end{aligned}$$

Here $[\pi_{mn}]$ is a "stochastic matrix" generating a Markov chain. Keep all other features of the problem as in exercise 5.3. The worker gets $n$ offers per period, where $n$ is now generated by a Markov chain so that the number of offers is possibly correlated over time.

**a.** Let $v(w, n)$ be the value of $E\sum_{t=0}^{\infty} \beta^t y_t$ for an unemployed worker who has received $n$ offers this period, the best of which is $w$. Formulate the Bellman equation for the worker's problem.

**b.** Show that the optimal policy is to set a reservation wage $\bar{w}(n)$ that depends on the number of offers received this period.

*Exercise 5.5* **Choosing the number of offers**

An unemployed worker must choose the number of offers $n$ to solicit. At a cost of $k(n)$ the worker receives $n$ offers this period. Here $k(n+1) > k(n)$ for $n \geq 1$. The number of offers $n$ must be chosen in advance at the beginning of the period and cannot be revised during the period. The worker wants to maximize $E\sum_{t=0}^{\infty}\beta^t y_t$. Here $y_t$ consists of $w$ each period she is employed but not searching, $[w - k(n)]$ the first period she is employed but searches for $n$ offers, and $[c - k(n)]$ each period she is unemployed but solicits and rejects $n$ offers. The offers are each independently drawn from $F(w)$. The worker who accepts an offer works forever at wage $w$.

Let $Q$ be the value of the problem for an unemployed worker who has not yet chosen the number of offers to solicit. Formulate the Bellman equation for this worker.

*Exercise 5.6* **Mortensen externality**

Two parties to a match (say, worker and firm) jointly draw a match parameter $\theta$ from a c.d.f. $F(\theta)$. Once matched, they stay matched forever, each one deriving a benefit of $\theta$ per period from the match. Each unmatched pair of agents can influence the number of offers received in a period in the following way. The worker receives $n$ offers per period, with $n = f(c_1 + c_2)$, where $c_1$ represents the resources the worker devotes to searching and $c_2$ represents the resources the typical firm devotes to searching. Symmetrically, the representative firm receives $n$ offers per period where $n = f(c_1 + c_2)$. (We shall define the situation so that firms and workers have the same reservation $\theta$ so that there is never unrequited love.) Both $c_1$ and $c_2$ must be chosen at the beginning of the period, prior to searching during the period. Firms and workers have the same preferences, given by the expected present value of the match parameter $\theta$, net of search costs. The discount factor $\beta$ is the same for worker and firm.

**a.** Consider a Nash equilibrium in which party $i$ chooses $c_i$, taking $c_j$, $j \neq i$, as given. Let $Q_i$ be the value for an unmatched agent of type $i$ before the level of $c_i$ has been chosen. Formulate the Bellman equation for agents of types 1 and 2.

**b.** Consider the social planning problem of choosing $c_1$ and $c_2$ sequentially so as to maximize the criterion of $\lambda$ times the utility of agent 1 plus $(1-\lambda)$ times the utility of agent 2, $0 < \lambda < 1$. Let $Q(\lambda)$ be the value for this problem for two unmatched agents before $c_1$ and $c_2$ have been chosen. Formulate the Bellman equation for this problem.

**c.** Comparing the results in a and b, argue that, in the Nash equilibrium, the optimal amount of resources has not been devoted to search.

*Exercise 5.7* **Variable labor supply**

An unemployed worker receives each period a wage offer $w$ drawn from the distribution $F(w)$. The worker has to choose whether to accept the job—and therefore to work forever—or to search for another offer and collect $c$ in unemployment compensation. The worker who decides to accept the job must choose the number of hours to work in each period. The worker chooses a strategy to maximize

$$E\sum_{t=0}^{\infty}\beta^t u(y_t, l_t), \qquad \text{where} \quad 0<\beta<1,$$

and $y_t = c$ if the worker is unemployed, and $y_t = w(1-l_t)$ if the worker is employed and works $(1-l_t)$ hours; $l_t$ is leisure with $0 \leq l_t \leq 1$.

Analyze the worker's problem. Argue that the optimal strategy has the reservation wage property. Show that the number of hours worked is the same in every period.

*Exercise 5.8* **Wage growth rate and the reservation wage**

An unemployed worker receives each period an offer to work for wage $w_t$ forever, where $w_t = w$ in the first period and $w_t = \phi^t w$ after $t$ periods on the job. Assume $\phi > 1$, that is, wages increase with tenure. The initial wage offer is drawn from a distribution $F(w)$ that is constant over time (entry-level wages are stationary); successive drawings across periods are independently and identically distributed.

The worker's objective function is to maximize

$$E\sum_{t=0}^{\infty}\beta^t y_t, \qquad \text{where} \quad 0<\beta<1,$$

and $y_t = w_t$ if the worker is employed and $y_t = c$ if the worker is unemployed, where $c$ is unemployment compensation. Let $v(w)$ be the optimal value of the objective function for an unemployed worker who has offer $w$ in hand. Write the Bellman equation for this problem. Argue that, if two economies differ only in the growth rate of wages of employed workers, say $\phi_1 > \phi_2$, the economy with the higher growth rate has the smaller reservation wage.

*Note:* Assume that $\phi_i\beta < 1$, $i = 1, 2$.

*Exercise 5.9* **Search with a finite horizon**

Consider a worker who lives two periods. In each period the worker, if unemployed, receives an offer of lifetime work at wage $w$, where $w$ is drawn from a distribution $F$. Wage offers are identically and independently distributed over time. The worker's objective is to maximize $E\{y_1 + \beta y_2\}$, where $y_t = w$ if the worker is employed and is equal to $c$—unemployment compensation—if the worker is not employed.

Analyze the worker's optimal decision rule. In particular, establish that the optimal strategy is to choose a reservation wage in each period and to accept any offer with a wage at least as high as the reservation wage and to reject offers below that level. Show that the reservation wage decreases over time.

*Exercise 5.10* **Finite horizon and mean-preserving spread**

Consider a worker who draws every period a job offer to work forever at wage $w$. Successive offers are independently and identically distributed drawings from a distribution $F_i(w)$, $i = 1, 2$. Assume that $F_1$ has been obtained from $F_2$ by a mean-preserving spread. The worker's objective is to maximize

$$E \sum_{t=0}^{T} \beta^t y_t, \qquad 0 < \beta < 1,$$

where $y_t = w$ if the worker has accepted employment at wage $w$ and is zero otherwise. Assume that both distributions, $F_1$ and $F_2$, share a common upper bound, $B$.

**a.** Show that the reservation wages of workers drawing from $F_1$ and $F_2$ coincide at $t = T$ and $t = T - 1$.

**b.** Argue that for $t \leq T - 2$ the reservation wage of the workers that sample wage offers from the distribution $F_1$ is higher than the reservation wage of the workers that sample from $F_2$.

**c.** Now introduce unemployment compensation: the worker who is unemployed collects $c$ dollars. Prove that the result in part a no longer holds; that is, the reservation wage of the workers that sample from $F_1$ is higher than the one corresponding to workers that sample from $F_2$ for $t = T - 1$.

*Exercise 5.11* **Pissarides' analysis of taxation and variable search intensity**

An unemployed worker receives each period a zero offer (or no offer) with probability $[1 - \pi(e)]$. With probability $\pi(e)$ the worker draws an offer $w$ from the distribution $F$. Here $e$ stands for effort—a measure of search intensity—and $\pi(e)$ is increasing in $e$. A worker who accepts a job offer can be fired with probability $\alpha$, $0 < \alpha < 1$. The worker chooses a strategy, that is, whether to accept an offer or not and how much effort to put into search when unemployed, to maximize

$$E \sum_{t=0}^{\infty} \beta^t y_t, \qquad 0 < \beta < 1,$$

where $y_t = w$ if the worker is employed with wage $w$ and $y_t = 1 - e + z$ if the worker spends $e$ units of leisure searching and does not accept a job. Here $z$ is unemployment compensation. For the worker who searched and accepted a job, $y_t = w - e - T(w)$; that is, in the first period the wage is net of search costs. Throughout, $T(w)$ is the amount paid in taxes when the worker is employed. We assume that $w - T(w)$ is increasing in $w$. Assume that $w - T(w) = 0$ for $w = 0$, that if $e = 0$, then $\pi(e) = 0$—that is, the worker gets no offers—and that $\pi'(e) > 0$, $\pi''(e) < 0$.

**a.** Analyze the worker's problem. Establish that the optimal strategy is to choose a reservation wage. Display the condition that describes the optimal choice of $e$, and show that the reservation wage is independent of $e$.

**b.** Assume that $T(w) = t(w - a)$ where $0 < t < 1$ and $a > 0$. Show that an increase in $a$ decreases the reservation wage and increases the level of effort, increasing the probability of accepting employment.

**c.** Show under what conditions a change in $t$ has the opposite effect.

*Exercise 5.12* **Search and nonhuman wealth**

An unemployed worker receives every period an offer to work forever at wage $w$, where $w$ is drawn from the distribution $F(w)$. Offers are independently and identically distributed. Every agent has another source of income, which we denote $\epsilon_t$, that may be regarded as nonhuman wealth. In every period all agents get a realization of $\epsilon_t$, which is independently and identically distributed over time, with distribution function $G(\epsilon)$. We also assume that $w_t$ and $\epsilon_t$ are independent. The objective of a worker is to maximize

$$E\sum_{t=0}^{\infty}\beta^t y_t, \qquad 0<\beta<1,$$

where $y_t = w+\phi\epsilon_t$ if the worker has accepted a job that pays $w$, and $y_t = c+\epsilon_t$ if the worker remains unemployed. We assume that $0<\phi<1$ to reflect the fact that an employed worker has less time to engage in the collection of nonhuman wealth. Assume $1>\text{prob}\{w\geq c+(1-\phi)\epsilon\}>0$.

Analyze the worker's problem. Write down the Bellman equation, and show that the reservation wage increases with the level of nonhuman wealth.

*Exercise 5.13* **Search and asset accumulation**

A worker receives, when unemployed, an offer to work forever at wage $w$, where $w$ is drawn from the distribution $F(w)$. Wage offers are identically and independently distributed over time. The worker maximizes

$$E\sum_{t=0}^{\infty}\beta^t u(c_t,l_t), \qquad 0<\beta<1,$$

where $c_t$ is consumption and $l_t$ is leisure. Assume $R_t$ is i.i.d. with distribution $H(R)$. The budget constraint is given by

$$a_{t+1}\leq R_t(a_t+w_t n_t-c_t)$$

and $l_t+n_t\leq 1$ if the worker has a job that pays $w_t$. If the worker is unemployed, the budget constraint is $a_{t+1}\leq R_t(a_t+z-c_t)$ and $l_t=1$. Here $z$ is unemployment compensation. It is assumed that $u(\cdot)$ is bounded and that $a_t$, the worker's asset position, cannot be negative. This assumption corresponds to a no-borrowing assumption. Write the Bellman equation for this problem.

*Exercise 5.14* **Temporary unemployment compensation**

Each period an unemployed worker draws one, and only one, offer to work forever at wage $w$. Wages are i.i.d. draws from the c.d.f. $F$, where $F(0)=0$ and $F(B)=1$. The worker seeks to maximize $E\sum_{t=0}^{\infty}\beta^t y_t$, where $y_t$ is the sum of the worker's wage and unemployment compensation, if any. The worker is entitled to unemployment compensation in the amount $\gamma>0$ only during the *first* period that she is unemployed. After one period on unemployment compensation, the worker receives none.

**a.** Write the Bellman equations for this problem. Prove that the worker's optimal policy is a time-varying reservation wage strategy.

**b.** Show how the worker's reservation wage varies with the duration of unemployment.

**c.** Show how the worker's "hazard of leaving unemployment" (i.e., the probability of accepting a job offer) varies with the duration of unemployment.

Now assume that the worker is also entitled to unemployment compensation if she quits a job. As before, the worker receives unemployment compensation in the amount of $\gamma$ during the first period of an unemployment spell, and zero during the remaining part of an unemployment spell. (To requalify for unemployment compensation, the worker must find a job and work for at least one period.) The timing of events is as follows. At the very beginning of a period, a worker who was employed in the previous period must decide whether or not to quit. The decision is irreversible; that is, a quitter cannot return to an old job. If the worker quits, she draws a new wage offer as described previously, and if she accepts the offer she immediately starts earning that wage without suffering any period of unemployment.

**d.** Write the Bellman equations for this problem. [*Hint*: At the very beginning of a period, let $v^e(w)$ denote the value of a worker who was employed in the previous period with wage $w$ (before any wage draw in the current period). Let $v_1^u(w')$ be the value of an unemployed worker who has drawn wage offer $w'$ and who is entitled to unemployment compensation, if she rejects the offer. Similarly, let $v_+^u(w')$ be the value of an unemployed worker who has drawn wage offer $w'$ but who is not eligible for unemployment compensation.]

**e.** Characterize the three reservation wages, $\bar{w}^e$, $\bar{w}_1^u$, and $\bar{w}_+^u$, associated with the value functions in part d. How are they related to $\gamma$? (*Hint*: Two of the reservation wages are straightforward to characterize, while the remaining one depends on the actual parameterization of the model.)

*Exercise 5.15* **Seasons, I**

An unemployed worker seeks to maximize $E\sum_{t=0}^{\infty}\beta^t y_t$, where $\beta \in (0,1)$, $y_t$ is her income at time $t$, and $E$ is the mathematical expectation operator. The person's income consists of one of two parts: unemployment compensation of $c$ that she receives each period she remains unemployed, or a fixed wage $w$ that the worker receives if employed. Once employed, the worker is employed forever with no chance of being fired. Every odd period (i.e., $t = 1, 3, 5, \ldots$) the worker receives one offer to work forever at a wage drawn from the c.d.f. $F(W) = \text{prob}(w \leq W)$. Assume that $F(0) = 0$ and $F(B) = 1$ for some $B > 0$. Successive draws from $F$ are independent. Every even period (i.e., $t = 0, 2, 4, \ldots$), the unemployed worker receives two offers to work forever at a wage drawn from $F$. Each of the two offers is drawn independently from $F$.

**a.** Formulate the Bellman equations for the unemployed person's problem.

**b.** Describe the form of the worker's optimal policy.

*Exercise 5.16* **Seasons, II**

Consider the following problem confronting an unemployed worker. The worker wants to maximize

$$E_0 \sum_0^\infty \beta^t y_t, \quad \beta \in (0, 1),$$

where $y_t = w_t$ in periods in which the worker is employed and $y_t = c$ in periods in which the worker is unemployed, where $w_t$ is a wage rate and $c$ is a constant level of unemployment compensation. At the start of each period, an unemployed worker receives one and only one offer to work at a wage $w$ drawn from a c.d.f. $F(W)$, where $F(0) = 0, F(B) = 1$ for some $B > 0$. Successive draws from $F$ are identically and independently distributed. There is no recall of past offers. Only unemployed workers receive wage offers. The wage is fixed as long as the worker remains in the job. The only way a worker can leave a job is if she is fired. At the *beginning* of each odd period ($t = 1, 3, \ldots$), a previously employed worker faces the probability of $\pi \in (0, 1)$ of being fired. If a worker is fired, she immediately receives a new draw of an offer to work at wage $w$. At each even period ($t = 0, 2, \ldots$), there is no chance of being fired.

**a.** Formulate a Bellman equation for the worker's problem.

**b.** Describe the form of the worker's optimal policy.

*Exercise 5.17* **Gittins indexes for beginners**

At the end of each period,[8] a worker can switch between two jobs, A and B, to begin the following period at a wage that will be drawn at the beginning of next period from a wage distribution specific to job A or B, and to the worker's history of past wage draws from jobs of either type A or type B. The worker must decide to stay or leave a job at the end of a period after his wage for this period on his current job has been received, but before knowing what his wage would be next period in either job. The wage at either job is described by a job-specific $n$-state Markov chain. Each period the worker works at either job A or job B. At the end of the period, before observing next period's wage on either job, he chooses which job to go to next period. We use lowercase letters $(i, j = 1, \ldots, n)$ to denote states for job A, and uppercase letters $(I, J = 1, \ldots n)$ for job B. There is no option of being unemployed.

Let $w_a(i)$ be the wage on job A when state $i$ occurs and $w_b(I)$ be the wage on job B when state $I$ occurs. Let $A = [A_{ij}]$ be the matrix of one-step transition probabilities between the states on job A, and let $B = [B_{ij}]$ be the matrix for job B. If the worker leaves a job and later decides to returns to it, he draws the wage for his first new period on the job from the conditional distribution determined by his last wage working at that job.

The worker's objective is to maximize the expected discounted value of his life-time earnings, $E_0 \sum_{t=0}^{\infty} \beta^t y_t$, where $\beta \in (0,1)$ is the discount factor, and where $y_t$ is his wage from whichever job he is working at in period $t$.

**a.** Consider a worker who has worked at both jobs before. Suppose that $w_a(i)$ was the last wage the worker receives on job A and $w_b(I)$ the last wage on job B. Write the Bellman equation for the worker.

**b.** Suppose that the worker is just entering the labor force. The first time he works at job A, the probability distribution for his initial wage is $\pi_a = (\pi_{a1}, \ldots, \pi_{an})$. Similarly, the probability distribution for his initial wage on job B is $\pi_b = (\pi_{b1}, \ldots, \pi_{bn})$ Formulate the decision problem for a new worker, who must decide which job to take initially. [*Hint*: Let $v_a(i)$ be the expected discounted present value of lifetime earnings for a worker who was last in state $i$ on job A and has never worked on job B; define $v_b(I)$ symmetrically.]

---

[8] See Gittins (1989) for more general versions of this problem.

*Exercise 5.18* **Jovanovic (1979b)**

An employed worker in the $t$th period of tenure on the current job receives a wage $w_t = x_t(1-\phi_t - s_t)$ where $x_t$ is job-specific human capital, $\phi_t \in (0,1)$ is the fraction of time that the worker spends investing in job-specific human capital, and $s_t \in (0,1)$ is the fraction of time that the worker spends searching for a new job offer. If the worker devotes $s_t$ to searching at $t$, then with probability $\pi(s_t) \in (0,1)$ at the beginning of $t+1$ the worker receives a new job offer to begin working at new job-specific capital level $\mu'$ drawn from the c.d.f. $F(\cdot)$. That is, searching for a new job offer promises the prospect of instantaneously reinitializing job-specific human capital at $\mu'$. Assume that $\pi'(s) > 0, \pi''(s) < 0$. While on a given job, job-specific human capital evolves according to

$$x_{t+1} = G(x_t, \phi_t) = g(x_t \phi_t) - \delta x_t,$$

where $g'(\cdot) > 0, g''(\cdot) < 0$, $\delta \in (0,1)$ is a depreciation rate, and $x_0 = \mu$ where $t$ is tenure on the job, and $\mu$ is the value of the "match" parameter drawn at the start of the current job. The worker is risk neutral and seeks to maximize $E_0 \sum_{\tau=0}^{\infty} \beta^\tau y_\tau$, where $y_\tau$ is his wage in period $\tau$.

**a.** Formulate the worker's Bellman equation.

**b.** Describe the worker's decision rule for deciding whether to accept an offer $\mu'$ at the beginning of next period.

**c.** Assume that $g(x\phi) = A(x\phi)^\alpha$ for $A > 0, \alpha \in (0,1)$. Assume that $\pi(s) = s^{.5}$. Assume that $F$ is a discrete $n$-valued distribution with probabilities $f_i$; for example, let $f_i = n^{-1}$. Write a Matlab program to solve the Bellman equation. Compute the optimal policies for $\phi, s$ and display them.

# 6
# *Recursive (Partial) Equilibrium*

## *An equilibrium concept*

This chapter formulates competitive and oligopolistic equilibria in some dynamic settings. Up to now, we have studied single-agent problems where components of the state vector not under the control of the agent were taken as given. In this chapter, we describe multiple-agent settings in which some of the components of the state vector that one agent takes as exogenous are determined by the decisions of other agents. We study partial equilibrium models, of a kind often applied in microeconomics.[1] We describe two closely related equilibrium concepts for such models: a rational expectations or recursive competitive equilibrium, and a Nash-Markov perfect equilibrium. The first equilibrium concept jointly restricts a Bellman equation and a transition law that is taken as given in that Bellman equation. The second equilibrium concept leads to pairs (in the duopoly case) or sets (in the oligopoly case) of Bellman equations and transition equations that are to be solved jointly by simultaneous backward induction.

Though the equilibrium concepts introduced in this chapter obviously transcend linear-quadratic setups, we choose to present them in the context of linear quadratic examples. These examples have the usual benefit associated with the optimal linear regulator that the Bellman equations remain tractable.

## *Example: adjustment costs*

This section describes a model of a competitive market with producers who face adjustment costs.[2] The model consists of $n$ identical firms whose profit function makes them want to forecast the aggregate output decisions of other

---

[1] For example, see Rosen and Topel (1988) and Rosen, Murphy, and Scheinkman (1994)

[2] The model is a version of one analyzed by Lucas and Prescott (1971) and Sargent (1987a). The recursive competitive equilibrium concept was used by Lucas and Prescott (1971) and described further by Prescott and Mehra (1980).

firms just like them in order to determine their own output. We assume that $n$ is a large number so that the output of any single firm has a negligible effect on aggregate output and, hence, firms are justified to treat the forecast of aggregate output as a constant when making their own output decisions. Thus, one of $n$ competitive firm sells output $y_t$ and chooses a production plan to maximize

$$\sum_{t=0}^{\infty} \beta^t R_t \tag{6.1}$$

where

$$R_t = p_t y_t - .5d(y_{t+1} - y_t)^2 \tag{6.2}$$

subject to $y_0$, a given initial condition. Here $\beta \in (0,1)$ is a discount factor, and $d > 0$ measures a cost of adjusting the rate of output. The firm is a price taker. The price $p_t$ lies on the demand curve

$$p_t = A_0 - A_1 Y_t \tag{6.3}$$

where $A_0 > 0, A_1 > 0$ and $Y_t$ is the marketwide level of output, being the sum of output of $n$ identical firms. The firm believes that marketwide output follows the law of motion

$$Y_{t+1} = H_0 + H_1 Y_t \equiv H(Y_t), \tag{6.4}$$

where $Y_0$ is a known initial condition. The belief parameters $H_0, H_1$ are to be among the equilibrium objects of the analysis, but for now we proceed on faith and take them as given. The firm observes $Y_t$ and $y_t$ at time $t$, when it chooses $y_{t+1}$. The adjustment costs $d(y_{t+1} - y_t)^2$ give the firm the incentive to forecast the market price.

Substituting equation (6.3) into equation (6.2) gives

$$R_t = (A_0 - A_1 Y_t) y_t - .5d(y_{t+1} - y_t)^2.$$

The firm's incentive to forecast the market price translates into an incentive to forecast the level of market output $Y$. We can write the Bellman equation for the firm as

$$v(y, Y) = \max_{y'} \left\{ A_0 y - A_1 y Y - .5d(y' - y)^2 + \beta v(y', Y') \right\}. \tag{6.5}$$

Here $'$ denotes next period's value of a variable. The Euler equation for the firm's problem is

$$-d(y' - y) + \beta v_y(y', Y') = 0. \tag{6.6}$$

Noting that for this problem the control is $y'$ and applying the Benveniste-Scheinkman formula from chapter 2 gives

$$v_y(y,Y) = A_0 - A_1 Y + d(y' - y).$$

Substituting this equation into equation (6.6) gives

$$-d(y_{t+1} - y_t) + \beta[A_0 - A_1 Y_{t+1} + d(y_{t+2} - y_{t+1})] = 0. \tag{6.7}$$

In the process of solving its Bellman equation, the firm sets an output path that satisfies equation (6.7), taking equation (6.4) as given, subject to the initial conditions $(y_0, Y_0)$ as well as an extra terminal condition. The terminal condition is

$$\lim_{t \to \infty} \beta^t y_t v_y(y_t, Y_t) = 0. \tag{6.8}$$

This is called the transversality condition and plays the role of a first-order necessary condition "at infinity." A solution of the firm's problem is a solution of the difference equation (6.7) subject to the given initial condition $y_0$ and the terminal condition (6.8). Solving the Bellman equation by backward induction automatically incorporates both equations (6.7) and (6.8).

The solution of the firm's Bellman equation is a policy function

$$y_{t+1} = h(y_t, Y_t). \tag{6.9}$$

Then with $n$ identical firms, setting $Y_t = n y_t$ makes the actual law of motion for output for the market

$$Y_{t+1} = n h(Y_t / n, Y_t). \tag{6.10}$$

Thus, when firms believe that the law of motion for marketwide output is equation (6.4), their optimizing behavior makes the actual law of motion equation (6.10).

A recursive competitive equilibrium equates the actual and perceived laws of motion (6.4) and (6.10). For this model, we adopt the following definition:

DEFINITION: A recursive competitive equilibrium[3] of the model with adjustment costs is a value function $v(y, Y)$, an optimal policy function $h(y, Y)$, and a law of motion $H(Y)$ such that

[3] This is also often called a rational expectations equilibrium.

a. Given $H$, $v(y,Y)$ satisfies the firm's Bellman equation and $h(y,Y)$ is the optimal policy function.

b. The law of motion $H$ satisfies $H(Y) = nh(Y/n, Y)$.

The firm's optimum problem induces a mapping $\mathcal{M}$ from a perceived law of motion for capital $H$ to an actual law of motion $\mathcal{M}(H)$. The mapping is summarized in equation (6.10). The $H$ component of a rational expectations equilibrium is a fixed point of the operator $\mathcal{M}$.

This equilibrium just defined is a special case of a recursive competitive equilibrium, to be defined more generally in the next section. How might we find an equilibrium? The next subsection shows a method that works in the present case and often works more generally. The method involves noting that the equilibrium solves an associated planning problem. For convenience, we'll assume from now on that the number of firms is one, while retaining the assumption of price-taking behavior.

*A planning problem*

Our solution strategy is to match the Euler equations of the market problem with those for a planning problem that can be solved as a single-agent dynamic programming problem. The optimal quantities from the planning problem are then the recursive competitive equilibrium quantities, and the equilibrium price can be coaxed from shadow prices for the planning problem.

To determine the planning problem, we first compute the sum of consumer and producer surplus at time $t$, defined as

$$S_t = S(Y_t, Y_{t+1}) = \int_0^{Y_t} (A_0 - A_1 x)\, d\, x - .5d(Y_{t+1} - Y_t)^2. \tag{6.11}$$

The first term is the area under the demand curve. The planning problem is to choose a production plan to maximize

$$\sum_{t=0}^{\infty} \beta^t S(Y_t, Y_{t-1}) \tag{6.12}$$

subject to an initial condition $Y_0$. The Bellman equation for the planning problem is

$$V(Y) = \max_{Y'} \left\{ A_0 Y - \frac{A_1}{2} Y^2 - .5d(Y' - Y)^2 + \beta V(Y') \right\}. \tag{6.13}$$

The Euler equation is

$$-d(Y' - Y) + \beta V'(Y') = 0. \tag{6.14}$$

Applying the Benveniste-Scheinkman formula gives

$$V'(Y) = A_0 - A_1 Y + d(Y' - Y). \tag{6.15}$$

Substituting this into equation (6.14) and rearranging gives

$$\beta A_0 + dY_t - [\beta A_1 + d(1+\beta)]Y_{t+1} + d\beta Y_{t+2} = 0 \tag{6.16}$$

Return to equation (6.7) and set $y_t = Y_t$ for all $t$. (Remember that we have set $n = 1$. When $n \neq 1$ we have to adjust pieces of the argument for $n$.) Notice that with $y_t = Y_t$, equations (6.16) and (6.7) are identical. Thus, a solution of the planning problem also is an equilibrium. Setting $y_t = Y_t$ in equation (6.7) amounts to dropping equation (6.4) and instead solving for the coefficients $H_0, H_1$ that make $y_t = Y_t$ true and that jointly solve equations (6.4) and (6.7).

It follows that for this example we can compute an equilibrium by forming the optimal linear regulator problem corresponding to the Bellman equation (6.13). The optimal policy function for this problem can be used to form the rational expectations $H(Y)$.[4]

## Recursive competitive equilibrium

The equilibrium concept of the previous section is widely used. Following Prescott and Mehra (1980), it is useful to define the equilibrium concept more generally as a *recursive competitive equilibrium.* Let $x$ be a vector of state variables under the control of a representative agent and let $X$ be the vector of those same variables chosen by "the market." Let $Z$ be a vector of other state variables chosen by "nature", that is, determined outside the model. The representative agent's problem is characterized by the Bellman equation

$$v(x, X, Z) = \max_u \{R(x, X, Z, u) + \beta v(x', X', Z')\} \tag{6.17}$$

[4] The method of this section was used by Lucas and Prescott (1971). It uses the connection between equilibrium and Pareto optimality expressed in the fundamental theorems of welfare economics. See Mas-Colell, Whinston, and Green (1995).

where $'$ denotes next period's value, and where the maximization is subject to the restrictions:

$$x' = g(x, X, Z, u) \tag{6.18}$$
$$X' = G(X, Z). \tag{6.19}$$
$$Z' = \zeta(Z) \tag{6.20}$$

Here $g$ describes the impact of the representative agent's controls $u$ on his state $x'$; $G$ and $\zeta$ describe his beliefs about the evolution of the aggregate state. The solution of the representative agent's problem is a decision rule

$$u = h(x, X, Z). \tag{6.21}$$

To make the representative agent representative, we impose $X = x$, but only "after" we have solved the agent's decision problem. Substituting equation (6.21) and $X = x_t$ into equation (6.18) gives the *actual* law of motion

$$X' = G_A(X, Z), \tag{6.22}$$

where $G_A(X, Z) \equiv g[X, X, Z, h(X, X, Z)]$. We are now ready to propose a definition:

DEFINITION: A *recursive competitive equilibrium* is a policy function $h$, an actual aggregate law of motion $G_A$, and a perceived aggregate law $G$ such that (a) Given $G$, $h$ solves the representative agent's optimization problem; and (b) $h$ implies that $G_A = G$.

This equilibrium concept is also sometimes called a *rational expectations equilibrium.* The equilibrium concept makes $G$ an outcome of the analysis. The functions giving the representative agent's expectations about the aggregate state variables contribute no free parameters and are *outcomes* of the analysis. In exercise 6.1, you are asked to implement this equilibrium concept.

## *Markov perfect equilibrium*

It is instructive to consider a dynamic model of duopoly. We now assume that the market has two firms. Each firm recognizes that its output decision will affect the aggregate output, and therefore influence the market price. That is, we drop the assumption of price-taking behavior.[5] The one-period return function of firm $i$ is

$$R_{it} = p_t y_{it} - .5d(y_{it+1} - y_{it})^2. \tag{6.23}$$

There is a demand curve

$$p_t = A_0 - A_1(y_{1t} + y_{2t}). \tag{6.24}$$

Substituting the demand curve into equation (6.23) lets us express the return as

$$R_{it} = A_0 y_{it} - A_1 y_{it}^2 - A_1 y_{it} y_{-i,t} - .5d(y_{it+1} - y_{it})^2, \tag{6.25}$$

where $y_{-i,t}$ denotes the output of the firm other than $i$. Firm $i$ chooses a decision rule that sets $y_{it+1}$ as a function of $(y_{it}, y_{-i,t})$ and that maximizes

$$\sum_{t=0}^{\infty} \beta^t R_{it}.$$

Temporarily assume that the maximizing decision rule is $y_{it+1} = f_i(y_{it}, y_{-i,t})$.
Given the function $f_{-i}$, the Bellman equation of firm $i$ is

$$v_i(y_{it}, y_{-i,t}) = \max_{y_{i,t+1}} \left\{R_{it} + \beta v_i(y_{it+1}, y_{-i,t+1})\right\}, \tag{6.26}$$

where the maximization is subject to the perceived decision rule of the other firm

$$y_{-i,t+1} = f_{-i}(y_{-i,t}, y_{it}). \tag{6.27}$$

Note the cross-reference between the two problems for $i = 1, 2$.
We now advance the following definition:

DEFINITION: A Markov perfect equilibrium is a pair of value functions $v_i$ and a pair of policy functions $f_i$ for $i = 1, 2$ such that

a. Given $f_{-i}, v_i$ satisfies the Bellman equation (6.26).

b. The policy function $f_i$ attains the right side of the Bellman equation (6.26).

[5] One consequence of departing from the price-taking framework is that the market outcome will no longer maximize welfare, measured as the sum of consumer and producer surplus. See exercise 6.4 for the case of a monopoly.

The adjective Markov denotes that the equilibrium decision rules depend only on the current values of the state variables $y_{it}$, not their histories. Perfect means that the equilibrium is constructed by backward induction and therefore builds in optimizing behavior for each firm for all conceivable future states, including many that are not realized by iterating forward on the pair of equilibrium strategies $f_i$.

*Computation*

If it exists, a Markov perfect equilibrium can be computed by iterating to convergence on the pair of Bellman equations (6.26). In particular, let $v_i^j, f_i^j$ be the value function and policy function for firm $i$ at the $j$th iteration. Then imagine constructing the iterates

$$v_i^{j+1}(y_{it}, y_{-i,t}) = \max_{y_{i,t+1}} \left\{ R_{it} + \beta v_i^j (y_{it+1}, y_{-i,t+1}) \right\}, \tag{6.28}$$

where the maximization is subject to

$$y_{-i,t+1} = f_{-i}^j (y_{-i,t}, y_{it}). \tag{6.29}$$

In general, these iterations are difficult.[6] In the next section, we describe how the calculations simplify for the case in which the return function is quadratic and the transition laws are linear.

## Linear Markov perfect equilibria

In this section, we show how the optimal linear regulator can be used to solve a model like that in the previous section. That was an example of a dynamic game. A dynamic game consists of these objects: (a) a list of players; (b) a list of dates and actions available to each player at each date; and (c) payoffs for each player expressed as functions of the actions taken by all players.

The optimal linear regulator is a good tool for formulating and solving dynamic games. The standard equilibrium concept—subgame perfection—in these games requires that each player's strategy be computed by backward induction.

[6] See Levhari and Mirman (1980) for how a Markov perfect equilibrium can be computed conveniently with logarithmic returns and Cobb-Douglas transition laws. Levhari and Mirman construct a model of fish and fishers.

This leads to an interrelated pair of Bellman equations. In linear-quadratic dynamic games, these "stacked Bellman equations" become "stacked Riccati equations" with a tractable mathematical structure. We now consider the following two-player, linear quadratic *dynamic game.* An $(n \times 1)$ state vector $x_t$ evolves according to a transition equation

$$x_{t+1} = A_t x_t + B_{1t} u_{1t} + B_{2t} u_{2t} \tag{6.30}$$

where $u_{jt}$ is a $(k_j \times 1)$ vector of controls of player $j$. We start with a finite horizon formulation, where $t_0$ is the initial date and $t_1$ is the terminal date for the common horizon of the two players. Player 1 maximizes

$$\sum_{t=t_0}^{t_1-1} \left( x_t^T R_1 x_t + u_{1t}^T Q_1 u_{1t} + u_{2t}^T S_1 u_{2t} \right) \tag{6.31}$$

where $R_1$ and $S_1$ are negative semidefinite and $Q_1$ is negative definite. Player 2 maximizes

$$\sum_{t=t_0}^{t_1-1} \left( x_t^T R_2 x_t + u_{2t}^T Q_2 u_{2t} + u_{1t}^T S_2 u_{1t} \right) \tag{6.32}$$

where $R_2$ and $S_2$ are negative semidefinite and $Q_2$ is negative definite.

We formulate a Markov perfect equilibrium as follows. Player $j$ employs linear decision rules

$$u_{jt} = -F_{jt} x_t, \quad t = t_0, \ldots, t_1 - 1$$

where $F_{jt}$ is a $(k_j \times n)$ matrix. Assume that player $i$ knows $\{F_{-i,t}; t = t_0, \ldots, t_1 - 1\}$. Then player 1's problem is to maximize expression (6.31) subject to the known law of motion (6.30) *and* the known control law $u_{2t} = -F_{2t} x_t$ of player 2. Symmetrically, player 2's problem is to maximize expression (6.32) subject to equation (6.30) and $u_{1t} = -F_{1t} x_t$. A Markov perfect equilibrium is a pair of sequences $\{F_{1t}, F_{2t}; t = t_0, t_0 + 1, \ldots, t_1 - 1\}$ such that $\{F_{1t}\}$ solves player 1's problem, given $\{F_{2t}\}$, and $\{F_{2t}\}$ solves player 2's problem, given $\{F_{1t}\}$. We have restricted each player's strategy to depend only on $x_t$, and not on the *history* $h_t = \{(x_s, u_{1s}, u_{2s}), s = t_0, \ldots, t\}$. This restriction on strategy spaces accounts for the adjective "Markov" in the phrase "Markov perfect equilibrium."

Player 1's problem is to maximize

$$\sum_{t=t_0}^{t_1-1} \left\{ x_t^T (R_1 + F_{2t}^T S_1 F_{2t}) x_t + u_{1t}^T Q_1 u_{1t} \right\}$$

subject to

$$x_{t+1} = (A_t - B_{2t}F_{2t})x_t + B_{1t}u_{1t}.$$

This is an optimal linear regulator problem, and it can be solved by working backward. Evidently, player 2's problem is also an optimal linear regulator problem.

The solution of player 1's problem is given by

$$F_{1t} = (B_{1t}^T P_{1t+1} B_{1t} + Q_1)^{-1} B_{1t}^T P_{1t+1}(A_t - B_{2t}F_{2t}) \tag{6.33}$$

$$t = t_0, t_0 + 1, \ldots, t_1 - 1$$

where $P_{1t}$ is the solution of the following matrix Riccati difference equation, with terminal condition $P_{1t_1} = 0$:

$$\begin{aligned} P_{1t} = {} & (A_t - B_{2t}F_{2t})^T P_{1t+1}(A_t - B_{2t}F_{2t}) + (R_1 + F_{2t}^T S_1 F_{2t}) \\ & -(A_t - B_{2t}F_{2t})^T P_{1t+1} B_{1t}(B_{1t}^T P_{1t+1} B_{1t} + Q_1)^{-1} B_{1t}^T P_{1t+1}(A_t - B_{2t}F_{2t}). \end{aligned} \tag{6.34}$$

The solution of player 2's problem is

$$F_{2t} = (B_{2t}^T P_{2t+1} B_{2t} + Q_2)^{-1} B_{2t}^T P_{2t+1}(A_t - B_{1t}F_{1t}) \tag{6.35}$$

where $P_{2t}$ solves the following matrix Riccati difference equation, with terminal condition $P_{2t_1} = 0$:

$$\begin{aligned} P_{2t} = {} & (A_t - B_{1t}F_{1t})^T P_{2t+1}(A_t - B_{1t}F_{1t}) + (R_2 + F_{1t}^T S_2 F_{1t}) \\ & -(A_t - B_{1t}F_{1t})^T P_{2t+1} B_{2t}(B_{2t}^T P_{2t+1} B_{2t} + Q_2)^{-1} B_{2t}^T P_{2t+1}(A_t - B_{1t}F_{1t}). \end{aligned} \tag{6.36}$$

The equilibrium sequences $\{F_{1t}, F_{2t}; t = t_0, t_0 + 1, \ldots, t_1 - 1\}$ can be calculated from the pair of coupled Riccati difference equations (6.34) and (6.36). In particular, we use equations (6.33), (6.34), (6.35), and (6.36) to "work backward" from time $t_1 - 1$. Notice that given $P_{1t+1}$ and $P_{2t+1}$, equations (6.33) and (6.35) are a system of $(k_2 \times n) + (k_1 \times n)$ *linear* equations in the $(k_2 \times n) + (k_1 \times n)$ unknowns in the matrices $F_{1t}$ and $F_{2t}$.

Notice how $j$'s control law $F_{jt}$ is a function of $\{F_{is}, s \geq t, i \neq j\}$. Thus, agent $i$'s choice of $\{F_{it}; t = t_0, \ldots, t_1 - 1\}$ influences agent $j$'s choice of control laws. However, in the Markov perfect equilibrium of this game, each agent is assumed to ignore the influence that his choice exerts on the other agent's choice.[7]

[7] In an equilibrium of a *Stackelberg* or *dominant player* game, the timing of moves is so altered relative to the present game that one of the agents called the leader takes into account the influence that his choices exert on the other agent's choices.

We often want to compute the solutions of such games for infinite horizons, in the hope that the decision rules $F_{it}$ settle down to be time invariant as $t_1 \to +\infty$. In practice, we usually fix $t_1$ and compute the equilibrium of an infinite horizon game by driving $t_0 \to -\infty$. Judd followed that procedure in the following example.

*An example*

This section describes the Markov perfect equilibrium of an infinite horizon linear quadratic game proposed by Kenneth Judd (1990). The equilibrium is computed by iterating to convergence on the pair of Riccati equations defined by the choice problems of two firms. Each firm solves a linear quadratic optimization problem, taking as given and known the sequence of linear decision rules used by the other player. The firms set prices and quantities of two goods interrelated through their demand curves. There is no uncertainty. Relevant variables are defined as follows:

$I_{it}$ = inventories of firm $i$ at beginning of $t$.
$q_{it}$ = production of firm $i$ during period $t$.
$p_{it}$ = price charged by firm $i$ during period $t$.
$S_{it}$ = sales made by firm $i$ during period $t$.
$E_{it}$ = costs of production of firm $i$ during period $t$.
$C_{it}$ = costs of carrying inventories for firm $i$ during $t$.

The firms' cost functions are

$$C_{it} = c_{i1} + c_{i2} I_{it} + .5 c_{i3} I_{it}^2$$
$$E_{it} = e_{i1} + e_{i2} q_{it} + .5 e_{i3} q_{it}^2$$

where $e_{ij}, c_{ij}$ are positive scalars.
Inventories obey the laws of motion

$$I_{i,t+1} = (1-\delta) I_{it} + q_{it} - S_{it}$$

Demand is governed by the linear schedule

$$S_t = d p_{it} + B$$

where $S_t = [\, S_{1t} \quad S_{2t} \,]'$, $d$ is a $(2 \times 2)$ negative definite matrix, and $B$ is a vector of constants. Firm $i$ maximizes the undiscounted sum

$$\lim_{T \to \infty} \frac{1}{T} \sum_{t=0}^{T} \left( p_{it} S_{it} - E_{it} - C_{it} \right)$$

by choosing a decision rule for price and quantity of the form

$$u_{it} = -F_i x_t$$

where $u_{it} = [\, p_{it} \quad q_{it} \,]'$, and the state is $x_t = [\, I_{1t} \quad I_{2t} \,]$.

In the web site for the book, we supply a Matlab program **nnash.m** that computes a Markov perfect equilibrium of the linear quadratic dynamic game in which player $i$ maximizes

$$\sum_{t=0}^{\infty} \{x_t' r_i x_t + 2x_t' w_i u_{it} + u_{it}' q_i u_{it} + u_{jt}' s_i u_{jt} + 2u_{jt}' m_i u_{it}\}$$

subject to the law of motion

$$x_{t+1} = a x_t + b_1 u_{1t} + b_2 u_{2t}$$

and a control law $u_{jt} = -f_j x_t$ for the other player; here $a$ is $n \times n$; $b_1$ is $n \times k_1$; $b_2$ is $n \times k_2$; $r_1$ is $n \times n$; $r_2$ is $n \times n$; $q_1$ is $k_1 \times k_1$; $q_2$ is $k_2 \times k_2$; $s_1$ is $k_2 \times k_2$; $s_2$ is $k_1 \times k_1$; $w_1$ is $n \times k_1$; $w_2$ is $n \times k_2$; $m_1$ is $k_2 \times k_1$; and $m_2$ is $k_1 \times k_2$. The equilibrium of Judd's model can be computed by filling in the matrices appropriately. A Matlab tutorial **judd.m** uses **nnash.m** to compute the equilibrium.

## Concluding remarks

This chapter has introduced two equilibrium concepts and illustrated how dynamic programming algorithms are embedded in each. For the linear models we have used as illustrations, the dynamic programs become optimal linear regulators, making it tractable to compute equilibria even for large state spaces. We chose to define these equilibria concepts in partial equilibrium settings that are more natural for microeconomic applications than for macroeconomic ones. In the next chapter, we use the recursive equilibrium concept to analyze a general equilibrium in an endowment economy. That setting serves as a natural starting point for addressing various macroeconomic issues.

## Exercises

These problems aim to teach about (1) mapping problems into recursive forms, (2) different equilibrium concepts, and (3) using Matlab. Computer programs are available from the web site for the book.[8]

*Exercise 6.1* **A competitive firm**

A competitive firm seeks to maximize

$$\sum_{t=0}^{\infty} \beta^t R_t \tag{1}$$

where $\beta \in (0,1)$, and time-$t$ revenue $R_t$ is

$$R_t = p_t y_t - .5d(y_{t+1} - y_t)^2, \quad d > 0, \tag{2}$$

where $p_t$ is the price of output, and $y_t$ is the time-$t$ output of the firm. Here $.5d(y_{t+1} - y_t)^2$ measures the firm's cost of adjusting its rate of output. The firm starts with a given initial level of output $y_0$. The price lies on the market demand curve

$$p_t = A_0 - A_1 Y_t, A_0, A_1 > 0 \tag{3}$$

where $Y_t$ is the market level of output, which the firm takes as exogenous, and which the firm believes follows the law of motion

$$Y_{t+1} = H_0 + H_1 Y_t, \tag{4}$$

with $Y_0$ as a fixed initial condition.

**a.** Formulate the Bellman equation for the firm's problem.

**b.** Formulate the firm's problem as a discounted optimal linear regulator problem, being careful to describe all of the objects needed. What is the *state* for the firm's problem?

---

[8] The web site is `ftp://zia.stanford.edu/pub/sargent/webdocs/matlab`.

**c.** Use the Matlab program `olrp.m` to solve the firm's problem for the following parameter values: $A_0 = 100, A_1 = .05, \beta = .95, d = 10, H_0 = 95.5$, and $H_1 = .95$. Express the solution of the firm's problem in the form

$$y_{t+1} = h_0 + h_1 y_t + h_2 Y_t, \tag{5}$$

giving values for the $h_j$'s.

**d.** If there were $n$ identical competitive firms all behaving according to equation (5), what would equation (5) imply for the *actual* law of motion (4) for the market supply $Y$?

**e.** Formulate the Euler equation for the firm's problem.

*Exercise 6.2* **Rational expectations**

Now assume that the firm in problem 1 is "representative." We implement this idea by setting $n = 1$. In equilibrium, we will require that $y_t = Y_t$, but we don't want to impose this condition at the stage that the firm is optimizing (because we want to retain competitive behavior). Define a rational expectations equilibrium to be a pair of numbers $H_0, H_1$ such that if the representative firm solves the problem ascribed to it in problem 1, then the firm's optimal behavior given by equation (5) implies that $y_t = Y_t \ \forall\, t \geq 0$.

**a.** Use the program that you wrote for exercise 6.1 to determine which if any of the following pairs $(H_0, H_1)$ is a rational expectations equilibrium: (i) (94.0888, .9211); (ii) (93.22, .9433), and (iii) (95.08187459215024, .95245906270392)?

**b.** Describe an iterative algorithm by which the program that you wrote for exercise 6.1 might be used to compute a rational expectations equilibrium. (You are not being asked actually to use the algorithm you are suggesting.)

*Exercise 6.3* **Maximizing welfare**

A planner seeks to maximize the welfare criterion

$$\sum_{t=0}^{\infty} \beta^t S_t, \tag{6}$$

where $S_t$ is "consumer surplus plus producer surplus" defined to be

$$S_t = S(Y_t, Y_{t+1}) = \int_0^{Y_t} (A_0 - A_1 x)\, d\,x - .5d(Y_{t+1} - Y_t)^2.$$

**a.** Formulate the planner's Bellman equation.

**b.** Formulate the planner's problem as an optimal linear regulator, and solve it using the Matlab program `olrp.m`. Represent the solution in the form $Y_{t+1} = s_0 + s_1 Y_t$.

**c.** Compare your answer in part b with your answer to part a of exercise 6.2.

*Exercise 6.4* **Monopoly**

A monopolist faces the industry demand curve (3) and chooses $Y_t$ to maximize $\sum_{t=0}^{\infty} \beta^t R_t$ where $R_t = p_t Y_t - .5d(Y_{t+1} - Y_t)^2$ and where $Y_0$ is given.

**a.** Formulate the firm's Bellman equation.

**b.** For the parameter values listed in exercise 6.1, formulate and solve the firm's problem using `olrp.m`.

**c.** Compare your answer in part b with the answer you obtained to part b of exercise 6.3.

*Exercise 6.5* **Duopoly**

An industry consists of two firms that jointly face the industry-wide demand curve (3), where now $Y_t = y_{1t} + y_{2t}$. Firm $i = 1, 2$ maximizes

$$\sum_{t=0}^{\infty} \beta^t R_{it} \tag{7}$$

where $R_{it} = p_t y_{it} - .5d(y_{i,t+1} - y_{it})^2$.

**a.** Define a Markov perfect equilibrium for this industry.

**b.** Formulate the Bellman equation for each firm.

**c.** Use the Matlab program `nash.m` to compute an equilibrium, assuming the parameter values listed in exercise 6.1.

*Exercise 6.6* **Self-control**

This is a model of a human who has time-inconsistent preferences, of a type proposed by Phelps and Pollak (1968) and used by Laibson (1994).[9] The human

[9] See Gul and Pesendorfer (2000) for a single-agent recursive representation of preferences exhibiting temptation and self-control.

lives from $t = 0, \ldots, T$. Think of the human as actually consisting of $T+1$ personalities, one for each period. Each personality is a distinct agent (i.e., a distinct utility function and constraint set). Personality $T$ has preferences ordered by $u(c_T)$ and personality $t < T$ has preferences that are ordered by

$$u(c_t) + \delta \sum_{j=1}^{T-t} \beta^j u(c_{t+j}), \tag{6.37}$$

where $u(\cdot)$ is a twice continuously differentiable, increasing and strictly concave function of consumption of a single good; $\beta \in (0,1)$, and $\delta \in (0,1]$. When $\delta < 1$, preferences of the sequence of personalities are time-inconsistent (that is, not recursive). At each $t$, let there be a savings technology described by

$$k_{t+1} + c_t \le f(k_t), \tag{6.38}$$

where $f$ is a production function with $f' > 0, f'' \le 0$.

**a.** Define a Markov perfect equilibrium for the $T+1$ personalities.

**b.** Argue that the Nash-Markov equilibrium can be computed by iterating on the following functional equations:

$$V_{j+1}(k) = \max_c \{u(c) + \beta\delta W_j(k')\} \tag{6.39a}$$

$$W_{j+1}(k) = u[c_{j+1}(k)] + \beta W_j[f(k) - c_{j+1}(k)] \tag{6.40}$$

where $c_{j+1}(k)$ is the maximizer of the right side of (6.39a) for $j+1$, starting from $W_0(k) = u[f(k)]$ . Here $W_j(k)$ is the value of $u(c_{T-j}) + \beta u(c_{T-j+1}) + \ldots + \beta^{T-j} u(c_T)$, taking the decision rules $c_h(k)$ as given for $h = 0, 1, \ldots, j$.

**c.** State the optimization problem of the time-0 person who is given the power to dictate the choices of all subsequent persons. Write the Bellman equations for this problem. The time zero person is said to have a commitment technology for "self-control" in this problem.

# 7
# Competitive Equilibrium with Complete Markets

## Time-0 versus sequential trading

This chapter describes the competitive equilibrium for a pure exchange infinite horizon economy with stochastic Markov endowments. This economy is a basic setting for studying risk sharing, asset pricing, and consumption. We describe two market structures: the Arrow-Debreu structure with complete markets in dated contingent claims all traded at time $0$, and a recursive structure with complete one-period Arrow securities. These two have different asset structures and timings of trades, but lead to identical allocations of consumption across people. Both are referred to as complete market economies. They allow more comprehensive sharing of risks than do the incomplete markets economies to be studied in chapters 13 and 14.

## The physical setting

### Preferences and endowments

Let $\pi(s'|s)$ be a Markov chain with initial distribution $\pi_0(s)$ given. That is, $\text{Prob}(s_{t+1} = s'|s_t = s) = \pi(s'|s)$ and $\text{Prob}(s_0 = s) = \pi_0(s)$. As we saw in chapter 1, the chain induces a sequence of probability measures $\pi(s^t)$ on histories $s^t = [s_t, s_{t-1}, \ldots, s_0]$ via the recursions

$$\pi(s^t) = \pi(s_t|s_{t-1})\pi(s_{t-1}|s_{t-2})\ldots\pi(s_1|s_0)\pi_0(s_0). \tag{7.1}$$

Formula (7.1) is the unconditional probability of $s^t$ when $s_0$ has not been realized. If $s_0$ has been observed, the appropriate distribution of $s^t$ is conditional on $s_0$, in which case we use

$$\pi(s^t|s_0) = \pi(s_t|s_{t-1})\pi(s_{t-1}|s_{t-2})\ldots\pi(s_1|s_0). \tag{7.2}$$

In the rest of this chapter, we shall assume that trading occurs after $s_0$ has been observed, so we shall use equation (7.2).[1] We let $\pi(s^t|s^\tau)$ denote the probability of state $s^t$ conditional on being in state $s^\tau$ at date $\tau$,

$$\pi(s^t|s^\tau) = \pi(s_t|s_{t-1})\pi(s_{t-1}|s_{t-2})\dots\pi(s_{\tau+1}|s_\tau). \tag{7.3}$$

Because of the Markov assumption, $\pi(s^t|s^\tau)$ does not depend on the history $s^{\tau-1}$.[2]

There are $I$ agents named $i=1,\dots,I$. Agent $i$ owns a stochastic endowment of one good $y_t^i = y^i(s_t)$ that depends on the realization of $s_t$. Note that $y_t^i$ is a time invariant function of $s_t$. The variable $s_t$ is publicly observable. Household $i$ purchases a history-dependent consumption plan $c^i = \{c_t^i(s^t)\}_{t=0}^\infty$. The household orders consumption streams by

$$U(c^i) = \sum_{t=0}^{\infty}\sum_{s^t} \beta^t u[c_t^i(s^t)]\pi(s^t|s_0) \tag{7.4}$$

where the right side is equal to $E_0 \sum_{t=0}^\infty \beta^t u(c_t^i)$, where $E_0$ is the mathematical expectation operator, conditioned on $s_0$. Here $u(c)$ is an increasing, twice continuously differentiable, strictly concave function of consumption $c$ of one good. The utility function satisfies the Inada condition

$$\lim_{c\downarrow 0} u'(c) = +\infty.$$

*Complete markets*

Households trade dated state-contingent claims to consumption. There is a complete set of securities. Trades occur at time $0$, after $s_0$ has been realized. The household can exchange claims on time-$t$ consumption, contingent on history $s^t$ at price $q_t^0(s^t)$. The superscript $0$ refers to the date at which trades occur, while

[1] Most of our formulas carry over to the case where trading occurs before $s_0$ has been realized; just replace equation (7.2) with equation (7.1).

[2] Note our slight abuse of the word "state" and the notation $\pi(x|z)$. Properly speaking, the state in period $t$ is $s_t$ but we do also refer to "state $s^t$ at time $t$" which is a vector containing the realization of states in all periods up and until time $t$. Further, $\pi(x|z) = \text{Prob}(s_{t+1}=x|s_t=z)$ when $x$ and $z$ are states in the word's proper meaning, but if $x$ is a vector of states with length $t>1$ and $z$ is a vector containing the last $\tau$ elements of vector $x$, where $t \geq \tau \geq 1$, then $\pi(x|z) = \text{Prob}(s^t = x|s^\tau = z)$.

the subscript $t$ refers to the date that deliveries are to be made. The household's budget constraint is

$$\sum_{t=0}^{\infty}\sum_{s^t} q_t^0(s^t)c_t^i(s^t) \leq \sum_{t=0}^{\infty}\sum_{s^t} q_t^0(s^t)y^i(s_t). \tag{7.5}$$

The household's problem is to choose $c^i$ to maximize expression (7.4) subject inequality (7.5). Here $q_t^0$ is the price of time $t$ consumption contingent on history $s^t$ at $t$ in terms of an abstract unit of account or numeraire. Underlying the *single* budget constraint (7.5) is thet fact that multilateral trades are possible through a clearing operation that keeps track of net claims.[3] All trades occur at time 0. After time 0, trades that were agreed to at 0 are executed.

For each consumer, there is a *single* budget constraint (7.5) to which we attach a Lagrange multiplier $\mu^i$. We obtain the first-order conditions for the household's problem:

$$\frac{\partial U(c^i)}{\partial c_t^i(s^t)} = \mu^i q_t^0(s^t). \tag{7.6}$$

The left side is the derivative of total utility to the time-$t$, state-$s^t$ component of consumption. Each individual has his own $\mu_i$, which is independent of time. Note also that with specification (7.4) of the utility functional, we have

$$\frac{\partial U(c^i)}{\partial c_t^i(s^t)} = \beta^t u'[c_t^i(s^t)]\pi(s^t|s_0). \tag{7.7}$$

This expression implies that equation (7.6) can be written

$$\beta^t u'[c_t^i(s^t)]\pi(s^t|s_0) = \mu^i q_t^0(s^t). \tag{7.8}$$

We use the following definitions:

DEFINITIONS: A *price system* is a sequence of functions $\{q_t^0(s^t)\}_{t=0}^{\infty}$. An *allocation* is a list of sequences of functions $c^i = \{c_t^i(s^t)\}_{t=0}^{\infty}$, one for each $i$. A *feasible allocation* satisfies

$$\sum_i y^i(s_t) \geq \sum_i c_t^i(s^t). \tag{7.9}$$

[3] In the language of modern payments systems, this is a system with net settlements, not gross settlements, of trades.

DEFINITION: A *competitive equilibrium* is a feasible allocation and price system such that the allocation solves each household's problem.

Notice that equation (7.8) implies

$$\frac{u'[c_t^i(s^t)]}{u'[c_t^j(s^t)]} = \frac{\mu^i}{\mu^j} \tag{7.10}$$

for all pairs $(i,j)$. Thus, ratios of marginal utilities between pairs of agents are constant across all states and dates.

An equilibrium allocation solves equations (7.10), (7.9), and (7.5). Note that equation (7.10) implies that

$$c_t^i(s^t) = u'^{-1}\left\{u'[c_t^1(s^t)]\frac{\mu^i}{\mu^1}\right\}. \tag{7.11}$$

Substituting this into equation (7.9) at equality gives

$$\sum_i u'^{-1}\left\{u'[c_t^1(s^t)]\frac{\mu^i}{\mu^1}\right\} = \sum_i y^i(s_t). \tag{7.12}$$

The right side of equation (7.12) depends only on the current $s_t$ and not the entire history $s^t$; therefore, the left side, and so $c_t^1(s^t)$, must also depend only on $s_t$. It follows from equation (7.11) that the equilibrium allocation $c_t^i(s^t)$ depends only on $s_t$ for each $i$. We summarize this analysis in the following proposition:

PROPOSITION 1: The competitive equilibrium allocation is not history dependent; $c_t^i(s^t) = \bar{c}^i(s_t)$.

*Equilibrium pricing function*

Suppose that $c^i$, $i = 1,\dots,I$ is an equilibrium allocation. Then the marginal condition (7.6) or (7.8) gives the price system $q_t^0(s^t)$ as a function of the allocation to household $i$, for any $i$. Note that the price system is a stochastic process. Because the units of the price system are arbitrary, one of the multipliers can be normalized at any positive value. We shall set $\mu^1 = u'[c^1(s_0)]$, so that $q_0^0(s_0) = 1$, putting the price system in units of time-0 goods.[4]

[4] This choice also implies that $\mu^i = u'[c^i(s_0)]$ for all $i$.

## Examples

### *Risk sharing*

Suppose that the one-period utility function is of the constant relative risk-aversion form

$$u(c) = (1-\gamma)^{-1} c^{1-\gamma}, \ \gamma > 0.$$

Then equation (7.10) implies

$$(c_t^i)^{-\gamma} = (c_t^j)^{-\gamma} \left(\frac{\mu^i}{\mu^j}\right)$$

or

$$c_t^i = c_t^j \left(\frac{\mu^i}{\mu^j}\right)^{-\frac{1}{\gamma}}. \tag{7.13}$$

Equation (7.13) states that time-$t$ elements of consumption allocations to distinct agents are constant fractions of one another. With a power utility function, it says that individual consumption is perfectly correlated with the aggregate endowment or aggregate consumption. The fractions assigned to each individual are independent of the realization of $s_t$. Thus, there is extensive cross-state and cross-time consumption smoothing. The constant fractions characterization of consumption comes from (1) complete markets and (2) a homothetic one-period utility function.

### *No aggregate uncertainty*

Let $s_t$ be a Markov process taking values on the unit interval $[0,1]$. There are two households, with $y_t^1 = s$ and $y_t^2 = 1-s$. Note that the aggregate endowment is $y_t = \sum_i y_t^i = 1$. We use a guess-and-verify method to find the equilibrium. Guess that the equilibrium allocation satisfies $c_t^i = c_0^i$ for all $t$ for $i = 1,2$. Then from equation (7.8),

$$q_t^0(s^t) = \beta^t \pi(s^t|s_0) \frac{u'(c_0^i)}{\mu^i}, \tag{7.14}$$

for all $t$ for $i = 1,2$. Household $i$'s budget constraint implies

$$\frac{u'(c_0^i)}{\mu^i} \sum_{t=0}^{\infty} \sum_{s^t} \beta^t \pi(s^t|s_0) \left[c_0^i - y^i(s_t)\right] = 0.$$

Solving this equation for $c_0^i$ gives

$$c_0^i = (1-\beta) \sum_{t=0}^{\infty} \sum_{s^t} \beta^t \pi(s^t|s_0) y^i(s_t). \tag{7.15}$$

Summing equation (7.15) verifies that $c_0^1 + c_0^2 = 1$.

*Periodic endowment processes*

Consider the special case of the previous example in which $s_t$ is a two-state Markov chain taking the two values $\{0,1\}$, with transition probabilities $\pi(1|0) = \pi(0|1) = 1$, $\pi(0|0) = \pi(1|1) = 0$. Suppose that $\pi_0(1) = 1$, so that the initial value of the shock $s_0$ and therefore of the endowment $y_0^1$ is $1$. Thus, the endowment processes are perfectly predictable sequences $(1,0,1,\ldots)$ for the first agent and $(0,1,0,\ldots)$ for the second agent. Let $\tilde{s}^t$ be the history of $(1,0,1,\ldots)$ up to $t$. Evidently, $\pi(\tilde{s}^t) = 1$, and the probability assigned to all other histories up to $t$ is zero. The equilibrium price system is then

$$q_t^0(s^t) = \begin{cases} \beta^t, & \text{if } s^t = \tilde{s}^t; \\ 0, & \text{otherwise} \ ; \end{cases}$$

when using the time-0 good as numeraire, $q_0^0(\tilde{s}_0) = 1$. From equation (7.15), we have

$$c_0^1 = (1-\beta) \sum_{j=0}^{\infty} \beta^{2j} = \frac{1}{1+\beta}, \tag{7.16a}$$

$$c_0^2 = (1-\beta)\beta \sum_{j=0}^{\infty} \beta^{2j} = \frac{\beta}{1+\beta}. \tag{7.16b}$$

Consumer 1 consumes more every period because he is richer by virtue of receiving his endowment earlier.

*Interpretation of trading arrangement*

In the competitive equilibrium, all trades occur at $t=0$ in one market. Deliveries occur after $t=0$, but no more trades. A vast credit system operates at $t=0$. An unspecified clearing system assures that condition (7.5) holds for each household $i$. A symptom of the once-and-for-all trading arrangement is that each household faces one budget constraint that accounts for all trades across dates and states.

In a subsequent section, we describe another trading arrangement with more trading dates but fewer securities at each date. As a prelude to that section, we describe some asset-pricing implications embedded in the model with time-0 trading.

## Primer on asset pricing

Many asset-pricing models assume complete markets and price assets by breaking an asset into a sequence of state-contingent claims, evaluating each component of that sequence at the relevant "state price deflator" $q_0^t$, then adding up those values. The asset being priced is viewed as redundant, in the sense that it offers a bundle of state-contingent dated claims, each component of which has already been priced by the market. While we shall devote a later chapter entirely to such asset-pricing theories, it is useful to give some pricing formulas at this point because they help illustrate the complete market competitive structure.

### Pricing redundant assets

Let $\{d(s_t)\}_{t=0}^{\infty}$ be a stream of claims on time $t$, state $s^t$ consumption, where $d(s_t)$ is a measurable function of $s_t$. The price of an asset entitling the owner to this stream must be

$$a_0^0 = \sum_{t=0}^{\infty} \sum_{s^t} q_t^0(s^t) d(s_t). \tag{7.17}$$

If this equation did not hold, someone could make unbounded profits by synthesizing this asset through purchases or sales of state-contingent dated commodities and then either buying or selling the composite asset. We shall elaborate this arbitrage argument below and in a later chapter on asset pricing.

### Riskless consol

As an example, consider the price of a *riskless consol*, that is, an asset offering to pay one unit of consumption for sure each period. Then $d_t(s_t) = 1$ for all $t$ and $s_t$, and the price of this asset is

$$\sum_{t=0}^{\infty} \sum_{s^t} q_t^0(s^t).$$

*Riskless strips*

As another example, consider a sequence of *strips* of returns on the riskless consol. The time- $t$ strip is just the return process $d_\tau = 1$ if $\tau = t \geq 0$, and 0 otherwise. Thus, the owner of the strip is entitled to just the time- $t$ interest payment. The value of the time- $t$ strip at time 0 is evidently

$$\sum_{s^t} q_t^0(s^t).$$

Compare this to the price of the consol reported earlier.

*Tail assets*

Return to the stream of dividends $\{d(s_t)\}_{t\geq 0}$ generated by the asset priced in equation (7.17). For $\tau \geq 1$, suppose that we strip off the first $\tau - 1$ periods of the dividend and want to get the time-0 value of the dividend stream $\{d(s_t)\}_{t\geq\tau}$. Specifically, we seek this asset value for each possible realization of $s^\tau$. Let $a_\tau^0(s^\tau)$ be the time-0 price of an asset that entitles the owner to dividend stream $\{d(s_t)\}_{t\geq\tau}$ if history $s^\tau$ is realized,

$$a_\tau^0(s^\tau) = \sum_{t\geq\tau} \sum_{\{\tilde{s}^t : \tilde{s}^\tau = s^\tau\}} q_t^0(\tilde{s}^t) d(\tilde{s}_t). \tag{7.18}$$

The units of the price are time-0 (state- $s_0$ ) goods per unit (the numeraire) so that $q_0^0(s_0) = 1$. To convert the price into units of time $\tau$, state $s^\tau$ consumption goods, divide by $q_\tau^0(s^\tau)$ to get

$$a_\tau^\tau(s^\tau) \equiv \frac{a_\tau^0(s^\tau)}{q_\tau^0(s^\tau)} = \sum_{t\geq\tau} \sum_{\{\tilde{s}^t : \tilde{s}^\tau = s^\tau\}} \frac{q_t^0(\tilde{s}^t)}{q_\tau^0(s^\tau)} d(\tilde{s}_t). \tag{7.19}$$

Notice that [5]

$$\begin{aligned} q_t^\tau(s^t) \equiv \frac{q_t^0(s^t)}{q_\tau^0(s^\tau)} &= \frac{\beta^t u'[c_t^i(s^t)]\pi(s^t)}{\beta^\tau u'[c_\tau^i(s^\tau)]\pi(s^\tau)} \\ &= \beta^{t-\tau} \frac{u'[c_t^i(s^t)]}{u'[c_\tau^i(s^\tau)]} \pi(s^t|s^\tau). \end{aligned} \tag{7.20}$$

Here $q_t^\tau(s^t)$ is the price of one unit of consumption delivered at time $t$, state $s^t$ in terms of the date- $\tau$, state- $s^\tau$ consumption good; $\pi(s^t|s^\tau)$ is the probability

[5] Because the marginal conditions hold for all consumers, this condition holds for all $i$.

of state $s^t$ conditional on being in state $s^\tau$ at date $\tau$, as given by (7.3). Thus, the price at $t$ for the "tail asset" is

$$a_\tau^\tau(s^\tau) = \sum_{t\geq\tau} \sum_{\{\tilde{s}^t:\tilde{s}^\tau=s^\tau\}} q_t^\tau(\tilde{s}^t)d(\tilde{s}_t). \tag{7.21}$$

When we want to create a time series of, say, equity prices, we use the "tail asset" pricing formula. An equity purchased at time $\tau$ entitles the owner to the dividends from time $\tau$ forward. Our formula (7.21) expresses the asset price in terms of prices with time $\tau$, state $s^\tau$ good as numeraire.

Notice how formula (7.20) takes the form of a pricing function for a complete markets economy with date- and state-contingent commodities, whose markets have been opened at date $\tau$. Recall Proposition 1 stating that the equilibrium consumption allocation is not history-dependent, which here implies that the relative price in equation (7.20) is also characterized by a lack of history-dependence in the following sense:

PROPOSITION 2: The equilibrium price of date-$t \geq 0$, state-$s^t$ consumption goods expressed in terms of date $\tau$ $(0 \leq \tau \leq t)$, state $s^\tau$ consumption goods is not history-dependent: $q_t^\tau(s^t) = q_k^j(\tilde{s}^k)$ for $j,k \geq 0$ such that $t-\tau = k-j$ and $[s_t, s_{t-1}, \ldots, s_\tau] = [\tilde{s}_k, \tilde{s}_{k-1}, \ldots, \tilde{s}_j]$.

This property of $q_t^\tau(s^t)$ will play a central role later when we study a recursive formulation of the model that only allows for sequential trades of one-period contingent claims.

*Pricing one period returns*

The one-period version of equation (7.20) is

$$q_{\tau+1}^\tau(s^{\tau+1}) = \beta\frac{u'(c_{\tau+1}^i)}{u'(c_\tau^i)}\pi(s_{\tau+1}|s_\tau).$$

The right side is the one-period *pricing kernel* at time $\tau$. If we want to find the price at time $\tau$ in state $s^\tau$ of a claim to a random payoff $\omega(s_{\tau+1})$, we use

$$p_\tau^\tau(s^\tau) = \sum_{s_{\tau+1}} q_{\tau+1}^\tau(s^{\tau+1})\omega(s_{\tau+1})$$

or

$$p_{\tau}^{\tau}(s^{\tau}) = E_{\tau}\left[\beta \frac{u'(c_{\tau+1})}{u'(c_{\tau})}\omega(s_{\tau+1})\right], \tag{7.22}$$

where $E_{\tau}$ is the conditional expectation operator. We have deleted the $i$ superscripts on consumption, with the understanding that equation (7.22) is true for any consumer $i$; we have also suppressed the dependence of $c_{\tau}$ on $s_{\tau}$, which is implicit.

Let $R_{\tau+1} \equiv \omega(s_{\tau+1})/p_{\tau}^{\tau}(s^{\tau})$ be the one-period gross *return* on the asset. Then for any asset, equation (7.22) implies

$$1 = E_{\tau}\left[\beta \frac{u'(c_{\tau+1})}{u'(c_{\tau})} R_{\tau+1}\right]. \tag{7.23}$$

The term $m_{\tau+1} \equiv \beta u'(c_{\tau+1})/u'(c_{\tau})$ functions as a *stochastic discount factor*. Like $R_{\tau+1}$, it is a random variable measurable with respect to $s_{\tau+1}$, given $s^{\tau}$. Equation (7.23) is a restriction on the conditional moments of returns and $m_{t+1}$. Applying the law of iterated expectations to equation (7.23) gives the unconditional moments restriction

$$1 = E\left[\beta \frac{u'(c_{\tau+1})}{u'(c_{\tau})} R_{\tau+1}\right]. \tag{7.24}$$

In the next section, we display another market structure in which the one-period pricing kernel $q_{t+1}^{t}$ also plays a decisive role. This structure has the celebrated one-period "Arrow securities," the sequential trading of which substitutes for the comprehensive trading of long horizon claims at time 0 envisioned previously.

## A recursive formulation: Arrow securities

This section describes an alternative market structure that preserves the equilibrium allocation from our competitive equilibrium. This setting also preserves the key one-period asset-pricing formula (7.22).

### Debt limits

In moving to the sequential formulation, we shall need to impose some restrictions on asset trades to prevent Ponzi schemes. We impose the weakest possible restrictions in this section. We'll synthesize restrictions that work by starting from the equilibrium allocation of the previous section (with time-0 markets), and find some state-by-state debt limits that suffice to support sequential trading. Often we'll refer to these weakest possible debt limits as the natural debt limits.

Let $q^t_\tau(s^\tau)$ be the Arrow-Debreu price, denominated in units of the date $t$, state $s^t$ consumption good. Consider the value of the tail of agent $i$'s endowment sequence at time $t$ in state $s^t$:

$$A^i_t(s^t) = \sum_{\tau=t}^{\infty} \sum_{\{\tilde{s}^\tau : \tilde{s}^t = s^t\}} q^t_\tau(\tilde{s}^\tau) y^i(\tilde{s}_\tau) \equiv \bar{A}^i(s_t), \tag{7.25}$$

where the definition of the function $\bar{A}^i(s_t)$ with $s_t$ as its sole argument invokes Proposition 2. Using formula (7.20), evidently $\bar{A}^i(s)$ solves the recursive equation

$$\bar{A}^i(s) = y^i(s) + \beta E_s \frac{u'[\bar{c}(s')]}{u'[\bar{c}(s)]} \bar{A}^i(s'), \tag{7.26}$$

where $E_s$ is the expectation conditioned on state $s$, $\bar{c}(s)$ is the consumption of any agent in the Arrow-Debreu complete markets equilibrium with time-0 trading, and $'$ denotes a next-period value. (Recall that Proposition 1 states that the equilibrium consumption allocation is not history-dependent.) We call $\bar{A}^i(s)$ the *natural debt limit* in state $s$. It is the value of the maximal amount that agent $i$ can repay starting from state $s$, assuming that his consumption is zero forever.

From now on, we shall require that households never promise to pay more than $\bar{A}(s')$ in any state $s'$ tomorrow, because it will not be feasible for them to repay more.

### *Arrow securities*

We build on an insight of Arrow (1964) that one-period securities are enough to implement complete markets, provided that new one-period markets are reopened for trading each period. Thus, at each date $t \geq 0$, trades occur in a set of claims to one-period-ahead state-contingent consumption. We describe a competitive equilibrium of this sequential trading economy. With a full array of these one-period-ahead claims, the sequential trading arrangement attains the same allocation as the competitive equilibrium that we described earlier.

Before turning to sequential trading, we ask the following question. In the preceding competitive equilibrium where all trading takes place at time 0, what is the implied wealth, other than the endowment, of household $i$ at time $t$ in state $s^t$? In period $t$, conditional on history $s^t$, we sum up the value of the household's purchased claims to current and future goods net of its outstanding liabilities. Since history $s^t$ is realized, all claims and liabilities contingent on another initial history can be discarded. For example, household $i$'s net claim to delivery of goods in a future period $\tau \geq t$, contingent on history $\tilde{s}^\tau$ such that $\tilde{s}^t = s^t$, is given by $[c_\tau^i(\tilde{s}^\tau) - y^i(\tilde{s}_\tau)]$. Thus, the household's wealth, or the value of all its current and future net claims, expressed in terms of the date $t$, state $s^t$ consumption good is

$$\Upsilon_t^i(s^t) = \sum_{\tau=t}^{\infty} \sum_{\{\tilde{s}^\tau : \tilde{s}^t = s^t\}} q_\tau^t(\tilde{s}^\tau) \left[c_\tau^i(\tilde{s}^\tau) - y^i(\tilde{s}_\tau)\right] \equiv \bar{\Upsilon}^i(s_t), \tag{7.27}$$

where the definition of the function $\bar{\Upsilon}^i(s_t)$ with $s_t$ as its sole argument invokes Propositions 1 and 2. Notice that feasibility constraint (7.9) at equality implies that

$$\sum_{i=1}^{I} \Upsilon_t^i(s^t) = 0, \qquad \forall t, s^t.$$

In a sequential trading economy, we assume there is a sequence of markets in one-period-ahead state-contingent claims to wealth or consumption. At each date $t \geq 0$, households trade claims to date $t+1$ consumption, whose payment is contingent on the realization of $s_{t+1}$. Let $\theta_t^i$ denote the claims to time $t$ consumption, other than its endowment, that household $i$ brings into time $t$. Suppose that $Q(s'|s)$ is a *pricing kernel* to be interpreted as follows: $Q(s_{t+1}|s_t)$ gives the price of one unit of time–$t+1$ consumption, contingent on the state at $t+1$ being $s_{t+1}$, given that the state at $t$ is $s_t$. Notice that we are guessing that this function exists and does not depend on $t$. The household faces a sequence of budget constraints for $t \geq 0$, where the time-$t$ budget constraint is

$$c_t^i + \sum_{s_{t+1}} \theta_{t+1}^i(s_{t+1}) Q(s_{t+1}|s_t) \leq y^i(s_t) + \theta_t^i. \tag{7.28}$$

Here it is understood that $\theta_{t+1}^i = \theta_{t+1}^i(s_{t+1})$, which depends on the realization of $s_{t+1}$ and the amount $\theta_{t+1}^i(s_{t+1})$ purchased at $t$. At time $t$, the household chooses $c_t^i$ and $\{\theta_{t+1}^i(s_{t+1})\}$, where $\{\theta_{t+1}^i(s_{t+1})\}$ is a vector of claims on time–$t+1$ consumption, one element of the vector for each value of the time–$t+1$

state $s_{t+1}$. To rule out Ponzi schemes, we impose the state-by-state borrowing constraints

$$-\theta^i_{t+1}(s_{t+1}) \leq \bar{A}^i(s_{t+1}), \tag{7.29}$$

where $\bar{A}^i$ are computed in equation (7.26).

Given that the pricing kernel $Q(s'|s)$ and the endowment $y^i(s)$ are functions of a Markov process $s$, we are motivated to seek a recursive solution to the household's optimization problem. The household $i$'s state at time $t$ is its wealth $\theta^i_t$ and the current realization $s_t$. We seek a pair of optimal policy functions $h^i(\theta, s)$, $g^i(\theta, s, s')$ such that the household's optimal decisions are

$$c^i_t = h^i(\theta^i_t, s_t), \tag{7.30a}$$

$$\theta^i_{t+1}(s_{t+1}) = g^i(\theta^i_t, s_t, s_{t+1}). \tag{7.30b}$$

Let $v^i(\theta, s)$ be the optimal value of household $i$'s problem starting from state $(\theta, s)$; $v^i(\theta, s)$ is the maximum expected discounted utility household $i$ with current wealth $\theta$ can attain in state $s$. The Bellman equation for the household's problem is

$$v^i(\theta, s) = \max_{c, \tilde{\theta}(s')} \left\{ u(c) + \beta \sum_{s'} v^i[\tilde{\theta}(s'), s'] \pi(s'|s) \right\} \tag{7.31}$$

where the maximization is subject to the following version of constraint (7.28):

$$c + \sum_{s'} \tilde{\theta}(s') Q(s'|s) \leq y^i(s) + \theta \tag{7.32}$$

and also

$$c \geq 0, \tag{7.33a}$$

$$-\tilde{\theta}(s') \leq \bar{A}^i(s'). \tag{7.33b}$$

Let the optimum decision rules be

$$c = h^i(\theta, s), \tag{7.34a}$$

$$\tilde{\theta}(s') = g^i(\theta, s, s'). \tag{7.34b}$$

Note that the solution of the Bellman equation implicitly depends on $Q(\cdot|\cdot)$ because it appears in the constraint (7.32).

DEFINITION: A *distribution of wealth* is a vector $\vec{\theta}_t = \{\theta^i_t\}_{i=1}^I$ satisfying $\sum_i \theta^i_t = 0$.

DEFINITION: A *recursive competitive equilibrium* is an initial distribution of wealth $\vec{\theta}_0$, a pricing kernel $Q(s'|s)$, sets of value functions $\{v^i(\theta, s)\}_{i=1}^I$ and decision rules $\{h^i(\theta, s), g^i(\theta, s, s')\}_{i=1}^I$ such that (a) for all $i$, given $\theta^i_0$ and the pricing kernel, the decision rules solve the household's problem; (b) for all realizations of $\{s_t\}_{t=0}^\infty$, the consumption and asset portfolios $\{\{c^i_t, \{\theta^i_{t+1}(s')\}_{s'}\}_i\}_t$ implied by the decision rules satisfy $\sum_i c^i_t = \sum_i y^i(s_t)$ and $\sum_i \theta^i_{t+1}(s') = 0$ for all $t$ and $s'$.

*Equivalence of allocations*

Use the first-order conditions for the problem on the right of equation (7.31) and the Benveniste-Scheinkman formula and rearrange to get

$$Q(s_{t+1}|s_t) = \frac{\beta u'(c^i_{t+1}) \pi(s_{t+1}|s_t)}{u'(c^i_t)}, \tag{7.35}$$

where it is understood that $c^i_t = h^i(\theta^i_t, s_t)$, $\theta^i_{t+1}(s_{t+1}) = g^i(\theta^i_t, s_t, s_{t+1})$.

By making an appropriate guess about the form of the pricing kernel, it is easy to show that a competitive equilibrium allocation of the complete markets model with time-0 trading is also a recursive competitive equilibrium allocation. Thus, take $q^0_t(s^t)$ as given from the Arrow-Debreu equilibrium and suppose that the pricing kernel makes the following recursion true:

$$q^0_{t+1}(s^{t+1}) = Q(s_{t+1}|s_t) q^0_t(s^t),$$

or

$$Q(s_{t+1}|s_t) = q^t_{t+1}(s^{t+1}). \tag{7.36}$$

Let $\{c^i_t(s^t)\}$ be a competitive equilibrium allocation. If equation (7.36) is satisfied, that allocation is a recursive competitive equilibrium allocation. To show this fact, take the household's first-order conditions for the competitive equilibrium economy from two successive periods and divide one by the other to get

$$\frac{\beta u'[c^i_{t+1}(s^{t+1})] \pi(s_{t+1}|s_t)}{u'[c^i_t(s^t)]} = \frac{q^0_{t+1}(s^{t+1})}{q^0_t(s^t)} = Q(s_{t+1}|s_t). \tag{7.37}$$

If the pricing kernel satisfies equation (7.36), this equation is equivalent with the first-order condition for the recursive competitive equilibrium economy (7.35). It remains for us to choose the initial wealth of the recursive equilibrium so that the recursive competitive equilibrium duplicates the competitive equilibrium allocation.

We conjecture that the initial wealth vector $\vec{\theta}_0$ of the sequential trading economy should be chosen to be the null vector. This is a natural conjecture, because it means that each household must rely on its own endowment stream to finance consumption, in the same way as households are constrained to finance their state-contingent purchases for the infinite future at time 0 in the Arrow-Debreu economy. To prove that the conjecture is correct, we must show that this particular initial wealth vector enables household $i$ to finance $\{c_t^i(s^t)\}$ and leaves no room for further increasing consumption in any period and state.

The proof proceeds by guessing that, at time $t \geq 0$ regardless of current state, household $i$ chooses an asset portfolio given by $\theta_{t+1}^i(s_{t+1}) = \bar{\Upsilon}^i(s_{t+1})$ for all $s_{t+1}$. If the history at date $t$ is $s^t$, the value of this asset portfolio expressed in terms of the date $t$, state $s^t$ consumption good is

$$\begin{aligned}\sum_{s_{t+1}} \theta_{t+1}^i(s_{t+1}) Q(s_{t+1}|s_t) &= \sum_{s_{t+1}} \bar{\Upsilon}^i(s_{t+1}) q_{t+1}^t(s^{t+1}) \\ &= \sum_{\tau=t+1}^{\infty} \sum_{\{\tilde{s}^\tau : \tilde{s}^t = s^t\}} q_\tau^t(\tilde{s}^\tau) \left[ c_\tau^i(\tilde{s}^\tau) - y^i(\tilde{s}_\tau) \right], \end{aligned} \tag{7.38}$$

where we have invoked expressions (7.27) and (7.36).[6] To demonstrate that household $i$ can afford this portfolio strategy, we now use budget constraint (7.28) to compute the implied consumption plan $\{\hat{c}_\tau^i(s^\tau)\}$. First, in the initial period $t = 0$ with $\theta_0^i = 0$, the substitution of equation (7.38) into budget constraint (7.28) at equality yields

$$\hat{c}_0^i(s_0) + \sum_{t=1}^{\infty} \sum_{s^t} q_t^0(s^t) \left[ c_t^i(s^t) - y^i(s_t) \right] = y^i(s_0) + 0.$$

[6] We have also used the following identities,

$$q_\tau^{t+1}(s^\tau) q_{t+1}^t(s^{t+1}) = \frac{q_\tau^0(s^\tau)}{q_{t+1}^0(s^{t+1})} \frac{q_{t+1}^0(s^{t+1})}{q_t^0(s^t)} = q_\tau^t(s^\tau) \text{ for } \tau > t.$$

This expression together with budget constraint (7.5) at equality imply $\hat{c}_0^i(s_0) = c_0^i(s_0)$. In other words, the proposed asset portfolio is affordable in period 0 and the associated consumption level is the same as in the competitive equilibrium of the Arrow-Debreu economy. In all consecutive future periods $t > 0$ and histories $s^t$, we replace $\theta_t^i$ in constraint (7.28) by $\bar{\Upsilon}^i(s_t)$ and after noticing that the value of the asset portfolio in (7.38) can be written as

$$\sum_{s_{t+1}} \theta_{t+1}^i(s_{t+1}) Q(s_{t+1}|s_t) = \bar{\Upsilon}^i(s_t) - \left[c_t^i(s^t) - y^i(s_t)\right], \tag{7.39}$$

it follows immediately from (7.28) that $\hat{c}_t^i(s^t) = c_t^i(s^t)$ for all periods and histories.

We have shown that the proposed portfolio strategy attains the same consumption plan as in the competitive equilibrium of the Arrow-Debreu economy, but what precludes household $i$ from further increasing current consumption by reducing some component of the asset portfolio? The answer lies in the debt limit restrictions to which the household must adhere. In particular, if the household wants to ensure that consumption plan $\{c_\tau^i(s^\tau)\}$ can be attained starting next period in all possible future states, the household should subtract the value of this commitment to future consumption from the natural debt limit in (7.25). Thus, the household is facing a self-imposed state-by-state borrowing constraint that is more restrictive than restriction (7.29): for any $s^{t+1}$,

$$-\theta_{t+1}^i(s_{t+1}) \leq \bar{A}^i(s_{t+1}) - \sum_{\tau=t+1}^{\infty} \sum_{\{\tilde{s}^\tau : \tilde{s}^{t+1} = s^{t+1}\}} q_\tau^{t+1}(\tilde{s}^\tau) c_\tau^i(\tilde{s}^\tau) = -\bar{\Upsilon}^i(s_{t+1}),$$

or

$$\theta_{t+1}^i(s_{t+1}) \geq \bar{\Upsilon}^i(s_{t+1}).$$

Hence, household $i$ does not want to increase consumption at time $t$ by reducing next period's wealth below $\bar{\Upsilon}^i(s_{t+1})$ because that would jeopardize the attainment of the preferred consumption plan satisfying first-order conditions (7.35) for all future periods and states.

We shall use the recursive competitive equilibrium concept extensively in our discussion of asset pricing in chapter 10.

## *j*-step pricing kernel

We are sometimes interested in the price at time $t$ of a claim to one unit of consumption at date $\tau > t$ contingent on the time-$\tau$ state being $s_\tau$, *regardless* of the particular history by which $s_\tau$ is reached at $\tau$. We let $Q_j(s'|s)$ denote the $j$-step pricing kernel to be interpreted as follows: $Q_j(s'|s)$ gives the price of one unit of consumption $j$ periods ahead, contingent on the state in that future period being $s'$, given that the current state is $s$. For example, $j = 1$ corresponds to the one-step pricing kernel $Q(s'|s)$.

With markets in all possible $j$-step ahead contingent claims, the counterpart to constraint (7.28), the household's budget constraint at time $t$, is

$$c_t^i + \sum_{j=1}^{\infty} \sum_{s_{t+j}} Q_j(s_{t+j}|s_t) z_{t,j}^i(s_{t+j}) \leq y^i(s_t) + \theta_t^i. \tag{7.40}$$

Here $z_{t,j}^i(s_{t+j})$ is household $i$'s holdings at the end of period $t$ of contingent claims that pay one unit of the consumption good $j$ periods ahead at date $t+j$, contingent on the state at date $t+j$ being $s_{t+j}$. The household's wealth in the next period depends on the chosen asset portfolio and the realization of $s_{t+1}$,

$$\theta_{t+1}^i(s_{t+1}) = z_{t,1}^i(s_{t+1}) + \sum_{j=2}^{\infty} \sum_{s_{t+j}} Q_{j-1}(s_{t+j}|s_{t+1}) z_{t,j}^i(s_{t+j}).$$

The realization of $s_{t+1}$ determines both which element of the vector of one-period ahead claims $\{z_{t,1}^i(s_{t+1})\}$ that pays off at time $t+1$, and the capital gains and losses inflicted on the holdings of longer horizon claims implied by equilibrium prices $Q_j(s_{t+j+1}|s_{t+1})$.

With respect to $z_{t,j}^i(s_{t+j})$ for $j > 1$, use the first-order condition for the problem on the right of (7.31) and the Benveniste-Scheinkman formula and rearrange to get

$$Q_j(s_{t+j}|s_t) = \sum_{s_{t+1}} \frac{\beta u'[c_{t+1}^i(s_{t+1})]\pi(s_{t+1}|s_t)}{u'(c_t^i)} Q_{j-1}(s_{t+j}|s_{t+1}). \tag{7.41}$$

This expression evaluated at the competitive equilibrium consumption allocation, characterizes two adjacent pricing kernels.[7] Together with first-order condition

[7] According to Proposition 1, the equilibrium consumption allocation is not history dependent, so that $(c_t^i, \{c_{t+1}^i(s_{t+1})\}_{s_{t+1}}) = (\bar{c}^i(s_t), \{\bar{c}^i(s_{t+1})\}_{s_{t+1}})$. Because marginal conditions hold for all households, the characterization of pricing kernels in (7.41) holds for any $i$.

(7.35), formula (7.41) implies that the kernels $Q_j, j = 2, 3, \ldots$ can be computed recursively:

$$Q_j(s_{t+j}|s_t) = \sum_{s_{t+1}} Q_1(s_{t+1}|s_t) Q_{j-1}(s_{t+j}|s_{t+1}). \tag{7.42}$$

*Arbitrage pricing*

It is useful briefly to describe how arbitrage pricing theory deduces restrictions on asset prices by manipulating budget sets with redundant assets.[8] We now present an arbitrage argument as an alternative way of deriving restriction (7.42) that was established above by using households' first-order conditions evaluated at the equilibrium consumption allocation. In addition to $j$-step-ahead contingent claims, we illustrate the arbitrage pricing theory by augmenting the trading opportunities in our Arrow securities economy by letting the consumer also trade an ex-dividend Lucas tree. Because markets are already complete, these additional assets are redundant. They have to be priced in to leave the budget set unaltered.[9]

Assume that at time $t$, in addition to purchasing a quantity $z_{t,j}(s_{t+j})$ of $j$-step-ahead claims paying one unit of consumption at time $t+j$ if the state takes value $s_{t+j}$ at time $t+j$, the consumer also purchases $N_{t+1}$ units of a stock or Lucas tree. Let the ex-dividend price of the tree at time-$t$ be $p(s_t)$. Next period, the tree pays a dividend $d(s_{t+1})$ depending on the state $s_{t+1}$. Ownership of the $N_{t+1}$ units of the tree at the beginning of $t+1$ entitles the consumer to a claim on $N_{t+1}[p(s_{t+1}) + d(s_{t+1})]$ units of time–$t+1$ consumption.[10] As before, let $\theta_t$ be the wealth of the consumer, apart from his endowment, $y(s_t)$. In this setting, the augmented version of constraint (7.40), the consumer's budget constraint, is

$$c_t + \sum_{j=1}^{\infty} \sum_{s_{t+j}} Q_j(s_{t+j}|s_t) z_{t,j}(s_{t+j}) + p(s_t) N_{t+1} \leq \theta_t + y(s_t) \tag{7.43a}$$

and

$$\begin{aligned} \theta_{t+1}(s_{t+1}) = {} & z_{t,1}(s_{t+1}) + [p(s_{t+1}) + d(s_{t+1})] N_{t+1} \\ & + \sum_{j=2}^{\infty} \sum_{s_{t+j}} Q_{j-1}(s_{t+j}|s_{t+1}) z_{t,j}(s_{t+j}). \end{aligned} \tag{7.43b}$$

---

[8] Arbitrage pricing theory was advocated by Stephen Ross (1976).

[9] That the additional assets are redundant follows from the fact that trading Arrow securities is sufficient to complete markets.

[10] We calculate the price of this asset using a different method in chapter 10.

Multiply equation (7.43b) by $Q_1(s_{t+1}|s_t)$, sum over $s_{t+1}$, solve for $\sum_{s_{t+1}} Q_1(s_{t+1}|s_t)z_1(s_t)$, and substitute this expression in (7.43a) to get

$$\begin{aligned} c_t + & \left\{ p(s_t) - \sum_{s_{t+1}} Q_1(s_{t+1}|s_t)[p(s_{t+1}) + d(s_{t+1})] \right\} N_{t+1} \\ & + \sum_{j=2}^{\infty} \sum_{s_{t+j}} \left\{ Q_j(s_{t+j}|s_t) - \sum_{s_{t+1}} Q_{j-1}(s_{t+j}|s_{t+1}) Q_1(s_{t+1}|s_t) \right\} z_{t,j}(s_{t+j}) \\ & + \sum_{s_{t+1}} Q_1(s_{t+1}|s_t)\theta_{t+1}(s_{+1}) \leq \theta_t + y(s_t). \end{aligned} \tag{7.44}$$

If the two terms in braces are not zero, the consumer can attain unbounded consumption and future wealth by purchasing or selling either the stock (if the first term in braces is not zero) or a state-contingent claim (if any of the terms in the second set of braces is not zero). Therefore, so long as the utility function has no satiation point, in any equilibrium, the terms in the braces must be zero. Thus we have the arbitrage pricing formulas

$$p(s_t) = \sum_{s_{t+1}} Q_1(s_{t+1}|s_t)[p(s_{t+1}) + d(s_{t+1})] \tag{7.45a}$$

$$Q_j(s_{t+j}|s_t) = \sum_{s_{t+1}} Q_{j-1}(s_{t+j}|s_{t+1}) Q_1(s_{t+1}|s_t). \tag{7.45b}$$

These are called arbitrage pricing formulas because if they were violated, there would exist an *arbitrage*. An arbitrage is defined as a risk-free transaction that earns positive profits.

## Consumption strips and the cost of business cycles

### Consumption strips

This section briefly describes ideas of Alvarez and Jermann (1999) and Lustig (2000). Their purpose is to link measures of the cost of business cycles with a risk premium for some assets. To this end, consider an endowment economy with a representative consumer endowed with a consumption process $c_t = c(s_t)$, where $s_t$ is Markov with transition probabilities $\pi(s'|s)$. Alvarez and Jermann define a one-period consumption strip as a claim to the random payoff $c_t$, sold at date $t-1$. The price in terms of time–$t-1$ consumption of this one-period consumption strip is

$$a_{t-1} = E_{t-1} m_t c_t, \tag{7.46}$$

where $m_t$ is the one-period stochastic discount factor

$$m_t = \frac{\beta u'(c_t)}{u'(c_{t-1})}. \tag{7.47}$$

Using the definition of a conditional covariance, equation (7.46) implies

$$a_{t-1} = E_{t-1} m_t E_{t-1} c_t + \mathrm{cov}_{t-1}(c_t, m_t), \tag{7.48}$$

where $\mathrm{cov}_{t-1}(c_t, m_t) < 0$. Note that the price of a one-period claim on $E_{t-1} c_t$ is simply

$$\tilde{a}_{t-1} = E_{t-1} m_t E_{t-1} c_t, \tag{7.49}$$

so that the negative covariance in equation (7.48) is a discount due to risk in the price of the risky claim on $c_t$ relative to the risk-free claim on a payout with the same mean. Define the multiplicative risk premium on the consumption strip as $(1+\mu_{t-1}) \equiv \tilde{a}_t / a_t$, which evidently equals

$$1 + \mu_{t-1} = \frac{E_{t-1} m_t E_{t-1} c_t}{E_{t-1} m_t c_t}. \tag{7.50}$$

*Link to business cycle costs*

The cost of business cycle as defined in chapter 3 does not link immediately to an asset-pricing calculation because it is intramarginal. Alvarez and Jermann (1999) and Hansen, Sargent, and Tallarini (1999) were interested in coaxing attitudes about the cost of business cycles from asset prices. Alvarez and Jermann designed a notion of the marginal costs of business cycles to match asset pricing. With the timing conventions of Lustig (2000), their concept of marginal cost corresponds to the risk premium in one-period consumption strips.

Alvarez and Jermann (1999) and Lustig (2000) define the total costs of business cycles in terms of a stochastic process of adjustments to consumption $\Omega_{t-1}$ constructed to satisfy

$$E_0 \sum_{t=0}^{\infty} \beta^t u[(1+\Omega_{t-1}) c_t] = E_0 \sum_{t=0}^{\infty} \beta^t u(E_{t-1} c_t).$$

The idea is to compensate the consumer for the one-period-ahead risk in consumption that he faces.

The time-$t$ component of the marginal cost of business cycles is defined as follows through a variational argument, taking the endowment as a benchmark. Let $\alpha \in (0,1)$ be a parameter to index consumption processes. Define $\Omega_{t-1}(\alpha)$ implicitly by means of

$$E_{t-1}u\{[1+\Omega_{t-1}(\alpha)]c_t\} = E_{t-1}u[\alpha E_{t-1}c_t + (1-\alpha)c_t]. \tag{7.51}$$

Differentiate equation (7.51) with respect to $\alpha$ and evaluate at $\alpha = 0$ to get

$$\Omega'_{t-1}(0) = \frac{E_{t-1}u'(c_t)(E_{t-1}c_t - c_{t-1})}{E_{t-1}c_t u'(c_t)}.$$

Multiply both numerator and denominator of the right side by $\beta/u'(c_{t-1})$ to get

$$\Omega'_{t-1}(0) = \frac{E_{t-1}m_t(E_{t-1}c_t - c_t)}{E_{t-1}m_t c_t}, \tag{7.52}$$

where we use $\Omega_{t-1}(0) = 0$. Rearranging gives

$$1 + \Omega'_{t-1}(0) = \frac{E_{t-1}m_t E_{t-1}c_t}{E_{t-1}m_t c_t}. \tag{7.53}$$

Comparing equation (7.53) with (7.50) shows that the marginal cost of business cycles equals the multiplicative risk premium on the one-period consumption strip. Thus, in this economy, the marginal cost of business cycles can be coaxed from asset market data.

## Gaussian asset pricing model

The theory of the preceding section is readily adapted to a setting in which the state of the economy evolves according to a continuous-state Markov process. We use such a version in chapter 10. Here we give a taste of how such an adaptation can be made by describing an economy in which the state follows a linear stochastic difference equation driven by a Gaussian disturbance. If we supplement this with the specification that preferences are quadratic, we get a setting in which asset prices can be calculated swiftly.

Suppose that the state evolves according to the stochastic difference equation

$$s_{t+1} = As_t + Cw_{t+1} \tag{7.54}$$

where $A$ is a matrix whose eigenvalues are bounded from above in modulus by $1/\sqrt{\beta}$ and $w_{t+1}$ is a Gaussian martingale difference sequence adapted to the history of $s_t$. Assume that $Ew_{t+1}w_{t+1} = I$. The conditional density of $s_{t+1}$ is Gaussian:

$$\pi(s_t|s_{t-1}) \sim \mathcal{N}(As_{t-1}, CC'). \tag{7.55}$$

More precisely,

$$\pi(s_t|s_{t-1}) = K \exp\left\{-.5(s_t - As_{t-1})(CC')^{-1}(s_t - As_{t-1})\right\}, \tag{7.56}$$

where $K = (2\pi)^{-k/2} \det(CC')^{-1/2}$ and $s_t$ is $k \times 1$. We also assume that $\pi_0(s_0)$ is Gaussian.[11]

If $\{c_t^i(s_t)\}_{t=0}^{\infty}$ is the equilibrium allocation to agent $i$, and the agent has preferences represented by (7.4), the equilibrium pricing function satisfies

$$q_t^0(s^t) = \frac{\beta^t u'[c_t^i(s_t)]\pi(s^t)}{u'[c_0^i(s_0)]}. \tag{7.57}$$

Once again, let $\{d_t(s_t)\}_{t=0}^{\infty}$ be a stream of claims to consumption. The time-0 price of the asset with this dividend stream is

$$a_0 = \sum_{t=0}^{\infty} \int_{s^t} q_t^0(s^t) d_t(s_t) d\, s^t.$$

Substituting equation (7.57) into the preceding equation gives

$$a_0 = \sum_t \int_{s^t} \beta^t \frac{u'[c_t^i(s_t)]}{u'[c_0^i(s_0)]} d_t(s_t)\pi(s^t) ds^t$$

or

$$a_0 = E \sum_{t=0}^{\infty} \beta^t \frac{u'[c_t(s_t)]}{u'[c_0(s_0)]} d_t(s_t). \tag{7.58}$$

[11] If $s_t$ is stationary, $\pi_0(s_0)$ can be specified to be the stationary distribution of the process.

This formula expresses the time-0 asset price as an inner product of a discounted marginal utility process and a dividend process.[12]

This formula becomes especially useful in the case that the one-period utility function $u(c)$ is quadratic, so that marginal utilities become linear, and that the dividend process $d_t$ is linear in $s_t$. In particular, assume that

$$u(c_t) = -.5(c_t - b)^2 \tag{7.59}$$
$$d_t = S_d s_t, \tag{7.60}$$

where $b > 0$ is a bliss level of consumption. Furthermore, assume that the equilibrium allocation to agent $i$ is

$$c_t^i = S_{ci} s_t, \tag{7.61}$$

where $S_{ci}$ is a vector conformable to $s_t$.

The utility function (7.59) implies that $u'(c_t^i) = b - c_t^i = b - S_{ci} s_t$. Suppose that unity is one element of the state space for $s_t$, so that we can express $b = S_b s_t$. Then $b - c_t = S_p s_t$, where $S_p = S_b - S_{ci}$, and the asset-pricing formula becomes

$$a_0 = \frac{E_0 \sum_{t=0}^{\infty} \beta^t s_t' S_p' S_d s_t}{S_p s_0}. \tag{7.62}$$

Thus, to price the asset, we have to evaluate the expectation of the sum of a discounted quadratic form in the state variable. This is easy to do by using results from chapter 1.

In chapter 1, we evaluated the conditional expectation of the geometric sum of the quadratic form

$$\alpha_0 = E_0 \sum_{t=0}^{\infty} \beta^t s_t' S_p' S_d s_t.$$

We found that it could be written in the form

$$\alpha_0 = s_0' \mu s_0 + \sigma, \tag{7.63}$$

where $\mu$ is an $(n \times n)$ matrix, and $\sigma$ is a scalar that satisfy

$$\begin{aligned} \mu &= S_p' S_d + \beta A' \mu A \\ \sigma &= \beta \sigma + \beta \text{ trace } \mu C C' \end{aligned} \tag{7.64}$$

[12] For two scalar stochastic processes $x, y$, the inner product is defined as $< x, y >= E \sum_{t=0}^{\infty} \beta^t x_t y_t$.

The first equation of (7.64) is a *discrete Lyapunov equation* in the square matrix $\mu$, and can be solved by using one of several algorithms.[13] After $\mu$ has been computed, the second equation can be solved for the scalar $\sigma$.

## Concluding remarks

The framework in this chapter serves much of macroeconomics either as foundation or benchmark. It is the foundation of extensive literatures on asset pricing and risk sharing. We describe the literature on asset pricing in more detail in chapter 10. The model also serves as benchmark, or point of departure, for a variety of models designed to confront observations that seem inconsistent with complete markets. In particular, for models with exogenously imposed incomplete markets, see chapters 13 on precautionary saving and 14 on incomplete markets. For models with endogenous incomplete markets, see chapter 15 on enforcement and information problems. For models of money, see chapters 17 and 18. To take monetary theory as an example, complete markets models dispose of any need for money because they contain an efficient multilateral trading mechanism, with such extensive netting of claims that no medium of exchange is required to facilitate bilateral exchanges. Any modern model of money introduces a friction that stops complete markets. Some monetary models (e.g., the cash-in-advance model of Lucas, 1981) impose minimal impediments to complete markets, to preserve many of the asset-pricing implications of complete markets models while also allowing classical monetary doctrines like the quantity theory of money. The shopping-time model of chapter 17 is constructed in a similar spirit. Other monetary models, such as the Townsend turnpike model of chapter 18 or the Kiyotaki-Wright search model of chapter 19, impose more extensive frictions on multilateral exchanges and leave the complete markets model farther behind. But before leaving the complete markets model, we'll put it hard to work in chapters 8, 9, and 10.

## Exercises

### *Exercise 7.1* **Existence of representative consumer**

Suppose households 1 and 2 have one-period utility functions $u(c^1)$ and $w(c^2)$, respectively, where $u$ and $w$ are both increasing, strictly concave, twice-differentiable

[13] The Matlab control toolkit has a program called `dlyap.m`; also see a program called `doublej.m`.

functions of a scalar consumption rate. Consider the Pareto problem:

$$\max_{\{c^1,c^2\}} \left[\theta u(c^1) + (1-\theta) w(c^2)\right]$$

subject to the constraint $c^1 + c^2 = c$. Show that the solution of this problem has the form of a concave utility function $v_\theta(c)$, which depends on the Pareto weight $\theta$.

The function $v_\theta(c)$ is the utility function of the *representative consumer.* Such a representative consumer always lurks within a complete markets competitive equilibrium even with heterogeneous preferences.

*Exercise 7.2* **Term structure of interest rates**

Consider an economy with a single consumer. There is one good in the economy, which arrives in the form of an exogenous endowment obeying[14]

$$y_{t+1} = \lambda_{t+1} y_t,$$

where $y_t$ is the endowment at time $t$ and $\{\lambda_{t+1}\}$ is governed by a two-state Markov chain with transition matrix

$$P = \begin{bmatrix} p_{11} & 1 - p_{11} \\ 1 - p_{22} & p_{22} \end{bmatrix},$$

and initial distribution $\pi_\lambda = [\,\pi_0 \quad 1 - \pi_0\,]$. The value of $\lambda_t$ is given by $\bar{\lambda}_1 = .98$ in state 1 and $\bar{\lambda}_2 = 1.03$ in state 2. Assume that the history of $y_s, \lambda_s$ up to $t$ is observed at time $t$. The consumer has endowment process $\{y_t\}$ and has preferences over consumption streams that are ordered by

$$E_0 \sum_{t=0}^{\infty} \beta^t u(c_t)$$

where $\beta \in (0,1)$ and $u(c) = \frac{c^{1-\gamma}}{1-\gamma}$, where $\gamma \geq 1$.

**a.** Define a competitive equilibrium, being careful to name all of the objects of which it consists.

**b.** Tell how to compute a competitive equilibrium.

---

[14] Such a specification was made by Mehra and Prescott (1985).

For the remainder of this problem, suppose that $p_{11} = .8, p_{22} = .85, \pi_0 = .5$, $\beta = .96$, and $\gamma = 2$. Suppose that the economy begins with $\lambda_0 = .98$ and $y_0 = 1$.

**c.** Compute the (unconditional) average growth rate of consumption, computed before having observed $\lambda_0$.

**d.** Compute the time-0 prices of three risk-free discount bonds, in particular, those promising to pay one unit of time-$j$ consumption for $j = 0, 1, 2$, respectively.

**e.** Compute the time-0 prices of three bonds, in particular, ones promising to pay one unit of time-$j$ consumption contingent on $\lambda_j = \bar{\lambda}_1$ for $j = 0, 1, 2$, respectively.

**f.** Compute the time-0 prices of three bonds, in particular, ones promising to pay one unit of time-$j$ consumption contingent on $\lambda_j = \bar{\lambda}_2$ for $j = 0, 1, 2$, respectively.

**g.** Compare the prices that you computed in parts d, e, and f.

*Exercise 7.3* An economy consists of two infinitely lived consumers named $i = 1, 2$. There is one nonstorable consumption good. Consumer $i$ consumes $c_t^i$ at time $t$. Consumer $i$ ranks consumption streams by

$$\sum_{t=0}^{\infty} \beta^t u(c_t^i),$$

where $\beta \in (0, 1)$ and $u(c)$ is increasing, strictly concave, and twice continuously differentiable. Consumer 1 is endowed with a stream of the consumption good $y_t^i = 1, 0, 0, 1, 0, 0, 1, \ldots$. Consumer 2 is endowed with a stream of the consumption good $0, 1, 1, 0, 1, 1, 0, \ldots$. Assume that there are complete markets with time-0 trading.

**a.** Define a competitive equilibrium.

**b.** Compute a competitive equilibrium.

**c.** Suppose that one of the consumers markets a derivative asset that promises to pay .05 units of consumption each period. What would the price of that asset be?

*Exercise 7.4* Consider a pure endowment economy with a single representative consumer; $\{c_t, d_t\}_{t=0}^{\infty}$ are the consumption and endowment processes, respectively. Feasible allocations satisfy

$$c_t \leq d_t.$$

The endowment process is described by [15]

$$d_{t+1} = \lambda_{t+1} d_t. \tag{1}$$

The growth rate $\lambda_{t+1}$ is described by a two-state Markov process with transition probabilities

$$P_{ij} = \text{Prob}(\lambda_{t+1} = \bar{\lambda}_j | \lambda_t = \bar{\lambda}_i).$$

Assume that

$$P = \begin{bmatrix} .8 & .2 \\ .1 & .9 \end{bmatrix},$$

and that

$$\bar{\lambda} = \begin{bmatrix} .97 \\ 1.03 \end{bmatrix}.$$

In addition, $\lambda_0 = .97$ and $d_0 = 1$ are both known at date 0. The consumer has preferences over consumption ordered by

$$E_0 \sum_{t=0}^{\infty} \beta^t \frac{c_t^{1-\gamma}}{1-\gamma},$$

where $E_0$ is the mathematical expectation operator, conditioned on information known at time 0, $\gamma = 2, \beta = .95$.

## Part I

At time 0, after $d_0$ and $\lambda_0$ are known, there are complete markets in date- and state-contingent claims. The market prices are denominated in units of time-0 consumption goods.

**a.** Define a competitive equilibrium, being careful to specify all the objects composing an equilibrium.

---

[15] See Mehra and Prescott (1985).

**b.** Compute the equilibrium price of a claim to one unit of consumption at date 5, denominated in units of time-0 consumption, contingent on the following history of growth rates: $(\lambda_1, \lambda_2, \ldots, \lambda_5) = (.97, .97, 1.03, .97, 1.03)$. Please give a numerical answer.

**c.** Compute the equilibrium price of a claim to one unit of consumption at date 5, denominated in units of time-0 consumption, contingent on the following history of growth rates: $(\lambda_1, \lambda_2, \ldots, \lambda_5) = (1.03, 1.03, 1.03, 1.03, .97)$.

**d.** Give a formula for the price at time 0 of a claim on the entire endowment sequence.

**e.** Give a formula for the price at time 0 of a claim on consumption in period 5, contingent on the growth rate $\lambda_5$ being .97 (regardless of the intervening growth rates).

**Part II**

Now assume a different market structure. Assume that at each date $t \geq 0$ there is a complete set of one-period forward Arrow securities.

**f.** Define a (recursive) competitive equilibrium with Arrow securities, being careful to define all of the objects that compose such an equilibrium.

**g.** For the representative consumer in this economy, for each state compute the "natural debt limits" that constrain state-contingent borrowing.

**h.** Compute a competitive equilibrium with Arrow securities. In particular, compute both the pricing kernel and the allocation.

**i.** An entrepreneur enters this economy and proposes to issue a new security each period, namely, a risk-free two-period bond. Such a bond issued in period $t$ promises to pay one unit of consumption at time $t+1$ for sure. Find the price of this new security in period $t$, contingent on $\lambda_t$.

*Exercise 7.5* **A periodic economy**

An economy consists of two consumers, named $i = 1, 2$. The economy exists in discrete time for periods $t \geq 0$. There is one good in the economy, which is not storable and arrives in the form of an endowment stream owned by each consumer. The endowments to consumers $i = 1, 2$ are

$$
\begin{aligned}
y_t^1 &= s_t \\
y_t^2 &= 1
\end{aligned} \tag{1}
$$

where $s_t$ is a random variable governed by a two-state Markov chain with values $s_t = \bar{s}_1 = 0$ or $s_t = \bar{s}_2 = 1$. The Markov chain has time-invariant transition probabilities denoted by $\pi(s_{t+1} = s'|s_t = s) = \pi(s'|s)$, and the probability distribution over the initial state is $\pi_0(s)$. The *aggregate endowment* at $t$ is $Y(s_t) = y_t^1 + y_t^2$.

Let $c^i$ denote the stochastic process of consumption for agent $i$. Household $i$ orders consumption streams according to

$$
U(c^i) = \sum_{t=0}^{\infty} \sum_{s^t} \beta^t \ln[c_t^i(s^t)] \pi(s^t), \tag{2}
$$

where $\pi(s^t)$ is the probability of the history $s^t = (s_0, s_1, \ldots, s_t)$.

**a.** Give a formula for $\pi(s^t)$.

**b.** Let $\theta \in (0, 1)$ be a Pareto weight on household 1. Consider the planning problem

$$
\max_{c^1, c^2} \left\{ \theta \ln(c^1) + (1 - \theta) \ln(c^2) \right\} \tag{3}
$$

where the maximization is subject to

$$
c_t^1(s^t) + c_t^2(s^t) \leq Y(s_t). \tag{4}
$$

Solve the Pareto problem, taking $\theta$ as a parameter.

**b.** Define a *competitive equilibrium* with history-dependent Arrow-Debreu securities traded once and for all at time 0. Be careful to define all of the objects that compose a competitive equilibrium.

**c.** Compute the competitive equilibrium price system (i.e., find the prices of all of the Arrow-Debreu securities).

**d.** Tell the relationship between the solutions (indexed by $\theta$) of the Pareto problem and the competitive equilibrium allocation. If you wish, refer to the two welfare theorems.

**e.** Briefly tell how you can compute the competitive equilibrium price system *before* you have figured out the competitive equilibrium allocation.

**f.** Now define a recursive competitive equilibrium with trading every period in one-period Arrow securities only. Describe all of the objects of which such an equilibrium is composed. (Please denominate the prices of one-period time–$t+1$ state-contingent Arrow securities in units of time-$t$ consumption.) Define the "natural borrowing limits" for each consumer in each state. Tell how to compute these natural borrowing limits.

**g.** Tell how to compute the prices of one-period Arrow securities. How many prices are there (i.e., how many numbers do you have to compute)? Compute all of these prices in the special case that $\beta = .95$ and $\pi(s_j|s_i) = P_{ij}$ where $P = \begin{bmatrix} .8 & .2 \\ .3 & .7 \end{bmatrix}$.

**h.** Within the one-period Arrow securities economy, a new asset is introduced. One of the households decides to market a one-period-ahead riskless claim to one unit of consumption (a one-period real bill). Compute the equilibrium prices of this security when $s_t = 0$ and when $s_t = 1$. Justify your formula for these prices in terms of first principles.

**i.** Within the one-period Arrow securities equilibrium, a new asset is introduced. One of the households decides to market a two-period-ahead riskless claim to one unit consumption (a two-period real bill). Compute the equilibrium prices of this security when $s_t = 0$ and when $s_t = 1$.

**j.** Within the one-period Arrow securities equilibrium, a new asset is introduced. One of the households decides at time $t$ to market five-period-ahead claims to consumption at $t+5$ contingent on the value of $s_{t+5}$. Compute the equilibrium prices of these securities when $s_t = 0$ and $s_t = 1$ and $s_{t+5} = 0$ and $s_{t+5} = 1$.

# 8
# *Overlapping Generations Models*

This chapter describes the pure-exchange overlapping generations model of Paul Samuelson (1958). Although later in the chapter we shall treat the model in a more plausible fashion, we choose to begin with an abstract presentation of the model, building on the previous chapter. Thus, we begin by treating the overlapping generations model as a special case of the chapter 7 general equilibrium model with complete markets and all trades occurring at time $0$. A peculiar type of heterogeneity across agents distinguishes the model. Each individual cares about consumption only at two adjacent dates, and the set of individuals who care about consumption at a particular date includes some who care about consumption one period earlier and others who care about consumption one period later. We shall study how this special preference and demographic pattern affects some of the outcomes of the chapter 7 model.

While it helps to reveal the fundamental structure, allowing complete markets with time-$0$ trading in an overlapping generations model strains credulity. The formalism envisions that equilibrium price and quantity sequences are set at time $0$, before the participants who are to execute the trades have been born. For that reason, most applied work with the overlapping generations model adopts a sequential trading arrangement, like the sequential trade in Arrow securities described in chapter 7. The sequential trading arrangement has all trades executed by agents living in the here and now. Nevertheless, equilibrium quantities and intertemporal prices are equivalent between these two trading arrangements. Therefore, analytical results found in one setting transfer to the other.

Later in the chapter, we use versions of the model with sequential trading to tell how the overlapping generations model provides a framework for thinking about equilibria with government debt and/or valued fiat currency, intergenerational transfers, and fiscal policy.

## *Endowments and preferences*

Time is discrete, starts at $t = 1$, and lasts forever, so $t = 1, 2, \ldots$. There is an infinity of agents named $i = 0, 1, \ldots$. There is a single good at each date. There is no uncertainty. Each agent has a strictly concave, twice continuously differentiable one-period utility function $u(c)$, which is strictly increasing in consumption $c$ of one good. Agent $i$ consumes a vector $c^i = \{c_t^i\}_{t=1}^\infty$ and has the special utility function

$$U^i(c^i) = u(c_i^i) + u(c_{i+1}^i), \quad i \geq 1, \tag{8.1a}$$
$$U^0(c^0) = u(c_1^0). \tag{8.1b}$$

Notice that agent $i$ only wants goods dated $i$ and $i+1$. The interpretation of equations (8.1) is that agent $i$ lives during periods $i$ and $i+1$ and wants to consume only when he is alive.

Each household has an endowment sequence $y^i$ satisfying $y_i^i \geq 0, y_{i+1}^i \geq 0, y_t^i = 0 \ \forall t \neq i$ or $i+1$. Thus, households are endowed with goods only when they are alive.

## *Time-*0 *trading*

We use the definition of competitive equilibrium from chapter 7. Thus, we temporarily suspend disbelief and proceed in the style of Debreu (1959) with time-0 trading. Specifically, we imagine that there is a "clearing house" at time 0 that posts prices and, at those prices, compiles aggregate demand and supply for goods in different periods. An equilibrium price vector makes markets for all periods $t \geq 2$ clear, but there may be excess supply in period 1; that is, the clearing house might end up with goods left over in period 1. Any such excess supply of goods in period 1 can be given to the initial old generation without any effects on the equilibrium price vector, since those old agents optimally consume all their wealth in period 1 and do not want to buy goods in future periods. The reason for our special treatment of period 1 will become clear as we proceed.

Thus, at date 0, there are complete markets in time-$t$ consumption goods with date-0 price $q_t^0$. A household's budget constraint is

$$\sum_{t=1}^{\infty} q_t^0 c_t^i \leq \sum_{t=1}^{\infty} q_t^0 y_t^i. \tag{8.2}$$

Letting $\mu^i$ be a multiplier attached to the household's budget constraint, the household's first-order conditions are

$$\begin{aligned} \mu^i q_i^0 &= u'(c_i^i), &(8.3a)\\ \mu^i q_{i+1}^0 &= u'(c_{i+1}^i), &(8.3b)\\ c_t^i &= 0 \text{ if } t \notin \{i, i+1\}. &(8.3c) \end{aligned}$$

Evidently an allocation is feasible if for all $t \geq 1$,

$$c_t^i + c_t^{i-1} \leq y_t^i + y_t^{i-1}. \tag{8.4}$$

DEFINITION: An allocation is *stationary* if $c_{i+1}^i = c_o, c_i^i = c_y \; \forall i \geq 1$.

Here the subscript $o$ denotes old and $y$ denotes young. Note that we do not require that $c_1^0 = c_o$. We call an equilibrium with a stationary allocation a *stationary equilibrium.*

*Example equilibrium*

Let $\epsilon \in (0, .5)$. The endowments are

$$\begin{aligned} y_i^i &= 1 - \epsilon, \; \forall i \geq 1,\\ y_{i+1}^i &= \epsilon, \; \forall i \geq 0, &(8.5)\\ y_t^i &= 0 \text{ otherwise.} \end{aligned}$$

This economy has many equilibria. We describe two stationary equilibria now, and later we shall describe some nonstationary equilibria. We can use a guess-and-verify method to confirm the following two equilibria.

1. Equilibrium H: a high-interest-rate equilibrium. Set $q_t^0 = 1 \; \forall t \geq 1$ and $c_i^i = c_{i+1}^i = .5$ for all $i \geq 1$ and $c_1^0 = \epsilon$. To verify that this is an equilibrium, notice that each household's first-order conditions are satisfied and that the allocation is feasible. There is extensive intergenerational trade, which occurs at time-0 at the equilibrium price vector $q_t^0$. Note that constraint (8.4) holds with equality for all $t \geq 2$ but with strict inequality for $t = 1$. Some of the $t = 1$ consumption good is left unconsumed.

2. Equilibrium L: a low-interest-rate equilibrium. Set $q_1^0 = 1$, $\frac{q_{t+1}^0}{q_t^0} = \frac{u'(\epsilon)}{u'(1-\epsilon)} = \alpha > 1$. Set $c_t^i = y_t^i$ for all $i, t$. This equilibrium is autarkic, with prices being set to eradicate all trade.

*Relation to the welfare theorems*

As we shall explain in more detail later, equilibrium H Pareto dominates Equilibrium L. In Equilibrium H every generation after the initial old one is better off and no generation is worse off than in Equilibrium L. The Equilibrium H allocation is peculiar because some of the time-1 good is not consumed. There is thus room to set up a giveaway program to the initial old that makes them better off and costs subsequent generations nothing. We shall see how the institution of fiat money accomplishes this purpose.

Equilibrium L is peculiar because it violates the implication of the first theorem of welfare economics, which asserts, under certain assumptions, that a competitive equilibrium allocation is Pareto optimal.[1] An assumption of the theorem is violated by Equilibrium L: the condition that the value of the aggregate endowment at the equilibrium prices is finite.[2]

*Nonstationary equilibria*

Our example economy has more equilibria. To construct all equilibria, we summarize preferences and consumption decisions in terms of an offer curve. We shall use a graphical apparatus proposed by David Gale (1973) and used further to good advantage by William Brock (1990).

DEFINITION: The household's *offer curve* is the locus of $(c_i^i, c_{i+1}^i)$ that solves

$$\max_{\{c_i^i, c_{i+1}^i\}} U(c^i)$$

subject to

$$c_i + \alpha_i c_{i+1}^i \leq y_i^i + \alpha_i y_{i+1}^i.$$

Here $\alpha_i \equiv \frac{q_{i+1}^0}{q_i^0}$, the reciprocal of the one-period gross rate of return from period $i$ to $i+1$, is treated as a parameter.

[1] See Mas-Collell, Whinston, and Green (1995) and Debreu (1954).

[2] Note that if the horizon of the economy were finite, then the counterpart of Equilibrium H would not exist and the allocation of the counterpart of Equilibrium L would be Pareto optimal.

Evidently, the offer curve solves the following pair of equations:

$$c_i^i + \alpha_i c_{i+1}^i = y_i^i + \alpha_i y_{i+1}^i \tag{8.6a}$$

$$\frac{u'(c_{i+1}^i)}{u'(c_i^i)} = \alpha_i \tag{8.6b}$$

for $\alpha_i > 0$. We denote the offer curve by

$$\psi(c_i^i, c_{i+1}^i) = 0.$$

The graphical construction of the offer curve is illustrated in Figure 8.1. We trace it out by varying $\alpha_i$ in the household's problem and reading tangency points between the household's indifference curve and the budget line. The resulting locus depends implicitly on the endowments and it lies above the indifference curve through the endowment point. By construction the following property is also true: at the intersection between the offer curve and a straight line through the endowment point, the straight line is tangent to an indifference curve.[3]

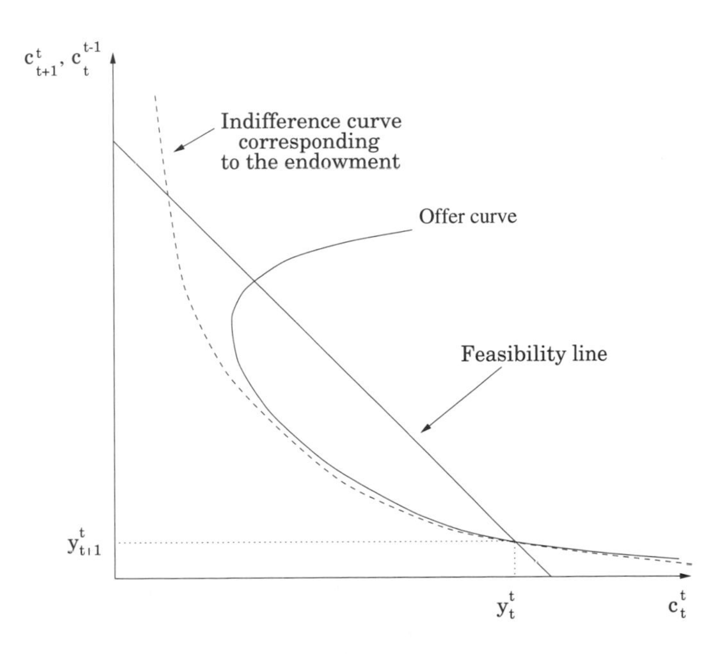


**Figure 8.1** The offer curve and feasibility line.

[3] Given our assumptions on preferences and endowments, the conscientious reader will find Figure 8.1 inaccurate since the offer curve fails to intersect the feasibility line at $c_t^t = c_{t+1}^t$, i.e., Equilibrium H above. The omission is motivated by expositional clarity when we next introduce additional objects in the graphs.

Following Gale (1973), we can use the offer curve and a straight line depicting feasibility in the $(c_i^i, c_i^{i-1})$ plane to construct a machine for computing equilibrium allocations and prices. In particular, we can use the following pair of difference equations to solve for an equilibrium allocation. For $i \geq 1$, the equations are[4]

$$\psi(c_i^i, c_{i+1}^i) = 0, \tag{8.7a}$$

$$c_i^i + c_i^{i-1} = y_i^i + y_i^{i-1}. \tag{8.7b}$$

After the allocation has been computed, the equilibrium price system can be computed from

$$q_i^0 = u'(c_i^i)$$

for all $i \geq 1$.

*Computing equilibria*

*Example 1* Gale's machine: A procedure for constructing an equilibrium is illustrated in Figure 8.2, which reproduces a version of a graph of David Gale (1973). Start with a proposed $c_1^0$, a time-1 allocation to the initial old. Then use the feasibility line to find $c_1^1$, the time-1 allocation to the initial young. This time-1 allocation is feasible, but the time-1 young will choose $c_1^1$ only if $(c_2^1, c_1^1)$ lies on the offer curve. Therefore, we choose $c_2^1$ from the point on the offer curve that cuts a vertical line through $c_1^1$. Then we proceed to find $c_2^2$ from the intersection of a horizontal line through $c_2^1$ and the feasiblity line. We continue recursively in this way, choosing $c_i^i$ as the intersection of the feasibility line with a horizontal line through $c_i^{i-1}$, then choosing $c_{i+1}^i$ as the intersection of a vertical line through $c_i^i$ and the offer curve. We can construct a sequence of $\alpha_i$'s from the slope of a straight line through the endowment point and the sequence of $(c_i^i, c_{i+1}^i)$ pairs that lie on the offer curve.

If the offer curve has the shape drawn in Figure 8.2, any $c_1^0$ between the upper and lower intersections of the offer curve and the feasibility line is an equilibrium setting of $c_1^0$. Each such $c_1^0$ is associated with a distinct $\alpha_i$ sequence, all but one of them converging to the *low*-interest-rate stationary equilibrium allocation and interest rate.

[4] By imposing equation (8.7b) with equality, we are implicitly possibly including a giveaway program to the initial old.

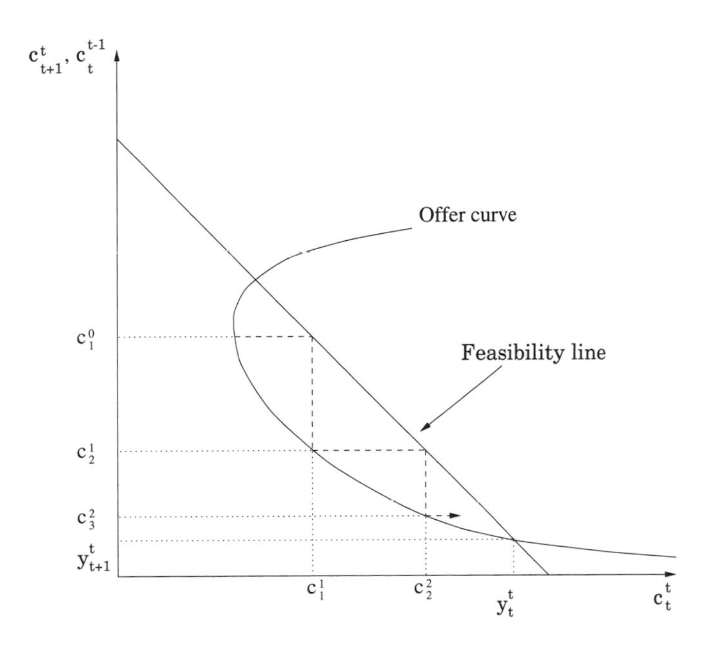


**Figure 8.2** A nonstationary equilibrium allocation.

*Example 2* Endowment at $+\infty$: Take the preference and endowment structure of the previous example and modify only one feature. Change the endowment of the initial old to be $y_1^0 = \epsilon$ *and* "$1-\epsilon$ units of consumption at $t=+\infty$," by which we mean that we take

$$\sum_t q_t^0 y_t^0 = q_1^0 \epsilon + (1-\epsilon) \lim_{t\to\infty} q_t^0 .$$

It is easy to verify that the only competitive equilibrium of the economy with this specification of endowments has $q_t^0 = 1\ \forall t \geq 1$, and thus $\alpha_t = 1\ \forall t \geq 1$. The reason is that all the "low-interest-rate" equilibria that we have described would assign an infinite value to the endowment of the initial old. Confronted with such prices, the initial old would demand unbounded consumption. Meeting such a demand is not feasible. Therefore, such a price system cannot be an equilibrium.

*Example 3* A Lucas tree: Take the preference and endowment structure to be the same as example 1 and modify only one feature. Endow the initial old with a "Lucas tree," namely, a claim to a constant stream of $d > 0$ units of consumption for each $t \geq 1$.[5] Thus, the budget constraint of the initial old person now becomes

$$q_1^0 c_1^0 = d \sum_{t=1}^{\infty} q_t^0 + q_1^0 y_1^0.$$

The offer curve of each young agent remains as before, but now the feasibility line is

$$c_i^i + c_i^{i-1} = y_i^i + y_i^{i-1} + d$$

for all $i \geq 1$. Note that young agents are now endowed below the feasibility line. From Figure 8.3, it seems that there are now two candidates for stationary equilibria, one with constant $\alpha < 1$, the other with constant $\alpha > 1$. The one with $\alpha < 1$ is associated with the steeper budget line in Figure 8.3. However, the candidate stationary equilibrium with $\alpha > 1$ cannot be an equilibrium for a reason similar to that encountered in example 2: for the price system associated with an $\alpha > 1$, the wealth of the initial old is unbounded, which cannot be true in an equilibrium; at such a price system, the initial old would be consuming an unbounded amount, which is not feasible. This argument rules out not only the stationary $\alpha > 1$ equilibrium but also all nonstationary candidate equilibria that converge to that constant $\alpha$. Therefore, there is a unique equilibrium; it is stationary and has $\alpha < 1$.

If we interpret the gross rate of return on the tree as $\alpha^{-1} = \frac{p+d}{p}$, where $p = \sum_{t=1}^{\infty} q_t^0 d$, we can compute that $p = \frac{d}{R-1}$ where $R = \alpha^{-1}$. Here $p$ is the price of the Lucas tree.

In terms of the logarithmic preference example, the difference equation (8.10) becomes modified to

$$\alpha_i = \frac{1+2d}{\epsilon} - \frac{\epsilon^{-1} - 1}{\alpha_{i-1}}. \tag{8.8}$$

*Example 4* Government expenditures: Take the preferences and endowments to be as in example 1 again, but now alter the feasibility condition to be

$$c_i^i + c_i^{i-1} + g = y_i^i + y_i^{i-1}$$

for all $i \geq 1$ where $g > 0$ is a positive level of government purchases. The "clearing house" is now looking for an equilibrium price vector such that this feasibility constraint is satisfied. We assume that government purchases do not

[5] This is a version of an example of Brock (1990).

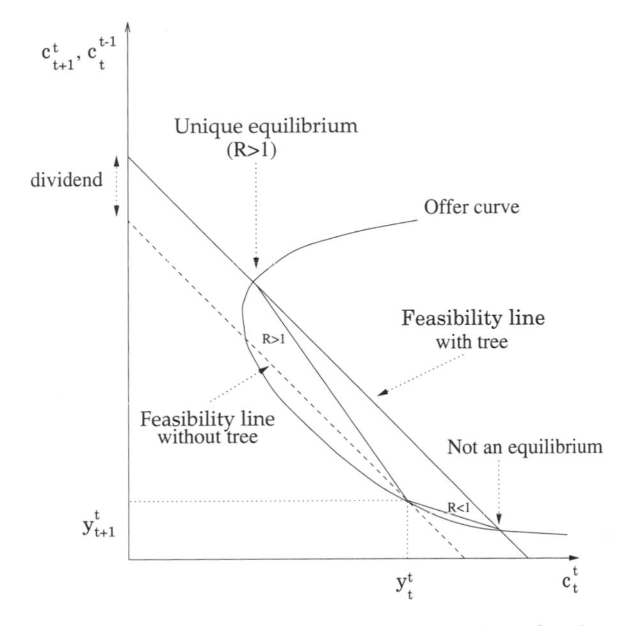


**Figure 8.3** Unique equilibrium with a fixed-dividend asset.

give utility. The offer curve and the feasibility line look as in Figure 8.4. Notice that the endowment point $(y^i_i, y^i_{i+1})$ now lies *outside* the relevant feasibility line. Formally, this graph looks like example 3, but with a "negative dividend $d$." Now there are two stationary equilibria with $\alpha > 1$, and a continuum of equilibria converging to the higher $\alpha$ equilibrium (the one with the lower slope $\alpha^{-1}$ of the associated budget line). Now the equilibria with $\alpha > 1$ cannot be ruled out by the argument in example 3 because no one's endowment sequence receives infinite value when $\alpha > 1$.

Later, we shall interpret this example in terms of a government financing a constant deficit through money creation, or through borrowing at a negative real net interest rate. We shall discuss this and other examples in a setting with sequential trading.

*Example 5* Log utility: Suppose that $u(c) = \ln c$ and that the endowment is given by equations (8.5). Then the offer curve is given by the recursive formulas $c^i_i = .5(1 - \epsilon + \alpha\epsilon), c^i_{i+1} = \alpha^{-1} c^i_i$. Let $\alpha_i$ be the gross rate of return facing the

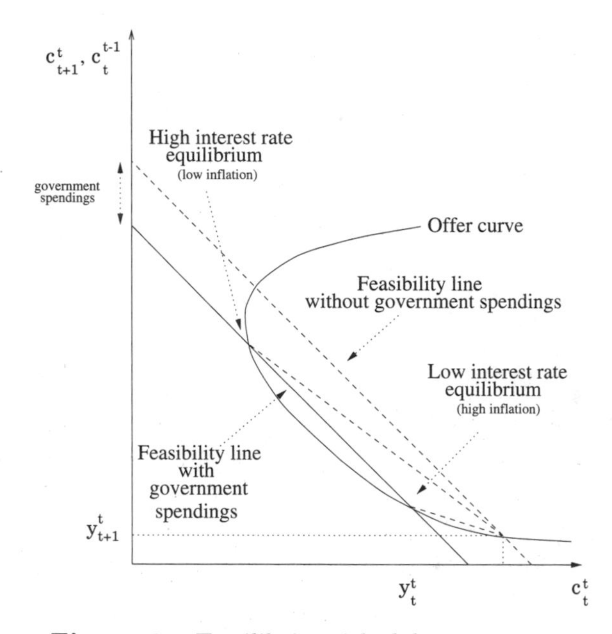


**Figure 8.4** Equilibria with debt- or money-financed government deficit finance.

young at $i$. Feasibility at $i$ and the offer curves then imply

$$\frac{1}{2\alpha_{i-1}}(1-\epsilon+\alpha_{i-1}\epsilon)+.5(1-\epsilon+\alpha_i\epsilon)=1. \tag{8.9}$$

This implies the difference equation

$$\alpha_i=\epsilon^{-1}-\frac{\epsilon^{-1}-1}{\alpha_{i-1}}. \tag{8.10}$$

See Figure 8.2. An equilibrium $\alpha_i$ sequence must satisfy equation (8.9) and have $\alpha_i>0$ for all $i$. Evidently, $\alpha_i=1$ for all $i\geq 1$ is an equilibrium $\alpha$ sequence. So is any $\alpha_i$ sequence satisfying equation (8.9) and $\alpha_1\geq 1$; $\alpha_1<1$ will not work because equation (8.9) implies that the tail of $\{\alpha_i\}$ is an unbounded negative sequence. The limiting value of $\alpha_i$ for any $\alpha_1>1$ is $\frac{1-\epsilon}{\epsilon}=u'(\epsilon)/u'(1-\epsilon)$, which is the interest factor associated with the stationary autarkic equilibrium. Notice that Figure 8.2 suggests that the stationary $\alpha_i=1$ equilibrium is not stable, while the autarkic equilibrium is.

## *Sequential trading*

We now alter the trading arrangement to bring us into line with standard presentations of the overlapping generations model. We leave behind the time-0, complete market trading arrangement and substitute for it sequential trading in which a durable asset, either government debt or money or claims on a Lucas tree, are passed from old to young. In addition to cross-generation transfers that are financed with voluntary exchanges of such assets, others can be financed by government tax and transfer programs.

## *Money*

We now describe Samuelson's (1958) version of the model. Trading occurs sequentially through a medium of exchange, an inconvertible (or "fiat") currency. In Samuelson's model, the preferences and endowments are as described previously, with one additional component of the endowment. At date $t = 1$, old agents are endowed in the aggregate with $M > 0$ units of intrinsically worthless currency. No one has promised to redeem the currency for goods. Although the currency is not "backed" by a convertibility promise, Samuelson showed that rational expectations that the currency will be valued in the future can give the currency value now. Currency is backed by expectations without promises: it is valued today because people expect it to be valued tomorrow.

For each date $t \geq 1$, young agents purchase $m_t^i$ units of currency at price of $1/p_t$ units of the time-$t$ consumption good. Here $p_t \geq 0$ is the time-$t$ price level. At each $t \geq 1$, each old agent exchanges his holdings of currency for the time-$t$ consumption good. The budget constraints of a young agent born in period $i \geq 1$ are

$$c_i^i + \frac{m_i^i}{p_i} \leq y_i^i, \tag{8.11}$$

$$c_{i+1}^i \leq \frac{m_i^i}{p_{i+1}} + y_{i+1}^i, \tag{8.12}$$

$$m_i^i \geq 0. \tag{8.13}$$

If $m_i^i \geq 0$, inequalities (8.11) and (8.12) imply

$$c_i^i + c_{i+1}^i \left(\frac{p_{i+1}}{p_i}\right) \leq y_i^i + y_{i+1}^i \left(\frac{p_{i+1}}{p_i}\right). \tag{8.14}$$

Note that this budget set is identical with equation (8.2), provided that we set

$$\frac{p_{i+1}}{p_i} = \alpha_i = \frac{q^0_{i+1}}{q^0_i}.$$

We use the following definitions:

DEFINITION: A nominal price sequence is a positive sequence $\{p_i\}_{i \geq 1}$.

DEFINITION: An equilibrium with valued fiat money is a feasible allocation and a nominal price sequence with $p_t < +\infty$ for all $t$ such that given the price sequence, the allocation solves the household's problem for each $i \geq 1$.

The qualification that $p_t < +\infty$ for all $t$ means that fiat money is valued.

*More computing equilibria*

Summarize the household's optimum problem with a saving function

$$y^i_i - c^i_i = s(\alpha_i; y^i_i, y^i_{i+1}). \tag{8.15}$$

Then the equilibrium conditions for the model are

$$\frac{M}{p_i} = s(\alpha_i; y^i_i, y^i_{i+1}) \tag{8.16a}$$

$$\alpha_i = \frac{p_{i+1}}{p_i}, \tag{8.16b}$$

where it is understood that $c^i_{i+1} = y^i_{i+1} + \frac{M}{p_{i+1}}$. To solve the model, we solve the difference equations (8.16) for $\{p_i\}^{\infty}_{i=1}$, then get the allocation from the household's budget constraints evaluated at equality at the equilibrium level of real balances. As an example, suppose that $u(c) = \ln(c)$, and that $(y^i_i, y^i_{i+1}) = (w_1, w_2)$ with $w_1 > w_2$. Then the saving function is $s(\alpha_i) = .5(w_1 - \alpha_i w_2)$. Then equation (8.16a) becomes

$$.5(w_1 - w_2 \frac{p_{t+1}}{p_t}) = \frac{M}{p_t}$$

or

$$p_t = 2M/w_1 + \left(\frac{w_2}{w_1}\right) p_{t+1}. \tag{8.17}$$

This is a difference equation whose solutions with a positive price level are

$$p_t = \frac{2M}{w_1(1 - \frac{w_2}{w_1})} + c\left(\frac{w_1}{w_2}\right)^t, \tag{8.18}$$

for any scalar $c > 0$.[6] The solution for $c = 0$ is the unique stationary solution. The solutions with $c > 0$ have uniformly higher price levels than the $c = 0$ solution, and have the value of currency going to zero.

*Equivalence of equilibria*

We briefly look back at the equilibria with time-0 trading and note that the equilibrium allocations are the same under time-0 and sequential trading. Thus, the following proposition asserts that with one adjustment, a competitive equilibrium allocation with time-0 trading is an equilibrium allocation in the fiat money economy (with sequential trading).

PROPOSITION: Let $\overline{c}^i$ denote a competitive equilibrium allocation (with time-0 trading) and suppose that it satisfies $\overline{c}_1^1 < y_1^1$. Then there exists an equilibrium (with sequential trading) of the monetary economy with allocation that satisfies $c_i^i = \overline{c}_i^i, c_{i+1}^i = \overline{c}_{i+1}^i$ for $i \geq 1$.

PROOF: Take the competitive equilibrium allocation and price system and form $\alpha_i = q_{i+1}^0/q_i^0$. Set $m_i^i/p_i = y_i^i - \overline{c}_i^i$. Set $m_i^i = M$ for all $i \geq 1$, and determine $p(1)$ from $\frac{M}{p(1)} = y_1^1 - \overline{c}_1^1$. This last equation determines a positive initial price level $p(1)$ provided that $y_1^1 - \overline{c}_1^1 > 0$. Determine subsequent price levels from $p_{i+1} = \alpha_i p_i$. Determine the allocation to the initial old from $c_1^0 = y_1^0 + \frac{M}{p_1} = y_1^0 + (y_1^1 - c_1^1)$. ∎

In the monetary equilibrium, time-$t$ real balances equal the per capita savings of the young and the per capita dissavings of the old. To be in a monetary equilibrium, both quantities must be positive for all $t \geq 1$.

A converse of the proposition is true.

PROPOSITION: Let $\overline{c}^i$ be an equilibrium allocation for the fiat money economy. Then there is a competitive equilibrium with time 0 trading with the same allocation, provided that the endowment of the initial old is augmented with a particular transfer from the "clearing house."

[6] See the appendix to chapter 1.

To verify this proposition, we have to construct the required transfer from the clearing house to the initial old. Evidently, it is $y_1^1 - \bar{c}_1^1$. We invite the reader to complete the rest of the proof.

## *Deficit finance*

For the rest of this chapter, we shall assume sequential trading. With sequential trading of fiat currency, this section reinterprets one of our earlier examples with time-0 trading, the example with government spending.

Consider the following overlapping generations model: The population is constant. At each date $t \geq 1$, $N$ identical young agents are endowed with $(y_t^t, y_{t+1}^t) = (w_1, w_2)$, where $w_1 > w_2 > 0$. A government levies lump-sum taxes of $\tau_1$ on each young agent and $\tau_2$ on each old agent alive at each $t \geq 1$. There are $N$ old people at time 1 each of whom is endowed with $w_2$ units of the consumption good and $M_0 > 0$ units of inconvertible perfectly durable fiat currency. The initial old have utility function $c_1^0$. The young have utility function $u(c_t^t) + u(c_{t+1}^t)$. For each date $t \geq 1$ the government augments the currency supply according to

$$M_t - M_{t-1} = p_t(g - \tau_1 - \tau_2), \tag{8.19}$$

where $g$ is a constant stream of government expenditures per capita and $0 < p_t \leq +\infty$ is the price level. If $p_t = +\infty$, we intend that equation (8.19) be interpreted as

$$g = \tau_1 + \tau_2. \tag{8.20}$$

For each $t \geq 1$, each young person's behavior is summarized by

$$s_t = f(R_t; \tau_1, \tau_2) = \arg\max_{s \geq 0} \left[u(w_1 - \tau_1 - s) + u(w_2 - \tau_2 + R_t s)\right]. \tag{8.21}$$

DEFINITION: An equilibrium with valued fiat currency is a pair of positive sequences $\{M_t, p_t\}$ such that (a) given the price level sequence, $M_t/p_t = f(R_t)$ (the dependence on $\tau_1, \tau_2$ being understood); (b) $R_t = p_t/p_{t+1}$; and (c) the government budget constraint (8.19) is satisfied for all $t \geq 1$.

The condition $f(R_t) = M_t/p_t$ can be written as $f(R_t) = M_{t-1}/p_t + (M_t - M_{t-1})/p_t$. The left side is the savings of the young. The first term on right side is the dissaving of the old (the real value of currency that they exchange for time-$t$ consumption). The second term on the right is the dissaving of the government (its deficit), which is the real value of the additional currency that the government prints at $t$ and uses to purchases time-$t$ goods from the young.

To compute an equilibrium, define $d = g - \tau_1 - \tau_2$ and write equation (8.19) as

$$\frac{M_t}{p_t} = \frac{M_{t-1}}{p_{t-1}} \frac{p_{t-1}}{p_t} + d$$

for $t \geq 2$ and

$$\frac{M_1}{p_1} = \frac{M_0}{p_1} + d$$

for $t = 1$. Substitute $M_t/p_t = f(R_t)$ into these equations to get

$$f(R_t) = f(R_{t-1})R_{t-1} + d \tag{8.22a}$$

for $t \geq 2$ and

$$f(R_1) = \frac{M_0}{p_1} + d. \tag{8.22b}$$

Given $p_1$, which determines an initial $R_1$ by means of equation (8.22b), equations (8.22) form an autonomous difference equation in $R_t$. This system can be solved using Figure 8.4.

### *Steady states and the Laffer curve*

Let's seek a stationary solution of equations (8.22), a quest that is rendered reasonable by the fact that $f(R_t)$ is time invariant (because the endowment and the tax patterns as well as the government deficit $d$ are time invariant). Guess that $R_t = R$ for $t \geq 1$. Then equations (8.22) become

$$f(R)(1 - R) = d, \tag{8.23a}$$

$$f(R) = \frac{M_0}{p_1} + d. \tag{8.23b}$$

For example, suppose that $u(c) = \ln(c)$. Then $f(R) = \frac{w_1 - \tau_1}{2} - \frac{w_2 - \tau_2}{2R}$. We have graphed $f(R)(1 - R)$ against $d$ in Figure 8.5. Notice that if there is one solution for equation (8.23a), then there are at least two.

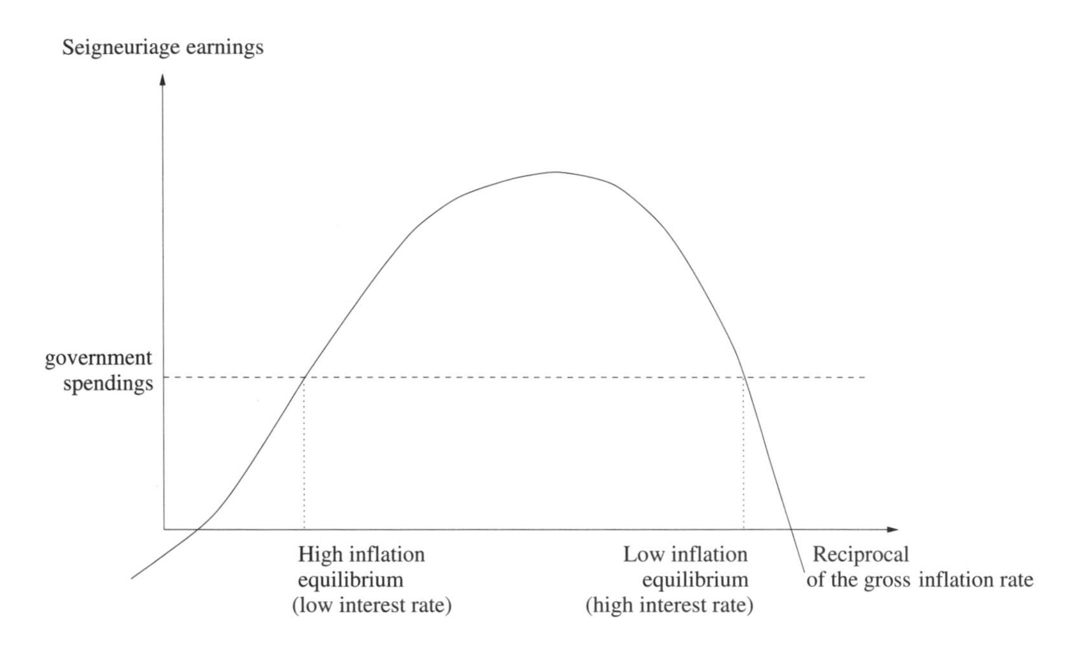


**Figure 8.5** The Laffer curve in revenues from the inflation tax.

Here $(1-R)$ can be interpreted as a tax rate on real balances, and $f(R)(1-R)$ is a Laffer curve for the inflation tax rate. The high-return (low-tax) $R = \overline{R}$ is associated with the good Laffer curve stationary equilibrium, and the low-return (high-tax) $R = \underline{R}$ comes with the bad Laffer curve stationary equilibrium. Once $R$ is determined, we can determine $p(1)$ from equation (8.23b).

Figure 8.5 is isomorphic with Figure 8.4. The saving rate function $f(R)$ can be deduced from the offer curve. Thus, a version of Figure 8.4 can be used to solve the difference equation (8.22a) graphically. If we do so, we discover a continuum of nonstationary solutions of equation (8.22a), all but one of which have $R_t \to \underline{R}$ as $t \to \infty$. Thus, the bad Laffer curve equilibrium is stable.

The stability of the bad Laffer curve equilibrium arises under perfect foresight dynamics. Bruno and Fischer (1990) and Marcet and Sargent (1989) analyze how the system behaves under two different types of adaptive dynamics. They find that either under a crude form of adaptive expectations or under a least squares learning scheme, $R_t$ converges to $\overline{R}$. This finding is comforting because the comparative dynamics are more plausible at $\overline{R}$ (larger deficits bring higher inflation). Furthermore, Marimon and Sunder (1993) present experimental evidence pointing toward the selection made by the adaptive dynamics. Marcet and Nicolini

(1999) build an adaptive model of several Latin American hyperinflations that rests on this selection.

## Equivalent setups

This section describes some alternative asset structures and trading arrangements that support the same equilibrium allocation. We take a model with a government deficit and show how it can be supported with sequential trading in government-indexed bonds, sequential trading in fiat currency, or time-0 trading in Arrow-Debreu dated securities.

### The economy

Consider an overlapping generations economy with one agent born at each $t \geq 1$ and an initial old person at $t = 1$. Young agents born at date $t$ have endowment pattern $(y_t^t, y_{t+1}^t)$ and the utility function described earlier. The initial old person is endowed with $M_0 > 0$ units of unbacked currency and $y_1^0$ units of the consumption good. There is a stream of per-young-person government purchases $\{g_t\}$.

DEFINITION: An equilibrium with money financed government deficits is a sequence $\{M_t, p_t\}_{t=0}^{\infty}$ with $0 < p_t < +\infty$ and $M_t > 0$ that satisfies (a) given $\{p_t\}$,

$$M_t = \arg\max_{\tilde{M} \geq 0} \left[ u(y_t^t - \tilde{M}/p_t) + u(y_{t+1}^t + \tilde{M}/p_{t+1}) \right]; \tag{8.24a}$$

and (b)

$$M_t - M_{t-1} = p_t g_t. \tag{8.24b}$$

Now consider a version of the same economy in which there is no currency but rather indexed government bonds. The demographics and endowments are identical with the preceding economy, but now each initial old person is endowed with $B_1$ units of a maturing bond, denominated in units of time-1 consumption good. In period $t$, the government sells new one-period bonds to the young to finance its purchases $g_t$ of time-$t$ goods and to pay off the one-period debt falling due at time $t$. Let $R_t > 0$ be the gross real one-period rate of return on government debt between $t$ and $t+1$.

DEFINITION: An equilibrium with bond-financed government deficits is a sequence $\{B_{t+1}, R_t\}_{t=0}^{\infty}$ that satisfies (a) given $\{R_t\}$,

$$B_{t+1} = \arg\max_{\tilde{B}} [u(y_t^t - \tilde{B}/R_t) + u(y_{t+1}^t + \tilde{B})]; \tag{8.25a}$$

and (b)

$$B_{t+1}/R_t = B_t + g_t, \tag{8.25b}$$

with $B_1 \geq 0$ given.

These two types of equilibria are isomorphic in the following sense: Take an equilibrium of the economy with money-financed deficits and transform it into an equilibrium of the economy with bond-financed deficits as follows: set $B_t = M_{t-1}/p_t, R_t = p_t/p_{t+1}$. It can be verified directly that these settings of bonds and interest rates, together with the original consumption allocation, form an equilibrium of the economy with bond-financed deficits.

Each of these two types of equilibria is evidently also isomorphic to the following equilibrium formulated with time-0 markets:

DEFINITION: Let $B_1^g$ represent claims to time-1 consumption owed by the government to the old at time 1. An equilibrium with time-0 trading is an initial level of government debt $B_1^g$, a price system $\{q_t^0\}_{t=1}^{\infty}$, and a sequence $\{s_t\}_{t=1}^{\infty}$ such that (a) given the price system,

$$s_t = \arg\max_{\tilde{s}} \left\{ u(y_t^t - \tilde{s}) + u\left[ y_{t+1}^t + \left( \frac{q_t^0}{q_{t+1}^0} \right) \tilde{s} \right] \right\};$$

and (b)

$$q_1^0 B_1^g + \sum_{t=1}^{\infty} q_t^0 g_t = 0. \tag{8.26}$$

Condition b is the Arrow-Debreu version of the government budget constraint. Condition a is the optimality condition for the intertemporal consumption decision of the young of generation $t$.

The government budget constraint in condition b can be represented recursively as

$$q_{t+1}^0 B_{t+1}^g = q_t^0 B_t^g + q_t^0 g_t. \tag{8.27}$$

If we solve equation (8.27) forward and impose $\lim_{t \to 0} q_{t+T}^0 B_{t+T}^g = 0$, we obtain the budget constraint (8.26) for $t = 1$. Condition (8.26) makes it evident that when $\sum_{t=1} q_t^0 g_t > 0$, $B_1^g < 0$, so that the government has negative net worth. This negative net worth corresponds to the unbacked claims that the market nevertheless values in the sequential trading versions of the model.

*Growth*

It is easy to extend these models to the case in which there is growth in the population. Let there be $N_t = nN_{t-1}$ identical young people at time $t$, with $n > 0$. For example, consider the economy with money-financed deficits. The total money supply is $N_t M_t$, and the government budget constraint is

$$N_t M_t - N_{t-1} M_{t-1} = N_t p_t g,$$

where $g$ is per-young-person government purchases. Dividing both sides of the budget constraint by $N_t$ and rearranging gives

$$\frac{M_t}{p_{t+1}} \frac{p_{t+1}}{p_t} = n^{-1} \frac{M_{t-1}}{p_t} + g. \tag{8.27c}$$

This equation replaces equation (8.24b) in the definition of an equilibrium with money-financed deficits. (Note that in a steady state $R = n$ is the high-interest-rate equilibrium.) Similarly, in the economy with bond-financed deficits, the government budget constraint would become

$$\frac{B_{t+1}}{R_t} = n^{-1} B_t + g_t.$$

It is also easy to modify things to permit the government to tax young and old people at $t$. In that case, with government bond finance the government budget constraint becomes

$$\frac{B_{t+1}}{R_t} = n^{-1} B_t + g_t - \tau_t^t - n^{-1} \tau_t^{t-1},$$

where $\tau_t^s$ is the time $t$ tax on a person born in period $s$.

## Optimality and the existence of monetary equilibria

Wallace (1980) discusses the connection between nonoptimality of the equilibrium without valued money and existence of monetary equilibria. Abstracting from his assumption of a storage technology, we study how the arguments apply to a pure endowment economy. The environment is as follows. At any date $t$, the population consists of $N_t$ young agents and $N_{t-1}$ old agents where $N_t = nN_{t-1}$ with $n > 0$. Each young person is endowed with $y_1 > 0$ goods, and an old person receives the endowment $y_2 > 0$. Preferences of a young agent at time $t$ are given by the utility function $u(c_t^t, c_{t+1}^t)$ which is twice differentiable with indifference curves that are convex to the origin. The two goods in the utility function are normal goods, and

$$\theta(c_1, c_2) \equiv u_1(c_1, c_2)/u_2(c_1, c_2),$$

the marginal rate of substitution function, approaches infinity as $c_2/c_1$ approaches infinity and approaches zero as $c_2/c_1$ approaches zero. The welfare of the initial old agents at time 1 is strictly increasing in $c_1^0$, and each one of them is endowed with $y_2$ goods and $m_0^0 > 0$ units of fiat money. Thus, the beginning-of-period aggregate nominal money balances in the initial period 1 are $M_0 = N_0 m_0^0$.

For all $t \geq 1$, $M_t$, the post-transfer time $t$ stock of fiat money, obeys $M_t = zM_{t-1}$ with $z > 0$. The time $t$ transfer (or tax), $(z-1)M_{t-1}$, is divided equally at time $t$ among the $N_{t-1}$ members of the current old generation. The transfers (or taxes) are fully anticipated and are viewed as lump-sum: they do not depend on consumption and saving behavior. The budget constraints of a young agent born in period $t$ are

$$c_t^t + \frac{m_t^t}{p_t} \leq y_1, \tag{8.28}$$

$$c_{t+1}^t \leq y_2 + \frac{m_t^t}{p_{t+1}} + \frac{(z-1)}{N_t} \frac{M_t}{p_{t+1}}, \tag{8.29}$$

$$m_t^t \geq 0, \tag{8.30}$$

where $p_t > 0$ is the time $t$ price level. In a nonmonetary equilibrium, the price level is infinite so the real value of both money holdings and transfers are zero. Since all members in a generation are identical, the nonmonetary equilibrium is autarky with a marginal rate of substitution equal to

$$\theta_{\text{aut}} \equiv \frac{u_1(y_1, y_2)}{u_2(y_1, y_2)}.$$

We ask two questions about this economy. Under what circumstances does a monetary equilibrium exist? And, when it exists, under what circumstances does it improve matters?

Let $\hat{m}_t$ denote the equilibrium real money balances of a young agent at time $t$, $\hat{m}_t \equiv M_t/(N_t p_t)$. Substitution of equilibrium money holdings into budget constraints (8.28) and (8.29) at equality yield $c_t^t = y_1 - \hat{m}_t$ and $c_{t+1}^t = y_2 + n\hat{m}_{t+1}$. In a monetary equilibrium, $\hat{m}_t > 0$ for all $t$ and the marginal rate of substitution $\theta(c_t^t, c_{t+1}^t)$ satisfies

$$\theta(y_1 - \hat{m}_t,\ y_2 + n\hat{m}_{t+1}) = \frac{p_t}{p_{t+1}} > \theta_{\text{aut}}, \quad \forall t \geq 1. \tag{8.31}$$

The equality part of (8.31) is the first-order condition for money holdings of an agent born in period $t$ evaluated at the equilibrium allocation. Since $c_t^t < y_1$ and $c_{t+1}^t > y_2$ in a monetary equilibrium, the inequality in (8.31) follows from the assumption that the two goods in the utility function are normal goods.

Another useful characterization of the equilibrium rate of return on money, $p_t/p_{t+1}$, can be obtained as follows. By the rule generating $M_t$ and the equilibrium condition $M_t/p_t = N_t\hat{m}_t$, we have for all $t$,

$$\frac{p_t}{p_{t+1}} = \frac{M_{t+1}}{zM_t}\frac{p_t}{p_{t+1}} = \frac{N_{t+1}\hat{m}_{t+1}}{zN_t\hat{m}_t} = \frac{n}{z}\frac{\hat{m}_{t+1}}{\hat{m}_t}. \tag{8.32}$$

We are now ready to address our first question, under what circumstances does a monetary equilibrium exist?

PROPOSITION: $\theta_{\text{aut}} z < n$ is necessary and sufficient for the existence of at least one monetary equilibrium.

PROOF: We first establish necessity. Suppose to the contrary that there is a monetary equilibrium and $\theta_{\text{aut}} z/n \geq 1$. Then, by the inequality part of (8.31) and expression (8.32), we have for all $t$,

$$\frac{\hat{m}_{t+1}}{\hat{m}_t} > \frac{z\theta_{\text{aut}}}{n} \geq 1. \tag{8.33}$$

If $z\theta_{\text{aut}}/n > 1$, the growth rate of $\hat{m}_t$ is bounded uniformly above one and, hence, the sequence $\{\hat{m}_t\}$ is unbounded which is inconsistent with an equilibrium because real money balances per capita cannot exceed the endowment $y_1$ of a young agent. If $z\theta_{\text{aut}}/n = 1$, the strictly increasing sequence $\{\hat{m}_t\}$ in (8.33) might not be unbounded but converge to some constant $\hat{m}_\infty$. According to

(8.31) and (8.32), the marginal rate of substitution will then converge to $n/z$ which by assumption is now equal to $\theta_{\text{aut}}$, the marginal rate of substitution in autarky. Thus, real balances must be zero in the limit which contradicts the existence of a strictly increasing sequence of positive real balances in (8.33).

To show sufficiency, we prove the existence of a unique equilibrium with constant per-capita real money balances when $\theta_{\text{aut}} z < n$. Substitute our candidate equilibrium, $\hat{m}_t = \hat{m}_{t+1} \equiv \hat{m}$, into (8.31) and (8.32), which yields two equilibrium conditions,

$$\theta(y_1 - \hat{m},\ y_2 + n\hat{m}) = \frac{n}{z} > \theta_{\text{aut}}.$$

The inequality part is satisfied under the parameter restriction of the proposition, and we only have to show the existence of $\hat{m} \in [0, y_1]$ that satisfies the equality part. But the existence (and uniqueness) of such a $\hat{m}$ is trivial. Note that the marginal rate of substitution on the left side of the equality is equal to $\theta_{\text{aut}}$ when $\hat{m} = 0$. Next, our assumptions on preferences imply that the marginal rate of substitution is strictly increasing in $\hat{m}$, and approaches infinity when $\hat{m}$ approaches $y_1$. ∎

The stationary monetary equilibrium in the proof will be referred to as the $\hat{m}$ equilibrium. In general, there are other nonstationary monetary equilibria when the parameter condition of the proposition is satisfied. For example, in the case of logarithmic preferences and a constant population, recall the continuum of equilibria indexed by the scalar $c > 0$ in expression (8.18). But here we choose to focus solely on the stationary $\hat{m}$ equilibrium, and its welfare implications. The $\hat{m}$ equilibrium will be compared to other feasible allocations using the Pareto criterion. Evidently, an allocation $C = \{c_1^0; (c_t^t, c_{t+1}^t), t \geq 1\}$ is feasible if

$$N_t c_t^t + N_{t-1} c_t^{t-1} \leq N_t y_1 + N_{t-1} y_2, \quad \forall t \geq 1,$$

or, equivalently,

$$n c_t^t + c_t^{t-1} \leq n y_1 + y_2, \quad \forall t \geq 1. \tag{8.34}$$

The definition of Pareto optimality is:

DEFINITION: A feasible allocation $C$ is Pareto optimal if there is no other feasible allocation $\tilde{C}$ such that

$$\tilde{c}_1^0 \geq c_1^0,$$
$$u(\tilde{c}_t^t, \tilde{c}_{t+1}^t) \geq u(c_t^t, c_{t+1}^t), \quad \forall t \geq 1,$$

and at least one of these weak inequalities holds with strict inequality.

We first examine under what circumstances the nonmonetary equilibrium (autarky) is Pareto optimal.

PROPOSITION: $\theta_{\text{aut}} \geq n$ is necessary and sufficient for the optimality of the nonmonetary equilibrium (autarky).

PROOF: To establish sufficiency, suppose to the contrary that there exists another feasible allocation $\tilde{C}$ that is Pareto superior to autarky and $\theta_{\text{aut}} \geq n$. Without loss of generality, assume that the allocation $\tilde{C}$ satisfies (8.34) with equality. (Given an allocation that is Pareto superior to autarky but that does not satisfy (8.34), one can easily construct another allocation that is Pareto superior to the given allocation, and hence to autarky.) Let period $t$ be the first period when this alternative allocation $\tilde{C}$ differs from the autarkic allocation. The requirement that the old generation in this period is not made worse off, $\tilde{c}_t^{t-1} \geq y_2$, implies that the first perturbation from the autarkic allocation must be $\tilde{c}_t^t < y_1$ with the subsequent implication that $\tilde{c}_{t+1}^t > y_2$. It follows that the consumption of young agents at time $t+1$ must also fall below $y_1$, and we define

$$\epsilon_{t+1} \equiv y_1 - \tilde{c}_{t+1}^{t+1} > 0. \tag{8.35}$$

Now, given $\tilde{c}_{t+1}^{t+1}$, we compute the smallest number $c_{t+2}^{t+1}$ that satisfies

$$u(\tilde{c}_{t+1}^{t+1}, c_{t+2}^{t+1}) \geq u(y_1, y_2).$$

Let $\bar{c}_{t+2}^{t+1}$ be the solution to this problem. Since the allocation $\tilde{C}$ is Pareto superior to autarky, we have $\tilde{c}_{t+2}^{t+1} \geq \bar{c}_{t+2}^{t+1}$. Before using this inequality, though, we want to derive a convenient expression for $\bar{c}_{t+2}^{t+1}$.

Consider the indifference curve of $u(c_1, c_2)$ that yields a fixed utility equal to $u(y_1, y_2)$. In general, along an indifference curve, $c_2 = h(c_1)$ where $h' = -u_1/u_2 = -\theta$ and $h'' > 0$. Therefore, applying the intermediate value theorem to $h$, we have

$$h(c_1) = h(y_1) + (y_1 - c_1)[-h'(y_1) + f(y_1 - c_1)], \tag{8.36}$$

where the function $f$ is strictly increasing and $f(0) = 0$.

Now since $(\tilde{c}_{t+1}^{t+1}, \bar{c}_{t+2}^{t+1})$ and $(y_1, y_2)$ are on the same indifference curve, we may use (8.35) and (8.36) to write

$$\bar{c}_{t+2}^{t+1} = y_2 + \epsilon_{t+1}[\theta_{\text{aut}} + f(\epsilon_{t+1})],$$

and after invoking $\tilde{c}_{t+2}^{t+1} \geq \bar{c}_{t+2}^{t+1}$, we have

$$\tilde{c}_{t+2}^{t+1} - y_2 \geq \epsilon_{t+1}[\theta_{\text{aut}} + f(\epsilon_{t+1})]. \tag{8.37}$$

Since $\tilde{C}$ satisfies (8.34) at equality, we also have

$$\epsilon_{t+2} \equiv y_1 - \tilde{c}_{t+2}^{t+2} = \frac{\tilde{c}_{t+2}^{t+1} - y_2}{n}. \tag{8.38}$$

Substitution of (8.37) into (8.38) yields

$$\epsilon_{t+2} \geq \epsilon_{t+1} \frac{\theta_{\text{aut}} + f(\epsilon_{t+1})}{n} > \epsilon_{t+1}, \tag{8.39}$$

where the strict inequality follows from $\theta_{\text{aut}} \geq n$ and $f(\epsilon_{t+1}) > 0$ (implied by $\epsilon_{t+1} > 0$). Continuing these computations of successive values of $\epsilon_{t+k}$ yields

$$\epsilon_{t+k} \geq \epsilon_{t+1} \prod_{j=1}^{k-1} \frac{\theta_{\text{aut}} + f(\epsilon_{t+j})}{n} > \epsilon_{t+1} \left[\frac{\theta_{\text{aut}} + f(\epsilon_{t+1})}{n}\right]^{k-1}, \quad \text{for } k > 2,$$

where the strict inequality follows from the fact that $\{\epsilon_{t+j}\}$ is a strictly increasing sequence. Thus the $\epsilon$ sequence is bounded below by a strictly increasing exponential and hence is unbounded. But such an unbounded sequence violates feasibility because $\epsilon$ cannot exceed the endowment $y_1$ of a young agent, it follows that we can rule out the existence of a Pareto superior allocation $\tilde{C}$, and conclude that $\theta_{\text{aut}} \geq n$ is a sufficient condition for the optimality of autarky.

To establish necessity, we prove the existence of an alternative feasible allocation $\hat{C}$ that is Pareto superior to autarky when $\theta_{\text{aut}} < n$. First, pick an $\epsilon > 0$ sufficiently small so that

$$\theta_{\text{aut}} + f(\epsilon) \leq n, \tag{8.40}$$

where $f$ is defined implicitly by equation (8.36). Second, set $\hat{c}_t^t = y_1 - \epsilon \equiv \hat{c}_1$, and

$$\hat{c}_{t+1}^t = y_2 + \epsilon[\theta_{\text{aut}} + f(\epsilon)] \equiv \hat{c}_2, \quad \forall t \geq 1. \tag{8.41}$$

That is, we have constructed a consumption bundle $(\hat{c}_1, \hat{c}_2)$ that lies on the same indifference curve as $(y_1, y_2)$, and from (8.40) and (8.41), we have

$$\hat{c}_2 \leq y_2 + n\epsilon,$$

which ensures that the condition for feasibility (8.34) is satisfied for $t \geq 2$. By setting $\hat{c}_1^0 = y_2 + n\epsilon$, feasibility is also satisfied in period 1 and the initial old generation is then strictly better off under the alternative allocation $\hat{C}$. ▮

With a constant nominal money supply, $z = 1$, the two propositions show that a monetary equilibrium exists if and only if the nonmonetary equilibrium is suboptimal. In that case, the following proposition establishes that the stationary $\hat{m}$ equilibrium is optimal.

PROPOSITION: Given $\theta_{\mathrm{aut}} z < n$, then $z \leq 1$ is necessary and sufficient for the optimality of the stationary monetary equilibrium $\hat{m}$.

PROOF: The class of feasible stationary allocations with $(c_t^t, c_{t+1}^t) = (c_1, c_2)$ for all $t \geq 1$, is given by

$$c_1 + \frac{c_2}{n} \leq y_1 + \frac{y_2}{n}, \tag{8.42}$$

i.e., the condition for feasibility in (8.34). It follows that the $\hat{m}$ equilibrium satisfies (8.42) at equality, and we denote the associated consumption allocation of an agent born at time $t \geq 1$ by $(\hat{c}_1, \hat{c}_2)$. It is also the case that $(\hat{c}_1, \hat{c}_2)$ maximizes an agent's utility subject to budget constraints (8.28) and (8.29). The consolidation of these two constraints yield

$$c_1 + \frac{z}{n} c_2 \leq y_1 + \frac{z}{n} y_2 + \frac{z}{n} \frac{(z-1)}{N_t} \frac{M_t}{p_{t+1}}, \tag{8.43}$$

where we have used the stationary rate or return in (8.32), $p_t/p_{t+1} = n/z$. After also invoking $zM_t = M_{t+1}$, $n = N_{t+1}/N_t$, and the equilibrium condition $M_{t+1}/(p_{t+1}N_{t+1}) = \hat{m}$, expression (8.43) simplifies to

$$c_1 + \frac{z}{n} c_2 \leq y_1 + \frac{z}{n} y_2 + (z-1)\hat{m}. \tag{8.44}$$

To prove the statement about necessity, Figure 8.6 depicts the two curves (8.42) and (8.44) when condition $z \leq 1$ fails to hold, i.e., we assume that $z > 1$.

The point that maximizes utility subject to (8.42) is denoted $(\overline{c}_1, \overline{c}_2)$. Transitivity of preferences and the fact that the slope of budget line (8.44) is flatter than that of (8.42) imply that $(\hat{c}_1, \hat{c}_2)$ lies southeast of $(\overline{c}_1, \overline{c}_2)$. By revealed preference, then, $(\overline{c}_1, \overline{c}_2)$ is preferred to $(\hat{c}_1, \hat{c}_2)$ and all generations born in period $t \geq 1$ are better off under the allocation $\overline{C}$. The initial old generation can also be made better off under this alternative allocation since it is feasible to strictly increase their consumption,

$$\overline{c}_1^0 = y_2 + n(y_1 - \overline{c}_1^1) > y_2 + n(y_1 - \hat{c}_1^1) = \hat{c}_1^0.$$

Thus, we have established that $z \leq 1$ is necessary for the optimality of the stationary monetary equilibrium $\hat{m}$.

To prove sufficiency, note that (8.31), (8.32) and $z \leq 1$ imply that

$$\theta(\hat{c}_1, \hat{c}_2) = \frac{n}{z} \geq n.$$

We can then construct an argument that is analogous to the sufficiency part of the proof to the preceding proposition. ▮

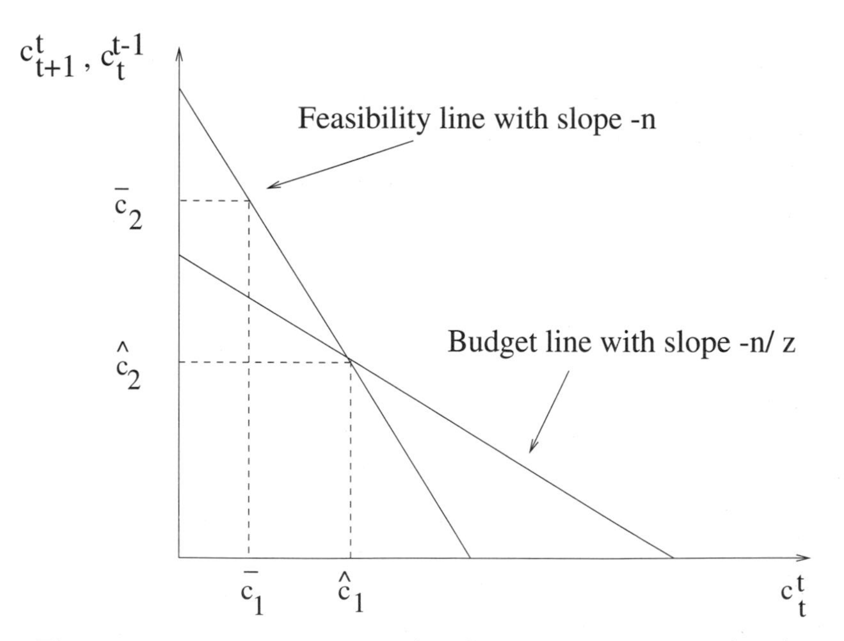


**Figure 8.6** The feasibility line (8.42) and the budget line (8.44) when $z > 1$. The consumption allocation in the monetary equilibrium is $(\hat{c}_1, \hat{c}_2)$, and the point that maximizes utility subject to the feasibility line is denoted $(\overline{c}_1, \overline{c}_2)$.

As pointed out by Wallace (1980), the proposition implies no connection between the path of the price level in a $\hat{m}$ equilibrium and the optimality of that equilibrium. Thus there may be an optimal monetary equilibrium with positive inflation — for example, if $\theta_{\text{aut}} < n < z \leq 1$ – and there may be a nonoptimal monetary equilibrium with a constant price level — for example, if $z = n > 1 > \theta_{\text{aut}}$. What counts is the nominal quantity of fiat money. The proposition suggests that the quantity of money should not be increased. In particular, if $z \leq 1$, then an optimal $\hat{m}$ equilibrium exists whenever the nonmonetary equilibrium is nonoptimal.

### *Balasko-Shell criterion for optimality*

For the case of constant population, Balasko and Shell (1980) have established a convenient general criterion for testing whether allocations are optimal. Balasko and Shell permit diversity among agents in terms of endowments $[w_t^{th}, w_{t+1}^{th}$ and utility functions $u^{th}(c_t^{th}, c_{t+1}^{th})$, where $w_s^{th}$ is the time $s$ endowment of an agent named $h$ who is born at $t$ and $c_s^{th}$ is the time $s$ consumption of agent named $h$ born at $t$. Balasko and Shell assume fixed populations of types $h$ over time. They impose several kinds of technical conditions that serve to rule out possible pathologies. The two main ones are these. First, they assume that indifference curves have neither flat parts nor kinks, and they also rule out indifference curves with flat parts or kinks as limits of sequences of indifference curves for given $h$ as $t \to \infty$. Second, they assume that the aggregate endowments $\sum_h (w_t^{th} + w_t^{t-1,h})$ are uniformly bounded from above and that there exists an $\epsilon > 0$ such that $w_t^{sh} > \epsilon$ for all $s, h$ and $t \in \{s, s+1\}$. They consider consumption allocations uniformly bounded away from the axes. With these conditions, Balasko and Shell consider the class of allocations in which all young agents at $t$ share a common marginal rate of substitution $1 + r_t \equiv u_1^{th}(c_t^{th}, c_{t+1}^{th}) / u_2^{th}(c_t^{th}, c_{t+1}^{th})$ and in which all of the endowments are consumed. Then Balasko and Shell show that an allocation is Pareto optimal if and only if

$$\sum_{t=1}^{\infty} \prod_{s=1}^{t} [1 + r_s] = +\infty, \tag{8.45}$$

that is, if and only if the infinite sum of $t$-period gross interest rates, $\prod_{s=1}^{t} [1+r_s]$, diverges.

The Balasko-Shell criterion for optimality succinctly summarizes the sense in which low-interest-rate economies are not optimal. We have already encountered repeated examples of the situation that, before an equilibrium with valued currency can exist, the equilibrium without valued currency must be a low-interest-rate economy in just the sense identified by Balasko-Shell's criterion, (8.45). Furthermore, by applying the Balasko-Shell criterion, (8.45), or by applying generalizations of it to allow for a positive net growth rate of population $n$, it can be shown that, among equilibria with valued currency, only equilibria with high rates of return on currency are optimal.

## Within generation heterogeneity

This section describes an overlapping generations model that has within-generation heterogeneity of endowments. We shall follow Sargent and Wallace (1982) and Smith (1988) and use this model as a vehicle for talking about some issues in monetary theory that require a setting in which government-issued currency coexists with and is a more or less good substitute for private IOU's.

We now assume that within each generation born at $t \geq 1$, there are $J$ groups agents. There is a constant number $N_j$ of group $j$ agents. Agents of group $j$ are endowed with $w_1(j)$ when young and $w_2(j)$ when old. The saving function of a household of group $j$ born at time $t$ solves the time $t$ version of problem (8.21). We denote this savings function $f(R_t, j)$. If we assume that all households of generation $t$ have preferences $U^t(c^t) = \ln c_t^t + \ln c_{t+1}^t$, the saving function is

$$f(R_t, j) = .5 \left( w_1(j) - \frac{w_2(j)}{R_t} \right).$$

At $t = 1$, there are old people who are endowed in the aggregate with $H = H(0)$ units of an inconvertible currency.

For example, assume that $J = 2$, that $(w_1(1), w_2(1)) = (\alpha, 0), (w_1(2), w_2(2)) = (0, \beta)$, where $\alpha > 0, \beta > 0$. The type 1 people are lenders, while the type 2 are borrowers. For the case of log preference we have the savings functions $f(R, 1) = \alpha/2, f(R, 2) = -\beta/(2R)$.

### Nonmonetary equilibrium

An equilibrium consists of sequences $(R, s_j$ of rates of return $R$ and savings rates for $j = 1, \ldots, J$ and $t \geq 1$that satisfy (1) $s_{tj} = f(R_t, j)$, and (2)

$\sum_{j=1}^{J} N_j f(R_t, j) = 0$. Condition (1) builds in household optimization; condition (2) says that aggregate net savings equals zero (borrowing equals lending).

For the case in which the endowments, preferences, and group sizes are constant across time, the interest rate is determined at the intersection of the aggregate savings function with the $R$ axis, depicted as $R_1$ in Figure 8.7. No intergenerational transfers occur in the nonmonetary equilibrium. The equilibrium consists of a sequence of separate two-period pure consumption loan economies of a type analyzed by Irving Fisher (1907).

*Monetary equilibrium*

In an equilibrium with valued fiat currency, at each date $t \geq 1$ the old receive goods from the young in exchange for the currency stock $H$. For any variable $x$, $x = \{x_t\}_{t=0}^{\infty}$. An equilibrium with valued fiat money is a set of sequences $R, p, s_j, \forall j$ such that (1) $p$ is a positive sequence, (2) $R_t = p_t/p_{t+1}$, (3) $s_{jt} = f(R_t, j)$, and (4) $\sum_{j=1}^{J} N_j f(R_t, j) = \frac{H}{p_t}$. Condition (1) states that currency is valued at all dates. Condition (2) states that currency and consumption loans are perfect substitutes. Condition (3) requires that savings decisions are optimal. Condition (4) equates the net savings of the young (the left side) to the net dissaving of the old (the right side). The old supply currency inelastically.

We can determine a stationary equilibrium graphically. A stationary equilibrium satisfies $p_t = p$ for all $t$, which implies $R = 1$ for all $t$. Thus, if it exists, a stationary equilibrium solves

$$\sum_{j=1}^{J} N_j f(1, j) = \frac{H}{p} \tag{8.46}$$

for a positive price level. See Figure 8.7. Evidently, a stationary monetary equilibrium exists if the net savings of the young are positive for $R = 1$.

For the special case of logarithmic preferences and two classes, the aggregate savings function of the young is time-invariant and equal to

$$\sum_j f(R, j) = .5(N_1 \alpha - N_2 \frac{\beta}{R}).$$

Note that the equilibrium condition (8.46) can be written

$$.5 N_1 \alpha = .5 N_2 \frac{\beta}{R} + \frac{H}{p}.$$

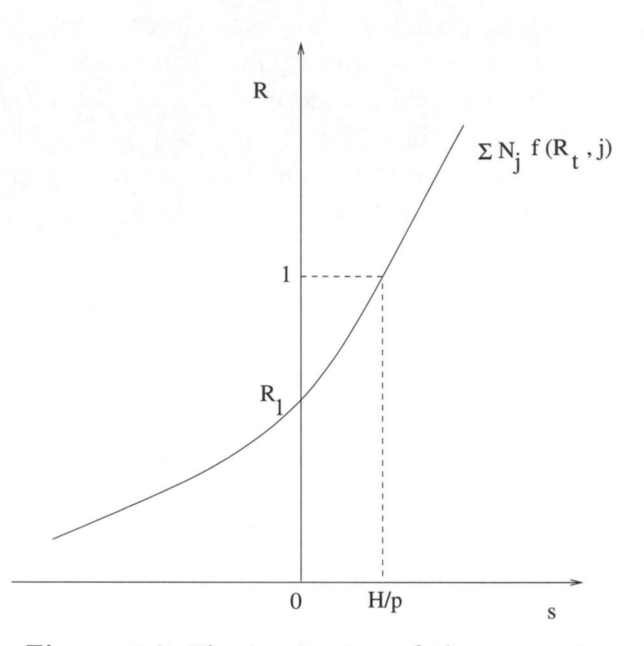


**Figure 8.7** The intersection of the aggregate savings function with a horizontal line at $R = 1$ determines a stationary equlibrium value of the price level, if positive.

The left side is the demand for savings or the demand for "currency" while the right side is the supply, consisting of privately issued IOU's (the first term) and government-issued currency. The right side is thus an abstract version of M1, which is a sum of privately issued IOU's (demand deposits) and government-issued reserves and currency.

*Nonstationary equilibria*

Mathematically, the equilibrium conditions for the model with log preferences and two groups have the same structure as the model analyzed previously in equations (8.17), (8.18), with simple reinterpretations of parameters. We leave it to the reader here and in an exercise to show that if there exists a stationary equilibrium with valued fiat currency, then there exists a continuum of equilibria with valued fiat currency, all but one of which have the real value of government currency approaching zero asymptotically. A linear difference equation like (8.17) supports this conclusion.

*The real bills doctrine*

In 19th century Europe and the early days of the Federal Reserve System in the U.S., central banks conducted open market operations not by purchasing government securities but by purchasing safe (risk-free) short-term private IOU's. We now introduce this old-fashioned type of open market operation. The government can issue additional currency each period. It uses the proceeds exclusively to purchase private IOU's (make loans to private agents). These open market operations are subject to the sequence of restrictions

$$L_t = R_{t-1} L_{t-1} + \frac{H_t - H_{t-1}}{p_t} \tag{8.47}$$

for $t \geq 1$ and $H_0 = H$ given, $L_0 = 0$. With the government injecting currency and purchasing loans in this way each period, the equilibrium condition in the loan market becomes

$$\sum_{j=1}^{J} N_j f(R_t, j) + L_t = \frac{H_{t-1}}{p_t} + \frac{H_t - H_{t-1}}{p_t} \tag{8.48}$$

where the first term on the right is the real dissaving of the old at $t$ (their real balances) and the second term is the real value of the new money printed by the monetary authority to finance purchases of private IOU's issued by the young at $t$. The left side is the net savings of the young plus the savings of the government.

Under several guises, the effects of open market operations like this have concerned monetary economists for centuries.[7] We can state the following proposition:

IRRELEVANCE OF OPEN MARKET OPERATIONS: Open market operations are irrelevant: any positive sequences $\{L_t, H_t\}_{t=0}^{\infty}$ that satisfy the constraint (8.47) are associated with the same equilibrium allocation, interest rate, and price level sequences.

PROOF: We can write the equilibrium condition (8.48) as

$$\sum_{j=1}^{J} N_j f(R_t, j) + L_t = \frac{H_t}{p_t}. \tag{8.49}$$

[7] One version of the issue concerned the effects on the price level of allowing banks to issue private bank notes. Notice that there is nothing in our setup that makes us take seriously that the notes $H_t$ are issued by the government. They could also be interpreted as being issued by a private bank.

For $t \geq 1$, iterating (8.47) once and using $R_{t-1} = \frac{p_{t-1}}{p_t}$ gives

$$L_t = R_{t-1} R_{t-2} L_{t-2} + \frac{H_t - H_{t-2}}{p_t}$$

Iterating back to time 0 and using $L_0 = 0$ gives

$$L_t = \frac{H_t - H_0}{p_t} \tag{8.50}$$

Substituting (8.50) into (8.49) gives

$$\sum_{j=1}^{J} N_j f(R_t, j) = \frac{H_0}{p_t}. \tag{8.51}$$

This is the same equilibrium condition in the economy with no open market operations, i.e., the economy with $L_t \equiv 0$ for all $t \geq 1$. Any price level and rate of return sequence that solves (8.51) also solves (8.48) for any $L_t$ sequence satisfying (8.47). ∎

This proposition captures some of the spirit of Adam Smith's real bills doctrine. According to that doctrine, if the government issues notes to purchase safe evidences of private indebtedness, it is not inflationary. Sargent and Wallace (1982) extend this discussion to settings in which the money market is separated from the credit market by some legal restrictions that inhibit intermediation. Then open market operations are no longer irrelevant because they can be used partially to undo the legal restrictions. Sargent and Wallace show how those legal restrictions can help stabilize the price level at a cost in terms of economic efficiency. Kahn and Roberds (1998) apply an extension of this setting to regulation issues for electronic payments systems.

## *Gift giving equilibrium*

Michihiro Kandori (1992) and Lones Smith (1992) have used ideas from the literature on reputation (see chapter 16) to study whether there exist history-dependent sequences of gifts that support an optimal allocation. Their idea is to set up the economy as a game played with a sequence of players. We briefly describe a gift-giving game for an overlapping generations economy in which voluntary intergenerational gifts supports an optimal allocation.

Suppose that the consumption of the initial old is

$$c_1^0 = y_1^0 + s_1^1$$

and the utility of each young agent is

$$u(y_i^i - s_i^i) + u(y_{i+1}^i + s_{i+1}^{i+1}), \quad i \geq 1 \tag{8.52}$$

where $s_i^i \geq 0$ is the gift from a young person at $i$. Suppose that the endowment pattern is $y_i^i = 1 - \epsilon, y_{i+1}^i = \epsilon$, where $\epsilon \in (0, .5)$.

Consider the following system of expectations, to which a young person chooses whether to conform:

$$s_i^i = \begin{cases} .5 - \epsilon & \text{if } v_i = \overline{v}; \\ 0 & \text{otherwise.} \end{cases} \tag{8.53a}$$

$$v_{i+1} = \begin{cases} \overline{v} & \text{if } v_i = \overline{v} \text{ and } s_i^i = .5 - \epsilon; \\ \underline{v} & \text{otherwise.} \end{cases} \tag{8.53b}$$

Here we are free to take $\overline{v} = 2u(.5)$ and $\underline{v} = u(1-\epsilon)+u(\epsilon)$. These are "promised utilities." We make them serve as "state variables" that summarize the history of intergenerational gift giving. To start things off, we need an initial value $v_1$. Equations (8.53) act as the transition laws that young agents face in choosing $\tilde{s}$ in (8.52).

An initial condition $v_1$ and the rule (8.53) is a system of expectations that tells the young person of each generation what he is expected to give. His gift is immediately handed over to an old person. A system of expectations is called an *equilibrium* if for each $i \geq 1$, each young agent chooses to conform.

We can immediately compute two equilibrium systems of expectations. The first one is the "autarky" equilibrium: give nothing yourself and expect all future generations to give nothing. To represent this equilibrium within equations (8.53), set $v_1 \neq \overline{v}$. It is easy to verify that each young person will confirm what is expected of him in this equilibrium. Given that future generations will not give, each young person chooses not to give.

For the second equilibrium, set $v_1 = \overline{v}$. Here each household chooses to give the expected amount, because failure to do so causes the next generation of young people not to give; whereas affirming the expectation to give passes that expectation along to the next generation, which affirms it in turn. Each of these equilibria is credible, in the sense of subgame perfection, to be studied extensively in chapter 16.

Narayana Kocherlakota (1998) has compared gift-giving and monetary equilibria in a variety of environments and has used the comparison to provide a precise sense in which money substitutes for memory.

## Concluding remarks

The overlapping generations model is a workhorse in analyses of public finance, welfare economics, and demographics. Diamond (1965) studies a version of the model with a neoclassical production function, and studies some fiscal policy issues within it. He shows that, depending on preference and productivity parameters, equilibria of the model can have too much capital; and that such capital overaccumulation can be corrected by having the government issue and perpetually roll over unbacked debt.[8]

Auerbach and Kotlikoff (1987) formulate a long-lived overlapping generations model with capital, labor, production, and various kinds of taxes. They use the model to study a host of fiscal issues. Rios-Rull (1994a) uses a calibrated overlapping generations growth model to examine the quantitative importance of market incompleteness for insuring against aggregate risk. See Attanasio (2000) for a review of theories and evidence about consumption within life-cycle models.

Several authors in a 1980 volume edited by John Kareken and Neil Wallace argued through example that the overlapping generations model is useful for analyzing a variety of issues in monetary economics. We refer to that volume, McCandless and Wallace (1992), Champ and Freeman (1994), Brock (1990), and Sargent (1987b) for a variety of applications of the overlapping generations model to issues in monetary economics.

## Exercises

*Exercise 8.1* At each date $t \geq 1$, an economy consists of overlapping generations of a constant number $N$ of two-period-lived agents. Young agents born in $t$ have preferences over consumption streams of a single good that are ordered by $u(c_t^t) + u(c_{t+1}^t)$, where $u(c) = c^{1-\gamma}/(1-\gamma)$, and where $c_t^i$ is the consumption of an agent born at $i$ in time $t$. It is understood that $\gamma > 0$, and that when $\gamma = 1$, $u(c) = \ln c$. Each young agent born at $t \geq 1$ has identical preferences and

[8] Abel, Mankiw, Summers, and Zeckhauser (1989) propose an empirical test of whether there is capital overaccumulation in the U.S. economy, and conclude that there is not.

endowment pattern $(w_1, w_2)$, where $w_1$ is the endowment when young and $w_2$ is the endowment when old. Assume $0 < w_2 < w_1$. In addition, there are some initial old agents at time 1 who are endowed with $w_2$ of the time-1 consumption good, and who order consumption streams by $c_1^0$. The initial old (i.e., the old at $t = 1$) are also endowed with $M$ units of unbacked fiat currency. The stock of currency is constant over time.

**a.** Find the saving function of a young agent.

**b.** Define an equilibrium with valued fiat currency.

**c.** Define a stationary equilibrium with valued fiat currency.

**d.** Compute a stationary equilibrium with valued fiat currency.

**e.** Describe how many equilibria with valued fiat currency there are. (You are not being asked to compute them.)

**f.** Compute the limiting value as $t \to +\infty$ of the rate of return on currency in each of the nonstationary equilibria with valued fiat currency. Justify your calculations.

*Exercise 8.2* Consider an economy with overlapping generations of a constant population of an even number $N$ of two-period-lived agents. New young agents are born at each date $t \geq 1$. Half of the young agents are endowed with $w_1$ when young and 0 when old. The other half are endowed with 0 when young and $w_2$ when old. Assume $0 < w_2 < w_1$. Preferences of all young agents are as in problem 1, with $\gamma = 1$. Half of the $N$ initial old are endowed with $w_2$ units of the consumption good and half are endowed with nothing. Each old person orders consumption streams by $c_1^0$. Each old person at $t = 1$ is endowed with $M$ units of unbacked fiat currency. No other generation is endowed with fiat currency. The stock of fiat currency is fixed over time.

**a.** Find the saving function of each of the two types of young person for $t \geq 1$.

**b.** Define an equilibrium without valued fiat currency. Compute all such equilibria.

**c.** Define an equilibrium with valued fiat currency.

**d.** Compute all the (nonstochastic) equilibria with valued fiat currency.

**e.** Argue that there is a unique stationary equilibrium with valued fiat currency.

**f.** How are the various equilibria with valued fiat currency ranked by the Pareto criterion?

*Exercise 8.3* Take the economy of exercise 8.1, but make one change. Endow the initial old with a tree that yields a constant dividend of $d > 0$ units of the consumption good for each $t \geq 1$.

**a.** Compute all the equilibria with valued fiat currency.

**b.** Compute all the equilibria without valued fiat currency.

**c.** If you want, you can answer both parts of this question in the context of the following particular numerical example: $w_1 = 10, w_2 = 5, d = .000001$.

*Exercise 8.4* Take the economy of exercise 8.1 and make the following two changes. First, assume that $\gamma = 1$. Second, assume that the number of young agents born at $t$ is $N(t) = nN(t-1)$, where $N(0) > 0$ is given and $n \geq 1$. Everything else about the economy remains the same.

**a.** Compute an equilibrium without valued fiat money.

**b.** Compute a stationary equilibrium with valued fiat money.

*Exercise 8.5* Consider an economy consisting of overlapping generations of two-period-lived consumers. At each date $t \geq 1$, there are born $N(t)$ identical young people each of whom is endowed with $w_1 > 0$ units of a single consumption good when young and $w_2 > 0$ units of the consumption good when old. Assume that $w_2 < w_1$. The consumption good is not storable. The population of young people is described by $N(t) = nN(t-1)$, where $n > 0$. Young people born at $t$ rank utility streams according to $\ln(c_t^t) + \ln(c_{t+1}^t)$ where $c_t^i$ is the consumption of the time-$t$ good of an agent born in $i$. In addition, there are $N(0)$ old people at time 1, each of whom is endowed with $w_2$ units of the time-1 consumption good. The old at $t = 1$ are also endowed with one unit of unbacked pieces of infinitely durable but intrinsically worthless pieces of paper called fiat money.

**a.** Define an equilibrium without valued fiat currency. Compute such an equilibrium.

**b.** Define an equilbrium with valued fiat currency.

**c.** Compute all equilibria with valued fiat currency.

**d.** Find the limiting rates of return on currency as $t \to +\infty$ in each of the equilibria that you found in part c. Compare them with the one-period interest rate in the equilibrium in part a.

**e.** Are the equilibria in part c ranked according to the Pareto criterion?

*Exercise 8.6* **Exchange rate determinacy**

The world consists of two economies, named $i = 1, 2$, which except for their governments' policies are "copies" of one another. At each date $t \geq 1$, there is a single consumption good, which is storable, but only for rich people. Each economy consists of overlapping generations of two-period-lived agents. For each $t \geq 1$, in economy $i$, $N$ poor people and $N$ rich people are born. Let $c_t^h(s), y_t^h(s)$ be the time $s$ (consumption, endowment) of a type-$h$ agent born at $t$. Poor agents are endowed $[y_t^h(t), y_t^h(t+1)] = (\alpha, 0)$; Rich agents are endowed $[y_t^h(t), y_t^h(t+1)] = (\beta, 0)$, where $\beta >> \alpha$. In each country, there are $2N$ initial old who are endowed in the aggregate with $H_i(0)$ units of an unbacked currency, and with $2N\epsilon$ units of the time-1 consumption good. For the rich people, storing $k$ units of the time-$t$ consumption good produces $Rk$ units of the time–$t+1$ consumption good, where $R > 1$ is a fixed gross rate of return on storage. Rich people can earn the rate of return $R$ either by storing goods or lending to either government by means of indexed bonds. We assume that poor people are prevented from storing capital or holding indexed government debt by the sort of denomination and intermediation restrictions described by Sargent and Wallace (1982).
For each $t \geq 1$, all young agents order consumption streams according to $\ln c_t^h(t) + \ln c_t^h(t+1)$.

For $t \geq 1$, the government of country $i$ finances a stream of purchases (to be thrown into the ocean) of $G_i(t)$ subject to the following budget constraint:

$$G_i(t) + RB_i(t-1) = B_i(t) + \frac{H_i(t) - H_i(t-1)}{p_i(t)} + T_i(t), \tag{1}$$

where $B_i(0) = 0$; $p_i(t)$ is the price level in country $i$; $T_i(t)$ are lump-sum taxes levied by the government on the *rich* young people at time $t$; $H_i(t)$ is the stock of $i$'s fiat currency at the end of period $t$; $B_i(t)$ is the stock of indexed government interest-bearing debt (held by the rich of either country). The government does not explicitly tax poor people, but might tax through an inflation tax. Each government levies a lump-sum tax of $T_i(t)/N$ on each young rich citizen of its own country.

Poor people in both countries are free to hold whichever currency they prefer. Rich people can hold debt of either government and can also store; storage and both government debts bear a constant gross rate of return $R$.

**a.** Define an *equilibrium* with valued fiat currencies (in both countries).

**b.** In a nonstochastic equilibrium, verify the following proposition: if an equilibrium exists in which both fiat currencies are valued, the exchange rate between the two currencies must be constant over time.

**c.** Suppose that government policy in each country is characterized by specified (exogenous) levels $G_i(t) = G_i, T_i(t) = T_i,\ B_i(t) = 0, \forall t \geq 1$. (The remaining elements of government policy adjust to satisfy the government budget constraints.) Assume that the exogenous components of policy have been set so that an equilibrium with two valued fiat currencies exists. Under this description of policy, show that the equilibrium exchange rate is indeterminate.

**d.** Suppose that government policy in each country is described as follows: $G_i(t) = G_i, T_i(t) = T_i, H_i(t+1) = H_i(1), B_i(t) = B_i(1)\ \forall t \geq 1$. Show that if there exists an equilibrium with two valued fiat currencies, the exchange rate is determinate.

**e.** Suppose that government policy in country 1 is specified in terms of exogenous levels of $s_1 = [H_1(t) - H_1(t-1)]/p_1(t)\ \forall t \geq 2$, and $G_1(t) = G_1\ \forall t \geq 1$. For country 2, government policy consists of exogenous levels of $B_2(t) = B_2(1), G_2(t) = G_2 \forall t \geq 1$. Show that if there exists an equilibrium with two valued fiat currencies, then the exchange rate is determinate.

### *Exercise 8.7* Credit controls

Consider the following overlapping generations model. At each date $t \geq 1$ there appear $N$ two-period-lived young people, said to be of generation $t$, who live and consume during periods $t$ and $(t+1)$. At time $t = 1$ there exist $N$ old people who are endowed with $H(0)$ units of paper "dollars," which they offer to supply inelastically to the young of generation 1 in exchange for goods. Let $p(t)$ be the price of the one good in the model, measured in dollars per time-$t$ good.

For each $t \geq 1$, $N/2$ members of generation $t$ are endowed with $y > 0$ units of the good at $t$ and 0 units at $(t+1)$, whereas the remaining $N/2$ members of generation $t$ are endowed with 0 units of the good at $t$ and $y > 0$ units when they are old. All members of all generations have the same utility function:

$$u[c_t^h(t), c_t^h(t+1)] = \ln c_t^h(t) + \ln c_t^h(t+1),$$

where $c_t^h(s)$ is the consumption of agent $h$ of generation $t$ in period $s$. The old at $t = 1$ simply maximize $c_0^h(1)$. The consumption good is nonstorable. The currency supply is constant through time, so $H(t) = H(0)$, $t \geq 1$.

**a.** Define a competitive equilibrium without valued currency for this model. Who trades what with whom?

**b.** In the equilibrium without valued fiat currency, compute competitive equilibrium values of the gross return on consumption loans, the consumption allocation of the old at $t = 1$, and that of the "borrowers" and "lenders" for $t \geq 1$.

**c.** Define a competitive equilibrium with valued currency. Who trades what with whom?

**d.** Prove that for this economy there does not exist a competitive equilibrium with valued currency.

**e.** Now suppose that the government imposes the restriction that $l_t^h(t)[1+r(t)] \geq -y/4$, where $l_t^h(t)[1+r(t)]$ represents claims on $(t+1)$–period consumption purchased (if positive) or sold (if negative) by household $h$ of generation $t$. This is a restriction on the amount of borrowing. For an equilibrium without valued currency, compute the consumption allocation and the gross rate of return on consumption loans.

**f.** In the setup of part e, show that there exists an equilibrium with valued currency in which the price level obeys the quantity theory equation $p(t) = qH(0)/N$. Find a formula for the undetermined coefficient $q$. Compute the consumption allocation and the equilibrium rate of return on consumption loans.

**g.** Are lenders better off in economy b or economy f? What about borrowers? What about the old of period 1 (generation 0)?

### *Exercise 8.8* Inside money and real bills

Consider the following overlapping generations model of two-period-lived people. At each date $t \geq 1$ there are born $N_1$ individuals of type 1 who are endowed with $y > 0$ units of the consumption good when they are young and zero units when they are old; there are also born $N_2$ individuals of type 2 who are endowed with zero units of the consumption good when they are young and $Y > 0$ units when they are old. The consumption good is nonstorable. At time $t = 1$, there are $N$ old people, all of the same type, each endowed with zero units of the consumption good and $H_0/N$ units of unbacked paper called "fiat currency." The populations of type 1 and 2 individuals, $N_1$ and $N_2$, remain constant for all $t \geq 1$. The young of each generation are identical in preferences and maximize the utility function $\ln c_t^h(t) + \ln c_t^h(t+1)$ where $c_t^h(s)$ is consumption in the $s$th period of a member $h$ of generation $t$.

**a.** Consider the equilibrium without valued currency (that is, the equilibrium in which there is no trade between generations). Let $[1 + r(t)]$ be the gross rate of return on consumption loans. Find a formula for $[1 + r(t)]$ as a function of $N_1, N_2, y$, and $Y$.

**b.** Suppose that $N_1, N_2, y$, and $Y$ are such that $[1+r(t)] > 1$ in the equilibrium without valued currency. Then prove that there can exist no quantity-theory-style equilibrium where fiat currency is valued and where the price level $p(t)$ obeys the quantity theory equation $p(t) = q \cdot H_0$, where $q$ is a positive constant and $p(t)$ is measured in units of currency per unit good.

**c.** Suppose that $N_1, N_2, y$, and $Y$ are such that in the nonvalued-currency equilibrium $1 + r(t) < 1$. Prove that there exists an equilibrium in which fiat currency is valued and that there obtains the quantity theory equation $p(t) = q \cdot H_0$, where $q$ is a constant. Construct an argument to show that the equilibrium with valued currency is not Pareto superior to the nonvalued-currency equilibrium.

**d.** Suppose that $N_1, N_2, y$, and $Y$ are such that, in the preceding nonvalued-currency economy, $[1 + r(t)] < 1$, there exists an equilibrium in which fiat currency is valued. Let $\bar{p}$ be the stationary equilibrium price level in that economy. Now consider an alternative economy, identical with the preceding one in all respects except for the following feature: a government each period purchases a constant amount $L_g$ of consumption loans and pays for them by issuing debt on itself, called "inside money" $M_I$, in the amount $M_I(t) = L_g \cdot p(t)$. The government never retires the inside money, using the proceeds of the loans to finance new purchases of consumption loans in subsequent periods. The quantity of outside money, or currency, remains $H_0$, whereas the "total high-power money" is now $H_0 + M_I(t)$.

(i) Show that in this economy there exists a valued-currency equilibrium in which the price level is constant over time at $p(t) = \bar{p}$, or equivalently, with $\bar{p} = qH_0$ where $q$ is defined in part c.

(ii) Explain why government purchases of private debt are not inflationary in this economy.

(iii) In standard macroeconomic models, once-and-for-all government open-market operations in private debt normally affect real variables and/or price level. What accounts for the difference between those models and the one in this exercise?

*Exercise 8.9* **Social security and the price level**

Consider an economy ("economy I") that consists of overlapping generations of two-period-lived people. At each date $t \geq 1$ there are born a constant number $N$ of young people, who desire to consume both when they are young, at $t$, and when they are old, at $(t+1)$. Each young person has the utility function $\ln c_t(t) + \ln c_t(t+1)$, where $c_s(t)$ is time-$t$ consumption of an agent born at $s$. For all dates $t \geq 1$, young people are endowed with $y > 0$ units of a single nonstorable consumption good when they are young and zero units when they are old. In addition, at time $t = 1$ there are $N$ old people endowed in the aggregate with $H$ units of unbacked fiat currency. Let $p(t)$ be the nominal price level at $t$, denominated in dollars per time-$t$ good.

**a.** Define and compute an equilibrium with valued fiat currency for this economy. Argue that it exists and is unique. Now consider a second economy ("economy II") that is identical to economy I except that economy II possesses a social security system. In particular, at each date $t \geq 1$, the government taxes $\tau > 0$ units of the time-$t$ consumption good away from each young person and at the same time gives $\tau$ units of the time-$t$ consumption good to each old person then alive.

**b.** Does economy II possess an equilibrium with valued fiat currency? Describe the restrictions on the parameter $\tau$, if any, that are needed to ensure the existence of such an equilibrium.

**c.** If an equilibrium with valued fiat currency exists, is it unique?

**d.** Consider the *stationary* equilibrium with valued fiat currency. Is it unique? Describe how the value of currency or price level would vary across economies with differences in the size of the social security system, as measured by $\tau$.

*Exercise 8.10* **Seignorage**

Consider an economy consisting of overlapping generations of two-period-lived agents. At each date $t \geq 1$, there are born $N_1$ "lenders" who are endowed with $\alpha > 0$ units of the single consumption good when they are young and zero units when they are old. At each date $t \geq 1$, there are also born $N_2$ "borrowers" who are endowed with zero units of the consumption good when they are young and $\beta > 0$ units when they are old. The good is nonstorable, and $N_1$ and $N_2$ are constant through time. The economy starts at time 1, at which time there are $N$ old people who are in the aggregate endowed with $H(0)$ units of unbacked, intrinsically worthless pieces of paper called dollars. Assume that $\alpha, \beta, N_1$, and $N_2$ are such that

$$\frac{N_2 \beta}{N_1 \alpha} < 1.$$

Assume that everyone has preferences

$$u[c_t^h(t), c_t^h(t+1)] = \ln c_t^h(t) + \ln c_t^h(t+1),$$

where $c_t^h(s)$ is consumption of time $s$ good of agent $h$ born at time $t$.

**a.** Compute the equilibrium interest rate on consumption loans in the equilibrium without valued currency.

**b.** Construct a *brief* argument to establish whether or not the equilibrium without valued currency is Pareto optimal.

The economy also contains a government that purchases and destroys $G_t$ units of the good in period $t$, $t \geq 1$. The government finances its purchases entirely by currency creation. That is, at time $t$,

$$G_t = \frac{H(t) - H(t-1)}{p(t)},$$

where $[H(t) - H(t-1)]$ is the additional dollars printed by the government at $t$ and $p(t)$ is the price level at $t$. The government is assumed to increase $H(t)$ according to

$$H(t) = zH(t-1), \qquad z \geq 1,$$

where $z$ is a constant for all time $t \geq 1$.

At time $t$, old people who carried $H(t-1)$ dollars between $(t-1)$ and $t$ offer these $H(t-1)$ dollars in exchange for time-$t$ goods. Also at $t$ the government offers $H(t) - H(t-1)$ dollars for goods, so that $H(t)$ is the total supply of dollars at time $t$, to be carried over by the young into time $(t+1)$.

**c.** Assume that $1/z > N_2\beta/N_1\alpha$. Show that under this assumption there exists a continuum of equilibria with valued currency.

**d.** Display the unique stationary equilibrium with valued currency in the form of a "quantity theory" equation. Compute the equilibrium rate of return on currency and consumption loans.

**e.** Argue that if $1/z < N_2\beta/N_1\alpha$, then there exists no valued-currency equilibrium. Interpret this result. (*Hint:* Look at the rate of return on consumption loans in the equilibrium without valued currency.)

**f.** Find the value of $z$ that *maximizes* the government's $G_t$ in a stationary equilibrium. Compare this with the largest value of $z$ that is compatible with the existence of a valued-currency equilibrium.

*Exercise 8.11* **Unpleasant monetarist arithmetic**

Consider an economy in which the aggregate demand for government currency for $t \geq 1$ is given by $[M(t)p(t)]^d = g[R_1(t)]$, where $R_1(t)$ is the gross rate of return on currency between $t$ and $(t+1)$, $M(t)$ is the stock of currency at $t$, and $p(t)$ is the value of currency in terms of goods at $t$ (the reciprocal of the price level). The function $g(R)$ satisfies

$$
\begin{aligned}
(1) \quad & g(R)(1-R) = h(R) > 0 \qquad \text{for } R \in (\underline{R}, 1), \\
& h(R) \leq 0 \qquad \text{for } R < \underline{R}, \qquad R \geq 1, \qquad \underline{R} > 0. \\
& h'(R) < 0 \qquad \text{for } R > R_m \\
& h'(R) > 0 \qquad \text{for } R < R_m \\
& h(R_m) > D, \qquad \text{where } D \text{ is a positive number to be defined shortly.}
\end{aligned}
$$

The government faces an infinitely elastic demand for its interest-bearing bonds at a constant-over-time gross rate of return $R_2 > 1$. The government finances a budget deficit $D$, defined as government purchases minus explicit taxes, that is constant over time. The government's budget constraint is

$$
(2) \qquad D = p(t)[M(t) - M(t-1)] + B(t) - B(t-1)R_2, \qquad t \geq 1,
$$

subject to $B(0) = 0, M(0) > 0$. In equilibrium,

$$
(3) \qquad M(t)p(t) = g[R_1(t)].
$$

The government is free to choose paths of $M(t)$ and $B(t)$, subject to equations (2) and (3).

**a.** Prove that, for $B(t) = 0$, for all $t > 0$, there exist two stationary equilibria for this model.

**b.** Show that there exist values of $B > 0$, such that there exist stationary equilibria with $B(t) = B$, $M(t)p(t) = Mp$.

**c.** Prove a version of the following proposition: among stationary equilibria, the lower the value of $B$, the lower the stationary rate of inflation consistent with equilibrium. (You will have to make an assumption about Laffer curve effects to obtain such a proposition.)

This problem displays some of the ideas used by Sargent and Wallace (1981). They argue that, under assumptions like those leading to the proposition stated in part c, the "looser" money is today [that is, the higher $M(1)$ and the lower $B(1)$], the lower the stationary inflation rate.

*Exercise 8.12* **Grandmont-Hall**

Consider a nonstochastic, one-good overlapping generations model consisting of two-period-lived young people born in each $t \geq 1$ and an initial group of old people at $t = 1$ who are endowed with $H(0) > 0$ units of unbacked currency at the beginning of period 1. The one good in the model is not storable. Let the aggregate first-period saving function of the young be time invariant and be denoted $f[1+r(t)]$ where $[1+r(t)]$ is the gross rate of return on consumption loans between $t$ and $(t+1)$. The saving function is assumed to satisfy $f(0) = -\infty$, $f'(1+r) > 0$, $f(1) > 0$.

Let the government pay interest on currency, starting in period 2 (to holders of currency between periods 1 and 2). The government pays interest on currency at a nominal rate of $[1+r(t)]p(t+1)/\bar{p}$, where $[1+r(t)]$ is the real gross rate of return on consumption loans, $p(t)$ is the price level at $t$, and $\bar{p}$ is a target price level chosen to satisfy

$$\bar{p} = H(0)/f(1).$$

The government finances its interest payments by printing new money, so that the government's budget constraint is

$$H(t+1) - H(t) = \left\{[1+r(t)]\frac{p(t+1)}{\bar{p}} - 1\right\} H(t), \qquad t \geq 1,$$

given $H(1) = H(0) > 0$. The gross rate of return on consumption loans in this economy is $1 + r(t)$. In equilibrium, $[1 + r(t)]$ must be at least as great as the real rate of return on currency

$$1 + r(t) \geq [1 + r(t)]p(t)/\bar{p} = [1 + r(t)]\frac{p(t+1)}{\bar{p}}\frac{p(t)}{p(t+1)}$$

with equality if currency is valued,

$$1 + r(t) = [1 + r(t)]p(t)/\bar{p}, \qquad 0 < p(t) < \infty.$$

The loan market-clearing condition in this economy is

$$f[1+r(t)] = H(t)/p(t).$$

**a.** Define an equilibrium.
**b.** Prove that there exists a unique monetary equilibrium in this economy and compute it.

*Exercise 8.13* **Bryant-Keynes-Wallace**

Consider an economy consisting of overlapping generations of two-period-lived agents. There is a constant population of $N$ young agents born at each date $t \geq 1$. There is a single consumption good that is not storable. Each agent born in $t \geq 1$ is endowed with $w_1$ units of the consumption good when young and with $w_2$ units when old, where $0 < w_2 < w_1$. Each agent born at $t \geq 1$ has identical preferences $\ln c_t^h(t) + \ln c_t^h(t+1)$, where $c_t^h(s)$ is time-$s$ consumption of agent $h$ born at time $t$. In addition, at time 1, there are alive $N$ old people who are endowed with $H(0)$ units of unbacked paper currency and who want to maximize their consumption of the time-1 good.

A government attempts to finance a constant level of government purchases $G(t) = G > 0$ for $t \geq 1$ by printing new base money. The government's budget constraint is

$$G = [H(t) - H(t-1)]/p(t),$$

where $p(t)$ is the price level at $t$, and $H(t)$ is the stock of currency carried over from $t$ to $(t+1)$ by agents born in $t$. Let $g = G/N$ be government purchases per young person. Assume that purchases $G(t)$ yield no utility to private agents.

**a.** Define a stationary equilibrium with valued fiat currency.
**b.** Prove that, for $g$ sufficiently small, there exists a stationary equilibrium with valued fiat currency.
**c.** Prove that, in general, if there exists one stationary equilibrium with valued fiat currency, with rate of return on currency $1 + r(t) = 1 + r_1$, then there exists at least one other stationary equilibrium with valued currency with $1 + r(t) = 1 + r_2 \neq 1 + r_1$.
**d.** Tell whether the equilibria described in parts b and c are Pareto optimal, among allocations among private agents of what is left after the government takes $G(t) = G$ each period. (A proof is not required here: an informal argument will suffice.)

Now let the government institute a forced saving program of the following form. At time 1, the government redeems the outstanding stock of currency

$H(0)$, exchanging it for government bonds. For $t \geq 1$, the government offers each young consumer the option of saving at least $F$ worth of time $t$ goods in the form of bonds bearing a constant rate of return $(1+r_2)$. A legal prohibition against private intermediation is instituted that prevents two or more private agents from sharing one of these bonds. The government's budget constraint for $t \geq 2$ is

$$G/N = B(t) - B(t-1)(1+r_2),$$

where $B(t) \geq F$. Here $B(t)$ is the saving of a young agent at $t$. At time $t=1$, the government's budget constraint is

$$G/N = B(1) - \frac{H(0)}{Np(1)},$$

where $p(1)$ is the price level at which the initial currency stock is redeemed at $t=1$. The government sets $F$ and $r_2$.

Consider stationary equilibria with $B(t) = B$ for $t \geq 1$ and $r_2$ and $F$ constant.

**e.** Prove that if $g$ is small enough for an equilibrium of the type described in part a to exist, then a stationary equilibrium with forced saving exists. (Either a graphical argument or an algebraic argument is sufficient.)

**f.** Given $g$, find the values of $F$ and $r_2$ that maximize the utility of a representative young agent for $t \geq 1$.

**g.** Is the equilibrium allocation associated with the values of $F$ and $(1+r_2)$ found in part f optimal among those allocations that give $G(t) = G$ to the government for all $t \geq 1$? (Here an informal argument will suffice.)

# 9
# *Ricardian Equivalence*

## *Borrowing limits and Ricardian equivalence*

This chapter studies whether the timing of taxes matters. Under some assumptions it does, and under others it does not. The Ricardian doctrine describes assumptions under which the timing of lump taxes does not matter. In this chapter, we will study how the timing of taxes interacts with restrictions on the ability of households to borrow. We study the issue in two equivalent settings: (1) an infinite horizon economy with an infinitely lived representative agent; and (2) an infinite horizon economy with a sequence of one-period-lived agents, each of whom cares about its immediate descendant. We assume that the interest rate is exogenously given. For example, the economy might be a small open economy that faces a given interest rate determined in the international capital market. Chapter 10 will describe a general equilibrium analysis of the Ricardian doctrine where the interest rate is determined within the model.

The key findings of the chapter are that in the infinite horizon model, Ricardian equivalence holds under what we earlier called the natural borrowing limit, but not under more stringent ones. The natural borrowing limit is the one that lets households borrow up to the capitalized value of their endowment sequences. These results have counterparts in the overlapping generations model, since that model is equivalent to an infinite horizon model with a no-borrowing constraint. In the overlapping generations model, the no-borrowing constraint translates into a requirement that bequests be nonnegative. Thus, in the overlapping generations model, the domain of the Ricardian proposition is restricted, at least relative to the infinite horizon model under the natural borrowing limit.

## *Infinitely lived–agent economy*

An economy consists of $N$ identical households each of whom orders a stream of consumption of a single good with preferences

$$\sum_{t=0}^{\infty} \beta^t u(c_t), \tag{9.1}$$

where $\beta \in (0,1)$ and $u(\cdot)$ is a strictly increasing, strictly concave, twice-differentiable one-period utility function. We impose the Inada condition

$$\lim_{c \downarrow 0} u'(c) = +\infty.$$

This condition is important because we will be stressing the feature that $c \geq 0$. There is no uncertainty. The household can invest in a single risk-free asset bearing a fixed gross one-period rate of return $R > 1$. The asset is either a risk-free loan to foreigners or to the government. At time $t$, the household faces the budget constraint

$$c_t + R^{-1} b_{t+1} \leq y_t + b_t, \tag{9.2}$$

where $b_0$ is given. Throughout this chapter, we assume that $R\beta = 1$. Here $\{y_t\}_{t=0}^{\infty}$ is a given nonstochastic nonnegative endowment sequence and $\sum_{t=0}^{\infty} \beta^t y_t < \infty$.

We shall investigate two alternative restrictions on asset holdings $\{b_t\}_{t=0}^{\infty}$. One is that $b_t \geq 0$ for all $t \geq 0$. This restriction states that the household can lend but not borrow. The alternative restriction permits the household to borrow, but only an amount that it is feasible to repay. To discover this amount, set $c_t = 0$ for all $t$ in formula (9.2) and solve forward for $b_t$ to get

$$\tilde{b}_t = -\sum_{j=0}^{\infty} R^{-j} y_{t+j}, \tag{9.3}$$

where we have ruled out Ponzi schemes by imposing the transversality condition

$$\lim_{T \to \infty} R^{-T} b_{T+1} = 0. \tag{9.4}$$

Following Aiyagari (1994), we call $\tilde{b}_t$ the *natural debt limit.* Even with $c_t = 0$, the consumer cannot repay more than $\tilde{b}_t$. Thus, our alternative restriction on assets is

$$b_t \geq \tilde{b}_t, \tag{9.5}$$

which is evidently weaker than $b_t \geq 0$.[1]

*Solution to consumption/savings decision*

Consider the household's problem of choosing $\{c_t, b_{t+1}\}_{t=0}^{\infty}$ to maximize expression (9.1) subject to (9.2) and $b_{t+1} \geq 0$ for all $t$. The first-order conditions for this problem are

$$u'(c_t) \geq \beta R u'(c_{t+1}), \quad \forall t \geq 0; \tag{9.6a}$$

and

$$u'(c_t) > \beta R u'(c_{t+1}) \quad \text{implies} \quad b_{t+1} = 0. \tag{9.6b}$$

Because $\beta R = 1$, these conditions and the constraint (9.2) imply that $c_{t+1} = c_t$ when $b_{t+1} > 0$; but when the consumer is borrowing constrained, $b_{t+1} = 0$ and $y_t + b_t = c_t < c_{t+1}$. The solution evidently depends on the $\{y_t\}$ path, as the following examples illustrate.

*Example 1* Assume $b_0 = 0$ and the endowment path $\{y_t\}_{t=0}^{\infty} = \{y_h, y_l, y_h, y_l, \ldots\}$, where $y_h > y_l > 0$. The present value of the household's endowment is

$$\sum_{t=0}^{\infty} \beta^t y_t = \sum_{t=0}^{\infty} \beta^{2t}(y_h + \beta y_l) = \frac{y_h + \beta y_l}{1 - \beta^2}.$$

The annuity value $\bar{c}$ that has the same present value as the endowment stream is given by

$$\frac{\bar{c}}{1-\beta} = \frac{y_h + \beta y_l}{1 - \beta^2}, \qquad \text{or} \quad \bar{c} = \frac{y_h + \beta y_l}{1 + \beta}.$$

The solution to the household's optimization problem is the constant consumption stream $c_t = \bar{c}$ for all $t \geq 0$, and using the budget constraint (9.2), we can back out the associated savings scheme; $b_{t+1} = (y_h - y_l)/(1 + \beta)$ for even $t$, and $b_{t+1} = 0$ for odd $t$. The consumer is never borrowing constrained.[2]

*Example 2* Assume $b_0 = 0$ and the endowment path $\{y_t\}_{t=0}^{\infty} = \{y_l, y_h, y_l, y_h, \ldots\}$, where $y_h > y_l > 0$. The solution is $c_0 = y_l$ and $b_1 = 0$, and from period 1 onward, the solution is the same as in example 1. Hence, the consumer is borrowing constrained the first period.[3]

---

[1] We encountered a more general version of equation (9.5) in chapter 7 when we discussed Arrow securities.

[2] Note $b_t = 0$ does not imply that the consumer is borrowing constrained. He is borrowing constrained if the Lagrange multiplier on the constraint $b_t \geq 0$ is not zero.

[3] Examples 1 and 2 illustrate a general result in chapter 13. Given a borrowing constraint and a non-stochastic endowment stream, the impact of the borrowing constraint will not vanish

*Example 3* Assume $b_0 = 0$ and $y_t = \lambda^t$ where $1 < \lambda < R$. Notice that $\lambda\beta < 1$. The solution with the borrowing constraint $b_t \geq 0$ is $c_t = \lambda^t, b_t = 0$ for all $t \geq 0$. The consumer is always borrowing constrained.

*Example 4* Assume the same $b_0$ and endowment sequence as in example 3, but now impose only the natural borrowing constraint (9.5). The present value of the household's endowment is

$$\sum_{t=0}^{\infty} \beta^t \lambda^t = \frac{1}{1-\lambda\beta}.$$

The household's budget constraint for each $t$ is satisfied at a constant consumption level $\hat{c}$ satisfying

$$\frac{\hat{c}}{1-\beta} = \frac{1}{1-\lambda\beta}, \qquad \text{or} \quad \hat{c} = \frac{1-\beta}{1-\lambda\beta}.$$

Substituting this consumption rate into formula (9.2) and solving forward gives

$$b_t = \frac{1-\lambda^t}{1-\beta\lambda}. \tag{9.7}$$

The consumer issues more and more debt as time passes, and uses his rising endowment to service it. The consumer's debt always satisfies the natural debt limit at $t$, namely, $\tilde{b}_t = -\lambda^t/(1-\beta\lambda)$.

*Example 5* Take the specification of example 4, but now impose $\lambda < 1$. Note that the solution (9.7) implies $b_t \geq 0$, so that the constant consumption path $c_t = \hat{c}$ in example 4 is now the solution even if the borrowing constraint $b_t \geq 0$ is imposed.

## *Government*

Add a government to the model. The government purchases a stream $\{g_t\}_{t=0}^{\infty}$ per household and imposes a stream of lump-sum taxes $\{\tau_t\}_{t=0}^{\infty}$ on the household, subject to the sequence of budget constraints

$$B_t + g_t = \tau_t + R^{-1} B_{t+1}, \tag{9.8}$$

---

until the household reaches the period with the highest annuity value of the remainder of the endowment stream.

where $B_t$ is one-period debt due at $t$, denominated in the time $t$ consumption good, that the government owes the households or foreign investors. Notice that we allow the government to borrow, even though in one of the preceding specifications, we did not permit the household to borrow. (If $B_t < 0$, the government lends to households or foreign investors.) Solving the government's budget constraint forward gives the intertemporal constraint

$$B_t = \sum_{j=0}^{\infty} R^{-j}(\tau_{t+j} - g_{t+j}) \tag{9.9}$$

for $t \geq 0$, where we have ruled out Ponzi schemes by imposing the transversality condition

$$\lim_{T\to\infty} R^{-T} B_{T+1} = 0.$$

*Effect on household*

We must now deduct $\tau_t$ from the household's endowment in (9.2),

$$c_t + R^{-1} b_{t+1} \leq y_t - \tau_t + b_t. \tag{9.10}$$

Solving this tax-adjusted budget constraint forward and invoking transversality condition (9.4) yield

$$b_t = \sum_{j=0}^{\infty} R^{-j}(c_{t+j} + \tau_{t+j} - y_{t+j}). \tag{9.11}$$

The natural debt limit is obtained by setting $c_t = 0$ for all $t$ in (9.11),

$$\tilde{b}_t = \sum_{j=0}^{\infty} R^{-j}(\tau_{t+j} - y_{t+j}). \tag{9.12}$$

Notice how taxes affect $\tilde{b}_t$ [compare equations (9.3) and (9.12)].

We use the following definition:

DEFINITION: Given initial conditions $(b_0, B_0)$, an *equilibrium* is a household plan $\{c_t, b_{t+1}\}$ and a government policy $\{g_t, \tau_t, B_{t+1}\}$ such that (a) the government plan satisfies the government budget constraint (9.8), and (b) given $\{\tau_t\}$, the household's plan is optimal.

We can now state a Ricardian proposition under the natural debt limit.

PROPOSITION 1: Suppose that the natural debt limit prevails. Given initial conditions $(b_0, B_0)$, let $\{\bar{c}_t, \bar{b}_{t+1}\}$ and $\{\bar{g}_t, \bar{\tau}_t, \bar{B}_{t+1}\}$ be an equilibrium. Consider any other tax policy $\{\hat{\tau}_t\}$ satisfying

$$\sum_{t=0}^{\infty} R^{-t}\hat{\tau}_t = \sum_{t=0}^{\infty} R^{-t}\bar{\tau}_t. \tag{9.13}$$

Then $\{\bar{c}_t, \hat{b}_{t+1}\}$ and $\{\bar{g}_t, \hat{\tau}_t, \hat{B}_{t+1}\}$ is also an equilibrium where

$$\hat{b}_t = \sum_{j=0}^{\infty} R^{-j}(\bar{c}_{t+j} + \hat{\tau}_{t+j} - y_{t+j}) \tag{9.14}$$

and

$$\hat{B}_t = \sum_{j=0}^{\infty} R^{-t}(\hat{\tau}_{t+j} - \bar{g}_{t+j}). \tag{9.15}$$

**Proof:** The first point of the proposition is that the same consumption plan $\{\bar{c}_t\}$, but adjusted borrowing plan $\{\hat{b}_{t+1}\}$, solve the household's optimum problem under the altered government tax scheme. Under the natural debt limit, the household in effect faces a single intertemporal budget constraint (9.11). At time $0$, the household can be thought of as choosing an optimal consumption plan subject to the single constraint,

$$b_0 = \sum_{t=0}^{\infty} R^{-t}(c_t - y_t) + \sum_{t=0}^{\infty} R^{-t}\tau_t.$$

Thus, the household's budget set, and therefore its optimal plan, does not depend on the timing of taxes, only their present value. The altered tax plan leaves the household's intertemporal budget set unaltered and therefore doesn't affect its optimal consumption plan. Next, we construct the adjusted borrowing plan $\{\hat{b}_{t+1}\}$ by solving the budget constraint (9.10) forward to obtain (9.14).[4] The adjusted borrowing plan satisfies trivially the (adjusted) natural debt limit in every period, since the consumption plan $\{\bar{c}_t\}$ is a nonnegative sequence.

---

[4] It is straightforward to verify that the adjusted borrowing plan $\{\hat{b}_{t+1}\}$ must satisfy the transversality condition (9.4). In any period $(k-1) \geq 0$, solving the budget constraint (9.10)

The second point of the proposition is that the altered government tax and borrowing plans continue to satisfy the government's budget constraint. In particular, we see that the government's budget set at time 0 does not depend on the timing of taxes, only their present value,

$$B_0 = \sum_{t=0}^{\infty} R^{-j} \tau_t - \sum_{t=0}^{\infty} R^{-j} g_t.$$

Thus, under the altered tax plan with an unchanged present value of taxes, the government can finance the same expenditure plan $\{\bar{g}_t\}$. The adjusted borrowing plan $\{\hat{B}_{t+1}\}$ is computed in a similar way as above to arrive at (9.15). ∎

This proposition depends on imposing the natural debt limit, which is weaker than the no-borrowing constraint on the household. Under the no-borrowing constraint, we require that the asset choice $b_{t+1}$ at time $t$ both satisfies budget constraint (9.10) and does not fall below zero. That is, under the no-borrowing constraint, we have to check more than just a single intertemporal budget constraint for the household at time 0. Changes in the timing of taxes that obey equation (9.13) evidently alter the right side of equation (9.10) and can, for example, cause a previously binding borrowing constraint no longer to be binding, and *vice versa*. Binding borrowing constraints in either the initial $\{\bar{\tau}_t\}$ equilibrium or

---

backward yields

$$b_k = \sum_{j=1}^{k} R^j \left[ y_{k-j} - \tau_{k-j} - c_{k-j} \right] + R^k b_0.$$

Evidently, the difference between $\bar{b}_k$ of the initial equilibrium and $\hat{b}_k$ is equal to

$$\bar{b}_k - \hat{b}_k = \sum_{j=1}^{k} R^j \left[ \hat{\tau}_{k-j} - \bar{\tau}_{k-j} \right],$$

and after multiplying both sides by $R^{1-k}$,

$$R^{1-k} \left( \bar{b}_k - \hat{b}_k \right) = R \sum_{t=0}^{k-1} R^{-t} \left[ \hat{\tau}_t - \bar{\tau}_t \right].$$

The limit of the right side is zero when $k$ goes to infinity due to condition (9.13), and hence, the fact that the equilibrium borrowing plan $\{\bar{b}_{t+1}\}$ satisfies transversality condition (9.4) implies that so must $\{\hat{b}_{t+1}\}$.

the new $\{\hat{\tau}_t\}$ equilibria eliminates a Ricardian proposition as general as Proposition 1. More restricted versions of the proposition evidently hold across restricted equivalence classes of taxes that do not alter when the borrowing constraints are binding across the two equilibria being compared.[5]

PROPOSITION 2: Consider an initial equilibrium with consumption path $\{\bar{c}_t\}$ in which $b_{t+1} > 0$ for all $t \geq 0$. Let $\{\bar{\tau}_t\}$ be the tax rate in the initial equilibrium, and let $\{\hat{\tau}_t\}$ be any other tax-rate sequence for which

$$\hat{b}_t = \sum_{j=0}^{\infty} R^{-j}(\bar{c}_{t+j} + \hat{\tau}_{t+j} - y_{t+j}) \geq 0$$

for all $t \geq 0$. Then $\{\bar{c}_t\}$ is also an equilibrium allocation for the $\{\hat{\tau}_t\}$ tax sequence.

We leave the proof of this proposition to the reader.

## *Linked generations interpretation*

Much of the preceding analysis with borrowing constraints applies to a setting with overlapping generations linked by a bequest motive. Assume that there is a sequence of one-period-lived agents. For each $t \geq 0$ there is a one-period-lived agent who values consumption and the utility of his direct descendant, a young person at time $t+1$. Preferences of a young person at $t$ are ordered by

$$u(c_t) + \beta V(b_{t+1}),$$

where $u(c)$ is the same utility function as in the previous section, $b_{t+1} \geq 0$ are bequests from the time-$t$ person to the time–$t+1$ person, and $V(b_{t+1})$ is the

[5] Note that the arguments in this chapter are cast in setting with an exogenous interest rate $R$ and a capital market that is outside of the model. When we discuss potential failures of Ricardian equivalence due to households facing no-borrowing constraints, we are also implicitly contemplating changes in the government's outside asset position. For example, consider an altered tax plan $\{\hat{\tau}_t\}$ that satisfies (9.13) and shifts taxes away from the future toward the present. A large enough change will definitely ensure that the government is a lender early on. But since the households are not allowed to become indebted, the government must lend abroad and we can show that Ricardian equivalence breaks down.

maximized utility function of a time–$t+1$ agent. The maximized utility function is defined recursively by

$$V(b_t) = \max_{c_t, b_{t+1}} \{u(c_t) + \beta V(b_{t+1})\} \tag{9.16}$$

where the maximization is subject to

$$c_t + R^{-1} b_{t+1} \le y_t - \tau_t + b_t \tag{9.17}$$

and $b_{t+1} \ge 0$. The constraint $b_{t+1} \ge 0$ requires that bequests cannot be negative. Notice that a person cares about his direct descendant, but not vice versa. We continue to assume that there is an infinitely lived government whose taxes and purchasing and borrowing strategies are as described in the previous section.

In consumption outcomes, this model is equivalent to the previous model with a no-borrowing constraint. Bequests here play the role of savings $b_{t+1}$ in the previous model. A positive savings condition $b_{t+1} > 0$ in the previous version of the model becomes an "operational bequest motive" in the overlapping generations model.

It follows that we can obtain a restricted Ricardian equivalence proposition, qualified as in Proposition 2. The qualification is that the initial equilibrium must have an operational bequest motive for all $t \ge 0$, and that the new tax policy must not be so different from the initial one that it renders the bequest motive inoperative.

## Concluding remarks

The arguments in this chapter were cast in a setting with an exogenous interest rate $R$ and a capital market that is outside of the model. When we discussed potential failures of Ricardian equivalence due to households facing no-borrowing constraints, we were also implicitly contemplating changes in the government's outside asset position. For example, consider an altered tax plan $\{\hat{\tau}_t\}$ that satisfies (9.13) and shifts taxes away from the future toward the present. A large enough change will definitely ensure that the government is a lender in early periods. But since the households are not allowed to become indebted, the government must lend abroad and we can show that Ricardian equivalence breaks down.

The readers might be able to anticipate the nature of the general equilibrium proof of Ricardian equivalence in chapter 10. First, private consumption and government expenditures must then be consistent with the aggregate endowment in each period, $c_t + g_t = y_t$, which implies that an altered tax plan cannot affect the consumption allocation as long as government expenditures are kept the same. Second, interest rates are determined by intertemporal marginal rates of substitution evaluated at the equilibrium consumption allocation, as studied in chapter 7. Hence, an unchanged consumption allocation implies that interest rates are also unchanged. Third, at those very interest rates, it can be shown that households would like to choose asset positions that exactly offset any changes in the government's asset holdings implied by an altered tax plan. For example, in the case of the tax change contemplated in the preceding paragraph, the households would demand loans exactly equal to the rise in government lending generated by budget surpluses in early periods. The households would use those loans to meet the higher taxes, and thereby, be able to finance an unchanged consumption plan.

The finding of Ricardian equivalence in the infinitely lived agent model is a useful starting point for identifying alternative assumptions under which the irrelevance result might fail to hold,[6] such as our imposition of borrowing constraints that are tighter than the "natural debt limit". Another deviation from the benchmark model is finitely lived agents, as analyzed by Diamond (1965) and Blanchard (1985). But as suggested by Barro (1974) and shown in this chapter, Ricardian equivalence will still continue to hold if agents are altruistic towards their descendants and there is an operational bequest motive. Bernheim and Bagwell (1988) take this argument to its extreme and formulate a model where all agents become interconnected because of linkages across dynastic families, which is shown to render neutral all redistributive policies including distortionary taxes. But in general, replacing lump sum taxes by distortionary taxes is a sure way to undo Ricardian equivalence, see e.g. Barsky, Mankiw and Zeldes (1986). We will return to the question of the timing of distortionary taxes in chapter 12.

---

[6] Seater (1993) reviews the theory and empirical evidence on Ricardian equivalence.

# 10
# Asset Pricing

## Introduction

Chapter 7 showed how an equilibrium price system for an economy with complete markets model could be used to determine the price of any redundant asset. That approach allowed us to price any asset whose payoff could be synthesized as a measurable function of the economy's state. We could use either the Arrow-Debreu time-0 prices or the prices of one-period Arrow securities to price redundant assets.

We shall use this complete markets approach again later in this chapter. However, we begin with another frequently used approach, one that does not require the assumption that there are complete markets. This approach spells out fewer aspects of the economy and assumes fewer markets, but nevertheless derives testable intertemporal restrictions on prices and returns of different assets, and also across those prices and returns and consumption allocations. This approach uses only the Euler equations for a maximizing consumer, and supplies stringent restrictions without specifying a complete general equilibrium model. In fact, the approach imposes only a *subset* of the restrictions that would be imposed in a complete markets model. As we shall see, even these restrictions have proved difficult to reconcile with the data, the equity premium being a widely discussed example.

Asset-pricing ideas have had diverse ramifications in macroeconomics. In this chapter, we describe some of these ideas, including the important Modigliani-Miller theorem asserting the irrelevance of firms' asset structures. We describe a closely related kind of Ricardian equivalence theorem. We describe various ways of representing the equity premium puzzle, and an idea of Mankiw (1986) that one day may help explain it.[1]

---

[1] See Duffie (1996) for a comprehensive treatment of discrete and continuous time asset pricing theories. See Campbell, Lo, and MacKinlay (1997) for a summary of recent work on empirical implementations.

## *Asset Euler equations*

We now describe the optimization problem of a single agent who has the opportunity to trade two assets. Following Hansen and Singleton (1983), the household's optimization by itself imposes ample restrictions on the co-movements of asset prices and the household's consumption. These restrictions remain true even if additional assets are made available to the agent, and so do not depend on specifying the market structure completely. Later we shall study a general equilibrium model with a large number of identical agents. Completing a general equilibrium model may impose additional restrictions, but will leave intact individual-specific versions of the ones to be derived here.

The agent has wealth $A_t > 0$ at time $t$ and wants to use this wealth to maximize expected lifetime utility,

$$E_t \sum_{j=0}^{\infty} \beta^t u(c_{t+j}), \qquad 0 < \beta < 1, \tag{10.1}$$

where $E_t$ denotes the mathematical expectation conditional on information known at time $t$, $\beta$ is a subjective discount factor, and $c_{t+j}$ is the agent's consumption in period $t+j$. The utility function $u(\cdot)$ is concave, strictly increasing, and twice continuously differentiable.

To finance future consumption, the agent can transfer wealth over time through bond and equity holdings. One-period bonds earn a risk-free real gross interest rate $R_t$, measured in units of time $t+1$ consumption good per time-$t$ consumption good. Let $L_t$ be gross payout on the agent's bond holdings between periods $t$ and $t+1$, payable in period $t+1$ with a present value of $R_t^{-1} L_t$ at time $t$. The variable $L_t$ is negative if the agent issues bonds and thereby borrows funds. The agent's holdings of equity shares between periods $t$ and $t+1$ are denoted $s_t$, where a negative number indicates a short position in shares. We impose the borrowing constraints $L_t \geq -b_b$ and $s_t \geq -b_s$, where $b_b \geq 0$ and $b_s \geq 0$.[2] A share of equity entitles the owner to its stochastic dividend stream $y_t$. Let $p_t$ be the share price in period $t$ net of that period's dividend. The budget constraint becomes

$$c_t + R_t^{-1} L_t + p_t s_t \leq A_t, \tag{10.2}$$

and next period's wealth is

$$A_{t+1} = L_t + (p_{t+1} + y_{t+1}) s_t. \tag{10.3}$$

[2] See chapters 7 and 14 for further discussions of natural and ad hoc borrowing constraints.

The stochastic dividend is the only source of exogenous fundamental uncertainty, with properties to be specified as needed later. The agent's maximization problem is then a dynamic programming problem with the state at $t$ being $A_t$ and current and past $y$,[3] and the controls being $L_t$ and $s_t$. At interior solutions, the Euler equations associated with controls $L_t$ and $s_t$ are

$$u'(c_t)R_t^{-1} = E_t\beta u'(c_{t+1}), \tag{10.4}$$

$$u'(c_t)p_t = E_t\beta(y_{t+1}+p_{t+1})u'(c_{t+1}). \tag{10.5}$$

These Euler equations give a number of insights into asset prices and consumption. Before turning to these, we first note that an optimal solution to the agent's maximization problem must also satisfy the following transversality conditions:[4]

$$\lim_{k\to\infty} E_t\beta^k u'(c_{t+k})R_{t+k}^{-1}L_{t+k} = 0, \tag{10.6}$$

$$\lim_{k\to\infty} E_t\beta^k u'(c_{t+k})p_{t+k}s_{t+k} = 0. \tag{10.7}$$

Heuristically, if any of the expressions in equations (10.6) and (10.7) were strictly positive, the agent would be overaccumulating assets so that a higher expected life-time utility could be achieved by, for example, increasing consumption today. The counterpart to such nonoptimality in a finite horizon model would be that the agent dies with positive asset holdings. For reasons like those in a finite horizon model, the agent would be happy if the two conditions (10.6) and (10.7) could be violated on the negative side. But the market would stop the agent from financing consumption by accumulating the debts that would be associated with such violations of (10.6) and (10.7) . No other agent would want to make those loans.

## Martingale theories of consumption and stock prices

In this section, we briefly recall some early theories of asset prices and consumption, each of which is derived by making special assumptions about either $R_t$ or $u'(c)$ in equations (10.4) and (10.5). These assumptions are too strong to be consistent with much empirical evidence, but they are instructive benchmarks.

[3] Current and past $y$'s enter as information variables. How many past $y$'s appear in the Bellman equation depends on the stochastic process for $y$.

[4] For a discussion of transversality conditions, see Benveniste and Scheinkman (1982) and Brock (1982).

First, suppose that the risk-free interest rate is constant over time, $R_t = R > 1$, for all $t$. Then equation (10.4) implies that

$$E_t u'(c_{t+1}) = (\beta R)^{-1} u'(c_t), \tag{10.8}$$

which is Robert Hall's (1978) result that the marginal utility of consumption follows a univariate linear first-order Markov process, so that no other variables in the information set help to predict (to Granger cause) $u'(c_{t+1})$, once lagged $u'(c_t)$ has been included.[5]

As an example, with the constant relative risk aversion utility function $u(c_t) = \frac{c_t^{1-\gamma}}{1-\gamma}$, equation (10.8) becomes

$$(\beta R)^{-1} = E_t \left( \frac{c_{t+1}}{c_t} \right)^{-\gamma}.$$

Using aggregate data, Hall tested implication (10.8) for the special case of quadratic utility by testing for the absence of Granger causality from other variables to $c_t$.

Efficient stock markets are sometimes construed to mean that the price of a stock ought to follow a martingale. Euler equation (10.5) shows that a number of simplifications must be made to get a martingale property for the stock price. We can transform the Euler equation

$$E_t \beta (y_{t+1} + p_{t+1}) \frac{u'(c_{t+1})}{u'(c_t)} = p_t.$$

by noting that for any two random variables $x, y$, we have the formula $E_t xy = E_t x E_t y + \text{cov}_t(x, y)$, where $\text{cov}_t(x, y) \equiv E_t(x - E_t x)(y - E_t y)$. This formula defines the conditional covariance $\text{cov}_t(x, y)$. Applying this formula in the preceding equation gives

$$\beta E_t(y_{t+1} + p_{t+1}) E_t \frac{u'(c_{t+1})}{u'(c_t)} + \beta \text{cov}_t \left[ (y_{t+1} + p_{t+1}), \frac{u'(c_{t+1})}{u'(c_t)} \right] = p_t. \tag{10.9}$$

[5] See Granger (1969) for his definition of causality. A random process $z_t$ is said *not* to cause a random process $x_t$ if $E(x_{t+1}|x_t, x_{t-1}, \ldots, z_t, z_{t-1}, \ldots) = E(x_{t+1}|x_t, x_{t-1}, \ldots)$. The absence of Granger causality can be tested in several ways. A direct way is to compute the two regressions mentioned in the preceding definition and test for their equality. An alternative test was described by Sims (1972).

To obtain a martingale theory of stock prices, it is necessary to assume, first, that $E_t u'(c_{t+1})/u'(c_t)$ is a constant, and second, that

$$\text{cov}_t \left[ (y_{t+1} + p_{t+1}), \frac{u'(c_{t+1})}{u'(c_t)} \right] = 0.$$

These conditions are obviously very restrictive and will only hold under very special circumstances. For example, a sufficient assumption is that agents are risk neutral so that $u(c_t)$ is linear in $c_t$ and $u'(c_t)$ becomes independent of $c_t$. In this case, equation (10.9) implies that

$$E_t \beta (y_{t+1} + p_{t+1}) = p_t. \tag{10.10}$$

Equation (10.10) states that, adjusted for dividends and discounting, the share price follows a first-order univariate Markov process and that no other variables Granger cause the share price. These implications have been tested extensively in the literature on efficient markets.[6]

We also note that the stochastic difference equation (10.10) has the class of solutions

$$p_t = E_t \sum_{j=1}^{\infty} \beta^j y_{t+j} + \xi_t \left( \frac{1}{\beta} \right)^t, \tag{10.11}$$

where $\xi_t$ is any random process that obeys $E_t \xi_{t+1} = \xi_t$ (that is, $\xi_t$ is a "martingale"). Equation (10.11) expresses the share price $p_t$ as the sum of discounted expected future dividends and a "bubble term" unrelated to any fundamentals. In the general equilibrium model that we will describe later, this bubble term always equals zero.

## *Equivalent martingale measure*

This section describes adjustments for risk and dividends that convert an asset price into a martingale. We return to the setting of chapter 7 and assume that the state $s_t$ that evolves according to a Markov chain with transition probabilities $\pi(s_{t+1}|s_t)$. Let an asset pay a stream of dividends $\{d(s_t)\}_{t \geq 0}$. The

---

[6] For a survey of this literature, see Fama (1976a). See Samuelson (1965) for the theory and Roll (1970) for an application to the term structure of interest rates.

cum-dividend[7] time- $t$ price of this asset, $a(s_t)$, can be expressed recursively as

$$a(s_t) = d(s_t) + \beta \sum_{s_{t+1}} \frac{u'[c^i_{t+1}(s_{t+1}]}{u'[c^i_t(s_t)]} a(s_{t+1}) \pi(s_{t+1}|s_t). \tag{10.12}$$

Notice that this equation can be written

$$a(s_t) = d(s_t) + R_t^{-1} \sum_{s_{t+1}} a(s_{t+1}) \tilde{\pi}(s_{t+1}|s_t) \tag{10.13}$$

or

$$a(s_t) = d(s_t) + R_t^{-1} \tilde{E}_t a(s_{t+1}),$$

where

$$R_t^{-1} = R_t^{-1}(s_t) \equiv \beta \sum_{s_{t+1}} \frac{u'[c^i_{t+1}(s_{t+1})]}{u'[c^i_t(s_t)]} \pi(s_{t+1}|s_t) \tag{10.14}$$

and $\tilde{E}$ is the mathematical expectation with respect to the distorted transition density

$$\tilde{\pi}(s_{t+1}|s_t) = R_t \beta \frac{u'[c^i_{t+1}(s_{t+1})]}{u'[c^i_t(s_t)]} \pi(s_{t+1}|s_t). \tag{10.15a}$$

Notice that $R_t^{-1}$ is the reciprocal of the gross one-period risk-free interest rate. The transformed transition probabilities are rendered probabilities—that is, made to sum to one—through the multiplication by $\beta R_t$ in equation (10.15a). The transformed or "twisted" transition measure $\tilde{\pi}(s_{t+1}|s_t)$ can be used to define the twisted measure

$$\tilde{\pi}_t(s^t) = \tilde{\pi}(s_t|s_{t-1}) \ldots \tilde{\pi}(s_1|s_0) \tilde{\pi}(s_0). \tag{10.15b}$$

For example,

$$\tilde{\pi}(s_{t+2}, s_{t+1}|s_t) = R_t(s_t) R_{t+1}(s_{t+1}) \beta^2 \frac{u'[c^i_{t+2}(s_{t+2})]}{u'[c^i_t(s_t)]} \pi(s_{t+2}|s_{t+1}) \pi(s_{t+1}|s_t).$$

The twisted measure $\tilde{\pi}_t(s^t)$ is called an *equivalent martingale measure.* We explain the meaning of the two adjectives. "Equivalent" means that $\tilde{\pi}$ assigns positive probability to any event that is assigned positive probability by $\pi$, and vice versa. The equivalence of $\pi$ and $\tilde{\pi}$ is guaranteed by the assumption that $u'(c) > 0$ in (10.15a).[8]

---

[7] Cum-dividend means that the person who owns the asset at the end of time $t$ is entitled to the time- $t$ dividend.

[8] The existence of an equivalent martingale measure implies both the existence of a *positive* stochastic discount factor (see the discussion of Hansen and Jagannathan bounds later in this chapter), and the absence of arbitrage opportunities; see Kreps (1979) and Duffie (1996).

We now turn to the adjective "martingale." To understand why this term is applied to (10.15a), consider the particular case of an asset with dividend stream $d_T = d(s_T)$ and $d_t = 0$ for $t < T$. Using the arguments in chapter 7 or iterating on equation (10.12), the cum-dividend price of this asset can be expressed as

$$a_T(s_T) = d(s_T) \tag{10.16a}$$

$$a_t(s_t) = E_{s_t} \beta^{T-t} \frac{u'[c_T^i(s_T)]}{u'[c_t^i(s_t)]} a_T(s_T), \tag{10.16b}$$

where $E_{s_t}$ denotes the conditional expectation under the $\pi$ probability measure. Now fix $t < T$ and define the "deflated" or "interest-adjusted" process

$$\tilde{a}_{t+j} = \frac{a_{t+j}}{R_t R_{t+1} \ldots R_{t+j-1}}, \tag{10.17}$$

for $j = 1, \ldots, T-t$. It follows directly from equations (10.16) and (10.15) that

$$\tilde{E}_t \tilde{a}_{t+j} = \tilde{a}_t(s_t) \tag{10.18}$$

where $\tilde{a}_t(s_t) = a(s_t) - d(s_t)$. Equation (10.18) asserts that relative to the twisted measure $\tilde{\pi}$, the interest-adjusted asset price is a martingale: using the twisted measure, the best prediction of the future interest-adjusted asset price is its current value.

Thus, when the equivalent martingale measure is used to price assets, we have so-called risk-neutral pricing. Notice that in equation (10.13) the adjustment for risk is absorbed into the twisted transition measure. We can write equation (10.18) as

$$\tilde{E}[a(s_{t+1})|s_t] = R_t[a(s_t) - d_t], \tag{10.19}$$

where $\tilde{E}$ is the expectation operator for the twisted transition measure. Equation (10.19) is another way of stating that, after adjusting for risk-free interest and dividends, the price of the asset is a *martingale* relative to the martingale equivalent measure.

Under the equivalent martingale measure, asset pricing reduces to calculating the conditional expectation of the stream of dividends that defines the asset. For example, consider a European call option written on the asset described earlier that is priced by equations (10.16). The owner of the call option has the right but not the obligation to the "asset" at time $T$ at a price $K$. The owner of the call will exercise this option only if $a_T \geq K$. The value at $T$ of the option is therefore $Y_T = \max(0, a_T - K) \equiv (a_T - K)^+$. The price of the option at $t < T$ is then

$$Y_t = \tilde{E}_t \left[ \frac{(a_T - K)^+}{R_t R_{t-1} \cdots R_{t+T-1}} \right]. \tag{10.20}$$

Black and Scholes (1973) used particular continuous time specification of $\tilde{\pi}$ that made it possible to solve equation (10.20) analytically for a function $Y_t$. Their solution is known as the Black-Scholes formula for option pricing.

## Equilibrium asset pricing

The preceding discussion of the Euler equations (10.4) and (10.5) leaves open how the economy, for example, generates the constant gross interest rate assumed in Hall's work. We now explore equilibrium asset pricing in a simple representative agent endowment economy, Lucas's asset-pricing model.[9] We imagine an economy consisting of a large number of identical agents with preferences as specified in expression (10.1). The only durable good in the economy is a set of identical "trees," one for each person in the economy. At the beginning of period $t$, each tree yields fruit or dividends in the amount $y_t$. The dividend $y_t$ is a function of $x_t$, which is assumed to be governed by a Markov process with a time-invariant transition probability distribution function given by $\text{prob}\{x_{t+1} \leq x' | x_t = x\} = F(x', x)$. The fruit is not storable, but the tree is perfectly durable. Each agent starts life at time zero with one tree.

All agents maximize expression (10.1) subject to the budget constraint (10.2)–(10.3) and transversality conditions (10.6)–(10.7). In an equilibrium, asset prices clear the markets. That is, the bond holdings of all agents sum to zero, and their total stock positions are equal to the aggregate number of shares. As a normalization, let there be one share per tree.

Due to the assumption that all agents are identical with respect to both preferences and endowments, we can work with a representative agent.[10] Lucas's model shares features with a variety of representative agent asset-pricing models. (See Brock, 1982, and Altug, 1989, for example.) These use versions of stochastic optimal growth models to generate allocations and price assets.

Such asset-pricing models can be constructed by the following steps:

---

[9] See Lucas (1978). Also see the important early work by Stephen LeRoy (1971, 1973). Breeden (1979) was an early work on the consumption-based capital-asset-pricing model.

[10] In chapter 7, we showed that some heterogeneity is also consistent with the notion of a representative agent.

1. Describe the preferences, technology, and endowments of a dynamic economy, then solve for the equilibrium intertemporal consumption allocation. Sometimes there is a particular planning problem whose solution equals the competitive allocation.

2. Set up a competitive market in some particular asset that represents a specific claim on future consumption goods. Permit agents to buy and sell at equilibrium asset prices subject to particular borrowing and short-sales constraints. Find an agent's Euler equation, analogous to equations (10.4) and (10.5), for this asset.

3. Equate the consumption that appears in the Euler equation derived in step 2 to the equilibrium consumption derived in step 1. This procedure will give the asset price at $t$ as a function of the state of the economy at $t$.

In our endowment economy, a planner that treats all agents the same would like to maximize $E_0 \sum_{t=0}^{\infty} \beta^t u(c_t)$ subject to $c_t \leq y_t$. Evidently the solution is to set $c_t$ equal to $y_t$. After substituting this consumption allocation into equations (10.4) and (10.5), we arrive at expressions for the risk-free interest rate and the share price,

$$u'(y_t) R_t^{-1} = E_t \beta u'(y_{t+1}), \tag{10.21}$$

$$u'(y_t) p_t = E_t \beta (y_{t+1} + p_{t+1}) u'(y_{t+1}). \tag{10.22}$$

## Stock prices without bubbles

Using recursions on equation (10.22) and the law of iterated expectations, which states that $E_t E_{t+1}(\cdot) = E_t(\cdot)$, we arrive at the following expression for the equilibrium share price,

$$u'(y_t) p_t = E_t \sum_{j=1}^{\infty} \beta^j u'(y_{t+j}) y_{t+j} + E_t \lim_{k \to \infty} \beta^k u'(y_{t+k}) p_{t+k}. \tag{10.23}$$

Moreover, equilibrium share prices have to be consistent with market clearing; that is, agents must be willing to hold their endowments of trees forever. It follows immediately that the last term in equation (10.23) must be zero. Suppose to the contrary that the term is strictly positive. That is, the marginal utility gain of selling shares, $u'(y_t)p_t$, exceeds the marginal utility loss of holding the asset forever and consuming the future stream of dividends, $E_t \sum_{j=1}^{\infty} \beta^j u'(y_{t+j}) y_{t+j}$. Thus, all agents would like to sell some of their shares and the price would be driven down. Analogously, if the last term in equation (10.23) were strictly negative, we would find that all agents would like to purchase more shares and the price would necessarily be driven up. We can therefore conclude that the equilibrium price must satisfy

$$p_t = E_t \sum_{j=1}^{\infty} \beta^j \frac{u'(y_{t+j})}{u'(y_t)} y_{t+j}, \tag{10.24}$$

which is a generalization of equation (10.11) in which the share price is an expected discounted stream of dividends but with time-varying and stochastic discount rates.

Note that asset bubbles could also have been ruled out by directly referring to transversality condition (10.7) and market clearing. In an equilibrium, the representative agent holds the per-capita outstanding number of shares. (We have assumed one tree per person and one share per tree.) After dividing transversality condition (10.7) by this constant time-invariant number of shares and replacing $c_{t+k}$ by equilibrium consumption $y_{t+k}$, we arrive at the implication that the last term in equation (10.23) must vanish.[11]

## Computing asset prices

We now turn to three examples in which it is easy to calculate an asset-pricing function by solving the expectational difference equation (10.22).

[11] Brock (1982) and Tirole (1982) use the transversality condition when proving that asset bubbles cannot exist in economies with a constant number of infinitely lived agents. However, Tirole (1985) shows that asset bubbles can exist in equilibria of overlapping generations models that are dynamically inefficient, that is, when the growth rate of the economy exceeds the equilibrium rate of return. O'Connell and Zeldes (1988) derive the same result for a dynamically inefficient economy with a growing number of infinitely lived agents. Abel, Mankiw, Summers, and Zeckhauser (1989) provide international evidence suggesting that dynamic inefficiency is not a problem in practice.

*Example 1: Logarithmic preferences*

Take the special case of equation (10.24) that emerges when $u(c_t) = \ln c_t$. Then equation (10.24) becomes

$$p_t = E_t \sum_{j=1}^{\infty} \beta^j y_{t+j}$$

or

$$p_t = \frac{\beta}{1-\beta} y_t. \tag{10.25}$$

Equation (10.25) is our asset-pricing function. It maps the state of the economy at $t$, $y_t$, into the price of a Lucas tree at $t$.

*Example 2: A finite-state version*

Mehra and Prescott (1985) consider a discrete state version of Lucas's one-kind-of-tree model. Let dividends assume the $n$ possible distinct values $[\sigma_1, \sigma_2, \ldots, \sigma_n]$. Let dividends evolve through time according to a Markov chain, with

$$\text{prob}\{y_{t+1} = \sigma_l | y_t = \sigma_k\} = P_{kl} > 0.$$

The $(n \times n)$ matrix $P$ with element $P_{kl}$ is called a stochastic matrix. The matrix satisfies $\sum_{l=1}^n P_{kl} = 1$ for each $k$. Express equation (10.22) of Lucas's model as

$$p_t u'(y_t) = \beta E_t p_{t+1} u'(y_{t+1}) + \beta E_t y_{t+1} u'(y_{t+1}). \tag{10.26}$$

Express the price at $t$ as a function of the state $\sigma_k$ at $t$, $p_t = p(\sigma_k)$. Define $p_t u'(y_t) = p(\sigma_k) u'(\sigma_k) \equiv v_k$, $k = 1, \ldots, n$. Also define $\alpha_k = \beta E_t y_{t+1} u'(y_{t+1}) = \beta \sum_{l=1}^n \sigma_l u'(\sigma_l) P_{kl}$. Then equation (10.26) can be expressed as

$$p(\sigma_k) u'(\sigma_k) = \beta \sum_{l=1}^n p(\sigma_l) u'(\sigma_l) P_{kl} + \beta \sum_{l=1}^n \sigma_l u'(\sigma_l) P_{kl}$$

or

$$v_k = \alpha_k + \beta \sum_{l=1}^n P_{kl} v_l,$$

or in matrix terms, $v = \alpha + \beta P v$, where $v$ and $\alpha$ are column vectors. The equation can be represented as $(I - \beta P) v = \alpha$. This equation has a unique solution given by[12]

$$v = (I - \beta P)^{-1}\alpha. \tag{10.27}$$

The price of the asset in state $\sigma_k$—call it $p_k$—can then be found from $p_k = v_k/[u'(\sigma_k)]$. Notice that equation (10.27) can be represented as

$$v = (I + \beta P + \beta^2 P^2 + \ldots)\alpha$$

or

$$p(\sigma_k) = p_k = \sum_l (I + \beta P + \beta^2 P^2 + \ldots)_{kl} \frac{\alpha_l}{u'(\sigma_k)},$$

where $(I + \beta P + \beta^2 P^2 + \ldots)_{kl}$ is the $(k,l)$ element of the matrix $(I + \beta P + \beta^2 P^2 + \ldots)$. We ask the reader to interpret this formula in terms of a geometric sum of expected future variables.

*Example 3: Asset pricing with growth*

Let's price a Lucas tree in a pure endowment economy with $c_t = d_t$ and $d_{t+1} = \lambda_{t+1} d_t$, where $\lambda_t$ is Markov with transition matrix $P$. Let $p_t$ be the ex dividend price of the Lucas tree. Assume the CRRA utility $u(c) = c^{1-\gamma}/(1-\gamma)$. Evidently, the price of the Lucas tree satisfies

$$p_t = E_t \left[ \beta \left( \frac{c_{t+1}}{c_t} \right)^{-\gamma} (p_{t+1} + d_{t+1}) \right].$$

Dividing both sides by $d_t$ and rearranging gives

$$\frac{p_t}{d_t} = E_t \left[ \beta (\lambda_{t+1})^{1-\gamma} \left( \frac{p_{t+1}}{d_{t+1}} + 1 \right) \right]$$

or

$$w_i = \beta \sum_j P_{ij} \lambda_j^{1-\gamma} (w_j + 1), \tag{10.28}$$

where $w_i$ represents the price-dividend ration. Equation (10.28) was used by Mehra and Prescott to compute equilibrium prices.

[12] Uniqueness follows from the fact that, because $P$ is a nonnegative matrix with row sums all equaling unity, the eigenvalue of maximum modulus $P$ has modulus unity. The maximum eigenvalue of $\beta P$ then has modulus $\beta$. (This point follows from Frobenius's theorem.) The implication is that $(I - \beta P)^{-1}$ exists and that the expansion $I + \beta P + \beta^2 P^2 + \ldots$ converges and equals $(I - \beta P)^{-1}$.

## The term structure of interest rates

We will now explore the term structure of interest rates by pricing bonds with different maturities.[13] We continue to assume that the time-$t$ state of the economy is the current dividend on a Lucas tree $y_t = x_t$, which is Markov with transition $F(x', x)$. The risk-free real gross return between periods $t$ and $t+j$ is denoted $R_{jt}$, measured in units of time–$(t+j)$ consumption good per time-$t$ consumption good. Thus, $R_{1t}$ replaces our earlier notation $R_t$ for the one-period gross interest rate. At the beginning of $t$, the return $R_{jt}$ is known with certainty and is risk free from the viewpoint of the agents. That is, at $t$, $R_{jt}^{-1}$ is the price of a perfectly sure claim to one unit of consumption at time $t+j$. For simplicity, we only consider such zero-coupon bonds, and the extra subscript $j$ on gross earnings $L_{jt}$ now indicates the date of maturity. The subscript $t$ still refers to the agent's decision to hold the asset between period $t$ and $t+1$.

As an example with one- and two-period safe bonds, the budget constraint and the law of motion for wealth in (10.2)–(10.3) are augmented as follows,

$$c_t + R_{1t}^{-1} L_{1t} + R_{2t}^{-1} L_{2t} + p_t s_{t+1} \leq A_t, \tag{10.29}$$

$$A_{t+1} = L_{1t} + R_{1t+1}^{-1} L_{2t} + (p_{t+1} + y_{t+1}) s_{t+1}. \tag{10.30}$$

Even though safe bonds represent sure claims to future consumption, these assets are subject to price risk prior to maturity. For example, two-period bonds from period $t$, $L_{2t}$, are traded at the price $R_{1t+1}^{-1}$ in period $t+1$, as shown in wealth expression (10.30). At time $t$, an agent who buys such assets and plans to sell them next period would be uncertain about the proceeds, since $R_{1t+1}^{-1}$ is not known at time $t$. The price $R_{1t+1}^{-1}$ follows from a simple arbitrage argument, since, in period $t+1$, these assets represent identical sure claims to time–$(t+2)$ consumption goods as newly issued one-period bonds in period $t+1$. The variable $L_{jt}$ should therefore be understood as the agent's net holdings between periods $t$ and $t+1$ of bonds that each pay one unit of consumption good at time $t+j$, without identifying when the bonds were initially issued.

Given wealth $A_t$ and current dividend $y_t = x_t$, let $v(A_t, x_t)$ be the optimal value of maximizing expression (10.1) subject to equations (10.29)–(10.30), the asset pricing function for trees $p_t = p(x_t)$, the stochastic process $F(x_{t+1}, x_t)$, and stochastic processes for $R_{1t}$ and $R_{2t}$. The Bellman equation can be written

---

[13] Dynamic asset-pricing theories for the term structure of interest rates have been developed by Cox, Ingersoll, and Ross (1985a, 1985b) and by LeRoy (1982).

as

$$v(A_t, x_t) = \max_{L_{1t}, L_{2t}, s_{t+1}} \left\{ u\left[A_t - R_{1t}^{-1} L_{1t} - R_{2t}^{-1} L_{2t} - p(x_t) s_{t+1}\right] \right.$$
$$\left. + \beta E_t v\left(L_{1t} + R_{1t+1}^{-1} L_{2t} + [p(x_{t+1}) + x_{t+1}] s_{t+1}, x_{t+1}\right)\right\},$$

where we have substituted for consumption $c_t$ and wealth $A_{t+1}$ from formulas (10.29) and (10.30), respectively. The first-order necessary conditions with respect to $L_{1t}$ and $L_{2t}$ are

$$u'(c_t) R_{1t}^{-1} = \beta E_t v_1\left(A_{t+1}, x_{t+1}\right), \tag{10.31}$$
$$u'(c_t) R_{2t}^{-1} = \beta E_t\left[v_1\left(A_{t+1}, x_{t+1}\right) R_{1t+1}^{-1}\right], \tag{10.32}$$

After invoking Benveniste and Scheinkman's result and equilibrium allocation $c_t = x_t$, we arrive at the following equilibrium rates of return,

$$R_{1t}^{-1} = \beta E_t\left[\frac{u'(x_{t+1})}{u'(x_t)}\right] \equiv R_1(x_t)^{-1}, \tag{10.33}$$

$$R_{2t}^{-1} = \beta E_t\left[\frac{u'(x_{t+1})}{u'(x_t)} R_{1t+1}^{-1}\right] = \beta^2 E_t\left[\frac{u'(x_{t+2})}{u'(x_t)}\right] \equiv R_2(x_t)^{-1}, \tag{10.34}$$

where the second equality in (10.34) is obtained by using (10.33) and the law of iterated expectations. Because of our Markov assumption, interest rates can be written as time-invariant functions of the economy's current state $x_t$. The general expression for the price at time $t$ of a bond that yields one unit of the consumption good in period $t+j$ is

$$R_{jt}^{-1} = \beta^j E_t\left[\frac{u'(x_{t+j})}{u'(x_t)}\right]. \tag{10.35}$$

The term structure of interest rates is commonly defined as the collection of yields to maturity for bonds with different dates of maturity. In the case of zero-coupon bonds, the yield to maturity is simply

$$\tilde{R}_{jt} \equiv R_{jt}^{1/j} = \beta^{-1}\left\{u'(x_t)\left[E_t u'(x_{t+j})\right]^{-1}\right\}^{1/j}. \tag{10.36}$$

As an example, let us assume that dividends are independently and identically distributed over time. The yields to maturity for a $j$-period bond and a $k$-period bond are then related as follows,

$$\tilde{R}_{jt} = \tilde{R}_{kt} \left\{ u'(x_t) \left[ Eu'(x) \right]^{-1} \right\}^{\frac{k-j}{kj}} .$$

The term structure of interest rates is therefore upward sloping whenever $u'(x_t)$ is less than $Eu'(x)$, that is, when consumption is relatively high today with a low marginal utility of consumption, and agents would like to save for the future. In an equilibrium, the short-term interest rate is therefore depressed if there is a diminishing marginal rate of physical transformation over time or, as in our model, there is no investment technology at all.

A classical theory of the term structure of interest rates is that long-term interest rates should be determined by expected future short-term interest rates. For example, the pure expectations theory hypothesizes that $R_{2t}^{-1} = R_{1t}^{-1} E_t R_{1t+1}^{-1}$. Let us examine if this relationship holds in our general equilibrium model. From equation (10.34) and by using equation (10.33), we obtain

$$\begin{aligned} R_{2t}^{-1} &= \beta E_t \left[ \frac{u'(x_{t+1})}{u'(x_t)} \right] E_t R_{1t+1}^{-1} + \text{cov}_t \left[ \beta \frac{u'(x_{t+1})}{u'(x_t)},\, R_{1t+1}^{-1} \right] \\ &= R_{1t}^{-1} E_t R_{1t+1}^{-1} + \text{cov}_t \left[ \beta \frac{u'(x_{t+1})}{u'(x_t)},\, R_{1t+1}^{-1} \right], \end{aligned} \tag{10.37}$$

which is a generalized version of the pure expectations theory, adjusted for the risk premium $\text{cov}_t[\beta u'(x_{t+1})/u'(x_t),\, R_{1t+1}^{-1}]$. The formula implies that the pure expectations theory holds only in special cases. One special case occurs when utility is linear in consumption, so that $u'(x_{t+1})/u'(x_t) = 1$. In this case, $R_{1t}$, given by equation (10.33), is a constant, equal to $\beta^{-1}$, and the covariance term is zero. A second special case occurs when there is no uncertainty, so that the covariance term is zero for that reason. These are the same conditions that suffice to eradicate the risk premium appearing in equation (10.9) and thereby sustain a martingale theory for a stock price.

## State-contingent prices

Thus far, this chapter has taken a different approach to asset pricing than we took in chapter 7. Recall that in chapter 7 we described two alternative complete markets models, one with once-and-for-all trading at time 0 of date- and history-contingent claims, the other with sequential trading of a complete set of one-period Arrow securities. After these state-contingent prices had been computed, we were able to price any asset whose payoffs were linear combinations of the basic state-contingent commodities, just by taking a weighted sum. That approach would work easily for the Lucas tree economy, which by its simple structure with a representative agent can readily be cast as an economy with complete markets. The pricing formulas that we derived in chapter 7 apply to the Lucas tree economy, adjusting only for the way we have altered the specification of the Markov process describing the state of the economy.

Thus, in chapter 7, we gave formulas for a pricing kernel for $j$-step-ahead state-contingent claims. In the notation of that chapter, we called $Q_j(s_{t+j}|s_t)$ the price when the time-$t$ state is $s_t$ of one unit of consumption in state $s_{t+j}$. In this chapter we have chosen to denote the state as $x_t$ and to let it be governed by a continuous-state Markov process. We now choose to use the notation $Q_j(x^j, x)$ to denote the $j$-step-ahead state-contingent price. We have the following version of the formula from chapter 7 for a $j$-period contingent claim

$$Q_j(x^j, x) = \beta^j \frac{u'(x^j)}{u'(x)} f^j(x^j, x), \tag{10.38}$$

where the $j$-step-ahead transition function obeys

$$f^j(x^j, x) = \int f(x^j, x^{j-1}) f^{j-1}(x^{j-1}, x) dx^{j-1}, \tag{10.39}$$

and

$$\text{prob}\{x_{t+j} \leq x^j | x_t = x\} = \int_{-\infty}^{x^j} f^j(w, x) dw.$$

In subsequent sections, we use the state-contingent prices to give expositions of several important ideas including the Modigliani-Miller theorem and a Ricardian theorem.

*Insurance premium*

We shall now use the contingent claims prices to construct a model of insurance. Let $q_\alpha(x)$ be the price in current consumption goods of a claim on one unit of consumption next period, contingent on the event that next period's dividends fall below $\alpha$. We think of the asset being priced as "crop insurance," a claim to consumption when next period's crops fall short of $\alpha$ per tree.

From the preceding section, we have

$$q_\alpha(x) = \beta \int_0^\alpha \frac{u'(x')}{u'(x)} f(x', x) dx'. \tag{10.40}$$

Upon noting that

$$\begin{aligned}\int_0^\alpha u'(x') f(x', x) dx' = & \text{prob}\{x_{t+1} \leq \alpha | x_t = x\} \cdot \\ & E\{u'(x_{t+1}) | x_{t+1} \leq \alpha, x_t = x\},\end{aligned}$$

we can represent the preceding equation as

$$\begin{aligned}q_\alpha(x) = & \frac{\beta}{u'(x_t)} \text{prob}\{x_{t+1} \leq \alpha | x_t = x\} \cdot \\ & E\{u'(x_{t+1}) | x_{t+1} \leq \alpha, x_t = x\}.\end{aligned} \tag{10.41}$$

Notice that, in the special case of risk neutrality $[u'(x)$ is a constant$]$, equation (10.41) collapses to

$$q_\alpha(x) = \beta \text{prob}\{x_{t+1} \leq \alpha | x_t = x\},$$

which is an intuitively plausible formula for the risk-neutral case. When $u'' < 0$ and $x_t \geq \alpha$, equation (10.41) implies that $q_\alpha(x) > \beta \text{prob}\{x_{t+1} \leq \alpha | x_t = x\}$ (because then $E\{u'(x_{t+1}) | x_{t+1} \leq \alpha, x_t = x\} > u'(x_t)$ for $x_t \geq \alpha$). In other words, when the representative consumer is risk averse ($u'' < 0$) and when $x_t \geq \alpha$, the price of crop insurance $q_\alpha(x)$ exceeds the "actuarially fair" price of $\beta \text{prob}\{x_{t+1} \leq \alpha | x_t = x\}$.

Another way to represent equation (10.40) that is perhaps more convenient for purposes of empirical testing is

$$1 = \frac{\beta}{u'(x_t)} E\left[u'(x_{t+1}) R_t(\alpha) | x_t\right] \tag{10.42}$$

where

$$R_t(\alpha) = \begin{cases} 0 & \text{if } x_{t+1} > \alpha \\ 1/q_\alpha(x_t) & \text{if } x_{t+1} \leq \alpha. \end{cases}$$

*Man-made uncertainty*

In addition to pricing assets with returns made risky by nature, we can use the model to price arbitrary man-made lotteries. Suppose that there is a market for one-period lottery tickets paying a stochastic prize $\omega$ in next period, and let $h(\omega, x', x)$ be a probability density for $\omega$, conditioned on $x'$ and $x$. The price of a lottery ticket in state $x$ is denoted $q_L(x)$. To obtain an equilibrium expression for this price, we follow the steps in the section "Equilibrium asset pricing" and include purchases of lottery tickets in the agent's budget constraint. (Quantities are negative if the agent is selling lottery tickets.) Then by reasoning similar to that leading to the arbitrage pricing formulas of chapter 7, we arrive at the lottery ticket price formula:

$$q_L(x) = \beta \int \int \frac{u'(x')}{u'(x)} \omega h(\omega, x', x) f(x', x) d\omega \, dx'. \tag{10.43}$$

Notice that if $\omega$ and $x'$ are independent, the integrals of equation (10.43) can be factored and, recalling equation (10.33), we obtain

$$q_L(x) = \beta \int \frac{u'(x')}{u'(x)} f(x', x) \, dx' \cdot \int \omega h(\omega, x', x) d\omega = R_1(x)^{-1} E\{\omega | x\}. \tag{10.44}$$

Thus, the price of a lottery ticket is the price of a sure claim to one unit of consumption next period, times the expected payoff on a lottery ticket. There is no risk premium, since in a competitive market no one is in a position to impose risk on anyone else, and no premium need be charged for risks not borne.

*The Modigliani-Miller theorem*

The Modigliani and Miller theorem[14] asserts circumstances under which the total value (stocks plus debt) of a firm is independent of the firm's financial structure, that is, the particular evidences of indebtedness or ownership that it issues. Following Hirshleifer (1966) and Stiglitz (1969), the Modigliani-Miller

[14] See Modigliani and Miller (1958).

theorem can be proved easily in a setting with complete state-contingent markets.

Suppose that an agent starts a firm at time $t$ with a tree as its sole asset, and then immediately sells the firm to the public by issuing $S$ number of shares and $B$ number of bonds as follows. Each bond promises to pay off $r$ per period, and $r$ is chosen so that $rB$ is less than all possible realizations of future crops $y_{t+j}$. After payments to bondholders, the owners of issued shares are entitled to the residual crop. Thus, the dividend of an issued share is equal to $(y_{t+j} - rB)/S$ in period $t+j$. Let $p_t^B$ and $p_t^S$ be the equilibrium prices of an issued bond and share, respectively, which can be obtained by using the contingent claims prices,

$$p_t^B = \sum_{j=1}^{\infty} \int rQ_j(x_{t+j}, x_t)dx_{t+j}, \tag{10.45}$$

$$p_t^S = \sum_{j=1}^{\infty} \int \frac{y_{t+j} - rB}{S} Q_j(x_{t+j}, x_t)dx_{t+j}. \tag{10.46}$$

The total value of issued bonds and shares is then

$$p_t^B B + p_t^S S = \sum_{j=1}^{\infty} \int y_{t+j} Q_j(x_{t+j}, x_t)dx_{t+j}, \tag{10.47}$$

which, by equations (10.24) and (10.38), is equal to the tree's initial value $p_t$. Equation (10.47) exhibits the Modigliani-Miller proposition that the value of the firm, that is, the total value of the firm's bonds and equities, is independent of the number of bonds $B$ outstanding. The total value of the firm is also independent of the coupon rate $r$.

The total value of the firm is independent of the financing scheme because the equilibrium prices of issued bonds and shares adjust to reflect the riskiness inherent in any mix of liabilities. To illustrate these equilibrium effects, let us assume that $u(c_t) = \ln c_t$ and $y_{t+j}$ is i.i.d. over time so that $E_t(y_{t+j}) = E(y)$, and $\frac{1}{y_{t+j}}$ is also i.i.d. for all $j \geq 1$. With logarithmic preferences, the price of a tree $p_t$ is given by equation (10.25), and the other two asset prices are now

$$p_t^B = \sum_{j=1}^{\infty} E_t \left[ r\beta^j \frac{u'(x_{t+j})}{u'(x_t)} \right] = \frac{\beta}{1-\beta} rE(y^{-1})y_t, \tag{10.48}$$

$$p_t^S = \sum_{j=1}^{\infty} E_t \left[ \frac{y_{t+j} - rB}{S} \beta^j \frac{u'(x_{t+j})}{u'(x_t)} \right]$$

$$= \frac{\beta}{1-\beta}\left[1 - rBE(y^{-1})\right]\frac{y_t}{S}, \qquad (10.49)$$

where we have used equations (10.45), (10.46), and (10.38) and $y_t = x_t$. (The expression $[1 - rBE(y^{-1})]$ is positive because $rB$ is less than the lowest possible realization of $y$.) As can be seen, the price of an issued share depends negatively on the number of bonds $B$ and the coupon $r$, and also the number of shares $S$. We now turn to the expected rates of return on different assets, which should be related to their riskiness. First, notice that, with our special assumptions, the expected capital gains on issued bonds and shares are all equal to that of the underlying tree asset,

$$E_t\left[\frac{p^B_{t+1}}{p^B_t}\right] = E_t\left[\frac{p^S_{t+1}}{p^S_t}\right] = E_t\left[\frac{p_{t+1}}{p_t}\right] = E_t\left[\frac{y_{t+1}}{y_t}\right]. \qquad (10.50)$$

It follows that any differences in expected total rates of return on assets must arise from the expected yields due to next period's dividends and coupons. Use equations (10.25), (10.48), and (10.49) to get

$$\begin{aligned}
\frac{r}{p^B_t} &= \left\{\left[1 - E_t(y_{t+1})E_t(y^{-1}_{t+1})\right] + E_t(y_{t+1})E_t(y^{-1}_{t+1})\right\}\frac{r}{p^B_t} \\
&= \frac{1 - E(y)E(y^{-1})}{E(y^{-1})p_t} + \frac{E_t(y_{t+1})}{p_t} < E_t\left[\frac{y_{t+1}}{p_t}\right], \qquad (10.51)
\end{aligned}$$

$$\begin{aligned}
E_t&\left[\frac{(y_{t+1} - rB)/S}{p^S_t}\right] \\
&= \left\{\left[1 - rBE(y^{-1})\right] + rBE(y^{-1})\right\}E_t\left[\frac{(y_{t+1} - rB)/S}{p^S_t}\right] \\
&= \frac{E_t(y_{t+1} - rB)}{p_t} + \frac{rBE(y^{-1})E_t(y_{t+1} - rB)}{\left[1 - rBE(y^{-1})\right]p_t} \\
&= \frac{E_t(y_{t+1})}{p_t} + \frac{rB\left[E(y^{-1})E(y) - 1\right]}{\left[1 - rBE(y^{-1})\right]p_t} > E_t\left[\frac{y_{t+1}}{p_t}\right], \qquad (10.52)
\end{aligned}$$

where the two inequalities follow from Jensen's inequality, which states that $E(y^{-1}) > [E(y)]^{-1}$ for any random variable $y$. Thus, from equations (10.50)-(10.52), we can conclude that the firm's bonds (shares) earn a lower (higher) expected rate of return as compared to the underlying asset. Moreover, equation (10.52) shows that the expected rate of return on the issued shares is positively related to payments to bondholders $rB$. In other words, equity owners demand a higher expected return from a more leveraged firm because of the greater risk borne.

## Government debt

### The Ricardian proposition

We now use a version of Lucas's tree model to describe the Ricardian proposition that tax financing and bond financing of a given stream of government expenditures are equivalent.[15] This proposition may be viewed as an application of the Modigliani-Miller theorem to government finance and obtains under circumstances in which the government is essentially like a firm in the constraints that it confronts with respect to its financing decisions.

We add to Lucas's model a government that spends current output according to a nonnegative stochastic process $\{g_t\}$ that satisfies $g_t < y_t$ for all $t$. The variable $g_t$ denotes per capita government expenditures at $t$. For analytical convenience we assume that $g_t$ is thrown away, giving no utility to private agents. The government finances its expenditures by issuing one-period debt that is permitted to be state contingent, and with a stream of lump-sum per capita taxes $\{\tau_t\}$, a stream that we assume is a stochastic process expressible at time $t$ as a function of $x_t \equiv \{y_t, g_t\}$ and any debt from last period. The state of the economy is now a vector including the dividend $y_t$ and government expenditures $g_t$. We assume that $y_t$ and $g_t$ are jointly described by a Markov process with transition density $f(x_{t+1}, x_t) = f(\{y_{t+1}, g_{t+1}\}, \{y_t, g_t\})$ where

$$\begin{aligned}&\text{prob}\{y_{t+1} \leq y', g_{t+1} \leq g' | y_t = y, g_t = g\}\\&= \int_0^{y'} \int_0^{g'} f(\{z, w\}, \{y, g\})\, dw\, dz.\end{aligned}$$

[15] An article by Robert Barro (1974) promoted strong interest in the Ricardian proposition. Barro described the proposition in a context distinct from the present one but closely related to it. Barro used an overlapping generations model but assumed altruistic agents who cared about their descendants. Restricting preferences to ensure an operative bequest motive, Barro described an overlapping generations structure that is equivalent with a model with an infinitely lived representative agent. See chapter 9 for more on Ricardian equivalence.

We can here apply the three steps outlined earlier to construct equilibrium prices. Since taxation is lump sum without any distortionary effects, the competitive equilibrium consumption allocation still equals that of a planning problem where all agents are assigned the same Pareto weight. Thus, the social planning problem for our purposes is to maximize $E_0 \sum_{t=0}^{\infty} \beta^t u(c_t)$ subject to $c_t \leq y_t - g_t$, whose solution is $c_t = y_t - g_t$. Proceeding as we did in earlier sections, the equilibrium share price, interest rates, and state-contingent claims prices are described by

$$p_t = E_t \sum_{j=1}^{\infty} \beta^j \frac{u'(y_{t+j} - g_{t+j})}{u'(y_t - g_t)} y_{t+j}, \tag{10.53}$$

$$R_{jt}^{-1} = \beta^j E_t \frac{u'(y_{t+j} - g_{t+j})}{u'(y_t - g_t)}, \tag{10.54}$$

$$Q_j(x_{t+j}, x_t) = \beta^j \frac{u'(y_{t+j} - g_{t+j})}{u'(y_t - g_t)} f^j(x_{t+j}, x_t), \tag{10.55}$$

where $f^j(x_{t+j}, x_t)$ is the $j$-step-ahead transition function that, for $j \geq 2$, obeys equation (10.39). Notice that these equilibrium prices are independent of the government's tax and debt policy. Our next step in showing Ricardian equivalence is to demonstrate that the private agents' budget sets are also invariant to government financing decisions.

Turning first to the government's budget constraint, we have

$$g_t = \tau_t + \int Q(x_{t+1}, x_t) b_t(x_{t+1}) dx_{t+1} - b_{t-1}(x_t), \tag{10.56}$$

where $b_t(x_{t+1})$ is the amount of $(t+1)$ goods that the government promises at $t$ to deliver, provided the economy is in state $x_{t+1}$ at $(t+1)$. If the government decides to issue only one-period risk-free debt, for example, we have $b_t(x_{t+1}) = b_t$ for all $x_{t+1}$, so that

$$\begin{aligned} \int Q(x_{t+1}, x_t) b_t(x_{t+1}) dx_{t+1} &= b_t \int Q(x_{t+1}, x_t) dx_{t+1} \\ &= b_t / R_{1t}. \end{aligned}$$

Equation (10.56) then becomes

$$g_t = \tau_t + b_t / R_{1t} - b_{t-1}. \tag{10.57}$$

Equation (10.57) is a standard form of the government's budget constraint under conditions of certainty.

If we write the budget constraint (10.56) in the form

$$b_{t-1}(x_t) = \tau_t - g_t + \int Q(x_{t+1}, x_t) b_t(x_{t+1}) dx_{t+1},$$

and iterate upon it to eliminate future $b_{t+j}(x_{t+j+1})$, we eventually find that[16]

$$b_{t-1}(x_t) = \tau_t - g_t + \sum_{j=1}^{\infty} \int \left[\tau_{t+j} - g_{t+j}\right] Q_j(x_{t+j}, x_t) dx_{t+j}, \tag{10.58}$$

as long as

$$\lim_{k\to\infty} \int \int Q_k(x_{t+k}, x_t) \cdot Q(x_{t+k+1}, x_{t+k}) b_{t+k}(x_{t+k+1}) dx_{t+k+1}\, dx_{t+k} = 0. \tag{10.59}$$

A strictly positive limit of equation (10.59) can be ruled out by using the transversality condition for a private agent's holdings of government bonds $b_t^d(x_{t+1})$. (The superscript $d$ stands for demand and distinguishes the variable from government's supply of bonds.) As in the case of private bonds in equation (10.6) and shares in equation (10.7), an individual would be overaccumulating assets unless

$$\lim_{k\to\infty} E_t \beta^k u'(c_{t+k}) \int Q(x_{t+k+1}, x_{t+k}) b_{t+k}^d(x_{t+k+1}) dx_{t+k+1} \leq 0. \tag{10.60}$$

After invoking equation (10.55) and equilibrium conditions $c_{t+k} = y_{t+k} - g_{t+k}$, $b_{t+k}^d(x_{t+k+1}) = b_{t+k}(x_{t+k+1})$, we see that the left-hand sides of equations (10.59) and (10.60) are equal except for a factor of $[u'(y_t - g_t)]^{-1}$ known at time $t$. Therefore, transversality condition (10.60) evaluated in an equilibrium ensures that the limit of expression (10.59) is nonpositive. Next, we simply assume away the case of a strictly negative limit of expression (10.59), since it would correspond to a rather uninteresting situation where the government accumulates

[16] Repeated substitution, exchange of orders of integration, and use of the expression for $j$-step-ahead contingent-claim-pricing functions are the steps used in deriving the present value budget constraint from the preceding equation.

"paper claims" against the private sector by setting taxes higher than needed for financial purposes. Thus, equation (10.58) states that the value of government debt maturing at time $t$ equals the present value of the stream of government surpluses.

We now turn to a private agent's budget constraint at time $t$,

$$\begin{aligned} c_t + \tau_t + p_t s_{t+1} + \int Q(x_{t+1}, x_t) b_t^d(x_{t+1}) dx_{t+1} \\ \leq (p_t + y_t) s_t + b_{t-1}^d(x_t). \end{aligned} \tag{10.61}$$

We multiply the corresponding budget constraint in period $t+1$ by $Q(x_{t+1}, x_t)$ and integrate over $x_{t+1}$. The resulting expression is substituted into equation (10.61) by eliminating the purchases of government bonds in period $t$. The two consolidated budget constraints become

$$\begin{aligned} & c_t + \tau_t + \int [c_{t+1} + \tau_{t+1}] Q(x_{t+1}, x_t) dx_{t+1} \\ & + \left\{ p_t - \int [p_{t+1} + y_{t+1}] Q(x_{t+1}, x_t) dx_{t+1} \right\} s_{t+1} + \int p_{t+1} s_{t+2} Q(x_{t+1}, x_t) dx_{t+1} \\ & + \int Q_2(x_{t+2}, x_t) b_{t+1}^d(x_{t+2}) dx_{t+2} \leq (p_t + y_t) s_t + b_{t-1}^d(x_t), \end{aligned} \tag{10.62}$$

where the expression in braces is zero by an arbitrage argument (or an Euler equation). When continuing the consolidation of all future budget constraints, we eventually find that

$$\begin{aligned} c_t + \tau_t + \sum_{j=1}^{\infty} \int [c_{t+j} + \tau_{t+j}] Q_j(x_{t+j}, x_t) dx_{t+j} \\ \leq (p_t + y_t) s_t + b_{t-1}^d(x_t), \end{aligned} \tag{10.63}$$

where we have imposed limits equal to zero for the two terms involving $s_{t+k+1}$ and $b_{t+k}^d(x_{t+k+1})$ when $k$ goes to infinity. The two terms vanish because of transversality conditions (10.7) and (10.60) and the reasoning in the preceding paragraph. Thus, equation (10.63) states that the present value of the stream of consumption and taxes cannot exceed the agent's initial wealth at time $t$.

Finally, we substitute the government's present value budget constraint (10.58) into that of the representative agent (10.63) by eliminating the present value of taxes. Thereafter, we invoke market-clearing conditions $s_t = 1$ and $b^d_{t-1}(x_t) = b_{t-1}(x_t)$ and we use the equilibrium expressions for prices (10.53) and (10.55) to express $p_t$ as the sum of all future dividends discounted by the $j$-step-ahead pricing kernel. The result is

$$c_t + \sum_{j=1}^{\infty} \int c_{t+j} Q_j(x_{t+j}, x_t) dx_{t+j} \leq y_t - g_t + \sum_{j=1}^{\infty} \int \left[ y_{t+j} - g_{t+j} \right] Q_j(x_{t+j}, x_t) dx_{t+j}. \tag{10.64}$$

Given that equilibrium prices $Q_j(x_{t+j}, x_t)$ have been shown to be independent of the government's tax and debt policy, the implication of formula (10.64) is that the representative agents' budget set is also invariant to government financing decisions. Having no effects on prices and private agents' budget constraints, taxes and government debt do not affect private consumption decisions.

We can summarize this discussion with the following proposition:

RICARDIAN PROPOSITION: Equilibrium consumption and state-contingent prices depend only on the stochastic process for output $y_t$ and government expenditure $g_t$. In particular, consumption and state-contingent prices are both independent of the stochastic process $\tau_t$ for taxes.

In this model, the choices of the time pattern of taxes and government bond issues have no effect on any "relevant" equilibrium price or quantity. (Some asset prices may be affected, however.) The reason is that, as indicated by equations (10.56) and (10.58), larger deficits $(g_t - \tau_t)$, accompanied by larger values of government debt $b_t(x_{t+1})$, now signal future government surpluses. The agents in this model accumulate these government bond holdings and expect to use their proceeds to pay off the very future taxes whose prospects support the value of the bonds. Notice also that, given the stochastic process for $(y_t, g_t)$, the way in which the government finances its deficits (or invests its surpluses) is irrelevant. Thus it does not matter whether it borrows using short-term, long-term, safe, or risky instruments. This irrelevance of financing is an application of the Modigliani-Miller theorem. Equation (10.58) may be interpreted as stating that the present value of the government is independent of such financing decisions.

The next section elaborates on the significance that future government surpluses in equation (10.58) are discounted with contingent claims prices and not the risk-free interest rate, even though the government may choose to issue only safe debt. This distinction is made clear by using equations (10.54) and (10.55) to rewrite equation (10.58) as follows,

$$b_{t-1}(x_t) = \tau_t - g_t + \sum_{j=1}^{\infty} \Bigg\{ R_{jt}^{-1} E_t[\tau_{t+j} - g_{t+j}] + \text{cov}_t \left[ \beta^j \frac{u'(y_{t+j} - g_{t+j})}{u'(y_t - g_t)}, \, \tau_{t+j} - g_{t+j} \right] \Bigg\}. \tag{10.65}$$

*No Ponzi schemes*

Bohn (1995) considers a nonstationary discrete-state-space version of Lucas's tree economy to demonstrate the importance of using a proper criterion when assessing long-run sustainability of fiscal policy, that is, determining whether the government's present-value budget constraint and the associated transversality condition are satisfied as in equations (10.58) and (10.59) of the earlier model. The present-value budget constraint says that any debt at time $t$ must be repaid with future surpluses because the transversality condition rules out Ponzi schemes—financial trading strategies that involve rolling over an initial debt with interest forever.

At each date $t$, there is now a finite set of possible states of nature, and $h_t$ is the history of all past realizations including the current one. Let $\pi(h_{t+j}|h_t)$ be the probability of a history $h_{t+j}$, conditional on history $h_t$ having been realized up until time $t$. The dividend of a tree in period $t$ is denoted $y(h_t) > 0$, and can depend on the whole history of states of nature. The stochastic process is such that a private agent's expected utility remains bounded for any fixed fraction $c \in (0,1]$ of the stream $y(h_t)$, implying

$$\lim_{j \to \infty} E_t \beta^j u'(c_{t+j}) c_{t+j} = 0 \tag{10.66}$$

for $c_t = c \cdot y(h_t)$.[17]

---

[17] Expected lifetime utility is bounded if the sequence of "remainders" converges to zero,

$$0 = \lim_{k \to \infty} E_t \sum_{j=k}^{\infty} \beta^j u(c_{t+j}) \geq \lim_{k \to \infty} E_t \sum_{j=k}^{\infty} \beta^j \left\{ u'(c_{t+j}) c_{t+j} \right\} \geq 0,$$

where the first inequality is implied by concavity of $u(\cdot)$. We obtain equation (10.66) because $u'(c_{t+j})c_{t+j}$ is positive at all dates.

Bohn (1995) examines the following government policy. Government spending is a fixed fraction $(1-c) = g_t/y_t$ of income. The government issues safe one-period debt so that the ratio of end-of-period debt to income is constant at some level $b = R_{1t}^{-1} b_t / y_t$. Given any initial debt, taxes can then be computed from budget constraint (10.57). It is intuitively clear that this policy can be sustained forever, but let us formally show that the government's transversality condition holds in any period $t$, given history $h_t$,

$$\lim_{j\to\infty} \sum_{h_{t+j+1}} Q(h_{t+j+1}|h_t) b_{t+j} = 0, \tag{10.67}$$

where $Q(h_{t+j}|h_t)$ is the price at $t$, given history $h_t$, of a unit of consumption good to be delivered in period $t+j$, contingent on the realization of history $h_{t+j}$. In an equilibrium, we have

$$Q(h_{t+j}|h_t) = \beta^j \frac{u'\left[c \cdot y(h_{t+j})\right]}{u'\left[c \cdot y(h_t)\right]} \pi(h_{t+j}|h_t). \tag{10.68}$$

After substituting equation (10.68), the debt policy, and $c_t = c \cdot y_t$ into the left-hand side of equation (10.67),

$$\begin{aligned}
&\lim_{j\to\infty} E_t \left[\beta^{j+1} \frac{u'(c_{t+j+1})}{u'(c_t)} R_{1,t+j}\, b \frac{c_{t+j}}{c}\right] \\
&= \lim_{j\to\infty} E_t E_{t+j} \left[\beta^j \frac{u'(c_{t+j})}{u'(c_t)} \beta \frac{u'(c_{t+j+1})}{u'(c_{t+j})} R_{1,t+j}\, b \frac{c_{t+j}}{c}\right] \\
&= \frac{b}{c\, u'(c_t)} \lim_{j\to\infty} E_t \left[\beta^j u'(c_{t+j}) c_{t+j}\right] \;=\; 0.
\end{aligned}$$

The first of these equalities invokes the law of iterated expectations; the second equality uses the equilibrium expression for the one-period interest rate, which is still given by expression (10.54); and the final equality follows from (10.66). Thus, we have shown that the government's transversality condition and therefore its present-value budget constraint are satisfied.

Bohn (1995) cautions us that this conclusion of fiscal sustainability might erroneously be rejected if we instead use the risk-free interest rate to compute present values. To derive expressions for the safe interest rate, we assume that preferences are given by the constant relative risk-aversion utility function $u(c_t) = (c_t^{1-\gamma} - 1)/(1-\gamma)$, and the dividend $y_t$ grows at the rate $\tilde{y}_t = y_t / y_{t-1}$

which is i.i.d. with mean $E(\tilde{y})$. Thus, risk-free interest rates given by equation (10.54) become

$$R_{jt}^{-1} = E_t \left[ \beta^j \left( \prod_{i=1}^{j} \tilde{y}_{t+i} \right)^{-\gamma} \right] = \prod_{i=1}^{j} E\left(\beta \tilde{y}^{-\gamma}\right) = R_1^{-j},$$

where $R_1$ is the time-invariant one-period risk-free interest rate. That is, the term structure of interest rates obeys the pure expectations theory, since interest rates are nonstochastic. [The analogue to expression (10.37) for this economy would therefore be one where the covariance term is zero.]

For the sake of the argument, we now compute the expected value of future government debt discounted at the safe interest rate and take the limit

$$\lim_{j\to\infty} E_t\left(\frac{b_{t+j}}{R_{j+1,t}}\right) = \lim_{j\to\infty} E_t\left(\frac{R_{1,t+j}\, b y_{t+j}}{R_{j+1,t}}\right) = \lim_{j\to\infty} E_t\left(\frac{R_1\, b y_t \prod_{i=1}^{j} \tilde{y}_{t+i}}{R_1^{j+1}}\right)$$

$$= b y_t \lim_{j\to\infty} \left[\frac{E(\tilde{y})}{R_1}\right]^j = \begin{cases} 0, & \text{if } R_1 > E(\tilde{y}); \\ b y_t, & \text{if } R_1 = E(\tilde{y}); \\ \infty, & \text{if } R_1 < E(\tilde{y}). \end{cases} \tag{10.69}$$

The limit is infinity if the expected growth rate of dividends $E(\tilde{y})$ exceeds the risk-free rate $R_1$. The level of the safe interest rate depends on risk aversion and on the variance of dividend growth. This dependence is best illustrated with an example. Suppose there are two possible states of dividend growth that are equally likely to occur with a mean of 1 percent, $E(\tilde{y}) - 1 = .01$, and let the subjective discount factor be $\beta = .98$. Figure 10.1 depicts the equilibrium interest rate $R_1$ as a function of the standard deviation of dividend growth and the coefficient of relative risk aversion $\gamma$. For $\gamma = 0$, agents are risk neutral, so the interest rate is given by $\beta^{-1} \approx 1.02$ regardless of the amount of uncertainty. When making agents risk averse by increasing $\gamma$, there are two opposing effects on the equilibrium interest rate. On the one hand, higher risk aversion implies also that agents are less willing to substitute consumption over time. Therefore, there is an upward pressure on the interest rate to make agents accept an upward-sloping consumption profile. This fact completely explains the positive relationship between $R_1$ and $\gamma$ when the standard deviation of growth is zero, that is, when deterministic growth is 1 percent. On the other hand, higher risk aversion in an uncertain environment means that agents attach a higher value to sure claims to future consumption, which tends to increase the bond price

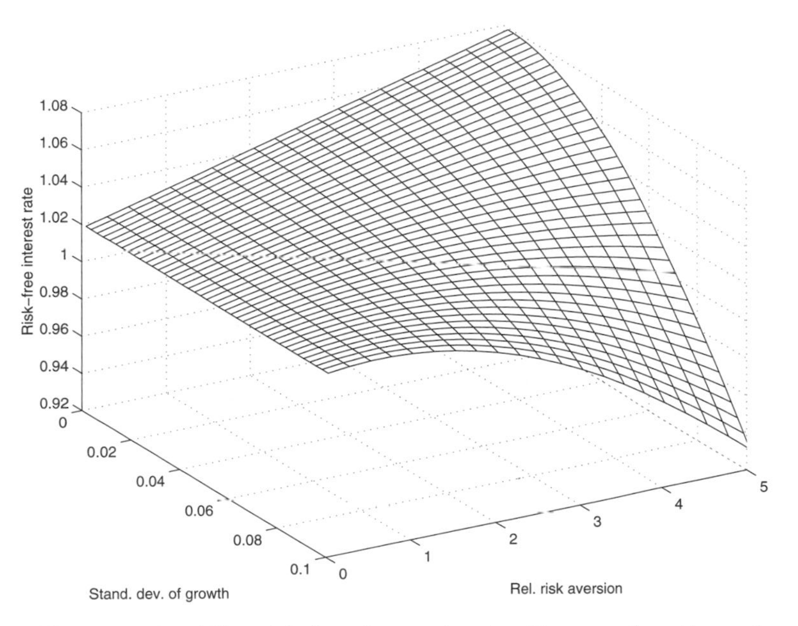


**Figure 10.1** The risk-free interest rate $R_1$ as a function of the coefficient of relative risk aversion $\gamma$ and the standard deviation of dividend growth. There are two states of dividend growth that are equally likely to occur with a mean of 1 percent, $E(\tilde{y}) - 1 = .01$, and the subjective discount factor is $\beta = .98$.

$R_1^{-1}$. As a result, Figure 10.1 shows how the risk-free interest $R_1$ falls below the expected gross growth rate of the economy when agents are sufficiently risk averse and the standard deviation of dividend growth is sufficiently large.[18]

If $R_1 \leq E(\tilde{y})$ so that the expected value of future debt discounted at the safe interest rate does not converge to zero in equation (10.69), it follows that the expected sum of all future government surpluses discounted at the safe interest rate in equation (10.65) falls short of the initial debt. In fact, our example is then associated with negative expected surpluses at all future horizons,

$$E_t\left(\tau_{t+j} - g_{t+j}\right) = E_t\left(b_{t+j-1} - b_{t+j}/R_{1,t+j}\right) = E_t\left[(R_1 - \tilde{y}_{t+j})by_{t+j-1}\right]$$

[18] A risk-free interest rate less than the growth rate would indicate dynamic inefficiency in a deterministic steady state but not necessarily in a stochastic economy. Our model here of an infinitely lived representative agent is dynamically efficient. For discussions of dynamic inefficiency, see Diamond (1965) and Romer (1996, chap. 2).

$$= [R_1 - E(\tilde{y})]\, b\, [E(\tilde{y})]^{j-1}\, y_t \begin{cases} > 0, & \text{if } R_1 > E(\tilde{y}); \\ = 0, & \text{if } R_1 = E(\tilde{y}); \\ < 0, & \text{if } R_1 < E(\tilde{y}); \end{cases} \tag{10.70}$$

where the first equality invokes budget constraint (10.57). Thus, for $R_1 \leq E(\tilde{y})$, the sum of covariance terms in equation (10.65) must be positive. The described debt policy also clearly has this implication where, for example, a low realization of $\tilde{y}_{t+j}$ implies a relatively high marginal utility of consumption and at the same time forces taxes up in order to maintain the targeted debt-income ratio in the face of a relatively low $y_{t+j}$.

As pointed out by Bohn (1995), this example illustrates the problem with empirical studies, such as Hamilton and Flavin (1986), Wilcox (1989), Hansen, Roberds, and Sargent (1991), Gali (1991), and Roberds (1996), which rely on safe interest rates as discount factors when assessing the sustainability of fiscal policy. Such an approach would only be justified if future government surpluses were uncorrelated with future marginal utilities so that the covariance terms in equation (10.65) would vanish. This condition is trivially true in a nonstochastic economy or if agents are risk neutral; otherwise, it is difficult, in practice, to imagine a tax and spending policy that is uncorrelated with the difference between aggregate income and government spending that determines the marginal utility of consumption.

## Interpretation of risk-aversion parameter

The next section will describe the equity premium puzzle. The equity premium depends on the consumer's willingness to bear risks, as determined by the curvature of a one-period utility function. To help understand why the measured equity premium is a puzzle it is important to interpret a parameter that measures curvature in terms of an experiment about choices between gambles. Economists' prejudice that reasonable values of the coefficient of relative risk aversion must be below 3 comes from such experiments.

The asset-pricing literature often uses the constant relative risk-aversion utility function

$$u(c) = (1-\gamma)^{-1} c^{1-\gamma}.$$

Note that

$$\gamma = \frac{-c u''(c)}{u'(c)},$$

which is the individual's coefficient of relative risk aversion.

We want to interpret the parameter $\gamma$ in terms of a preference for avoiding risk. Following Pratt (1964), consider offering two alternatives to a consumer who starts off with risk-free consumption level $c$: he can receive $c - \pi$ with certainty or a lottery paying $c - y$ with probability .5 and $c + y$ with probability .5. For given values of $y$ and $c$, we want to find the value of $\pi = \pi(y, c)$ that leaves the consumer indifferent between these two choices. That is, we want to find the function $\pi(y, c)$ that solves

$$u[c - \pi(y, c)] = .5u(c + y) + .5u(c - y). \tag{10.71}$$

For given values of $c, y$, we can solve the nonlinear equation (10.71) for $\pi$.

Alternatively, for small values of $y$, we can appeal to Pratt's local argument. Taking a Taylor series expansion of $u(c - \pi)$ gives[19]

$$u(c - \pi) = u(c) - \pi u'(c) + O(\pi^2). \tag{10.72}$$

Taking a Taylor series expansion of $u(c + \tilde{y})$ gives

$$u(c + \tilde{y}) = u(c) + \tilde{y}u'(c) + \frac{1}{2}\tilde{y}^2 u''(c) + O(\tilde{y}^3), \tag{10.73}$$

where $\tilde{y}$ is the random variable that takes value $y$ with probability .5 and $-y$ with probability .5. Taking expectations on both sides gives

$$Eu(c + \tilde{y}) = u(c) + \frac{1}{2}y^2 u''(c) + o(y^2). \tag{10.74}$$

Equating formulas (10.72) and (10.74) and ignoring the higher order terms gives

$$\pi(y, c) \approx \frac{1}{2}y^2 \left[\frac{-u''(c)}{u'(c)}\right].$$

For the constant relative risk-aversion utility function, we have

$$\pi(y, c) \approx \frac{1}{2}y^2 \frac{\gamma}{c}.$$

[19] Here $O(\cdot)$ means terms of order at most $(\cdot)$, while $o(\cdot)$ means terms of smaller order than $(\cdot)$.

This can be expressed as

$$\pi/y = \frac{1}{2}\gamma\,(y/c)\,. \tag{10.75}$$

The left side is the premium that the consumer is willing to pay to avoid a fair bet of size $y$; the right side is one-half $\gamma$ times the ratio of the size of the bet $y$ to his initial consumption level $c$.

**Table 10.1** Risk premium $\pi(y,c)$ for various values of $y$ and $\gamma$

| $\gamma \backslash y$ | 10 | 100 | 1,000 | 5,000 |
|---|---|---|---|---|
| 2 | .02 | .2 | 20 | 500 |
| 5 | .05 | 5 | 50 | 1,217 |
| 10 | .1 | 1 | 100 | 2,212 |

Following Cochrane (1997), think of confronting someone with initial consumption of $50,000 per year with a 50–50 chance of winning or losing $y$ dollars. How much would the person be willing to pay to avoid that risk? For $c = 50,000$, we calculated $\pi$ from equation (10.71) for values of $y = 10, 100, 1,000, 5,000$. See Table 10.1. A common reaction to these premiums is that for values of $\gamma$ even as high as 5, they are too big. This result is one important source of macroeconomists' prejudice that $\gamma$ should not be much higher than 2 or 3.

## *The equity premium puzzle*

Mehra and Prescott (1985) describe an empirical problem for the representative agent model of this chapter. For plausible parameterizations of the utility function, the model cannot explain the large differential in average yields on relatively riskless bonds and risky equity in the U.S. data over the ninety-year period 1889–1978, as depicted in Table 10.2. The average real yield on the Standard and Poor 500 index was 7 percent, while the average yield on short-term debt was only 1 percent. As pointed out by Kocherlakota (1996a), the theory is qualitatively correct in predicting a positive equity premium, but it fails quantitatively because stocks are not sufficiently riskier than bonds to rationalize a spread of 6 percentage points.[20]

[20] For recent reviews and possible resolutions of the equity premium puzzle, see Aiyagari (1993), Kocherlakota (1996a), and Cochrane (1997).

**Table 10.2** Summary statistics for U.S. annual data, 1889-1978

| | **Mean** | **Variance-Covariance** | | |
|---|---|---|---|---|
| | | $1 + r^s_{t+1}$ | $1 + r^b_{t+1}$ | $c_{t+1}/c_t$ |
| $1 + r^s_{t+1}$ | 1.070 | 0.0274 | 0.00104 | 0.00219 |
| $1 + r^b_{t+1}$ | 1.010 | | 0.00308 | −0.000193 |
| $c_{t+1}/c_t$ | 1.018 | | | 0.00127 |

The quantity $1 + r^s_{t+1}$ is the real return to stocks, $1 + r^b_{t+1}$ is the real return to relatively riskless bonds, and $c_{t+1}/c_t$ is the growth rate of per capita real consumption of nondurables and services.

Source: Kocherlakota (1996a, Table 1), who uses the same data as Mehra and Prescott (1985).

Rather than calibrating a general equilibrium model as in Mehra and Prescott (1985), we proceed in the fashion of Hansen and Singleton (1983) and demonstrate the equity premium puzzle by studying unconditional averages of Euler equations under assumptions of lognormal returns. Let the real rates of return on stocks and bonds between periods $t$ and $t+1$ be denoted $1 + r^s_{t+1}$ and $1 + r^b_{t+1}$, respectively. In our Lucas tree model, these numbers would be given by $1 + r^s_{t+1} = (y_{t+1} + p_{t+1})/p_t$ and $1 + r^b_{t+1} = R_{1t}$. Concerning the real rate of return on bonds, we now use time subscript $t+1$ to allow for uncertainty at time $t$ about its realization. Since the numbers in Table 10.2 are computed on the basis of nominal bonds, real bond yields are subject to inflation uncertainty. To allow for such uncertainty and to switch notation, we rewrite Euler equations (10.4) and (10.5) as

$$1 = \beta E_t \left[ (1 + r^i_{t+1}) \frac{u'(c_{t+1})}{u'(c_t)} \right], \qquad \text{for } i = s, b. \tag{10.76}$$

Departing from our earlier general equilibrium approach, we now postulate exogenous stochastic processes for both endowments (consumption) and rates of return,

$$\frac{c_{t+1}}{c_t} = \bar{c}_{\triangle} \exp\left\{\epsilon_{c,t+1} - \sigma_c^2/2\right\}, \tag{10.77}$$

$$1 + r^i_{t+1} = (1 + \bar{r}^i) \exp\left\{\epsilon_{i,t+1} - \sigma_i^2/2\right\}, \qquad \text{for } i = s, b; \tag{10.78}$$

where exp is the exponential function and $\{\epsilon_{c,t+1}, \epsilon_{s,t+1}, \epsilon_{b,t+1}\}$ are jointly normally distributed with zero means and variances $\{\sigma_c^2, \sigma_s^2, \sigma_b^2\}$. Thus, the logarithm of consumption growth and the logarithms of rates of return are log-normal distribution jointly normally distributed. When the logarithm of a variable is normally distributed with some mean $\mu$ and variance $\sigma^2$, the formula for the mean of the untransformed variable is $\exp(\mu + \sigma^2/2)$. Thus, the mean of consumption growth, and the means of real yields on stocks and bonds are here equal to $\bar{c}_\triangle$, $1+\bar{r}^s$, and $1+\bar{r}^b$, respectively.

As in the previous section, preferences are assumed to be given by the constant relative risk-aversion utility function $u(c_t) = (c_t^{1-\gamma} - 1)/(1-\gamma)$. After substituting this utility function and the stochastic processes (10.77) and (10.78) into equation (10.76), we take unconditional expectations of equation (10.76). By the law of iterated expectations, the result is

$$\begin{aligned} 1 &= \beta E\left[(1+r_{t+1}^i)\left(\frac{c_{t+1}}{c_t}\right)^{-\gamma}\right], \\ &= \beta(1+\bar{r}^i)\bar{c}_\triangle^{-\gamma} E\left\{\exp\left[\epsilon_{i,t+1} - \sigma_i^2/2 - \gamma\left(\epsilon_{c,t+1} - \sigma_c^2/2\right)\right]\right\} \\ &= \beta(1+\bar{r}^i)\bar{c}_\triangle^{-\gamma} \exp\left[(1+\gamma)\gamma\sigma_c^2/2 - \gamma\,\mathrm{cov}(\epsilon_i, \epsilon_c)\right], \\ &\quad \text{for} \quad i = s, b; \end{aligned} \tag{10.79}$$

where the second equality follows from the expression in braces being lognormally distributed and the application of the preceding formula for computing its mean. Taking logarithms of equation (10.79) yields

$$\log(1+\bar{r}^i) = -\log(\beta) + \gamma\log(\bar{c}_\triangle) - (1+\gamma)\gamma\sigma_c^2/2 + \gamma\,\mathrm{cov}(\epsilon_i, \epsilon_c), \quad \text{for} \quad i = s, b. \tag{10.80}$$

It is informative to interpret equation (10.80) for the risk-free interest rate in the model of the section on Bohn's model, under the auxiliary assumption of log-normally distributed dividend growth so that equilibrium consumption growth is given by equation (10.77). Since interest rates are time invariant, we have $\mathrm{cov}(\epsilon_b, \epsilon_c) = 0$. In the case of risk-neutral agents ($\gamma = 0$), equation (10.80) has the familiar implication that the interest rate is equal to the inverse of the subjective discount factor $\beta$ regardless of any uncertainty. In the case of deterministic growth ($\sigma_c^2 = 0$), the second term of equation (10.80) says that the safe interest rate is positively related to the coefficient of relative risk aversion $\gamma$, as we also found in the example of Figure 10.1. Likewise, the downward pressure on the interest rate due to uncertainty in Figure 10.1 shows up as the third term of equation (10.80). Since the term involves the square of $\gamma$, the safe interest rate must eventually be a decreasing function of the coefficient of relative risk aversion when $\sigma_c^2 > 0$.

We now turn to the equity premium by taking the difference between the expressions for the rates of return on stocks and bonds, as given by equation (10.80),

$$\log(1 + \bar{r}^s) - \log(1 + \bar{r}^b) = \gamma \left[ \text{cov}(\epsilon_s, \epsilon_c) - \text{cov}(\epsilon_b, \epsilon_c) \right]. \tag{10.81}$$

Using the approximation $\log(1 + r) \approx r$, and noting that the covariance between consumption growth and real yields on bonds in Table 10.2 is virtually zero, we can write the theory's interpretation of the historical equity premium as

$$\bar{r}^s - \bar{r}^b \approx \gamma \, \text{cov}(\epsilon_s, \epsilon_c). \tag{10.82}$$

After approximating $\text{cov}(\epsilon_s, \epsilon_c)$ with the covariance between consumption growth and real yields on stocks in Table 10.2, equation (10.82) states that an equity premium of 6 percent would require a $\gamma$ of 27. Kocherlakota (1996a, p. 52) summarizes the prevailing view that "a vast majority of economists believe that values of $[\gamma]$ above ten (or, for that matter, above five) imply highly implausible behavior on the part of individuals." That statement is a reference to the argument of Pratt, described in the preceding section. This constitutes the equity premium puzzle. Mehra and Prescott (1985) and Weil (1989) point out that an additional part of the puzzle relates to the low historical mean of the riskless rate of return. According to equation (10.80) for bonds, a high $\gamma$ is needed to rationalize an average risk-free rate of only 1 percent given historical consumption data and the standard assumption that $\beta$ is less than one.[21]

## *Market price of risk*

Gallant, Hansen, and Tauchen (1990) and Hansen and Jagannathan (1991) interpret the equity premium puzzle in terms of the high "market price of risk" implied by time-series data on asset returns. The market price of risk is defined

[21] For $\beta < 0.99$, equation (10.80) for bonds with data from Table 10.1 produces a coefficient of relative risk aversion of at least 27. If we use the lower variance of the growth rate of U.S. consumption in post–World War II data, the implied $\gamma$ exceeds 200 as noted by Aiyagari (1993).

in terms of asset prices and their one-period payoffs. Let $q_t$ be the time-$t$ price of an asset bearing a one-period payoff $p_{t+1}$. A household's Euler equation for holdings of this asset can be represented as

$$q_t = E_t(m_{t+1}p_{t+1}) \tag{10.83}$$

where $m_{t+1} = \frac{\beta u'(c_{t+1})}{u'(c_t)}$ serves as a stochastic discount factor for discounting the stochastic return $p_{t+1}$. Using the definition of a conditional covariance, equation (10.83) can be written[22]

$$q_t = E_t m_{t+1} E_t p_{t+1} + \text{cov}_t(m_{t+1}, p_{t+1}).$$

Applying the Cauchy-Schwarz inequality[23] to the covariance term in the preceding equation gives

$$\frac{q_t}{E_t m_{t+1}} \geq E_t p_{t+1} - \left(\frac{\text{std}_t m_{t+1}}{E_t m_{t+1}}\right) \text{std}_t p_{t+1}, \tag{10.84}$$

where $\text{std}_t$ denotes a conditional standard deviation. In expression (10.84), the term $\left(\frac{\text{std}_t m_{t+1}}{E_t m_{t+1}}\right)$ is called the market price of risk. According to expression (10.84), it provides an estimate of the rate at which the price of a security falls with an increase in the conditional standard deviation of its payoff.

Gallant, Hansen, and Tauchen (1990) and Hansen and Jagannathan (1991) used asset prices and returns alone to estimate the market price of risk, without imposing the link to consumption data implied by any particular specification of a stochastic discount factor. Their version of the equity premium puzzle is that the market price of risk implied by the asset market data alone is much higher than can be reconciled with the aggregate consumption data, say, with a specification that $m_{t+1} = \beta\left(\frac{c_{t+1}}{c_t}\right)^{-\gamma}$. Aggregate consumption is not volatile enough to make the standard deviation of the object high enough for the reasonable values of $\gamma$ that we have discussed.

In the next section, we describe how Hansen and Jagannathan coaxed evidence about the market price of risk from asset prices and one-period returns.

[22] Note that $E_t m_{t+1}$ is the reciprocal of the gross one-period risk-free return.

[23] The Cauchy-Schwarz inequality is $\frac{|\text{cov}_t(m_{t+1}, p_{t+1})|}{\text{std}_t m_{t+1} \text{std}_t p_{t+1}} \leq 1$.

## Hansen-Jagannathan bounds

Our earlier exposition of the equity premium puzzle based on the lognormal specification of returns was highly parametric, being tied to particular specifications of preferences and the distribution of asset returns. Hansen and Jagannathan (1991) described a nonparametric way of summarizing the equity premium puzzle. Their work can be regarded as substantially generalizing Robert Shiller's and Stephen LeRoy's earlier work on variance bounds to handle stochastic discount factors.[24] We present one of Hansen and Jagannathan's bounds.

Hansen and Jagannathan are interested in restricting asset prices possibly in more general settings than we have studied so far. We have described a theory that prices assets in terms of a particular "stochastic discount factor," defined as $m_{t+1} = \beta \frac{u'(c_{t+1})}{u'(c_t)}$. The theory asserted that the price at $t$ of an asset with one-period payoff $p_{t+1}$ is $E_t m_{t+1} p_{t+1}$. Hansen and Jagannathan were interested in more general models, in which the stochastic discount factor could assume other forms.

Following Hansen and Jagannathan, let $x^j$ be a random payoff on a security. Let there be $J$ basic securities, so $j = 1, \ldots, J$. Thus, let $x \in I\!R^J$ be a random vector of payoffs on the basic securities. Assume that the $J \times J$ matrix $Exx'$ exists. Also assume that a $J \times 1$ vector $q$ of prices on the basic securities is observed, where the $j$th component of $q$ is the price of the $j$th component of the payoff vector $x_j$. Consider forming portfolios of the primitive securities. We want to determine the relationship of the prices of portfolios to the prices of the basic securities from which they have been formed. With this in mind, let $c \in I\!R^J$ be a vector of portfolio weights. The return on a portfolio with weights $c$ is $c \cdot x$.

Define the space of payouts attainable from portfolios of the basic securities:

$$P \equiv \left\{ p : p = c \cdot x \text{ for some } c \in I\!R^J \right\}.$$

We want to price portfolios, that is, payouts, in $P$. We seek a price functional $\pi$ mapping $P$ into $I\!R$: $\pi : P \to I\!R$. Because $q$ is observed, we insist that $q = \pi(x)$, that is, $q_j = \pi(x_j)$.

Note that $\pi(c \cdot x)$ is the value of a portfolio costing $c \cdot q$. The *law of one price* asserts that the value of a portfolio equals what it costs:

$$c \cdot q = \pi(c \cdot x).$$

---

[24] See Hansen's (1982a) early call for such a generalization.

The law of one price states that the pricing functional $\pi$ is linear on $P$.

An aspect of the law of one price is that $\pi(c \cdot x)$ depends on $c \cdot x$, not on $c$. If any other portfolio has return $c \cdot x$, it should also be priced at $\pi(c \cdot x)$. Thus, two portfolios with the same payoff have the same price:

$$\pi(c_1 \cdot x) = \pi(c_2 \cdot x) \text{ if } c_1 \cdot x = c_2 \cdot x.$$

If the $x$'s are *returns*, then $q = \mathbf{1}$, the unit vector, and

$$\pi(c \cdot x) = c \cdot \mathbf{1}.$$

*Inner product representation of the pricing kernel*

If $y$ is a scalar random variable, $E(yx)$ is the vector whose $j$th component is $E(yx_j)$. The cross-moments $E(yx)$ are called the inner product of $x$ and $y$. According to the Riesz representation theorem, a linear functional can be represented as the inner product of the random payoff with *some* scalar random variable $y$. This random variable is called a stochastic discount factor. Thus, a *stochastic discount factor* is a scalar random variable $y$ that makes the following equation true:

$$\pi(p) = E(yp) \; \forall p \in P. \tag{10.85}$$

For example, the vector of prices of the primitive securities, $q$, satisfies

$$q = E(yx). \tag{10.86}$$

Because it implies that the pricing functional is linear, the law of one price implies that there exists a stochastic discount factor. In fact, there exist many stochastic discount factors. Hansen and Jagannathan sought to characterize admissible discount factors.

Note

$$\text{cov}(y, p) = E(yp) - E(y)E(p),$$

which implies that the price functional can be represented as

$$\pi(p) = E(y)E(p) + \text{cov}(y, p).$$

This expresses the price of a portfolio as the expected return times the expected discount factor plus the covariance between the return and the discount factor.

Notice that the expected discount factor is simply the price of a sure scalar payoff of unity:

$$\pi(1) = E(y).$$

The linearity of the pricing functional leaves open the possibility that prices of some portfolios are negative. This would open up arbitrage opportunities. David Kreps (1979) showed that the principle that the price system should offer *no arbitrage* opportunities requires that the stochastic discount factor be strictly positive. For most of this section, we shall not impose the principle of no arbitrage, just the law of one price. Thus, we do not require stochastic discount factors to be positive.

### *Classes of stochastic discount factors*

In previous sections we constructed structural models of the stochastic discount factor. In particular, for the stochastic discount factor, our theories typically advocated

$$y = m_t \equiv \frac{\beta u'(c_{t+1})}{u'(c_t)}, \tag{10.87}$$

the intertemporal substitution of consumption today for consumption tomorrow. For a particular utility function, this specification leads to a parametric form of the stochastic discount factor that depends on the random consumption of a particular consumer or set of consumers.

Hansen and Jagannathan want to approach the data with a *class* of stochastic discount factors. To begin, Hansen and Jagannathan note that one candidate for a stochastic discount factor is

$$y^* = x'(Exx')^{-1}q. \tag{10.88}$$

This can be verified directly, by substituting into equation (10.86) and verifying that $q = E(y^* x)$.

Besides equation (10.88), many other stochastic discount factors work, in the sense of pricing the random returns $x$ correctly, that is, recovering $q$ as their price. It can be verified directly that any other $y$ that satisfies

$$y = y^* + e$$

is also a stochastic discount factor, where $e$ is orthogonal to $x^*$. Let $\mathcal{Y}$ be the space of all stochastic discount factors.

*A Hansen-Jagannathan bound*

Given data on $q$ and the distribution of returns $x$, Hansen and Jagannathan wanted to infer properties of $y$ while imposing no more structure than linearity of the pricing functional (the law of one price). Imposing only this, they constructed bounds on the first and second moments of stochastic discount factors $y$ that are consistent with a given distribution of payoffs on a set of primitive securities. For $y \in \mathcal{Y}$, here is how they constructed one of their bounds:

Let $y$ be an unobserved stochastic discount factor. Though $y$ is unobservable, we can represent it in terms of the population linear regression[25]

$$y = a + x'b + e \tag{10.89}$$

where $e$ is orthogonal to $x$ and

$$\begin{aligned} b &= [\text{cov}(x, x)]^{-1} \text{cov}(x, y) \\ a &= Ey - Ex'b. \end{aligned}$$

Here $\text{cov}(x, x) = E(xx)' - E(x)E(x)'$. We have data that allow us to estimate the second-moment matrix of $x$, but no data on $y$ and therefore on $\text{cov}(x, y)$. But we do have data on $q$, the vector of security prices. So Hansen and Jagannathan proceeded indirectly to use the data on $q, x$ to infer something about $y$. Notice that $q = E(yx)$ implies $\text{cov}(x, y) = q - E(y)E(x)$. Therefore

$$b = [\text{cov}(x, x)]^{-1} [q - E(y)E(x)]. \tag{10.90}$$

Thus, *given* a guess about $E(y)$, asset returns and prices can be used to estimate $b$. Because the residuals in equation (10.89) are orthogonal to $x$,

$$\text{var}(y) = \text{var}(x'b) + \text{var}(e).$$

Therefore

$$[\text{var}(x'b)]^{.5} \leq \sigma(y), \tag{10.91}$$

where $\sigma(y)$ denotes the standard deviation of the random variable $y$. This is the lower bound on the standard deviation of all[26] stochastic discount factors with

[25] See chapter 1 for the definition and construction of a population linear regression.

[26] The stochastic discount factors are not necessarily positive. Hansen and Jagannathan (1991) derive another bound that imposes positivity.

prespecified mean $E(y)$. For various specifications, Hansen and Jagannathan used expressions (10.90) and (10.91) to compute the bound on $\sigma(y)$ as a function of $E(y)$, tracing out a frontier of admissible stochastic discount factors in terms of their means and standard deviations.

Here are two such specifications. First, recall that a (gross) return for an asset with price $q$ and payoff $x$ is defined as $z = x/q$. A return is risk free if $z$ is constant (not random). Then note that if there is an asset with risk-free return $z^{RF} \in x$, it follows that $E(yz^{RF}) = z^{RF}Ey = 1$, and therefore $Ey$ is a known constant. Then there is only one point on the frontier that is of interest, the one with the known $E(y)$. If there is no risk-free asset, we can calculate a different bound for every specified value of $E(y)$.

Second, take a case where $E(y)$ is not known because there is no risk-free payout in the set of returns. Suppose, for example, that the data set consists of "excess returns." Let $x^s$ be a return on a stock portfolio and $x^b$ be a return on a risk-free bond. Let $z = x^s - x^b$ be the excess return. Then

$$E[yz] = 0.$$

Thus, for an excess return, $q = 0$, so formula (10.90) becomes[27]

$$b = -[\text{cov}(z,z)]^{-1}E(y)E(z).$$

Then

$$\text{var}(z'b) = E(y)^2 E(z)'[\text{cov}(z,z)^{-1}]E(z).$$

Therefore, the Hansen-Jagannathan bound becomes

$$\sigma(y) \geq [E(z)'\text{cov}(z,z)^{-1}E(z)]^{.5}E(y). \tag{10.92}$$

In the special case of a scalar excess return, (10.92) becomes

$$\frac{\sigma(y)}{E(y)} \geq \frac{E(z)}{\sigma(z)}. \tag{10.93}$$

The left side, the ratio of the standard deviation of the discount factor to its mean, is called the *market price of risk.* Thus, the bound (10.93) says that the market price of risk is at least $\frac{E(z)}{\sigma(z)}$. The ratio $\frac{E(z)}{\sigma(z)}$ thus determines a straight-line frontier in the $[E(y), \sigma(y)]$ plane above which the stochastic discount factor must reside.

---

[27] This formula follows from $\text{var}(b'z) = b'\text{cov}(z,z)b$.

For a set of returns, $q = \mathbf{1}$ and equation (10.90) becomes

$$b = [\text{cov}(x,x)]^{-1}[\mathbf{1} - E(y)E(x)]. \tag{10.94}$$

The bound is computed by solving equation (10.94) and

$$\sqrt{b'\text{cov}(x,x)b} \leq \sigma(y). \tag{10.95}$$

In more detail, we compute the bound for various values of $E(y)$ by using equation (10.94) to compute $b$, then using that $b$ in expression (10.95) to compute the lower bound on $\sigma(y)$.

Cochrane and Hansen (1992) used data on two returns, the real return on a value-weighted NYSE stock return and the real return on U.S. Treasury bills. They used the excess return of stocks over Treasury bills to compute bound (10.92) and both returns to compute equation (10.94). The bound (10.94) is a parabola, while formula (10.92) is a straight line in the $[E(y), \sigma(y)]$ plane.

### *The Mehra-Prescott data*

In exercise 10.1, we ask you to calculate the Hansen-Jagannathan bounds for the annual U.S. time series studied by Mehra and Prescott. Figures 10.2 and 10.3 describe the basic data and the bounds that you should find.[28]

Figure 10.2 plots annual gross real returns on stocks and bills in the United States for 1889 to 1979, and Figure 10.3 plots the annual gross rate of consumption growth. Notice the extensive variability around the mean returns of (1.01, 1.069) apparent in Figure 10.2.

Figure 10.4 plots the Hansen-Jagannathan bounds for these data, obtained by treating the sample second moments as population moments in the preceding formulas. For $\beta = .99$, we have also plotted the mean and standard deviation of the candidate stochastic discount factor $\beta\lambda_t^{-\gamma}$, where $\lambda_t$ is the gross rate of consumption growth and $\gamma$ is the coefficient of relative risk aversion. Figure 10.4 plots the mean and standard deviation of candidate discount factors for $\gamma = 0$ (the square), $\gamma = 7.5$ (the circle), $\gamma = 15$ (the diamond), and $\gamma = 22.5$ (the triangle). Notice that it takes a high value of $\gamma$ to bring the stochastic discount factor within the bounds for these data. This is Hansen and Jagannathan's statement of the equity premium puzzle.

---

[28] These bounds were computed using the Matlab programs `hjbnd1.m` and `hjbnd2.m`.

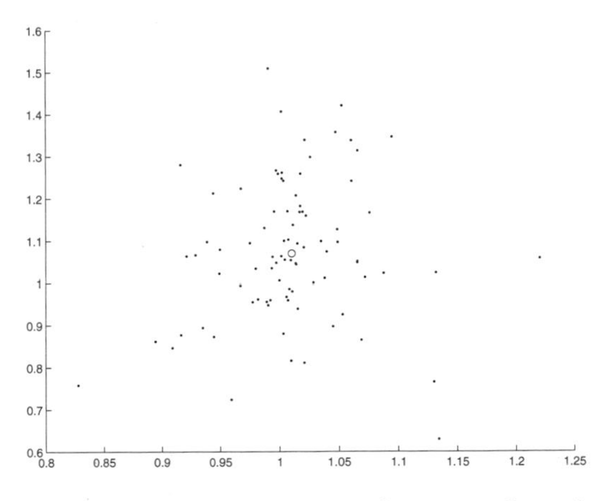


**Figure 10.2** Scatter plot of gross real stock returns ($y$ axis) against real Treasury bill return ($x$-axis), annual data 1889–1979. The circle denotes the means, (1.010, 1.069).

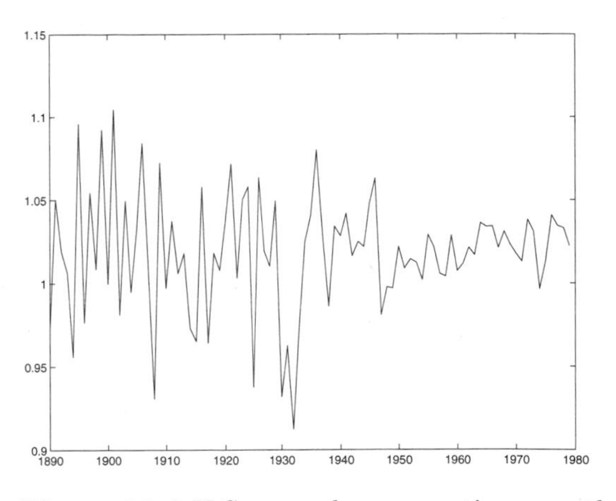


**Figure 10.3** U.S. annual consumption growth, 1889–1979.

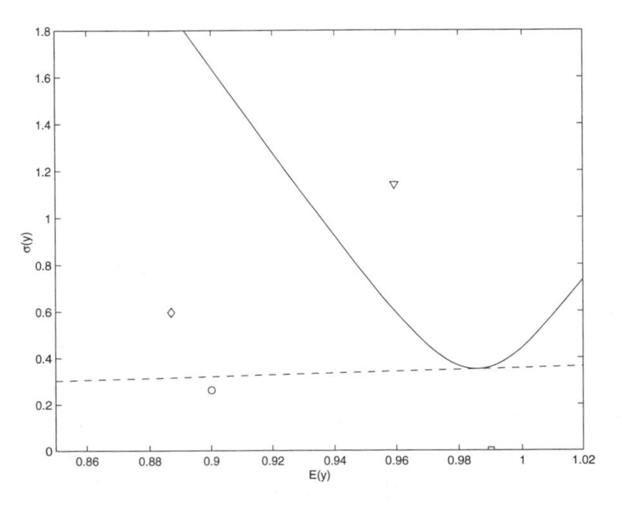


**Figure 10.4** Hansen-Jagannathan bounds for excess return of stock over bills (dotted line) and the stock and bill returns (solid line), U.S. annual data, 1889–1979.

## Factor models

In the two previous sections we have seen the equity premium puzzle that follows upon imposing that the stochastic discount factor is taken as $\lambda_t^{-\gamma}$, where $\lambda_t$ is the gross growth rate of consumption between $t$ and $t+1$ and $\gamma$ is the coefficient of relative risk aversion. In response to this puzzle, or empirical failure, researchers have resorted to "factor models." These preserve the law of one price and often the no-arbitrage principle, but they abandon the link between the stochastic discount factor and the consumption process. They posit a model-free process for the stochastic discount factor, and use the overidentifying restrictions from the household's Euler equations from a set of $N$ returns $R_{it+1}, i = 1, \ldots, N$, to let the data tell what the factors are.

Thus, suppose that we have a time series of data on returns $R_{i,t+1}$. The Euler equations are

$$E_t M_{t+1} R_{it+1} = 1, \tag{10.96}$$

for some stochastic discount factor $M_{t+1}$ that is unobserved by the econometrician. Posit the model

$$\log(M_{t+1}) = \alpha_0 + \sum_{j=1}^{k} \alpha_j f_{jt+1} \tag{10.97}$$

where the $k$ factors $f_{jt}$ are governed by the stochastic processes

$$f_{jt+1} = \beta_{j0} + \sum_{h=1}^{m} \beta_{jh} f_{j,t+1-h} + a_{jt+1}, \tag{10.98}$$

where $a_{jt+1}$ is a Gaussian error process with specified covariance matrix. This model keeps $M_{t+1}$ positive. The factors $f_{jt+1}$ may or may not be observed. Whether they are observed can influence the econometric procedures that are feasible. If we substitute equations (10.97) and (10.98) into equation (10.96) we obtain the $N$ sets of moment restrictions

$$E_t\Big\{\exp\Big[\alpha_0 + \sum_{j=1}^{k} \alpha_j \big(\beta_{j0} + \sum_{h=1}^{m} \beta_{jh} f_{j,t+1-h} + a_{jt+1}\big)\Big] R_{it+1}\Big\} = 1. \tag{10.99}$$

If current and lagged values of the factors $f_{jt}$ are observed, these conditions can be used to estimate the coefficients $\alpha_j, \beta_{jh}$ by the generalized method of moments. If the factors are not observed, by making the further assumption that the logs of returns are jointly normally distributed and by exploiting the assumption that the errors $a_{jt}$ are Gaussian, analytic solutions for $R_{i,t+1}$ as a function of current and lagged values of the $k$ factors can be attained, and these can be used to form a likelihood function.[29]

This structure is known as an affine factor model. The term "affine" describes the function (10.98) (linear plus a constant). This kind of model has been used extensively to study the term structure of interest rates. There the returns are taken to be a vector of one-period holding-period yields on bonds of different maturities.[30]

## *Heterogeneity and incomplete markets*

As Hansen and Jagannathan (1991) and the preceding analysis of the log-linear model both indicate, the equity premium reflects restrictions across returns and consumption imposed by Euler equations. These restrictions do not assume complete markets. A complete markets assumption might enter indirectly, to justify using aggregate consumption growth to measure the intertemporal rate of substitution.

The equity premium puzzle is that data on asset returns and aggregate consumption say that the equity premium is much larger than is predicted by Euler equations for asset holdings with a plausible coefficient of relative risk aversion $\gamma$. Gregory Mankiw (1986) posited a pattern of systematically varying spreads across individual's intertemporal rates of substitution that could magnify the theoretical equity premium. Mankiw's mechanism requires (a) incomplete markets, (b) a precautionary savings motive, in the sense of convex marginal utilities of consumption, and (c) a negative covariance between the cross-sectional variance of consumption and the aggregate level of consumption. To magnify the quantitative importance of Mankiw's mechanism, it helps if there are (d) highly persistent endowment processes.

---

[29] Sometimes even if the factors are unobserved, it is possible to deduce good enough estimates of them to proceed as though they are observed. Thus, in their empirical term structure model, Chen and Scott (1993) and Dai and Singleton (forthcoming) set the number of factors $k$ equal to the number of yields studied. Letting $R_t$ be the $k \times 1$ vector of yields and $f_t$ the $k \times 1$ vector of factors, they can solve equation (10.96) for an expression of the form $R = g_0 + g_1 f_t$ from which Chen and Scott could deduce $f_t = g_1^{-1}(R_t - g_0)$ to get observable factors. See Gong and Remolona (1997) for a discrete-time affine term-structure model.

[30] See Piazzesi (2000) for an ambitious factor model of the term structure where some of the factors are interpreted in terms of a monetary policy authority's rule for setting a short rate.

We shall study incomplete markets and precautionary savings models in chapters 13 and 14. But here is a sketch of Mankiw's idea: Consider a heterogeneous consumer economy. Let $M(g_{it})$ be the stochastic discount factor associated with consumer $i$, say, $M(g_{it}) = \beta g_{it}^{-\gamma}$, where $\beta$ is a constant discount factor, $g_{it}$ is consumer $i$'s gross growth rate of consumption, and $\gamma$ is the coefficient of relative risk aversion in a CRRA utility function. Here $M(g_{it})$ is consumer $i$'s intertemporal rate of substitution between consumption at $t-1$ and consumption at $t$ when the random growth rate $c_{i,t}/c_{i,t-1} = g_{it}$. With complete markets, $M(g_{it}) = M(g_{jt})$ for all $i, j$. This equality follows from the household's first-order conditions with complete markets (see Rubinstein, 1974). However, with incomplete markets, the $M(g_{it})$'s need not be equal across consumers. Mankiw used this fact to magnify the theoretical value of the equity premium.

Mankiw considered the consequences of time variation in the cross-section distribution of personal stochastic discount factors $M(g_{it})$. Mankiw assumed an incomplete market setting in which for each household $i$, the following Euler equations held, say, for a risk-free return $R_{ft}$ from $t-1$ to $t$ and an excess return of stocks over a bond, $R_{xt}$:

$$ER_{ft}M(g_{it}) = 1 \tag{10.100a}$$
$$ER_{xt}M(g_{it}) = 0. \tag{10.100b}$$

Consumers share the same function $M$, but the gross rate of consumption growth varies across households. The cross-section distribution of $M$ across households varies across time.[31] Thus, assume $\text{Prob}(g_{it} \leq G) = F_t(G)$ and define the first moment of the cross-sectional distribution at time $t$ as $\mu_{1t} = \int g F_t(dg)$. Also define higher moments $\mu_{jt}$ of $g_{it}$, $\mu_{jt} = \int g^j F_t(dg)$. Following Mankiw (1986) and Cogley (1999), use the second-order Taylor series approximation

$$M(g_{it}) \approx M(\mu_{1t}) + M'(\mu_{1t})(g_{it} - \mu_{1t}) + \frac{1}{2}M''(\mu_{1t})(g_{it} - \mu_{1t})^2. \tag{10.101}$$

Substitute equation (10.101) into equations (10.100) to get

$$E\left\{R_{ft}\left[M(\mu_{1t}) + M'(\mu_{1t})(g_{it} - \mu_{1t}) + \frac{1}{2}M''(\mu_{1t})(g_{it} - \mu_{1t})^2\right]\right\} = 1.$$

[31] For a setting in which the cross section of $M_{it}$'s varies over time, see the model of Krusell and Smith (1998) described in chapter 14.

Because $M'(\mu_{1t})$ is nonstochastic, using the law of iterated expectations gives $M'(\mu_{1t})E[R_{ft}(g_{it}-\mu_{1t})] = M'(\mu_{1t})EE_t[R_{ft}(g_{it}-\mu_{1t})] = M'(\mu_{1t})E[R_{ft}E_t(g_{it}-\mu_{1t})] = 0$ and so

$$ER_t[M(\mu_{1t}) + \frac{1}{2}M''(\mu_{1t})\mu_{2t}] = 1 \tag{10.102}$$

for any return $R_t$, where $\mu_{2t} = E(g_{it}-\mu_{1t})^2$. For the risk-free return $R_{ft}$, equation (10.102) implies

$$ER_{ft} = \frac{1}{E[M(\mu_{1t}) + \frac{1}{2}M''(\mu_{1t})\mu_{2t}]}.$$

For an excess return, the counterpart to equation (10.102) is

$$ER_{xt}[M(\mu_{1t}) + \frac{1}{2}M''(\mu_{1t})\mu_{2t}] = 0. \tag{10.103}$$

Thus, for an excess return $R_{xt}$, equation (10.103) and the definition of a covariance imply

$$E(R_{xt}) = \frac{-\text{cov}[R_{xt}, M(\mu_{1t}) + \frac{1}{2}M''(\mu_{1t})\mu_{2t}]}{E[M(\mu_{1t}) + \frac{1}{2}M''(\mu_{1t})\mu_{2t}]}. \tag{10.104}$$

When $M''(\mu_{1t})\mu_{2t} = 0$, equation (10.104) collapses to a version of the standard formula for the equity premium in a representative agent model. When $M''(\mu_{1t}) > 0$ [that is, when marginal utility is *convex* and when the variance $\mu_{2t}$ of the cross section of distribution of $M(g_{it})$'s covaries *inversely* with the excess return], the expected excess return is *higher*. Thus, variations in the cross-section heterogeneity of stochastic discount factors can potentially boost the equity premium under three conditions: (a) convexity of the marginal utility of consumption, which implies that $M'' > 0$; (b) an inverse correlation between excess returns and the cross-section second moment of the cross-section distribution of $M(g_{it})$; and (c) sufficient dispersion in the cross-section distribution of $M(g_{it})$ to make the covariance large in absolute magnitude.

The third aspect is relevant because in many incomplete markets settings, households can achieve much risk sharing and intertemporal consumption smoothing by frequently trading a small number of assets (sometimes only one asset). See the Bewley models of chapter 14. In Bewley models, households each have an idiosyncratic endowment process that follows an identically distributed but household-specific Markov process. Households use purchases of an asset to smooth out endowment fluctuations. Their ability to do so depends on the rate of return of the asset and the *persistence* of their endowment shocks. Broadly speaking, the more persistent are the endowment shocks, the more difficult it is to self-insure, and therefore the larger is the cross-section variation in $M(g_{it})$ that emerges. Thus, higher persistence in the endowment shock process enhances the mechanism described by Mankiw.

Constantinides and Duffie (1996)[32] set down a general equilibrium with incomplete markets incorporating Mankiw's mechanism. Their general equilibrium generates the volatility of the cross-section distribution of stochastic discount factors as well as the negative covariation between excess returns and the cross-section dispersion of stochastic discount factors required to activate Mankiw's mechanism. An important ingredient of Constantinides and Duffie's example is that each household's endowment process is very persistent (a random walk).

Storesletten, Telmer, and Yaron (1998) are pursuing ideas from Mankiw and Constantinides and Duffie by using evidence from the PSID to estimate the persistence of endowment shocks. They use a different econometric specification than that of Heaton and Lucas (1996), who found limited persistence in endowments from the PSID data, limited enough to shut down Mankiw's mechanism. Cogley (1999) checked the contribution of the covariance term in equation (10.104) using data from the use Consumer Expenditure Survey, and found what he interpreted as weak support for the idea. The cross-section covariance found by Cogley has the correct sign but is not very large.

## Concluding remarks

Chapter 7 studied asset pricing within a complete markets setting and introduced some arbitrage pricing arguments. This chapter has given more applications of arbitrage pricing arguments, for example, in deriving Modigliani-Miller and Ricardian irrelevance theorems. We have gone beyond chapter 7 in studying how, in the spirit of Hansen and Singleton (1983), consumer optimization alone puts restrictions on asset returns and consumption, without requiring complete markets or a fully spelled out general equilibrium model. At various points in this chapter, we have alluded to incomplete markets models. In chapters 14 and 15, we describe other ingredients of such models.

---

[32] Also see Attanasio and Weber (1993) for important elements of the argument of this section.

## Exercises

*Exercise 10.1* **Hansen-Jagannathan bounds**

Consider the following annual data for annual gross returns on U.S. stocks and U.S. Treasury bills from 1890 to 1979. These are the data used by Mehra and Prescott. The mean returns are $\mu = [1.07 \quad 1.02]$ and the covariance matrix of returns is $\begin{bmatrix} .0274 & .00104 \\ .00104 & .00308 \end{bmatrix}$.

**a.** For data on the excess return of stocks over bonds, compute Hansen and Jagannathan's bound on the stochastic discount factor $y$. Plot the bound for $E(y)$ on the interval $[.9, 1.02]$.

**b.** Using data on both returns, compute and plot the bound for $E(y)$ on the interval $[.9, 1.02]$. Plot this bound on the same figure as you used in part a.

**c.** On the textbook's web page (ftp://zia.stanford.edu/pub/sargent/webdocs/matlab), there is a Matlab file epdata.m with Kydland and Prescott's time series. The series epdata(:,4) is the annual growth rate of aggregate consumption $c_t/c_{t-1}$. Assume that $\beta = .99$ and that $m_t = \beta u'(c_t)/u'(c_{t-1})$, where $u(\cdot)$ is the CRRA utility function. For the three values of $\gamma = 0, 5, 10$, compute the standard deviation and mean of $m_t$ and plot them on the same figure as in part b. What do you infer from where the points lie?

*Exercise 10.2* **The term structure and regime switching**, donated by Rodolfo Manuelli

Consider a pure exchange economy where the stochastic process for consumption is given by,

$$c_{t+1} = c_t \exp[\alpha_0 - \alpha_1 s_t + \varepsilon_{t+1}],$$

where

(i) $\alpha_0 > 0$, $\alpha_1 > 0$, and $\alpha_0 - \alpha_1 > 0$.

(ii) $\varepsilon_t$ is a sequence of i.i.d. random variables distributed $N(\mu, \tau^2)$. Note: Given this specification, it follows that $E[e^{\varepsilon}] = \exp[\mu + \tau^2/2]$.

(iii) $s_t$ is a Markov process independent from $\varepsilon_t$ that can take only two values, $\{0, 1\}$. The transition probability matrix is completely summarized by

$$\text{Prob}[s_{t+1} = 1 | s_t = 1] = \pi(1),$$
$$\text{Prob}[s_{t+1} = 0 | s_t = 0] = \pi(0).$$

(iv) The information set at time $t, \Omega_t$, contains $\{c_{t-j}, s_{t-j}, \varepsilon_{t-j}; j \geq 0\}$.

There is a large number of individuals with the following utility function

$$U = E_0 \sum_{t=0}^{\infty} \beta^t u(c_t),$$

where $u(c) = c^{(1-\sigma)}/(1-\sigma)$. Assume that $\sigma > 0$ and $0 < \beta < 1$. As usual, $\sigma = 1$ corresponds to the log utility function.

**a.** Compute the "short-term" (one-period) interest rate.

**b.** Compute the "long-term" (two-period) interest rate measured in the same time units as the rate you computed in a. (That is, take the appropriate square root.)

**c.** Note that the log of the rate of growth of consumption is given by

$$\log(c_{t+1}) - \log(c_t) = \alpha_0 - \alpha_1 s_t + \varepsilon_{t+1}.$$

Thus, the conditional expectation of this growth rate is just $\alpha_0 - \alpha_1 s_t + \mu$. Note that when $s_t = 0$, growth is high and, when $s_t = 1$, growth is low. Thus, loosely speaking, we can identify $s_t = 0$ with the peak of the cycle (or good times) and $s_t = 1$ with the trough of the cycle (or bad times). Assume $\mu > 0$. Go as far as you can describing the implications of this model for the cyclical behavior of the term structure of interest rates.

**d.** Are short term rates pro- or countercyclical?

**e.** Are long rates pro- or countercyclical? If you cannot give a definite answer to this question, find conditions under which they are either pro- or countercyclical, and interpret your conditions in terms of the "permanence" (you get to define this) of the cycle.

*Exercise 10.3* **Growth slowdowns and stock market crashes**, donated by Rodolfo Manuelli[33]

Consider a simple one-tree pure exchange economy. The only source of consumption is the fruit that grows on the tree. This fruit is called dividends by the tribe inhabiting this island. The stochastic process for dividend $d_t$ is described as follows: If $d_t$ is not equal to $d_{t-1}$, then $d_{t+1} = \gamma d_t$ with probability $\pi$, and $d_{t+1} = d_t$ with probability $(1 - \pi)$. If in any pair of periods $j$ and $j+1$, $d_j = d_{j+1}$, then for all $t > j$, $d_t = d_j$. In words, the process – if not stopped – grows at a rate $\gamma$ in every period. However, once it stops growing for one period, it remains constant forever on. Let $d_0$ equal one.

Preferences over stochastic processes for consumption are given by

$$U = E_0 \sum_{t=0}^{\infty} \beta^t u(c_t),$$

where $u(c) = c^{(1-\sigma)}/(1-\sigma)$. Assume that $\sigma > 0$, $0 < \beta < 1$, $\gamma > 1$, and $\beta\gamma^{(1-\sigma)} < 1$.

**a.** Define a competitive equilibrium in which shares to this tree are traded.

**b.** Display the equilibrium process for the price of shares in this tree $p_t$ as a function of the history of dividends. Is the price process a Markov process in the sense that it depends just on the last period's dividends?

**c.** Let $T$ be the first time in which $d_{T-1} = d_T = \gamma^{(T-1)}$. Is $p_{T-1} > p_T$? Show conditions under which this is true. What is the economic intuition for this result? What does it say about stock market declines or crashes?

**d.** If this model is correct, what does it say about the behavior of the aggregate value of the stock market in economies that switched from high to low growth (e.g., Japan)?

[33] See also Joseph Zeira (1999).

*Exercise 10.4* **The term structure and consumption**, donated by Rodolfo Manuelli

Consider an economy populated by a large number of identical households. The (common) utility function is

$$\sum_{t=0}^{\infty} \beta^t u(c_t),$$

where $0 < \beta < 1$, and $u(x) = x^{1-\theta)}/(1-\theta)$, for some $\theta > 0$. (If $\theta = 1$, the utility is logarithmic.) Each household owns one tree. Thus, the number of households and trees coincide. The amount of consumption that grows in a tree satisfies

$$c_{t+1} = c^* c_t^{\varphi} \varepsilon_{t+1},$$

where $0 < \varphi < 1$, and $\varepsilon_t$ is a sequence of i.i.d. log normal random variables with mean one, and variance $\sigma^2$. Assume that, in addition to shares in trees, in this economy bonds of all maturities are traded.

**a.** Define a competitive equilibrium.

**b.** Go as far as you can calculating the term structure of interest rates, $\tilde{R}_{jt}$, for $j = 1, 2, \ldots$.

**c.** Economist A argues that economic theory predicts that the variance of the log of short-term interest rates (say one-period) is always lower than the variance of long-term interest rates, because short rates are "riskier." Do you agree? Justify your answer.

**d.** Economist B claims that short-term interest rates, i.e., $j = 1$, are "more responsive" to the state of the economy, i.e., $c_t$, than are long-term interest rates, i.e., $j$ large. Do you agree? Justify your answer.

**e.** Economist C claims that the Fed should lower interest rates because whenever interest rates are low, consumption is high. Do you agree? Justify your answer.

**f.** Economist D claims that in economies in which output (consumption in our case) is very persistent ($\varphi \approx 1$), changes in output (consumption) do not affect interest rates. Do you agree? Justify your answer and, if possible, provide economic intuition for your argument.

# 11
# *Economic Growth*

## *Introduction*

This chapter describes basic nonstochastic models of sustained economic growth. We begin by describing a benchmark exogenous growth model, where sustained growth is driven by exogenous growth in labor productivity. Then we turn our attention to several endogenous growth models, where sustained growth of labor productivity is somehow *chosen* by the households in the economy. We describe several models that differ in whether the equilibrium market economy matches what a benevolent planner would choose. Where the market outcome doesn't match the planner's outcome, there can be room for welfare-improving government interventions. The objective of the chapter is to shed light on the mechanisms at work in different models. We try to facilitate comparison by using the same production function for most of our discussion while changing the meaning of one of its arguments.

Paul Romer's work has been an impetus to the revived interest in the theory of economic growth. In the spirit of Arrow's (1962) model of learning by doing, Romer (1986) presents an endogenous growth model where the accumulation of capital (or knowledge) is associated with a positive externality on the available technology. The aggregate of all agents' holdings of capital is positively related to the level of technology, which in turn interacts with individual agents' savings decisions and thereby determines the economy's growth rate. Thus, the households in this economy are *choosing* how fast the economy is growing but do so in an unintentional way. The competitive equilibrium growth rate falls short of the socially optimal one.

Another approach to generating endogenous growth is to assume that all production factors are reproducible. Following Uzawa (1965), Lucas (1988) formulates a model with accumulation of both physical and human capital. The joint accumulation of all inputs ensures that growth will not come to a halt even though each individual factor in the final-good production function is subject to diminishing returns. In the absence of externalities, the growth rate in the competitive equilibrium coincides in this model with the social optimum.

Romer (1987) constructs a model where agents can choose to engage in research that produces technological improvements. Each invention represents a technology for producing a new type of intermediate input that can be used in the production of final goods without affecting the marginal product of existing intermediate inputs. The introduction of new inputs enables the economy to experience sustained growth even though each intermediate input taken separately is subject to diminishing returns. In a decentralized equilibrium, private agents will only expend resources on research if they are granted property rights over their inventions. Under the assumption of infinitely lived patents, Romer solves for a monopolistically competitive equilibrium that exhibits the classic tension between static and dynamic efficiency. Patents and the associated market power are necessary for there to be research and new inventions in a decentralized equilibrium, while the efficient production of existing intermediate inputs would require marginal-cost pricing, that is, the abolition of granted patents. The monopolistically competitive equilibrium is characterized by a smaller supply of each intermediate input and a lower growth rate as compared to the socially optimal outcome.

Finally, we revisit the question of when nonreproducible factors may not pose an obstacle to growth. Rebelo (1991) shows that even if there are nonreproducible factors in fixed supply in a neoclassical growth model, sustained growth is possible if there is a "core" of capital goods that is produced without the direct or indirect use of the nonreproducible factors. Because of the ever-increasing relative scarcity of a nonreproducible factor, Rebelo finds that its price increases over time relative to a reproducible factor. Romer (1990) assumes that research requires the input of labor and not only goods as in his earlier model (1987). Now, if labor is in fixed supply and workers' innate productivity is constant, it follows immediately that growth must asymptotically come to an halt. To make sustained growth feasible, we can take a cue from our earlier discussion. One modeling strategy would be to introduce an externality that enhances researchers' productivity, and an alternative approach would be to assume that researchers can accumulate human capital. Romer adopts the first type of assumption, and we find it instructive to focus on its role in overcoming a barrier to growth that nonreproducible labor would otherwise pose.

## *The economy*

The economy has a constant population of a large number of identical agents who order consumption streams $\{c_t\}_{t=0}^{\infty}$ according to

$$\sum_{t=0}^{\infty} \beta^t u(c_t), \quad \text{with } \beta \in (0,1) \text{ and } u(c) = \frac{c^{1-\sigma} - 1}{1-\sigma} \quad \text{for } \sigma \in [0, \infty), \tag{11.1}$$

and $\sigma = 1$ is taken to be logarithmic utility.[1] Lowercase letters for quantities, such as $c_t$ for consumption, are used to denote individual variables, and upper case letters stand for aggregate quantities.

For most part of our discussion of economic growth, the production function takes the form

$$F(K_t, X_t) = X_t f(\hat{K}_t), \quad \text{where } \hat{K}_t \equiv \frac{K_t}{X_t}. \tag{11.2}$$

That is, the production function $F(K, X)$ exhibits constant returns to scale in its two arguments, which via Euler's theorem on linearly homogeneous functions implies

$$F(K, X) = F_1(K, X)K + F_2(K, X)X, \tag{11.3}$$

where $F_i(K, X)$ is the derivative with respect to the $i$th argument [and $F_{ii}(K, X)$ will be used to denote the second derivative with respect to the $i$th argument]. The input $K_t$ is physical capital with a rate of depreciation equal to $\delta$. New capital can be created by transforming one unit of output into one unit of capital. Past investments are reversible. It follows that the relative price of capital in terms of the consumption good must always be equal to one. The second argument $X_t$ captures the contribution of labor. Its precise meaning will differ among the various setups that we will examine.

We assume that the production function satisfies standard assumptions of positive but diminishing marginal products,

$$F_i(K, X) > 0, \quad F_{ii}(K, X) < 0, \quad \text{for } i = 1, 2;$$

and the Inada conditions,

$$\begin{aligned} \lim_{K \to 0} F_1(K, X) &= \lim_{X \to 0} F_2(K, X) = \infty, \\ \lim_{K \to \infty} F_1(K, X) &= \lim_{X \to \infty} F_2(K, X) = 0, \end{aligned}$$

[1] By virtue of L'Hôpital's rule, the limit of $(c^{1-\sigma} - 1)/(1-\sigma)$ is $\log(c)$ as $\sigma$ goes to one.

which imply

$$\lim_{\hat{K}\to 0} f'(\hat{K}) = \infty, \qquad \lim_{\hat{K}\to\infty} f'(\hat{K}) = 0. \tag{11.4}$$

We will also make use of the mathematical fact that a linearly homogeneous function $F(K, X)$ has first derivatives $F_i(K, X)$ homogeneous of degree 0; thus, the first derivatives are only functions of the ratio $\hat{K}$. In particular, we have

$$F_1(K, X) = \frac{\partial X f(K/X)}{\partial K} = f'(\hat{K}), \tag{11.5a}$$

$$F_2(K, X) = \frac{\partial X f(K/X)}{\partial X} = f(\hat{K}) - f'(\hat{K})\hat{K}. \tag{11.5b}$$

*Balanced growth path*

We seek additional technological assumptions to generate market outcomes with steady-state growth of consumption at a constant rate $1 + \mu = c_{t+1}/c_t$. The literature uses the term "balanced growth path" to denote a situation where all endogenous variables grow at constant (but possibly different) rates. Along such a steady-state growth path (and during any transition toward the steady state), the return to physical capital must be such that households are willing to hold the economy's capital stock.

In a competitive equilibrium where firms rent capital from the agents, the rental payment $r_t$ is equal to the marginal product of capital,

$$r_t = F_1(K_t, X_t) = f'(\hat{K}_t). \tag{11.6}$$

Households maximize utility given by equation (11.1) subject to the sequence of budget constraints

$$c_t + k_{t+1} = r_t k_t + (1 - \delta)k_t + \chi_t, \tag{11.7}$$

where $\chi_t$ stands for labor-related budget terms. The first-order condition with respect to $k_{t+1}$ is

$$u'(c_t) = \beta u'(c_{t+1})\,(r_{t+1} + 1 - \delta). \tag{11.8}$$

After using equations (11.1) and (11.6) in equation (11.8), we arrive at the following equilibrium condition:

$$\left(\frac{c_{t+1}}{c_t}\right)^{\sigma} = \beta[f'(\hat{K}_{t+1}) + 1 - \delta]. \tag{11.9}$$

We see that a constant consumption growth rate on the left-hand side is sustained in an equilibrium by a constant rate of return on the right-hand side. It was also for this reason that we chose the class of utility functions in equation (11.1) that exhibits a constant intertemporal elasticity of substitution. These preferences allow for balanced growth paths.[2]

Equation (11.9) makes clear that capital accumulation alone cannot sustain steady-state consumption growth when the labor input $X_t$ is constant over time, $X_t = L$. Given the second Inada condition in equations (11.4), the limit of the right-hand side of equation (11.9) is $\beta(1-\delta)$ when $\hat{K}$ approaches infinity. The steady state with a constant labor input must therefore be a constant consumption level and a capital-labor ratio $\hat{K}^\star$ given by

$$f'(\hat{K}^\star) = \beta^{-1} - (1-\delta). \tag{11.10}$$

In chapter 2 we derived a closed-form solution for the transition dynamics toward such a steady state in the case of logarithmic utility, a Cobb-Douglas production function, and $\delta = 1$.

## Exogenous growth

As in Solow's (1956) classical article, the simplest way to ensure steady-state consumption growth is to postulate exogenous labor-augmenting technological change at the constant rate $1 + \mu \geq 1$,

$$X_t = A_t L, \qquad \text{with } A_t = (1+\mu)A_{t-1},$$

where $L$ is a fixed stock of labor. Our conjecture is then that both consumption and physical capital will grow at that same rate $1+\mu$ along a balanced growth path. The same growth rate of $K_t$ and $A_t$ implies that the ratio $\hat{K}$ and, therefore, the marginal product of capital remain constant in the steady state. A time-invariant rate of return is in turn consistent with households choosing a constant growth rate of consumption, given the assumption of isoelastic preferences.

---

[2] To ensure well-defined maximization problems, a maintained assumption throughout the chapter is that parameters are such that any derived consumption growth rate $1+\mu$ yields finite lifetime utility; i.e., the implicit restriction on parameter values is that $\beta(1+\mu)^{1-\sigma} < 1$. To see that this condition is needed, substitute the consumption sequence $\{c_t\}_{t=0}^{\infty} = \{(1+\mu)^t c_0\}_{t=0}^{\infty}$ into equation (11.1).

Evaluating equation (11.9) at a steady state, the optimal ratio $\hat{K}^\star$ is given by

$$(1+\mu)^\sigma = \beta[f'(\hat{K}^\star)+1-\delta]. \tag{11.11}$$

While the steady-state consumption growth rate is exogenously given by $1+\mu$, the endogenous steady-state ratio $\hat{K}^\star$ is such that the implied rate of return on capital induces the agents to choose a consumption growth rate of $1+\mu$. As can be seen, a higher degree of patience (a larger $\beta$), a higher willingness intertemporally to substitute (a lower $\sigma$) and a more durable capital stock (a lower $\delta$) each yield a higher ratio $\hat{K}^\star$, and therefore more output (and consumption) at a point in time; but the growth rate remains fixed at the rate of exogenous labor-augmenting technological change. It is straightforward to verify that the competitive equilibrium outcome is Pareto optimal, since the private return to capital coincides with the social return.

Physical capital is compensated according to equation (11.6), and labor is also paid its marginal product in a competitive equilibrium,

$$w_t = F_2(K_t, X_t)\,\frac{\mathrm{d}\,X_t}{\mathrm{d}\,L} = F_2(K_t, X_t)\,A_t. \tag{11.12}$$

So, by equation (11.3), we have

$$r_t K_t + w_t L = F(K_t, A_t L).$$

Factor payments are equal to total production, which is the standard result of a competitive equilibrium with constant-returns-to-scale technologies. However, it is interesting to note that if $A_t$ were a separate production factor, there could not exist a competitive equilibrium, since factor payments based on marginal products would exceed total production. In other words, the dilemma would then be that the production function $F(K_t, A_t L)$ exhibits increasing returns to scale in the three "inputs" $K_t$, $A_t$, and $L$, which is not compatible with the existence of a competitive equilibrium. This problem is to be kept in mind as we now turn to one way to endogenize economic growth.

## *Externality from spillovers*

Inspired by Arrow's (1962) paper on learning by doing, Romer (1986) suggests that economic growth can be endogenized by assuming that technology grows because of aggregate spillovers coming from firms' production activities.

The problem alluded to in the previous section that a competitive equilibrium fails to exist in the presence of increasing returns to scale is avoided by letting technological advancement be external to firms.[3] As an illustration, we assume that firms face a fixed labor productivity that is proportional to the current economy-wide average of physical capital per worker.[4] In particular,

$$X_t = \bar{K}_t L, \qquad \text{where } \bar{K}_t = \frac{K_t}{L}.$$

The competitive rental rate of capital is still given by equation (11.6) but we now trivially have $\hat{K}_t = 1$, so equilibrium condition (11.9) becomes

$$\left(\frac{c_{t+1}}{c_t}\right)^{\sigma} = \beta[f'(1) + 1 - \delta]. \tag{11.13}$$

Note first that this economy has no transition dynamics toward a steady state. Regardless of the initial capital stock, equation (11.13) determines a time-invariant growth rate. To ensure a positive growth rate, we require the parameter restriction $\beta[f'(1) + 1 - \delta] \geq 1$. A second critical property of the model is that the economy's growth rate is now a function of preference and technology parameters.

The competitive equilibrium is no longer Pareto optimal, since the private return on capital falls short of the social rate of return, with the latter return given by

$$\frac{\mathrm{d}\, F\left(K_t, \frac{K_t}{L} L\right)}{\mathrm{d}\, K_t} = F_1(K_t, K_t) + F_2(K_t, K_t) = f(1), \tag{11.14}$$

where the last equality follows from equations (11.5). This higher social rate of return enters a planner's first-order condition, which then also implies a higher optimal consumption growth rate,

$$\left(\frac{c_{t+1}}{c_t}\right)^{\sigma} = \beta[f(1) + 1 - \delta]. \tag{11.15}$$

Let us reconsider the suboptimality of the decentralized competitive equilibrium. Since the agents and the planner share the same objective of maximizing

---

[3] Arrow (1962) focuses on learning from experience that is assumed to get embodied in capital goods, while Romer (1986) postulates spillover effects of firms' investments in knowledge. In both analyses, the productivity of a given firm is a function of an aggregate state variable, either the economy's stock of physical capital or stock of knowledge.

[4] This specific formulation of spillovers is analyzed in a rarely cited paper by Frankel (1962).

utility, we are left with exploring differences in their constraints. For a given sequence of the spillover $\{\bar{K}_t\}_{t=0}^{\infty}$, the production function $F(k_t, \bar{K}_t l_t)$ exhibits constant returns to scale in $k_t$ and $l_t$. So, once again, factor payments in a competitive equilibrium will be equal to total output, and optimal firm size is indeterminate. Therefore, we can consider a representative agent with one unit of labor endowment who runs his own production technology, taking the spillover effect as given. His resource constraint becomes

$$c_t + k_{t+1} = F(k_t, \bar{K}_t) + (1-\delta)k_t = \bar{K}_t f\left(\frac{k_t}{\bar{K}_t}\right) + (1-\delta)k_t,$$

and the private gross rate of return on capital is equal to $f'(k_t/\bar{K}_t) + 1 - \delta$. After invoking the equilibrium condition $k_t = \bar{K}_t$, we arrive at the competitive equilibrium return on capital $f'(1) + 1 - \delta$ that appears in equation (11.13). In contrast, the planner maximizes utility subject to a resource constraint where the spillover effect is internalized,

$$C_t + K_{t+1} = F\left(K_t, \frac{K_t}{L}L\right) + (1-\delta)K_t = [f(1) + 1 - \delta]K_t.$$

## *All factors reproducible*

### *One-sector model*

An alternative approach to generating endogenous growth is to assume that all factors of production are producible. Remaining within a one-sector economy, we now assume that human capital $X_t$ can be produced in the same way as physical capital but rates of depreciation might differ. Let $\delta_X$ and $\delta_K$ be the rates of depreciation of human capital and physical capital, respectively.

The competitive equilibrium wage is equal to the marginal product of human capital

$$w_t = F_2(K_t, X_t). \tag{11.16}$$

Households maximize utility subject to budget constraint (11.7) where the term $\chi_t$ is now given by

$$\chi_t = w_t x_t + (1-\delta_X)x_t - x_{t+1}.$$

The first-order condition with respect to human capital becomes

$$u'(c_t) = \beta u'(c_{t+1})\,(w_{t+1} + 1 - \delta_X). \tag{11.17}$$

Since both equations (11.8) and (11.17) must hold, the rates of return on the two assets have to obey

$$F_1(K_{t+1}, X_{t+1}) - \delta_K = F_2(K_{t+1}, X_{t+1}) - \delta_X,$$

and after invoking equations (11.5),

$$f(\hat{K}_{t+1}) - (1 + \hat{K}_{t+1})f'(\hat{K}_{t+1}) \;=\; \delta_X - \delta_K, \tag{11.18}$$

which uniquely determines a time-invariant competitive equilibrium ratio $\hat{K}^\star$, as a function solely of depreciation rates and parameters of the production function.[5]

After solving for $f'(\hat{K}^\star)$ from equation (11.18) and substituting into equation (11.9), we arrive at an expression for the equilibrium growth rate

$$\left(\frac{c_{t+1}}{c_t}\right)^\sigma = \beta\left[\frac{f(\hat{K}^\star)}{1+\hat{K}^\star} + 1 - \frac{\delta_X + \hat{K}^\star\delta_K}{1+\hat{K}^\star}\right]. \tag{11.19}$$

As in the previous model with an externality, the economy here is void of any transition dynamics toward a steady state. But this implication hinges now critically upon investments being reversible so that the initial stocks of physical capital and human capital are inconsequential. In contrast to the previous model, the present competitive equilibrium is Pareto optimal because there is no longer any discrepancy between private and social rates of return.[6]

---

[5] The left side of equation (11.18) is strictly increasing, since the derivative with respect to $\hat{K}$ is $-(1+\hat{K})f''(\hat{K}) > 0$. Thus, there can only be one solution to equation (11.18) and existence is guaranteed because the left-hand side ranges from minus infinity to plus infinity. The limit of the left-hand side when $\hat{K}$ approaches zero is $f(0) - \lim_{\hat{K}\to 0} f'(\hat{K})$, which is equal to minus infinity by equations (11.4) and the fact that $f(0) = 0$. [Barro and Sala-i-Martin (1995) show that the Inada conditions and constant returns to scale imply that all production factors are essential, i.e., $f(0) = 0$.] To establish that the left side of equation (11.18) approaches plus infinity when $\hat{K}$ goes to infinity, we can define the function $g$ as $F(K,X) = Kg(\hat{X})$ where $\hat{X} \equiv X/K$ and derive an alternative expression for the left-hand side of equation (11.18), $(1+\hat{X})g'(\hat{X}) - g(\hat{X})$, for which we take the limit when $\hat{X}$ goes to zero.

[6] It is instructive to compare the present model with two producible factors, $F(K,X)$, to the previous setup with one producible factor and an externality, $\tilde{F}(K,X)$ with $X = \bar{K}L$. Suppose the present technology is such that $\hat{K}^\star = 1$ and $\delta_K = \delta_X$, and the two different setups are equally productive; i.e., we assume that $F(K,X) = \tilde{F}(2K, 2X)$, which implies $f(\hat{K}) = 2\tilde{f}(\hat{K})$. We can then verify that the present competitive equilibrium growth rate in equation (11.19) is the same as the planner's solution for the previous setup in equation (11.15).

The problem of optimal taxation with commitment (see chapter 12) is studied for this model of endogenous growth by Jones, Manuelli, and Rossi (1993), who adopt the assumption of irreversible investments.

*Two-sector model*

Following Uzawa (1965), Lucas (1988) explores endogenous growth in a two-sector model with all factors being producible. The resource constraint in the goods sector is

$$C_t + K_{t+1} = K_t^{\alpha}(\phi_t X_t)^{1-\alpha} + (1-\delta)K_t, \tag{11.20a}$$

and the linear technology for accumulating additional human capital is

$$X_{t+1} - X_t = A(1-\phi_t)X_t, \tag{11.20b}$$

where $\phi_t \in [0,1]$ is the fraction of human capital employed in the goods sector, and $(1-\phi_t)$ is devoted to human capital accumulation. (Lucas provides an alternative interpretation that we will discuss later.)

We seek a balanced growth path where consumption, physical capital, and human capital grow at constant rates (but not necessarily the same ones) and the fraction $\phi$ stays constant over time. Let $1+\mu$ be the growth rate of consumption, and equilibrium condition (11.9) becomes

$$(1+\mu)^{\sigma} \;=\; \beta\left(\alpha K_t^{\alpha-1}\left[\phi X_t\right]^{1-\alpha} + 1 - \delta\right). \tag{11.21}$$

That is, along the balanced growth path, the marginal product of physical capital must be constant. With the assumed Cobb-Douglas technology, the marginal product of capital is proportional to the average product, so that by dividing equation (11.20a) through by $K_t$ and applying equation (11.21) we obtain

$$\frac{C_t}{K_t} + \frac{K_{t+1}}{K_t} = \frac{(1+\mu)^{\sigma}\beta^{-1} - (1-\alpha)(1-\delta)}{\alpha}. \tag{11.22}$$

By definition of a balanced growth path, $K_{t+1}/K_t$ is constant, so equation (11.22) implies that $C_t/K_t$ is constant; that is, the capital stock must grow at the same rate as consumption.

Substituting $K_t = (1+\mu)K_{t-1}$ into equation (11.21),

$$(1+\mu)^{\sigma} - \beta(1-\delta) = \beta\alpha\left[(1+\mu)K_{t-1}\right]^{\alpha-1}\left[\phi X_t\right]^{1-\alpha},$$

and dividing by the similarly rearranged equation (11.21) for period $t-1$, we arrive at

$$1 = (1+\mu)^{\alpha-1}\left[\frac{X_t}{X_{t-1}}\right]^{1-\alpha},$$

which directly implies that human capital must also grow at the rate $1+\mu$ along a balanced growth path. Moreover, by equation (11.20b), the growth rate is

$$1+\mu = 1 + A(1-\phi), \tag{11.23}$$

so it remains to determine the steady-state value of $\phi$.

The equilibrium value of $\phi$ has to be such that a unit of human capital receives the same factor payment in both sectors; that is, the marginal products of human capital must be the same,

$$p_t A = (1-\alpha)K_t^{\alpha}\left[\phi X_t\right]^{-\alpha},$$

where $p_t$ is the relative price of human capital in terms of the composite consumption/capital good. Since the ratio $K_t/X_t$ is constant along a balanced growth path, it follows that the price $p_t$ must also be constant over time. Finally, the remaining equilibrium condition is that the rates of return on human and physical capital are equal,

$$\frac{p_t(1+A)}{p_{t-1}} = \alpha K_t^{\alpha-1}\left[\phi X_t\right]^{1-\alpha} + 1 - \delta,$$

and after invoking a constant steady-state price of human capital and equilibrium condition (11.21), we obtain

$$1+\mu = \left[\beta(1+A)\right]^{1/\sigma}. \tag{11.24}$$

Thus, the growth rate is positive as long as $\beta(1+A) \geq 1$, but feasibility requires also that solution (11.24) falls below $1+A$ which is the maximum growth rate of human capital in equation (11.20b). This parameter restriction, $[\beta(1+A)]^{1/\sigma} < (1+A)$, also ensures that the growth rate in equation (11.24) yields finite lifetime utility.

As in the one-sector model, there is no discrepancy between private and social rates of return, so the competitive equilibrium is Pareto optimal. Lucas (1988) does allow for an externality (in the spirit of our earlier section) where the economy-wide average of human capital per worker enters the production function in the goods sector, but as he notes, the externality is not needed to generate endogenous growth.

Lucas provides an alternative interpretation of the technologies in equations (11.20). Each worker is assumed to be endowed with one unit of time. The time spent in the goods sector is denoted $\phi_t$, which is multiplied by the agent's human capital $x_t$ to arrive at the efficiency units of labor supplied. The remaining time is spent in the education sector with a constant marginal productivity of $Ax_t$ additional units of human capital acquired. Even though Lucas's interpretation does introduce a nonreproducible factor in form of a time endowment, the multiplicative specification makes the model identical to an economy with only two factors that are both reproducible. One section ahead we will study a setup with a nonreproducible factor that has some nontrivial implications.

## *Research and monopolistic competition*

Building on Dixit and Stiglitz's (1977) formulation of the demand for differentiated goods and the extension to differentiated inputs in production by Ethier (1982), Romer (1987) studies an economy with an aggregate resource constraint of the following type:

$$C_t + \int_0^{A_{t+1}} Z_{t+1}(i)\,\mathrm{d}i + (A_{t+1} - A_t)\kappa = L^{1-\alpha} \int_0^{A_t} Z_t(i)^\alpha\,\mathrm{d}i, \tag{11.25}$$

where one unit of the intermediate input $Z_{t+1}(i)$ can be produced from one unit of output at time $t$, and $Z_{t+1}(i)$ is used in production in the following period $t+1$. The continuous range of inputs at time $t$, $i \in [0, A_t]$, can be augmented for next period's production function at the constant marginal cost $\kappa$.

In the allocations that we are about to study, the quantity of an intermediate input will be the same across all existing types, $Z_t(i) = Z_t$ for $i \in [0, A_t]$. The resource constraint (11.25) can then be written as

$$C_t + A_{t+1}Z_{t+1} + (A_{t+1} - A_t)\kappa = L^{1-\alpha}\, A_t Z_t^\alpha. \tag{11.26}$$

If $A_t$ were constant over time, say, let $A_t = 1$ for all $t$, we would just have a parametric example of an economy yielding a no-growth steady state given by equation (11.10) with $\delta = 1$. Hence, growth can only be sustained by allocating resources to a continuous expansion of the range of inputs. But this approach poses a barrier to the existence of a competitive equilibrium, since the production relationship $L^{1-\alpha} A_t Z_t^\alpha$ exhibits increasing returns to scale in its three "inputs." Following Judd's (1985a) treatment of patents in a dynamic setting of Dixit and Stiglitz's (1977) model of monopolistic competition, Romer (1987) assumes that an inventor of a new intermediate input obtains an infinitely lived patent on that design. As the sole supplier of an input, the inventor can recoup the investment cost $\kappa$ by setting a price of the input above its marginal cost.

*Monopolistic competition outcome*

The final-goods sector is still assumed to be characterized by perfect competition because it exhibits constant returns to scale in the labor input $L$ and the existing continuous range of intermediate inputs $Z_t(i)$. Thus, a competitive outcome prescribes that each input is paid its marginal product,

$$w_t = (1-\alpha)L^{-\alpha} \int_0^{A_t} Z_t(i)^\alpha \, di, \tag{11.27}$$

$$p_t(i) = \alpha L^{1-\alpha} Z_t(i)^{\alpha-1}, \tag{11.28}$$

where $p_t(i)$ is the price of intermediate input $i$ at time $t$ in terms of the final good.

Let $1+R_m$ be the steady-state interest rate along the balanced growth path that we are seeking. In order to find the equilibrium invention rate of new inputs, we first compute the profits from producing and selling an existing input $i$. The profit at time $t$ is equal to

$$\pi_t(i) = [p_t(i) - (1+R_m)]\, Z_t(i), \tag{11.29}$$

where the cost of supplying one unit of the input $i$ is one unit of the final good acquired in the previous period; that is, the cost is the intertemporal price $1+R_m$. The first-order condition of maximizing the profit in equation (11.29) is the familiar expression that the monopoly price $p_t(i)$ should be set as a markup above marginal cost, $1+R_m$, and the markup is inversely related to the absolute value of the demand elasticity of input $i$, $|\epsilon_t(i)|$;

$$p_t(i) = \frac{1+R_m}{1+\epsilon_t(i)^{-1}}, \tag{11.30}$$

$$\epsilon_t(i) = \left[\frac{\partial\, p_t(i)}{\partial\, Z_t(i)} \frac{Z_t(i)}{p_t(i)}\right]^{-1} < 0.$$

The constant marginal cost, $1 + R_m$, and the constant-elasticity demand curve (11.28), $\epsilon_t(i) = -(1-\alpha)^{-1}$, yield a time-invariant monopoly price which substituted into demand curve (11.28) results in a time-invariant equilibrium quantity of input $i$:

$$p_t(i) = \frac{1 + R_m}{\alpha}, \tag{11.31a}$$

$$Z_t(i) = \left(\frac{\alpha^2}{1 + R_m}\right)^{1/(1-\alpha)} L \equiv Z_m. \tag{11.31b}$$

By substituting equation (11.31) into equation (11.29), we obtain an input producer's steady-state profit flow,

$$\pi_t(i) = (1-\alpha)\alpha^{1/(1-\alpha)} \left(\frac{\alpha}{1 + R_m}\right)^{\alpha/(1-\alpha)} L \equiv \Omega_m(R_m). \tag{11.32}$$

In an equilibrium with free entry, the cost $\kappa$ of inventing a new input must be equal to the discounted stream of future profits associated with being the sole supplier of that input,

$$\sum_{t=1}^{\infty} (1 + R_m)^{-t} \Omega_m(R_m) = \frac{\Omega_m(R_m)}{R_m}; \tag{11.33}$$

that is,

$$R_m \kappa = \Omega_m(R_m). \tag{11.34}$$

The profit function $\Omega_m(R)$ is positive, strictly decreasing in $R$, and convex, as depicted in Figure 11.1. It follows that there exists a unique intersection between $\Omega(R)$ and $R\kappa$ that determines $R_m$. Using the corresponding version of equilibrium condition (11.9), the computed interest rate $R_m$ characterizes a balanced growth path with

$$\left(\frac{c_{t+1}}{c_t}\right)^{\sigma} = \beta(1 + R_m), \tag{11.35}$$

as long as $1 + R_m \geq \beta^{-1}$; that is, the technology must be sufficiently productive relative to the agents' degree of impatience.[7] It is straightforward to verify that

[7] If the computed value $1 + R_m$ falls short of $\beta^{-1}$, the technology does not present sufficient private incentives for new inventions, so the range of intermediate inputs stays constant over time, and the equilibrium interest rate equals $\beta^{-1}$.

the range of inputs must grow at the same rate as consumption in a steady state. After substituting the constant quantity $Z_m$ into resource constraint (11.26) and dividing by $A_t$, we see that a constant $A_{t+1}/A_t$ implies that $C_t/A_t$ stays constant; that is, the range of inputs must grow at the same rate as consumption.

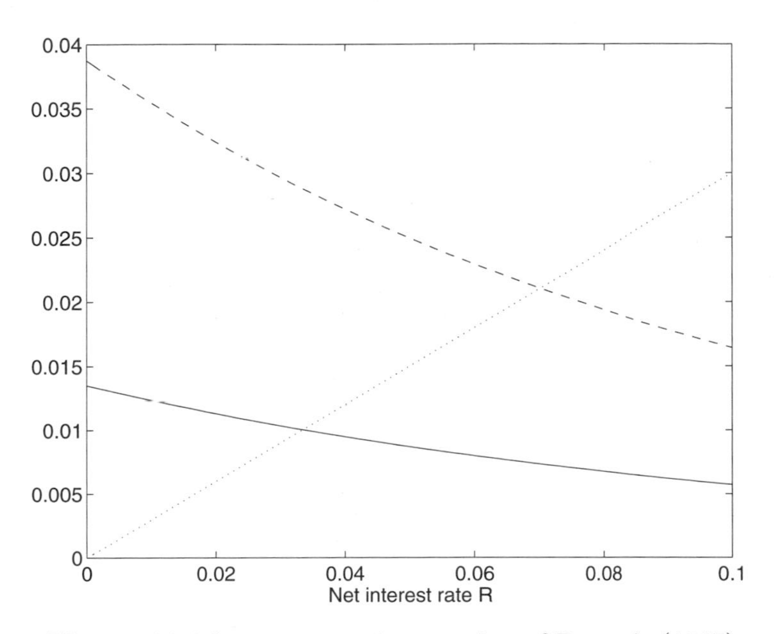


**Figure 11.1** Interest rates in a version of Romer's (1987) model of research and monopolistic competition. The dotted line is the linear relationship $\kappa R$, while the solid and dashed curves depict $\Omega_m(R)$ and $\Omega_s(R)$, respectively. The intersection between $\kappa R$ and $\Omega_m(R)$ $[\Omega_s(R)]$ determines the interest rate along a balanced growth path for the laissez-faire economy (planner allocation), as long as $R \geq \beta^{-1} - 1$. The parameterization is $\alpha = 0.9$, $\kappa = 0.3$, and $L = 1$.

Note that the solution to equation (11.34) exhibits positive scale effects where a larger labor force $L$ implies a higher interest rate and therefore a higher growth rate in equation (11.35). The reason is that a larger economy enables input producers to profit from a larger sales volume in equation (11.31b), which spurs more inventions until the discounted stream of profits of an input is driven down to the invention cost $\kappa$ by means of the higher equilibrium interest rate. In other words, it is less costly for a larger economy to expand its range of inputs because the cost of an additional input is smaller in per capita terms.

*Planner solution*

Let $1+R_s$ be the social rate of interest along an optimal balanced growth path. We analyze the planner problem in two steps. First, we establish that the socially optimal supply of an input $i$ is the same across all existing inputs and constant over time. Second, we derive $1+R_s$ and the implied optimal growth rate of consumption.

For a given social interest rate $1+R_s$ and a range of inputs $[0, A_t]$, the planner would choose the quantities of intermediate inputs that maximize

$$L^{1-\alpha}\int_0^{A_t} Z_t(i)^{\alpha}\,\mathrm{d}i - (1+R_s)\int_0^{A_t} Z_t(i)\,\mathrm{d}i,$$

with the following first-order condition with respect to $Z_t(i)$,

$$Z_t(i) = \left(\frac{\alpha}{1+R_s}\right)^{1/(1-\alpha)} L \equiv Z_s. \tag{11.36}$$

Thus, the quantity of an intermediate input is the same across all inputs and constant over time. Hence, the planner's problem is simplified to one where utility function (11.1) is maximized subject to resource constraint (11.26) with quantities of intermediate inputs given by equation (11.36). The first-order condition with respect to $A_{t+1}$ is then

$$\left(\frac{c_{t+1}}{c_t}\right)^{\sigma} = \beta\frac{L^{1-\alpha}Z_s^{\alpha}+\kappa}{Z_s+\kappa} = \beta(1+R_s), \tag{11.37}$$

where the last equality merely invokes the definition of $1+R_s$ as the social marginal rate of intertemporal substitution, $\beta^{-1}(c_{t+1}/c_t)^{\sigma}$. After substituting equation (11.36) into equation (11.37) and rearranging the last equality, we obtain

$$R_s\kappa = (1-\alpha)\left(\frac{\alpha}{1+R_s}\right)^{\alpha/(1-\alpha)} L \equiv \Omega_s(R_s). \tag{11.38}$$

The solution to this equation, $1+R_s$, is depicted in Figure 11.1, and existence is guaranteed in the same way as in the case of $1+R_m$.

We conclude that the social rate of return $1+R_s$ and, therefore, the optimal growth rate exceed the laissez-faire outcome, since the function $\Omega_s(R)$ lies above the function $\Omega_m(R)$,

$$\Omega_m(R) = \alpha^{1/(1-\alpha)}\Omega_s(R). \tag{11.39}$$

We can also show that the laissez-faire supply of an input falls short of the socially optimal one,

$$Z_m < Z_s \qquad \Longleftrightarrow \qquad \alpha\frac{1+R_s}{1+R_m} < 1. \tag{11.40}$$

To establish condition (11.40), divide equation (11.31b) by equation (11.36). Thus, the laissez-faire equilibrium is characterized by a smaller supply of each intermediate input and a lower growth rate as compared to the socially optimal outcome. These inefficiencies reflect the fact that suppliers of intermediate inputs do not internalize the full contribution of their inventions and so their monopolistic pricing results in less than socially efficient quantities of inputs.

## *Growth in spite of nonreproducible factors*

### *"Core" of capital goods produced without nonreproducible inputs*

It is not necessary that all factors be producible in order to experience sustained growth through factor accumulation in the neoclassical framework. Instead, Rebelo (1991) shows that the critical requirement for perpetual growth is the existence of a "core" of capital goods that is produced with constant returns technologies and without the direct or indirect use of nonreproducible factors. Here we will study the simplest version of his model with a single capital good that is produced without any input of the economy's constant labor endowment. Jones and Manuelli (1990) provide a general discussion of convex models of economic growth and highlight the crucial feature that the rate of return to accumulated capital must remain bounded above the inverse of the subjective discount factor in spite of any nonreproducible factors in production.

Rebelo (1991) analyzes the competitive equilibrium for the following technology,

$$C_t = L^{1-\alpha}\,(\phi_t K_t)^{\alpha}, \tag{11.41a}$$

$$I_t = A(1-\phi_t)K_t, \tag{11.41b}$$

$$K_{t+1} = (1-\delta)K_t + I_t, \tag{11.41c}$$

where $\phi_t \in [0,1]$ is the fraction of capital employed in the consumption goods sector, and $(1-\phi_t)$ is employed in the linear technology producing investment goods $I_t$. In a competitive equilibrium, the rental price of capital $r_t$ (in terms of consumption goods) is equal to the marginal product of capital, which then has to be the same across the two sectors (as long as they both are operating),

$$r_t = \alpha L^{1-\alpha} \left(\phi_t K_t\right)^{\alpha-1} = p_t A, \tag{11.42}$$

where $p_t$ is the relative price of capital in terms of consumption goods.

Along a steady-state growth path with a constant $\phi$, we can compute the growth rate of capital by substituting equation (11.41b) into equation (11.41c) and dividing by $K_t$,

$$\frac{K_{t+1}}{K_t} = (1-\delta) + A(1-\phi) \equiv 1 + \rho(\phi). \tag{11.43}$$

Given the growth rate of capital, $1+\rho(\phi)$, it is straightforward to compute other rates of change

$$\frac{p_{t+1}}{p_t} = [1+\rho(\phi)]^{\alpha-1}, \tag{11.44a}$$

$$\frac{C_{t+1}}{C_t} = \frac{p_{t+1} I_{t+1}}{p_t I_t} = \frac{p_{t+1} K_{t+1}}{p_t K_t} = [1+\rho(\phi)]^{\alpha}. \tag{11.44b}$$

Since the values of investment goods and the capital stock in terms of consumption goods grow at the same rate as consumption, $[1+\rho(\phi)]^{\alpha}$, this common rate is also the steady-state growth rate of the economy's net income, measured as $C_t + p_t I_t - \delta p_t K_t$.

Agents maximize utility given by condition (11.1) subject to budget constraint (11.7) modified to incorporate the relative price $p_t$,

$$c_t + p_t k_{t+1} = r_t k_t + (1-\delta) p_t k_t + \chi_t. \tag{11.45}$$

The first-order condition with respect to capital is

$$\left(\frac{c_{t+1}}{c_t}\right)^{\sigma} = \beta \frac{(1-\delta)p_{t+1} + r_{t+1}}{p_t}. \tag{11.46}$$

After substituting $r_{t+1} = p_{t+1} A$ from equation (11.42) and steady-state rates of change from equation (11.44) into equation (11.46), we arrive at the following equilibrium condition:

$$[1+\rho(\phi)]^{1-\alpha(1-\sigma)} = \beta(1-\delta+A). \tag{11.47}$$

Thus, the growth rate of capital and, therefore, the growth rate of consumption are positive as long as

$$\beta(1 - \delta + A) \geq 1. \tag{11.48a}$$

Moreover, the maintained assumption of this chapter that parameters are such that derived growth rates yield finite lifetime utility, $\beta(c_{t+1}/c_t)^{1-\sigma} < 1$, imposes here the parameter restriction $\beta[\beta(1-\delta+A)]^{\alpha(1-\sigma)/[1-\alpha(1-\sigma)]} < 1$ which can be simplified to read

$$\beta(1 - \delta + A)^{\alpha(1-\sigma)} < 1. \tag{11.48b}$$

Given that conditions (11.48) are satisfied, there is a unique equilibrium value of $\phi$ because the left side of equation (11.47) is monotonically decreasing in $\phi \in [0, 1]$ and it is strictly greater (smaller) than the right side for $\phi = 0$ ($\phi = 1$). The outcome is socially efficient because private and social rates of return are the same, as in the previous models with all factors reproducible.

*Research labor enjoying an externality*

Romer's (1987) model includes labor as a fixed nonreproducible factor, but similar to the last section, an important assumption is that this nonreproducible factor is not used in the production of inventions that expand the input variety (which constitutes a kind of reproducible capital in that model). In his sequel, Romer (1990) assumes that the input variety $A_t$ is expanded through the effort of researchers rather than the resource cost $\kappa$ in terms of final goods. Suppose that we specify this new invention technology as

$$A_{t+1} - A_t = \eta(1 - \phi_t)L,$$

where $(1 - \phi_t)$ is the fraction of the labor force employed in the research sector (and $\phi_t$ is working in the final-goods sector). After dividing by $A_t$, it becomes clear that this formulation cannot support sustained growth, since new inventions bounded from above by $\eta L$ must become a smaller fraction of any growing range $A_t$. Romer solves this problem by assuming that researchers' productivity grows with the range of inputs (i.e., an externality as discussed previously),

$$A_{t+1} - A_t = \eta A_t(1 - \phi_t)L,$$

so the growth rate of $A_t$ is

$$\frac{A_{t+1}}{A_t} = 1 + \eta(1 - \phi_t)L. \tag{11.49}$$

When seeking a balanced growth path with a constant $\phi$, we can use the earlier derivations, since the optimization problem of monopolistic input producers is the same as before. After replacing $L$ in equations (11.31b) and (11.32) by $\phi L$, the steady-state supply of an input and the profit flow of an input producer are

$$Z_m = \left(\frac{\alpha^2}{1+R_m}\right)^{1/(1-\alpha)} \phi L, \tag{11.50a}$$

$$\Omega_m(R_m) = (1-\alpha)\alpha^{1/(1-\alpha)} \left(\frac{\alpha}{1+R_m}\right)^{\alpha/(1-\alpha)} \phi L. \tag{11.50b}$$

In an equilibrium, agents must be indifferent between earning the wage in the final-goods sector equal to the marginal product of labor and being a researcher who expands the range of inputs by $\eta A_t$ and receives the associated discounted stream of profits in equation (11.33):

$$(1-\alpha)\,(\phi L)^{-\alpha}\, A_t Z_m^{\alpha} = \eta A_t \frac{\Omega_m(R_m)}{R_m}.$$

The substitution of equation (11.50) into this expression yields

$$\phi = \frac{R_m}{\alpha \eta L}, \tag{11.51}$$

which used in equation (11.49) determines the growth rate of the input range,

$$\frac{A_{t+1}}{A_t} = 1 + \eta L - \frac{R_m}{\alpha}. \tag{11.52}$$

Thus, the maximum feasible growth rate in equation (11.49), that is, $1+\eta L$ with $\phi = 0$, requires an interest rate $R_m = 0$, while the growth vanishes as $R_m$ approaches $\alpha\eta L$.

As previously, we can show that both consumption and the input range must grow at the same rate along a balanced growth path. It then remains to determine which consumption growth rate given by equation (11.52), is supported by Euler equation (11.35);

$$1 + \eta L - \frac{R_m}{\alpha} = \left[\beta(1+R_m)\right]^{1/\sigma}. \tag{11.53}$$

The left side of equation (11.53) is monotonically decreasing in $R_m$, and the right side is increasing. It is also trivially true that the left-hand side is strictly greater than the right-hand side for $R_m = 0$. Thus, a unique solution exists as long as the technology is sufficiently productive, in the sense that $\beta(1 + \alpha\eta L) > 1$. This parameter restriction ensures that the left side of equation (11.53) is strictly less than the right side at the interest rate $R_m = \alpha\eta L$ corresponding to a situation with zero growth, since no labor is allocated to the research sector, $\phi = 1$.

Equation (11.53) shows that this alternative model of research shares the scale implications described earlier; that is, a larger economy in terms of $L$ has a higher equilibrium interest rate and therefore a higher growth rate. It can also be shown that the laissez-faire outcome continues to produce a smaller quantity of each input and yield a lower growth rate than what is socially optimal. An additional source of underinvestment is now that agents who invent new inputs do not take into account that their inventions will increase the productivity of all future researchers.

## Concluding comments

This chapter has focused on the mechanical workings of endogenous growth models with only limited references to the motivation behind assumptions. For example, we have examined how externalities might enter models to overcome the onset of diminishing returns from nonreproducible factors without referring too much to the authors' interpretation of those externalities. The formalism of models is of course silent on why the assumptions are made, but the conceptual ideas behind can hold valuable insights. In the last setup, Paul Romer argues that input designs represent excludable factors in the monopolists' production of inputs but the input variety $A$ is also an aggregate stock of knowledge that enters as a nonexcludable factor in the production of new inventions. That is, the patent holder of an input type has the sole right to produce and sell that particular input, but she cannot stop inventors from studying the input design and learning knowledge that helps to invent new inputs. This multiple use of an input design hints at the nonrival nature of ideas and technology, i.e., a nonrival object has the property that its use by one person in no way limits its use by another. Romer (1990, p. S75) emphasizes this fundamental nature of technology and its implication; "If a nonrival good has productive value, then output cannot be a constant-returns-to-scale function of all its inputs taken together. The standard replication argument used to justify homogeneity of degree one does not apply because it is not necessary to replicate nonrival inputs." Thus, an endogenous growth model that is driven by technological change must be one where the advancement enters the economy as an externality or the assumption of perfect competition must be abandoned. Besides technological change, an alternative approach in the endogenous growth literature is to assume that all production factors are reproducible, or that there is a "core" of capital goods produced without the direct or indirect use of nonreproducible factors.

As we have seen, much of the effort in the endogenous growth literature is geared toward finding the proper technology specification. Even though growth is an endogenous outcome in these models, its manifestation hinges ultimately upon technology assumptions. In the case of the last setup, as pointed out by Romer (1990, p. S84), "Linearity in $A$ is what makes unbounded growth possible, and in this sense, unbounded growth is more like an assumption than a result of the model." It follows that various implications of the analyses stand and fall with the assumptions on technology. For example, the preceding model of research and monopolistic competition implies that the laissez-faire economy grows at a slower rate than the social optimum, but Benassy (1998) shows how this result can be overturned if the production function for final goods on the right side of equation (11.25) is multiplied by the input range raised to some power $\nu$, $A_t^{\nu}$. It then becomes possible that the laissez-faire growth rate exceeds the socially optimal rate because the new production function disentangles input producers' market power, determined by the parameter $\alpha$, and the economy's returns to specialization, which is here also related to the parameter $\nu$.

Segerstrom, Anant, and Dinopoulos (1990), Grossman and Helpman (1991), and Aghion and Howitt (1992) provide early attempts to explore endogenous growth arising from technologies that allow for product improvements and, therefore, product obsolescence. These models open the possibility that the laissez-faire growth rate is excessive because of a *business-stealing* effect where agents fail to internalize the fact that their inventions exert a negative effect on incumbent producers. Similar to the models of research by Romer (1987, 1990) covered in this chapter, these other technologies exhibit scale effects so that increases in the resources devoted to research imply faster economic growth. Charles Jones (1995), Young (1998), and Segerstrom (1998) criticize this feature and propose assumptions on technology that do not give rise to scale effects.

## *Exercises*

*Exercise 11.1* **Government spending and investment**, donated by Rodolfo Manuelli

Consider the following economy. There is a representative agent who has preferences given by

$$\sum_{t=0}^{\infty} \beta^t u(c_t),$$

where the function $u$ is differentiable, increasing, and strictly concave. The technology in this economy is given by

$$\begin{aligned} c_t + x_t + g_t &\leq f(k_t, g_t), \\ k_{t+1} &\leq (1-\delta)k_t + x_t, \\ (c_t, k_{t+1}, x_t) &\geq (0,0,0), \end{aligned}$$

and the initial condition $k_0 > 0$, given. Here $k_t$ and $g_t$ are capital per worker and government spending per worker. The function $f$ is assumed to be strictly concave, increasing in each argument, twice differentiable, and such that the partial derivative with respect to both arguments converge to zero as the quantity of them grows without bound.

**a.** Describe a set of equations that characterize an interior solution to the planner's problem when the planner can choose the sequence of government spending.

**b.** Describe the steady state for the "general" specification of this economy. If necessary, make assumptions to guarantee that such a steady state exists.

**c.** Go as far as you can describing how the steady-state levels of capital per worker and government spending per worker change as a function of the discount factor.

**d.** Assume that the technology level can vary. More precisely, assume that the production function is given by $f(k,g,z) = zk^{\alpha}g^{\eta}$, where $0 < \alpha < 1$, $0 < \eta < 1$, and $\alpha + \eta < 1$. Go as far as you can describing how the investment/GDP ratio and the government spending/GDP ratio vary with the technology level $z$ at the steady state.

*Exercise 11.2* **Productivity and employment**, donated by Rodolfo Manuelli

Consider a basic growth economy with one modification. Instead of assuming that the labor supply is fixed at one, we include leisure in the utility function. To simplify, we consider the total endowment of time to be one. With this modification, preferences and technology are given by

$$\sum_{t=0}^{\infty} \beta^t u(c_t, 1 - n_t),$$

$$c_t + x_t + g_t \le z f(k_t, n_t),$$

$$k_{t+1} \le (1 - \delta) k_t + x_t.$$

In this setting, $n_t$ is the number of hours worked by the representative household at time $t$. The rest of the time, $1 - n_t$, is consumed as leisure. The functions $u$ and $f$ are assumed to be strictly increasing in each argument, concave, and twice differentiable. In addition, $f$ is such that the marginal product of capital converges to zero as the capital stock goes to infinity for any given value of labor, $n$.

**a.** Describe the steady state of this economy. If necessary, make additional assumptions to guarantee that it exists and is unique. If you make additional assumptions, go as far as you can giving an economic interpretation of them.

**b.** Assume that $f(k, n) = k^\alpha n^{1-\alpha}$ and $u(c, 1 - n) = \left[c^\mu (1 - n)^{1-\mu}\right]^{1-\sigma} / (1 - \sigma)$. What is the effect of changes in the technology (say increases in $z$) upon employment and output per capita?

**c.** Consider next an increase in $g$. Are there conditions under which an increase in $g$ will result in an increase in the steady-state $k/n$ ratio? How about an increase in the steady-state level of output per capita? Go as far as you can giving an economic interpretation of these conditions. [Try to do this for general $f(k, n)$ functions – with the appropriate convexity assumptions – but if this proves too hard, use the Cobb-Douglas specification.]

*Exercise 11.3* **Vintage capital and cycles**, dontated by Rodolfo Manuelli

Consider a standard one sector optimal growth model with only one difference: If $k_{t+1}$ new units of capital are built at time $t$, these units remain fully productive (i.e. they do not depreciate) until time $t+2$, at which point they disappear. Thus, the technology is given by

$$c_t + k_{t+1} \leq zf(k_t + k_{t-1}).$$

**a.** Formulate the optimal growth problem.

**b.** Show that, under standard conditions, a steady state exists and is unique.

**c.** A researcher claims that with the unusual depreciation pattern, it is possible that the economy displays cycles. By this he means that, instead of a steady state, the economy will converge to a period two sequence like $(c^o, c^e, c^o, c^e, \ldots)$ and $(k^o, k^e, k^o, k^e, \ldots)$, where $c^o$ $(k^o)$ indicates consumption (investment) in odd periods, and $c^e$ $(k^e)$ indicates consumption (investment) in even periods. Go as far as you can determining whether this can happen. If it is possible, try to provide an example.

*Exercise 11.4* **Excess capacity**, donated by Rodolfo Manuelli

In the standard growth model, there is no room for varying the rate of utilization of capital. In this problem you explore how the nature of the solution is changed when variable rates of capital utilization are allowed.

As in the standard model, there is a representative agent with preferences given by

$$\sum_{t=0}^{\infty} \beta^t u(c_t), \quad 0 < \beta < 1.$$

It is assumed that $u$ is strictly increasing, concave, and twice differentiable. Output depends on the actual number of machines used at time $t$, $\kappa_t$. Thus, the aggregate resource constraint is

$$c_t + x_t \leq zf(\kappa_t),$$

where the function $f$ is strictly increasing, concave, and twice differentiable. In addition, $f$ is such that the marginal product of capital converges to zero as the stock goes to infinity. Capital that is not used does not depreciate. Thus, capital accumulation satisfies

$$k_{t+1} \leq (1-\delta)\kappa_t + (k_t - \kappa_t) + x_t,$$

where we require that the number of machines used, $\kappa_t$, is no greater than the number of machines available, $k_t$, or $k_t \geq \kappa_t$. This specification captures the idea that if some machines are not used, $k_t - \kappa_t > 0$, they do not depreciate.

**a.** Describe the planner's problem and analyze, as thoroughly as you can, the first order conditions. Discuss your results.

**b.** Describe the steady state of this economy. If necessary, make additional assumptions to guarantee that it exists and is unique. If you make additional assumptions, go as far as you can giving an economic interpretation of them.

**c.** What is the optimal level of capacity utilization in this economy in the steady state?

**d.** Is this model consistent with the view that cross country differences in output per capita are associated with differences in capacity utilization?

*Exercise 11.5* **Heterogeneity and growth**, donated by Rodolfo Manuelli

Consider an economy populated by a large number of households indexed by $i$. The utility function of household $i$ is

$$\sum_{t=0}^{\infty} \beta^t u_i(c_{it}),$$

where $0 < \beta < 1$, and $u_i$ is differentiable, increasing and strictly concave. Note that although we allow the utility function to be "household specific," all households share the same discount factor. All households are endowed with one unit of labor that is supplied inelastically.

Assume that in this economy capital markets are perfect and that households start with initial capital given by $k_{i0} > 0$. Let total capital in the economy at time $t$ be denoted $k_t$ and assume that total labor is normalized to $1$.

Assume that there are a large number of firms that produce output using capital and labor. Each firm has a production function given by $F(k, n)$ which is increasing, differentiable, concave and homogeneous of degree one. Firms maximize the present discounted value of profits. Assume that initial ownership of firms is uniformly distributed across households.

**a.** Define a competitive equilibrium.

**b.** i) Economist A argues that the steady state of this economy is unique and independent of the $u_i$ functions, while B says that without knowledge of the $u_i$ functions it is impossible to calculate the steady-state interest rate.

ii) Economist A says that if $k_0$ is the steady-state aggregate stock of capital, then the pattern of "consumption inequality" will mirror exactly the pattern of "initial capital inequality" (i.e., $k_{i0}$), even though capital markets are perfect. Economist B argues that for all $k_0$, in the long run, per capita consumption will be the same for all households.

Discuss i) and ii) and justify your answer. Be as formal as you can.

**c.** Assume that the economy is at the steady state. Describe the effects of the following three policies.

i) At time zero, capital redistributed across households (i.e., some people must surrender capital and others get their capital).

ii) Half of the households are required to pay a lump sum tax. The proceeds of the tax are used to finance a transfer program to the other half of the population.

iii) Two thirds of the households are required to pay a lump sum tax. The proceeds of the tax are used to finance the purchase of a public good, say $g$, which does not enter in either preferences or technology.

*Exercise 11.6* **Taxes and growth**, donated by Rodolfo Manuelli

Consider a simple two-planner economy. The first planner picks "tax rates," $\tau_t$, and makes transfers to the representative agent, $v_t$. The second planner takes the tax rates and the transfers as given. That is, even though we know the connection between tax rates and transfers, the second planner does not, he/she takes the sequence of tax rates and transfers as given and beyond his/her control when solving for the optimal allocation. Thus the problem faced by the second planner (the only one we will analyze for now) is

$$\max \sum_{t=0}^{\infty} \beta^t u(c_t)$$

subject to

$$c_t + x_t + g_t - v_t \leq (1 - \tau_t) f(k_t),$$
$$k_{t+1} \leq (1 - \delta) k_t + x_t,$$
$$(c_t, k_{t+1}, x_t) \geq (0, 0, 0),$$

and the initial condition $k_0 > 0$, given. The functions $u$ and $f$ are assumed to be strictly increasing, concave, and twice differentiable. In addition, $f$ is such that the marginal product of capital converges to zero as the capital stock goes to infinity.

**a.** Assume that $0 < \tau_t = \tau < 1$, that is, the tax rate is constant. Assume that $v_t = \tau f(k_t)$ (remember that we know this, but the planner takes $v_t$ as given at the time he/she maximizes). Show that there exists a steady state, and that for any initial condition $k_0 > 0$ the economy converges to the steady state.

**b.** Assume now that the economy has reached the steady state you analyzed in a. The first planner decides to change the tax rate to $0 < \tau' < \tau$. (Of course, the first planner and we know that this will result in a change in $v_t$; however, the second planner — the one that maximizes — acts as if $v_t$ is a given sequence that is independent of his/her decisions.) Describe the new steady state as well as the dynamic path followed by the economy to reach this new steady state. Be as precise as you can about consumption, investment and output.

**c.** Consider now a competitive economy in which households — but not firms — pay income tax at rate $\tau_t$ on both labor and capital income. In addition, each household receives a transfer, $v_t$, that it takes to be given and independent of its own actions. Let the aggregate per capita capital stock be $k_t$. Then, balanced budget on the part of the government implies $v_t = \tau_t(r_t k_t + w_t, n_t)$, where $r_t$ and $w_t$ are the rental prices of capital and labor, respectively. Assume that the production function is $F(k, n)$, with $F$ homogeneous of degree one, concave and "nice." Go as far as you can describing the impact of the change described in b upon the equilibrium interest rate.

# 12
# *Optimal Taxation with Commitment*

## *Introduction*

This chapter formulates dynamic optimal taxation as a Ramsey problem. The government's goal is to maximize agents' welfare subject to raising set revenues through distortionary taxation. When designing an optimal policy, the government takes into account the equilibrium reactions by consumers and firms to the tax system. We first study a nonstochastic economy, then a stochastic economy.

The model is a competitive equilibrium version of the basic neoclassical growth model with a government that finances an exogenous stream of government purchases. In the simplest version, the production factors are raw labor and physical capital on which the government levies distorting flat-rate taxes. The problem is to determine the optimal settings over time for the two tax rates. In a nonstochastic economy, Chamley (1986) and Judd (1985b) show in related settings that if an equilibrium has an asymptotic steady state, then the optimal policy is eventually to set the tax rate on capital to zero. This remarkable result asserts that capital income taxation serves neither efficiency nor redistributive purposes in the long run. This conclusion is robust to whether the government can issue debt or must run a balanced budget in each period. However, the optimal capital tax may be different from zero if the tax system is incomplete. Following an approach similar to one taken by Correia (1996), this chapter studies the case of an additional fixed production factor that cannot be taxed by the government.

In a stochastic version of the model, we find indeterminacy of state-contingent debt and capital taxes. There are infinitely many plans that implement a given competitive allocation. For example, two alternative extreme cases are (1) that the government issues risk-free bonds and lets the capital tax rate depend on the current state, or (2) that it fixes the capital tax rate one period ahead and lets debt be state-contingent. While the state-by-state capital tax rates cannot be pinned down, the optimal Ramsey plan does determine the present market value of next period's tax payments across states of nature. After dividing by the present market value of capital income, we have a measure that we call the *ex ante capital tax rate.* If there exists a stationary Ramsey allocation, Zhu (1992) shows that there are two possible outcomes. For some utility functions, the Ramsey plan prescribes a zero ex ante capital tax rate, which can be implemented by setting a zero tax on capital income. But, except for special classes of preferences, Zhu concludes that the ex ante capital tax rate should vary around zero, in the sense that there is a positive measure of states with positive tax rates and a positive measure of states with negative tax rates. Chari, Christiano, and Kehoe (1994) perform numerical simulations and conclude that there is a quantitative presumption that the ex ante capital tax rate is approximately zero.

Returning to a nonstochastic setup, Jones, Manuelli, and Rossi (1997) augment the model by allowing for human capital accumulation. They make the particular assumption about the technology for human capital accumulation that it is linearly homogeneous in a stock of human capital and a flow of inputs coming from current output. Under this special constant returns assumption, they show that a zero limiting tax applies also to labor income; that is, the return to human capital should not be taxed in the limit. Instead, the government should resort to a consumption tax. But even this consumption tax, and therefore all taxes, should be zero in the limit for a particular class of preferences where it is optimal for the government under a transition period to amass so many claims on the private economy that the interest earnings suffice to finance government expenditures. While these successive results on optimal taxation require ever more stringent assumptions, a zero capital tax in a nonstochastic steady state is an immediate implication of a standard constant-returns-to-scale production technology, competitive markets, and a complete set of flat-rate taxes.

Throughout the chapter we maintain the assumption that the government can commit to future tax rates.

## A nonstochastic economy

There is an infinitely lived representative household living in a single-good economy. The household likes consumption, leisure streams $\{c_t, \ell_t\}_{t=0}^{\infty}$ that give higher values of

$$\sum_{t=0}^{\infty} \beta^t u(c_t, \ell_t), \ \ \beta \in (0,1) \tag{12.1}$$

where $u$ is increasing, strictly concave, and three times continuously differentiable in $c$ and $\ell$. The household is endowed with one unit of time that can be used for leisure $\ell_t$ and labor $n_t$;

$$\ell_t + n_t = 1. \tag{12.2}$$

The single good is produced with labor $n_t$ and capital $k_t$ as inputs. The output can be consumed by households, used by the government, or used to augment the capital stock. The technology is described by

$$c_t + g_t + k_{t+1} = F(k_t, n_t) + (1-\delta)k_t, \tag{12.3}$$

where $\delta \in (0,1)$ is the rate at which capital depreciates, and $\{g_t\}_{t=0}^{\infty}$ is an exogenous sequence of government purchases. We assume a standard concave production function $F(k,n)$ that exhibits constant returns to scale. By Euler's theorem, linear homogeneity of $F$ implies

$$F(k,n) = F_k k + F_n n. \tag{12.4}$$

Let $u_c$ be the derivative of $u(c_t, \ell_t)$ with respect to consumption; $u_\ell$ is the derivative with respect to $\ell$. We use $u_c(t)$ and $F_k(t)$ and so on to denote the time-$t$ values of the indicated objects, evaluated at an allocation to be understood from the context.

*Government*

The government finances its stream of purchases $\{g_t\}_{t=0}^{\infty}$ by levying flat-rate, time-varying taxes on earnings from capital at rate $\tau_t^k$ and from labor at rate $\tau_t^n$. The government might also trade one-period bonds, which would suffice to accomplish any intertemporal trade in a world without uncertainty. Let $b_t$ be government indebtedness to the private sector, denominated in time $t$-goods, maturing at the beginning of period $t$. The government's budget constraint becomes

$$g_t = \tau_t^k r_t k_t + \tau_t^n w_t n_t + \frac{b_{t+1}}{R_t} - b_t, \tag{12.5}$$

where $r_t$ and $w_t$ are the market-determined rental rate of capital and the wage rate for labor, respectively, denominated in units of time $t$ goods, and $R_t$ is the gross rate of return on one-period bonds held from $t$ to $t+1$. Interest earnings on bonds are assumed to be tax exempt; this assumption is innocuous for bond exchanges between the government and the private sector.

*Households*

The representative household maximizes expression (12.1) subject to the following sequence of budget constraints:

$$c_t + k_{t+1} + \frac{b_{t+1}}{R_t} = (1 - \tau_t^n) w_t n_t + (1 - \tau_t^k) r_t k_t + (1 - \delta) k_t + b_t. \tag{12.6}$$

With $\beta^t \lambda_t$ as the Lagrange multiplier on the time-$t$ budget constraint, the first-order conditions are

$$\begin{aligned} c_t\colon &\quad u_c(t) = \lambda_t, && (12.7)\\ n_t\colon &\quad u_\ell(t) = \lambda_t (1 - \tau_t^n) w_t, && (12.8)\\ k_{t+1}\colon &\quad \lambda_t = \beta \lambda_{t+1} \left[ (1 - \tau_{t+1}^k) r_{t+1} + 1 - \delta \right], && (12.9)\\ b_{t+1}\colon &\quad \lambda_t \frac{1}{R_t} = \beta \lambda_{t+1}. && (12.10) \end{aligned}$$

Substituting equation (12.7) into equations (12.8) and (12.9), we obtain

$$\begin{aligned} u_\ell(t) &= u_c(t)(1 - \tau_t^n) w_t, && (12.11a)\\ u_c(t) &= \beta u_c(t+1) \left[ (1 - \tau_{t+1}^k) r_{t+1} + 1 - \delta \right]. && (12.11b) \end{aligned}$$

Moreover, equations (12.9) and (12.10) imply

$$R_t = (1 - \tau_{t+1}^k) r_{t+1} + 1 - \delta, \tag{12.12}$$

which is a condition not involving any quantities that the household is free to adjust. Because only one financial asset is needed to accomplish all intertemporal trades in a world without uncertainty, condition (12.12) constitutes an arbitrage condition for trades in capital and bonds that ensures that these two assets have the same rate of return. The same condition can be obtained by consolidating two consecutive budget constraints; constraint (12.6) and its counterpart for time $t+1$ can be merged by eliminating the common quantity $b_{t+1}$,

$$\begin{aligned} c_t + \frac{c_{t+1}}{R_t} + \frac{k_{t+2}}{R_t} + \frac{b_{t+2}}{R_t R_{t+1}} &= (1 - \tau_t^n) w_t n_t + \frac{(1 - \tau_{t+1}^n) w_{t+1} n_{t+1}}{R_t}\\ &+ \left[ \frac{(1 - \tau_{t+1}^k) r_{t+1} + 1 - \delta}{R_t} - 1 \right] k_{t+1} + (1 - \tau_t^k) r_t k_t + (1 - \delta) k_t + b_t. \end{aligned} \tag{12.13}$$

where the left side is the use of funds, and the right side measures the resources at the household's disposal. If the term multiplying $k_{t+1}$ is not zero, the household can make its budget set unbounded by either buying an arbitrarily large $k_{t+1}$ when $(1-\tau_{t+1}^k)r_{t+1}+1-\delta > R_t$, or, in the opposite case, selling capital short with an arbitrarily large negative $k_{t+1}$. In such arbitrage transactions, the household would finance purchases of capital or invest the proceeds from short sales in the bond market between periods $t$ and $t+1$. Thus, to ensure the existence of a competitive equilibrium with bounded budget sets, condition (12.12) must hold.

If we continue the process of recursively using successive budget constraints to eliminate successive $b_{t+j}$ terms, begun in equation (12.13), we arrive at the household's present-value budget constraint,

$$\sum_{t=0}^{\infty}\left(\prod_{i=1}^{t} R_i^{-1}\right) c_t = \sum_{t=0}^{\infty}\left(\prod_{i=1}^{t} R_i^{-1}\right)(1-\tau_t^n)w_t n_t + \left[(1-\tau_0^k)r_0 + 1 - \delta\right] k_0 + b_0, \tag{12.14}$$

where we have imposed the transversality conditions

$$\lim_{T\to\infty}\left(\prod_{i=0}^{T-1} R_i^{-1}\right) k_{T+1} = 0, \tag{12.15}$$

$$\lim_{T\to\infty}\left(\prod_{i=0}^{T-1} R_i^{-1}\right) \frac{b_{T+1}}{R_T} = 0. \tag{12.16}$$

As discussed in chapter 10, the household would not like to violate these transversality conditions by choosing $k_{t+1}$ or $b_{t+1}$ to be larger, because alternative feasible allocations with higher consumption in finite time would yield higher lifetime utility. A consumption/savings plan that made either expression negative would not be possible because the household would not find any agents willing to be on the lending side of the implied transactions.

### *Firms*

In each period, the representative firm takes $(r_t, w_t)$ as given and rents capital and labor from households to maximize profits,

$$\Pi = F(k_t, n_t) - r_t k_t - w_t n_t. \tag{12.17}$$

The first-order conditions for this problem are

$$r_t = F_k(t), \tag{12.18a}$$

$$w_t = F_n(t). \tag{12.18b}$$

In other words, inputs should be employed until the marginal product of the last unit is equal to its rental price. With constant returns to scale, we get the standard result that pure profits are zero and the size of an individual firm is indeterminate.

An alternative way of establishing the equilibrium conditions for the rental price of capital and the wage rate for labor is as follows. Substituting equation (12.4) in equation (12.17) yields

$$\Pi = [F_k(t) - r_t]k_t + [F_n(t) - w_t]n_t.$$

If the firm's profits are to be nonnegative and finite, the terms multiplying $k_t$ and $n_t$ must be zero; that is, condition (12.18) must hold. Thus, these conditions imply that in any equilibrium, $\Pi = 0$.

## The Ramsey problem

### Definitions

We shall use symbols without subscripts to denote the one-sided infinite sequence for the corresponding variable, e.g., $c \equiv \{c_t\}_{t=0}^{\infty}$.

DEFINITION: A *feasible allocation* is a sequence $(k, c, \ell, g)$ that satisfies equation (12.3).

DEFINITION: A *price system* is a 3-tuple of nonnegative bounded sequences $(w, r, R)$.

DEFINITION: A *government policy* is a 4-tuple of sequences $(g, \tau^k, \tau^n, b)$.

DEFINITION: A *competitive equilibrium* is a feasible allocation, a price system, and a government policy such that (a) given the price system and the government policy, the allocation solves both the firm's problem and the household's problem; and (b) given the allocation and the price system, the government policy satisfies the sequence of government budget constraints (12.5).

There are many competitive equilibria, indexed by different government policies. This multiplicity motivates the Ramsey problem.

DEFINITION: Given $k_0, b_0$, the *Ramsey problem* is to choose a competitive equilibrium that maximizes expression (12.1).

To make the Ramsey problem interesting, we always impose a restriction on $\tau_0^k$, for example, by taking it as given at a small number, say, $0$. This approach rules out taxing the initial capital stock, a so-called capital levy that would constitute a lump-sum tax, since $k_0$ is in fixed supply. Further, one often imposes other restrictions on $\tau_t^k, t \geq 1$, namely, that they be bounded above by some arbitrarily given numbers. These bounds play an important role in shaping the near-term temporal properties of the optimal tax plan, as discussed by Chamley (1986) and explored in computational work by Jones, Manuelli, and Rossi (1993). In the analysis that follows, we shall impose the bound on $\tau_t^k$ only for $t = 0$.[1]

## Zero capital tax

Following Chamley (1986), we formulate the Ramsey problem as if the government chooses the after-tax rental rate of capital $\tilde{r}_t$, and the after-tax wage rate $\tilde{w}_t$;

$$\begin{aligned}\tilde{r}_t &\equiv (1 - \tau_t^k) r_t, \\ \tilde{w}_t &\equiv (1 - \tau_t^n) w_t.\end{aligned}$$

Using equations (12.18) and (12.4), Chamley expresses government tax revenues as

$$\begin{aligned}\tau_t^k r_t k_t + \tau_t^n w_t n_t &= (r_t - \tilde{r}_t) k_t + (w_t - \tilde{w}_t) n_t \\ &= F_k(t) k_t + F_n(t) n_t - \tilde{r}_t k_t - \tilde{w}_t n_t \\ &= F(k_t, n_t) - \tilde{r}_t k_t - \tilde{w}_t n_t.\end{aligned}$$

After substituting this expression into equation (12.5), we have incorporated the firm's first-order conditions into the government's budget constraint. The government's policy choice is also constrained by the aggregate resource constraint

[1] According to our assumption on the technology in equation (12.3), capital is reversible and can be transformed back into the consumption good. Thus, the capital stock is a fixed factor for only one period at a time, so $\tau_0^k$ is the only tax that we need to restrict to ensure an interesting Ramsey problem.

(12.3) and the household's first-order conditions (12.11). The Ramsey problem in Lagrangian form becomes

$$
\begin{aligned}
L = \sum_{t=0}^{\infty} \beta^t \{ & u(c_t, 1-n_t) \\
& + \Psi_t \left[ F(k_t, n_t) - \tilde{r}_t k_t - \tilde{w}_t n_t + \frac{b_{t+1}}{R_t} - b_t - g_t \right] \\
& + \theta_t \left[ F(k_t, n_t) + (1-\delta) k_t - c_t - g_t - k_{t+1} \right] \\
& + \mu_{1t} \left[ u_\ell(t) - u_c(t) \tilde{w}_t \right] \\
& + \mu_{2t} \left[ u_c(t) - \beta u_c(t+1) \left( \tilde{r}_{t+1} + 1 - \delta \right) \right] \}, \qquad (12.19)
\end{aligned}
$$

where $R_t = \tilde{r}_{t+1} + 1 - \delta$, as given by equation (12.12). Note that the household's budget constraint is not explicitly included because it is redundant when the government satisfies its budget constraint and the resource constraint holds.

The first-order condition with respect to $k_{t+1}$ is

$$
\theta_t = \beta \left\{ \Psi_{t+1} \left[ F_k(t+1) - \tilde{r}_{t+1} \right] + \theta_{t+1} \left[ F_k(t+1) + 1 - \delta \right] \right\}. \qquad (12.20)
$$

The equation has a straightforward interpretation. A marginal increment of capital investment in period $t$ increases the quantity of available goods at time $t+1$ by the amount $[F_k(t+1) + 1 - \delta]$, which has a social marginal value $\theta_{t+1}$. In addition, there is an increase in tax revenues equal to $[F_k(t+1) - \tilde{r}_{t+1}]$, which enables the government to reduce its debt or other taxes by the same amount. The reduction of the "excess burden" equals $\Psi_{t+1}[F_k(t+1) - \tilde{r}_{t+1}]$. The sum of these two effects in period $t+1$ is discounted back by the discount factor $\beta$, and is equal to the social marginal value of the initial investment good in period $t$, given by $\theta_t$.

Suppose that government expenditures stay constant after some period $T$, and assume that the solution to the Ramsey problem converges to a steady state; that is, all endogenous variables remain constant. Using equation (12.18a), the steady-state version of equation (12.20) is

$$
\theta = \beta \left[ \Psi \left( r - \tilde{r} \right) + \theta \left( r + 1 - \delta \right) \right]. \qquad (12.21)
$$

Now with a constant consumption stream, the steady-state version of the household's optimality condition for the choice of capital in equation (12.11b) is

$$
1 - \beta \left( \tilde{r} + 1 - \delta \right). \qquad (12.22)
$$

A substitution of equation (12.22) into equation (12.21) yields

$$(\theta + \Psi)\,(r - \tilde{r}) = 0. \tag{12.23}$$

Since the marginal social value of goods, $\theta$, is strictly positive and the marginal social value of reducing government debt or taxes, $\Psi$, is nonnegative, it follows that $r$ must be equal to $\tilde{r}$, so that $\tau^k = 0$. This analysis establishes the following celebrated result, versions of which were attained by Chamley (1986) and Judd (1985b).

PROPOSITION 1: If there exists a steady-state Ramsey allocation, the associated limiting tax rate on capital is zero.

Note that the conclusion is robust to whether the government can issue debt or must run a balanced budget in each period. In the latter case, we just set $b_t$ and $b_{t+1}$ equal to zero in equation (12.19), and nothing changes in the derivation of $\tau^k = 0$. We emphasize this fact so that we can eliminate the possibility that zero capital taxes are optimal because all tax revenues are zero asymptotically. That is, a government trading in the bond market *could* choose to amass claims on the private economy so that eventually the interest earnings suffice to finance the stream of government expenditures. We will examine a situation later when such a policy would actually be the optimal one.

## Limits to redistribution

The optimality of a limiting zero capital tax extends also to an economy with heterogeneous agents, as mentioned by Chamley (1986) and explored in depth by Judd (1985b). We assume a finite number of different classes of agents, $N$, and for simplicity, let each class be the same size. The consumption, labor supply, and capital stock of the representative agent in class $i$ are denoted $c_t^i$, $n_t^i$, and $k_t^i$, respectively. The utility function might also depend on the class, $u^i(c_t^i, 1 - n_t^i)$, but the discount factor is assumed to be identical across all agents.

The government can make positive class-specific lump-sum transfers $S_t^i \geq 0$, but there are no lump-sum taxes. As before, the government must rely on flat-rate taxes on earnings from capital and labor. We assume that the government has a social welfare function that is a positively weighted average of individual utilities, with the weight $\alpha^i \geq 0$ on class $i$. Without affecting the limiting result for the capital tax, the government is here assumed to run a balanced budget. The Lagrangian of the government's optimization problem becomes

$$
\begin{aligned}
L = \sum_{t=0}^{\infty} \beta^t \Bigg\{ & \sum_{i=1}^{N} \alpha^i u^i(c_t^i, 1 - n_t^i) \\
& + \Psi_t \left[ F(k_t, n_t) - \tilde{r}_t k_t - \tilde{w}_t n_t - g_t - S_t \right] \\
& + \theta_t \left[ F(k_t, n_t) + (1 - \delta) k_t - c_t - g_t - k_{t+1} \right] \\
& + \sum_{i=1}^{N} \epsilon_t^i \left[ \tilde{w}_t n_t^i + \tilde{r}_t k_t^i + (1 - \delta) k_t^i + S_t^i - c_t^i - k_{t+1}^i \right] \\
& + \sum_{i=1}^{N} \mu_{1t}^i \left[ u_\ell^i(t) - u_c^i(t) \tilde{w}_t \right] \\
& + \sum_{i=1}^{N} \mu_{2t}^i \left[ u_c^i(t) - \beta u_c^i(t+1) \left( \tilde{r}_{t+1} + 1 - \delta \right) \right] \Bigg\}, \qquad (12.24)
\end{aligned}
$$

where $x_t \equiv \sum_{i=1}^{N} x_t^i$, for $x = c, n, k, S$. Here we have to include the budget constraints and the first-order conditions for each class of agents.

The social marginal value of an increment in the capital stock depends now on whose capital stock is augmented. The Ramsey problem's first-order condition with respect to $k_{t+1}^i$ is

$$
\begin{aligned}
\theta_t + \epsilon_t^i = \beta \{ & \Psi_{t+1} \left[ F_k(t+1) - \tilde{r}_{t+1} \right] \\
& + \theta_{t+1} \left[ F_k(t+1) + 1 - \delta \right] + \epsilon_{t+1}^i \left( \tilde{r}_{t+1} + 1 - \delta \right) \}. \qquad (12.25)
\end{aligned}
$$

If an asymptotic steady state exists in equilibrium, the time-invariant version of this condition becomes

$$
\theta + \epsilon^i \left[ 1 - \beta \left( \tilde{r} + 1 - \delta \right) \right] = \beta \left[ \Psi \left( r - \tilde{r} \right) + \theta \left( r + 1 - \delta \right) \right]. \qquad (12.26)
$$

Since the steady-state condition (12.22) holds for each individual household, the term multiplying $\epsilon^i$ is zero, and we can once again deduce condition (12.23), asserting that the limiting capital tax must be zero in any convergent Pareto-efficient tax program.

Judd (1985b) discusses one extreme version of heterogeneity with two classes of agents. Agents of class 1 are workers who do not save, so their budget constraint is

$$c_t^1 = \tilde{w}_t n_t^1 + S_t^1.$$

Agents of class 2 are capitalists who do not work, so their budget constraint is

$$c_t^2 + k_{t+1}^2 = \tilde{r}_t k_t^2 + (1-\delta) k_t^2 + S_t^2.$$

Since this setup is also covered by the preceding analysis, a limiting zero capital tax remains optimal if there is a steady state. This fact implies, for example, that if the government only values the welfare of workers ($\alpha^1 > \alpha^2 = 0$), there will not be any redistribution in the limit, and government expenditures will be financed solely by levying wage taxes on workers.

It is important to keep in mind that the results only pertain to the limiting steady state. Our analysis says nothing about how much redistribution is accomplished in the transition period.

## *Primal approach to the Ramsey problem*

In the formulation of the Ramsey problem in expression (12.19), Chamley reduced a pair of taxes $(\tau_t^k, \tau_t^n)$ and a pair of prices $(r_t, w_t)$ to just one pair of numbers $(\tilde{r}_t, \tilde{w}_t)$ by utilizing the firm's first-order conditions and equilibrium outcomes in factor markets. In a similar spirit, we will now eliminate all prices and taxes so that the government can be thought of as directly choosing a feasible allocation, subject to constraints that ensure the existence of prices and taxes such that the chosen allocation is consistent with the optimization behavior of households and firms. This primal approach to the Ramsey problem, as opposed to the dual approach in which tax rates are viewed as governmental decision variables, is used in Lucas and Stokey's (1983) analysis of an economy without capital. Here we will follow the setup of Jones, Manuelli, and Rossi (1997).

As a comparison to the formulation in equation (12.19), we will now only consider the case when the government is free to trade in the bond market. The constraints with Lagrange multipliers $\Psi_t$ can therefore be replaced with a single present-value budget constraint for either the government or the representative household. (One of them is redundant, since we are also imposing the aggregate resource constraint.) It turns out that the problem simplifies nicely if we choose the present-value budget constraint of the household (12.14), in which future capital stocks have been eliminated with the use of arbitrage conditions. For convenience, we repeat the household's present-value budget constraint here:

$$\sum_{t=0}^{\infty} q_t^0 c_t = \sum_{t=0}^{\infty} q_t^0 (1-\tau_t^n) w_t n_t + \left[(1-\tau_0^k) r_0 + 1 - \delta\right] k_0 + b_0, \tag{12.27}$$

In equation (12.27), $q_t^0$ is the Arrow-Debreu price given by

$$q_t^0 = \prod_{i=1}^{t} R_i^{-1} \tag{12.28}$$

with the numeraire $q_0^0 = 1$. The last two constraints in expression (12.19) will be used to replace prices $q_t^0$ and $(1-\tau_t^n)w_t$ in equation (12.27) with the household's marginal rates of substitution.

A stepwise summary of the primal approach is as follows:

1. Obtain the first-order conditions of the household's and the firm's problems, as well as any arbitrage pricing conditions. Solve these conditions for $\{q_t^0, r_t, w_t, \tau_t^k, \tau_t^n\}_{t=0}^{\infty}$ as functions of the allocation $\{c_t, n_t, k_{t+1}\}_{t=0}^{\infty}$.
2. Substitute these expressions for taxes and prices in terms of the allocation into the household's present-value budget constraint. This is an intertemporal constraint involving only the allocation.
3. Solve for the Ramsey allocation by maximizing expression (12.1) subject to equation (12.3) and the "adjusted budget constraint" derived in step 2.
4. After the Ramsey allocation is solved, use the formulas from step 1 to find taxes and prices.

*Constructing the Ramsey plan*

We now carry out the steps outlined in the preceding list.

*Step 1.* Let $\lambda$ be a Lagrange multiplier on the household's budget constraint (12.27). The first-order conditions for the household's problem are

$$\begin{aligned} c_t: &\quad \beta^t u_c(t) - \lambda q_t^0 = 0, \\ n_t: &\quad -\beta^t u_\ell(t) + \lambda q_t^0 (1-\tau_t^n) w_t = 0. \end{aligned}$$

With the numeraire $q_0^0 = 1$, these conditions imply

$$q_t^0 = \beta^t \frac{u_c(t)}{u_c(0)}, \tag{12.29a}$$

$$(1 - \tau_t^n) w_t = \frac{u_\ell(t)}{u_c(t)}. \tag{12.29b}$$

As before, we can derive the arbitrage condition (12.12), which now reads

$$\frac{q_t^0}{q_{t+1}^0} = (1 - \tau_{t+1}^k) r_{t+1} + 1 - \delta. \tag{12.30}$$

Profit maximization and factor market equilibrium imply equations (12.18).

*Step 2.* Substitute equations (12.29) and $r_0 = F_k(0)$ into equation (12.27), so that we can write the household's budget constraint as

$$\sum_{t=0}^{\infty} \beta^t [u_c(t) c_t - u_\ell(t) n_t] - A = 0, \tag{12.31}$$

where $A$ is given by

$$A = A(c_0, n_0, \tau_0^k) = u_c(0) \left\{ [(1 - \tau_0^k) F_k(0) + 1 - \delta] k_0 + b_0 \right\}. \tag{12.32}$$

*Step 3.* The Ramsey problem is to maximize expression (12.1) subject to equation (12.31) and the feasibility constraint (12.3). As before, we proceed by assuming that government expenditures are small enough that the problem has a convex constraint set and that we can approach it using Lagrangian methods. In particular, let $\Phi$ be a Lagrange multiplier on equation (12.31) and define

$$V(c_t, n_t, \Phi) = u(c_t, 1 - n_t) + \Phi \left[ u_c(t) c_t - u_\ell(t) n_t \right]. \tag{12.33}$$

Then form the Lagrangian

$$J = \sum_{t=0}^{\infty} \beta^t \{ V(c_t, n_t, \Phi) + \theta_t [F(k_t, n_t) + (1 - \delta) k_t - c_t - g_t - k_{t+1}] \} - \Phi A, \tag{12.34}$$

where $\{\theta_t\}_{t=0}^{\infty}$ is a sequence of Lagrange multipliers. For given $k_0$ and $b_0$, we fix $\tau_0^k$ and maximize $J$ with respect to $\{c_t, n_t, k_{t+1}\}_{t=0}^{\infty}$. First-order conditions for this problem are[2]

$$\begin{aligned}
c_t\colon\ & V_c(t) = \theta_t, \quad t \geq 1 \\
n_t\colon\ & V_n(t) = -\theta_t F_n(t), \quad t \geq 1 \\
k_{t+1}\colon\ & \theta_t = \beta\theta_{t+1}[F_k(t+1) + 1 - \delta], \quad t \geq 0 \\
c_0\colon\ & V_c(0) = \theta_0 + \Phi A_c, \\
n_0\colon\ & V_n(0) = -\theta_0 F_n(0) + \Phi A_n.
\end{aligned}$$

These conditions become

$$V_c(t) = \beta V_c(t+1)[F_k(t+1) + 1 - \delta], \quad t \geq 1 \tag{12.35a}$$

$$V_n(t) = -V_c(t) F_n(t), \quad t \geq 1 \tag{12.35b}$$

$$V_n(0) = [\Phi A_c - V_c(0)]\, F_n(0) + \Phi A_n. \tag{12.35c}$$

To these we add equations (12.3) and (12.31), which we repeat here for convenience:

$$c_t + g_t + k_{t+1} = F(k_t, n_t) + (1-\delta)k_t, \tag{12.36a}$$

$$\sum_{t=0}^{\infty} \beta^t [u_c(t) c_t - u_\ell(t) n_t] - A = 0. \tag{12.36b}$$

We seek an allocation $\{c_t, n_t, k_{t+1}\}_{t=0}^{\infty}$, and a multiplier $\Phi$ that satisfies the system of difference equations formed by equations (12.35)–(12.36).[3]

*Step 4:* After an allocation has been found, obtain $q_t^0$ from equation (12.29a), $r_t$ from equation (12.18a), $w_t$ from equation (12.18b), $\tau_t^n$ from equation (12.29b), and finally $\tau_t^k$ from equation (12.30).

---

[2] When comparing the first-order condition for $k_{t+1}$ to the earlier one in equation (12.20), obtained under Chamley's alternative formulation of the Ramsey problem, note that the Lagrange multiplier $\theta_t$ is different across formulations. Specifically, the present specification of the objective function $V$ subsumes parts of the household's present-value budget constraint. To bring out this difference, a more informative notation would be to write $V_j(t, \Phi)$ for $j = c, n$ rather than just $V_j(t)$.

[3] This nonlinear system of equations can be solved iteratively. First, fix $\Phi$, and solve equations (12.35) and (12.36a) for an allocation. Then check the "adjusted budget constraint" (12.36b), and increase or decrease $\Phi$ depending on whether the budget constraint is binding or slack.

*Revisiting a zero capital tax*

Consider the special case in which there is a $T \geq 0$ for which $g_t = g$ for all $t \geq T$. Assume that there exists a solution to the Ramsey problem, and that it converges to a time-invariant allocation, so that $c, n$, and $k$ are constant after some time. Then because $V_c(t)$ converges to a constant, the stationary version of equation (12.35a) implies equations

$$1 = \beta(F_k + 1 - \delta). \tag{12.37}$$

Now because $c_t$ is constant in the limit, equation (12.29a) implies that $(q_t^0 / q_{t+1}^0) \to \beta^{-1}$ as $t \to \infty$. Then the arbitrage condition for capital (12.30) becomes

$$1 = \beta[(1 - \tau^k)F_k + 1 - \delta], \tag{12.38}$$

The two equalities, (12.37) and (12.38), imply that $\tau_k = 0$.

## Taxation of initial capital

Thus far, we have set $\tau_0^k$ at zero (or some other small fixed number). Now suppose that the government is free to choose $\tau_0^k$. The derivative of $J$ in equation (12.34) with respect to $\tau_0^k$ is

$$\frac{\partial J}{\partial \tau_0^k} = \Phi u_c(0) F_k(0) k_0, \tag{12.39}$$

which is strictly positive for all $\tau_0^k$ as long as $\Phi > 0$. The nonnegative Lagrange multiplier $\Phi$ measures the utility costs of raising government revenues through distorting taxes. Without distortionary taxation, a competitive equilibrium would attain the first-best outcome for the representative household, and $\Phi$ would be equal to zero, so that the household's (or equivalently, by Walras' Law, the government's) present-value budget constraint would not exert any additional constraining effect on welfare maximization beyond what is present in the economy's technology. In contrast, when the government has to use some of the tax rates $\{\tau_t^n, \tau_{t+1}^k\}_{t=0}^{\infty}$, the multiplier $\Phi$ is strictly positive and reflects the welfare cost of the distorted margins, implicit in the present-value budget constraint (12.36b), which govern the household's optimization behavior.

By raising $\tau_0^k$ and thereby increasing the revenues from lump-sum taxation of $k_0$, the government reduces its future need to rely on distortionary taxation, and hence the value of $\Phi$ falls. In fact, the ultimate implication of condition (12.39) is that the government should set $\tau_0^k$ high enough to drive $\Phi$ down to zero. In other words, the government should raise *all* revenues through a time-0 capital levy, then lend the proceeds to the private sector and finance government expenditures by using the interest from the loan; this would enable the government to set $\tau_t^n = 0$ for all $t \geq 0$ and $\tau_t^k = 0$ for all $t \geq 1$.[4]

## *Nonzero capital tax due to incomplete taxation*

The result that the limiting capital tax should be set equal to zero hinges on a complete set of flat-rate taxes. The effects of incomplete taxation is illustrated with an example by Correia (1996), who introduces an additional production factor, $z_t$, in fixed supply, $z_t = Z$, which cannot be taxed, $\tau_t^z = 0$.

The new production function $F(k_t, n_t, z_t)$ is assumed to exhibit constant returns to scale in all of its inputs. Profit maximization implies that the rental price of the new factor is equal to its marginal product,

$$p_t^z = F_z(t).$$

The only change to the household's present-value budget constraint (12.27) is that a stream of revenues is added to the right-hand side,

$$\sum_{t=0}^{\infty} q_t^0 p_t^z Z.$$

Following our scheme of constructing the Ramsey plan, step 2 yields the following "adjusted budget constraint" of the household,

$$\sum_{t=0}^{\infty} \beta^t \left\{ u_c(t)[c_t - F_z(t)Z] - u_\ell(t)n_t \right\} - A = 0, \tag{12.40}$$

[4] The scheme may involve $\tau_0^k > 1$ for high values of $\{g_t\}_{t=0}^{\infty}$ and $b_0$. However, such a scheme cannot be implemented if the household is able to avoid the tax liability by not renting out its capital stock at time $0$. The government would then be constrained to choose $\tau_0^k \leq 1$.

In the rest of the chapter, we do not impose that $\tau_t^k \leq 1$. If we were to do so, the extra constraint in the Ramsey problem would be

$$u_c(t) \geq \beta(1-\delta)u_c(t+1),$$

which can be seen after substituting equation (12.29a) into equation (12.30).

where $A$ remains defined by equation (12.32). In step 3 we formulate

$$V(c_t, n_t, k_t, \Phi) = u(c_t, 1 - n_t) + \Phi \left\{ u_c(t)[c_t - F_z(t)Z] - u_\ell(t)n_t \right\}. \tag{12.41}$$

In contrast to equation (12.33), $k_t$ enters now as an argument in $V$ because of the presence of the marginal product of the factor $Z$ (but we have chosen to suppress the quantity $Z$ itself, since it is in fixed supply).

Except for these changes of the functions $F$ and $V$, the Lagrangian of the Ramsey problem is the same as equation (12.34). The first-order condition with respect to $k_{t+1}$ is

$$\theta_t = \beta V_k(t+1) + \beta\theta_{t+1}[F_k(t+1) + 1 - \delta]. \tag{12.42}$$

Assuming the existence of a steady state, the stationary version of equation (12.42) reads

$$1 = \beta(F_k + 1 - \delta) + \beta\frac{V_k}{\theta}. \tag{12.43}$$

Condition (12.43) and the arbitrage condition for capital (12.38) imply an optimal value for $\tau^k$,

$$\tau^k = \frac{V_k}{\theta F_k} = \frac{\Phi u_c Z}{\theta F_k} F_{zk}.$$

As discussed earlier, $\Phi > 0$ in a second-best solution with distortionary taxation, so the limiting tax rate on capital is zero only if $F_{zk} = 0$. Moreover, the sign of $\tau^k$ depends on the direction of the effect of capital on the marginal product of the untaxed factor $Z$. If $k$ and $Z$ are complements, the limiting capital tax is positive, and it is negative in the case where the two factors are substitutes.

Other examples of a nonzero limiting capital tax are presented by Stiglitz (1987) and Jones, Manuelli, and Rossi (1997), who assume that two types of labor must be taxed at the same tax rate. Once again, the incompleteness of the tax system makes the optimal capital tax dependent on how capital affects the marginal products of the other factors.

## A stochastic economy

We now turn to optimal taxation in a stochastic version of our economy. With the notation of chapter 7, we follow the setups of Zhu (1992) and Chari, Christiano, and Kehoe (1994). The stochastic event $s_t$ at time $t$ constitutes an exogenous shock both to the production function $F(\cdot, \cdot, s_t)$ and to government purchases $g(s_t)$. We use the history of events $s^t$ as a commodity space; $c(s^t)$, $\ell(s^t)$, and $n(s^t)$ are the household's consumption, leisure, and labor at time $t$ given history $s^t$; and $k(s^t)$ denotes the capital stock carried over to next period $t+1$. Following our earlier convention, $u_c(s^t)$ and $F_k(s^t)$ and so on denote the values of the indicated objects at time $t$ for history $s^t$, evaluated at an allocation to be understood from the context.

The household's preferences are given by

$$\sum_{t=0}^{\infty} \sum_{s^t} \beta^t \pi(s^t) u[c(s^t), \ell(s^t)]. \tag{12.44}$$

The production function is still constant returns to scale in labor and capital. Feasibility requires that

$$c(s^t) + g(s_t) + k(s^t) = F[k(s^{t-1}), n(s^t), s_t] + (1-\delta)k(s^{t-1}). \tag{12.45}$$

*Government*

Given history $s^t$ at time $t$, the government finances its exogenous purchase $g(s_t)$ and any debt obligation by levying flat-rate taxes on earnings from capital at rate $\tau^k(s^t)$ and from labor at rate $\tau^n(s^t)$, and by issuing state-contingent debt. Let $b(s_{t+1}|s^t)$ be government indebtedness to the private sector at the beginning of period $t+1$ if event $s_{t+1}$ is realized. This state-contingent asset is traded in period $t$ at the price $p(s_{t+1}|s^t)$, in terms of time-$t$ goods. The government's budget constraint becomes

$$\begin{aligned} g(s_t) =& \tau^k(s^t) r(s^t) k(s^{t-1}) + \tau^n(s^t) w(s^t) n(s^t) \\ &+ \sum_{s_{t+1}} p(s_{t+1}|s^t) b(s_{t+1}|s^t) - b(s_t|s^{t-1}), \end{aligned} \tag{12.46}$$

where $r(s^t)$ and $w(s^t)$ are the market-determined rental rate of capital and the wage rate for labor, respectively.

*Households*

The representative household maximizes expression (12.44) subject to the following sequence of budget constraints:

$$c(s^t) + k(s^t) + \sum_{s_{t+1}} p(s_{t+1}|s^t) b(s_{t+1}|s^t)$$

$$= \left[1-\tau^k(s^t)\right] r(s^t)k(s^{t-1}) + \left[1-\tau^n(s^t)\right] w(s^t)n(s^t) \\ + (1-\delta)k(s^{t-1}) + b(s_t|s^{t-1}) \quad \forall t. \tag{12.47}$$

The first-order conditions for this problem imply

$$\frac{u_\ell(s^t)}{u_c(s^t)} = [1-\tau^n(s^t)]w(s^t), \tag{12.48a}$$

$$p(s_{t+1}|s^t) = \beta \frac{\pi(s^{t+1})}{\pi(s^t)} \frac{u_c(s^{t+1})}{u_c(s^t)}, \tag{12.48b}$$

$$u_c(s^t) = \beta E_t \left\{ u_c(s^{t+1}) \left[(1-\tau^k(s^{t+1}))r(s^{t+1}) + 1 - \delta\right]\right\}, \tag{12.48c}$$

where $E_t$ is the mathematical expectation conditional upon information available at time $t$,

$$E_t x(s^{t+1}) \equiv \sum_{s_{t+1}} \pi(s_{t+1}|s^t) x(s^{t+1}) = \sum_{s_{t+1}} \frac{\pi(s^{t+1})}{\pi(s^t)} x(s^{t+1}).$$

Similar to arbitrage condition (12.12) in the nonstochastic economy, conditions (12.48b) and (12.48c) imply

$$1 = \sum_{s_{t+1}} p(s_{t+1}|s^t) \left\{[1-\tau^k(s^{t+1})]r(s^{t+1}) + 1 - \delta\right\}. \tag{12.49}$$

And once again, this arbitrage condition can also be obtained by consolidating the budget constraints of two consecutive periods. Multiply the time–$t+1$ version of equation (12.47) by $p(s_{t+1}|s^t)$ and sum over all realizations $s_{t+1}$. The resulting expression can be substituted into equation (12.47) by eliminating $\sum_{s_{t+1}} p(s_{t+1}|s^t) b(s_{t+1}|s^t)$. Then, to rule out arbitrage transactions in capital and state-contingent assets, the term multiplying $k(s^t)$ must be zero; this approach amounts to imposing condition (12.49). Similar arbitrage arguments were made in chapters 7 and 10.

As before, by repeated substitution of one-period budget constraints, we can obtain the household's present-value budget constraint:

$$\sum_{t=0}^{\infty} \sum_{s^t} q_t^0(s^t)c(s^t) = \sum_{t=0}^{\infty} \sum_{s^t} q_t^0(s^t)[1-\tau^n(s^t)]w(s^t)n(s^t) \\ + \left[(1-\tau_0^k)r_0 + 1 - \delta\right] k_0 + b_0, \tag{12.50}$$

where we denote time-0 variables by the time subscript 0. The price system $q_t^0(s^t)$ is given by the following formula, versions of which were displayed in chapter 7:

$$q_{t+1}^0(s^{t+1}) = p(s_{t+1}|s^t)q_t^0(s^t) = \beta^{t+1}\pi(s^{t+1})\frac{u_c(s^{t+1})}{u_c(s^0)}. \tag{12.51}$$

Alternatively, equilibrium price (12.51) can be computed from the first-order conditions when maximizing expression (12.44) subject to equation (12.50) (and choosing the numeraire $q_0^0 = 1$). Furthermore, the arbitrage condition (12.49) can be expressed as

$$q_t^0(s^t) = \sum_{s_{t+1}} q_{t+1}^0(s^{t+1})\left\{[1-\tau^k(s^{t+1})]r(s^{t+1}) + 1 - \delta\right\}. \tag{12.52}$$

In deriving the present-value budget constraint (12.50), we imposed two transversality conditions that specify that for any infinite history $s^\infty$

$$\lim q_t^0(s^t)k(s^t) = 0, \tag{12.53a}$$

$$\lim \sum_{s_{t+1}} q_{t+1}^0(s^{t+1})b(s_{t+1}|s^t) = 0, \tag{12.53b}$$

where the limits are taken over sequences of histories $s^t$ contained in the infinite history $s^\infty$.

*Firms*

The static maximization problem of the representative firm remains the same as before. Thus, in a competitive equilibrium, production factors are paid their marginal products,

$$r(s^t) = F_k(s^t), \tag{12.54a}$$
$$w(s^t) = F_n(s^t). \tag{12.54b}$$

## *Indeterminacy of state-contingent debt and capital taxes*

Consider a feasible government policy $\{g(s_t),\ \tau^k(s^t),\ \tau^n(s^t),\ b(s_{t+1}|s^t);\ \forall s^t,\ s_{t+1}\}_{t\geq 0}$ with an associated competitive allocation $\{c(s^t), n(s^t), k(s^t); \forall s^t\}_{t\geq 0}$. We will now show that there are infinitely many plans for state-contingent debt and capital taxes that can implement this particular competitive allocation. But first, by way of contrast, note that the labor tax is uniquely determined by equations (12.48a) and (12.54b).

The intuition for the indeterminacy of state-contingent debt and capital taxes can be gained from the household's first-order condition (12.48c), which states that capital tax rates affect the household's intertemporal allocation by changing the present market value of after-tax returns on capital. If a different set of capital taxes induces the same present market value of after-tax returns on capital, then they will also be consistent with the same competitive allocation. Of course, it remains to be shown whether the change of capital tax receipts in different states can be offset by restructuring the government's issue of state-contingent debt. Zhu (1992) suggests how such feasible alternative policies can be constructed.

Let $\{\epsilon(s^t); \forall s^t\}_{t\geq 0}$ be a random process such that

$$E_t u_c(s^{t+1})\epsilon(s^{t+1})r(s^{t+1}) = 0. \tag{12.55}$$

We can then construct an alternative policy for capital taxes and state-contingent debt, $\{\hat\tau^k(s^t), \hat b(s_{t+1}|s^t); \forall s^t, s_{t+1}\}_{t\geq 0}$, as follows:

$$\hat\tau_0^k = \tau_0^k, \tag{12.56a}$$

$$\hat\tau^k(s^{t+1}) = \tau^k(s^{t+1}) + \epsilon(s^{t+1}), \qquad t\geq 0 \tag{12.56b}$$

$$\hat b(s_{t+1}|s^t) = b(s_{t+1}|s^t) - \epsilon(s^{t+1})r(s^{t+1})k(s^t). \qquad t\geq 0 \tag{12.56c}$$

Compared to the original fiscal policy, we can verify that this alternative policy does not change the following:

1. the household's intertemporal consumption choice, governed by first-order condition (12.48c).
2. the present market value of all government debt issued at time $t$, when discounted with the equilibrium expression for $p(s_{t+1}|s^t)$ in equation (12.48b).
3. the government's revenue from capital taxation net of maturing government debt in any state $s^{t+1}$.

Thus, the alternative policy is feasible and leaves the competitive allocation unchanged.

Since there are infinitely many ways of constructing sequences of random variables $\{\epsilon(s^t)\}$ that satisfy equation (12.55), it follows that the competitive allocation can be implemented by many different plans for capital taxes and state-contingent debt. It is instructive to consider two special cases where there is no uncertainty one period ahead about one of the two policy instruments. We first take the case of risk-free one-period bonds. In period $t$, the government issues bonds that promise to pay off $\bar{b}_{t+1}(s^t)$ at time $t+1$ with certainty. Let the amount of bonds be such that their present market value is the same as that for the original fiscal plan,

$$\sum_{s_{t+1}} p(s_{t+1}|s^t)\bar{b}_{t+1}(s^t) = \sum_{s_{t+1}} p(s_{t+1}|s^t) b(s_{t+1}|s^t).$$

After invoking the equilibrium expression for prices (12.48b), we can solve for the constant $\bar{b}_{t+1}(s^t)$,

$$\bar{b}_{t+1}(s^t) = \frac{E_t u_c(s^{t+1}) b(s_{t+1}|s^t)}{E_t u_c(s^{t+1})}. \tag{12.57}$$

The change in capital taxes needed to offset this shift to risk-free bonds is then implied by equation (12.56c),

$$\epsilon(s^{t+1}) = \frac{b(s_{t+1}|s^t) - \bar{b}_{t+1}(s^t)}{r(s^{t+1})k(s^t)}. \tag{12.58}$$

We can check that equations (12.57) and (12.58) describe a permissible policy by substituting these expressions into equation (12.55) and verifying that the restriction is indeed satisfied.

Next, we examine a policy where the capital tax is not contingent on the realization of the current state but is already set in the previous period. Let $\bar{\tau}_{t+1}(s^t)$ be the capital tax rate in period $t+1$, conditional on information at time $t$. We choose $\bar{\tau}_{t+1}(s^t)$ such that the household's first-order condition (12.48c) is unaffected,

$$\begin{aligned} &E_t \left\{ u_c(s^{t+1}) \left[ (1 - \bar{\tau}^k_{t+1}(s^t)) r(s^{t+1}) + 1 - \delta \right] \right\} \\ &= E_t \left\{ u_c(s^{t+1}) \left[ (1 - \tau^k(s^{t+1})) r(s^{t+1}) + 1 - \delta \right] \right\}, \end{aligned}$$

which gives

$$\bar{\tau}^k_{t+1}(s^t) = \frac{E_t u_c(s^{t+1}) \tau^k(s^{t+1}) r(s^{t+1})}{E_t u_c(s^{t+1}) r(s^{t+1})}. \tag{12.59}$$

Thus, the alternative policy in equations (12.56) with capital taxes known one period in advance is accomplished by setting

$$\epsilon_{t+1}(s^{t+1}) = \bar{\tau}^k_{t+1}(s^t) - \tau^k(s^{t+1}).$$

## The Ramsey plan under uncertainty

We now ask what competitive allocation should be chosen by a benevolent government; that is, we solve the Ramsey problem for the stochastic economy. The computational strategy is in principle the same as the preceding one for a nonstochastic economy.

Step 1, in which we use private first-order conditions to solve for prices and taxes in terms of the allocation, has already been accomplished with equations (12.48a), (12.51), (12.52) and (12.54). In step 2, we use these expressions to eliminate prices and taxes from the household's present-value budget constraint (12.50), which leaves us with

$$\sum_{t=0}^{\infty}\sum_{s^t}\beta^t\pi(s^t)[u_c(s^t)c(s^t) - u_\ell(s^t)n(s^t)] - A = 0, \tag{12.60}$$

where $A$ is still given by equation (12.32). Proceeding to step 3, we define

$$\begin{aligned} V\left[c(s^t), n(s^t), \Phi\right] &= u[c(s^t), 1 - n(s^t)] \\ &\quad + \Phi\left[u_c(s^t)c(s^t) - u_\ell(s^t)n(s^t)\right], \end{aligned} \tag{12.61}$$

where $\Phi$ is a Lagrange multiplier on equation (12.60). Then form the Lagrangian

$$\begin{aligned} J = \sum_{t=0}^{\infty}\sum_{s^t}\beta^t\pi(s^t)\Big\{V[c(s^t), n(s^t), \Phi] + \theta(s^t)\left[F\left(k(s^{t-1}), n(s^t), s_t\right)\right. \\ \left. +(1-\delta)k(s^{t-1}) - c(s^t) - g(s_t) - k(s^t)\right]\Big\} - \Phi A, \end{aligned} \tag{12.62}$$

where $\{\theta(s^t); \forall s^t\}_{t\geq 0}$ is a sequence of Lagrange multipliers. For given $k_0$ and $b_0$, we fix $\tau^k_0$ and maximize $J$ with respect to $\{c(s^t), n(s^t), k(s^t); \forall s^t\}_{t\geq 0}$.

The first-order conditions of the Ramsey problem are

$$\begin{aligned} c(s^t)&: \; V_c(s^t) = \theta(s^t), && t \geq 1; \\ n(s^t)&: \; V_n(s^t) = -\theta(s^t)F_n(s^t), && t \geq 1; \\ k(s^t)&: \; \theta(s^t) = \beta\sum_{s_{t+1}}\pi(s_{t+1}|s^t)\theta(s^{t+1})[F_k(s^{t+1}) + 1 - \delta], && t \geq 0; \end{aligned}$$

where we have left out the conditions for $c_0$ and $n_0$, which are different because they include terms related to the initial stocks of capital and bonds. The first-order conditions of the problem imply, for $t \geq 1$,

$$V_c(s^t) = \beta E_t V_c(s^{t+1})[F_k(s^{t+1}) + 1 - \delta], \tag{12.63a}$$

$$V_n(s^t) = -V_c(s^t) F_n(s^t). \tag{12.63b}$$

These expressions reveal an interesting property of the Ramsey allocation. If the stochastic process on $s$ follows a Markov process, equations (12.63) suggest that the allocations from period 1 onward can be described by time-invariant allocation rules $c(s,k)$, $n(s,k)$, and $k'(s,k)$.[5]

## Ex ante capital tax varies around zero

In a nonstochastic economy, we proved that the optimal limiting capital tax is zero if the equilibrium converges to a steady state. The analogue to a steady state in a stochastic economy is a stationary equilibrium. Therefore, we now assume that the process on $s$ follows a Markov process with transition probabilities $\tilde{\pi}(s'|s) \equiv \text{Prob}(s_{t+1} = s'|s_t = s)$. As noted in the previous section, this assumption implies that the allocation rules are time-invariant functions of $(k, s)$. If the economy converges to a stationary equilibrium, the stochastic process $\{k_t, s_t\}$ is a stationary, ergodic Markov process on the compact set $[0, \bar{k}] \times \mathbf{S}$ where $\mathbf{S}$ is a finite set of possible realizations for $s_t$, and $\bar{k}$ is an upper bound on the capital stock.[6]

Because of the indeterminacy of state-contingent government debt and capital taxes, it is not possible to pose any questions about the stationary distribution of realized capital tax rates but we can study the *ex ante capital tax rate* defined as

$$\bar{\tau}^k_{t+1}(s^t) = \frac{\sum_{s_{t+1}} p(s_{t+1}|s^t)\tau^k(s^{t+1})r(s^{t+1})}{\sum_{s_{t+1}} p(s_{t+1}|s^t)r(s^{t+1})}. \tag{12.64}$$

[5] As a reminder, we might want to include the constant $\Phi$ as an explicit argument in $c(s,k)$, $n(s,k)$, and $k'(s,k)$, to emphasize that the second-best allocation depends critically on the extent to which the government has to resort to distortionary taxation, as captured by the Lagrange multiplier $\Phi$ on the household's present-value budget constraint.

[6] An upper bound on the capital stock can be constructed as follows,

$$\bar{k} = \max\{\bar{k}(s) : F[\bar{k}(s), 1, s] = \delta\bar{k}(s); s \in \mathbf{S}\}.$$

That is, the ex ante capital tax rate is the ratio of present market value of taxes on capital income over the present market value of capital income. After invoking the equilibrium price of equation (12.48b), we see that this expression is identical to equation (12.59). Recall that equation (12.59) provided one resolution to the indeterminacy of the Ramsey plan by pinning down a unique fixed capital tax rate for period $t+1$ conditional on information at time $t$. It follows that the alternative interpretation of $\bar{\tau}^k_{t+1}(s^t)$ in equation (12.64) as the ex ante capital tax rate offers a unique measure for the multiplicity of capital tax schedules under the Ramsey plan. Moreover, it is quite intuitive that one way for the government to tax away, in present-value terms, a fraction $\bar{\tau}^k_{t+1}(s^t)$ of next period's capital income is to set a constant tax rate exactly equal to that number.

Let $P^\infty(\cdot)$ be the probability measure over the outcomes in such a stationary equilibrium. We now state the proposition of Zhu (1992) that the ex ante capital tax rate in a stationary equilibrium is either equal to zero or varies around zero.

PROPOSITION 2: If there exists a stationary Ramsey allocation, the ex ante capital tax rate is such that either

(a) $P^\infty(\bar{\tau}^k_t = 0) = 1$ or $P^\infty(\bar{\tau}^k_t > 0) > 0$ and $P^\infty(\bar{\tau}^k_t < 0) > 0$; or

(b) $P^\infty(\bar{\tau}^k_t = 0) = 1$ if and only if $P^\infty[V_c(c_t, n_t, \Phi)/u_c(c_t, \ell_t) = \Lambda] = 1$ for some constant $\Lambda$.

A sketch of the proof is provided in the next subsection. Let us here just add that the two possibilities with respect to the ex ante capital tax rate are not vacuous. One class of utilities that imply $P^\infty(\bar{\tau}^k_t = 0) = 1$, is given by

$$u(c, \ell) = \frac{c^{1-\sigma}}{1-\sigma} + v(\ell),$$

for which the ratio $V_c(c_t, n_t, \Phi)/u_c(c_t, \ell_t)$ is equal to $[1 + \Phi(1-\sigma)]$; that is, some constant $\Lambda$ as required by Proposition 2. Chari, Christiano, and Kehoe (1994) solve numerically for Ramsey plans when the preferences do not satisfy this condition. In their simulations, the ex ante tax on capital income remains approximately equal to zero.

To revisit Chamley (1986) and Judd's (1985b) result on the optimality of a zero capital tax in a nonstochastic economy, it is trivially true that the ratio $V_c(c_t, n_t, \Phi)/u_c(c_t, \ell_t)$ is constant in a steady state. In a stationary equilibrium of a stochastic economy, Proposition 2 qualifies this result as follows: For some utility functions, the Ramsey plan prescribes a zero ex ante capital tax rate, which can be implemented by setting a zero tax on capital income. But, except for such special classes of preferences, Proposition 2 states that the ex ante capital tax rate should vary around zero, in the sense that $P^\infty(\bar\tau_t^k > 0) > 0$ and $P^\infty(\bar\tau_t^k < 0) > 0$.

*Sketch of the proof of Proposition 2*

Note from equation (12.64) that $\bar\tau_{t+1}^k(s^t) \geq (\leq)\ 0$ if and only if

$$\sum_{s_{t+1}} p(s_{t+1}|s^t)\tau^k(s^{t+1})r(s^{t+1}) \geq (\leq)\ 0,$$

which, together with equation (12.49), implies

$$1 \leq (\geq) \sum_{s_{t+1}} p(s_{t+1}|s^t)\left[r(s^{t+1}) + 1 - \delta\right].$$

Substituting equations (12.48b) and (12.54a) into this expression yields

$$u_c(s^t) \leq (\geq) \beta E_t u_c(s^{t+1})\left[F_k(s^{t+1}) + 1 - \delta\right], \tag{12.65}$$

if and only if $\bar\tau_{t+1}^k(s^t) \geq (\leq)\ 0$.

Define

$$H(s^t) \equiv \frac{V_c(s^t)}{u_c(s^t)}. \tag{12.66}$$

Using equation (12.63a), we have

$$u_c(s^t)H(s^t) = \beta E_t u_c(s^{t+1})H(s^{t+1})[F_k(s^{t+1}) + 1 - \delta]. \tag{12.67}$$

By formulas (12.65) and (12.67), $\bar\tau_{t+1}^k(s^t) \geq (\leq)\ 0$ if and only if

$$H(s^t) \geq (\leq) = \frac{E_t\omega(s^{t+1})H(s^{t+1})}{E_t\omega(s^{t+1})}, \tag{12.68}$$

where $\omega(s^{t+1}) \equiv u_c(s^{t+1})[F_k(s^{t+1}) + 1 - \delta]$.

Since a stationary Ramsey equilibrium has time-invariant allocation rules $c(s,k)$, $n(s,k)$, and $k'(s,k)$, it follows that $\bar\tau_{t+1}^k(s^t)$, $H(s^t)$, and $\omega(s^t)$ can also be expressed as functions of $(s,k)$. The stationary version of expression (12.68) with transition probabilities $\tilde\pi(s'|s)$ becomes

$$\begin{aligned} &\bar\tau^k(s,k) \geq (\leq)\ 0 \quad \text{if and only if} \\ &H(s,k) > (<) = \frac{\sum_{s'} \tilde\pi(s'|s)\omega[s', k'(k,s)]H[s', k'(k,s)]}{\sum_{s'} \tilde\pi(s'|s)\omega[s', k'(k,s)]} \equiv \Gamma H(s,k). \end{aligned} \tag{12.69}$$

Note that the operator $\Gamma$ is a weighted average of $H[s', k'(k,s)]$ and that it has the property that $\Gamma H^* = H^*$ for any constant $H^*$.

Under some regularity conditions, $H(s,k)$ attains a minimum $H^-$ and a maximum $H^+$ in the stationary equilibrium. That is, there exist equilibrium states $(s^-, k^-)$ and $(s^+, k^+)$ such that

$$P^\infty \left[H(s,k) \geq H^-\right] = 1, \tag{12.70a}$$

$$P^\infty \left[H(s,k) \leq H^+\right] = 1, \tag{12.70b}$$

where $H^- = H(s^-, k^-)$ and $H^+ = H(s^+, k^+)$. We will now show that if

$$P^\infty \left[H(s,k) \geq \Gamma H(s,k)\right] = 1, \tag{12.71a}$$

or,

$$P^\infty \left[H(s,k) \leq \Gamma H(s,k)\right] = 1, \tag{12.71b}$$

then, there must exist a constant $H^*$ such that

$$P^\infty \left[H(s,k) = H^*\right] = 1. \tag{12.71c}$$

First, take equation (12.71a) and consider the state $(s,k) = (s^-, k^-)$ that is associated with a set of possible states in the next period, $\{s', k'(s,k); \forall s' \in \mathbf{S}\}$. By equation (12.70a), $H(s',k') \geq H^-$, and since $H(s,k) = H^-$, condition (12.71a) implies that $H(s',k') = H^-$. Now, we can repeat the same argument for each one of $(s',k')$, and thereafter for the equilibrium states that they in turn map into, and so on. Thus, using the ergodicity of $\{s_t, k_t\}$, we obtain equation (12.71c) with $H^* = H^-$. A similar reasoning can be applied to equation (12.71b), but we now use $(s,k) = (s^+, k^+)$ and equation (12.70b) to show that equation (12.71c) is implied.

By the correspondence in expression (12.69) we have established part (a) of Proposition 2. Part (b) follows also after recalling definition (12.66); that is, the constant $H^*$ in equation (12.71c) is the sought-after $\Lambda$.

## Examples of labor tax smoothing

To gain some insight into optimal tax policies, we consider a few examples of government expenditures to be financed in a simplified model that abstracts from physical capital. The technology is now described by

$$c(s^t) + g(s_t) = n(s^t). \tag{12.72}$$

Since one unit of labor yields one unit of output, the competitive equilibrium wage will trivially be $w(s^t) = 1$. The model is otherwise identical to the previous framework. The household's present-value budget constraint is given by equation (12.50) when deleting the part involving physical capital. Prices and taxes are expressed in terms of the allocation by conditions (12.48a) and (12.51). After using these expressions to eliminate prices and taxes, the "adjusted budget constraint," equation (12.60), becomes

$$\sum_{t=0}^{\infty} \sum_{s^t} \beta^t \pi(s^t)[u_c(s^t)c(s^t) - u_\ell(s^t)n(s^t)] - u_c(s^0)b_0 = 0. \tag{12.73}$$

We then form the Lagrangian in the same way as before. After writing out the derivatives $V_c(s^t)$ and $V_n(s^t)$, the first-order conditions of this Ramsey problem are

$$\begin{aligned}
c(s^t)\!:\quad & (1+\Phi)u_c(s^t) + \Phi\left[u_{cc}(s^t)c(s^t) - u_{\ell c}(s^t)n(s^t)\right] \\
& \qquad - \theta(s^t) = 0, \qquad\qquad t \geq 1; \quad (12.74a)\\
n(s^t)\!:\quad & -(1+\Phi)u_\ell(s^t) - \Phi\left[u_{c\ell}(s^t)c(s^t) - u_{\ell\ell}(s^t)n(s^t)\right] \\
& \qquad + \theta(s^t) = 0, \qquad\qquad t \geq 1; \quad (12.74b)\\
c(s^0)\!:\quad & (1+\Phi)u_c(s^0) + \Phi\left[u_{cc}(s^0)c(s^0) - u_{\ell c}(s^0)n(s^0)\right] \\
& \qquad - \theta(s^0) - \Phi u_{cc}(s^0)b_0 = 0; \quad (12.74c)\\
n(s^0)\!:\quad & -(1+\Phi)u_\ell(s^0) - \Phi\left[u_{c\ell}(s^0)c(s^0) - u_{\ell\ell}(s^0)n(s^0)\right] \\
& \qquad + \theta(s^0) - \Phi u_{c\ell}(s^0)b_0 = 0. \quad (12.74d)
\end{aligned}$$

This very model is analyzed by Lucas and Stokey (1983), who also address time consistency of the optimal fiscal policy.[7]

The following preliminary calculations will be useful in shedding light on optimal tax policies for a few examples of government expenditure streams. First,

---

[7] The optimal tax policy is in general time inconsistent as studied in chapter 16 and indicated by the preceding discussion about taxation of initial capital. However, Lucas and Stokey (1983) show that the optimal policy in the present model without physical capital is time consistent as long as the government can issue debt at all maturities (not just one-period debt as in our formulation). It is then possible for the government to choose a restructured debt so that the initial optimal allocation, $\{c(s^t), n(s^t); \forall s^t\}_{t\geq 0}$, remains the solution to next period's reoptimization of the remainder problem. By induction, the same is true in all later periods. Persson, Persson, and Svensson (1988) extend the argument to the maturity structure of both real and *nominal* bonds in a monetary economy.

substitute equations (12.48a) and (12.72) into equation (12.73) to get

$$\sum_{t=0}^{\infty} \sum_{s^t} \beta^t \pi(s^t) u_c(s^t) \left[\tau^n(s^t) n(s^t) - g(s_t)\right] - u_c(s^0) b_0 = 0. \tag{12.75}$$

Then multiplying equation (12.74a) by $c(s^t)$ and equation (12.74b) by $n(s^t)$ and summing, we find

$$\begin{aligned}
&(1+\Phi)\left[c(s^t)u_c(s^t) - n(s^t)u_\ell(s^t)\right] \\
&+ \Phi\left[c(s^t)^2 u_{cc}(s^t) - 2n(s^t)c(s^t)u_{\ell c}(s^t) + n(s^t)^2 u_{\ell\ell}(s^t)\right] \\
&- \theta(s^t)\left[c(s^t) - n(s^t)\right] = 0, \qquad t \geq 1.
\end{aligned} \tag{12.76a}$$

Similarly, multiplying equation (12.74c) by $[c(s^0) - b_0]$ and equations (12.74d) by $n(s^0)$ and summing, we obtain

$$\begin{aligned}
&(1+\Phi)\left\{\left[c(s^0) - b_0\right] u_c(s^0) - n(s^0) u_\ell(s^0)\right\} \\
&+ \Phi\left\{\left[c(s^0) - b_0\right]^2 u_{cc}(s^0) - 2n(s^0)\left[c(s^0) - b_0\right] u_{\ell c}(s^0) + n(s^0)^2 u_{\ell\ell}(s^0)\right\} \\
&- \theta(s^0)\left[c(s^0) - b_0 - n(s^0)\right] = 0.
\end{aligned} \tag{12.76b}$$

Note that since the utility function is strictly concave, the quadratic term in equation (12.76) is negative.[8] Finally, multiplying equation (12.76a) by $\beta^t \pi(s^t)$, summing over $t$ and $s^t$, and adding equation (12.76b), we find that

$$(1+\Phi)\left(\sum_{t=0}^{\infty} \sum_{s^t} \beta^t \pi(s^t) \left[c(s^t)u_c(s^t) - n(s^t)u_\ell(s^t)\right] - u_c(s^0) b_0\right)$$

[8] To see that the quadratic term in equation (12.76a) is negative, complete the square by adding and subtracting the quantity $n^2 u_{\ell c}^2 / u_{cc}$ (where we have suppressed the argument $s^t$),

$$\begin{aligned}
&c^2 u_{cc} - 2nc u_{\ell c} + n^2 u_{\ell\ell} + n^2 \frac{u_{\ell c}^2}{u_{cc}} - n^2 \frac{u_{\ell c}^2}{u_{cc}} \\
&= u_{cc}\left(c^2 - 2nc\frac{u_{\ell c}}{u_{cc}} + n^2 \frac{u_{\ell c}^2}{u_{cc}^2}\right) + \left(u_{\ell\ell} - \frac{u_{\ell c}^2}{u_{cc}}\right) n^2 \\
&= u_{cc}\left(c - \frac{u_{\ell c}}{u_{cc}} n\right)^2 + \frac{u_{cc}u_{\ell\ell} - u_{\ell c}^2}{u_{cc}} n^2.
\end{aligned}$$

Since the conditions for a strictly concave $u$ are $u_{cc} < 0$ and $u_{cc}u_{\ell\ell} - u_{\ell c}^2 > 0$, it follows immediately that the quadratic term in equation (12.76a) is negative. The same argument applies to the quadratic term in equation (12.76b).

$$+\Phi Q-\sum_{t=0}^{\infty}\sum_{s^t}\beta^t\pi(s^t)\theta(s^t)\left[c(s^t)-n(s^t)\right]+\theta(s^0)b_0=0,$$

where $Q$ is the sum of negative (quadratic) terms. Using equations (12.73) and (12.72), we arrive at

$$\Phi Q+\sum_{t=0}^{\infty}\sum_{s^t}\beta^t\pi(s^t)\theta(s^t)g(s_t)+\theta(s^0)b_0=0. \tag{12.77}$$

Expression (12.77) furthers our understanding of the Lagrange multiplier $\Phi$ that assigns a marginal shadow price to imposing the household's present-value budget constraint, in excess of the shadow value associated with the economy's resource constraint as captured by the multipliers $\{\theta(s^t);\forall s^t\}_{t\geq 0}$. Let us first examine under what circumstances the Lagrange multiplier $\Phi$ is equal to zero. Setting $\Phi=0$ in equations (12.74) and (12.77) yields

$$u_c(s^t)=u_\ell(s^t)=\theta(s^t), \qquad t\geq 0; \tag{12.78}$$

and, thus,

$$\sum_{t=0}^{\infty}\sum_{s^t}\beta^t\pi(s^t)u_c(s^t)g(s_t)+u_c(s^0)b_0=0.$$

Dividing this expression by $u_c(s^0)$ and using equation (12.51), we find that

$$\sum_{t=0}^{\infty}\sum_{s^t}q_t^0(s^t)g(s_t)=-b_0.$$

In other words, when the government's initial claims $-b_0$ against the private sector equal the present-value of all future government expenditures, the Lagrange multiplier $\Phi$ is zero; that is, the household's present-value budget does not exert any additional constraining effect on welfare maximization than what is not already present in the economy's technology. The reason then is that the government does not have to resort to any distortionary taxation, as can be seen from conditions (12.48a) and (12.78), which imply $\tau^n(s^t)=0$. If the government's initial claims against the private sector were to exceed the present value of future government expenditures, a trivial implication would be that the government would like to hand back this excess financial wealth as lump-sum transfers to the households, and our argument here with $\Phi = 0$ would remain applicable. In the opposite case, when the present value of all government expenditures exceeds the value of any initial claims against the private sector, the Lagrange multiplier $\Phi > 0$. For example, suppose $b_0 = 0$ and there is some $g(s_t) > 0$. After recalling that $Q < 0$ and $\theta(s^t) > 0$, it follows from equation (12.77) that $\Phi > 0$.

Following Lucas and Stokey (1983), we now turn to three examples of government expenditure streams to shed light on optimal tax policies. Throughout we assume that $b_0 = 0$.

*Example 1: $g_t = g$ for all $t$.*

Using resource constraints (12.72) and $\ell_t + n_t = 1$ in first-order conditions (12.74), we find that

$$\begin{aligned}
&(1+\Phi)u_c(c_t, 1-c_t-g) \\
&\quad + \Phi\left[c_t u_{cc}(c_t, 1-c_t-g) - (c_t+g)u_{\ell c}(c_t, 1-c_t-g)\right] \\
=&(1+\Phi)u_\ell(c_t, 1-c_t-g) \\
&\quad + \Phi\left[c_t u_{c\ell}(c_t, 1-c_t-g) - (c_t+g)u_{\ell\ell}(c_t, 1-c_t-g)\right].
\end{aligned} \tag{12.79}$$

Since this condition is the same in every period, we conclude that the optimal allocation is constant over time: $(c_t, n_t) = (\hat{c}, \hat{n})$ for $t \geq 0$. It then follows from condition (12.48a) that the tax rate required to implement the optimal allocation is also constant over time: $\tau_t^n = \hat{\tau}^n$, for $t \geq 0$. Consequently, equation (12.75) implies that the government budget is balanced in each period.

The function of government debt issues in this economy is to smooth distortions over time. Since government expenditures are already smooth in this example, they are optimally financed from contemporaneous taxes. Nothing can be gained from using debt to change the timing of tax collection.

*Example 2: $g_t = 0$ for $t \neq T$, and $g_T > 0$.*

Setting $g = 0$ in expression (12.79), the optimal allocation $(c_t, n_t) = (\hat{c}, \hat{n})$ is constant for $t \neq T$, and consequently, from condition (12.48a), the tax rate is also constant over these periods, $\tau_t^n = \hat{\tau}^n$ for $t \neq T$. Using equations (12.76), we can study tax revenues. Recall that $c_t - n_t = 0$ for $t \neq T$ and that $b_0 = 0$. Thus, the last term in equations (12.76) drops out. Since $\Phi > 0$, the second (quadratic) term is negative, so the first term must be positive. Since $(1+\Phi) > 0$, this fact implies

$$0 < \hat{c} - \frac{u_\ell}{u_c}\hat{n} = \hat{c} - (1-\hat{\tau}^n)\hat{n} = \hat{\tau}^n\hat{n},$$

where the first equality invokes condition (12.48a). We conclude that tax revenue is positive for $t \neq T$. For period $T$, the last term in equation (12.76), $\theta_T g_T$, is positive. Therefore, the sign of the first term is indeterminate: labor may be either taxed or subsidized in period $T$.

This example is a stark illustration of tax smoothing where debt is used to redistribute tax distortions over time. With the same tax revenues in all periods before and after time $T$, the optimal debt policy is as follows: In each period $t = 0, 1, \ldots, T-1$, the government runs a surplus, using it to buy bonds issued by the private sector. In period $T$, the expenditure $g_T$ is met by selling all of these bonds, possibly levying a tax on current labor income, and issuing new bonds that are rolled over forever. The interest payments on that constant outstanding government debt are equal to the constant tax revenue for $t \neq T$, $\hat{\tau}^n \hat{n}$. Thus, the tax distortion is the same in all periods around period $T$, regardless of the proximity to the date $T$. This symmetry was first noted by Barro (1979).

*Example 3:* $g_t = 0$ *for* $t \neq T$*, and* $g_T$ *is stochastic.*

We assume that $g_T = g > 0$ with probability $\alpha$ and $g_T = 0$ with probability $1 - \alpha$. As in the previous example, there is an optimal constant allocation $(c_t, n_t) = (\hat{c}, \hat{n})$ for all periods $t \neq T$ (although the optimum values of $\hat{c}$ and $\hat{n}$ will not, in general, be the same as in example 2). In addition, equation (12.79) implies that $(c_T, n_T) = (\hat{c}, \hat{n})$ if $g_T = 0$. The argument in example 2 shows that tax revenue is positive in all these states. Consequently, debt issues are as follows.

In each period $t = 0, 1, \ldots, T-2$, the government runs a surplus, using it to buy bonds issued by the private sector. The critical difference from example 2 occurs in period $T-1$ when the government now sells all these bonds and uses the proceeds plus current labor tax revenue to buy one-period contingent bonds that only pay off in the next period if $g_T = g$ and otherwise have no value. Moreover, the government buys additional such contingent claims by going short in noncontingent claims. As in example 2, the noncontingent government debt will be rolled over forever with interest payments equal to $\hat{\tau}^n \hat{n}$, but here it is issued one period earlier. If $g_T = 0$ in the next period, the government clearly satisfies its intertemporal budget constraint. In the case $g_T = g$, the construction of our Ramsey equilibrium ensures that the payoff on the government's holdings of contingent claims against the private sector is equal to $g$ plus interest payments of $\hat{\tau}^n \hat{n}$ on government debt net of any current labor tax/subsidy in period $T$. In periods $T+1, T+2, \ldots$, the situation is as in example 2, regardless of whether $g_T = 0$ or $g_T = g$.

This is another example of tax smoothing over time where the tax distortion is the same in all periods around time $T$. It also demonstrates the risk-spreading aspects of fiscal policy under uncertainty. In effect, the government in period $T-1$ buys insurance from the private sector against the event that $g_T = g$.

Lucas and Stokey (1983) draw three lessons from such examples. The first is built into the model at the outset: budget balance in a present-value sense must be respected. In a stationary economy, fiscal policies that have occasional deficits necessarily have offsetting surpluses at other dates. Thus, in the examples with erratic government expenditures, good times are associated with budget surpluses. Second, in the face of erratic government spending, the role of government debt is to smooth tax distortions over time, and it is clear that no general, welfare-economic case can be developed for budget balance on a continual basis. Third, the contingent-claim character of government debt is important for an optimal policy.[9]

Finally, we take a look at the value of contingent government debt in our earlier model with physical capital. By multiplying equation (12.47) by $p(s_t|s^{t-1})$ and summing over $s_t$, we express the household's budget constraint for period $t$ in terms of time $t-1$ values,

$$\begin{aligned} k(s^{t-1}) &+ \sum_{s_t} p(s_t|s^{t-1}) b(s_t|s^{t-1}) \\ &= \sum_{s_t} p(s_t|s^{t-1}) \left\{c(s^t) - \left[1 - \tau^n(s^t)\right] w(s^t) n(s^t)\right\} \\ &+ \sum_{s_t} p(s_t|s^{t-1}) \Big[k(s^t) + \sum_{s_{t+1}} p(s_{t+1}|s^t) b(s_{t+1}|s^t)\Big], \end{aligned} \tag{12.80}$$

where the unit coefficient on $k(s^{t-1})$ is obtained by invoking conditions (12.48b) and (12.48c). Expression (12.80) states that the household's ownership of capital and contingent debt at the end of period $t-1$ is equal to the present value of next period's contingent purchases of goods and financial assets net of labor earnings. We can eliminate the terms in braces by using next period's version of equation

[9] Marcet, Sargent, and Seppälä (1996) offer a qualification to the importance of state-contingent government debt in the model by Lucas and Stokey (1983). In numerical simulations, they explore Ramsey outcomes under the assumption that contingent claims cannot be traded. They find that the incomplete-markets Ramsey allocation is very close to the complete-markets Ramsey allocation. This closeness comes from the Ramsey policy's use of self-insurance through risk-free borrowing and lending with households. Compare to our chapter 14 on heterogeneous agents and how they can overcome market incompleteness through self-insurance.

(12.80). After invoking transversality conditions (12.53), continued substitutions yield

$$\sum_{s_t} p(s_t|s^{t-1}) b(s_t|s^{t-1})$$
$$= \sum_{j=t}^{\infty} \sum_{s^j} \beta^j \pi(s^j|s^{t-1}) \frac{u_c(s^j)c(s^j) - u_\ell(s^j)n(s^j)}{u_c(s^{t-1})} - k(s^{t-1}), \qquad (12.81)$$

where we have invoked conditions (12.48a) and (12.48b). Suppose now that the stochastic process on $s$ follows a Markov process. Then recall from earlier that the allocations from period 1 onward can then be described by time-invariant allocation rules with the current state $s$ and capital stock $k$ as arguments. Thus, equation (12.81) implies that the end-of-period government debt is also a function of the state vector $(s, k)$, since the current state fully determines the end-of-period capital stock and is the only information needed to form conditional expectations of future states. Putting together the lessons of this section with earlier ones, reliance on state-contingent debt and/or state-contingent capital taxes enables the government to avoid any lingering effects on indebtedness from past shocks to government expenditures and past productivity shocks that affected labor tax revenues.

## *Zero tax on human capital*

Returning to the nonstochastic model, Jones, Manuelli, and Rossi (1997) show that the optimality of a limiting zero tax also applies to labor income in a model with human capital, $h_t$, as long as the technology for accumulating human capital is constant returns to scale in the stock of human capital and goods used (not including raw labor).

We postulate the following human capital technology,

$$h_{t+1} = (1 - \delta_h)h_t + H(x_{ht}, h_t, n_{ht}), \qquad (12.82)$$

where $\delta_h \in (0, 1)$ is the rate at which human capital depreciates. The function $H$ describes how new human capital is created with the input of a market good $x_{ht}$, the stock of human capital $h_t$, and raw labor $n_{ht}$. Human capital is in turn used to produce "efficiency units" of labor $e_t$,

$$e_t = M(x_{mt}, h_t, n_{mt}), \qquad (12.83)$$

where $x_{mt}$ and $n_{mt}$ are the market good and raw labor used in the process. We assume that both $H$ and $M$ are homogeneous of degree one in market goods $(x_{jt}, j = h, m)$ and human capital $(h_t)$, and twice continuously differentiable with strictly decreasing (but everywhere positive) marginal products of all factors.

The number of efficiency units of labor $e_t$ replaces our earlier argument for labor in the production function, $F(k_t, e_t)$. The household's preferences are still described by expression (12.1), with leisure $\ell_t = 1 - n_{ht} - n_{mt}$. The economy's aggregate resource constraint is

$$c_t + g_t + k_{t+1} + x_{mt} + x_{ht} = F\left[k_t, M(x_{mt}, h_t, n_{mt})\right] + (1-\delta)k_t. \quad (12.84)$$

The household's present-value budget constraint is

$$\begin{aligned}\sum_{t=0}^{\infty} q_t^0(1+\tau_t^c)c_t = \sum_{t=0}^{\infty} q_t^0 &\left[(1-\tau_t^n)w_t e_t - (1+\tau_t^m)x_{mt} - x_{ht}\right] \\ &+ \left[(1-\tau_0^k)r_0 + 1 - \delta\right] k_0 + b_0, \end{aligned} \quad (12.85)$$

where we have added $\tau_t^c$ and $\tau_t^m$ to the set of tax instruments, to enhance the government's ability to control various margins. Substitute equation (12.83) into equation (12.85), and let $\lambda$ be the Lagrange multiplier on this budget constraint, while $\alpha_t$ denotes the Lagrange multiplier on equation (12.82). The household's first-order conditions are then

$$\begin{aligned}
c_t\!: &\quad \beta^t u_c(t) - \lambda q_t^0(1+\tau_t^c) = 0, && (12.86a)\\
n_{mt}\!: &\quad -\beta^t u_\ell(t) + \lambda q_t^0(1-\tau_t^n)w_t M_n(t) = 0, && (12.86b)\\
n_{ht}\!: &\quad -\beta^t u_\ell(t) + \alpha_t H_n(t) = 0, && (12.86c)\\
x_{mt}\!: &\quad \lambda q_t^0\left[(1-\tau_t^n)w_t M_x(t) - (1+\tau_t^m)\right] = 0, && (12.86d)\\
x_{ht}\!: &\quad -\lambda q_t^0 + \alpha_t H_x(t) = 0, && (12.86e)\\
h_{t+1}\!: &\quad -\alpha_t + \lambda q_{t+1}^0(1-\tau_{t+1}^n)w_{t+1}M_h(t+1) \\
&\qquad + \alpha_{t+1}\left[1-\delta_h + H_h(t+1)\right] = 0. && (12.86f)
\end{aligned}$$

Substituting equation (12.86e) into equation (12.86f) yields

$$\frac{q_t^0}{H_x(t)} = q_{t+1}^0\left[\frac{1-\delta_h+H_h(t+1)}{H_x(t+1)} + (1-\tau_{t+1}^n)w_{t+1}M_h(t+1)\right]. \quad (12.87)$$

We now use the household's first-order conditions to simplify the sum on the right-hand side of the present-value constraint (12.85). First, note that homogeneity of $H$ implies that equation (12.82) can be written as

$$h_{t+1} = (1-\delta_h)h_t + H_x(t)x_{ht} + H_h(t)h_t.$$

Solve for $x_{ht}$ with this expression, use $M$ from equation (12.83) for $e_t$, and substitute into the sum on the right side of equation (12.85), which then becomes

$$\sum_{t=0}^{\infty} q_t^0 \Big\{ (1-\tau_t^n) w_t M_x(t) x_{mt} + (1-\tau_t^n) w_t M_h(t) h_t \\ -(1+\tau_t^m) x_{mt} - \frac{h_{t+1} - [1-\delta_h + H_h(t)]h_t}{H_x(t)}. \Big\}$$

Here we have also invoked the homogeneity of $M$. First-order condition (12.86d) implies that the term multiplying $x_{mt}$ is zero, $[(1-\tau_t^n)w_t M_x(t) - (1+\tau_t^m)] = 0$. After rearranging, we are left with

$$\left[\frac{1-\delta_h + H_h(0)}{H_x(0)} + (1-\tau_0^n) w_0 M_h(0)\right] h_0 \\ - \sum_{t=1}^{\infty} h_t \left\{ \frac{q_{t-1}^0}{H_x(t-1)} - q_t^0 \left[\frac{1-\delta_h + H_h(t)}{H_x(t)} + (1-\tau_t^n) w_t M_h(t)\right]\right\}. \tag{12.88}$$

However, the term in braces is zero by first-order condition (12.87), so the sum on the right side of equation (12.85) simplifies to the very first term in this expression.

Following our standard scheme of constructing the Ramsey plan, a few more manipulations of the household's first-order conditions are needed to solve for prices and taxes in terms of the allocation. We first assume that $\tau_0^c = \tau_0^k = \tau_0^n = \tau_0^m = 0$. If the numeraire is $q_0^0 = 1$, then condition (12.86a) implies

$$q_t^0 = \beta^t \frac{u_c(t)}{u_c(0)} \frac{1}{1+\tau_t^c}, \tag{12.89a}$$

From equations (12.86b) and (12.89a) and $w_t = F_e(t)$, we obtain

$$(1+\tau_t^c)\frac{u_\ell(t)}{u_c(t)} = (1-\tau_t^n) F_e(t) M_n(t), \tag{12.89b}$$

and, by equations (12.86c), (12.86e), and (12.89a),

$$(1+\tau_t^c)\frac{u_\ell(t)}{u_c(t)} = \frac{H_n(t)}{H_x(t)}, \tag{12.89c}$$

and equation (12.86d) with $w_t = F_e(t)$ yields

$$1+\tau_t^m = (1-\tau_t^n)F_e(t)M_x(t). \tag{12.89d}$$

For a given allocation, expressions (12.89) allow us to recover prices and taxes in a recursive fashion: (12.89c) defines $\tau_t^c$ and (12.89a) can be used to compute $q_t^0$, (12.89b) sets $\tau_t^n$, and (12.89d) pins down $\tau_t^m$.

There is only one task remaining to complete our strategy of determining prices and taxes that achieve any allocation. The additional condition (12.87) characterizes the household's intertemporal choice of human capital, which imposes still another constraint on the price $q_t^0$ and the tax $\tau_t^n$. Our determination of $\tau_t^n$ in equation (12.89b) can be thought of as manipulating the margin that the household faces in its static choice of supplying effective labor $e_t$, but the tax rate also affects the household's dynamic choice of human capital $h_t$. Thus, in the Ramsey problem, we will have to impose the extra constraint that the allocation is consistent with the same $\tau_t^n$ entering both equations (12.89b) and (12.87). To find an expression for this extra constraint, solve for $(1-\tau_t^n)$ from equation (12.89b) and a lagged version of equation (12.87), which are then set equal to each other. We eliminate the price $q_t^0$ by using equations (12.89a) and (12.89c), and the final constraint becomes

$$u_\ell(t-1)H_n(t) = \beta u_\ell(t)H_n(t-1)\left[1-\delta_h + H_h(t) + H_n(t)\frac{M_h(t)}{M_n(t)}\right]. \tag{12.90}$$

Proceeding to step 2 in constructing the Ramsey plan, we use condition (12.89a) to eliminate $q_t^0(1+\tau_t^c)$ in the household's budget constraint (12.85). After also invoking the simplified expression (12.88) for the sum on the right-hand of (12.85), the household's "adjusted budget constraint" can be written as

$$\sum_{t=0}^{\infty}\beta^t u_c(t)c_t - \tilde{A} = 0, \tag{12.91}$$

where $\tilde{A}$ is given by

$$\begin{aligned}\tilde{A} &= \tilde{A}(c_0, n_{m0}, n_{h0}, x_{m0}, x_{h0}) \\ &= u_c(0)\left\{\left[\frac{1-\delta_h + H_h(0)}{H_x(0)} + F_e(0)M_h(0)\right]h_0 + [F_k(0)+1-\delta_k]k_0 + b_0\right\}.\end{aligned}$$

In step 3, we define

$$V(c_t, n_{mt}, n_{ht}, \Phi) = u(c_t, 1 - n_{mt} - n_{ht}) + \Phi u_c(t) c_t, \tag{12.92}$$

and formulate a Lagrangian,

$$\begin{aligned} J = \sum_{t=0}^{\infty} \beta^t \{ & V(c_t, n_{mt}, n_{ht}, \Phi) \\ & + \theta_t \{ F[k_t, M(x_{mt}, h_t, n_{mt})] + (1-\delta) k_t - c_t - g_t - k_{t+1} - x_{mt} - x_{ht} \} \\ & + \nu_t [(1 - \delta_h) h_t + H(x_{ht}, h_t, n_{ht}) - h_{t+1}] \} \; - \; \Phi \tilde{A}. \end{aligned} \tag{12.93}$$

This formulation would correspond to the Ramsey problem if it were not for the missing constraint (12.90). Following Jones, Manuelli, and Rossi (1997), we will solve for the first-order conditions associated with equation (12.93), and when it is evaluated at a steady state, we can verify that constraint (12.90) is satisfied even though it has not been imposed. Thus, if both the problem in expression (12.93) and the proper Ramsey problem with constraint (12.90) converge to a unique steady state, they will converge to the same steady state.

The first-order conditions for equation (12.93) evaluated at the steady state are

$$c: \quad V_c = \theta \tag{12.94a}$$

$$n_m: \quad V_{n_m} = -\theta F_e M_n \tag{12.94b}$$

$$n_h: \quad V_{n_h} = -\nu H_n \tag{12.94c}$$

$$x_m: \quad 1 = F_e M_x \tag{12.94d}$$

$$x_h: \quad \theta = \nu H_x \tag{12.94e}$$

$$h: \quad 1 = \beta \left( 1 - \delta_h + H_h + \frac{\theta}{\nu} F_e M_h \right) \tag{12.94f}$$

$$k: \quad 1 = \beta (1 - \delta_k + F_k). \tag{12.94f}$$

Note that $V_{n_m} = V_{n_h}$, so by conditions (12.94b) and (12.94c),

$$\frac{\theta}{\nu} = \frac{H_n}{F_e M_n}, \tag{12.95}$$

which we substitute into equation (12.94f),

$$1 = \beta \left( 1 - \delta_h + H_h + H_n \frac{M_h}{M_n} \right). \tag{12.96}$$

Condition (12.96) coincides with constraint (12.90), evaluated in a steady state. In other words, we have confirmed that the problem (12.93) and the proper Ramsey problem with constraint (12.90) share the same steady state, under the maintained assumption that both problems converge to a unique steady state.

What is the optimal $\tau^n$? The substitution of equation (12.94e) into equation (12.95) yields

$$H_x = \frac{H_n}{F_e M_n}. \tag{12.97}$$

The household's first-order conditions (12.89b) and (12.89c) imply in a steady state that

$$(1 - \tau^n) H_x = \frac{H_n}{F_e M_n}. \tag{12.98}$$

It follows immediately from equations (12.97) and (12.98) that $\tau^n = 0$. Given $\tau^n = 0$, conditions (12.89d) and (12.94d) imply $\tau^m = 0$. We conclude that in the present model neither labor nor capital should be taxed in the limit.

## Should all taxes be zero?

The optimal steady-state tax policy of the model in the previous section is to set $\tau^k = \tau^n = \tau^m = 0$. However, in general, this implies $\tau^c \neq 0$. To see this point, use equation (12.89b) and $\tau^n = 0$ to get

$$1 + \tau^c = \frac{u_c}{u_\ell} F_e M_n. \tag{12.99}$$

From equations (12.94a) and (12.94b)

$$F_e M_n = -\frac{V_{n_m}}{V_c} = \frac{u_\ell + \Phi u_{c\ell} c}{u_c + \Phi(u_c + u_{cc} c)}. \tag{12.100}$$

Hence,

$$1 + \tau^c = \frac{u_c u_\ell + \Phi u_c u_{c\ell} c}{u_c u_\ell + \Phi(u_c u_\ell + u_{cc} u_\ell c)}. \tag{12.101}$$

As discussed earlier, a first-best solution without distortionary taxation has $\Phi = 0$, so $\tau^c$ should trivially be set equal to zero. In a second-best solution, $\Phi > 0$ and we get $\tau^c = 0$ if and only if

$$u_c u_{c\ell} c = u_c u_\ell + u_{cc} u_\ell c, \tag{12.102}$$

which is in general not satisfied. However, Jones, Manuelli, and Rossi (1997) point out one interesting class of utility functions that is consistent with equation (12.102),

$$u(c,\ell) = \begin{cases} \frac{c^{1-\sigma}}{1-\sigma} v(\ell), & \text{if } \sigma > 0, \sigma \neq 1 \\ \ln(c) + v(\ell). & \text{if } \sigma = 1; \end{cases}$$

If a steady state exists, the optimal solution for these preferences is eventually to set all taxes equal to zero. It follows that the optimal plan involves collecting tax revenues in excess of expenditures in the initial periods. When the government has amassed claims against the private sector so large that the interest earnings suffice to finance $g$, all taxes are set equal to zero. Since the steady-state interest rate is $R = \beta^{-1}$, we can use the government's budget constraint (12.5) to find the corresponding value of government indebtedness,

$$b = \frac{\beta}{\beta - 1} g < 0.$$

## Concluding remarks

Perhaps the most startling finding of this chapter is that the optimal steady-state tax on physical capital in a nonstochastic economy is equal to *zero*. The result that capital should not be taxed in the steady state is robust to whether or not the government must balance its budget in each period and to any redistributional concerns arising from a social welfare function. As a stark illustration, Judd's (1985b) example demonstrates that the result holds when the government is constrained to run a balanced budget and it only cares about the workers who are exogenously constrained to not hold any assets. Thus, the capital owners who are assumed not to work will be exempt from taxation in the steady state, and the government will finance its expenditures solely by levying wage taxes on the group of agents that it cares about.

It is instructive to consider Jones, Manuelli, and Rossi's (1997) extension of the no-tax result to labor income, or more precisely human capital. They ask rhetorically, Is physical capital special? We are inclined to answer yes to this question for the following reason. The zero tax on human capital is derived in a model where the production of both human capital and "efficiency units" of labor are constant returns to scale in the stock of human capital and the use of final goods but not raw labor that otherwise enters as an input in the production functions. These assumptions explain why the stream of future labor income in the household's present-value budget constraint in equation (12.85) is reduced to the first term in equation (12.88), which is the value of the household's human capital at time 0. Thus, the functional forms have made raw labor disappear as an object for taxation in future periods. Or in the words of Jones, Manuelli, and Rossi (1997, p. 103 and 99), "Our zero tax results are driven by zero profit conditions. Zero profits follow from the assumption of linearity in the accumulation technologies. Since the activity 'capital income' and the activity 'labor income' display constant returns to scale in reproducible factors, their 'profits' cannot enter the budget constraint in equilibrium." But for alternative production functions that make the endowment of raw labor reappear, the optimal labor tax would not be zero. It is for this reason that we think physical capital is special because the zero-tax result arises with the minimal assumptions of the standard neoclassical growth model, while the zero-tax result on labor income requires that raw labor vanish from the agents' present-value budget constraints.[10]

The weaknesses of our optimal steady-state tax analysis are that it says nothing about how long it takes to reach the zero tax on capital income and how taxes and any redistributive transfers are set during the transition period. These questions will have to be studied numerically as was done by Chari, Christiano, and Kehoe (1994), though their paper does not involve any redistributional concerns because of the assumption of a representative agent. Domeij and Heathcote (2000) construct a model with heterogeneous agents and incomplete insurance markets to study the welfare implications of eliminating capital income taxation. Using earnings and wealth data from the United States, they calibrate a stochastic process for labor earnings that implies a wealth distribution of asset holdings resembling the empirical one. Setting initial tax rates equal to estimates of present taxes in the United States, they study the effects of an unexpected policy reform that sets the capital tax permanently equal to zero and raises the labor tax to maintain long-run budget balance. They find that a majority of households prefers the status quo to the tax reform because of the distributional implications. This example illustrates the importance of a well-designed tax and transfer policy in the transition to a new steady state. In addition, as shown by Aiyagari (1995), the optimal capital tax in an heterogeneous-agent model with

[10] One special case of Jones, Manuelli, and Rossi's (1997) framework with its zero-tax result for labor is Lucas's (1988) endogenous growth model studied in chapter 11. Recall our alternative interpretation of that model as one without any nonreproducible raw labor but just two reproducible factors: physical and human capital. No wonder that raw labor in Lucas's model does not affect the optimal labor tax, since the model can equally well be thought of as an economy without raw labor.

incomplete insurance markets is actually positive, even in the long run. A positive capital tax is used to counter the tendency of such an economy to overaccumulate capital because of too much precautionary saving. We say more about these heterogeneous-agent models in chapter 14.

An assumption maintained throughout the chapter has been that the government can commit to future tax rates when solving the Ramsey problem at time 0. As noted earlier, taxing the capital stock at time 0 amounts to lump-sum taxation and therefore disposes of distortionary taxation. It follows that a government without a commitment technology would be tempted in future periods to renege on its promises and levy a confiscatory tax on capital. An interesting question arises: Can there exist a reputational mechanism that replaces the assumption of a commitment technology? That is, can an announced policy be sustained in an equilibrium because the government *wants* to preserve its reputation? These issues will be studied in chapter 16.

## Exercises

*Exercise 12.1* **A small open economy** (Razin and Sadka, 1995)

Consider the nonstochastic model with capital and labor in this chapter, but assume that the economy is a small open economy that cannot affect the international rental rate on capital, $r_t^*$. Domestic firms can rent any amount of capital at this price, and the households and the government can choose to go short or long in the international capital market at this rental price. There is no labor mobility across countries. We retain the assumption that the government levies a tax $\tau_t^n$ on households' labor income but households no longer have to pay taxes on their capital income. Instead, the government levies a tax $\hat{\tau}_t^k$ on domestic firms' rental payments to capital regardless of the capital's origin (domestic or foreign). Thus, a domestic firm faces a total cost of $(1+\hat{\tau}_t^k)r_t^*$ on a unit of capital rented in period $t$.

**a.** Solve for the optimal capital tax $\hat{\tau}_t^k$.

**b.** Compare the optimal tax policy of this small open economy to that of the closed economy of this chapter.

*Exercise 12.2* **Consumption taxes**

Consider the nonstochastic model with capital and labor in this chapter, but instead of labor and capital taxation assume that the government sets labor and consumption taxes, $\{\tau_t^n, \tau_t^c\}$. Thus, the household's present-value budget constraint is now given by

$$\sum_{t=0}^{\infty} q_t^0 (1+\tau_t^c) c_t = \sum_{t=0}^{\infty} q_t^0 (1-\tau_t^n) w_t n_t + [r_0 + 1 - \delta] k_0 + b_0.$$

**a.** Solve for the Ramsey plan.

**b.** Suppose that the solution to the Ramsey problem converges to a steady state. Characterize the optimal limiting sequence of consumption taxes.

**c.** In the case of capital taxation, we imposed an exogenous upper bound on $\tau_0^k$. Explain why a similar exogenous restriction on $\tau_0^c$ is needed to ensure an interesting Ramsey problem. (Hint: Explore the implications of setting $\tau_t^c = \tau^c$ and $\tau_t^n = -\tau^c$ for all $t \geq 0$, where $\tau^c$ is a large positive number.)

*Exercise 12.3* **Specific utility function** (Chamley, 1986)

Consider the nonstochastic model with capital and labor in this chapter, and assume that the period utility function in equation (12.1) is given by

$$u(c_t, \ell_t) = \frac{c_t^{1-\sigma}}{1-\sigma} + v(\ell_t),$$

where $\sigma > 0$. When $\sigma$ is equal to one, the term $c_t^{1-\sigma}/(1-\sigma)$ is replaced by $\log(c_t)$.

**a.** Show that the optimal tax policy in this economy is to set capital taxes equal to zero in period 2 and from thereon, i.e., $\tau_t^k = 0$ for $t \geq 2$. (Hint: Given the preference specification, evaluate and compare equations (12.30) and (12.35a).)

**b.** Suppose there is uncertainty in the economy as in the stochastic model with capital and labor in this chapter. Derive the optimal *ex ante capital tax rate* for $t \geq 2$.

*Exercise 12.4* **Two labor inputs** (Jones, Manuelli, and Rossi, 1997)

Consider the nonstochastic model with capital and labor in this chapter, but assume that there are two labor inputs, $n_{1t}$ and $n_{2t}$, entering the production function, $F(k_t, n_{1t}, n_{2t})$. The household's period utility function is still given by $u(c_t, \ell_t)$ where leisure is now equal to

$$\ell_t = 1 - n_{1t} - n_{2t}.$$

Let $\tau_{it}^n$ be the flat-rate tax at time $t$ on wage earnings from labor $n_{it}$, for $i = 1, 2$, and $\tau_t^k$ denotes the tax on earnings from capital.

**a.** Solve for the Ramsey plan. What is the relationship between the optimal tax rates $\tau_{1t}^n$ and $\tau_{2t}^n$ for $t \geq 1$? Explain why your answer is different for period $t = 0$. As an example, assume that $k$ and $n_1$ are complements while $k$ and $n_2$ are substitutes.

We now assume that the period utility function is given by $u(c_t, \ell_{1t}, \ell_{2t})$ where

$$\ell_{1t} = 1 - n_{1t}, \qquad \text{and} \qquad \ell_{2t} = 1 - n_{2t}.$$

Further, the government is now constrained to set the same tax rate on both types of labor, i.e., $\tau_{1t}^n = \tau_{2t}^n$ for all $t \geq 0$.

**b.** Solve for the Ramsey plan. (Hint: Using the household's first-order conditions, we see that the restriction $\tau_{1t}^n = \tau_{2t}^n$ can be incorporated into the Ramsey problem by adding the constraint $u_{\ell_1}(t) F_{n_2}(t) = u_{\ell_2}(t) F_{n_1}(t)$.)

**c.** Suppose that the solution to the Ramsey problem converges to a steady state where the constraint that the two labor taxes should be equal is binding. Show that the limiting capital tax is not zero unless $F_{n_1} F_{n_2 k} = F_{n_2} F_{n_1 k}$.

# 13
# Self-Insurance

## Introduction

This chapter describes a version of what is sometimes called a savings problem (e.g., Chamberlain and Wilson, 1984). An agent wants to maximize the expected discounted sum of a concave function of one-period consumption rates, as in chapter 7. However, the agent is cut off from all insurance markets and almost all asset markets. The consumer can only purchase nonnegative amounts of a single risk-free asset. The absence of insurance opportunities induces the consumer to adjust his asset holdings to acquire "self-insurance."

This model is interesting to us partly as a benchmark to compare both with the complete markets model of chapter 7 and also with some of the recursive contracts models of chapter 15, where information and enforcement problems restrict allocations relative to chapter 7, but nevertheless permit more insurance than is allowed in this chapter. A generalization of the single-agent model of this chapter will also be an important component of the incomplete markets models of chapter 14. Finally, the chapter provides our first brush with the powerful martingale convergence theorem.

## The economy

An agent orders consumption streams according to

$$E_0 \sum_{t=0}^{\infty} \beta^t u(c_t), \tag{13.1}$$

where $\beta \in (0,1)$, and $u(c)$ is a strictly increasing, strictly concave, twice continuously differentiable function of the consumption of a single good $c$. The agent is endowed with an infinite random sequence $\{y_t\}_{t=0}^{\infty}$ of the good. Each period, the endowment takes one of a finite number of values, indexed by $s \in \mathbf{S}$. Elements of the sequence of endowments are independently and identically distributed with $\text{Prob}(y = y_s) = \Pi_s, \Pi_s \geq 0$, and $\sum_{s \in \mathbf{S}} \Pi_s = 1$. There are no insurance markets.

The agent can hold nonnegative amounts of a single risk-free asset that has a rate of return $r$ where $(1+r)\beta = 1$. Let $a_t \geq 0$ be the agent's assets at the beginning of period $t$ including the current realization of the income process. The agent faces the sequence of budget constraints

$$a_{t+1} = (1+r)(a_t - c_t) + y_{t+1} \,. \tag{13.2}$$

where $0 \leq c_t \leq a_t$, with $a_0$ given. The constraint that $c_t \leq a_t$ is the constraint that holdings of the asset at the end of the period (which equal $a_t - y_t$) must be nonnegative. The constraint $c_t \geq 0$ is either imposed or comes from an Inada condition $\lim_{c \downarrow 0} u'(c) = +\infty$.

The Bellman equation for an agent with $a > 0$ is

$$\begin{aligned} V(a) &= \max_c \left\{ u(c) + \sum_{s=1}^{S} \beta \Pi_s V\Big[(1+r)(a-c) + y_s\Big] \right\}, \\ &\text{subject to} \quad 0 \leq c \leq a \,, \end{aligned} \tag{13.3}$$

where $y_s$ is the income realization in state $s \in \mathbf{S}$. The value function $V(a)$ inherits the basic properties of $u(c)$; that is, $V(y)$ is increasing, strictly concave, and differentiable.

"Self-insurance" occurs when the agent uses savings to insure himself against income fluctuations. On the one hand, in response to low income realizations, an agent can draw down his savings and avoid temporary large drops in consumption. On the other hand, high income realizations can partly be saved for poor outcomes in the future. We are interested in the long-run properties of an optimal "self-insurance" scheme. Will the agent's future consumption settle down around some level $\bar{c}$?[1] Or will the agent eventually become impoverished?[2] Following the analysis of Chamberlain and Wilson (1984) and Sotomayor (1984), we will show that neither outcome occurs, but that the consumption level will rather converge to infinity.

---

[1] As will occur in the model of social insurance without commitment, to be analyzed in chapter 15.

[2] As in the case of social insurance with asymmetric information, to be analyzed in chapter 15.

## *Nonstochastic endowment*

Without uncertainty the question of insurance is of course remote. However, it is instructive to study the optimal consumption decisions of an agent with an uneven income stream who faces a borrowing constraint. Along the optimal path, it must be true that either

(a) $c^*_{t-1} = c^*_t$ ; or
(b) $c^*_{t-1} < c^*_t$ and $c^*_{t-1} = a^*_{t-1}$, and hence $a^*_t = y_t$ .

According to conditions (a) and (b), $c_{t-1}$ can never exceed $c_t$. The reason is that a declining consumption sequence can be improved by cutting a marginal unit of consumption at time $t-1$ with a utility loss of $u'(c_{t-1})$ and increasing consumption at time $t$ by the saving plus interest with a discounted utility gain of $\beta(1+r)u'(c_t) = u'(c_t) > u'(c_{t-1})$, where the inequality follows from the strict concavity of $u(c)$ and $c_{t-1} > c_t$. A symmetrical argument rules out $c_{t-1} < c_t$ as long as the nonnegativity constraint on savings is not binding; that is, an agent would choose to cut his savings to make $c_{t-1}$ equal to $c_t$ as in condition (a). Therefore, consumption increases from one period to another as in condition (b) only for a constrained agent with zero savings, $a^*_{t-1} - c^*_{t-1} = 0$. It follows that next period's assets are then equal to next period's income, $a^*_t = y_t$ .

Suppose that an agent arrives in period $t$ with zero savings, and he knows that the borrowing constraint will never bind again. He would then find it optimal to choose the highest sustainable constant consumption level, given by the annuity value of the income process starting from period $t$,

$$x_t \equiv \frac{r}{1+r} \sum_{j=t}^{\infty} (1+r)^{t-j} y_j \,.$$

In the optimization problem under certainty, we can show that the impact of the borrowing constraint will not vanish until an agent reaches the period with the highest annuity value of the remainder of the income process, as stated in the following proposition.

PROPOSITION: Given a borrowing constraint and a nonstochastic endowment stream, the limit of the nondecreasing optimal consumption path is

$$\bar{c} \equiv \lim_{t \to \infty} c^*_t = \sup_t x_t \equiv \bar{x} \,. \tag{13.4}$$

PROOF: We will first show that $\bar{c} \leq \bar{x}$. Suppose to the contrary that $\bar{c} > \bar{x}$. Then conditions (a) and (b) imply that there is a $t$ such that $a_t^* = y_t$ and $c_j^* > x_t$ for all $j \geq t$. Therefore, there is a $\tau$ sufficiently large so that

$$0 < \sum_{j=t}^{\tau} (1+r)^{t-j} (c_j^* - y_j) = (1+r)^{t-\tau} (c_\tau^* - a_\tau^*),$$

where the equality uses $a_t^* = y_t$ and successive iterations on budget constraint (13.2). The implication that $c_\tau^* > a_\tau^*$ constitutes a contradiction because it violates the constraint that savings are nonnegative in optimization problem (13.3).

To show that $\bar{c} \geq \bar{x}$, suppose to the contrary that $\bar{c} < \bar{x}$. Then there is an $x_t$ such that $c_j^* < x_t$ for all $j \geq t$, and hence

$$\sum_{j=t}^{\infty} (1+r)^{t-j} c_j^* < \sum_{j=t}^{\infty} (1+r)^{t-j} x_t = \sum_{j=t}^{\infty} (1+r)^{t-j} y_j \leq a_t^* + \sum_{j=t+1}^{\infty} (1+r)^{t-j} y_j,$$

where the last weak inequality uses $a_t^* \geq y_t$. Therefore, there is an $\epsilon > 0$ and $\hat{\tau} > t$ such that for all $\tau > \hat{\tau}$,

$$\sum_{j=t}^{\tau} (1+r)^{t-j} c_j^* < a_t^* + \sum_{j=t+1}^{\tau} (1+r)^{t-j} y_j - \epsilon,$$

and after invoking budget constraint (13.2) repeatedly,

$$(1+r)^{t-\tau} c_\tau^* < (1+r)^{t-\tau} a_\tau^* - \epsilon,$$

or, equivalently,

$$c_\tau^* < a_\tau^* - (1+r)^{\tau-t} \epsilon.$$

We can then construct an alternative feasible consumption sequence $\{c_j^\epsilon\}$ such that $c_j^\epsilon = c_j^*$ for $j \neq \hat{\tau}$ and $c_j^\epsilon = c_j^* + \epsilon$ for $j = \hat{\tau}$. The fact that this alternative sequence yields higher utility establishes the contradiction. ▮

Thus, we have shown that under certainty the optimal consumption sequence converges to a finite limit as long as the discounted value of future income is bounded. Surprisingly enough, that result is overturned when there is uncertainty.

## *Stochastic endowment process*

With uncertain endowments, the first-order condition to the optimization problem (13.3) is

$$u'(c) \geq \sum_{s=1}^{S} \beta(1+r)\Pi_s V'\Big[(1+r)(a-c)+y_s\Big], \tag{13.5}$$

with equality if the nonnegativity constraint on savings is not binding. The Benveniste-Scheinkman formula implies $u'(c) = V'(a)$, so the first-order condition can also be written as

$$V'(a) \geq \sum_{s=1}^{S} \beta(1+r)\Pi_s V'(a'_s), \tag{13.6}$$

where $a'_s$ is next period's assets if the income shock is $y_s$. Since $\beta^{-1} = (1+r)$, $V'(a)$ is a nonnegative supermartingale. By a theorem of Doob (1953, p. 324), $V'(a)$ must then converge almost surely. The limiting value of $V'(a)$ must be zero based on the following argument: Suppose to the contrary that $V'(a)$ converges to a strictly positive limit. That supposition implies that $a$ converges to a finite positive value. But this implication is immediately contradicted by budget constraint (13.2) where assets are equal to the value of the past period's savings including interest, and a stochastic income $y_s$. The random nature of $y_s$ contradicts a finite limit for $a$. Instead, $V'(a)$ must converge to zero, implying that assets converge to infinity. (We return to this result in chapter 14 on incomplete market models.)

Though assets converge to infinity, they do not increase monotonically. Since assets are used for self-insurance, we would expect that low income realizations are associated with reductions in assets. To show this point, suppose to the contrary that even the lowest income realization $y_1$ is associated with nondecreasing assets; that is, $(1+r)(a-c)+y_1 \geq a$. Then we have

$$V'\Big[(1+r)(a-c)+y_1\Big] \leq V'(a) = \sum_{s=1}^{S} \Pi_s V'\Big[(1+r)(a-c)+y_s\Big], \tag{13.7}$$

where the last equality is first-order condition (13.6) when the nonnegativity constraint on savings is not binding and after using $\beta^{-1} = (1+r)$. Since $V'[(1+$

$r)(a-c)+y_s] \le V'[(1+r)(a-c)+y_1]$ for all $s \in \mathbf{S}$, expression (13.7) implies that the derivatives of $V$ evaluated at different asset values are equal to each other, an implication that is contradicted by the strict concavity of $V$.

The fact that assets converge to infinity means that the individual's consumption also converges to infinity. To shed some light on the optimality of such an ever-increasing consumption level, we now adopt the assumption that $u''' > 0$ and study first-order condition (13.5) at equality along an asset accumulation path. After invoking the Benveniste-Scheinkman formula, the first-order condition can be rewritten as

$$u'(c) = \sum_{s=1}^{S} \beta(1+r)\Pi_s u'(c'_s) = \sum_{s=1}^{S} \Pi_s u'(c'_s), \tag{13.8}$$

where $c'_s$ is next period's consumption if the income shock is $y_s$, and the last equality uses $(1+r) = \beta^{-1}$. As in any utility-maximizing trade-off, the marginal rate of substitution is equal to the marginal rate of transformation. When the gross interest rate is equal to the inverse of the discount factor, it is optimal to set today's marginal utility of consumption equal to next period's expected marginal utility of consumption.

It is important to recognize that the individual will never find it optimal to choose a time-invariant consumption level for the indefinite future. Suppose to the contrary that the individual at time $t$ were to choose a constant consumption level for all future periods. The maximum constant consumption level that would be sustainable under all conceivable future income realizations is the annuity value of his current assets $a_t$ and a stream of future incomes all equal to the lowest income realization. But whenever there is a future period with a higher income realization, we can use an argument similar to our earlier construction of the sequence $\{c_j^\epsilon\}$ in the case of certainty to show that the initial time-invariant consumption level does not maximize the agent's utility. It follows that future consumption levels will vary with income realizations, and by Jensen's inequality, first-order condition (13.8) implies

$$c < \sum_{s=1}^{S} \Pi_s c'_s . \tag{13.9}$$

In other words, the curvature of $u(c)$ satisfying $u' > 0$, $u'' < 0$ and $u''' > 0$, means that a particular absolute decline in consumption is not only more costly in utility terms than the gain of an identical absolute increase in consumption but the former is also associated with a larger rise in marginal utility as compared to the drop in marginal utility of the latter. To set today's marginal utility of consumption equal to next period's expected marginal utility of consumption, the agent must therefore balance future states with expected declines in consumption with so much higher expected increases in consumption for other states. Of course, when next period arrives with its optimal consumption (which is on average higher than last period's consumption), the same argument applies again. That is, the process exhibits a "ratchet effect" by which consumption tends toward ever higher levels.

## Concluding remarks

This chapter has maintained the assumption that $\beta(1+r)=1$, which is a very important ingredient in delivering the divergence toward infinity of the agent's asset and consumption level. Chamberlain and Wilson (1984) study a much more general version of the model where they relax this condition.

Chapter 14 will put together continua of agents facing generalizations of the savings problems in order to build some incomplete markets models. The models of that chapter will determine the interest rate $1+r$ as an equilibrium object. In a stationary equilibrium without aggregate uncertainty, the findings of the present chapter anticipate that the equilibrium interest rate in those models must fall short of the subjective rate of discounting $\beta^{-1}$. In a production economy with physical capital, this result implies that the marginal product of capital will be less than the one that would prevail in a complete markets world when the stationary interest rate is given by $\beta^{-1}$. In other words, an incomplete markets economy is characterized by overaccumulation of capital that drives down the interest rate below $\beta^{-1}$, which in turn chokes the desire to accumulate an infinite amount of assets that agents would have had if the interest rate had been equal to $\beta^{-1}$.

Chapter 15 will consider several models in which the condition $\beta(1+r)=1$ is maintained. The assumption will be that a social planner has access to risk-free loans outside the economy and seeks to maximize agents' welfare subject to enforcement and/or information problems. The environment is once again assumed to be stationary without aggregate uncertainty, so in the absence of enforcement and information problems the social planner would just redistribute the economy's resources in each period without any intertemporal trade with the outside world. But when agents are free to leave the economy with their endowment streams and forever live in autarky, the optimal solution prescribes that the social planner amass sufficient outside claims so that each agent is granted a constant consumption stream in the limit, at a level that weakly dominates autarky for all realizations of an agent's endowment. In the case of asymmetric information where the social planner can only induce agents to tell the truth by manipulating promises of future utilities, we obtain a conclusion that is diametrically opposite to the self-insurance outcome of the present chapter. Instead of consumption approaching infinity in the limit, the optimal solution has all agents' consumption approaching its lower bound.

# 14
# *Incomplete Markets Models*

## *Introduction*

In the complete markets model of chapter 7, the optimal consumption allocation is not history dependent: the allocation depends on the current value of the Markov state variable only. This outcome reflects the comprehensive opportunities to insure risks that markets provide. This chapter and the next describe settings with more impediments to exchanging risks. These reduced opportunities make allocations history dependent. In this chapter, the history dependence is encoded in the dependence of a household's consumption on the household's current asset holdings. In the next chapter, history dependence is encoded in the dependence of the consumption allocation on a continuation value promised by a planner or principal.

The present chapter describes a particular type of incomplete markets model. The models have a large number of ex ante identical but ex post heterogeneous agents who trade a single security. For most of this chapter, we study models with no aggregate uncertainty and no variation of an aggregate state variable over time (so macroeconomic time series variation is absent). But there is much uncertainty at the individual level. Households' only option is to "self-insure" by managing a stock of a single asset to buffer their consumption against adverse shocks. We study several models that differ mainly with respect to the particular asset that is the vehicle for self-insurance, for example, fiat currency or capital.

The tools for constructing these models are discrete-state discounted dynamic programming—used to formulate and solve problems of the individuals; and Markov chains—used to compute a stationary wealth distribution. The models produce a stationary wealth distribution that is determined simultaneously with various aggregates that are defined as means across corresponding individual-level variables.

We begin by recalling our discrete state formulation of a single-agent infinite horizon savings problem. We then describe several economies in which households face some version of this infinite horizon saving problem, and where some of the prices taken parametrically in each household's problem are determined by the *average* behavior of all households.[1]

This class of models was invented by Bewley (1977, 1980, 1983, 1986) partly to study a set of classic issues in monetary theory. The second half of this chapter joins that enterprise by using the model to represent inside and outside money, a free banking regime, a subtle limit to the scope of Friedman's optimal quantity of money, a model of international exchange rate indeterminacy, and some related issues. The chapter closes by describing some recent work of Krusell and Smith (1998) designed to extend the domain of such models to include a time-varying stochastic aggregate state variable. As we shall see, this innovation makes the state of the household's problem include the time-$t$ cross-section distribution of wealth, an immense object.

Researchers have used calibrated versions of Bewley models to give quantitative answers to questions including the welfare costs of inflation (İmrohoroğlu, 1992), the risk-sharing benefits of unfunded social security systems (İmrohoroğlu, İmrohoroğlu, and Joines ,1995), the benefits of insuring unemployed people (Hansen and İmrohoroğlu, 1992), and the welfare costs of taxing capital (Aiyagari, 1995).

## A savings problem

Recall the discrete state saving problem described in chapter 3. The household's labor income at time $t$, $s_t$, evolves according to an $m$-state Markov chain with transition matrix $\mathcal{P}$. If the realization of the process at $t$ is $\bar{s}_i$, then at time $t$ the household receives labor income $w\bar{s}_i$. Thus, employment opportunities determine the labor income process. We shall sometimes assume that $m$ is 2, and that $s_t$ takes the value 0 in an unemployed state and 1 in an employed state.

We constrain holdings of a single asset to a grid $\mathcal{A} = [0 < a_1 < a_2 < \ldots < a_n]$. For given values of $(w, r)$ and given initial values $(a_0, s_0)$ the household chooses

[1] Most of the heterogeneous agent models in this chapter have been arranged to shut down aggregate variations over time, to avoid the "curse of dimensionality" that comes into play in formulating the household's dynamic programming problem when there is an aggregate state variable. But we also describe a model of Krusell and Smith (1998) that has an aggregate state variable.

a policy for $\{a_{t+1}\}_{t=0}^{\infty}$ to maximize

$$E_0 \sum_{t=0}^{\infty} \beta^t u(c_t), \tag{14.1}$$

subject to

$$\begin{aligned} c_t + a_{t+1} &= (1+r)a_t + w s_t \\ a_{t+1} &\in \mathcal{A} \end{aligned} \tag{14.2}$$

where $\beta \in (0,1)$ is a discount factor; $u(c)$ is a strictly increasing, strictly concave, twice continuously differentiable one-period utility function satisfying the Inada condition $\lim_{c \downarrow 0} u'(c) = +\infty$; and $\beta(1+r) < 1$.[2]

The Bellman equation, for each $i \in [1, \ldots, m]$ and each $h \in [1, \ldots, n]$, is

$$v(a_h, \bar{s}_i) = \max_{a' \in \mathcal{A}} \{u[(1+r)a_h + w\bar{s}_i - a'] + \beta \sum_{j=1}^{m} \mathcal{P}(i,j) v(a', \bar{s}_j)\}, \tag{14.3}$$

where $a'$ is next period's value of asset holdings. Here $v(a,s)$ is the optimal value of the objective function, starting from asset-employment state $(a,s)$. Note that the grid $\mathcal{A}$ incorporates upper and lower limits on the quantity that can be borrowed (i.e., the amount of the asset that can be issued). The upper bound on $\mathcal{A}$ is restrictive. In some of our theortical discussion to follow, it will be important to dispense with that upper bound.

In chapter 13, we described how to solve equation (14.3) for a value function $v(a,s)$ and an associated policy function $a' = g(a,s)$ mapping this period's $(a,s)$ pair into an optimal choice of assets to carry into next period.

### *Wealth-employment distributions*

Define the unconditional distribution of $(a_t, s_t)$ pairs, $\lambda_t(a,s) = \text{Prob}(a_t = a, s_t = s)$. The exogenous Markov chain $\mathcal{P}$ on $s$ and the optimal policy function $a' = g(a,s)$ induce a law of motion for the distribution $\lambda_t$, namely,

$$\begin{aligned} \text{Prob}(s_{t+1} = s', a_{t+1} = a') = \sum_{a_t} \sum_{s_t} &\text{Prob}(a_{t+1} = a' | a_t = a, s_t = s) \\ &\cdot \text{Prob}(s_{t+1} = s' | s_t = s) \cdot \text{Prob}(a_t = a, s_t = s), \end{aligned}$$

[2] The Inada condition makes consumption nonnegative, and this fact plays a role in justifying the natural debt limit below.

or

$$\lambda_{t+1}(a',s') = \sum_a \sum_s \lambda_t(a,s)\text{Prob}(s_{t+1}=s'|s_t=s)\cdot \mathcal{I}(a',s,a),$$

where we define the indicator function $\mathcal{I}(a',a,s)=1$ if $a'=g(a,s)$, and 0 otherwise.[3] The indicator function $\mathcal{I}(a',a,s)=1$ identifies the time-$t$ states $a,s$ that are sent into $a'$ at time $t+1$. The preceding equation can be expressed as

$$\lambda_{t+1}(a',s') = \sum_s \sum_{\{a:a'=g(a,s)\}} \lambda_t(a,s)\mathcal{P}(s,s'). \tag{14.4}$$

A time-invariant distribution $\lambda$ that solves equation (14.4) (i.e., one for which $\lambda_{t+1}=\lambda_t$) is called a *stationary distribution*. One way to compute a stationary distribution is to iterate to convergence on equation (14.4). An alternative is to create a Markov chain that describes the solution of the optimum problem, then to compute an invariant distribution from a left eigenvector associated with a unit eigenvalue of the stochastic matrix (see chapter 1).

To deduce this Markov chain, we map the pair $(a,s)$ of vectors into a single-state vector $x$ as follows. For $i=1,\ldots,n$, $h=1,\ldots,m$, let the $j$th element of $x$ be the *pair* $(a_i,s_h)$, where $j=(i-1)m+h$. Thus, we denote $x' = [(a_1,s_1),(a_1,s_2),\ldots,(a_1,s_m),(a_2,s_1),\ldots,(a_2,s_m),\ldots,(a_n,s_1),\ldots,(a_n,s_m)]$. The optimal policy function $a'=g(a,s)$ and the Markov chain $\mathcal{P}$ on $s$ induce a Markov chain on $x_t$ via the formula

$$\begin{aligned}\text{Prob}[(a_{t+1}=a',s_{t+1}=s')|(a_t=a,s_t=s)] &= \text{Prob}(a_{t+1}=a'|a_t=a,s_t=s)\\ &\cdot \text{Prob}(s_{t+1}=s'|s_t=s) = \mathcal{I}(a',a,s)\mathcal{P}(s,s'),\end{aligned}$$

where $\mathcal{I}(a',a,s)=1$ is defined as above. This formula defines an $N\times N$ matrix $P$, where $N=n\cdot m$. This is the Markov chain on the household's state vector $x$.[4]

Suppose that the Markov chain associated with $P$ is asymptotically stationary and has a unique invariant distribution $\pi_\infty$. Typically, all states in the Markov chain will be recurrent, and the individual will occasionally revisit each state. For long samples, the distribution $\pi_\infty$ tells the fraction of time that the household

[3] This construction exploits the fact that the optimal policy is a deterministic function of the state, which comes from the concavity of the objective function and the convexity of the constraint set.

[4] Various Matlab programs to be described later in this chapter create the Markov chain for the joint $(a,s)$ state.

spends in each state. We can "unstack" the state vector $x$ and use $\pi_\infty$ to deduce the stationary probability measure $\lambda(a_i, s_h)$ over $(a, s)$ pairs, where

$$\lambda(a_i, s_h) = \text{Prob}(a_t = a_i, s_t = s_h) = \pi_\infty(j),$$

and where $\pi_\infty(j)$ is the $j$th component of the vector $\pi_\infty$, and $j = (i-1)m + h$.

*Reinterpretation of the distribution $\lambda$*

The solution of the household's optimum saving problem induces a stationary distribution $\lambda(a, s)$ that tells the fraction of time that an infinitely lived agent spends in state $(a, s)$. We want to reinterpret $\lambda(a, s)$. Thus, let $(a, s)$ index the state of a particular household at a particular time period $t$, and assume that there is a probability distribution of households over state $(a, s)$. We start the economy at time $t = 0$ with a distribution $\lambda(a, s)$ of households that we want to repeat itself over time. The models in this chapter arrange the initial distribution and other things so that the *distribution* of agents over individual state variables $(a, s)$ remains constant over time even though the state of the individual household is a stochastic process. We shall study several models of this type.

*Example 1: A pure credit model*

Mark Huggett (1993) studied a pure exchange economy. Each of a continuum of households has access to a centralized loan market in which it can borrow or lend at a constant net risk-free interest rate of $r$. Each household's endowment is governed by the Markov chain $(\mathcal{P}, \bar{s})$. The household can either borrow or lend at a constant risk-free rate. However, total borrowings cannot exceed $\phi > 0$, where $\phi$ is a parameter set by Huggett. A household's setting of next period's level of assets is restricted to the discrete set $\mathcal{A} = [a_1, \dots, a_m]$, where the lower bound on assets $a_1 = -\phi$. Later we'll discuss alternative ways to set $\phi$, and how it relates to a natural borrowing limit.

The solution of the household's problem is a policy function $a' = g(a, s)$ that induces a stationary distribution $\lambda(a, s)$ over states. Huggett uses the following definition:

DEFINITION: Given $\phi$, a *stationary equilibrium* is an interest rate $r$, a policy function $g(a, s)$, and a stationary distribution $\lambda(a, s)$ for which

a. The policy function $g(a, s)$ solves the household's optimum problem.

b. The stationary distribution $\lambda(a,s)$ is induced by $(\mathcal{P},\bar{s})$ and $g(a,s)$.

c. The loan market clears

$$\sum_{a,s} \lambda(a,s) g(a,s) = 0.$$

*Equilibrium computation*

Huggett computed equilibria by using an iterative algorithm. He fixed an $r = r_j$ for $j = 0$, and for that $r$ solved the household's problem for a policy function $g_j(a,s)$ and an associated stationary distribution $\lambda_j(a,s)$. Then he checked to see whether the loan market clears at $r_j$ by computing

$$\sum_{a,s} \lambda_j(a,s) g(a,s) = e_j^*.$$

If $e_j^* > 0$, Huggett raised $r_{j+1}$ above $r_j$ and recomputed excess demand, continuing these iterations until he found an $r$ at which excess demand for loans is zero.

*Example 2: A model with capital*

The next model was created by Rao Aiyagari (1994). He used a version of the saving problem in an economy with many agents and interpreted the single asset as homogeneous physical capital, denoted $k$. The capital holdings of a household evolve according to

$$k_{t+1} = (1-\delta) k_t + x_t$$

where $\delta \in (0,1)$ is a depreciation rate and $x_t$ is gross investment. The household's consumption is constrained by

$$c_t + x_t = \tilde{r} k_t + w s_t,$$

where $\tilde{r}$ is the rental rate on capital and $w$ is a competitive wage, to be determined later. The preceding two equations can be combined to become

$$c_t + k_{t+1} = (1 + \tilde{r} - \delta) k_t + w s_t,$$

which agrees with equation (14.2) if we take $a_t \equiv k_t$ and $r \equiv \tilde{r} - \delta$.

There is a large number of households with identical preferences (14.1) whose distribution across $(k,s)$ pairs is given by $\lambda(k,s)$, and whose average behavior determines $(w,r)$ as follows: Households are identical in their preferences, the Markov processes governing their employment opportunities, and the prices that they face. However, they differ in their histories $s_0^t = \{s_h\}_{h=0}^t$ of employment opportunities, and therefore in the capital that they have accumulated. Each household has its own history $s_0^t$ as well as its own initial capital $k_0$. The productivity processes are assumed to be independent across households. The behavior of the collection of these households determines the wage and interest rate $(w,r)$.

Assume an initial distribution *across* households of $\lambda(k,s)$. The average level of capital per household $K$ satisfies

$$K = \sum_{k,s} \lambda(k,s) g(k,s),$$

where $k' = g(k,s)$. Assuming that we start from the invariant distribution, the average level of employment is

$$N = \xi_\infty' \bar{s},$$

where $\xi_\infty$ is the invariant distribution associated with $\mathcal{P}$ and $\bar{s}$ is the exogenously specified vector of individual employment rates. The average employment rate is exogenous to the model, but the average level of capital is endogenous.

There is an aggregate production function whose arguments are the average levels of capital and employment. The production function determines the rental rates on capital and labor from the marginal conditions

$$\begin{aligned} w &= \partial F(K,N)/\partial N \\ \tilde{r} &= \partial F(K,N)/\partial K \end{aligned}$$

where $F(K,N) = AK^\alpha N^{1-\alpha}$ and $\alpha \in (0,1)$.

We now have identified all of the objects in terms of which a stationary equilibrium is defined.

Definition of Equilibrium: A *stationary equilibrium* is a policy function $g(k,s)$, a probability distribution $\lambda(k,s)$, and positive real numbers $(K,\tilde{r},w)$ such that

a. The prices $(w,r)$ satisfy

$$\begin{aligned} w &= \partial F(K,N)/\partial N \\ r &= \partial F(K,N)/\partial K - \delta. \end{aligned} \tag{14.5}$$

b. The policy function $g(k,s)$ solves the household's optimum problem.

c. The probability distribution $\lambda(k,s)$ is a stationary distribution associated with $[g(k,s),\mathcal{P}]$; that is, it satisfies

$$\lambda(k',s') = \sum_{s} \sum_{\{k:k'=g(k,s)\}} \lambda(k,s)\mathcal{P}(s,s').$$

d. The average value of $K$ is implied by the average the households' decisions

$$K = \sum_{k,s} \lambda(k,s) g(k,s).$$

*Computation of equilibrium*

Aiyagari computed an equilibrium of the model by defining a mapping from $K \in I\!R$ into $I\!R$, with the property that a fixed point of the mapping is an equilibrium $K$. Here is an algorithm for finding a fixed point:

1. For fixed value of $K = K_j$ with $j = 0$, compute $(w, r)$ from equation (14.5), then solve the household's optimum problem. Use the optimal policy $g_j(k,s)$ to deduce an associated stationary distribution $\lambda_j(k,s)$.

2. Compute the average value of capital associated with $\lambda_j(k,s)$, namely,

$$K_j^* = \sum_{k,s} \lambda_j(k,s) g_j(k,s).$$

3. For a fixed "relaxation parameter" $\xi \in (0,1)$, compute a new estimate of $K$ from method[5]

$$K_{j+1} = \xi K_j + (1-\xi) K_j^*.$$

4. Iterate on this scheme to convergence.

Later, we shall display some computed examples of equilibria of both Huggett's model and Aiyagari's model. But first we shall analyze some features of both models more formally.

[5] By setting $\xi < 1$, the relaxation method often converges to a fixed point in cases in which direct iteration (i.e., setting $\xi = 0$) fails to converge.

## *Unification and further analysis*

We can display salient features of several models by using a graphical apparatus of Aiyagari (1994). We shall show relationships among several models that have identical household sectors but make different assumptions about the single asset being traded.

For convenience, recall the basic savings problem. The household's objective is to maximize

$$E_0 \sum_{t=0}^{\infty} \beta^t u(c_t) \tag{14.6a}$$

$$c_t + a_{t+1} = w s_t + (1+r) a_t \tag{14.6b}$$

subject to the borrowing constraint

$$a_{t+1} \geq -\phi. \tag{14.6c}$$

We now temporarily suppose that $a_{t+1}$ can take any real value exceeding $-\phi$. Thus, we now suppose that $a_t \in [-\phi, +\infty)$. We occasionally find it useful to express the discount factor $\beta \in (0,1)$ in terms of a discount *rate* $\rho$ as $\beta = \frac{1}{1+\rho}$. In equation (14.6b), $w$ is sometimes a given function $\psi(r)$ of the net interest rate $r$.

## *Digression: the nonstochastic savings problem*

It is useful briefly to recall the nonstochastic version of the savings problem when $\beta(1+r) < 1$. To get the nonstochastic savings problem, assume that $s_t$ is fixed at some positive level $s$. Associated with the household's maximum problem is the Lagrangian

$$L = \sum_{t=0}^{\infty} \beta^t \left\{ u(c_t) + \theta_t \left[ (1+r) a_t + w s - c_t - a_{t+1} \right] \right\}, \tag{14.7}$$

where $\{\theta_t\}_{t=0}^{\infty}$ is a sequence of nonnegative Lagrange multipliers on the budget constraint. The first-order conditions for this problem are

$$u'(c_t) \geq \beta(1+r) u'(c_{t+1}), \quad = \text{if } a_{t+1} > -\phi. \tag{14.8}$$

When $a_{t+1} > -\phi$, the first-order condition implies

$$u'(c_{t+1}) = \frac{1}{\beta(1+r)} u'(c_t), \tag{14.9}$$

which because $\beta(1+r) < 1$ in turn implies that $u'(c_{t+1}) > u'(c_t)$ and $c_{t+1} < c_t$. Thus, consumption is declining during periods when the household is not borrowing constrained. Further, the household will eventually be borrowing constrained.

We can compute the steady level of consumption when the household eventually becomes permanently stuck at the borrowing constraint. Set $a_{t+1} = a_t = -\phi$. This step gives

$$c_t = \bar{c} = ws - r\phi. \tag{14.10}$$

This is the level of labor income left after paying the net interest on the debt. The household would like to shift consumption from tomorrow to today but can't.

If we solve the budget constraint forward, we obtain the present-value budget constraint

$$a_0 = (1+r)^{-1} \sum_{t=0}^{\infty} (1+r)^{-t} (c_t - ws). \tag{14.11}$$

Thus, when $\beta(1+r) < 1$, the household's consumption plan can be found from solving equations (14.11), (14.10), and (14.9) for an initial $c_0$ and a date $T$ after which the debt limit is binding and $c_t$ is constant.

Equation (14.10) implies that if consumption is required to be nonnegative,[6] then the debt limit must satisfy

$$\phi \leq \frac{ws}{r}. \tag{14.12}$$

We call the right side the *natural debt limit.* If $\phi < \frac{ws}{r}$, we say that there is an *ad hoc* debt limit.

We have deduced that when $\beta(1+r) < 1$, if a steady-state level exists, consumption is given by equation (14.10) and assets by $a_t = -\phi$.

Now turn to the case that $\beta(1+r) = 1$. Here equation (14.9) implies that $c_{t+1} = c_t$ and the budget constraint implies $c_t = ws + ra$ and $a_{t+1} = a_t = a_0$. So when $\beta(1+r) = 1$, *any* $a_0$ is a stationary value of $a$. It is optimal forever to roll over the initial asset level.

---

[6] Consumption must be nonnegative, for example, if we impose the Inada condition discussed earlier.

In summary, in the deterministic case, the steady-state demand for assets is $-\phi$ when $(1+r) < \beta^{-1}$ (i.e., when $r < \rho$); and it equals $a_0$ when $r = \rho$. Letting the steady-state level be $\bar{a}$, we have

$$\bar{a} = \begin{cases} -\phi, & \text{if } r < \rho; \\ a_0, & \text{if } r = \rho, \end{cases}$$

where $\beta = 1 + \rho^{-1}$. When $r = \rho$, we say that $\bar{a}$ is indeterminate.

## Borrowing limits: "natural" and "ad hoc"

We return to the stochastic case and take up the issue of debt limits. Imposing $c_t \geq 0$ implies the emergence of what Aiyagari calls a "natural" debt limit. Thus, imposing $c_t \geq 0$ and solving equation (14.6b) forward gives

$$a_t \geq -\frac{1}{1+r} \sum_{j=0}^{\infty} w s_{t+j} (1+r)^{-j}. \tag{14.13}$$

Since the right side is a random variable, not known at $t$, we have to supplement equation (14.13) to obtain the borrowing constraint. One possible approach is to replace the right side of equation (14.13) with its conditional expectation, and to require equation (14.13) to hold in expected value. But this expected value formulation is incompatible with the notion that the loan is risk free, and that the household can repay it for sure. If we insist that equation (14.13) hold almost surely, for all $t \geq 0$, then we obtain the constraint that emerges by replacing $s_t$ with $\min s \equiv s_1$, which yields

$$a_t \geq -\frac{s_1 w}{r}. \tag{14.14}$$

Aiyagari (1994) calls this the "natural debt limit." To accommodate possibly more stringent debt limits, beyond those dictated by the notion that it is feasible to repay the debt for sure, Aiyagari specifies the debt limit as

$$a_t \geq -\phi, \tag{14.15}$$

where

$$\phi = \min\left[b, \frac{s_1 w}{r}\right], \tag{14.16}$$

and $b > 0$ is an arbitrary parameter defining an "ad hoc" debt limit.

*A candidate for a single state variable*

For the special case in which $s$ is i.i.d., Aiyagari showed how to cast the model in terms of a single state variable. To synthesize a single state variable, note that the "disposable resources" available to be allocated at $t$ are $z_t = ws_t + (1+r)a_t + \phi$ or

$$z_t = ws_t + (1+r)\hat{a}_t - r\phi$$

where $\hat{a}_t \equiv a_t + \phi$. In terms of the single state variable $z_t$, the household's budget set can be represented recursively as

$$c_t + \hat{a}_{t+1} \leq z_t \tag{14.17a}$$
$$z_{t+1} = ws_{t+1} + (1+r)\hat{a}_{t+1} - r\phi \tag{14.17b}$$

where we must have $\hat{a}_{t+1} \geq 0$. The Bellman equation is

$$v(z_t, s_t) = \max_{\hat{a}_{t+1} \geq 0} \left\{ u(z_t - \hat{a}_{t+1}) + \beta E v(z_{t+1}, s_{t+1}) \right\}. \tag{14.18}$$

Here $s_t$ appears in the state vector purely as an information variable for predicting the employment component $s_{t+1}$ of next period's disposable resources $z_{t+1}$, conditional on the choice of $\hat{a}_{t+1}$ made this period. Therefore, it disappears from both the value function and the decision rule in the i.i.d. case.

More generally, with a serially correlated state, associated with the solution of the Bellman equation is a policy function

$$\hat{a}_{t+1} = A(z_t, s_t). \tag{14.19}$$

*Supermartingale convergence again*

From equation (14.17a), optimal consumption satisfies $c_t = z_t - A(z_t, s_t)$. The optimal policy obeys the Euler inequality:

$$u'(c_t) \geq \beta(1+r)E_t u'(c_{t+1}), \quad = \text{ if } \hat{a}_{t+1} > 0. \tag{14.20}$$

We can use equation (14.20) to deduce significant aspects of the limiting behavior of mean assets as a function of $r$. Following Chamberlain and Wilson (1984) and others, to deduce the effect of $r$ on the mean of assets, we analyze the limiting behavior of consumption implied by the Euler inequality (14.20). Define

$$M_t = \beta^t (1+r)^t u'(c_t) \geq 0.$$

Then $M_{t+1} - M_t = \beta^t (1+r)^t [\beta(1+r)u'(c_{t+1}) - u'(c_t)]$. Equation (14.20) can be written

$$E_t(M_{t+1} - M_t) \leq 0, \tag{14.21}$$

which asserts that $M_t$ is a supermartingale. Because $M_t$ is nonnegative, the supermartingale convergence theorem applies. It asserts that $M_t$ converges almost surely to a nonnegative random variable $\bar{M}$: $M_t \rightarrow_{\text{a.s.}} \bar{M}$.

It is interesting to consider three cases: (1) $\beta(1+r) > 1$; (2) $\beta(1+r) < 1$, and (3) $\beta(1+r) = 1$. In case 1, the fact that $M_t$ converges implies that $u'(c_t)$ converges to zero almost surely. If $u(\cdot)$ is unbounded (has no satiation point), this fact then implies that $c_t \rightarrow +\infty$ and that the consumer's asset holdings must be diverging to $+\infty$. Chamberlain and Wilson (1984) show that such results also characterize the borderline case (3). In case 2, convergence of $M_t$ leaves open the possibility that $u'(c)$ does not converge a.s., that it remains finite and continues to vary randomly. Indeed, when $\beta(1+r) < 1$, the average level of assets remains finite, and so does the level of consumption.

It is easier to analyze the borderline case $\beta(1+r) = 1$ in the special case that the employment process is independently and identically distributed, meaning that the stochastic matrix $\mathcal{P}$ has identical rows. In this case, $s_t$ provides no information about $z_{t+1}$, and so $s_t$ can be dropped as an argument of both $v(\cdot)$ and $A(\cdot)$. For the case in which $s_t$ is i.i.d., Aiyagari (1994) uses the following argument by contradiction to show that if $\beta(1+r) = 1$, then $z_t$ diverges to $+\infty$. Assume that there is some upper limit $z_{\max}$ such that $z_{t+1} \leq z_{\max} = ws_{\max} + (1+r)A(z_{\max}) - r\phi$. Then when $\beta(1+r) = 1$, the strict concavity of the value function, the Benveniste-Scheinkman formula, and equation (14.20) imply

$$\begin{aligned} v'(z_{\max}) &\geq E_t v'\left[ws_{t+1} + (1+r)A(z_{\max}) - r\phi\right] \\ &> v'\left[ws_{\max} + (1+r)A(z_{\max}) - r\phi\right] = v'(z_{\max}), \end{aligned}$$

which is a contradiction.

## *Average assets as function of* $r$

In the next several sections we use versions of a graph of Aiyagari (1994) to analyze several models. The graph plots the average level of assets as a function of $r$. In the model with capital, the graph is constructed to incorporate the equilibrium dependence of the wage $w$ on $r$. In models without capital, like Huggett's, the wage is fixed. We shall focus on situations where $\beta(1+r) < 1$. We consider cases where the optimal decision rule $A(z_t, s_t)$ and the Markov chain for $s$ induce a Markov chain jointly for assets and $s$ that has a unique invariant distribution. For fixed $r$, let $Ea(r)$ denote the mean level of assets $a$ and $E\hat{a}(r) = Ea(r) + \phi$ be the mean level of $a + \phi$, where the mean is taken with respect to the invariant distribution. Here it is understood that $Ea(r)$ is a function of $\phi$; when we want to make the dependence explicit we write $Ea(r;\phi)$. Also, as we have said, where the single asset is capital, it is appropriate to make the wage $w$ a function of $r$. This approach incorporates the way different values of $r$ affect average capital, the marginal product of labor, and therefore the wage.

The preceding analysis applying supermartingale convergence implies that as $\beta(1+r)$ goes to 1 from below (i.e., $r$ goes to $\rho$ from below), $Ea(r)$ diverges to $+\infty$. This feature is reflected in the shape of the $Ea(r)$ curve in Figure 14.1.[7]

Figure 14.1 assumes that the wage $w$ is fixed in drawing the $Ea(r)$ curve. Later, we will discuss how to draw a similar curve, making $w$ adjust as the function of $r$ that is induced by the marginal productivity conditions for positive values of $K$. For now, we just assume that $w$ is fixed at the value equal to the marginal product of labor when $K = K_1$, the equilibrium level of capital in the model. The equilibrium interest rate is determined at the intersection of the $Ea(r)$ curve with the marginal productivity of capital curve. Notice that the equilibrium interest rate $r$ is lower than $\rho$, its value in the nonstochastic version of the model, and that the equilibrium value of capital $K_1$ exceeds the equilibrium value $K_0$ (determined by the marginal productivity of capital at $r = \rho$ in the nonstochastic version of the model.)

For a pure credit version of the model like Huggett's, but the same $Ea(r)$ curve, the equilibrium interest rate is determined by the intersection of the $Ea(r)$ curve with the $r$ axis.

For the purpose of comparing some of the models that follow, it is useful to note the following aspect of the dependence of $Ea(0)$ on $\phi$:

[7] As discussed in Aiyagari (1994), $Ea(r)$ need not be a monotonically increasing function of $r$, especially because $w$ can be a function of $r$.

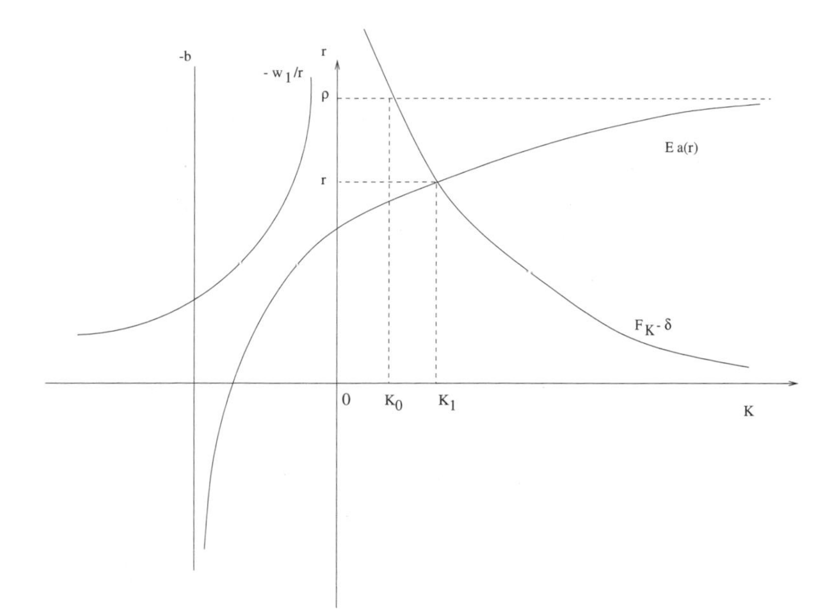


**Figure 14.1** Demand for capital and determination of interest rate. The $Ea(r)$ curve is constructed for a fixed wage that equals the marginal product of labor at level of capital $K_1$. In the nonstochastic version of the model with capital, the equilibrium interest rate and capital stock are $(\rho, K_0)$, while in the stochastic version they are $(r, K_1)$. For a version of the model without capital in which $w$ is fixed at this same fixed wage, the equilibrium interest rate in Huggett's pure credit economy occurs at the intersection of the $Ea(r)$ curve with the $r$ axis.

PROPOSITION 1: When $r = 0$, the optimal rule $\hat{a}_{t+1} = A(z_t, s_t)$ is independent of $\phi$. That is, for $\phi > 0$, $Ea(0;\phi) = Ea(0;0) - \phi$.

*Proof:* It is sufficient to note that when $r = 0$, $\phi$ disappears from the right side of equation (14.17b). Therefore, the optimal rule $\hat{a}_{t+1} = A(z_t, s_t)$ does not depend on $\phi$ when $r = 0$. More explicitly, when $r = 0$, add $\phi$ to both sides of the household's budget constraint to get

$$(a_{t+1} + \phi) + c_t \leq (a_t + \phi) + w s_t.$$

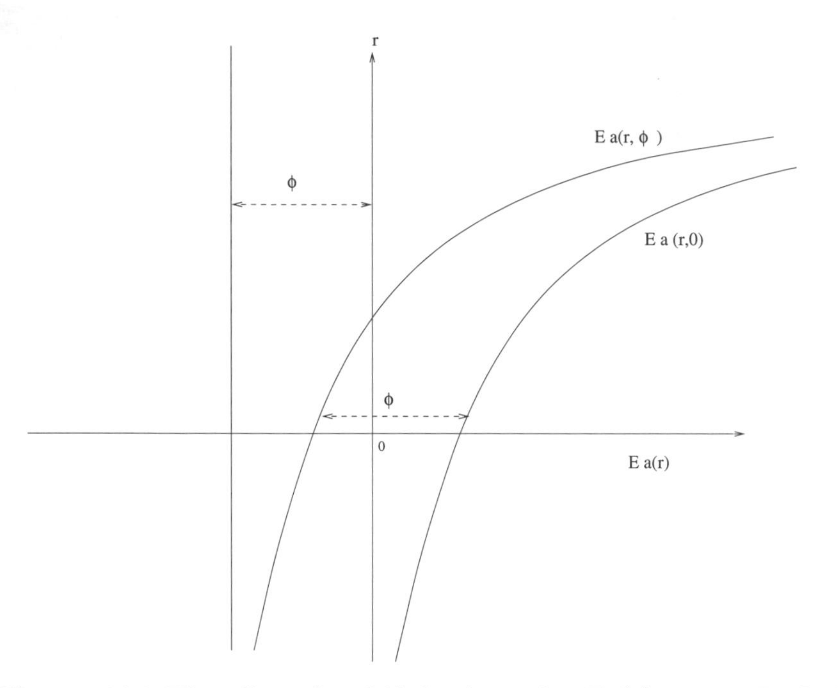


**Figure 14.2** The effect of a shift in $\phi$ on the $Ea(r)$ curve. Both $Ea(r)$ curves are drawn assuming that the wage is fixed.

If the household's problem with $\phi = 0$ is solved by the decision rule $a_{t+1} = g(a_t, z_t)$, then the household's problem with $\phi > 0$ is solved with the same decision rule evaluated at $a_{t+1} + \phi = g(a_t + \phi, z_t)$. ∎

Thus, it follows that at $r = 0$, an increase in $\phi$ displaces the $Ea(r)$ curve to the left by the same amount. See Figure 14.2. We shall use this result to analyze several models.

In the following sections, we use a version Figure 14.1 to compute equilibria of various models. For models without capital, the figure is drawn assuming that the wage is fixed. Typically, the $Ea(r)$ curve will have the same shape as Figure 14.1. In Huggett's model, the equilibrium interest rate is determined by the intersection of the $Ea(r)$ curve with the $r$-axis, reflecting that thc asset (pure consumption loans) is available in zero net supply. In some models with money, a perfect substitute for consumption loans (fiat currency) creates positive net supply.

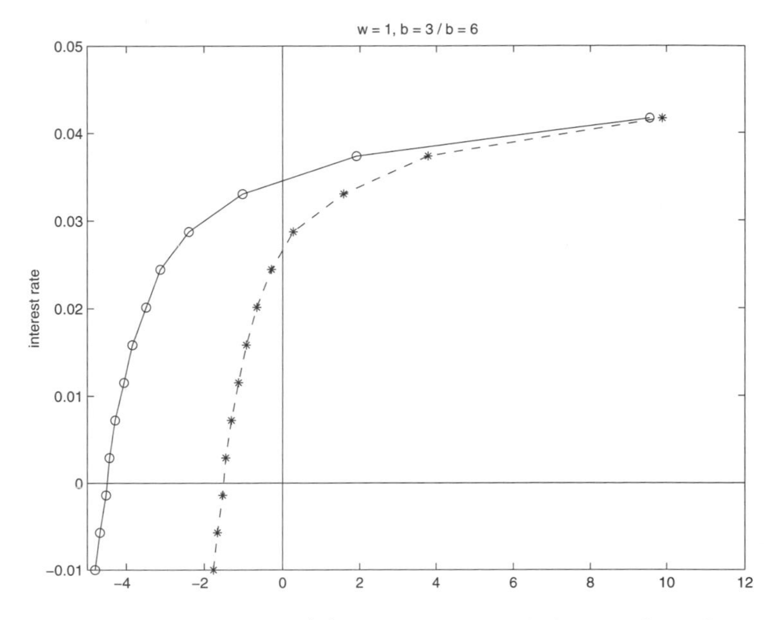


**Figure 14.3** Two $Ea(r)$ curves, one with $b = 6$, the other with $b = 3$, with $w$ fixed at $w = 1$.

## Computed examples

We used some Matlab programs that solve discrete-state dynamic programming problems to compute some examples.[8] We discretized the space of assets from $-\phi$ to a parameter $a_{\max} = 16$ with step size .2.

The utility function is $u(c) = \mu^{-1} c^{\mu}$, with $\mu = 3$. We set $\beta = .96$. We used two specifications of the Markov process for $s$. First, we used Tauchen's (1986) method to get a discrete-state Markov chain to approximate a first-order autoregressive process

$$\log s_t = \rho \log s_{t-1} + u_t,$$

[8] The Matlab programs used to compute the $Ea(r)$ functions are `bewley99.m`, `bewley99v2.m`, `aiyagari2.m`, `bewleyplot.m`, and `bewleyplot2.m`. The program `markovapprox.m` implements Tauchen's method for approximating a continuous autoregressive process with a Markov chain. A program `markov.m` simulates a Markov chain. The programs can be downloaded via anonymous ftp from `ftp://zia.stanford.edu/pub/sargent/webdocs/matlab`.

where $u_t$ is a sequence of i.i.d. Gaussian random variables. We set $\rho = .2$ and the standard deviation of $u_t$ equal to $.4\sqrt{(1-\rho)^2}$. We used Tauchen's method with $N = 7$ being the number of points in the grid for $s$.

For the second specification, we assumed that $s$ is i.i.d. with mean 1.0903. For this case, we compared two settings for the variance: .22 and .68.

Figures 14.3 and 14.5 plot the $Ea(r)$ curves for these various specifications. Figure 14.3 plots $Ea(r)$ for the first case of serially correlated $s$. The two $E[a(r)]$ curves correspond to two distinct settings of the ad hoc debt constraint. One is for $b = 3$, the other for $b = 6$. Figure 14.4 plots the invariant distribution of asset holdings for the case in which $b = 3$ and the interest rate is determined at the intersection of the $Ea(r)$ curve and the $r$ axis.

Figure 14.5 summarizes a precautionary savings experiment for the i.i.d. specification of $s$. Two $Ea(r)$ curves are plotted. For each, we set the ad hoc debt limit $b = 0$. The $Ea(r)$ curve further to the right is the one for the higher variance of the endowment shock $s$. Thus, a larger variance in the random shock causes increased savings.

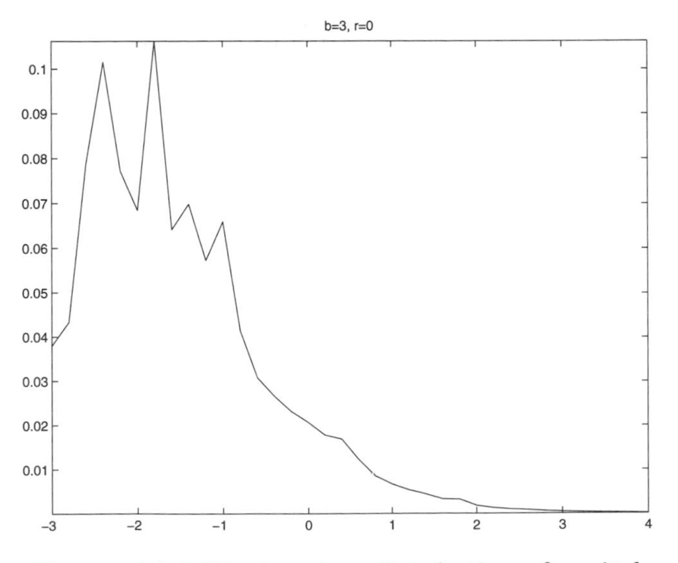


**Figure 14.4** The invariant distribution of capital when $b = 3$.

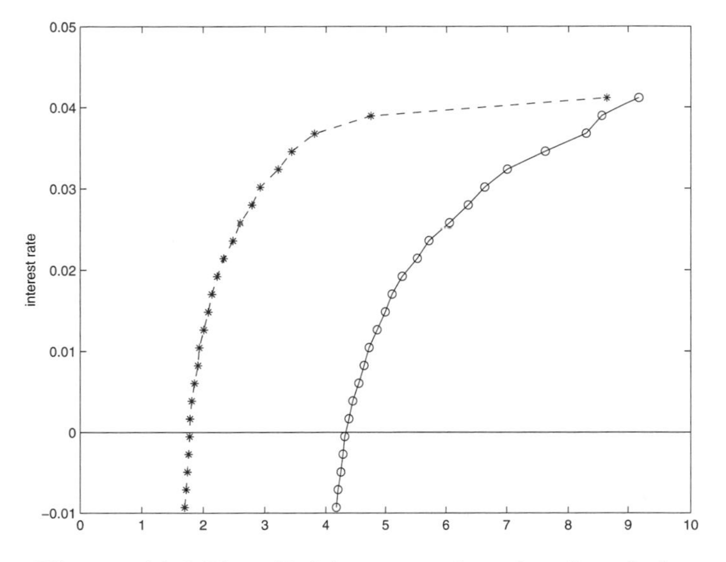


**Figure 14.5** Two $Ea(r)$ curves when $b = 0$ and the endowment shock $s$ is i.i.d. but with different variances; the curve with circles belongs to the economy with the higher variance.

Keep these graphs in mind as we turn to analyze some particular models in more detail.

## *Several models*

We consider several models in which a continuum of households faces the same problem. Their behavior generates the asset demand function $Ea(r;\phi)$. The models share the same family of $Ea(r;\phi)$ curves, but differ in their settings of $\phi$ and in their interpretations of the supply of the asset. The models are (1) Aiyagari's (1994, 1995) model in which the risk-free asset is either physical capital or private IOUs, with physical capital being the net supply of the asset; (2) Huggett's model (1993), where the asset is private IOUs, available in zero net supply; (3) Bewley's model of fiat currency; (4) modifications of Bewley's model to permit an inflation tax; and (5) modifications of Bewley's model to pay interest on currency, either explicitly or implicitly through deflation.

*Optimal stationary allocation*

Because there is no aggregate risk and the aggregate endowment is constant, a stationary optimal allocation would have consumption constant over time for each household. Each household's consumption plan would have constant consumption over time. The implicit risk-free interest rate associated with such an allocation would be $r = \rho$. In the version of the model with capital, the stationary aggregate capital stock solves

$$F_K(K, N) - \delta = \rho. \tag{14.22}$$

Equation (14.22) restricts the stationary optimal capital stock in the nonstochastic optimal growth model of Cass (1965) and Koopmans (1965). The stationary level of capital is $K_0$ in Figure 14.1, depicted as the ordinate of the intersection of the marginal productivity net of depreciation curve with a horizontal line $r = \rho$. As we saw before, the horizontal line at $r = \rho$ acts as a "long-run" demand curve for savings for a nonstochastic version of the savings problem. The stationary optimal allocation matches the one produced by a nonstochastic growth model. We shall use the risk-free interest rate $r = \rho$ as a benchmark against which to compare some alternative incomplete market allocations. Aiyagari's (1994) model replaces the horizontal line $r = \rho$ with an upward sloping curve $Ea(r)$, causing the stationary equilibrium interest rate to fall and the capital stock to rise relative to the risk-free model.

## *A model with capital and private IOUs*

Figure 14.1 can be used to depict the equilibrium of Aiyagari's model described previously. The single asset is capital. There is an aggregate production function $Y = F(K, N)$, and $w = F_N(K, N)$, $r + \delta = F_K(K, N)$. We can invert the marginal condition for capital to deduce a downward-sloping curve $K = K(r)$. This is drawn as the curve labeled $F_K - \delta$ in Figure 14.1. We can use the marginal productivity conditions to deduce a factor price frontier $w = \psi(r)$. For fixed $r$, we use $w = \psi(r)$ as the wage in the savings problem and then deduce $Ea(r)$. We want the equilibrium $r$ to satisfy

$$Ea(r) = K(r). \tag{14.23}$$

The equilibrium interest rate occurs at the intersection of $Ea(r)$ with the $F_K - \delta$ curve. See Figure 14.1.[9]

[9] Recall that Figure 14.1 was drawn for a fixed wage $w$, fixed at the value equal to the marginal product of labor when $K = K_1$. Thus, the new version of Figure 14.1 that incorporates

It follows from the shape of the curves that the equilibrium capital stock $K_1$ exceeds $K_0$, the capital stock required at the given level of total labor to make the interest rate equal $\rho$. There is capital overaccumulation in the stochastic version of the model.

## Private IOUs only

It is easy to compute the equilibrium of Mark Huggett's (1993) model with Figure 14.1. We recall that in Huggett's model, the one asset consists of risk-free loans issued by other households. There are no "outside" assets. This fits the basic model with $a_t$ being the quantity of loans owed to the individual at the beginning of $t$. The equilibrium condition is

$$Ea(r, \phi) = 0, \tag{14.24}$$

which is depicted as the intersection of the $Ea(r)$ curve in Figure 14.1 with the $r$-axis. There is a family of such curves, one for each value of the "ad hoc" debt limit. Relaxing the ad hoc debt limit (by driving $b \to +\infty$) sends the equilibrium interest rate upward toward the intersection of the furthest to the left $Ea(r)$ curve, the one that is associated with the natural debt limit, with the $r$-axis.

### *Limitation of what credit can achieve*

The equilibrium condition (14.24) and the condition that $\lim_{r \nearrow \rho} Ea(r) = +\infty$ imply that the equilibrium value of $r$ is less than $\rho$, for all values of the debt limit respecting the natural debt limit. This outcome supports the following conclusion:

PROPOSITION 2: (Suboptimality of equilibrium with credit) The equilibrium interest rate associated with the "natural debt limit" is the highest one that Huggett's model can support. This interest rate falls short of $\rho$, the interest rate that would prevail in a complete market world.[10]

---

$w = \psi(r)$ has a new curve $Ea(r)$ that intersects the $F_K - \delta$ curve at the same point $(r_1, K_1)$ as the old curve $Ea(r)$ with the fixed wage. Further, the new $Ea(r)$ curve would not be defined for negative values of $K$.

[10] Huggett used the model to study how tightening the ad hoc debt limit parameter $b$ could be used to reduce the risk-free rate enough below $\rho$ to explain the "risk-free rate" puzzle.

*Inside money interpretation*

Huggett's can be viewed as a model of pure "inside money," or of circulating private IOUs. Every person is a "banker" in this setting, entitled to issue "notes" or evidences of indebtedness, subject to the debt limit (14.15). A household has issued notes whenever $a_{t+1} < 0$.

There are several ways to think about the "clearing" of notes imposed by equation (14.24). Here is one: In period $t$, trading occurs in subperiods as follows: First, households realize their $s_t$. Second, some households who choose to set $a_{t+1} < a_t \leq 0$ issue new IOUs in the amount $-a_{t+1} + a_t$. Other households with $a_t < 0$ may decide to set $a_{t+1} \geq 0$, meaning that they want to "redeem" their outstanding notes and possibly acquire notes issued by others. Third, households go to the market and exchange goods for notes. Fourth, notes are "cleared" or "netted out" in a centralized clearing house: positive holdings of notes issued by others are used to retire possibly negative initial holdings of one's own notes. If a person holds positive amounts of notes issued by others, some of these are used to retire any of his own notes outstanding. This clearing operation leaves each person with a particular $a_{t+1}$ to carry into the next period, with no owner of IOUs also being in the position of having some notes outstanding.

There are other ways to interpret the trading arrangement in terms of circulating notes that implement multilateral long-term lending among corresponding "banks": notes issued by individual A and owned by B are "honored" or redeemed by individual C by being exchanged for goods.[11] In a different setting, Kocherlakota (1996b) and Kocherlakota and Wallace (1998) describe such trading mechanisms.

We now turn to a model with some "outside money."

*Bewley's basic model of fiat money*

This version of the model is set up to generate a demand for fiat money, an inconvertible currency supplied in a fixed nominal amount by the government (an entity outside the model). Individuals can hold currency, but not issue it. To map the individual's problem into problem (14.6), we let $m_{t+1}/p = a_{t+1}, b = \phi = 0$, where $m_{t+1}$ is the individual's holding of currency from $t$ to $t+1$, and $p$ is a constant price level. With a constant price level, $r = 0$. With $b = \phi = 0$, $\hat{a}_t = a_t$. Currency is the only asset that can be held. The fixed supply of currency

[11] It is even possible to tell versions of this story in which notes issued by one individual or group of individuals are "extinguished" by another.

is $M$. The condition for a stationary equilibrium is

$$Ea(0) = \frac{M}{p}. \tag{14.25}$$

This is a version of the quantity theory of money and is to be solved for $p$.

Since $r = 0$, we need *some* ad hoc borrowing constraint (i.e., $b < \infty$) to make this model have a stationary equilibrium. If we relax the borrowing constraint from $b = 0$ to permit some borrowing (letting $b > 0$), the $Ea(r)$ curve shifts to the left, causing $Ea(0)$ to fall and the stationary price level to rise.

Let $\bar{m} = Ea(0, \phi = 0)$ be the solution of equation (14.25) when $\phi = 0$. Proposition 1 tells how to construct a set of stationary equilibria, indexed by $\phi \in (0, \bar{m})$, which have identical allocations but different price levels. Given an initial stationary equilibrium with $\phi = 0$ and a price level satisfying equation (14.25), we construct the equilibrium for $\phi \in (0, \bar{m})$ by setting $\hat{a}_t$ for the new equilibrium equal to $\hat{a}_t$ for the old equilibrium for each person for each period.

This set of equilibria highlights how expanding the amount of "inside money," by substituting for "outside" money, causes the value of outside money (currency) to fall. The construction also indicates that if we set $\phi > \bar{m}$, then there exists no stationary monetary equilibrium with a finite positive price level. For $\phi > \bar{m}$, $Ea(0) < 0$ indicating a force for the interest rate to rise and for private IOUs to dominate currency in rate of return and to drive it out of the model. This outcome leads us to consider proposals to get currency back into the model by paying interest on it. Before we do, let's consider some situations more often observed, where a government raises revenues by an inflation tax.

## *A model of seigniorage*

The household side of the model is described in the previous section; we continue to summarize this in a stationary demand function $Ea(r)$. We suppose that $\phi = 0$, so individuals cannot borrow. But now the government augments the nominal supply of currency over time to finance a fixed aggregate flow of real purchases $G$. The government budget constraint at $t \geq 0$ is

$$M_{t+1} = M_t + p_t G, \tag{14.26}$$

which for $t \geq 1$ can be expressed

$$\frac{M_{t+1}}{p_t} = \frac{M_t}{p_{t-1}} \left( \frac{p_{t-1}}{p_t} \right) + G.$$

We shall seek a stationary equilibrium with $\frac{p_{t-1}}{p_t} = (1+r)$ for $t \geq 1$ and $\frac{M_{t+1}}{p_t} = \bar{a}$ for $t \geq 0$. These guesses make the previous equation become

$$\bar{a} = \frac{G}{-r}. \tag{14.27}$$

For $G > 0$, this is a rectangular hyperbola in the southeast quadrant. A stationary equilibrium value of $r$ is determined at the intersection of this curve with $Ea(r)$ (see Figure 14.6). Evidently, when $G > 0$, the equilibrium net interest rate $r < 0$; $-r$ can be regarded as an inflation tax. Notice that if there is one equilibrium value, there is typically more than one. This is a symptom of the Laffer curve present in this model. Typically if a stationary equilibrium exists, there are at least two stationary inflation rates that finance the government budget. This conclusion follows from the fact that both curves in Figure 14.6 have positive slopes.

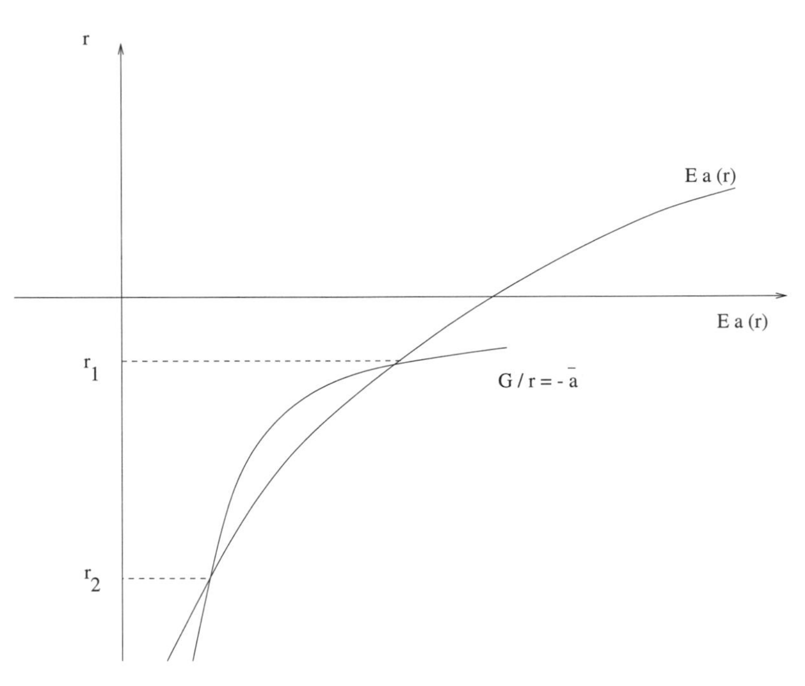


**Figure 14.6** Two stationary equilibrium rates of return on currency that finance the constant government deficit $G$.

After $r$ is determined, the initial price level can be determined by the time-0 version of the government budget constraint (14.26), namely,

$$\bar{a} = M_0/p_0 + G.$$

This is the version of the quantity theory of money that prevails in this model. An increase in $M_0$ increases $p_0$ and all subsequent prices proportionately.

Since there are generally multiple stationary equilibrium inflation rates, which one should we select? We recommend choosing the one with the highest rate of return to currency, that is, the lowest inflation tax. This selection gives "classical" comparative statics: increasing $G$ causes $r$ to fall. In distinct but related settings, Marcet and Sargent (1989) and Bruno and Fischer (1990) give learning procedures that select the same equilibrium we have recommended. Marimon and Sunder (1993) describe experiments with human subjects that they interpret as supporting this selection.

Note the effects of alterations in the debt limit $\phi$ on the inflation rate. Raising $\phi$ causes the $Ea(r)$ curve to shift to the left, and *lowers* $r$. It is even possible for such an increase in $\phi$ to cause all stationary equilibria to vanish. This experiment indicates why governments intent on raising seigniorage might want to restrict private borrowing. See Bryant and Wallace (1984) for an extensive theoretical elaboration of this and related points. See Sargent and Velde (1995) for a practical example from the French Revolution.

## *Exchange rate indeterminacy*

We can adapt the preceding model to display a version of Kareken and Wallace's (1980) theory of exchange rate indeterminacy. Consider a model consisting of two countries, each of which is a Bewley economy with stationary money demand function $Ea_i(r)$ in country $i$. The same single consumption good is available in each country. Residents of both countries are free to hold the currency of either country. Households of either country are indifferent between the two currencies as long as their rates of return are equal. Let $p_{it}$ be the price level in country $i$, and let $p_{1t} = e_t p_{2t}$ define the time-$t$ exchange rate $e_t$. The gross return on currency $i$ between $t-1$ and $t$ is $(1+r) = \left(\frac{p_{i,t-1}}{p_{i,t}}\right)$ for $i = 1, 2$. Equality of rates of return implies $e_t = e_{t-1}$ for all $t$ and therefore $p_{1,t} = ep_{2,t}$ for all $t$, where $e$ is a *constant* exchange rate to be determined.

Each of the two countries finances a fixed expenditure level $G_i$ by printing its own currency. Let $\bar{a}_i$ be the stationary level of real balances in country $i$'s currency. Stationary versions of the two countries' budget constraints are

$$\bar{a}_1 = \bar{a}_1(1+r) + G_1 \tag{14.28}$$

$$\bar{a}_2 = \bar{a}_2(1+r) + G_2 \tag{14.29}$$

Sum these to get

$$\bar{a}_1 + \bar{a}_2 = \frac{(G_1 + G_2)}{-r}.$$

Setting this curve against $Ea_1(r) + Ea_2(r)$ determines a stationary equilibrium rate of return $r$. To determine the initial price level and exchange rate, we use the time-0 budget constraints of the two governments. The time-0 budget constraint for country $i$ is

$$\frac{M_{i,1}}{p_{i,0}} = \frac{M_{i,0}}{p_{i,0}} + G_i$$

or

$$\bar{a}_i = \frac{M_{i,0}}{p_{i,0}} + G_i. \tag{14.30}$$

Add these and use $p_{1,0} = ep_{2,0}$ to get

$$(\bar{a}_1 + \bar{a}_2) - (G_1 + G_2) = \frac{M_{1,0} + eM_{2,0}}{p_{1,0}}.$$

This is one equation in two variables $(e, p_{1,0})$. If there is a solution for some $e \in (0, +\infty)$, then there is a solution for any other $e \in (0, +\infty)$. In this sense, the equilibrium exchange rate is indeterminate.

Equation (14.30) is a quantity theory of money stated in terms of the initial "world money supply" $M_{1,0} + eM_{2,0}$.

*Interest on currency*

Bewley (1980, 1983) studied whether Friedman's recommendation to pay interest on currency could improve outcomes in a stationary equilibrium, and possibly even support an optimal allocation. He found that when $\beta < 1$, Friedman's rule could improve things but could not implement an optimal allocation for reasons we now describe.

As in the earlier fiat money model, there is one asset, fiat currency, issued by a government. Households cannot borrow $(b = 0)$. The consumer's budget constraint is

$$m_{t+1} + p_t c_t \leq (1 + \tilde{r}) m_t + p_t w s_t - \tau p_t$$

where $m_{t+1} \geq 0$ is currency carried over from $t$ to $t+1$, $p_t$ is the price level at $t$, $\tilde{r}$ is nominal interest on currency paid by the government, and $\tau$ is a real lump-sum tax. This tax is used to finance the interest payments on currency. The government's budget constraint at $t$ is

$$M_{t+1} = M_t + \tilde{r} M_t - \tau p_t,$$

where $M_t$ is the nominal stock of currency per person at the beginning of $t$.

There are two versions of this model: one where the government pays explicit interest, while keeping the nominal stock of currency fixed; another where the government pays no explicit interest, but varies the stock of currency to pay interest through deflation.

For each setting, we can show that paying interest on currency, where currency holdings continue to obey $m_t \geq 0$, can be viewed as a device for weakening the impact of this nonnegativity constraint. We establish this point for each setting by showing that the household's problem is isomorphic with Aiyagari's problem of expressions (14.6), (14.15), and (14.16).

*Explicit interest*

In the first setting, the government leaves the money supply fixed, setting $M_{t+1} = M_t \ \forall t$, and undertakes to support a constant price level. These settings make the government budget constraint imply

$$\tau = \tilde{r} M / p.$$

Substituting this into the household's budget constraint and rearranging gives

$$\frac{m_{t+1}}{p} + c_t \leq \frac{m_t}{p}(1 + \tilde{r}) + w s_t - \tilde{r}\frac{M}{p}$$

where the choice of currency is subject to $m_{t+1} \geq 0$. With appropriate transformations of variables, this matches Aiyagari's setup of expressions (14.6), (14.15), and (14.16). In particular, take $r = \tilde{r}$, $\phi = \frac{M}{p}$, $\frac{m_{t+1}}{p} = \hat{a}_{t+1} \geq 0$. With these choices, the solution of the household's saving problem living in an economy with aggregate real balances of $\frac{M}{p}$ and with nominal interest $\tilde{r}$ on currency can be read from the solution of the savings problem with the real interest rate $\tilde{r}$ and a borrowing constraint parameter $\phi \equiv \frac{M}{p}$. Let the solution of this problem be given by the policy function $a_{t+1} = g(a, s; r, \phi)$. Because we have set $\frac{m_{t+1}}{p} = \hat{a}_{t+1} \equiv a_{t+1} + \frac{M}{p}$, the condition that the supply of real balances equals the demand $E\frac{m_{t+1}}{p} = \frac{M}{p}$ is equivalent with $E\hat{a}(r) = \phi$. Note that because

$a_t = \hat{a}_t - \phi$, the equilibrium can also be expressed as $Ea(r) = 0$, where as usual $Ea(r)$ is the average of $a$ computed with respect to the invariant distribution $\lambda(a, s)$.

The preceding argument shows that an equilibrium of the money economy with $m_{t+1} \geq 0$, equilibrium real balances $\frac{M}{p}$, and explicit interest on currency $r$ therefore is isomorphic to a pure credit economy with borrowing constraint $\phi = \frac{M}{p}$. We formalize this conclusion in the following proposition:

PROPOSITION 3: A stationary equilibrium with interest on currency financed by lump-sum taxation has the same allocation and interest rate as an equilibrium of Huggett's free banking model for debt limit $\phi$ equaling the equilibrium real balances from the monetary economy.

To compute an equilibrium with interest on currency, we use a "backsolving" method.[12] Thus, even though the spirit of the model is that the government names $\tilde{r} = r$ and commits itself to set the lump-sum tax needed to finance interest payments on whatever $\frac{M}{p}$ emerges, we can compute the equilibrium by naming $\frac{M}{p}$ *first*, then finding an $r$ that makes things work. In particular, we use the following steps:

1. Set $\phi$ to satisfy $0 \leq \phi \leq \frac{ws_1}{r}$. (We will elaborate on the upper bound in the next section.) Compute real balances and therefore $p$ by solving $\frac{M}{p} = \phi$.

2. Find $r$ from $E\hat{a}(r) = \frac{M}{p}$ or $Ea(r) = 0$.

3. Compute the equilibrium tax rate from the government budget constraint $\tau = r\frac{M}{p}$.

This construction finds a constant tax that satisfies the government budget constraint and that supports a level of real balances in the interval $0 \leq \frac{M}{p} \leq \frac{ws_1}{r}$. Evidently, the largest level of real balances that can be supported in equilibrium is the one associated with the natural debt limit. The levels of interest rates that are associated with monetary equilibria are in the range $0 \leq r \leq r_{FB}$ where $Ea(r_{FB}) = 0$ and $r_{FB}$ is the equilibrium interest rate in the pure credit economy (i.e., Huggett's model) under the natural debt limit.

[12] See Sims (1989) and Diaz-Giménez, Prescott, Fitgerald, and Alvarez (1992) for an explanation and application of backsolving.

*The upper bound on* $\frac{M}{p}$

To interpret the upper bound on attainable $\frac{M}{p}$, note that the government's budget constraint and the budget constraint of a household with zero real balances imply that $\tau = r\frac{M}{p} \leq ws$ for all realizations of $s$. Assume that the stationary distribution of real balances has a positive fraction of agents with real balances arbitrarily close to zero. Let the distribution of employment shocks $s$ be such that a positive fraction of these low-wealth consumers receive income $ws_1$ at any time. Then for it to be feasible for the lowest wealth consumers to pay their lump-sum taxes, we must have $\tau \equiv \frac{rM}{p} \leq ws_1$ or $\frac{M}{p} \leq \frac{ws_1}{r}$.

In a figure like Figure 14.1 or 14.2, the equilibrium real interest rate $r$ can be read from the intersection of the $Ea(r)$ curve and the $r$-axis. Think of a graph with two $Ea(r)$ curves, one with the "natural debt limit" $\phi = \frac{s_1 w}{r}$, the other one with an "ad hoc" debt limit $\phi = \min[b, \frac{s_1 w}{r}]$ shifted to the right. The highest interest rate that can be supported by an interest on currency policy is evidently determined by the point where the $Ea(r)$ curve for the "natural" debt limit passes through the $r$-axis. This is higher than the equilibrium interest rate associated with any of the ad hoc debt limits, but must be below $\rho$. Note that $\rho$ is the interest rate associated with the "optimal quantity of money." Thus, we have Aiyagari's (1994) graphical version of Bewley's (1983) result that the optimal quantity of money (Friedman's rule) cannot be implemented in this setting.

We summarize this discussion with a proposition:

PROPOSITION 4: *Free Banking and Friedman's Rule* The highest interest rate that can be supported by paying interest on currency equals that associated with the pure credit (i.e., the pure inside money) model with the natural debt limit.

If $\rho > 0$, Friedman's rule—to pay real interest on currency at the rate $\rho$—cannot be implemented in this model. The most that can be achieved by paying interest on currency is to eradicate the restriction that prevents households from issuing currency in competition with the government and to implement the free banking outcome.

*A very special case*

Levine and Zame (1999) have studied a special limiting case of the preceding model in which the free banking equilibrium, which we have seen is equivalent to the best stationary equilibrium with interest on currency, is optimal. They attain this special case as the limit of a sequence of economies with $\rho \downarrow 0$. Heuristically, under the natural debt limits, the $Ea(r)$ curves converge to a horizontal line at $r = 0$. At the limit $\rho = 0$, the argument leading to Proposition 4 allows for the optimal $r = \rho$ equilibrium.

*Implicit interest through inflation*

There is another arrangement equivalent to paying explicit interest on currency. Here the government aspires to pay interest through deflation, but abstains from paying explicit interest. This purpose is accomplished by setting $\tilde{r} = 0$ and $\tau p_t = -gM_t$, where it is intended that the outcome will be $(1+r)^{-1} = (1+g)$, with $g < 0$. The government budget constraint becomes $M_{t+1} = M_t(1+g)$. This can be written

$$\frac{M_{t+1}}{p_t} = \frac{M_t}{p_t}\frac{p_{t-1}}{p_t}(1+g).$$

We seek a steady state with constant real balances and inverse of the gross inflation rate $\frac{p_{t-1}}{p_t} = (1+r)$. Such a steady state implies that the preceding equation gives $(1+r) = (1+g)^{-1}$, as desired. The implied lump-sum tax rate is $\tau = -\frac{M_t}{p_{t-1}}(1+r)g$. Using $(1+r) = (1+g)^{-1}$, this can be expressed

$$\tau = \frac{M_t}{p_{t-1}}r.$$

The household's budget constraint with taxes set in this way becomes

$$c_t + \frac{m_{t+1}}{p_t} \leq \frac{m_t}{p_{t-1}}(1+r) + ws_t - \frac{M_t}{p_{t-1}}r \tag{14.31}$$

This matches Aiyagari's setup with $\frac{M_t}{p_{t-1}} = \phi$.

With these matches the steady-state equilibrium is determined just as though explicit interest were paid on currency. The intersection of the $Ea(r)$ curve with the $r$-axis determines the real interest rate. Given the parameter $b$ setting the debt limit, the interest rate equals that for the economy with explicit interest on currency.

## Precautionary savings

As we have seen in the production economy with idiosyncratic labor income shocks, the steady-state capital stock is larger when agents have no access to insurance markets as compared to the capital stock in a complete-markets economy. The "excessive" accumulation of capital can be thought of as the economy's aggregate amount of *precautionary savings*—a point emphasized by Huggett and Ospina (2000). The precautionary demand for savings is usually described as the extra savings caused by future income being random rather than determinate. In our general-equilibrium analysis, the finding of strictly positive precautionary savings depends only on agents being risk averse; that is, the second derivative of the utility function is negative. It is useful to contrast this finding to the partial-equilibrium results on precautionary savings.

In a partial-equilibrium savings problem, it has been known since Leland (1968) and Sandmo (1970) that precautionary savings in response to risk are associated with convexity of the marginal utility function, or a positive third derivative of the utility function. In a two-period model, the intuition can be obtained from the Euler equation, assuming an interior solution with respect to consumption:

$$u'[(1+r)a_0 + w_0 - a_1] = \beta(1+r)E_0 u'[(1+r)a_1 + w_1],$$

where $1+r$ is the gross interest rate, $w_t$ is labor income (endowment) in period $t = 0, 1$; $a_0$ are initial assets and $a_1$ is the optimal amount of savings between periods 0 and 1. Now compare the optimal choice of $a_1$ in two economies where next period's labor income $w_1$ is either determinate and equal to $\bar{w}_1$, or random with a mean value of $\bar{w}_1$. Let $a_1^n$ and $a_1^s$ denote the optimal choice of savings in the nonstochastic and stochastic economy, respectively, that satisfy the Euler equations:

$$\begin{aligned} u'[(1+r)a_0 + w_0 - a_1^n] &= \beta(1+r)u'[(1+r)a_1^n + \bar{w}_1] \\ u'[(1+r)a_0 + w_0 - a_1^s] &= \beta(1+r)E_0 u'[(1+r)a_1^s + w_1] \\ &> \beta(1+r)u'[(1+r)a_1^s + \bar{w}_1], \end{aligned}$$

where the strict inequality is implied by Jensen's inequality under the assumption that $u''' > 0$. It follows immediately from these expressions that the optimal asset level is strictly greater in the stochastic economy as compared to the nonstochastic economy, $a_1^s > a_1^n$.

Versions of precautionary savings have been analyzed by Miller (1974), Sibley (1975), Zeldes (1989), Caballero (1990), Kimball (1990, 1993), and Carroll and Kimball (1996), just to mention a few other studies in a vast literature. Using numerical methods for a finite-horizon savings problem and assuming a constant relative risk-aversion utility function, Zeldes (1989) found that introducing labor income uncertainty made the optimal consumption function concave in assets. That is, the marginal propensity to consume out of assets or transitory income declines with the level of assets. In contrast, without uncertainty and when $\beta(1+r)=1$ (as assumed by Zeldes), the marginal propensity to consume depends only on the number of periods left to live, and is neither a function of the agent's asset level nor the present-value of lifetime wealth.[13] Here we briefly summarize Carroll and Kimball's (1996) analytical explanation for the concavity of the consumption function that income uncertainty seemed to induce.

In a finite-horizon model where both the interest rate and endowment are stochastic processes, Carroll and Kimball cast their argument in terms of the class of hyperbolic absolute risk-aversion (HARA) one-period utility functions. These are defined by $\frac{u'''u'}{u''^2}=k$ for some number $k$. To induce precautionary savings, it must be true that $k>0$. Most commonly used utility functions are of the HARA class: quadratic utility has $k=0$, constant absolute risk-aversion (CARA) corresponds to $k=1$, and constant relative risk-aversion (CRRA) utility functions satisify $k>1$.

Carroll and Kimball show that if $k>0$, then consumption is a concave function of wealth. Moreover, except for some special cases, they show that the consumption function is *strictly* concave; that is, the marginal propensity to consume out of wealth declines with increases in wealth. The exceptions to strict concavity include two well-known cases: CARA utility if all of the risk is to labor income (no rate-of-return risk), and CRRA utility if all of the risk is rate-of-return risk (no labor-income risk).

In the course of the proof, Carroll and Kimball generalize the result of Sibley (1975) that a positive third derivative of the utility function is inherited by the value function. For there to be precautionary savings, the third derivative of the value function with respect to assets must be *positive*; that is, the marginal

[13] When $\beta(1+r)=1$ and there are $T$ periods left to live in a nonstochastic economy, consumption smoothing prescibes a constant consumption level $c$ given by $\sum_{t=0}^{T-1}\frac{c}{(1+r)^t}=\Omega$, which implies $c=\frac{r}{1+r}\left[1-\frac{1}{(1+r)^T}\right]^{-1}\Omega\equiv \text{MPC}_T\,\Omega$, where $\Omega$ is the agent's current assets plus the present value of her future labor income. Hence, the marginal propensity to consume out of an additional unit of assets or transitory income, $\text{MPC}_T$, is only a function of the time horizon $T$.

utility of assets must be a convex function of assets. The case of quadratic one-period utility is an example where there is no precautionary saving. Off corners, the value function is quadratic, and the third derivative of the value function is zero.[14]

Where precautionary saving occurs, and where the marginal utility of consumption is always positive, the derivative of the value function becomes approximately linear for large asset levels, and the consumption function becomes approximately linear.[15] This feature of the consumption function plays a decisive role in governing the behavior of a model of Krusell and Smith (1998), to which we now turn.

## *Models with fluctuating aggregate variables*

That the aggregate equilibrium state variables are constant helps makes the preceding models tractable. This section describes a way to extend such models to situations with time-varying stochastic aggregate state variables.[16]

Krusell and Smith (1998) modified Aiyagari's (1994) model by adding an aggregate state variable $z$, a technology shock that follows a Markov process. Each household continues to receive an idiosyncratic labor-endowment shock $s$ that averages to the same constant value for each value of the aggregate shock $z$. The aggregate shock causes the size of the state of the economy to expand dramatically because every household's wealth will depend on the history of the *aggregate* shock $z$, call it $z^t$, as well as the history of the household-specific shock $s^t$. That makes the joint histories of $z^t, s^t$ correlated across households, which in turn makes the cross-section distribution of $(k, s)$ vary randomly over time. Therefore, the interest rate and wage will also vary randomly over time.

One way to specify the state is to include the cross-section distribution $\lambda(k, s)$ *each period* among the state variables. Thus, the state includes a cross-section probability distribution of (capital, employment) pairs. In addition, a description of a recursive competitive equilibrium must include a law of motion mapping today's distribution $\lambda(k, s)$ into tomorrow's distribution.

---

[14] In linear quadratic models, decision rules for consumption and asset accumulation are independent of the variances of innovations to exogenous income processes. See Weil (1993).

[15] Roughly speaking, this follows from applying the Benveniste-Scheinkman formula and noting that, where $v$ is the value function, $v''$ is increasing in savings and $v''$ is bounded.

[16] See Duffie, Geanakoplos, Mas-Colell, and McLennan (1994) for a general formulation and equilibrium existence theorem for such models. See Marcet and Singleton (1999) for a computational strategy for incomplete markets models with a finite number of heterogeneous agents.

*Aiyagari's model again*

To prepare the way for Krusell and Smith's way of handling such a model, we recall the structure of Aiyagari's model. The household's Bellman equation in Aiyagari's model is

$$v(k,s) = \max_{c,k'} \{u(c) + \beta E[v(k',s')|s]\} \tag{14.32}$$

where the maximization is subject to

$$c + k' = \tilde{r}k + ws + (1-\delta)k, \tag{14.33}$$

and the prices $\tilde{r}$ and $w$ are fixed numbers satisfying

$$\tilde{r} = \tilde{r}(K,N) = \alpha \left(\frac{K}{N}\right)^{\alpha-1} \tag{14.34a}$$

$$w = w(K,N) = (1-\alpha)\left(\frac{K}{N}\right)^{\alpha}. \tag{14.34b}$$

Recall that aggregate capital and labor $K, N$ are the average values of $k, s$ computed from

$$K = \int k\lambda(k,s)dkds \tag{14.35}$$

$$N = \int s\lambda(k,s)dkds. \tag{14.36}$$

Here we are following Aiyagari by assuming a Cobb-Douglas aggregate production function. The definition of a stationary equilibrium requires that $\lambda(k,s)$ be the stationary distribution of $(k,s)$ across households induced by the decision rule that attains the right side of equation (14.32).

*Krusell and Smith's extension*

Krusell and Smith (1998) modify Aiyagari's model by adding an aggregate productivity shock $z$ to the price equations, emanating from the presence of $z$ in the production function. The shock $z$ is governed by an exogenous Markov process. Now the state must include $\lambda$ and $z$ too, so the household's Bellman equation becomes

$$v(k,s;\lambda,z) = \max_{c,k'} \{u(c) + \beta E[v(k',s';\lambda',z')|(s,z,\lambda)]\} \tag{14.37}$$

where the maximization is subject to

$$c + k' = \tilde{r}(K, N, z)k + w(K, N, z)s + (1 - \delta)k \tag{14.38a}$$

$$\tilde{r} = \tilde{r}(K, N, z) = z\alpha \left(\frac{K}{N}\right)^{\alpha - 1} \tag{14.38b}$$

$$w = w(K, N, z) = z(1 - \alpha) \left(\frac{K}{N}\right)^{\alpha} \tag{14.38c}$$

$$\lambda' = H(\lambda, z) \tag{14.38d}$$

where $(K, N)$ is a stochastic processes determined from[17]

$$K_t = \int k\lambda_t(k, s)dkds \tag{14.39}$$

$$N_t = \int s\lambda_t(k, s)dkds. \tag{14.40}$$

Here $\lambda_t(k, s)$ is the distribution of $k, s$ across households at time $t$. The *distribution* is itself a random function disturbed by the aggregate shock $z_t$.

The Bellman equation and the pricing functions induce the household to want to forecast the average capital stock $K$, in order to forecast future prices. That desire makes the household want to forecast the cross-section distribution of asset capital. To do so it consults the law of motion (14.38d).

DEFINITION: A recursive competitive equilibrium is a pair of price functions $\tilde{r}, w$, a value function and decision rule $k' = f(k, s; \lambda, z)$, and a law of motion $H$ for $\lambda(k, s)$ such that (a) given the price functions and $H$, the value function solves the Bellman equation (14.37) and the optimal decision rule is $f$; and (b) the decision rule $f$ and the Markov processes for $s$ and $z$ imply that today's distribution $\lambda(k, s)$ is matched into tomorrow's $\lambda'(k, s)$ by $H$.

The curse of dimensionality makes an equilibrium difficult to compute. Krusell and Smith propose a way to approximate an equilibrium using simulations. First, they characterize the distribution $\lambda(k, s)$ by a finite set of moments of capital $m = (m_1, \ldots, m_I)$. They assume a parametric functional form for $H$ mapping

---

[17] In our simplified formulation, $N$ is actually constant over time. But in Krusell and Smith's model, $N$ too can be a stochastic process, because leisure is in the one-period utility function.

today's $m$ into next period's value $m'$. They assume a form that can be conveniently estimated using nonlinear or linear least squares. They assume initial values for the parameters of $H$. Given $H$, they use numerical dynamic programming to solve the Bellman equation

$$v(k, s; m, z) = \max_{c,k'} \{u(c) + \beta E[v(k', s'; m, z')|(s, z, \lambda)]\}$$

subject to the assumed law of motion $H$ for $m$. They take the solution of this problem and draw a single long realization from the Markov process for $\{z_t\}$, say, of length $T$. For that realization of $z$, they then simulate paths of $\{k_t, s_t\}$ of length $T$ for a large number $M$ of households. They assemble these $M$ simulations into a history of $T$ empirical cross-section distributions $\lambda_t(k, s)$. They use these cross-section distributions to compute a time series of length $T$ of the moments $m(t)$. They use this sample and nonlinear least squares to estimate the transition function $H$. They go back to the beginning of the procedure, use this new guess at $H$, and continue, iterating to convergence of the function $H$.

Krusell and Smith compare the aggregate time series $K_t, N_t, \tilde{r}_t, w_t$ from this model with a corresponding representative agent (or complete markets) model. They find that the statistics for the aggregate quantities and prices for the two types of models are very close. Krusell and Smith interpret this result in terms of an "approximate aggregation" that follows from two properties of their parameterized model. First, consumption as a function of wealth is concave but close to linear for moderate-to-high wealth levels. Second, most of the saving is done by the high-wealth people. These two properties mean that fluctuations in the distribution of wealth have only a small effect on the aggregate amount saved and invested. Thus, distribution effects are small. Also, for these high-wealth people, self-insurance works quite well, so aggregate consumption is not much lower than it would be for the complete markets economy.

Krusell and Smith compare the distributions of wealth from their model to the U.S. data. Relative to the data, the model with a constant discount factor generates too few very poor people and too many rich people. Krusell and Smith modify the model by making the discount factor an exogenous stochastic process. The discount factor switches occasionally between two values. Krusell and Smith find that a modest difference between two discount factors can bring the model's wealth distribution much closer to the data. Patient people become wealthier; impatient people eventually become poorer.

## *Concluding remarks*

The models in this chapter pursue some of the adjustments that households make when their preferences and endowments give a motive to insure but markets provide limit opportunities to do so. We have studied settings where households' savings occurs through a single risk-free asset. Households use the asset to "self-insure," by making intertemporal adjustments of the asset holdings to smooth their consumption. Their consumption rates at a given date become a function of their asset holdings, which in turn depend on the histories of their endowments. In pure exchange versions of the model, the equilibrium allocation becomes individual-history specific, in contrast to the history-independence of the corresponding complete markets model.

The models of this chapter arbitrarily shut down or allow markets without explanation. The market structure is imposed, its consequences then analyzed. In the next chapter, we study a class of models for similar environments that, like the models of this chapter, make consumption allocations history dependent. But the spirit of the models in the next chapter differs from those in this chapter in requiring that the trading structure be more firmly motivated by the environment. In particular, the models in the next chapter posit a particular reason that complete markets do not exist, based in enforcement or information problems, and then study how risk sharing among people can best be arranged.

## *Exercises*

*Exercise 14.1* **Stochastic discount factor** (Bewley-Krusell-Smith)

A household has preferences over consumption of a single good ordered by a value function defined recursively according to $v(\beta_t, a_t, s_t) = u(c_t) + \beta_t E_t v(\beta_{t+1}, a_{t+1}, s_{t+1})$, where $\beta_t \in (0,1)$ is the time-$t$ value of a discount factor, and $a_t$ is time-$t$ holding of a single asset. Here $v$ is the discounted utility for a consumer with asset holding $a_t$, discount factor $\beta_t$, and employment state $s_t$. The discount factor evolves according to a three-state Markov chain with transition probabilities $P_{i,j} = \text{Prob}(\beta_{t+1} = \bar{\beta}_j | \beta_t = \bar{\beta}_i)$. The discount factor and employment state at $t$ are both known. The household faces the sequence of budget constraints

$$a_{t+1} + c_t \leq (1+r)a_t + w s_t$$

where $s_t$ evolves according to an $n$-state Markov chain with transition matrix $\mathcal{P}$. The household faces the borrowing constraint $a_{t+1} \geq -\phi$ for all $t$.

Formulate Bellman equations for the household's problem. Describe an algorithm for solving the Bellman equations. *Hint:* Form three coupled Bellman equations.

*Exercise 14.2* **Mobility costs** (Bertola)

A worker seeks to maximize $E\sum_{t=0}^{\infty}\beta^t u(c_t)$, where $\beta \in (0,1)$ and $u(c) = \frac{c^{1-\sigma}}{(1-\sigma)}$, and $E$ is the expectation operator. Each period, the worker supplies one unit of labor inelastically (there is no unemployment) and either $w^g$ or $w^b$, where $w^g > w^b$. A new "job" starts off paying $w^g$ the first period. Thereafter, a job earns a wage governed by the two-state Markov process governing transition between good and bad wages on all jobs; the transition matrix is $\begin{bmatrix} p & (1-p) \\ (1-p) & p \end{bmatrix}$. A new (well-paying) job is always available, but the worker must pay mobility cost $m > 0$ to change jobs. The mobility cost is paid at the beginning of the period that a worker decides to move. The worker's period-$t$ budget constraint is

$$A_{t+1} + c_t + mI_t \le RA_t + w_t,$$

where $R$ is a gross interest rate on assets, $c_t$ is consumption at $t$, $m > 0$ is moving costs, $I_t$ is an indicator equaling 1 if the worker moves in period $t$, zero otherwise, and $w_t$ is the wage. Assume that $A_0 > 0$ is given and that the worker faces the no-borrowing constraint, $A_t \ge 0$ for all $t$.

**a.** Formulate the Bellman equation for the worker.

**b.** Write a Matlab program to solve the worker's Bellman equation. Show the optimal decision rules computed for the following parameter values: $m = .9, p = .8, R = 1.02, \beta = .95, w^g = 1.4, w^b = 1, \sigma = 4$. Use a range of assets levels of $[0,3]$. Describe how the decision to move depends on wealth.

**c.** Compute the Markov chain governing the transition of the individual's state $(A, w)$. If it exists, compute the invariant distribution.

**d.** In the fashion of Bewley, use the invariant distribution computed in part **c** to describe the distribution of wealth across a large number of workers all facing this same optimum problem.

*Exercise 14.3* **Unemployment**

There is a continuum of workers with identical probabilities $\lambda$ of being fired each period when they are employed. With probability $\mu \in (0,1)$, each unemployed worker receives one offer to work at wage $w$ drawn from the cumulative distribution function $F(w)$. If he accepts the offer, the worker receives the offered wage each period until he is fired. With probability $1-\mu$, an unemployed worker receives no offer this period. The probability $\mu$ is determined by the function $\mu = f(U)$, where $U$ is the unemployment rate, and $f'(U) < 0, f(0) = 1, f(1) = 0$. A worker's utility is given by $E\sum_{t=0}^{\infty} \beta^t y_t$, where $\beta \in (0,1)$ and $y_t$ is income in period $t$, which equals the wage if employed and zero otherwise. There is no unemployment compensation. Each worker regards $U$ as fixed and constant over time in making his decisions.

**a.** For fixed $U$, write the Bellman equation for the worker. Argue that his optimal policy has the reservation wage property.

**b.** Given the typical worker's policy (i.e., his reservation wage), display a difference equation for the unemployment rate. Show that a stationary unemployment rate must satisfy

$$\lambda(1-U) = f(U)\big[1 - F(\bar{w})\big]U,$$

where $\bar{w}$ is the reservation wage.

**c.** Define a *stationary equilibrium.*

**d.** Describe how to compute a stationary equilibrium. You don't actually have to compute it.

*Exercise 14.4* **Asset insurance**

Consider the following setup. There is a continuum of households who maximize

$$E\sum_{t=0}^{\infty} \beta^t u(c_t),$$

subject to

$$c_t + k_{t+1} + \tau \le y + \max(x_t, g)k_t^{\alpha}, \quad c_t \ge 0,\ k_{t+1} \ge 0,\ t \ge 0,$$

where $y > 0$ is a constant level of income not derived from capital, $\alpha \in (0,1)$, $\tau$ is a fixed lump sum tax, $k_t$ is the capital held at the beginning of $t$, $g \le 1$ is an "investment insurance" parameter set by the government, and $x_t$ is a stochastic household-specific gross rate of return on capital. We assume that $x_t$ is governed by a two-state Markov process with stochastic matrix $\mathcal{P}$, which takes on the two values $\bar{x}_1 > 1$ and $\bar{x}_2 < 1$. When the bad investment return occurs, $(x_t = \bar{x}_2)$, the government supplements the household's return by $\max(0, g - \bar{x}_2)$.

The household-specific randomness is distributed identically and independently across households. Except for paying taxes and possibly receiving insurance payments from the government, households have no interactions with one another; there are no markets.

Given the government policy parameters $\tau, g$, the household's Bellman equation is

$$v(k, x) = \max_{k'}\{u\left[\max(x, g)k^\alpha - k' - \tau\right] + \beta \sum_{x'} v(k', x')\mathcal{P}(x, x')\}.$$

The solution of this problem is attained by a decision rule

$$k' = G(k, x),$$

that induces a stationary distribution $\lambda(k, x)$ of agents across states $(k, x)$.

The average (or per capita) physical output of the economy is

$$Y = \sum_k \sum_x (x \times k^\alpha)\lambda(k, x).$$

The average return on capital to households, *including* the investment insurance, is

$$\nu = \sum_k \bar{x}_1 k^\alpha \lambda(k, x_1) + \max(g, \bar{x}_2) \sum_k k^\alpha \lambda(k, x_2),$$

which states that the government pays out insurance to all households for which $g > \bar{x}_2$.

Define a stationary equilibrium.

# 15
# *Optimal Social Insurance*

## *Insurance with recursive contracts*

This chapter studies how to insure when there are incentive or commitment problems. We leave the competitive incomplete market setting of the previous chapter and study a planner who designs an efficient contract to supply insurance, subject to incentive constraints imposed by limited ability to enforce or to observe agents' actions or incomes.

This chapter pursues two themes, one substantive, the other technical. The substantive theme is a trade-off between efficiency and incentives that emerges in settings with private information or lack of commitment. As we shall see, one way to cope with incentive problems is to manipulate the future wealth distribution, and thereby sacrifice some insurance. Thus, the models contain theories of the evolution of distributions of income and wealth that are alternatives to the ones described in chapter 14 on incomplete markets.

The technical theme is how incentive problems can partially be overcome with contracts that use memory and promises, and how memory can be encoded recursively. Contracts are constrained to issue rewards that depend on the history of publicly observable outcomes. Histories are large-dimensional objects. Analytical progress accelerated after the discovery of Spear and Srivastava (1987), Thomas and Worrall (1988), Abreu, Pearce, and Stacchetti (1990), Phelan and Townsend (1991), and others that the dimension of the state can be held down by using an accounting system cast solely in terms of a "promised value," a one-dimensional object that can summarize the relevant aspects of an agent's history. Throughout this chapter, working with promised values permits formulating the contracting problem recursively.

We study three basic models. The first is a model of Kocherlakota (1996b) that focuses solely on commitment problems, but has only public information. The second is a model of Thomas and Worrall (1990). It has an incentive problem coming from private information, but assumes away commitment problems. The third model, by Atkeson (1991), combines both commitment and information problems. We also study a version of Shavell and Weiss's (1979) and Hopenhayn and Nicolini's (1997) model of unemployment insurance.

## Social insurance without commitment

This section describes a simplified version of a contract design problem studied by Kocherlakota (1996b).[1] The environment is like ones in the incomplete markets competitive equilibrium models of chapter 14, but the allocation mechanism differs. There is an enforcement problem. A moneylender allocates goods to households in ways designed to get them to share resources voluntarily.

Thus, imagine a village with a large number of ex ante identical households. Each household has preferences over consumption streams that are ordered by

$$E\sum_{t=0}^{\infty}\beta^t u(c_t), \tag{15.1}$$

where $u(c)$ is an increasing, strictly concave, and twice continuously differentiable function, and $\beta \in (0,1)$ is a discount factor. Each household receives a stochastic endowment stream $\{y_t\}_{t=0}^{\infty}$, where for each $t \geq 0$, $y_t$ is independently and identically distributed according to the discrete probability distribution $\text{Prob}(y_t = y_s) = \Pi_s$, where $s \in \{1,2,\ldots,S\} \equiv \mathbf{S}$ and $y_{s+1} > y_s$. The consumption good is not storable. At time $t \geq 1$, the household has experienced a history of endowments $h_t = (y_t, y_{t-1}, \ldots, y_0)$. For now we shall assume that the endowment processes are i.i.d. across time and across households.

Following a tradition started by Green (1987), the only person in the village who has access to a "storage technology" or to a risk-free loan market outside the village is the "moneylender," also sometimes called the "planner." The moneylender can borrow or lend at the constant risk-free gross interest rate of $\beta^{-1}$ and is in the business of borrowing from and lending to each villager. The villagers cannot borrow or lend with one another, and can only deal with the moneylender. Furthermore, we assume that the moneylender is committed to honor his promises. The villagers are not committed to the contract. The villagers are free to walk away from their arrangement with the moneylender at any time. They must be

[1] Kocherlakota studies a closed system in which there are two households who receive stochastic endowments that sum to a constant amount each period, and each of whom must be induced each period not to walk away from the arrangement to live in autarky.

induced not to by the structure of the contract with the moneylender. This is a model of "one-sided commitment" in which the contract must be "self-enforcing" on the villager.

The moneylender (or planner) designs a contract. A contract is a sequence of functions $c_t = f_t(h_t)$ for $t \geq 0$. The sequence of functions $\{f_t\}$ assigns a history-dependent consumption stream $c_t = f_t(h_t)$ to the household. The contract specifies that each period the villager contributes his time-$t$ endowment $y_t$ to the moneylender who then returns $c_t$ to the villager. From this arrangement, the moneylender earns an expected present value

$$P = E \sum_{t=0}^{\infty} \beta^t (y_t - c_t). \tag{15.2}$$

By plugging the associated consumption process into expression (15.1), we find that the contract gives the villager a prescribed present value of $v = E \sum_{t=0}^{\infty} \beta^t u(c_t)$.

The contract must be "self-enforcing" in the face of lack of commitment to the contract by the household. At any point in time, the household is free to walk away from the contract and simply consume its endowment stream. However, if the household walks away from the contract, it must live in autarky evermore. The ex ante value associated with consuming the endowment stream, to be called the autarky value, is

$$v_{\rm aut} = E \sum_{t=0}^{\infty} \beta^t u(y_t).$$

At any point in time, having observed his current-period endowment, the household can guarantee itself a present value of utility of $u(y_t) + \beta v_{\rm aut}$ by consuming its own endowment. The planner's contract must offer him more than this amount at every possible history and every date. Thus, the contract must satisfy

$$u[f_t(h_t)] + \beta E_t \sum_{j=1}^{\infty} \beta^{j-1} u[f_{t+j}(h_{t+j})] \geq u(y_t) + \beta v_{\rm aut}, \tag{15.3}$$

for all $t \geq 0$ and all histories $h_t$. Equation (15.3) is called the participation constraint for the borrower. It is interpreted as a constraint on the contract. A contract that satisfies equation (15.3) is said to be *sustainable*.

A difficulty of problems with constraints like equation (15.3) is that the dimension of the argument $h_t$ of the contract is so large and grows so rapidly with $t$. Fortunately, a recursive formulation of history-dependent contracts applies.

The goal is to represent the sequence of functions $\{f_t\}$ recursively by finding a state variable $v_t$ such that the contract takes the form

$$\begin{aligned} c_t &= g(v_t, y_t) \\ v_{t+1} &= \ell(v_t, y_t). \end{aligned}$$

Here $g$ and $\ell$ are time-invariant functions. By iterating the $\ell(\cdot)$ function $t$ times, one obtains

$$v_t = m(v_0; y_{t-1}, \ldots, y_0).$$

Thus, $v_t$ summarizes histories of endowments. We shall see that a state variable $v_t$ that works in several related contexts is a *promised discounted future value* $v_t$. We can treat the promised value $v$ as a state variable, then formulate a functional equation for a planner. The planner delivers a prescribed value $v$ by delivering a state-dependent current consumption $c$ and a promised value starting tomorrow, say $v'$, where $c$ and $v'$ each depend on the current endowment $y_t$ and the promise $v$. The planner aims to deliver $v$ in a way that maximizes his profits (15.2) to be recorded as a function of $v$ in a value function $P(v)$. Using dynamic programming, we can develop a functional equation for $P(v)$. Evidently, $P$ must be a decreasing function of $v$ because the higher is the consumption stream of the villager, the lower must be the profits of the moneylender.

Each period, the household must be induced to turn over the time-$t$ endowment $y_t$ to the planner, who invests it outside the village at a constant one-period gross interest rate of $\beta^{-1}$. The moneylender delivers a state-contingent consumption stream to the household that keeps the household participating in the arrangement after every history. He wants to do so in the "most efficient," that is, profit-maximizing, way. Let $P(v)$ be the expected present value of the "profit stream" $\{y_t - c_t\}$ for a planner who delivers value $v$ in the optimal way. The optimum value $P(v)$ obeys the functional equation

$$P(v) = \max_{\{c_s, w_s\}} \sum_{s=1}^{S} \Pi_s [(y_s - c_s) + \beta P(w_s)] \tag{15.4}$$

where the maximization is subject to the constraints

$$\sum_{s=1}^{S} \Pi_s [u(c_s) + \beta w_s] \geq v \tag{15.5}$$

$$u(c_s) + \beta w_s \geq u(y_s) + \beta v_{\text{aut}}, \quad s = 1, \ldots, S \tag{15.6}$$

$$c_s \in [c_{\min}, c_{\max}] \tag{15.7}$$
$$w_s \in [v_{\text{aut}}, \bar{v}]. \tag{15.8}$$

Here $w_s$ is the promised value with which the consumer enters next period, given that $y = y_s$ this period; $[c_{\min}, c_{\max}]$ is a bounded set to which we restrict the choice of $c_t$ each period. Constraint (15.5) is the promise-keeping constraint. It requires that at least the promised value $v$ be delivered. Constraints (15.6), one for each state $s$, are the participation constraints.

The constraint set is convex. The one-period return function in equation (15.4) is concave. The value function $P(v)$ that solves equation (15.4) is concave. Form the Lagrangian

$$\begin{aligned} L = & \sum_{s=1}^{S} \Pi_s[(y_s - c_s) + \beta P(w_s)] \\ & + \mu \left\{ \sum_{s=1}^{S} \Pi_s[u(c_s) + \beta w_s] - v \right\} \\ & + \sum_{s=1}^{S} \lambda_s \{u(c_s) + \beta w_s - [u(y_s) + \beta v_{\text{aut}}]\}. \end{aligned} \tag{15.9}$$

For each $v$ and for $s = 1, \ldots, S$, the first-order conditions with respect to $c_s, w_s$, respectively, are

$$(\lambda_s + \mu \Pi_s) u'(c_s) = \Pi_s \tag{15.10}$$
$$\lambda_s + \mu \Pi_s = -\Pi_s P'(w_s). \tag{15.11}$$

By the envelope theorem, if $P$ is differentiable, then $P'(v) = -\mu$; $P(v)$ is evidently decreasing in $v$, and is concave. Thus, $P'(v)$ becomes more and more negative as $v$ increases.

Equations (15.10) and (15.11) imply the following relationship between $c_s, w_s$:

$$u'(c_s) = -P'(w_s)^{-1}. \tag{15.12}$$

This condition simply states that the household's marginal rate of substitution between $c_s$ and $w_s$, given by $u'(c_s)/\beta$, should be equal to the planner's marginal rate of transformation as given by $-1/[\beta P'(w_s)]$. The concavity of $P$ and $u$ means that equation (15.12) traces out a positively sloped curve in the $c, w$

plane, as depicted in Figure 15.1. We can interpret this condition as making $c_s$ a function of $w_s$. To complete the optimal contract, it will be enough to find how $w_s$ depends on the promised value $v$ and the income state $y_s$.

How $w_s$ depends on $v$ depends on which of two mutually exclusive and exhaustive sets of states $s, v$ falls into after the realization of $y_s$: those in which the participation constraint (15.6) binds (i.e., states in which $\lambda_s > 0$) and those in which it does not (i.e., states in which $\lambda_s = 0$). Condition (15.11) can be written $P'(w_s) = P'(v) - \lambda_s/\Pi_s$. We shall analyze what happens in these two types of states, those in which $\lambda_s > 0$, and those in which $\lambda_s = 0$.

*States Where* $\lambda_s > 0$

When $\lambda_s > 0$, this fact implies that $P'(w_s) < P'(v)$, which in turn implies, by the concavity of $P$, that $w_s > v$. Further, the participation constraint at equality implies that $c_s \leq y_s$ (because $w_s \geq v_{\rm aut}$). Taken together, these results say that when the participation constraint (15.6) binds, the planner induces the household to consume less than its endowment today by raising its continuation value.

When $\lambda_s > 0$, $c_s$ and $w_s$ are determined by solving the two equations

$$u(c_s) + \beta w_s = u(y_s) + \beta v_{\rm aut} \tag{15.13}$$

$$u'(c_s) = -P'(w_s)^{-1}. \tag{15.14}$$

The participation constraint holds with equality. Notice that these equations are independent of $v$. This property is a key to understanding the form of the optimal contract. Let us denote the solutions of equations (15.13) and (15.14) by

$$c_s = g_1(y_s) \tag{15.15}$$

$$w_s = \ell_1(y_s). \tag{15.16}$$

*States Where* $\lambda_s = 0$

When the participation constraint does not bind, $\lambda_s = 0$ and first-order condition (15.11) imply that $P'(v) = P'(w_s)$, which implies that $w_s = v$. Therefore, from (15.14), we can write $u'(c_s) = -P'(v)^{-1}$, so that consumption in state $s$ depends on promised utility $v$ but not on the endowment in state $s$. Thus, when the participation constraint does not bind, the planner awards

$$c_s = g_2(v) \tag{15.17}$$

$$w_s = v, \tag{15.18}$$

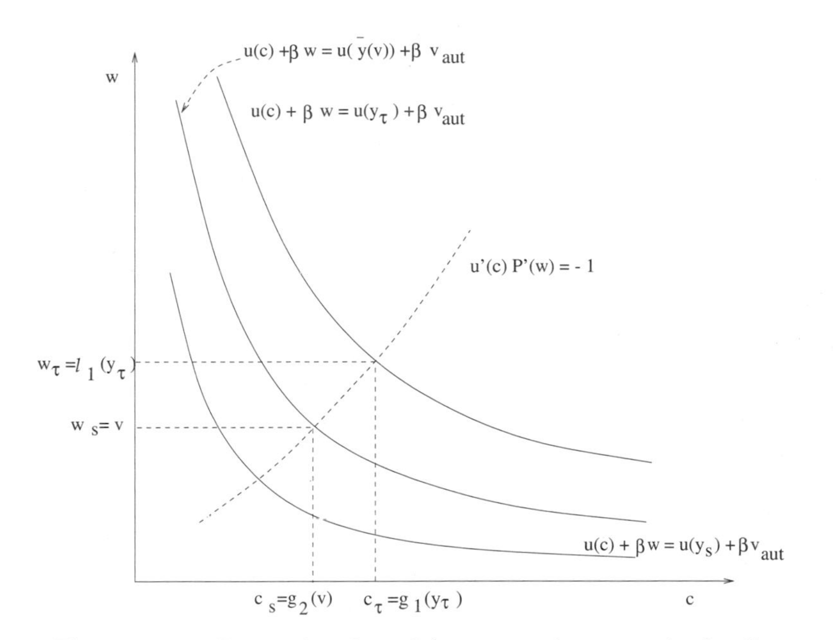


**Figure 15.1** Determination of (consumption, promised utility). Higher realizations of $y_s$ are associated with higher indifference curves $u(c) + \beta w = u(y_s) + \beta v_{\rm aut}$ . For a given $v$, there is a threshold level $\bar{y}(v)$ above which the participation constraint is binding and below which the planner awards a constant level of consumption, depending on $v$, and maintains the same promised value $w_s = v$. The cutoff level $\bar{y}(v)$ is determined by the indifference curve going through the intersection of a horizontal line at level $v$ with the "expansion path" $u'(c_s)P'(w_s) = -1$.

where $g_2(v)$ solves $u'[g_2(v)] = P'(v)^{-1}$.

*The Optimal Contract*

Combining the branches of the policy functions for the cases where the participation constraint does and does not bind, we obtain

$$c = \max\{g_1(y), g_2(v)\} \tag{15.19}$$
$$w = \max\{\ell_1(y), v\}. \tag{15.20}$$

The nature of the optimal policy is displayed graphically in Figures 15.1 and 15.2. To interpret the graphs, it is useful to study equations (15.13) and (15.14) for the case in which $w_s = v$. By setting $w_s = v$, we can solve these equations for a "cutoff value," call it $\bar{y}(v)$, such that the participation constraint binds only when $y_s \geq \bar{y}(v)$. To find $\bar{y}(v)$, we first solve equation (15.14) for the value $c_s$ associated with $v$ for those states in which the participation constraint is not binding:

$$u'[g_2(v)] = -P'(v)^{-1},$$

and then substitute this value into equation (15.13) to solve for $\bar{y}(v)$:

$$u[\bar{y}(v)] = u[g_2(v)] + \beta(v - v_{\text{aut}}). \tag{15.21}$$

By the concavity of $P$, the cutoff value $\bar{y}(v)$ is increasing in $v$.

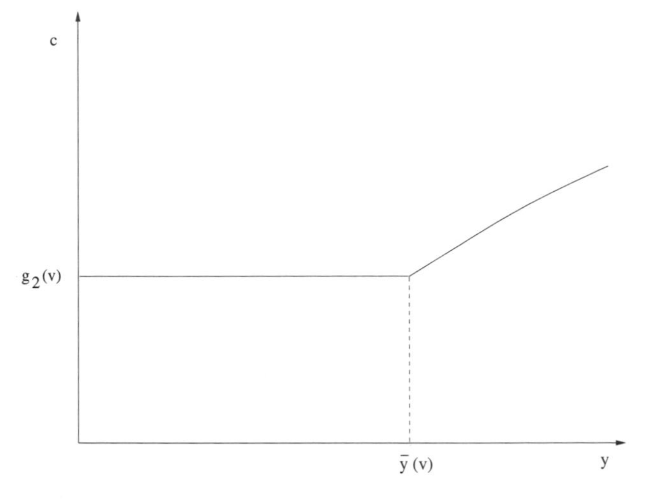


**Figure 15.2** The shape of consumption as a function of realized endowment, when the promised initial value is $v$.

Associated with a given level of $v_t \in (v_{\text{aut}}, \bar{v})$, there are two numbers $g_2(v_t), \bar{y}(v_t)$ such that if $y_t \leq \bar{y}(v_t)$ the planner offers the household $c_t = g_2(v_t)$ and leaves the promised utility unaltered, $v_{t+1} = v_t$. The planner is thus insuring against the states $y_s \leq \bar{y}(v_t)$ at time $t$. If $y_t > \bar{y}(v_t)$, the participation constraint is binding, prompting the planner to induce the household to surrender some of its current-period endowment in exchange for a raised promised utility $v_{t+1} > v_t$. Promised values never decrease, stay constant for low-$y$ states $y_s < \bar{y}(v_t)$, and increase in high-endowment states that threaten to violate the participation constraint. Consumption stays constant during periods when the participation constraint fails to bind and increases during periods when it threatens to bind. Thus, a household that realizes the highest endowment $y_S$ is permanently awarded the highest consumption level with an associated promised value $\bar{v}$ as given by

$$u[g_2(\bar{v})] + \beta\bar{v} = u(y_S) + \beta v_{\text{aut}}.$$

Notice that $u(y_1) < Eu(y)$ implies that $u(c) + \beta w = u(y_1) + \beta v_{\text{aut}} < v_{\text{aut}}$. This means that with $v = v_{\text{aut}}$, the participation constraint is not binding when $y = y_1$. Thus the lowest indifference curve in Figure 15.1, that indexed by $u(y_1) + \beta v_{\text{aut}}$, is above $(c, w) = [g_2(v_{\text{aut}}), v_{\text{aut}}]$, which graphically expresses that the participation constraint is not binding at endowment level $y_1$. The optimal contract trades off consumption against continuation values only for sufficiently higher realizations of $y$.

*Many households*

Consider a large village in which a moneylender faces a continuum of such households. At the beginning of time $t = 0$, before the realization of $y_0$, the moneylender offers each household $v_{\text{aut}}$ (or maybe just a small amount more). As time unfolds, the moneylender executes the contract for each household. A society of such households would experience a "fanning out" of the distribution of consumption across households for a while, to be followed by an eventual "fanning in" as the cross-sectional distribution of consumption asymptotically becomes concentrated at the single point $g_2(\bar{v})$ computed earlier (i.e., the minimum $c$ such that the participation constraint will never again be binding). Notice that the moneylender would on average, across villagers, be saving early on, investing the proceeds at the gross interest rate $\beta^{-1}$. Later he would be using the interest on his account outside the village to finance payments to the villagers.

## A Lagrangian method

Marcet and Marimon (1992, 1999) have proposed an approach that applies to most of the contract design problems of this chapter. They form a Lagrangian and use the Lagrange multipliers on incentive constraints to keep track of promises. Their approach extends work of Kydland and Prescott (1980) and is related to

Hansen, Epple, and Roberds' (1985) formulation for linear quadratic environments. We can illustrate the method in the context of the preceding model.

Marcet and Marimon's approach would be to formulate the problem directly in the space of stochastic processes (i.e., random sequences) and to form a Lagrangian for the moneylender. The contract specifies a stochastic process for consumption obeying the following constraints:

$$u(c_t) + E_t \sum_{j=1}^{\infty} \beta^j u(c_{t+j}) \geq u(y_t) + \beta v_{\rm aut} \qquad \forall t \geq 0, \tag{15.22a}$$

$$E_{-1} \sum_{t=0}^{\infty} \beta^t u(c_t) \geq v. \tag{15.22b}$$

Here $v$ is the initial promised value to be delivered to the villager starting in period 0. Equation (15.22a) gives the participation constraints.

The moneylender's Lagrangian is

$$\begin{aligned} J = E_{-1} \sum_{t=0}^{\infty} \beta^t \Big\{ (y_t - c_t) + \alpha_t \Big[ E_t \sum_{j=0}^{\infty} \beta^j u(c_{t+j}) - \big[ u(y_t) + \beta v_{\rm aut} \big] \Big] \Big\} \\ + \phi \Big[ E_{-1} \sum_{t=0}^{\infty} \beta^t u(c_t) - v \Big], \end{aligned} \tag{15.23}$$

where $\{\alpha_t\}_{t=0}^{\infty}$ is a stochastic process of nonnegative Lagrange multipliers on the participation constraint of the villager and $\phi$ is the strictly positive multiplier on the promise-keeping constraint; that is, the moneylender must deliver on the initial promise $v$. It is useful to transform the Lagrangian by making use of the following equality, which is a version of the "partial summation formula of Abel" (see Apostol, 1975, p. 194):

$$\sum_{t=0}^{\infty} \beta^t \alpha_t \sum_{j=0}^{\infty} \beta^j u(c_{t+j}) = \sum_{t=0}^{\infty} \beta^t \mu_t u(c_t), \tag{15.24}$$

where

$$\mu_t = \mu_{t-1} + \alpha_t, \qquad \text{with} \quad \mu_{-1} = 0. \tag{15.25}$$

Formula (15.24) can be verified directly. If we substitute formula (15.24) into formula (15.23) and use the law of iterated expectations to justify $E_0 E_t(\cdot) =$

$E_0(\cdot)$, we obtain

$$J = E_{-1} \sum_{t=0}^{\infty} \beta^t \{(y_t - c_t) + (\mu_t + \phi) u(c_t) \\ -(\mu_t - \mu_{t-1}) [u(y_t) + \beta v_{\text{aut}}]\} - \phi v. \tag{15.26}$$

For a given value $v$, we seek a saddle point: a maximum with respect to $\{c_t\}$, a minimum with respect to $\{\mu_t\}$ and $\phi$. The first-order condition with respect to $c_t$ is

$$u'(c_t) = \frac{1}{\mu_t + \phi}, \tag{15.27a}$$

which is a version of equation (15.14). We also have the complementary slackness conditions

$$u(c_t) + E_t \sum_{j=1}^{\infty} \beta^j u(c_{t+j}) - [u(y_t) + \beta v_{\text{aut}}] \geq 0, \qquad = 0 \text{ if } \alpha_t > 0; \tag{15.27b}$$

$$E_{-1} \sum_{t=0}^{\infty} \beta^t u(c_t) - v = 0. \tag{15.27c}$$

Equation (15.27) together with the transition law (15.25) characterizes the solution to the moneylender's maximization problem. To explore the optimal time profile of the consumption process, we now consider some period $t \geq 0$ when $(y_t, \mu_{t-1}, \phi)$ are known. First, we tentatively try the solution $\alpha_t = 0$ (i.e., the participation constraint is not binding). Equation (15.25) instructs us then to set $\mu_t = \mu_{t-1}$, which by first-order condition (15.27a) implies that $c_t = c_{t-1}$. If this outcome satisfies participation constraint (15.27b), we have our solution for period $t$. If not, it signifies that the participation constraint binds. In other words, the solution has $\alpha_t > 0$ and $c_t > c_{t-1}$. Thus, equations (15.25) and (15.27a) immediately show us that $c_t$ is a nondecreasing random sequence, that $c_t$ stays constant when the participation constraint is not binding, and that it rises when the participation constraint binds.

The numerical computation of a solution to equation (15.26) is complicated by the fact that slackness conditions (15.27b) and (15.27c) involve conditional expectations of future endogenous variables $\{c_{t+j}\}$. Marcet and Marimon (1992) handle this complication by resorting to the parameterized expectation approach; that is, they replace the conditional expectation by a parameterized function of the state variables.[2] For our particular problem, we can recursively solve for the optimal consumption process by using the fact that the villager will ultimately receive a constant welfare level equal to $u(y_S) + \beta v_{\rm aut}$, contingent on having experienced the maximum endowment $y_S$.

First, given an endowment history $h_t$, we define the following continuation value $w_s$:

$$w_s \equiv E\Big[\sum_{\tau=0}^{\infty} \beta^\tau u(c_{t+\tau+1}) \;\Big|\; \max\{h_t\} = y_s\Big];$$

That is, $w_s$ is the continuation value of a household whose highest endowment so far has been $y_s$. As noted previously, a realization of the maximum endowment $y_S$ is associated with a consumption level $c_S$ and a continuation value $w_S$ such that

$$u(c_S) + \beta w_S = u(y_S) + \beta v_{\rm aut} \tag{15.28a}$$

$$w_S = \frac{u(c_S)}{1-\beta}, \tag{15.28b}$$

where the latter expression uses the fact that participation constraints will never be strictly binding again, which implies a constant consumption level from there on. All other continuation values and consumption levels can then be computed recursively using the formulas

$$u(c_s) + \beta w_s = u(y_s) + \beta v_{\rm aut}, \tag{15.29a}$$

$$w_s = \sum_{t=0}^{\infty} \beta^t \Big\{ \Big[\sum_{k=1}^{s} \Pi_k\Big]^{t+1} u(c_s) + \Big[\sum_{k=1}^{s} \Pi_k\Big]^t \sum_{j=s+1}^{S} \Pi_j w_j \Big\}. \tag{15.29b}$$

Solving for $u(c_s)$ from equation (15.29b) and substituting into equation (15.29a) yields

$$w_s = \Big[\sum_{k=1}^{s} \Pi_k\Big] [u(y_s) + \beta v_{\rm aut}] + \sum_{j=s+1}^{S} \Pi_j w_j. \tag{15.30}$$

Recall that $y_s > y_{s-1}$, so equation (15.30) implies that the values $\{w_s\}$ are increasing in $s$. Moreover, since the optimal contract is constrained to deliver lifetime utility $v$ by condition (15.27c), it is only $w_s \geq v$ that are relevant for our

[2] For details on the implementation of the parameterized expectation approach in a simple growth model, see den Haan and Marcet (1990).

optimal contract design. The initial consumption level $c$ of a contract promising to deliver $v$ is obtained as follows: Select the maximum continuation $w_s$ that is less than or equal to $v$, and replace $(w_s, c_s)$ by $(v, c)$ in the corresponding equation (15.29b) [or equation (15.28b) if $v \geq w_S$].

Marcet and Marimon (1992, 1999) describe a variety of other examples using the Lagrangian method. See Kehoe and Perri (1998) for an application to an international trade model.

## A closed system

Kocherlakota (1996) studied a model with no moneylender. There are equal numbers of two types of households. Each of the households has the preferences, endowments, and autarkic utility possibilities described previously. However, the endowments of the two types of households are perfectly negatively correlated. Whenever a household of type 1 receives $y_s$, a household of type 2 receives $1-y_s$. We assume that $y_s \in [0,1]$, that the distribution of $y_t$ is i.i.d. over time, and that the distribution of $y_t$ is identical to that of $1-y_t$.[3] Also, now the planner does not have access to outside funds or borrowing and lending opportunities, but is confined simply to reallocating consumption goods between the two types of households. This limitation leads to two participation constraints. At time $t$, the type 1 household receives endowment $y_t$ and consumption $c_t$, while the type 2 household receives $1-y_t$ and $1-c_t$.

Kocherlakota formulated the contract design problem as a dynamic program, where the state of the system is the value $v$ assigned to a type 1 agent. Let $P(v)$ denote the expected discounted utility of a type 2 agent when a type 1 agent is awarded promised utility $v$. Then the Bellman equation is

$$P(v) = \max_{\{c_s, w_s\}} \sum_{s=1}^{S} \Pi_s \{u(1-c_s) + \beta P(w_s)\} \tag{15.31}$$

subject to

$$\sum_{s=1}^{S} \Pi_s [u(c_s) + \beta w_s] \geq v, \tag{15.32}$$

$$u(c_s) + \beta w_s \geq u(y_s) + \beta v_{\text{aut}}, \quad s = 1, \ldots, S \tag{15.33}$$

[3] This last assumption is made simply to make $v_{\text{aut}}$ the reservation value for both types of consumer; it is easy to relax this assumption.

$$u(1-c_s)+\beta P(w_s)\geq u(1-y_s)+\beta v_{\rm aut}, \quad s=1,\ldots,S \tag{15.34}$$

$$c_s \in [0,1] \tag{15.35}$$

$$w_s \in [v_{\rm aut}, v_{\rm max}]. \tag{15.36}$$

Here expression (15.32) is the promise-keeping constraint; expression (15.33) is the participation constraint for the type 1 agent; and expression (15.34) is the participation constraint for the type 2 agent. The set of feasible $c_s$ is given by expression (15.35), and expression (15.36) imposes bounds on $w_s$. The type 1 agent cannot be awarded a promised utility below the autarkic level $v_{\rm aut}$, and there is an upper bound $v_{\rm max}$ above which the other agent, the type 2 agent, cannot be induced to participate.

Attaching Lagrange multipliers $\mu, \lambda_s, \theta_s$ to expressions (15.32), (15.33), and (15.34), and assuming that $P(v)$ is differentiable, the first-order conditions at an interior solution are

$$c_s: \quad -(\Pi_s+\theta_s)u'(1-c_s)+(\lambda_s+\Pi_s\mu)u'(c_s)=0 \tag{15.37}$$

$$w_s: \quad (\Pi_s+\theta_s)P'(w_s)+(\lambda_s+\Pi_s\mu)=0. \tag{15.38}$$

By the envelope theorem,

$$P'(v)=-\mu. \tag{15.39}$$

Equations (15.37) and (15.38) imply the following relationship between $c_s$ and $w_s$:

$$-\frac{u'(1-c_s)}{u'(c_s)}=P'(w_s). \tag{15.40}$$

Similar to equation (15.14), condition (15.40) characterizes efficient trade-offs but this time it equalizes the type 1 and type 2 agents' marginal rates of substitution between $c_s$ and $w_s$, that is, $u'(c_s)/\beta=-u'(1-c_s)/[\beta P'(w_s)]$. The concavity of $P$ and $u$ means that equation (15.40) traces out a positively sloped contract curve in the $c,w$ plane, as depicted in Figure 15.3.

For a given $v$, at most one of the participation constraints (15.33) and (15.34) can bind at any state. Hence, there are three regions of interest for any given $v$:

1. Neither participation constraint binds. When $\lambda_s=\theta_s=0$, the $(c_s,w_s)$ pair is determined as the solution of

$$u'(1-c_s)=\mu u'(c_s)$$
$$-\mu=P'(w_s)=P'(v),$$

so that this is a situation in which consumption is independent of the endowment, and promises are not changed for either consumer. The first equation can be expressed $u'(1-c_s)/u'(c_s) = \mu$, and traces out the contract curve as we vary $\mu$. Here $\mu$ serves as a "temporary relative Pareto weight" for weighting the two consumers.

2. The participation constraint of a type 1 person binds ($\lambda_s > 0$), but $\theta_s = 0$. Thus, conditions (15.38) and (15.39) imply

$$P'(w_s) = P'(v) - \frac{\lambda_s}{\Pi_s}.$$

Since $P'(w_s) < P'(v) < 0$, it follows that the type 1 agent's promised value is raised, $w_s > v$, and by equation (15.40), so is his consumption level $c_s$. In other words, the contract raises both $c_s$ and $w_s$ to induce the type 1 agent to surrender some of his endowment to the planner, who transfers it to the type 2 agent. Since $P(w_s)$ is decreasing in $w_s$, the planner *reduces* both promised utility and current consumption to the type 2 agent. The type 2 agent accepts the reduction in promised utility because his current endowment is so low. The optimal $(c_s, w_s)$ pair is determined by solving equation (15.40) and the participation constraint of a type 1 agent at equality.

3. The participation constraint of a type 2 person binds ($\theta_s > 0$), but $\lambda_s = 0$. Thus, conditions (15.38) and (15.39) imply

$$P'(w_s) = \frac{\Pi_s}{\Pi_s + \theta_s} P'(v).$$

In this case the planner must lower both $w_s$ and $c_s$ from their previous period values in order to induce the type 2 agent to part with some of its endowment this period. The type 2 agent's promised utility $P(w_s)$ is raised. The optimal $(c_s, w_s)$ pair is determined by solving equation (15.40) and the participation constraint of the type 2 agent at equality.

Given $v$, the first case prevails in "intermediate levels" of $y_t$'s, the second in high-$y_t$ states, and the third in low-$y_t$ states. The optimal contract expresses $c_s, w_s$ each as nondecreasing functions of $y_s$, with the properties that there are two numbers, $\tilde{y}(v) \leq \hat{y}(v)$, each increasing functions of $v$, such that $c_s, w_s$ are each constant for $y_s \in [\tilde{y}(v), \hat{y}(v)]$, and increasing in $y_s$ otherwise. Thus, given

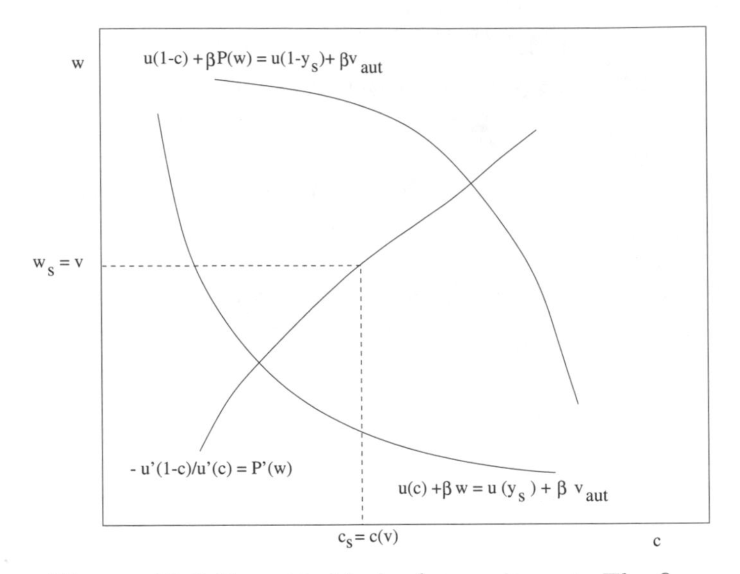


**Figure 15.3** Two-sided lack of commitment. The figure depicts "case 1" in which $v, y_s$ are such that neither household's participation constraint is binding. The two indifference curves depict $(c, w)$ and $[1-c, P(w)]$ pairs that leave households of type 1 and type 2, respectively, indifferent between remaining in the contract and withdrawing to autarky. The location of these indifference curves depends on $y_s$; higher values of $y_s$ shift both of these indifference curves toward the northeast. Given $v$, higher values of $y_s$ make it more likely that the household of type 1's participation constraint will be binding.

$v$, the planner insures the houschold fully over interior values $y_s \in [\tilde{y}(v), \hat{y}(v)]$ while risk sharing is less than complete outside of this range.

*Many households*

A village consisting of many such pairs of households could eventually converge to a stationary distribution of consumptions and continuation values, much like that of one of the incomplete markets models of chapter 14. The key feature of the two-sided lack-of-commitment model that permits such behavior is that it breaks the monotone increasing behavior of the continuation value displayed by the one-sided lack-of-commitment model.

## Endogenous borrowing constraints

Following some suggestions of Kehoe and Levine (1993), Alvarez and Jermann (1999) use a setup like Kocherlakota's as a model of complete markets with endogenous borrowing constraints. For expositional simplicity, we let $y^i(y)$ denote the endowment of a household of type $i$ when a household of type 1 receives $y$. Recall the earlier assumption that $[y^1(y), y^2(y)] = (y, 1-y)$. The state of the economy is the current endowment realization $y$ and the beginning-of-period asset holdings $A = (A_1, A_2)$, where $A_i$ is the asset holding of a household of type $i$ and $A_1 + A_2 = 0$. The assets are one-period Arrow securities, and there is a complete set of such markets. In particular, let $q(y'|y, A)$ be the price of one unit of consumption in state $y'$ tomorrow given state $(y, A)$ today. A household of type $i$ with beginning-of-period assets $a$ can purchase and sell these securities subject to the budget constraint

$$c + \sum_{y'} q(y'|y, A) a(y') \leq y^i(y) + a, \tag{15.41}$$

where $a(y')$ is the quantity purchased (if positive) or sold (if negative) of Arrow securities that pay one unit of consumption tomorrow if $y'$ is realized, and also subject to the borrowing constraints

$$a(y') \geq B^i(y', A'). \tag{15.42}$$

Notice that the borrowing constraints are history dependent as encoded by tomorrow's asset distribution, and there is one constraint for each state $(y', A')$.

The Bellman equation for the household in the decentralized economy is

$$V^i(a, y, A) = \max_{c, \{a(y')\}_{y' \in Y}} \left\{ u(c) + \beta \sum_{y'} V^i[a(y'), y', A'] \Pi(y') \right\}$$

subject to the budget constraint (15.41), the borrowing constraints (15.42), and the equilibrium law of motion for the asset distribution, $A' = \Gamma(y', y, A)$. Here we are using $\Pi(y')$ to denote the probability of income realization $y'$.

Alvarez and Jermann define a competitive equilibrium with borrowing constraints in a standard way, with the qualification that among the equilibrium objects are the borrowing constraints $B^i(y', A')$, themselves a stochastic process that the households take as given. Alvarez and Jermann show how to choose the borrowing constraints to make the allocation that solves the planning problem also be an equilibrium allocation. They do so by construction, identifying the elements of the borrowing constraints that are binding from having identified the states in the planning problem where one or another agent's participation constraint is binding.

It is easy for Alvarez and Jermann to compute the equilibrium pricing kernel from the allocation that solves the planning problem. The pricing kernel satisfies

$$q(y'|y,A) = \max_{i=1,2} \beta \frac{u'[c^i(A_i', y', A']}{u'[c^i(A_i, y, A)]} \Pi(y'),$$

where $c^i(a, y, A)$ is the consumption decision rule of a household of type $i$ with beginning-of-period assets $a$. Thus the intertemporal marginal rate of substitution of the agent whose participation constraint (or borrowing constraint) is not binding determines the pricing kernel. Alvarez and Jermann study how these asset prices behave as they vary the discount factor and the stochastic process for $y$. Also see Kocherlakota (1996a) and Kehoe and Levine (1993).

## *Social insurance with asymmetric information*

The environment of Kocherlakota has a commitment problem, because agents are free to choose autarky each period; but there is no information problem. We now study a contract design problem where the incentive problem comes not from a commitment problem, but instead from asymmetric information. As before, the planner can borrow or lend outside the village at the constant risk-free gross interest rate of $\beta^{-1}$, and each household's income $y_t$ is independently and identically distributed across time and across households. However, we now assume that both the planner and households can credibly enter into enduring and binding contracts. At the beginning of time, let $v^o$ be the expected lifetime utility that the planner promises to deliver to a household. The initial promise $v^o$ could presumably not be less than $v_{\rm aut}$, since a household would not accept a contract that gives a lower utility as compared to remaining in autarky. We defer discussing how $v^o$ is determined until the end of the section. The other new assumption here is that households have private information about their own income, and the planner can see neither their income nor their consumption. It follows that any insurance payments between the planner and a household must be based on the household's own reports about income realizations. An incentive-compatible contract makes households choose to report their incomes truthfully.

Our analysis follows the work by Thomas and Worrall (1990), who make a few additional assumptions about the preferences in expression (15.1): $u : (a, \infty) \rightarrow \mathbf{R}$ is twice continuously differentiable with $\sup u(c) < \infty$, $\inf u(c) = -\infty$, $\lim_{c \rightarrow a} u'(c) = \infty$. Thomas and Worrall also use the following special assumption:

CONDITION A: $-u''/u'$ is nonincreasing.

This is a sufficient condition to make the value function concave, as we will discuss. The roles of the other restrictions on preferences will also be revealed.

The efficient insurance contract again solves a dynamic programming problem. A planner maximizes expected discounted profits, $P(v)$, where $v$ is the household's promised utility from last period. The planner's current payment to the household, denoted $b$ (repayments from the household register as negative numbers), is a function of the state variable $v$ and the household's reported current income $y$. Let $b_s$ and $w_s$ be the payment and future utility awarded to the household if it reports income $y_s$. The optimum value function $P(v)$ obeys the functional equation

$$P(v) = \max_{\{b_s, w_s\}} \sum_{s=1}^{S} \Pi_s[-b_s + \beta P(w_s)] \tag{15.43}$$

where the maximization is subject to the constraints

$$\sum_{s=1}^{S} \Pi_s \left[u(y_s + b_s) + \beta w_s\right] = v \tag{15.44}$$

$$C_{s,k} \equiv u(y_s + b_s) + \beta w_s - \left[u(y_s + b_k) + \beta w_k\right] \geq 0, \quad s, k \in \mathbf{S} \tag{15.45}$$

$$b_s \in [a - y_s, \infty], \quad s \in \mathbf{S} \tag{15.46}$$

$$w_s \in [-\infty, v_{\max}], \quad s \in \mathbf{S} \tag{15.47}$$

where $v_{\max} = \sup u(c)/(1 - \beta)$. Equation (15.44) is the "promise-keeping" constraint guaranteeing that the promised utility $v$ is delivered. Note that the earlier weak inequality in (15.5) is replaced by an equality. The planner cannot award a higher utility than $v$ because it might then violate incentive-compatibility constraint for telling the truth in earlier periods. Constraint set (15.45) ensures that the households have no incentive to lie. Here $s$ is the actual income state, and $k$ is the reported income state. Note also that there are no "participation constraints" like expression (15.6) in the Kocherlakota model, an absence that reflects the assumption that both parties are committed to the contract.

It is instructive to establish bounds for the value function $P(v)$. Consider first a contract that pays a constant amount $\bar{b}$ in all periods, where $\bar{b}$ satisfies $\sum_{s=1}^{S} \Pi_s u(y_s + \bar{b})/(1-\beta) = v$. It is trivially incentive compatible and delivers the promised utility $v$. Therefore, the discounted profits from this contract, $-\bar{b}/(1-\beta)$, provide a lower bound to $P(v)$. However, $P(v)$ cannot exceed the value of the unconstrained first-best contract that pays $\bar{c} - y_s$ in all periods, where $\bar{c}$ satisfies $\sum_{s=1}^{S} \Pi_s u(\bar{c})/(1-\beta) = v$. Thus, the value function is bounded by

$$-\bar{b}(v)/(1-\beta) \;\leq\; P(v) \;\leq\; \sum_{s=1}^{S} \Pi_s [y_s - \bar{c}(v)]/(1-\beta). \tag{15.48}$$

The bounds are depicted in Figure 15.4, which also illustrates a few other properties of $P(v)$. Since $\lim_{c \to a} u'(c) = \infty$, it becomes very cheap for the planner to increase the promised utility when the current promise is very low, that is, $\lim_{v \to -\infty} P'(v) = 0$. The situation is the opposite when the household's promised utility is close to the upper bound $v_{\max}$ where the household has a low marginal utility of additional consumption, which implies that both $\lim_{v \to v_{\max}} P'(v) = -\infty$ and $\lim_{v \to v_{\max}} P(v) = -\infty$.

### *Efficiency implies $b_{s-1} \geq b_s, w_{s-1} \leq w_s$*

An incentive-compatible contract must satisfy $b_{s-1} \geq b_s$ and $w_{s-1} \leq w_s$. This requirement can be seen by adding the "downward constraint" $C_{s,s-1} \geq 0$ and the "upward constraint" $C_{s-1,s} \geq 0$ to get

$$u(y_s + b_s) \;-\; u(y_{s-1} + b_s) \;\geq\; u(y_s + b_{s-1}) \;-\; u(y_{s-1} + b_{s-1}),$$

where the concavity of $u(c)$ implies $b_s \leq b_{s-1}$. It then follows directly from $C_{s,s-1} \geq 0$ that $w_s \geq w_{s-1}$. In other words, a household reporting a low income receives a higher transfer from the planner in exchange for a lower future utility.

### *Local upward and downward constraints are enough*

Constraint set (15.45) can be simplified. We can show that if the local downward constraints $C_{s,s-1} \geq 0$ and upward constraints $C_{s,s+1} \geq 0$ hold for each $s \in \mathbf{S}$, then the global constraints $C_{s,k} \geq 0$ hold for each $s, k \in \mathbf{S}$. The argument

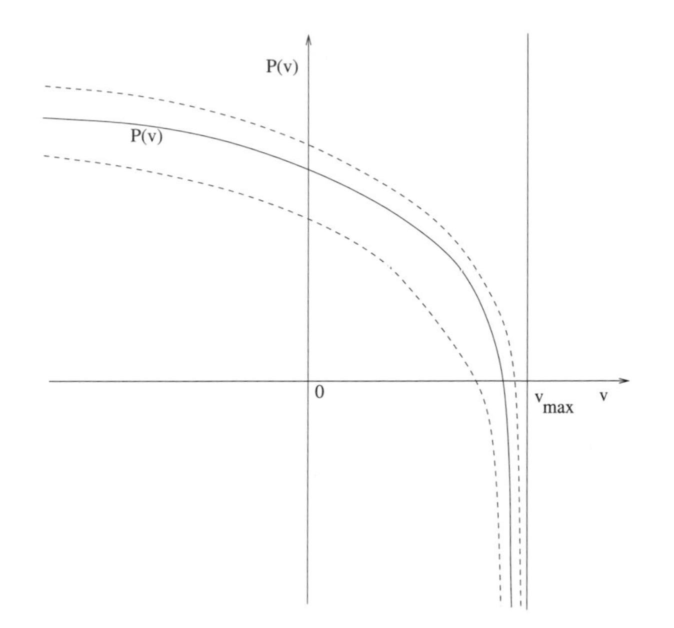


**Figure 15.4** Value function $P(v)$ and the two dashed curves depict the bounds on the value function. The vertical solid line indicates $v_{\max} = \sup u(c)/(1-\beta)$.

goes as follows: Suppose we know that the downward constraint $C_{s,k} \geq 0$ holds for some $s > k$,

$$u(y_s + b_s) + \beta w_s \ \geq \ u(y_s + b_k) + \beta w_k \, . \tag{15.49}$$

From above we know that $b_s \leq b_k$, so the concavity of $u(c)$ implies

$$u(y_{s+1} + b_s) \ - \ u(y_s + b_s) \ \geq \ u(y_{s+1} + b_k) \ - \ u(y_s + b_k) \, . \tag{15.50}$$

By adding expressions (15.49) and (15.50) and using the local downward constraint $C_{s+1,s} \geq 0$, we arrive at

$$u(y_{s+1} + b_{s+1}) + \beta w_{s+1} \ \geq \ u(y_{s+1} + b_k) + \beta w_k$$

that is, we have shown that the downward constraint $C_{s+1,k} \geq 0$ holds. In this recursive fashion we can verify that all global downward constraints are satisfied when the local downward constraints hold. A symmetric reasoning applies to the upward constraints. Starting from any upward constraint $C_{k,s} \geq 0$ with $k < s$, we can show that the local upward constraint $C_{k-1,k} \geq 0$ implies that the upward constraint $C_{k-1,s} \geq 0$ must also hold, and so forth.

*Concavity of P*

Thus far, we have not appealed to the concavity of the value function, but henceforth we shall have to. Thomas and Worrall showed that under Condition A, $P$ is concave.

PROPOSITION: The value function $P(v)$ is concave.

We recommend just skimming the following proof on first reading:

PROOF: Let $T(P)$ be the operator associated with the right side of equation (15.43). We would compute the optimum value function by iterating to convergence on $T$. We want to show that $T$ maps strictly concave $P$ to strictly concave function $T(P)$. Thomas and Worrall use the following argument:

Let $P_{k-1}(v)$ be the $k-1$ iterate on $T$. Assume that $P_{k-1}(v)$ is strictly concave. We want to show that $P_k$ is strictly concave. Consider any $v^o$ and $v'$ with associated contracts $(b_s^o, w_s^o)_{s\in S}$, $(b_s', w_s')_{s\in S}$. Let $w_s^* = \delta w_s^o + (1-\delta)w_s'$ and define $b_s^*$ by $u(b_s^* + y_s) = \delta u(b_s^o + y_s) + (1-\delta)u(b_s' + y_s)$ where $\delta \in (0,1)$. Therefore, $(b_s^*, w_s^*)_{s\in S}$ gives the borrower a utility that is the weighted average of the two utilities, and gives the lender no less than the average utility $\delta P_k(v^o)+(1-\delta)P_k(v')$. Then $C_{s,s-1}^* = \delta C_{s,s-1}^o + (1-\delta)C_{s,s-1}' + [\delta u(b_{s-1}^o + y_s) + (1-\delta)u(b_{s-1}' + y_s) - u(b_{s-1}^* + y_s)]$. Because the downward constraints $C_{s,s-1}^o$ and $C_{s,s-1}'$ are satisfied, and because the third term is nonnegative under Condition A, the downward incentive constraints $C_{s,s-1}^* \geq 0$ are satisfied. However, $(b_s^*, w_s^*)_{s\in S}$ may violate the upward incentive constraints. But Thomas and Worrall construct a new contract from $(b_s^*, w_s^*)_{s\in S}$ that is incentive compatible and that offers both the lender and the borrower no less utility. Thus, keep $w_1$ fixed and reduce $w_2$ until $C_{2,1} = 0$ or $w_2 = w_1$. Then reduce $w_3$ in the same way, and so on. Add the constant necessary to leave $\sum_s \Pi_s w_s$ constant. This step will not make the lender worse off, by the concavity of $P_{k-1}v$. Now if $w_2 = w_1$, which implies $b_2^* > b_1^*$, reduce $b_2$ until $C_{2,1} = 0$, and proceed in the same way for $b_3$, and so on. Since $b_s + y_s > b_{s-1} + y_{s-1}$, adding a constant to each $b_s$ to leave $\sum_s \Pi_s b_s$ constant cannot make the borrowers worse off. So in this new contract, $C_{s,s-1} = 0$ and $b_{s-1} \geq b_s$. Thus, the upward constraints also hold. Strict concavity of $P_k(v)$ then follows because it is not possible to have both $b_s^o = b_s'$ and $w_s^o = w_s'$ for all

$s \in S$ and $v^o \neq v'$, so the contract $(b_s^*, w_s^*)$ yields the lender strictly more than $\delta P_k(v^o) + (1-\delta)P_k(v')$. To complete the induction argument, note that starting from $P_0(v) = 0$, $P_1(v)$ is strictly concave. Therefore, $\lim_{k=\infty} P_k(v)$ is concave. ∎

We will now turn to some properties of the optimal allocation that require strict concavity of the value function. Thomas and Worrall derive these results for the finite horizon problem with value function $P_k(v)$, which is strictly concave by the preceding proposition. In order for us to stay with the infinite horizon value function $P(v)$, we make the following assumption that the limit, $\lim_{k=\infty} P_k(v)$, is not only concave but strictly concave.

ASSUMPTION: The value function $P(v)$ is strictly concave.

Concerning the following main result that all households become impoverished in the limit, Thomas and Worrall provide proof that only requires concavity of $P(v)$ as established in the preceding proposition.

*Local downward constraints always bind*

At the optimal solution, the local downward incentive constraints always bind, while the local upward constraints never do. That is, a household is always indifferent about reporting that its income was actually a little lower than it was but would never want to report that its income was in fact higher. To see that the downward constraints must be binding, suppose to the contrary that $C_{k,k-1} > 0$ for some $k \in \mathbf{S}$. Since $b_k \leq b_{k-1}$, it must then be the case that $w_k > w_{k-1}$. Consider changing $\{b_s, w_s; s \in \mathbf{S}\}$ as follows. Keep $w_1$ fixed, and if necessary reduce $w_2$ until $C_{2,1} = 0$. Next reduce $w_3$ until $C_{3,2} = 0$, and so on, until $C_{s,s-1} = 0$ for all $s \in \mathbf{S}$. (Note that any reductions cumulate when moving up the sequence of constraints.) Thereafter, add the necessary constant to each $w_s$ to leave the overall expected value of future promises unchanged, $\sum_{s=1}^{S} \Pi_s w_s$. The new contract offers the household the same utility and is incentive compatible because $b_s \leq b_{s-1}$ and $C_{s,s-1} = 0$ together imply $C_{s-1,s} > 0$; that is, the local upward constraints hold with strict inequality. At the same time, since the mean of promised values is unchanged and the differences $(w_s - w_{s-1})$ have either been left the same or reduced, the strict concavity of the value function $P(v)$ implies that the planner's profits have increased. That is, we have reversed a mean-preserving spread in the continuation values $w$. Because $P(v)$ is strictly concave,

$\sum_{s\in\mathbf{S}} \Pi_s P(w_s)$ rises and therefore $P(v)$ rises. Thus, the original contract with a nonbinding local downward constraint could not have been an optimal solution.

*Coinsurance*

The optimal contract is characterized by coinsurance: both the household's utility and the planner's profits increase with a higher income realization:

$$u(y_s + b_s) + \beta w_s \; > \; u(y_{s-1} + b_{s-1}) + \beta w_{s-1} \tag{15.51}$$
$$-b_s + \beta P(w_s) \; \geq \; -b_{s-1} + \beta P(w_{s-1}) \, . \tag{15.52}$$

The higher utility of the household in expression (15.51) follows trivially from the downward incentive-compatibility constraint $C_{s,s-1} = 0$. Concerning the planner's profits in expression (15.52), suppose to the contrary that $-b_s + \beta P(w_s) < -b_{s-1} + \beta P(w_{s-1})$. Then replacing $(b_s, w_s)$ in the contract by $(b_{s-1}, w_{s-1})$ raises the planner's profits but leaves the household's utility unchanged because $C_{s,s-1} = 0$, and the change is also incentive compatible. Thus, an optimal contract must be such that the planner's profits weakly increase in the household's income realization.

*$P'(v)$ is a martingale*

If we let $\lambda$ and $\mu_s$, $s = 2, \ldots, S$, be the multipliers associated with the constraints (15.44) and $C_{s,s-1} \geq 0$, $s = 2, \ldots, S$, the first-order necessary conditions with respect to $b_s$ and $w_s$, $s \in \mathbf{S}$, are

$$\Pi_s \Big[ 1 - \lambda \, u'(y_s + b_s) \Big] \; = \; \mu_s \, u'(y_s + b_s) \; - \; \mu_{s+1} \, u'(y_{s+1} + b_s) \tag{15.53}$$
$$\Pi_s \Big[ P'(w_s) \; + \; \lambda \Big] \; = \; \mu_{s+1} \; - \; \mu_s \, , \tag{15.54}$$

for $s \in \mathbf{S}$, where $\mu_1 = \mu_{S+1} = 0$. (There are no constraints corresponding to $\mu_1$ and $\mu_{S+1}$.) From the envelope condition,

$$P'(v) \; = \; -\lambda \, . \tag{15.55}$$

Summing equation (15.54) over $s \in \mathbf{S}$ and using equation (15.55) yields

$$\sum_{s=1}^{S} \Pi_s \, P'(w_s) \; = \; P'(v) \, . \tag{15.56}$$

*Spreading continuation values*

An efficient contract requires that the promised future utility falls (rises) when the household reports the lowest (highest) income realization, that is, that $w_1 < v < w_S$. To show that $w_S > v$, suppose to the contrary that $w_S \leq v$. Since $w_S \geq w_s$ for all $s \in \mathbf{S}$ and $P(v)$ is strictly concave, equation (15.56) implies that $w_s = v$ for all $s \in \mathbf{S}$. The substitution of equation (15.55) into equation (15.54) then yields a zero left-hand side of equation (15.54). Moreover, the right-hand side of equation (15.54) is equal to $\mu_2$ when $s = 1$ and $-\mu_S$ when $s = S$, so we can successively unravel from the constraint set (15.54) that $\mu_s = 0$ for all $s \in \mathbf{S}$. Turning to equation (15.53), it follows that the marginal utility of consumption is equalized across income realizations, $u'(y_s + b_s) = \lambda^{-1}$ for all $s \in \mathbf{S}$. Such a consumption smoothing requires $b_{s-1} > b_s$, but from incentive compatibility, $w_{s-1} = w_s$ implies $b_{s-1} = b_s$, a contradiction. We conclude that an efficient contract must have $w_S > v$. A symmetric argument establishes $w_1 < v$.

It is understandable that the planner must spread out promises to future utility, since otherwise it would be impossible to provide any insurance in the form of contingent payments today. How the planner balances the delivery of utility today as compared to future utilities is characterized by equation (15.56). To understand this expression, consider having the planner increase the household's promised utility $v$ by one unit. One way of doing so is to increase every $w_s$ by an increment $1/\beta$ while keeping every $b_s$ constant. Such a change preserves incentive compatibility at an expected discounted cost to the planner of $\sum_{s=1}^{S} \Pi_s P'(w_s)$. By the envelope theorem, this is locally as good a way to increase $v$ as any other, and its cost is therefore equal to $P'(v)$; that is, we obtain expression (15.56). In other words, given a planner's obligation to deliver utility $v$ to the agent, it is cost-efficient to choose a balance between today's contingent deliveries of goods, $\{b_s\}$, and the bundle of future utilities, $\{w_s\}$, such that the expected marginal cost of next period's promises, $\sum_{s=1}^{S} \Pi_s P'(w_s)$, is equal to the marginal cost of the current obligation, $P'(v)$. There is no intertemporal price involved in this trade-off, since any interest earnings on postponed payments are just sufficient to compensate the agent for his own subjective rate of discounting, $(1+r) = \beta^{-1}$.

*Martingale convergence and poverty*

Expression (15.56) has an intriguing implication for the long-run tendency of a household's promised future utility. Recall that $\lim_{v \to -\infty} P'(v) = 0$ and

$\lim_{v \to v_{\max}} P'(v) = -\infty$, so $P'(v)$ in expression (15.56) is a nonpositive martingale. By a theorem of Doob (1953, p. 324), $P'(v)$ then converges almost surely. We can show that $P'(v)$ must converge to 0, so that $v$ converges to $-\infty$. Suppose to the contrary that $P'(v)$ converges to a nonzero limit, which implies that $v$ converges to a finite limit. However, this assumption contradicts our earlier result where future $w_s$ are always spread out to aid incentive compatibility. The contradiction is only avoided for $v$ converging to $-\infty$; that is, the limit of $P'(v)$ must be zero.

The result that all households become impoverished in the limit can be understood from the concavity of $P(v)$. First, if there were no asymmetric information, the least expensive way of delivering lifetime utility $v$ would be to assign the household a constant consumption stream, given by the upper bound on the value function in expression (15.48). On the one hand, the concavity of $P(v)$ and standard intertemporal considerations favor a time-invariant consumption stream. But the presence of asymmetric information makes it necessary for the planner to vary promises of future utility to induce truth telling, which is costly due to the concavity of $P(v)$. For example, as pointed out by Thomas and Worrall, if $S = 2$ the cost of spreading $w_1$ and $w_2$ an equal small amount $\epsilon$ either side of their average value $\bar{w}$, is approximately $-0.5\epsilon^2 P''(\bar{w})$.[4] In general, we cannot say how this cost differs for any two values of $\bar{w}$, but it follows from the properties of $P(v)$ at its endpoints that $\lim_{v \to -\infty} P''(v) = 0$, and $\lim_{v \to v_{\max}} P''(v) = -\infty$. Thus, the cost of spreading promised values goes to zero at one endpoint and to infinity at the other endpoint. Therefore, on the other hand, the concavity of $P(v)$ and incentive compatibility considerations favor a downward drift in future utilities and, consequently, consumption. That is, the ideal time-invariant consumption level is abandoned in favor of an expected consumption path tilted toward the present because of incentive problems.

Finally, concerning the initial utility level $v^o$, one possibility is that this value is determined in competition between insurance providers. If there are no costs associated with administering contracts, $v^o$ would then be implicitly determined

[4] The expected discounted profits of providing promised values $w_1 = \bar{w} - \epsilon$ and $w_2 = \bar{w} + \epsilon$ with equal probabilities, can be approximated with a Taylor series expansion around $\bar{w}$,

$$\sum_{s=1}^{2} \frac{1}{2} P(w_s) \approx \sum_{s=1}^{2} \frac{1}{2} \left[ P(\bar{w}) + (w_s - \bar{w}) P'(\bar{w}) + \frac{(w_s - \bar{w})^2}{2} P''(\bar{w}) \right] = P(\bar{w}) + \frac{\epsilon^2}{2} P''(\bar{w}).$$

by the zero-profit condition, $P(v^o) = 0$. It remains important that such a contract is enforceable because, as we have seen, the household will eventually want to return to autarky. However, since the contract is the solution to a dynamic programming problem where the continuation of the contract is always efficient at every date, the insurer and the household will never mutually agree to renegotiate the contract.

### *Extension to general equilibrium*

Atkeson and Lucas (1992) provide examples of closed economies where the constrained efficient allocation also has each household's expected utility converging to the minimum level with probability one. Here the social planner chooses the incentive-compatible allocation for all agents subject to a constraint that the total consumption handed out in each period to the population of households cannot exceed some constant endowment level. Households are assumed to experience unobserved idiosyncratic taste shocks $\epsilon$ that are i.i.d. over time and households. The taste shock enters multiplicatively into preferences that take either the logarithmic form $u(c,\epsilon) = \epsilon \log(c)$, the constant relative risk aversion (CRRA) form $u(c,\epsilon) = \epsilon c^\gamma/\gamma$, $\gamma < 1$, $\gamma \neq 0$, or the constant absolute risk aversion (CARA) form $u(c,\epsilon) = -\epsilon \exp(-\gamma c)$, $\gamma > 0$. The assumption of a utility function belonging to these preference families greatly simplifies the analytics of the evolution of the wealth distribution. Atkeson and Lucas show that the general equilibrium analysis of this model yields an efficient allocation that delivers an ever-increasing fraction of resources to an ever-diminishing fraction of the economy's population.

### *Comparison with self-insurance*

We have just seen how in the Thomas and Worrall model, the planner responds to the incentive problem created by the consumer's private information by putting a downward tilt into temporal consumption profiles. It is useful to recall how in the "savings problem" of chapters 13 and 14, the martingale convergence theorem was used to show that the consumption profile acquired an upward tilt coming from the motive of the consumer to self-insure.

## *Optimal unemployment compensation*

This section describes a model of optimal unemployment compensation along the lines of Shavell and Weiss (1979) and Hopenhayn and Nicolini (1997). We shall use the techniques of Hopenhayn and Nicolini to analyze a model closer to Shavell and Weiss's. An unemployed worker orders stochastic processes of consumption, search effort $\{c_t, a_t\}_{t=0}^{\infty}$ according to

$$E \sum_{t=0}^{\infty} \beta^t \left[u(c_t) - a_t\right] \tag{15.57}$$

where $\beta \in (0,1)$ and $u(c)$ is strictly increasing, twice differentiable, and strictly concave. It is required that $c_t \geq 0$ and $a_t \geq 0$. All jobs are alike and pay wage $w > 0$ units of the consumption good each period forever. Once a worker has found a job, he is beyond the grasp of the unemployment insurance agency.[5] The probability of finding a job is $p(a)$ where $p$ is an increasing and strictly concave and twice differentiable function of $a$, satisfying $p(a) \in [0,1]$ for $a \geq 0$, $p(0) = 0$. The consumption good is nonstorable. The unemployed person cannot borrow or lend and holds no assets. If the unemployed worker is to do any consumption smoothing, it has to be through the unemployment insurance agency.

*The autarky problem*

As one benchmark, we first study the fate of the unemployed worker who has no access to unemployment insurance. Suppose the worker is on his own and has no access to unemployment insurance. Because employment is an absorbing state for the worker, we work backward from that state. Let $V^e$ be the expected sum of discounted utility of an employed worker. Evidently,

$$V^e = \frac{u(w)}{(1-\beta)}. \tag{15.58}$$

Now let $V^u$ be the expected present value of utility for an unemployed worker who chooses optimally. The Bellman equation is

$$V^u = \max_{a \geq 0} \left\{u(0) - a + \beta\left[p(a)V^e + (1-p(a))V^u\right]\right\}. \tag{15.59}$$

The first-order condition for this problem is

$$\beta p'(a)\left[V^e - V^u\right] \leq 1, \tag{15.60}$$

[5] This is Shavell and Weiss's assumption, but not Hopenhayn and Nicolini's. Hopenhayn and Nicolini allow the unemployment insurance agency to impose a permanent per-period history-dependent tax on previously unemployed workers.

with equality if $a > 0$. Since there is no state variable in this infinite horizon problem, there is a time-invariant optimal search intensity $a$ and an associated value of being unemployed $V^u$, which we can denote $V_{\rm aut}$.

Equations (15.59) and (15.60) form the basis for an iterative algorithm for computing $V^u$. Let $V_j^u$ be the estimate of $V^u$ at the $j$th iteration. Use this value in equation (15.60) and solve for an estimate of effort $a_j$. Use this value in equation (15.59) to compute $V_{j+1}^u$. Iterate to convergence.

*Unemployment insurance with full information*

As another benchmark, we study the provision of insurance with full information. The insurance agency can observe and control the unemployed person's consumption and search effort. The agency wants to design an unemployment insurance contract to give the unemployed worker discounted expected value $V > V_{\rm aut}$. The planner wants to deliver value $V$ in the most efficient way, meaning the way that minimizes discounted costs, using $\beta$ as the discount factor. We formulate the optimal insurance problem recursively. Let $C(V)$ be the expected discounted costs of giving the worker value $V$. The cost function is strictly convex because a higher $V$ implies a lower marginal utility of the worker; that is, additional expected "utils" can only be granted to the worker at an increasing marginal cost in terms of the consumption good. Given $V$, the planner assigns first-period pair $(c,a)$ and promised continuation value $V^u$, should the worker be unlucky and not find a job. Both of these objects will be chosen to be functions of $V$ and to satisfy the Bellman equation

$$C(V) = \min_{c,a,V^u} \left\{c + \beta[1-p(a)]C(V^u)\right\}, \tag{15.61}$$

where the maximization is subject to

$$V \leq u(c) - a + \beta\left\{p(a)V^e + [1-p(a)]V^u\right\}. \tag{15.62}$$

Here $V^e$ is given by equation (15.58), which reflects the assumption that once the worker is employed, he is beyond the reach of the unemployment insurance agency. The right side of the Bellman equation is attained by policy functions $c = c(V), a = a(V)$, and $V^U = V^U(V)$. Equation (15.62) is a "promise-keeping" constraint, and asserts that the 3-tuple $(c,a,V^u)$ attains $V$. Let $\theta$ be the multiplier on constraint (15.62). At an interior solution, the first-order conditions with respect to $c, a$, and $V^u$, respectively, are

$$\theta = \frac{1}{u'(c)}, \tag{15.63a}$$

$$C(V^u) = \theta \left[ \frac{1}{\beta p'(a)} - (V^e - V^u) \right], \tag{15.63b}$$

$$C'(V^u) = \theta. \tag{15.63c}$$

The envelope condition $C'(V) = \theta$ and equation (15.63c) imply that $C'(V^u) = C'(V)$. Convexity of $C$ then implies that $V^u = V$. Applied repeatedly over time, $V^u = V$ makes the continuation value remain constant during the entire spell of unemployment. Equation (15.63a) determines $c$, and equation (15.63b) determines $a$, both as functions of the promised $V$. That $V^u = V$ then implies that $c$ and $a$ are held constant during the unemployment spell. Thus, the worker's consumption is "fully smoothed" during the unemployment spell. However, because $V^u$ need not equal $V^e$, the worker's consumption is not necessarily smoothed across states of employment and unemployment.

Relative to autarky, the agency delivers a higher $V^u$ by both increasing the unemployed worker's consumption $c$ and decreasing his search effort $a$. However, it is important to notice that the prescribed search effort is larger than what the worker would have chosen on his own if being promised consumption level $c$ while unemployed. This follows from equations (15.63a) and (15.63b) and the fact that the insurance scheme is costly, $C(V^u) > 0$, which imply $[\beta p'(a)]^{-1} > (V^e - V^u)$. According to the worker's first-order condition (15.60), this relationship should hold with equality for a positive search effort. The worker would therefore like to establish such an equality by lowering $a$, thereby lowering the term $[\beta p'(a)]^{-1}$ [which also lowers $(V^e - V^u)$ when the value of being unemployed $V^u$ increases]. If an equality is established before $a$ reaches zero, this would be the worker's preferred search effort; otherwise the worker would find it optimal to set $a = 0$ and live on $c$ indefinitely. In other words, since the worker does not take the cost of the insurance scheme into account, he would choose a search effort below the socially optimal one. Thus, the efficient contract hinges critically on the agency's ability to control both the unemployed worker's consumption and his search effort.

### *Unemployment insurance with asymmetric information*

Following Shavell and Weiss (1979) and Hopenhayn and Nicolini (1997), now assume that the unemployment insurance agency cannot observe or enforce $a$, though it can observe and control $c$. The worker is free to choose $a$, which puts expression (15.60) back in the picture. Given any contract, the individual will choose search effort according to the first-order condition (15.60). This fact leads the insurance agency to design the unemployment insurance contract to respect this restriction. Thus, the recursive contract design problem is now to minimize equation (15.61) subject to expression (15.62) and the incentive constraint (15.60).

Since the restrictions (15.60) and (15.62) are not linear and generally do not define a convex set, it becomes difficult to provide conditions under which the solution to the dynamic programming problem results in a convex function $C(V)$. As discussed in appendix A of this chapter, this complication can be handled by convexifying the constraint set through the introduction of lotteries. However, a common finding is that optimal plans do not involve lotteries, because convexity of the constraint set is a sufficient but not necessary condition for convexity of the cost function. Following Hopenhayn and Nicolini (1997), we therefore proceed under the assumption that $C(V)$ is strictly convex in order to characterize the optimal solution.

Let $\eta$ be the multiplier on constraint (15.60), while $\theta$ continues to denote the multiplier on constraint (15.62). At an interior solution, the first-order conditions with respect to $c, a$, and $V^u$, respectively, are[6]

$$\theta = \frac{1}{u'(c)}, \tag{15.64a}$$

$$\begin{aligned} C(V^u) &= \theta \left[ \frac{1}{\beta p'(a)} - (V^e - V^u) \right] - \eta \frac{p''(a)}{p'(a)} (V^e - V^u) \\ &= -\eta \frac{p''(a)}{p'(a)} (V^e - V^u), \end{aligned} \tag{15.64b}$$

$$C'(V^u) = \theta - \eta \frac{p'(a)}{1 - p(a)}. \tag{15.64c}$$

where the second equality in equation (15.64b) follows from strict equality of the incentive constraint (15.60) when $a > 0$. As long as the insurance scheme is associated with costs, $C(V^u) > 0$, first-order condition (15.64b) implies that the multiplier $\eta$ is strictly positive. The first-order condition (15.64c) and the envelope condition $C'(V) = \theta$ together allow us to conclude that $C'(V^u) < C'(V)$. Convexity of C then implies that $V^u < V$. After we have also used equation (15.64a), it follows that consumption of the unemployed worker must decrease over time in order to provide him with the proper incentives.

---

[6] Hopenhayn and Nicolini let the insurance agency also choose $V^e$, the continuation value from $V$, if the worker finds a job. This approach reflects their assumption that the agency can tax a previously unemployed worker after he becomes employed.

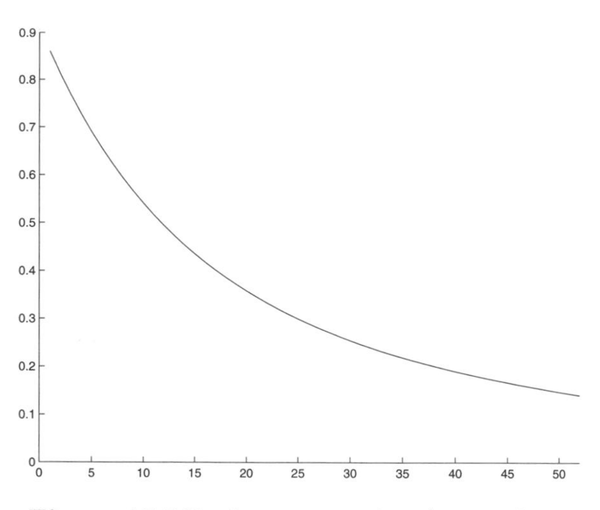


**Figure 15.5** Replacement ratio $c/w$ as a function of duration of unemployment in Shavell-Weiss model.

*Computed example*

For parameters chosen by Hopenhayn and Nicolini, Figure 15.5 displays the replacement ratio $c/w$ as a function of the duration of the unemployment spell.[7] This schedule was computed by finding the optimal policy functions

$$\begin{aligned} V_{t+1}^u &= f(V_t^u) \\ c_t &= g(V_t^u). \end{aligned}$$

and iterating on them, starting from some initial $V_0^u > V_{\rm aut}$, where $V_{\rm aut}$ is the autarky level for an unemployed worker. Notice how the replacement ratio declines with duration. Figure 15.5 sets $V_0^u$ at 16,942, a number that has to be interpreted in the context of Hopenhayn and Nicolini's parameter settings.

We computed these numbers using the parametric version studied by Hopenhayn and Nicolini. We shall briefly describe the computational strategy that we

[7] This figure was computed using the Matlab programs `hugo.m`, `hugo1a.m`, `hugofoc1.m`, `valhugo.m`. These are available in the subdirectory `hugo`, which contains a readme file. These programs were composed by various members of Economics 233 at Stanford in 1998, especially Eva Nagypal, Laura Veldkamp, and Chao Wei.

used, and shall ask the reader to try it out in exercise 15.2. The basic computational strategy is to iterate to convergence on the Bellman equation (15.61), subject to expressions (15.60) and (15.62).

Hopenhayn and Nicolini chose parameterizations and parameters as follows: They interpreted one period as one week, which led them to set $\beta = .999$. They took $u(c) = \frac{c^{(1-\sigma)}}{1-\sigma}$ and set $\sigma = .5$. They set the wage $w = 100$ and specified the hazard function to be $p(a) = 1 - \exp(-ra)$, with $r$ chosen to give a hazard rate $p(a^*) = .1$, where $a^*$ is the optimal search effort under autarky. We shall use these same settings.

In exercise 15.2 the reader is asked to solve the Bellman equation numerically. In doing so, it is useful to note that there is a natural upper bound to the set of continuation values $V^u$. To compute it, represent condition (15.60) as

$$V^u \geq V^e - [\beta p'(a)]^{-1},$$

with equality if $a > 0$. If there is zero search effort, then $V^u > V^e - [\beta p'(0)]^{-1}$. Therefore, to rule out zero search effort we require

$$V^u \leq V^e - [\beta p'(0)]^{-1}.$$

[Remember that $p''(a) < 0$.] This step gives us our upper bound for $V^u$.

To formulate the Bellman equation numerically, we suggest using the constraints to eliminate $c$ and $a$ as choice variables, thereby reducing the Bellman equation to a minimization over the one choice variable $V^u$. First express the promise-keeping constraint (15.62) as $u(c) \geq V + a - \beta\{p(a)V^e + [1 - p(a)]V^u\}$. For the preceding utility function, whenever the right side of this inequality is negative, then this promise-keeping constraint is not binding and can be satisfied with $c = 0$. This observation allows us to write

$$c = u^{-1}\left(\max\left\{0, V + a - \beta[p(a)V^e + (1 - p(a))V^u]\right\}\right). \tag{15.65}$$

Similarly, solving the inequality (15.60) for $a$, using the assumed functional form for $p(a)$, leads to

$$a = \max\left\{0, \frac{\log[r\beta(V^e - V^u)]}{r}\right\}. \tag{15.66}$$

Formulas (15.65) and (15.66) express $(c, a)$ as functions of $V$ and the continuation value $V^u$. Using these functions allows us to write the Bellman equation in $C(V)$ as

$$C(V) = \min_{V^u}\left\{c + \beta[1 - p(a)]C(V^u)\right\} \tag{15.67}$$

where $c$ and $a$ are given by equations (15.65) and (15.66).

*Interpretations*

We computed Figure 15.5 using the computational strategy described in chapter 3. The substantial downward slope in the replacement ratio in Figure 15.5 comes entirely from the incentive constraints on the planner's problem. We saw earlier that without private information, the planner would smooth consumption across unemployment states, so that the replacement ratio would be constant. In Figure 15.5, the rate at which the replacement ratio falls is designed to promote search effort, especially early in the unemployment spell. There is a "carrot and stick" aspect to the replacement schedule that we shall see again when we study credible government policy.

It is worth emphasizing that the sole motive for declining benefits over the unemployment spell is to provide the unemployed workers with the proper incentives, and is not any "punishment" for wrong behavior. In fact, the solution to the design problem ensures that the unemployed workers will act according to the optimal social contract, and falling consumption of the unlucky ones with long unemployment spells is just a necessary "price" to be paid for the common good of providing proper incentives.

*Extensions*

Hopenhayn and Nicolini studied the situation in which the planner could tax the worker after he becomes employed, where the tax could depend on the duration of unemployment. They showed that giving the planner this additional instrument substantially decreased the rate at which the replacement ratio falls during a spell of unemployment. Instead, the planner makes use of a more potent tool: a *permanent* bonus or tax after the worker becomes employed. With high discount factors, this tax or bonus is potent because it endures. In exercise 15.3, we ask the reader to set up the functional equation for Hopenhayn and Nicolini's model.

In Hopenhayn and Nicolini's model, employment is an absorbing state and there are no incentive problems once a job is found. There are not multiple spells of unemployment. Wang and Williamson (1996) built a model in which there can be multiple unemployment spells, and in which there is also an incentive problem on the job. As in Hopenhayn and Nicolini's model, search effort affects the probability of finding a job. In addition, while on a job, effort affects the probability that the job ends and the worker becomes unemployed again. Each job pays the same wage. In Wang and Williamson's setup, the contract makes the promised value keep track of the duration and number of spells of employment as well as of the number and duration of spells of unemployment. There is one contract that transcends employment and unemployment.

Rui Zhao (1999) builds a model that modifies and extends features of Wang and Williamson's. In her model, effort on the job affects output as well as the probability that the job will end. In response to this added feature, the contract responds to the history of outputs while on the job, and makes the compensation at the beginning of an unemployment spell vary directly with the compensation attained on the previous job. Again, a single contract governs the employment relation and the unemployment insurance arrangement. Zhao uses her model to help understand why unemployment insurance systems often feature a "replacement ratio" that gives more unemployment insurance payments to workers who had higher wages in their prior jobs.

## *Lending with moral hazard*

Andrew Atkeson (1991) composed a model designed to explain how, in defiance of the pattern predicted by complete markets models, low output realizations in various countries in the mid-1980s prompted international lenders to ask those countries for net repayments. A complete markets model would have net flows to a borrower during periods of bad endowment shocks. Atkeson's idea was that information and enforcement problems could produce the observed outcome. Thus, Atkeson's model combines two features of the models we have seen earlier in this chapter: incentive problems from private information and participation constraints coming from enforcement problems.

Atkeson showed that the optimal contract the remarkable feature that the job of handling enforcement and information problems is done completely by a repayment schedule without any direct manipulation of continuation values. Continuation values respond only by updating a single state variable – a measure of resources available to the borrower – that appears in the optimum value function, which in turn is affected only through the repayment schedule. Once this state variable is taken into account, promised values do not appear as independently manipulated state variables.[8]

Atkeson's model brings together several features. He studies a "borrower" who by himself is situated like a planner in a stochastic growth model, with the only

[8] To understand how Atkeson achieves this outcome, the reader should also digest the approach described in the following chapter.

vehicle for saving being a stochastic investment technology. Atkeson adds the possibility that the planner can also borrow subject to both participation and information constraints.

A borrower lives for $t = 0, 1, 2, \ldots$. He begins life with $Q_0$ units of a single good. At each date $t \geq 0$, the borrower has access to an investment technology. If $I_t \geq 0$ units of the good are invested at $t$, $Y_{t+1} = f(I_t, \varepsilon_{t+1})$ units of time $t+1$ goods are available, where $\varepsilon_{t+1}$ is an i.i.d. random variable. Let $g(Y_{t+1}, I_t)$ be the probability density of $Y_{t+1}$ conditioned on $I_t$. It is assumed that increased investment shifts the distribution of returns toward higher returns.

The borrower has preferences over consumption streams ordered by

$$U = (1-\delta) E_0 \sum_{t=0}^{\infty} \delta^t u(c_t) \tag{15.68}$$

where $\delta \in (0,1)$ and $u(\cdot)$ is increasing, strictly concave, and twice continuously differentiable.

Atkeson used various technical conditions to render his model tractable. He assumed that for each investment $I$, $g(Y, I)$ has finite support $(Y_1, \ldots, Y_n)$ with $Y_n > Y_{n-1} > \ldots > Y_1$. He assumed that $g(Y_i, I) > 0$ for all values of $I$ and all states $Y_i$, making it impossible precisely to infer $I$ from $Y$. He further assumed that the distribution $g(Y, I)$ is given by the convex combination of two underlying distributions $g_0(Y)$ and $g_1(Y)$ as follows:

$$g(Y, I) = \lambda(I) g_0(Y) + [1 - \lambda(I)] g_1(Y), \tag{15.69}$$

where $g_0(Y_i)/g_1(Y_i)$ is monotone and increasing in $i$, $0 \leq \lambda(I) \leq 1$, $\lambda'(I) > 0$, and $\lambda''(I) \leq 0$ for all $I$. Note that

$$g_I(Y, I) = \lambda'(I)[g_0(Y) - g_1(Y)],$$

where $g_I$ denotes the derivative with respect to $I$. Moreover, the assumption that increased investment shifts the distribution of returns toward higher returns implies

$$\sum_i Y_i \left[g_0(Y_i) - g_1(Y_i)\right] > 0. \tag{15.70}$$

We shall consider the borrower's choices in three environments: (1) autarky, (2) lending from risk-neutral lenders under complete observability of the borrower's choices and complete enforcement, and (3) lending under incomplete observability and limited enforcement. Environment 3 is Atkeson's. We can use environments 1 and 2 to construct bounds on the value function for performing computations described in an appendix.

*Autarky*

Suppose that there are no lenders. Thus the "borrower" is just an isolated household endowed with the technology. The household chooses $(c_t, I_t)$ to maximize expression (15.68) subject to

$$\begin{aligned} c_t + I_t &\leq Q_t \\ Q_{t+1} &= Y_{t+1}. \end{aligned}$$

The optimal value function $U(Q)$ for this problem satisfies the Bellman equation

$$U(Q) = \max_{Q \geq I \geq 0} \left\{ (1-\delta)\, u(Q - I) + \delta \sum_{Q'} U\big(Q'\big) g(Q', I) \right\}. \tag{15.71}$$

The first-order condition for $I$ is

$$-(1-\delta) u'(c) + \delta \sum_{Q'} U(Q') g_I(Q', I) = 0 \tag{15.72}$$

for $0 < I < Q$. This first-order condition implicitly defines a rule for accumulating capital under autarky.

## Investment with full insurance

We now consider an environment in which in addition to investing $I$ in the technology, the borrower can issue Arrow securities at a vector of prices $q(Y', I)$, where we let $'$ denote next period's values, and $d(Y')$ the quantity of one-period Arrow securities issued by the borrower; $d(Y')$ is the number of units of next period's consumption good that the borrower promises to deliver. We shall assume that the Arrow securities are priced by risk-neutral investors who also have one-period discount factor $\delta$. As in chapter 7, we formulate the borrower's budget constraints recursively as

$$c - \sum_{Y'} q(Y', I) d(Y') + I \leq Q \tag{15.73a}$$

$$Q' = Y' - d(Y'). \tag{15.73b}$$

Let $W(Q)$ be the optimal value for a borrower with goods $Q$. The borrower's Bellman equation is

$$\begin{aligned} W(Q) = \max_{c,I,d(Y')} \Big\{ &(1-\delta)u(c) + \delta \sum_{Y'} W[Y' - d(Y')] g(Y', I) \\ &+ \lambda [Q - c + \sum_{Y'} q(Y', I) d(Y') - I] \Big\}, \end{aligned} \tag{15.74}$$

where $\lambda$ is a Lagrange multiplier on expression (15.73a). First-order conditions with respect to $c, I, d(Y')$, respectively, are

$$c: \quad (1-\delta)u'(c) - \lambda = 0, \tag{15.75a}$$

$$I: \quad \delta \sum_{Y'} W[Y' - d(Y')] g_I(Y', I) + \lambda q_I(Y', I) d(Y') - \lambda = 0, \tag{15.75b}$$

$$d(Y'): \quad -\delta W'[Y' - d(Y')] g(Y', I) + \lambda q(Y', I) = 0. \tag{15.75c}$$

Letting risk-neutral lenders determine the price of Arrow securities implies that

$$q(Y', I) = \delta g(Y', I), \tag{15.76}$$

which in turn implies that the gross one-period risk-free interest rate is $\delta^{-1}$. At these prices for Arrow securities, it is profitable to invest in the stochastic technology until the expected rate of return on the marginal unit of investment is driven down to $\delta^{-1}$;

$$\sum_{Y'} [Y' - d(Y')] g_I(Y', I) = \delta^{-1}, \tag{15.77}$$

and after invoking equation (15.69)

$$\lambda'(I) \sum_{Y'} [Y' - d(Y')] \left( g_0(Y') - g_1(Y') \right) = \delta^{-1}.$$

This condition uniquely determines the investment level $I$, since the left side is decreasing in $I$ and must eventually approach zero because of the upper bound on $\lambda(I)$. (The investment level is strictly positive as long as the left-hand side exceeds $\delta^{-1}$ when $I = 0$.)

The first-order condition (15.75c) and the Benveniste-Scheinkman condition, $W'(Q') = (1-\delta)u'(c')$, imply the consumption-smoothing result $c' = c$. This in turn implies, via the status of $Q$ as the state variable in the Bellman equation, that $Q' = Q$. Thus, the solution has $I$ constant over time at a level determined by equation (15.77), and $c$ and the functions $d(Y')$ satisfying

$$c + I = Q + \sum_{Y'} q(Y', I) d(Y') \tag{15.78a}$$

$$d(Y') = Y' - Q \tag{15.78b}$$

The borrower borrows a constant $\sum_{Y'} q(Y', I) d(Y')$ each period, invests the same $I$ each period, and makes high repayments when $Y'$ is high and low repayments when $Y'$ is low. This is the standard full-insurance solution.

We now turn to Atkeson's setting where the borrower does better than under autarky but worse than with the loan contract under perfect enforcement and observable investment. Atkeson found a contract with value $V(Q)$ for which $U(Q) \leq V(Q) \leq W(Q)$. We shall want to compute $W(Q)$ and $U(Q)$ in order to compute the value of the borrower under the more restricted contract.

## *Limited commitment and unobserved investment*

Atkeson designed an optimal recursive contract that copes with two impediments to risk sharing: (1) moral hazard, that is, hidden action: the lender cannot observe the borrower's action $I_t$ that affects the probability distribution of returns $Y_{t+1}$; and (2) one-sided limited commitment: the borrower is free to default on the contract and can choose to revert to autarky at any state.

Each period, the borrower confronts a two-period-lived, risk-neutral lender who is endowed with $M > 0$ in each period of his life. Each lender can lend or borrow at a risk-free gross interest rate of $\delta^{-1}$ and must earn an expected return of at least $\delta^{-1}$ if he is to lend to the borrower. The lender is also willing to *borrow* at this same expected rate of return. The lender can lend up to $M$ units of consumption to the borrower in the first period of his life, and could *repay* (if the borrower lends) up to $M$ units of consumption in the second period of his life. The lender lends $b_t \leq M$ units to the borrower and gets a state-contingent repayment $d(Y_{t+1})$, where $-M \leq d(Y_{t+1})$, in the second period of his life. That the repayment is state-contingent lets the lender insure the borrower.

A lender is willing to make a one-period loan to the borrower, but only if the loan contract assures repayment. The borrower will fulfill the contract only if he wants. The lender observes $Q$, but observes neither $C$ nor $I$. Next period, the lender can observe $Y_{t+1}$. He bases the repayment on that observation.

Where $c_t + I_t - b_t = Q_t$, Atkeson's optimal recursive contract takes the form

$$d_{t+1} = d\left(Y_{t+1}, Q_t\right) \tag{15.79a}$$
$$Q_{t+1} = Y_{t+1} - d_{t+1} \tag{15.79b}$$
$$b_t = b(Q_t). \tag{15.79c}$$

The repayment schedule $d(Y_{t+1}, Q_t)$ depends only on observables and is designed to recognize the limited-commitment and moral-hazard problems.

Notice how $Q_t$ is the only state variable in the contract. Atkeson uses the apparatus of Abreu, Pearce, and Stacchetti (1990), to be discussed in chapter 16, to show that the state can be taken to be $Q_t$, and that it is not necessary to keep track of the history of past $Q$'s. Atkeson obtains the following Bellman equation. Let $V(Q)$ be the optimum value of a borrower in state $Q$ under the optimal contract. Let $A = (c, I, b, d(Y'))$, all to be chosen as functions of $Q$. The Bellman equation is

$$V(Q) = \max_{A}\Big\{(1-\delta)\, u\,(c) + \delta \sum_{Y'} V\,[Y' - d\,(Y', Q)]\, g\,(Y', I)\Big\} \tag{15.80a}$$

subject to

$$c + I - b \le Q,\ b \le M,\ -d(Y', Q) \le M,\ c \ge 0,\ I \ge 0 \tag{15.80b}$$
$$b \le \delta \sum_{Y'} d(Y')\, g\,(Y', I) \tag{15.80c}$$
$$V\,[Y' - d\,(Y')] \ge U\,(Y') \tag{15.80d}$$

$$I = \arg\max_{\tilde{I}\,\epsilon[0, Q+b]}\Big\{(1-\delta)\ u\ (Q + b - \tilde{I}) + \delta \sum_{Y'} V\,\left[Y' - d\,(Y', Q)\right] g\,(Y', \tilde{I})\Big\} \tag{15.80e}$$

Condition (15.80b) is feasibility. Condition (15.80c) is a rationality constraint for lenders: it requires that the expected value of the payoffs exceed the risk-free interest rate $\delta^{-1}$ available to the lenders. Condition (15.80d) says that in every state tomorrow, the borrower must want to comply with the contract; thus the value of affirming the contract (the left side) must be at least as great as the value of autarky. Condition (15.80e) states that the borrower chooses $I$ to maximize his expected utility under the contract.

There are many value functions $V(Q)$ and associated contracts $b(Q), d(Y', Q)$ that satisfy conditions (15.80). Because we want the optimal contract, we want the $V(Q)$ that is the largest (hopefully, pointwise). The usual strategy of iterating on the Bellman equation, starting from an arbitrary guess $V^0(Q)$, say, 0, will not work in this case because high candidate continuation values $V(Q')$ are needed to support good current-period outcomes. But a modified version of the usual iterative strategy does work, which is to make sure that we start with a large enough initial guess at the continuation value function $V^0(Q')$. Atkeson (1988, 1991) verified that the optimal contract can be constructed by iterating to convergence on conditions (15.80), provided that the iterations begin from a large enough initial value function $V^0(Q)$. (See the appendix for a computational exercise using Akkeson's iterative stategy.) He adapted ideas from Abreu, Pearce, and Stacchetti (1990) to show this result.[9] In the next subsection, we shall form a Lagrangian in which the role of continuation values is explicitly accounted for.

*Binding participation constraint*

Atkeson motivated his work as an effort to explain why countries often experience capital outflows in the very-low-income periods in which they would be borrowing *more* in a complete markets setting. The optimal contract associated with conditions (15.80) has the feature that Atkeson sought: the borrower makes net repayments $d_t > b_t$ in states with low output realizations.

Atkeson establishes this property using the following argument. First, he assumes the following condition about the outcomes:

ASSUMPTION: For the optimum contract

$$\sum_i d_i \left[g_0(Y_i) - g_1(Y_i)\right] \geq 0. \tag{15.81}$$

This makes the value of repayments increasing in investment.

Atkeson assumes conditions (15.81) and (15.69) to justify using the first-order condition for the right side of equation (15.80e) to characterize the investment decision. The first-order condition for investment is

$$-(1-\delta)u'(Q+b-I) + \delta \sum_i V(Y_i - d_i) g_I(Y_i, I) = 0.$$

[9] See chapter 16 for some work with the Abreu, Pearce, and Stacchetti structure, and for how, with history dependence, dynamic programming principles direct attention to *sets* of continuation value functions. The need to handle a set of continuation values appropriately is why Atkeson must initiate his iterations from a sufficiently high initial value function.

*A property of the repayment schedule*

To deduce a key property of the repayment schedule, we will follow Atkeson by introducing a continuation value $\tilde{V}$ as an additional choice variable in a programming problem for the contract design. Atkeson shows how (15.80) can be viewed as the outcome of a more elementary programming problem in which the contract designer chooses the continuation value function from a set of permissible values.[10] Following Atkeson, let $U_d(Y_i) \equiv \tilde{V}(Y_i - d(Y_i))$ where $\tilde{V}(Y_i - d(Y_i))$ is a continuation value function to be chosen by the author of the contract. Atkeson shows that we can regard the contract author choosing a continuation value function along with the elements of $A$, but that in the end it will be optimal for him to choose the continuation values to satisfy the Bellman equation (15.80). For the purposes of deriving a feature of the optimal repayment schedule, we follow Atkeson and regard the $U_d(Y_i)$'s as choice variables. They must satisfy $U_d(Y_i) \leq V(Y_i - d_i)$, where $V(Y_i - d_i)$ satisfies the Bellman equation (15.80). Let $\mu$ be a vector of multipliers and form the Lagrangian

$$\begin{aligned} J(A, U_d, \mu) = {} & (1-\delta)u(c) + \delta \sum_i U_d(Y_i) g(Y_i, I) \\ & + \mu_1 (Q + b - c - I) \\ & + \mu_2 \Big[ \delta \sum_i d_i g(Y_i, I) - b \Big] \\ & + \delta \sum_i \mu_3(Y_i) g(Y_i, I) \big[ U_d(Y_i) - U(Y_i) \big] \\ & + \mu_4 \Big[ -(1-\delta) u'(Q + b - I) + \delta \sum_i U_d(Y_i) g_I(Y_i, I) \Big] \\ & + \delta \sum_i \mu_5(Y_i) g(Y_i, I) \big[ V(Y_i - d_i) - U_d(Y_i) \big], \end{aligned} \tag{15.82}$$

where the $\mu_j$'s are nonnegative Lagrange multipliers. To investigate the consequences of when the participation constraint is binding, rearrange the first-order condition with respect to $U_d(Y_i)$ to get

$$1 + \mu_4 \frac{g_I(Y_i, I)}{g(Y_i, I)} = \mu_5(Y_i) - \mu_3(Y_i), \tag{15.83}$$

[10] See Atkeson (1991) and chapter 16.

where $g_I/g = \lambda'(I)\left[\frac{g_0(Y_i)-g_1(Y_i)}{g(Y,I)}\right]$, which is negative for low $Y_i$ and positive for high $Y_i$. All the multipliers are nonnegative. Then evidently when the left side of equation (15.83) is *negative*, we must have $\mu_3(Y_i) > 0$, so that condition (15.80d) is binding and $U_d(Y_i) = U(Y_i)$. Therefore, $V(Y_i - d_i) = U(Y_i)$ for states with $\mu_3(Y_i) > 0$. In states $Y_i$ where $\mu_3(Y_i) > 0$, new loans $b'$ cannot exceed repayments $d_i = d(Y_i)$, which follows from the programs that define $V(\cdot)$ and $U(\cdot)$.

Thus, in low output realizations states where condition (15.80d) is binding, the borrower experiences a "capital outflow." This outcome obviously differs from the complete markets contract (15.78b), which provides a "capital inflow" to the lender in low output states. That the pair of functions $b_t = b(Q_t)$, $d_t = d(Y_t, Q_{t-1})$ forming the optimal contract specifies repayments in those distressed states is how the contract provides the incentive for the borrower to make investment decisions that reduce the likelihood that $Y_t, Q_t, Q_{t-1}$ will enter the region that triggers those capital outflows under distress.

We remind the reader of the remarkable feature of Atkeson's contract that the repayment schedule and the state variable $Q$ 'do all the work.' Atkeson's contract manages to encode all history dependence in an extremely economical fashion. In the end, there is no need, as occurred in the problems that we studied earlier in this chapter, to add a promised value as an independent state variable.

## Concluding remarks

The idea of using promised values as a state variable has made it possible to use dynamic programming to study problems with history dependence. In this chapter we have studied how using a promised value as a state variable helps to study optimal risk-sharing arrangements when there are incentive problems coming from limited enforcement or limited information or both. We have referred to various papers, such as Zhao (1999), that extend some of the models studied in this chapter. We also recommend Bond and Park (1998), who study the gradualism of some trade liberalization agreements between large and small countries in terms of a model where the agreement manipulates a country's continuation valuation to manage incentive problems. Krueger (1999) uses a model with participation constraints and Arrow securities to confront some empirical puzzles about consumption. The following chapter uses dynamic programming with promised values as state variables to study "reputational macroeconomics."

There are two appendices to this chapter. The first describes two more models, with the aim partly of describing a useful device for convexifying constraint sets in recursive contracts problems. The second describes a possible computational strategy for Atkeson's model.

## Appendix A: Historical development

### Spear and Srivastava

Spear and Srivastava (1987) introduced the following recursive formulation of an infinitely repeated, discounted repeated principal-agent problem: A *principal* owns a technology that produces output $q_t$ at time $t$, where $q_t$ is determined by a family of c.d.f.'s $F(q_t|a_t)$, and $a_t$ is an action taken at the beginning of $t$ by an *agent* who operates the technology. The principal has access to an outside loan market with constant risk-free gross interest rate $\beta^{-1}$. The agent has preferences over consumption streams ordered by

$$E_0 \sum_{t=0}^{\infty} \beta^t u(c_t, a_t).$$

The principal is risk neutral and offers a contract to the agent designed to maximize

$$E_0 \sum_{t=0}^{\infty} \beta^t \{q_t - c_t\}$$

where $c_t$ is the payment from the principal to the agent at $t$.

### Timing

Let $w$ denote the discounted utility promised to the agent at the beginning of the period. Given $w$, the principal selects three functions $a(w)$, $c(w,q)$, and $\tilde{w}(w,q)$ determining the current action $a_t = a(w_t)$, the current consumption $c = c(w_t, q_t)$, and a promised utility $w_{t+1} = \tilde{w}(w_t, q_t)$. The choice of the three functions $a(w)$, $c(w,q)$, and $\tilde{w}(w,q)$ must satisfy the following two sets of constraints:

$$w = \int \{u[c(w,q), a(w)] + \beta \tilde{w}(w,q)\}\, dF[q|a(w)] \tag{15.84}$$

and

$$\int \{u[c(w,q),a(w)] + \beta\tilde{w}(w,q)\}\ dF[q|a(w)]$$
$$\geq \int \{u[c(w,q),\hat{a}] + \beta\tilde{w}(w,q)\}dF(q|\hat{a})\ = \forall \quad \hat{a} \in A. \tag{15.85}$$

Equation (15.84) requires the contract to deliver the promised level of discounted utility. Equation (15.85) is the *incentive compatibility* constraint requiring the agent to want to deliver the amount of effort called for in the contract. Let $v(w)$ be the value to the principal associated with promising discounted utility $w$ to the agent. The principal's Bellman equation is

$$v(w) = \max_{a,c,\tilde{w}} \{q - c(w,q) + \beta\ v[\tilde{w}(w,q)]\}\ dF[q|a(w)] \tag{15.86}$$

where the maximization is over functions $a(w)$, $c(w,q)$, and $\tilde{w}(w,q)$ and is subject to the constraints (15.84) and (15.85). This value function $v(w)$ and the associated optimum policy functions are to be solved by iterating on the Bellman equation (15.86).

*Use of lotteries*

In various implementations of this approach, a difficulty can be that the constraint set fails to be convex as a consequence of the structure of the incentive constraints. This problem has been overcome by Phelan and Townsend (1991) by convexifying the constraint set through randomization. Thus, Phelan and Townsend simplify the problem by extending the principal's choice to the space of lotteries over actions $a$ and outcomes $c, w'$. To introduce Phelan and Townsend's formulation, let $P(q|a)$ be a family of discrete probability distributions over discrete spaces of outputs and actions $Q, A$; and imagine that consumption and values are also constrained to lie in discrete spaces $C, W$, respectively. Phelan and Townsend instruct the principal to choose a probability distribution $\Pi(a,q,c,w')$ subject first to the constraint that for all fixed $(\bar{a},\bar{q})$

$$\sum_{C\times W} \Pi(\bar{a},\bar{q},c,w') = P(\bar{q}|\bar{a}) \sum_{Q\times C\times W} \Pi(\bar{a},q,c,w') \tag{15.87a}$$

$$\Pi(a,q,c,w') \geq 0 \tag{15.87b}$$

$$\sum_{A\times Q\times C\times W} \Pi(a,q,c,w') = 1. \tag{15.87c}$$

Equation (15.87a) simply states that $\text{Prob}(\bar{a},\bar{q}) = \text{Prob}(\bar{q}|\bar{a})\text{Prob}(\bar{a})$. The remaining pieces of (15.87) just require that "probabilities are probabilities." The counterpart of Spear-Srivastava's equation (15.84) is

$$w = \sum_{A\times Q\times C\times W} \{u(c,a)+\beta w'\}\ \Pi(a,q,c,w'). \tag{15.88}$$

The counterpart to Spear-Srivastava's equation (15.85) for each $a, \hat{a}$ is

$$\sum_{Q\times C\times W} \{u(c,a)+\beta w'\}\ \Pi(c,w'|q,a)P(q|a)$$
$$\geq \sum_{Q\times C\times W} \{u(c,a)+\beta w'\}\ \Pi(c,w'|q,a)P(q|\hat{a}).$$

Here $\Pi(c,w'|q,a)P(q|\hat{a})$ is the probability of $(c,w',q)$ if the agent claims to be working $a$ but is actually working $\hat{a}$. Express

$$\Pi(c,w'|q,a)P(q|\hat{a}) =$$
$$\Pi(c,w'|q,a)P(q|a)\ \frac{P(q|\hat{a})}{P(q|a)} = \Pi(c,w',q|a)\ \cdot\ \frac{P(q|\hat{a})}{P(q|a)}.$$

To write the incentive constraint as

$$\sum_{Q\times C\times W} \{u(c,a)+\beta w'\}\Pi(c,w',q|a)$$
$$\geq \sum_{Q\times C\times W} \{u(c,\hat{a})+\beta w'\}\ \Pi(c,w',q|\hat{a})\ \cdot\ \frac{P(q|\hat{a})}{P(q|a)}.$$

Multiplying both sides by the unconditional probability $P(a)$ gives expression (15.89).

$$\sum_{Q\times C\times W} \{u(c,a)+\beta w'\}\ \Pi(a,q,c,w')$$
$$\geq \sum_{Q\times C\times W} \{u(c,\hat{a})+\beta w'\}\ \frac{P(q|\hat{a})}{P(q|a)}\ \Pi(a,q,c,w') \tag{15.89}$$

The Bellman equation for the principal's problem is

$$v(w) = \max_{\Pi}\{(q-c)+\beta v(w')\}\Pi(a,q,c,w'), \tag{15.90}$$

where the maximization is over the probabilities $\Pi(a,q,c,w')$ subject to equations (15.87), (15.88), and (15.89). The problem on the right side of equation (15.90) is a linear programming problem. Think of each of $(a,q,c,w')$ being constrained to a discrete grid of points. Then, for example, the term $(q-c)+\beta v(w')$ on the right side of equation (15.90) can be represented as a *fixed* vector that multiplies a vectorized version of the probabilities $\Pi(a,q,c,w')$. Similarly, each of the constraints (15.87), (15.88), and (15.89) can be represented as a linear inequality in the choice variables, the probabilities $\Pi$. Phelan and Townsend compute solutions of these linear programs to iterate on the Bellman equation (15.90). Note that at each step of the iteration on the Bellman equation, there is one linear program to be solved for each point $w$ in the space of grid values for $W$.

In practice, Phelan and Townsend have found that lotteries are often redundant in the sense that most of the $\Pi(a,q,c,w')$'s are zero, and a few are one.

## Appendix B: Computations for Atkeson's model

It is instructive to compute a numerical example of the optimal contract. We follow Atkeson's (1988) work with the following numerical example. Assume $u(c) = 2c^{.5}, \lambda(I) = \left(\frac{I}{Y_n+2M}\right)^{.5}, g_i(Y_j) = \frac{\exp^{-\alpha_i Y_j}}{\sum_{k=1}^{n}\exp^{-\alpha_i Y_k}}$ with $n=5, Y_1 = 100, Y_n = 200, M = 100, \alpha_1 = \alpha_2 = -.5, \delta = .9$. Here is a version of Atkeson's numerical algorithm.

1. First solve the Bellman equation (15.71) and (15.72) for the autarky value $U(Q)$. Use a polynomial for the value function.[11]

2. Solve the Bellman equation for the full-insurance setting for the value function $W(Q)$ as follows. First, solve equation (15.77) for $I$. Then solve equation (15.78b) for $d(Y') = Y' - Q$ and compute $c = c(Q)$ from (15.78a). Since $c$ is constant, $W(Q) = u[c(Q)]$.

Now solve the Bellman equation for the contract with limited commitment and unobserved action. First, approximate $V(Q)$ by a polynomial, using the method

[11] We recommend the Schumaker shape-preserving spline mentioned in chapter 3 and described by Judd (1998).

described in chapter 3. Next, iterate on the Bellman equation, starting from initial value function $V^0(Q) = W(Q)$ computed earlier. As Atkeson shows, it is important to start with a value function *above* $V(Q)$. We know that $W(Q) \geq V(Q)$.

Use the following steps:

1. Let $V^j(Q)$ be the value function at the $j$th iteration. Let $d$ be the vector $[\, d_1 \quad \ldots \quad d_n \,]'$. Define

$$X(d) = \sum_i V^j(Y_i - d_i)[g_0(Y_i) - g_1(Y_i)]. \tag{15.91}$$

The first-order condition for the borrower's problem (15.80e) is

$$-(1-\delta)u'(Q+b-I) + \delta\lambda'(I)X \geq 0, \quad = 0 \text{ if } I > 0.$$

Given a candidate continuation value function $V^j$, a value $Q$, and $b, d_1, \ldots, d_n$, solve the borrower's first-order condition for a function

$$I = f(b, d_1, \ldots, d_n; Q).$$

Evidently, when $X(d) < 0$, $I = 0$. From equation (15.91) and the particular example,

$$I = f(b, d; Q) = \frac{\delta^2 (Y_n + 2M) X(d)^2}{4(1-\delta)^2 + \delta^2 (Y_n + 2M) X(d)^2} (Q + b). \tag{15.92}$$

Summarize this equation in a Matlab function.

2. Use equation (15.92) and the constraint (15.80c) at equality to form

$$b = \delta \sum_i d_i g[Y_i, f(b, d)].$$

Solve this equation for a new function

$$b = m(d). \tag{15.93}$$

3. Write one step on the Bellman equation as

$$
\begin{aligned}
V^{j+1}(Q) = \max_{d} \Big\{ & (1-\delta)u\big[Q + m(d) - f(m(d),d)\big] \\
& + \delta \sum_i V^j(Y_i - d_i) g\big[Y_i, f(m(d),d)\big] \\
& - \sum_i \theta_i \Big[\max\big(0, U(Y_i) - V^j(Y_i - d_i)\big)\Big] \\
& - \sum_i \eta_i \max[0, -d_i - M] - \eta_0 \max[0, m(d) - M] \Big\},
\end{aligned} \tag{15.94}
$$

where $V^j(Q)$ is the value function at the $j$th iteration, and $\theta_i > 0, \eta_i$ are positive penalty parameters designed to enforce the participation constraints (15.80d) and the restrictions on the size of borrowing and repayments. The idea is to set the $\theta_i$'s and $\eta_i$'s large enough to assure that $d$ is set so that constraint (15.80d) is satisfied for all $i$.

## Exercises

*Exercise 15.1* **Lagrangian method with two-sided no commitment**

Consider the model of Kocherlakota with two-sided lack of commitment. There are two consumers, each having preferences $E_0 \sum_{t=0}^{\infty} \beta^t u[c_i(t)]$, where $u$ is increasing, twice differentiable, and strictly concave, and where $c_i(t)$ is the consumption of consumer $i$. The good is not storable, and the consumption allocation must satisfy $c_1(t) + c_2(t) \leq 1$. In period $t$, consumer 1 receives an endowment of $y_t \in [0,1]$, and consumer 2 receives an endowment of $1 - y_t$. Assume that $y_t$ is i.i.d. over time and is distributed according to the discrete distribution $\text{Prob}(y_t = y_s) = \Pi_s$. At the start of each period, after the realization of $y_s$ but before consumption has occurred, each consumer is free to walk away from the loan contract.

**a.** Find expressions for the expected value of autarky, before the state $y_s$ is revealed, for consumers of each type. (*Note:* These need not be equal.)

**b.** Using the Lagrangian method, formulate the contract design problem of finding an optimal allocation that for each history respects feasibility and the participation constraints of the two types of consumers.

**c.** Use the Lagrangian method to characterize the optimal contract as completely as you can.

*Exercise 15.2* **Optimal unemployment compensation**

**a.** Write a program to compute the autarky solution, and use it to reproduce Hopenhayn and Nicolini's calibration of $r$, as described in text.

**b.** Use your calibration from part a. Write a program to compute the optimum value function $C(V)$ for the insurance design problem with incomplete information. Use the program to form versions of Hopenhayn and Nicolini's table 1, column 4 for three different initial values of $V$, chosen by you to belong to the set $(V_{\rm aut}, V^e)$.

*Exercise 15.3* **Taxation after employment**

Show how the functional equation (15.61), (15.62) would be modified if the planner were permitted to tax workers after they became employed.

*Exercise 15.4* **A model of Dixit, Grossman, and Gul**

For each date $t \geq 0$, two political parties divide a "pie" of fixed size $1$. Party 1 receives a sequence of shares $y = \{y_t\}_{t\geq 0}$ and has utility function $E\sum_{t=0}^{\infty} \beta^t U(y_t)$, where $\beta \in (0,1)$, $E$ is the mathematical expectation operator, and $U(\cdot)$ is an increasing, strictly concave, twice differentiable period utility function. Party 2 receives share $1-y_t$ and has utility function $E\sum_{t=0}^{\infty} \beta^t U(1-y_t)$. A state variable $X_t$ is governed by a Markov process; $X$ resides in one of $K$ states. There is a partition $S_1, S_2$ of the state space. If $X_t \in S_1$, party 1 chooses the division $y_t, 1-y_t$, where $y_t$ is the share of party $1$. If $X_t \in S_2$, party 2 chooses the division. At each point in time, each party has the option of choosing "autarky," in which case its share is 1 when it is in power and zero when it is not in power.

Formulate the optimal history-dependent sharing rule as a recursive contract. Formulate the Bellman equation. *Hint:* Let $V[u_0(x), x]$ be the optimal value for party 1 in state $x$ when party 2 is promised value $u_0(x)$.

*Exercise 15.5* **Two-state numerical example of social insurance**

Consider an endowment economy populated by a large number of individuals with identical preferences,

$$E\sum_{t=0}^{\infty} \beta^t u(c_t) = E\sum_{t=0}^{\infty} \beta^t \left(4c_t - \frac{c_t^2}{2}\right), \qquad \text{with } \beta = 0.8.$$

With respect to endowments, the individuals are divided into two types of equal size. All individuals of a particular type receive 0 goods with probability 0.5 and 2 goods with probability 0.5 in any given period. The endowments of the two types of individuals are perfectly negatively correlated so that the per capita endowment is always 1 good in every period.

The social planner attaches the same welfare weight to all individuals. Without access to outside funds or borrowing and lending opportunities, the social planner seeks to provide insurance by simply reallocating goods between the two types of individuals. The design of the social insurance contract is constrained by a lack of commitment on behalf of the individuals. The individuals are free to walk away from any social arrangement, but they must then live in autarky evermore.

**a.** Compute the optimal insurance contract when the social planner lacks memory; that is, transfers in any given period can be a function only of the current endowment realization.

**b.** Can the insurance contract in part a be improved if we allow for history-dependent transfers?

**c.** Explain how the optimal contract changes when the parameter $\beta$ goes to one. Explain how the optimal contract changes when the parameter $\beta$ goes to zero.

*Exercise 15.6* **Optimal unemployment compensation with unobservable wage offers**

Consider an unemployed person with preferences given by

$$E \sum_{t=0}^{\infty} \beta^t u(c_t)\,,$$

where $\beta \in (0,1)$ is a subjective discount factor, $c_t \geq 0$ is consumption at time $t$, and the utility function $u(c)$ is strictly increasing, twice differentiable, and strictly concave. Each period the worker draws one offer $w$ from a uniform wage distribution on the domain $[w_L, w_H]$ with $0 \leq w_L < w_H < \infty$. Let the cumulative density function be denoted $F(x) = \text{prob}\{w \leq x\}$, and denote its density by $f$, which is constant on the domain $[w_L, w_H]$. After the worker has accepted a wage offer $w$, he receives the wage $w$ per period forever. He is then beyond the grasp of the unemployment insurance agency. During the unemployment spell, any consumption smoothing has to be done through the unemployment insurance agency because the worker holds no assets and cannot borrow or lend.

**a.** Characterize the worker's optimal reservation wage when he is entitled to a time-invariant unemployment compensation $b$ of indefinite duration.

**b.** Characterize the optimal unemployment compensation scheme under full information. That is, we assume that the insurance agency can observe and control the unemployed worker's consumption and reservation wage.

**c.** Characterize the optimal unemployment compensation scheme under asymmetric information where the insurance agency cannot observe wage offers, though it can observe and control the unemployed worker's consumption. Discuss the optimal time profile of the unemployed worker's consumption level.

# 16
# Credible Government Policies

## Introduction

In situations with strategic interactions, the *timing* of choices matters.[1] Kydland and Prescott (1977) opened the modern discussion of time consistency in macroeconomics with some examples that show how outcomes differ in otherwise identical economies when the assumptions about the timing of government policy choices are altered. In particular, they compared a timing protocol in which a government determines its (possibly state-contingent) policies once and for all at the beginning of the economy with one in which the government chooses sequentially. Because outcomes are worse when the government chooses sequentially, Kydland and Prescott's examples indicate the value to a government of having access to a commitment technology that binds it not to choose sequentially.

Much subsequent work on time consistency has focused on the extent to which it is possible for a reputation to substitute for a commitment technology, even when the government chooses sequentially.[2] The issue is whether incentives and expectations can be arranged so that a government has a reputation that it protects by adhering to an expected pattern of behavior, and that is replaced by a worse reputation if it does not behave as expected. The literature focuses on whether there exist rational expectations that can induce a government through fear of loss of a good reputation to behave as it would under a commitment technology, even when it chooses sequentially.

Reputational models of government policy exploit ideas from dynamic programming. These models arrange a setting in which a government each period faces choices whose consequences consist of a first-period return and a reputation

---

[1] Consider two extensive-form versions of the "battle of the sexes" game described by Kreps (1990), one in which the man chooses first, the other in which the woman chooses first. Backward induction recovers different outcomes in these two different games. Though they share the same choice sets and payoffs, these are different games.

[2] Barro and Gordon (1983a, 1983b) are early contributors to this literature. See Kenneth Rogoff (1989) for a survey.

to begin next period. Because these models impose rational expectations, any reputation that the government is assigned at the beginning of next period must be one that the government will *want* to preserve. The literature has been driven by its internal logic to study the set of *all* possible reputations that it could ever be in the government's interests to protect.

This chapter applies an apparatus of Abreu, Pearce, and Stacchetti (1986, 1990) to reputational equilibria in a class of macroeconomic models. The economic model is identical to one that has been used by Chari, Kehoe, and Prescott (1989) and Stokey (1989, 1991) to exhibit what Chari and Kehoe (1990) call sustainable government policies and what Stokey calls credible public policies. The literature on sustainable or credible government policies in macroeconomics adapts ideas from the literature on repeated games so that they can be applied in contexts in which a single agent (a government) behaves strategically, and in which the remaining agents' behavior can be summarized as a "competitive equilibrium" that responds nonstrategically to the government's choices.[3]

Abreu, Pearce, and Stacchetti exploit ideas from dynamic programming. This chapter closely follow Stacchetti (1991), who applies Abreu, Pearce, and Stacchetti (1986, 1990) to a more general class of models than that treated here.[4]

## The one-period economy

There is a continuum of households, each of whom chooses an action $\xi \in X$. There is a government that chooses an action $y \in Y$. The sets $X$ and $Y$ are compact sets. The average level of $\xi$ across households is denoted $x \in X$. The utility of a particular household when it chooses $\xi$, when the average household setting is $x$, and when the government chooses $y$ is $u(\xi, x, y)$. The payoff function $u(\xi, x, y)$ is strictly concave and continuously differentiable.

### Competitive equilibrium

For given levels of $y$ and $x$, the representative household faces the problem $\max_{\xi \in X} u(\xi, x, y)$. Let the solution be a function $\xi = f(x, y)$. When a household thinks that the government's setting is $y$ and believes that the average level of other households' setting is $x$, it acts to set $\xi = f(x, y)$. Because all households

[3] For descriptions of theories of credible government policy see Chari and Kehoe (1990), Stokey (1989, 1991), Rogoff (1989), and Chari, Kehoe, and Prescott (1989). For applications of the framework of Abreu, Pearce, and Stacchetti, see Chang (1998), Phelan and Stacchetti (1999).

[4] Stacchetti also studies a class of setups in which the private sector observes only a noise-ridden signal of the government's actions.

are alike, this fact implies that the actual level of $x$ is $f(x, y)$. For expectations about the average to be consistent with the average outcome, we require that $\xi = x$, or $x = f(x, y)$. We use the following definitions:

DEFINITION 1: A *competitive equilibrium* or a *rational expectations equilibrium* is an $x \in X$ that satisfies $x = f(x, y)$.

Note that a competitive equilibrium satisfies $u(x, x, y) = \max_{\xi \in X} u(\xi, x, y)$.

For each $y \in Y$, let $x = h(y)$ denote the corresponding competitive equilibrium. We adopt this definition:

DEFINITION 2: The set of competitive equilibria is $C = \{(x, y) \mid u(x, x, y) = \max_{\xi \in X} u(\xi, x, y)\}$, or equivalently $C = \{(x, y) \mid x = h(y)\}$.

*The Ramsey problem*

The following timing of actions underlies a *Ramsey plan.* First, the government selects a $y \in Y$. Then knowing the setting for $y$, the aggregate of households "responds" with a competitive equilibrium. The government evaluates policies $y \in Y$ with the payoff function $u(x, x, y)$; that is, the government is benevolent.

In making its choice of $y$, the government has to "forecast" how the economy will respond. The government correctly forecasts that the economy will respond to $y$ with a competitive equilibrium, $x = h(y)$. We use these definitions:

DEFINITION 3: The *Ramsey problem* for the government is $\max_{y \in Y} u[h(y), h(y), y]$, or equivalently $\max_{(x,y) \in C} u(x, x, y)$.

DEFINITION 4: The policy that attains the maximum of the Ramsey problem is denoted $y^R$, and $(y^R, x^R)$ where $x^R = h(y^R)$ is called the *Ramsey outcome* or *Ramsey plan.*

Two remarks about the Ramsey problem are in order. First, the Ramsey outcome is typically inferior to the "dictatorial outcome" that solves the unrestricted problem $\max_{x \in X, y \in Y} u(x, x, y)$, because the restriction $(x, y) \in C$ is in general binding. Second, the timing of actions is important. The formulation of the Ramsey problem assumes that the government has a technology that permits it to choose first, and then not to reconsider its action.

If the government were granted the opportunity to reconsider its plan *after* households had chosen $x^R$, it would in general want to deviate from the setting $y = y^R$ because usually there exists an $\alpha \neq y^R$ for which $u(x^R, x^R, \alpha) > u(x^R, x^R, y^R)$. The "time consistency problem" is about the incentive to deviate from the Ramsey plan if the government were given a chance to react *after* households have set $x = x^R$. In this one-shot setting, to support the Ramsey plan requires a set of protocols that forces the government to choose first and then to walk away from the table.

*Nash equilibrium*

Consider an alternative setting in which households also face a "forecasting problem." Households forecast that, given $x$, the government will set $y$ to solve $\max_{y \in Y} u(x, x, y)$. We use the following:

DEFINITION 5: A *Nash equilibrium* $(x^N, y^N)$ satisfies
(1) $(x^N, y^N) \in C$
(2) Given $x^N$, $u(x^N, x^N, y^N) = \max_{\eta \in Y} u(x^N, x^N, \eta)$

Condition (1) asserts that $x^N = h(y^N)$, or that the economy "responds" to $y^N$ with a competitive equilibrium. In other words, condition (1) says that given $(x^N, y^N)$, each individual household wants to set $\xi = x^N$; that is, it has no incentive to deviate from $x^N$. Condition (2) asserts that given $x^N$, the government chooses a policy $y^N$ from which it has no incentive to deviate.[5]

We can use the solution of the problem in condition (2) to define the government's *best response* function $y = H(x)$. The definition of a Nash equilibrium can be phrased as a pair $(x, y) \in C$ such that $y = H(x)$.

There are two "timings of choices" for which a Nash equilibrium is a natural equilibrium concept. One is where households "choose first," forecasting that the government will respond to the aggregate outcome $x$ by setting $y = H(x)$. Another is where the government and all households choose simultaneously, in which case the Nash equilibrium $(x^N, y^N)$ depicts a situation in which everyone has "rational expectations": given that each household expects the aggregate variables to be $(x^N, y^N)$, each household responds in a way to make $x = x^N$;

[5] Much of the language of this chapter is borrowed from game theory, but the object under study is not a game, because we do not specify all of the objects that formally define a game. In particular, we do not specify the payoffs to all agents for all feasible choices. We only specify the payoffs $u(\xi, x, y)$ where each agent chooses the *same* value of $\xi$.

and given that the government expects that $x = x^N$, it responds by setting $y = y^N$.

We let $v^N = u(x^N, x^N, y^N)$ and $v^R = u(x^R, x^R, y^R)$. Note that $v^N \leq v^R$. Because of the additional constraint embedded in the Nash equilibrium, outcomes are ordered according to

$$v^N \leq \max_{\{(x,y)\in C:\, y=H(x)\}} u(x,x,y) \leq \max_{(x,y)\in C} u(x,x,y) = v^R .$$

## Examples of economies

To illustrate these concepts, we consider two models: a taxation example with a fully specified economy and a reduced-form model with discrete choice sets.

### Taxation example

There is a continuum of households with preferences over leisure $\ell$, private consumption $c$, and per capita government expenditures $g$. The utility function is

$$U(\ell, c, g) = \ell + \log(\alpha + c) + \log(\alpha + g), \qquad \alpha \in (0,\, 1/2).$$

Each household is endowed with one unit of time that can be devoted to leisure or work. The production technology is linear in labor, and the economy's resource constraint is

$$\bar{c} + g = 1 - \bar{\ell},$$

where $\bar{c}$ and $\bar{\ell}$ are the average levels of private consumption and leisure, respectively.

A benevolent government that maximizes the welfare of the representative household would choose $\ell = 0$ and $c = g = 1/2$. This "dictatorial outcome" yields welfare $W^d = 2\log(\alpha + 1/2)$.

Here we will focus on competitive equilibria where the government finances its expenditures by levying a flat-rate tax $\tau$ on labor income. The household's budget constraint becomes $c = (1-\tau)(1-\ell)$. Given a government policy $(\tau, g)$, an individual household's optimal decision rule for leisure is

$$\ell(\tau) = \begin{cases} \dfrac{\alpha}{1-\tau} & \text{if } \tau \in [0,\, 1-\alpha]; \\ 1 & \text{if } \tau > 1-\alpha. \end{cases}$$

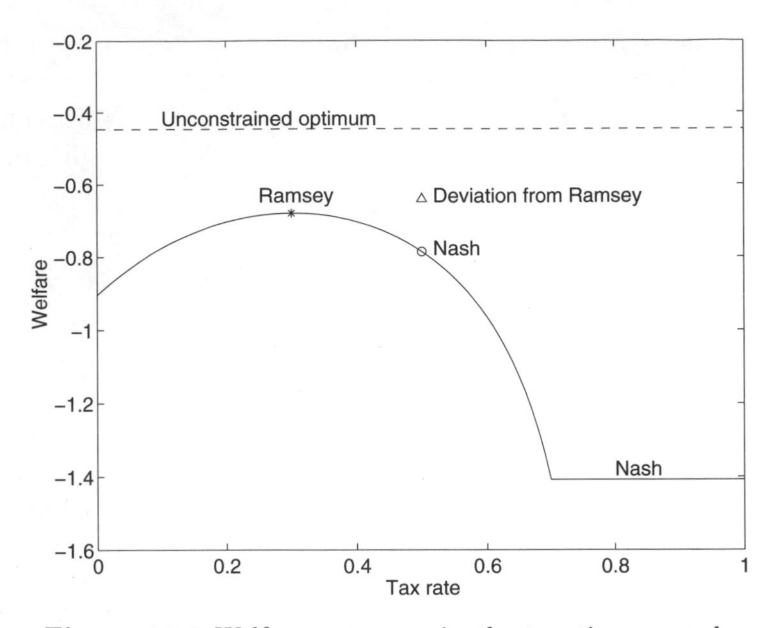


**Figure 16.1** Welfare outcomes in the taxation example. The solid portion of the curve depicts the set of competitive equilibria, $W^c(\tau)$. The set of Nash equilibria is the horizontal portion of the solid curve and the equilibrium at $\tau = 1/2$. The Ramsey outcome is marked with an asterisk. The "time inconsistency problem" is indicated with the triangle showing the outcome if the government were able to reset $\tau$ after households had chosen the Ramsey labor supply. The dashed line describes the welfare level at the unconstrained optimum, $W^d$. The graph sets $\alpha = 0.3$.

Due to the linear technology and the fact that government expenditures enter additively in the utility function, the household's decision rule $\ell(\tau)$ is also the equilibrium value of individual leisure at a given tax rate $\tau$. Imposing government budget balance, $g = \tau(1-\ell)$, the representative household's welfare in a competitive equilibrium is indexed by $\tau$ and equal to

$$W^c(\tau) = \ell(\tau) + \log\{\alpha + (1-\tau)[1-\ell(\tau)]\} + \log\{\alpha + \tau[1-\ell(\tau)]\}.$$

The Ramsey allocation is the solution to $\max_\tau W^c(\tau)$. The government's problem in a Nash equilibrium is $\max_\tau\{\ell + \log[\alpha + (1-\tau)(1-\ell)] + \log[\alpha + \tau(1-\ell)]\}$. If $\ell > 0$, the optimizer is $\tau = .5$. There is a continuum of Nash equilibria indexed by $\tau \in [1-\alpha,\, 1]$ where agents choose not to work, and consequently $c = g = 0$. The only Nash equilibrium with production is $\tau = 1/2$ with welfare level $W^c(1/2)$.

This conclusion follows directly from the fact that the government's best response is $\tau = 1/2$ for any $\ell < 1$. These outcomes are numerically illustrated in Figure 16.1. Here the time inconsistency problem surfaces in the government's incentive to reset the tax rate $\tau$, if offered the choice, after the household has set its labor supply.

The objects of the general setup in the preceding section can be mapped into the present taxation example as follows: $\xi = \ell$, $x = \bar{\ell}$, $X = [0, 1]$, $y = \tau$, $Y = [0, 1]$, $u(\xi, x, y) = \xi + \log[\alpha + (1-y)(1-\xi)] + \log[\alpha + y(1-x)]$, $f(x, y) = \ell(y)$, $h(y) = \ell(y)$, and $H(x) = 1/2$ if $x < 1$; and $H(x) \in [0, 1]$ if $x = 1$.

*Reduced-form example with discrete choice sets*

Consider a reduced-form economy with $X = \{x_L, x_H\}$ and $Y = \{y_L, y_H\}$, in which $u(x, x, y)$ assume the values given in Table 16.1. Assume that values of $u(\xi, x, y)$ for $\xi \neq x$ are such that the values with asterisks for $\xi = x$ are competitive equilibria. In particular, we might assume that

$$u(\xi, x_i, y_j) = 0 \quad \text{when } \xi \neq x_i \text{ and } i = j$$
$$u(\xi, x_i, y_j) = 20 \quad \text{when } \xi \neq x_i \text{ and } i \neq j.$$

These payoffs imply that $u(x_L, x_L, y_L) > u(x_H, x_L, y_L)$ (i.e., $3 > 0$); and $u(x_H, x_H, y_H) > u(x_L, x_H, y_H)$ (i.e., $10 > 0$). Therefore $(x_L, x_L, y_L)$ and $(x_H, x_H, y_H)$ are competitive equilibria. Also, $u(x_H, x_H, y_L) < u(x_L, x_H, y_L)$ (i.e., $12 < 20$), so the dictatorial outcome cannot be supported as a competitive equilibrium.

**Table 16.1** One-period payoffs to the government–household [values of $u(x_i, x_i, y_j)$].

| | $x_L$ | $x_H$ |
|---|---|---|
| $y_L$ | 3* | 12 |
| $y_H$ | 1 | 10* |

* Denotes $(x, y) \in C$.

The Ramsey outcome is $(x_H, y_H)$; the Nash equilibrium outcome is $(x_L, y_L)$.

Figure 16.2 depicts a timing of choices that supports the Ramsey outcome for this example. The government chooses first, then walks away. The Ramsey outcome $(x_H, y_H)$ is the competitive equilibrium yielding the highest value of $u(x, x, y)$.

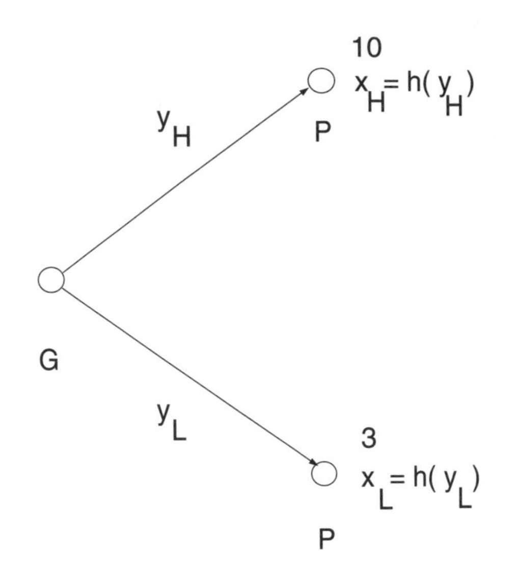


**Figure 16.2** Timing of choices that supports Ramsey outcome. Here $P$ and $G$ denote nodes at which the public and the government, respectively, choose. The government has a commitment technology that binds it to "choose first." The government chooses the $y \in Y$ that maximizes $u[h(y), h(y), y]$, where $x = h(y)$ is the function mapping government actions into equilibrium values of $x$.

Figure 16.3 diagrams a timing of choices that supports the Nash equilibrium. Recall that by definition every Nash equilibrium outcome has to be a competitive equilibrium outcome. We denote competitive equilibrium pairs $(x, y)$ with asterisks. The government sector chooses after knowing that the private sector has set $x$, and chooses $y$ to maximize $u(x, x, y)$. With this timing, if the private sector chooses $x = x_H$, the government has an incentive to set $y = y_L$, a setting of $y$ that does not support $x_H$ as a Nash equilibrium. The unique Nash equilibrium is $(x_L, y_L)$, which gives a lower utility $u(x, x, y)$ than does the competitive equilibrium $(x_H, y_H)$.

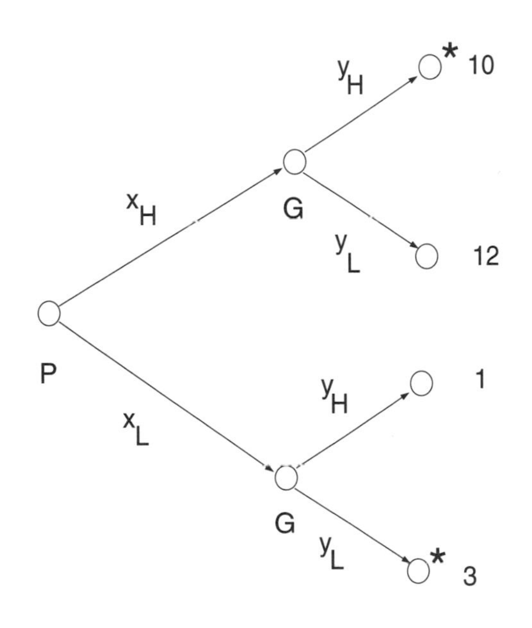


**Figure 16.3** Timing of actions in a Nash equilibrium in which the private sector acts first. Here $G$ denotes a node at which the government chooses and $P$ denotes a node at which the public chooses. The private sector sets $x \in X$ before knowing the government's setting of $y \in Y$. Competitive equilibrium pairs $(x, y)$ are denoted with an asterisk. The unique Nash equilibrium is $(x_L, y_L)$.

## *Reputational mechanisms: General idea*

What happens if the one-period economy is repeated infinitely? We shall study economies repeated an infinite number of times, because we want to study stationary situations in which a government sustains a Ramsey outcome forever. In a finitely repeated economy, the government will certainly behave opportunistically the last period, implying that nothing better than a Nash outcome can be supported the last period. In a finite horizon economy with a unique Nash equilibrium, we won't be able to sustain anything better than a Nash equilibrium outcome for *any* earlier period. If there are multiple Nash equilibria, it is sometimes possible to sustain a better than Nash equilibrium outcome for a while in a finite horizon economy.[6] Let the government have the recursive utility function $v = (1-\delta)u(x,x,y) + \delta v'$ where $\delta \in (0,1)$ and $v'$ is next period's value of $v$. Because the sets $(X,Y)$ from which $(x,y)$ are drawn simply repeat themselves period after period, there is no purely physical state variable in the economy. We might seek a "reputational state variable" $r$ with law of motion $r' = g(x,y,r)$, and a function $x(r)$ describing the market's one-period response to the government's reputational state variable such that the government's problem at each $t \geq 1$ becomes to maximize over choice of $(x,y) \in C$ the criterion $v(r) = (1-\delta)u[x(r),x(r),y] + \delta v(r')$ subject to the law of motion for reputation $r' = g(x,y,r)$. We seek a reputational function $g(x,y,r)$, a function $x(r)$ giving the market's response to the reputational state variable, and an initial condition on $r_0$, the government's initial reputation, that can induce the government repeatedly to choose a better outcome than the Nash equilibrium $(x^N, y^N)$. Though the government chooses sequentially (i.e., each period) and within each period chooses simultaneously or after "the market" has set $x$ for the period, the hope is that $x(r)$ and $g(x,y,r)$ can be chosen in a way to induce the government to improve upon repetition of the Nash outcome.

If we are free to choose the reputational function $g$ *arbitrarily*, that is, without regard to the government's motives, then it is easy to make this idea work: we can craft the law of motion of the reputation variable and the market's dependence on it to induce any behavior that we might wish to support from the government. However, the aim of the literature has been to find specifications of $g(x,y,r)$ and $x(r)$ that are consistent with optimizing behavior by the government and rational expectations. We have to set up the model so that the government wants to fulfill its reputation and so that the market is not repeatedly fooled by false assessments of the government's reputation.

A reputational variable is peculiar in that it is both "backward looking" and "forward looking." Its backward-looking character comes from its role in encoding historical behavior. Its forward-looking behavior comes from how its value portrays future payoffs to the government. We are about to study the ingenious machinery of Abreu, Pearce, and Stacchetti, which exploits these aspects of a

[6] See Exercise 16.1, which uses an idea of Benoit and Krishna (1985).

reputational variable. They will show us how the ideal reputational variable is a "promised value."

*Dynamic programming*

Abreu, Pearce, and Stacchetti (1986, 1990) used dynamic programming. To implement their approach, we would define a strategy profile as a pair of contingency plans, one each for the private sector and the government, mapping the observed history of the economy into first-period outcomes $(x, y)$. An *equilibrium* is a strategy profile that delivers a competitive equilibrium $(x, y)$ in every period and that is optimal for the government at each $t \geq 1$ and for every possible history of the economy up to that $t$.

To characterize equilibria à la Abreu, Pearce, and Stacchetti, we formulate a dynamic programming problem that restricts the government's strategy. For each $t \geq 1$, the government's strategy instructs it to choose a first-period action $y \in Y$ that determines a first-period competitive equilibrium $(x, y) \in C$ and a value for its problem for next period $v_1$, a so-called continuation value if it adheres to its strategy. The government's choice of $y$ gives it current value $v$ of

$$v = (1 - \delta) u(x, x, y) + \delta v_1, \tag{16.1a}$$

where $(x, y) \in C$, $v_1$ is a continuation value and the choice of $y$ is subject to the incentive constraint

$$v \geq (1 - \delta) u(x, x, \eta) + \delta v_2, \quad \forall \eta \in Y, \tag{16.1b}$$

or equivalently

$$v \geq (1 - \delta) u\big[x, x, H(x)\big] + \delta v_2.$$

In the incentive constraint, $v_2$ is a value that the government attains next period if it deviates from the first-period outcome $(x, y)$ prescribed by the strategy.

The preceding equation is a Bellman equation that maps *pairs* of values $(v_1, v_2)$ into values $v$ [and first-period outcomes $(x, y)$]. Figure 16.4 illustrates this mapping for the infinitely repeated version of the taxation example. Given a pair $(v_1, v_2)$, the solid curve depicts $v$ in equation (16.1a), and the dashed curve describes the right side of the incentive constraint (16.1b). The solid curve above the dashed curve identifies tax rates and competitive equilibria that are incentive compatible at the given continuation values $(v_1, v_2)$. As can be seen, tax rates below 18 percent cannot be sustained in this numerical example.

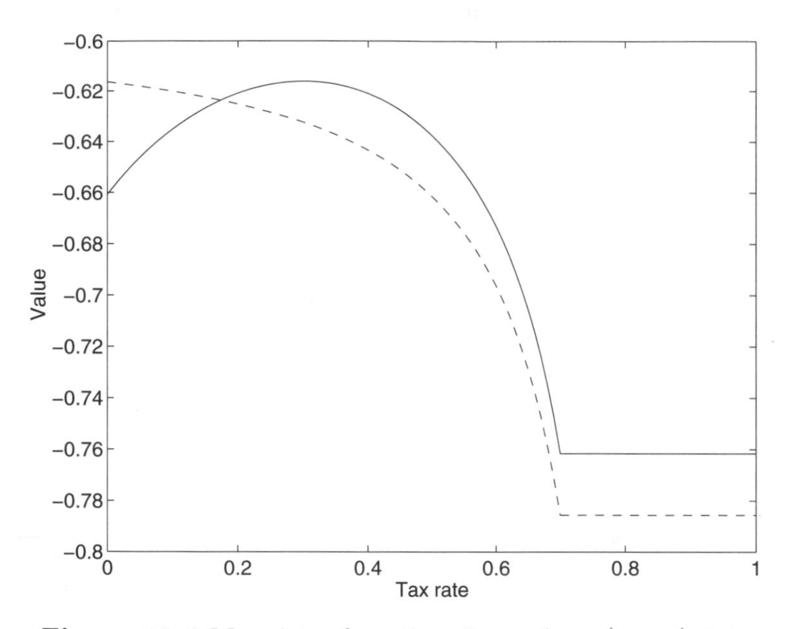


**Figure 16.4** Mapping of continuation values $(v_1, v_2)$ into values $v$ in the infinitely repeated version of the taxation example. The solid curve depicts $v = (1-\delta)u[\ell(\tau), \ell(\tau), \tau] + \delta v_1$. The dashed curve is the right side of the incentive constraint, $v \geq (1-\delta)u\{\ell(\tau), \ell(\tau), H[\ell(\tau)]\} + \delta v_2$, where $H$ is the government's best response function. The solid curve above the dashed curve shows competitive equilibria that are incentive compatible for continuation values $(v_1, v_2)$. The parameterization is $\alpha = 0.3$ and $\delta = 0.8$, and the continuation values are chosen as $(v_1, v_2) = (-0.6, -0.63)$.

Abreu, Pearce, and Stacchetti's approach is to calculate the set of equilibrium values by iterating on equation (16.1). Notice that as we vary $(v_1, v_2)$, this Bellman equation maps out a set of values $v \in V$. Abreu, Pearce, and Stacchetti extend the equation to one that maps *sets* of values into sets of values, and they define iterations that determine sets of equilibrium values and equilibrium strategies. The remainder of the chapter fills in some of the details of Abreu, Pearce, and Stacchetti's dynamic programming formulation.

## The infinitely repeated economy

Consider an economy that repeats the preceding one-period economy forever. At each $t \geq 1$, each household chooses $\xi_t \in X$, with the result that the average $x_t \in X$; the government chooses $y_t \in Y$. We use the notation

$(\vec{x}, \vec{y}) = \{(x_t, y_t)\}_{t=1}^{\infty}$, $\vec{\xi} = \{\xi_t\}_{t=1}^{\infty}$. To denote the *history* of $(x_t, y_t)$ up to $t$ we use the notation $x^t = \{x_s\}_{s=1}^{t}$, $y^t = \{y_s\}_{s=1}^{t}$. These histories live in the spaces $X^t$ and $Y^t$, respectively, where $X^t = X \times \cdots \times X$, the Cartesian product of $X$ taken $t$ times, and $Y^t$ is the Cartesian product of $Y$ taken $t$ times.

For the repeated economy, each household and the government, respectively, evaluate paths $(\vec{\xi}, \vec{x}, \vec{y})$ according to

$$V_h(\vec{\xi}, \vec{x}, \vec{y}) = \frac{(1-\delta)}{\delta} \sum_{t=1}^{\infty} \delta^t u(\xi_t, x_t, y_t), \tag{16.2a}$$

$$V_g(\vec{x}, \vec{y}) = \frac{(1-\delta)}{\delta} \sum_{t=1}^{\infty} \delta^t\, r(x_t, y_t), \tag{16.2b}$$

where $r(x_t, y_t) \equiv u(x_t, x_t, y_t)$ and $0 < \delta < 1$. (Note that we have not defined the government's payoff when $\xi_t \neq x_t$.) A *pure strategy* is defined as a sequence of functions, the $t$th element of which maps the history $(x^{t-1}, y^{t-1})$ observed at the beginning of $t$ into an action at $t$. In particular, for the aggregate of households, a strategy is a sequence $\sigma^h = \{\sigma_t^h\}_{t=1}^{\infty}$ such that

$$\begin{aligned} &\sigma_1^h \in X \\ &\sigma_t^h : X^{t-1} \times Y^{t-1} \to X \qquad \text{for each} \quad t \geq 2\,. \end{aligned}$$

Similarly, for the government, a strategy $\sigma^g = \{\sigma_t^g\}_{t=1}^{\infty}$ is a sequence such that

$$\begin{aligned} &\sigma_1^g \in Y \\ &\sigma_t^g : X^{t-1} \times Y^{t-1} \to Y \qquad \text{for each} \quad t \geq 2. \end{aligned}$$

We let $\sigma_t = (\sigma_t^h, \sigma_t^g)$ be the $t$th component of the *strategy profile*, which is a pair of functions mapping $X^{t-1} \times Y^{t-1} \to X \times Y$. Note that since there is "no history" at $t = 1$, at $t = 1$ a strategy profile is just a point from the set $X \times Y$.

The strategy profile $\sigma = (\sigma^g, \sigma^h)$ evidently recursively generates a trajectory of outcomes, which we denote $\{[x(\sigma)_t, y(\sigma)_t]\}_{t=1}^{\infty}$:

$$\begin{aligned} \left[x(\sigma)_1, y(\sigma)_1\right] &= (\sigma_1^h, \sigma_1^g) \\ \left[x(\sigma)_t,\, y(\sigma)_t\right] &= \sigma_t\left[x(\sigma)^{t-1},\, y(\sigma)^{t-1}\right]. \end{aligned}$$

The value to the government of a strategy profile $\sigma = (\sigma^h, \sigma^g)$ is the value of the trajectory that it generates

$$V_g(\sigma) = V_g\left[\vec{x}(\sigma),\, \vec{y}(\sigma)\right].$$

*Recursive formulation*

Since the value of a path $(\xi, x, y)$ in equation (16.2a) or (16.2b) is additively separable in its one-period returns, we can express the value in a recursive form with a current period and a "continuation economy." In particular, the value to the government of an outcome sequence $(x, y)$ can be represented

$$\begin{aligned} V_h(\vec{\xi}, \vec{x}, \vec{y}) &= (1-\delta)\, u(\xi_1, x_1, y_1) + \delta V_h(\{\xi_t\}_{t=2}^{\infty}, \{x_t\}_{t=2}^{\infty}, \ \{y_t\}_{t=2}^{\infty}) \\ V_g(\vec{x}, \vec{y}) &= (1-\delta)\, r(x_1, y_1) + \delta V_g(\{x_t\}_{t=2}^{\infty}, \ \{y_t\}_{t=2}^{\infty}) \end{aligned} \tag{16.3}$$

and the value for a household can similarly be represented recursively. Notice also that a strategy profile $\sigma$ induces a strategy profile for the continuation economy, as follows: We let $\sigma|_{(x^t,y^t)}$ denote the strategy profile for a continuation economy whose first period is $t+1$, and which is initiated after history $(x^t, y^t)$ has been observed; here $(\sigma|_{(x^t,y^t)})_s$ is the $s$th component of $(\sigma|_{(x^t,y^t)})$, which for $s \geq 2$ is a function that maps $X^{s-1} \times Y^{s-1}$ into $X \times Y$, and for $s = 1$ is a point in $X \times Y$. Thus, after a first-period outcome pair $(x_1, y_1)$, strategy $\sigma$ induces the continuation strategy

$$\begin{aligned} (\sigma|_{(x_1,y_1)})_{s+1}\,(\nu^s, \eta^s) &= \sigma_{s+2}\,(x_1, \nu_1, \ldots, \nu_s, y_1, \eta_1, \ldots, \eta_s) \\ &\qquad \text{for all } \ (\nu^s, \eta^s) \in X^s \times Y^s\ , \quad \forall s \geq 0. \end{aligned}$$

It might be helpful to write out a few terms for $s = 0, 1, \ldots$:

$$\begin{aligned} (\sigma|_{(x_1,y_1)})_1 &= \sigma_2(x_1, y_1) = (\nu_1, \eta_1) \\ (\sigma|_{(x_1,y_1)})_2(\nu_1, \eta_1) &= \sigma_3(x_1, \nu_1, y_1, \eta_1) = (\nu_2, \eta_2) \\ (\sigma|_{(x_1,y_1)})_3(\nu_1, \nu_2, \eta_1, \eta_2) &= \sigma_4(x_1, \nu_1, \nu_2, y_1, \eta_1, \eta_2) = (\nu_3, \eta_3) \end{aligned}$$

More generally, define the continuation strategy

$$\begin{aligned} (\sigma|_{(x^t,y^t)})_1 &= \sigma_{t+1}(x^t, y^t) \\ (\sigma|_{(x^t,y^t)})_{s+1}\,(\nu^s, \eta^s) &= \\ &\sigma_{t+s+1}\,(x_1, \ldots, x_t, \nu_1, \ldots, \nu_s; y_1, \ldots, y_t, \eta_1, \ldots, \eta_s) \\ &\qquad \text{for all } \ s \geq 1 \quad \text{and all } \ (\nu^s, \eta^s) \in X^s \times Y^s\ . \end{aligned}$$

Here $(\sigma|_{(x^t,y^t)})_{s+1}\,(\nu^s, \eta^s)$ is the induced strategy pair to apply in the $(s+1)$th period of the continuation economy. This equation says we attain this strategy by shifting the original strategy forward $t$ periods and evaluating it at history $(x_1, \ldots, x_t, \nu_1, \ldots, \nu_s; y_1, \ldots, y_t, \eta_1, \ldots, \eta_s)$ for the *original* economy. Figure 16.5 depicts the unfolding of choices over time in such an economy.

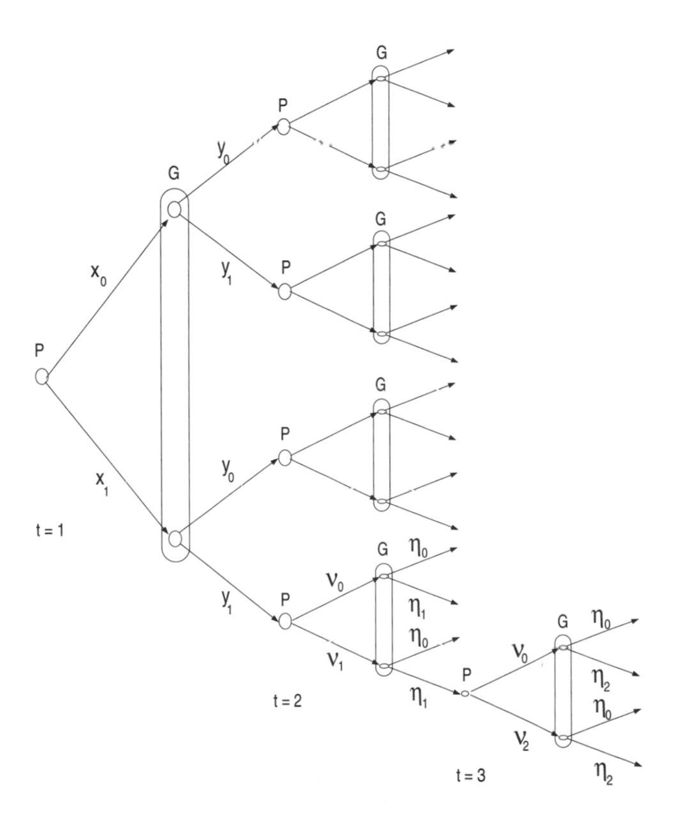


**Figure 16.5** An infinite horizon economy. The government and the public choose simultaneously. Only the first three periods of the economy are shown. Each period the economy repeats itself. The outcome trajectory is $(x_1, \nu_1, \nu_2; y_1, \eta_1, \eta_2)$.

In terms of the continuation strategy $\sigma|_{(x_1,y_1)}$, from equation (16.3) we know that $V_g(\sigma)$ can be represented as

$$V_g(\sigma) = (1 - \delta)r(x_1, y_1) + \delta V_g(\sigma|_{(x_1,y_1)}) \tag{16.4}$$

Representation (16.4) decomposes the value to the government of strategy profile $\sigma$ into a one-period return and the continuation value $V_g(\sigma|_{(x_1,y_1)})$ associated with the continuation strategy $\sigma|_{(x_1,y_1)}$.

Any sequences $(x, y)$ in equation (16.3) or any strategy profile $\sigma$ in equation (16.4) can be assigned a value. But these objects will only be of interest if they can constitute an equilibrium outcome. The recursive structure of the economy motivates the following equilibrium concept.

## *Subgame perfect equilibrium (SPE)*

DEFINITION 6: A strategy profile $\sigma = (\sigma^h, \sigma^g)$ is a *subgame perfect equilibrium* of the infinitely repeated economy if for each $t \geq 1$ and each history $(x^{t-1}, y^{t-1}) \in X^{t-1} \times Y^{t-1}$,

(1) The outcome $x_t = \sigma_t^h\left(x^{t-1}, y^{t-1}\right)$ is a competitive equilibrium, given that $y_t = \sigma_t^g\left(x^{t-1}, y^{t-1}\right)$.

(2) For each $\eta \in Y$

$$\begin{aligned}(1-\delta)\, r(x_t, y_t) + \delta V_g(\sigma|_{(x^t, y^t)}) \\ \geq (1-\delta)\, r(x_t, \eta) + \delta V_g(\sigma|_{(x^t; y^{t-1}, \eta)})\ .\end{aligned}$$

Requirement (1) attributes a theory of forecasting government behavior to members of the public, in particular, that they use the time-$t$ component $\sigma_t^g$ of the government's strategy and information available at the end of period $t-1$ to forecast the government's behavior at $t$. Condition (1) also asserts that a competitive equilibrium appropriate to the public's forecast value for $y_t$ is the market outcome at time $t$. Requirement (2) says that at each point in time and following each history, the government has no incentive to deviate from the first-period outcome called for by its strategy $\sigma^g$; that is, the government always has the incentive to behave as the public is expecting. Notice that in the formulation of condition (2), the government *contemplates* setting its time-$t$ choice $\eta_t$ at something other than the value forecast by the public, but confronts consequences of its choices that deter it ever from choosing an $\eta_t$ that fails to confirm the public's expectations of it.

The definition implies that for each $t \geq 2$ and each $(x^{t-1}, y^{t-1}) \in X^{t-1} \times Y^{t-1}$, the continuation strategy $\sigma|_{(x^{t-1}, y^{t-1})}$ is itself a subgame perfect equilibrium. We state this formally for the case $t = 2$.

PROPOSITION 1: Assume that $\sigma$ is a subgame perfect equilibrium. Then for all $(\nu, \eta) \in X \times Y$, $\sigma|_{(\nu, \eta)}$ is a subgame perfect equilibrium.

PROOF: Write out requirements (1) and (2) of Definition 6, which the continuation strategy $\sigma|_{(\nu,\eta)}$ must satisfy to qualify as a subgame perfect equilibrium. In particular, for all $s \geq 1$ and for all $(x^{s-1}, y^{s-1}) \in X^{s-1} \times Y^{s-1}$, we require

$$(x_s, y_s) \in C, \tag{16.5}$$

where $x_s = \sigma^h|_{(\nu,\eta)}(x^{s-1}, y^{s-1}), y_s = \sigma^g|_{(\nu,\eta)}(x^{s-1}, y^{s-1})$. We also require that for all $\tilde{\eta} \in Y$,

$$\begin{aligned}&(1-\delta)r(x_s, y_s) + \delta V_g(\sigma|_{(\eta, x^s; \nu, y^s)})\\&\geq (1-\delta)r(x_s, \tilde{\eta}) + \delta V_g(\sigma|_{(\nu, x^s; \eta, y^{s-1}, \tilde{\eta})})\end{aligned} \tag{16.6}$$

Notice that requirements (1) and (2) of Definition 6 for $t = 2, 3, \ldots$ imply expressions (16.5) and (16.6) for $s = 1, 2, \ldots$. ∎

The statement that $\sigma|_{(\nu,\eta)}$ is subgame perfect for all $(\nu, \eta) \in X \times Y$ assures that $\sigma$ is "almost" subgame perfect. To assure that $\sigma$ itself is a subgame perfect equilibrium, we must add two requirements to the subgame perfection of $\sigma|_{(\nu,\eta)}$ for all $(\nu, \eta) \in X \times Y$: first, that the $t = 1$ outcome pair $(x_1, y_1)$ is a competitive equilibrium, and second, that the government's choice of $y_1$ satisfies the time-1 version of the incentive constraint (2) in Definition 6.

This reasoning leads us to the following important lemma:

LEMMA: Consider a strategy profile $\sigma$, and let the associated first-period outcome be given by $x = \sigma_1^h, y = \sigma_1^g$. The profile $\sigma$ is a subgame perfect equilibrium if and only if

(1) for each $(\nu, \eta) \in X \times Y, \quad \sigma|_{(\nu,\eta)}$ is a subgame perfect equilibrium.
(2) $(x, y)$ is a competitive equilibrium.
(3) $\forall\, \eta \in Y$, $(1-\delta)\, r(x, y) + \delta\, V_g(\sigma|_{(x,y)}) \geq (1-\delta)\, r(x, \eta) + \delta V_g(\sigma|_{(x,\eta)})$.

PROOF: First, prove the "if" part. Property (1) of the lemma and properties (16.5) and (16.6) of Proposition 1 show that requirements (1) and (2) of Definition 6 are satisfied for $t \geq 2$. Properties (2) and (3) of the lemma imply that requirements (1) and (2) of Definition 6 hold for $t = 1$.

Second, prove the "only if" part. Part (1) of the lemma follows from Proposition 1. Parts (2) and (3) of the lemma follow from requirements (1) and (2) of Definition 6 for $t = 1$. ∎

The lemma is very important because it characterizes subgame perfect equilibria in terms of a first-period competitive equilibrium outcome pair $(x, y)$, and a *pair* of continuation values: a value $V_g(\sigma|_{(x,y)})$ to be "paid" to the government next period if it now adheres to the first-period pair $(x,y)$, and a value $V_g(\sigma|_{(x,\eta)})$, $\eta \neq y$, to be paid to the government if it now deviates from $(x,y)$. Each of these values has to be "drawn" from the set of values $V_g(\sigma)$ associated with subgame perfect equilibrium $\sigma$. Note that the continuation value for a deviation must also be associated with a subgame perfect equilibrium in order for that outcome to be a "credible punishment." We now illustrate this construction.

## Examples of SPE

### Infinite repetition of one-period Nash equilibrium

It is easy to verify that the following strategy profile $\sigma^N = (\sigma^h, \sigma^g)$ forms a subgame perfect equilibrium:

$$\begin{aligned} \sigma_t^h &= x^N \qquad \forall\, t\,, \quad \forall\, (x^{t-1}, y^{t-1}); \\ \sigma_t^g &= y^N \qquad \forall\, t\,, \quad \forall\, (x^{t-1}, y^{t-1}). \end{aligned}$$

These strategies instruct the households and the government to choose the static Nash equilibrium outcomes for all periods for all histories. Evidently, for these strategies $V_g(\sigma^N) = v^N = r(x^N, y^N)$. Furthermore, for these strategies the continuation value $V_g(\sigma|_{(x^t; y^{t-1}, \eta)}) = v^N$ for all outcomes $\eta \in Y$. These strategies satisfy requirement (1) of Definition 6 because $(x^N, y^N)$ is a competitive equilibrium. The strategies satisfy requirement (2) because $r(x^N, y^N) = \max_{y \in Y} r(x^N, y)$ and because the continuation value $V_g(\sigma) = v^N$ is independent of the action chosen by the government in the first period. In this subgame perfect equilibrium, $\sigma_t^N = \{\sigma_t^h, \sigma_t^g\} = (x^N, y^N)$ for all $t$ and for all $(x^{t-1}, y^{t-1})$, and the value $V_g(\sigma^N)$ and the continuation values for each history $(x^t, y^t)$, $V_g(\sigma^N|_{(x^t, y^t)})$, all equal $v^N$.

It is useful to look at this subgame perfect equilibrium in terms of the lemma. To verify that $\sigma^N$ is a subgame perfect equilibrium using the lemma, we work with the first-period outcome pair $(x^N, y^N)$ and the pair of values $V_g(\sigma|_{(x^N, y^N)}) = v^N, V_g(\sigma|_{(x,\eta)}) = v^N$, where $v^N = r(x^N, y^N)$. With these settings, we proceed by verifying that $(x^N, y^N)$ and $v^N$ satisfy requirements (1), (2), and (3) of the lemma.

*Supporting better outcomes with trigger strategies*

The idea behind a trigger strategy is that the public might have a system of expectations about the government's behavior that induces the government to choose a better than Nash outcome $(\tilde{x}, \tilde{y})$. The public expects that as long as the government chooses $\tilde{y}$, it will continue to do so in the future; but once the government deviates from this choice, the public expects that it will choose $y^N$ thereafter, prompting the public (really "the market") to react with $x^N = h(y^N)$. This system of expectations confronts the government with the prospect of being "punished by the market's expectations" if it chooses to deviate from $\tilde{y}$.

To formalize this idea, we shall use the subgame perfect equilibrium $\sigma^N$ as a continuation strategy and the value $v^N$ as a continuation value in part (2) of Definition 6 of a subgame equilibrium (for $\eta \neq y_t$); then by working backward one step, we shall try to construct *another* subgame perfect equilibrium [with first-period outcome $(\tilde{x}, \tilde{y}) \neq (x^N, y^N)$]. In particular, for our new subgame perfect equilibrium we propose to set

$$\begin{gathered} \tilde{\sigma}_1 = (\tilde{x}, \tilde{y}) \\ \tilde{\sigma}|_{(x,y)} = \begin{cases} \tilde{\sigma} & \text{if } (x,y) = (\tilde{x}, \tilde{y}) \\ \sigma^N & \text{if } (x,y) \neq (\tilde{x}, \tilde{y}) \end{cases} \end{gathered} \tag{16.7}$$

where $(\tilde{x}, \tilde{y})$ is a competitive equilibrium that satisfies the following particular case of part (2):

$$(1-\delta)\, r(\tilde{x}, \tilde{y}) + \delta \tilde{v} \geq (1-\delta)\, r(\tilde{x}, \eta) + \delta v^N , \tag{16.8}$$

for all $\eta \in Y$, where $\tilde{v} = V_g\Big(\{x_t = \tilde{x}\}_{t=1}^{\infty},\ \{y_t = \tilde{y}\}_{t=1}^{\infty}\Big) = r(\tilde{x}, \tilde{y})$. Inequality (16.8) is equivalent with

$$\max_{\eta \in Y} r(\tilde{x}, \eta) - r(\tilde{x}, \tilde{y}) \leq \frac{\delta}{1-\delta}\, (\tilde{v} - v^N). \tag{16.9}$$

For any $(\tilde{x}, \tilde{y}) \in C$ that satisfies expression (16.9) with $\tilde{v} = r(\tilde{x}, \tilde{y})$, equation (16.7) is a subgame perfect equilibrium with value $\tilde{v}$.

If $(\tilde{x}, \tilde{y}) = (x^R, y^R)$ satisfies inequality (16.9) with $\tilde{v} = r(x^R, y^R)$, then repetition of the Ramsey outcome $(x^R, y^R)$ is supportable by a subgame perfect equilibrium of the form (16.7).

This construction uses the following objects:

1. A proposed first-period equilibrium $(\tilde{x}, \tilde{y}) \in C$;
2. A subgame perfect equilibrium $\sigma^2$ with value $V_g(\sigma^2)$, which is used to synthesize the continuation strategy in the event that the first-period outcome does not equal $(\tilde{x}, \tilde{y})$, via the relationship $\tilde{\sigma}|_{(x,y)} = \sigma^2$, if $(x, y) \neq (\tilde{x}, \tilde{y})$. In the preceding example, $\sigma^2 = \sigma^N$ and $V_g(\sigma^2) = v^N$.
3. A subgame perfect equilibrium $\sigma^1$, with value $V_g(\sigma^1)$, used to define the continuation value to be assigned after first-period outcome $(\tilde{x}, \tilde{y})$ and the continuation strategy $\tilde{\sigma}|_{(\tilde{x},\tilde{y})} = \sigma^1$. In the preceding example, $\sigma^1 = \tilde{\sigma}$, which is defined recursively (and self-referentially) via equation (16.7).
4. A candidate for a new equilibrium $\tilde{\sigma}$, defined in object 3, and a corresponding value $V_g(\tilde{\sigma})$. In the example, $V_g(\tilde{\sigma}) = r(\tilde{x}, \tilde{y})$.

In the preceding example, objects 3 and 4 are equated, and this fact constitutes a special feature.

Note how we have used the lemma in verifying that $\tilde{\sigma}$ is a subgame perfect equilibrium. We start with the subgame perfect equilibrium $\sigma^N$ with associated value $v^N$. We guess a first-period outcome pair $(\tilde{x}, \tilde{y})$ and a value $\tilde{v}$ for a new subgame perfect equilibrium, where $\tilde{v} = r(\tilde{x}, \tilde{y})$. Then we verify requirements (2) and (3) of the lemma with $(v^N, \tilde{v})$ as continuation values and $(\tilde{x}, \tilde{y})$ as first-period outcomes.

## *Values of all SPE*

The role played by the lemma in analyzing our two examples hints at the central role that it plays in the methods that Abreu, Pearce, and Stacchetti have developed for describing and computing values for *all* the subgame perfect equilibria for setups like ours. APS build on the way that the lemma characterizes subgame perfect equilibrium in terms of a first-period equilibrium, along with a pair of continuation values, each element of which is itself a value associated with a different subgame perfect equilibrium.

The lemma directs Abreu, Pearce and Stacchetti's attention away from the *set of strategy profiles* and toward the *set of values* $V_g(\sigma)$ associated with those profiles. They define the set $V$ of values associated with subgame perfect equilibria:

$$V = \{V_g(\sigma) \mid \sigma \text{ is a subgame perfect equilibrium}\}.$$

Evidently, $V \subset I\!R$. From the lemma, for a given competitive equilibrium $(x, y) \in C$, there exists a subgame perfect equilibrium $\sigma$ for which $x = \sigma_1^h, y = \sigma_1^g$ if and only if there exist two values $v_1, v_2 \in V$ such that

$$(1-\delta)\, r(x,y) + \delta v_1 \geq (1-\delta)\, r(x,\eta) + \delta v_2 \quad \forall\, \eta \in Y. \tag{16.10}$$

Let $\sigma^1$ and $\sigma^2$ be subgame perfect equilibria for which $v_1 = V_g(\sigma^1)$, $v_2 = V_g(\sigma^2)$. The subgame perfect equilibrium $\sigma$ that supports $(x,y) = (\sigma_1^h, \sigma_1^g)$ is completed by specifying $\sigma|_{(x,y)} = \sigma^1$ and $\sigma|_{(\nu,\eta)} = \sigma^2$ if $(\nu,\eta) \neq (x,y)$.

This construction produces out of two values $v_1, v_2 \in V$ a subgame perfect equilibrium $\sigma$ with value $v \in V$ given by

$$v = (1-\delta)\, r(x,y) + \delta v_1 \,.$$

Thus, the construction maps pairs $(v_1, v_2)$ into a strategy $\sigma$, a first-period competitive equilibrium $(x,y)$, and a value $v = V_g(\sigma)$.

APS characterize subgame perfect equilibria by studying a mapping from pairs of continuation values $v_1, v_2 \in V$ into values $v \in V$. They use the following definitions:

DEFINITION 7: Let $W \subset I\!R$. A 4-tuple $(x, y, w_1, w_2)$ is said to be *admissible with respect to* $W$ if $(x,y) \in C$, $w_1, w_2 \in W \times W$, and

$$(1-\delta)\, r(x,y) + \delta w_1 \geq (1-\delta)\, r(x,\eta) + \delta w_2\,, \quad \forall\, \eta \in Y. \tag{16.11}$$

Notice that when $W \subset V$, the admissible 4-tuple $(x, y, w_1, w_2)$ determines a subgame perfect equilibrium with strategy profile

$$\sigma_1 = (x,y),\ \sigma|_{(x,y)} = \sigma^1,\ \sigma|_{(\nu,\eta)} = \sigma^2 \ \text{ for } (\nu,\eta) \neq (x,y)$$

where $w_1 = V_g(\sigma^1)$, $w_2 = V_g(\sigma^2)$. The value of the equilibrium is $V_g(\sigma) = w = (1-\delta)\, r(x,y) + \delta w_1$.

Abreu, Pearce, and Stacchetti use admissible 4-tuples to generate new candidate values.

DEFINITION 8: For each set $W \subset I\!R$, let $B(W)$ be the set of possible values $w = (1-\delta)\, r(x,y) + \delta w_1$ associated with admissible tuples $(x, y, w_1, w_2)$.

The operator $B(W)$ maps sets of values into sets of values. Think of $W$ as a set of potential continuation values and $B(W)$ as the set of values that they support. By the definition of admissibility it immediately follows that the operator $B$ is *monotone.*

PROPERTY (monotonicity of $B$): If $W \subseteq W' \subseteq R$, then $B(W) \subseteq B(W')$.

PROOF: It can be verified directly from the definition of admissible 4-tuples that if $w \in B(W)$, then $w \in B(W')$: simply use the $(w_1, w_2)$ pair that supports $w \in B(W)$ to support $w \in B(W')$. ▮

It can also be verified that $B(\cdot)$ maps compact sets $W$ into compact sets $B(W)$.

The self-referential character of subgame perfect equilibria is exploited in the following definition:

DEFINITION 9: The set $W$ is said to be *self-generating* if $W \subseteq B(W)$.

The set $V$ of subgame perfect equilibrium values $V_g(\sigma)$ is self-generating, by virtue of the lemma. Thus, we can write $V \subseteq B(V)$. APS show that $V$ is the *largest* self-generating set. The key to showing this point is the following theorem:[7]

THEOREM 1 (Self-Generation): If $W \subset I\!R$ is bounded and self-generating, then $B(W) \subseteq V$.

The proof proceeds by taking a point $w \in W \subseteq B(W)$ and constructing a subgame perfect equilibrium with value $w$.

PROOF: Assume $W \subseteq B(W)$. Choose an element $w \in B(W)$ and transform it as follows into a subgame perfect equilibrium:

*Step 1.* Because $w \in B(W)$, we know that there exist outcomes $(x, y)$ and values $w_1$ and $w_2$ that satisfy

$$\begin{aligned} w = (1-\delta)\, r(x, y) + \delta w_1 \geq (1-\delta)\, r(x, \eta) + \delta w_2 \quad \forall \eta \in Y \\ (x, y) \in C \\ w_1, w_2 \in W \times W. \end{aligned}$$

[7] The *unbounded* set $I\!R$ (the extended real line) is self-generating but not meaningful. It is self-generating because any value $v \in I\!R$ can be supported because there are no limits on the continuation values. It is not meaningful because most points in $I\!R$ are values that cannot be attained with *any* strategy profile.

Set $\sigma_1 = (x, y)$.

*Step 2.* Since $w_1 \in W \subseteq B(W)$, there exist outcomes $(\tilde{x}, \tilde{y})$ and values $(\tilde{w}_1, \tilde{w}_2) \in W$ that satisfy

$$w_1 = (1-\delta)\, r(\tilde{x}, \tilde{y}) + \delta \tilde{w}_1 \geq (1-\delta)\, r(\tilde{x}, \eta) + \delta \tilde{w}_2, \quad \forall\, \eta \in Y$$
$$(\tilde{x}, \tilde{y}) \in C.$$

Set the first-period outcome in period 2 (the outcome to be played *given* that $y$ was chosen in period 1) equal to $(\tilde{x}, \tilde{y})$; that is, set $(\sigma|_{(x,y)})_1 = (\tilde{x}, \tilde{y})$.

Continuing in this way, for each $w \in B(W)$, we can create a sequence of continuation values $w_1, \tilde{w}_1, \tilde{\tilde{w}}_1, \ldots$ and a corresponding sequence of first-period outcomes $(x, y)$, $(\tilde{x}, \tilde{y})$, $(\tilde{\tilde{x}}, \tilde{\tilde{y}})$.

At each stage in this construction, policies are *unimprovable*, which means that given the continuation values, one-period deviations from the prescribed policies are not optimal. It follows that the strategy profile is optimal. By construction $V_g(\sigma) = w$. ∎

Collecting results, we know that

1. $V \subseteq B(V)$ (by the lemma).
2. If $W \subseteq B(W)$, then $B(W) \subseteq V$ (by self-generation).
3. $B$ is monotone and maps compact sets into compact sets.

Facts 1 and 2 imply that $V = B(V)$, so that the set of equilibrium values is a "fixed point" of $B$, in particular, the largest bounded fixed point.

Monotonicity of $B$ and the fact that it maps compact sets into compact sets provides us with an algorithm for computing the set $V$, namely, to start with a set $W_0$ for which $V \subseteq B(W_0) \subseteq W_0$, and to iterate to convergence on $B$. In more detail, we use the following steps:

1. Start with a set $W_0 = [\underline{w}_0, \bar{w}_0]$ that we know is bigger than $V$, and for which $B(W_0) \subseteq W_0$. It will always work to set $\bar{w}_0 = \max_{(x,y)\in C} r(x, y)$, $\underline{w}_0 = \min_{(x,y)\in C} r(x, y)$.
2. Compute the boundaries of the set $B(W_0) = [\underline{w}_1, \bar{w}_1]$. The value $\bar{w}_1$ solves the problem

$$\bar{w}_1 = \max_{(x,y)\in C} (1-\delta)\, r(x, y) + \delta \bar{w}_0$$

subject to

$$(1-\delta)\, r(x, y) + \delta \bar{w}_0 \geq (1-\delta)\, r(x, \eta) + \delta \underline{w}_0 \quad \text{for all } \eta \in Y$$

The value $\underline{w}_1$ solves the problem

$$\underline{w}_1 = \min_{(x,y)\in C;\, (w_1,w_2)\in[\underline{w}_0,\bar{w}_0]^2} (1-\delta)\, r(x,y) + \delta w_1$$

subject to

$$(1-\delta)\, r(x,y) + \delta w_1 \geq (1-\delta)\, r(x,\eta) + \delta w_2 \quad \forall\, \eta \in Y.$$

With $(\underline{w}_0, \bar{w}_0)$ chosen as before, it will be true that $B(W_0) \subseteq W_0$.

3. Having constructed $W_1 = B(W_0) \subseteq W_0$, continue to iterate, producing a decreasing sequence of compact sets $W_{j+1} = B(W_j) \subseteq W_j$. Iterate until the sets converge.

## Self-enforcing SPE

The subgame perfect equilibrium with the *worst* value $v \in V$ has a remarkable property, namely, that it is "self-enforcing." We use the following definition:

DEFINITION 10: A subgame perfect equilibrium $\sigma$ with first-period outcome $(\tilde{x}, \tilde{y})$ is said to be *self-enforcing* if

$$\sigma|_{(x,y)} = \sigma \qquad \text{if } (x,y) \neq (\tilde{x},\tilde{y}). \tag{16.12}$$

A strategy profile satisfying equation (16.12) is called self-enforcing because after a one-shot deviation the prescribed punishment is simply to restart the equilibrium.

The value $\underline{v}$ associated with the worst subgame perfect equilibrium $\sigma$ satisfies

$$\underline{v} = \min_{x,y,v} \left\{ (1-\delta)\, r(x,y) + \delta v \right\} \tag{16.13}$$

where the minimization is subject to $(x,y) \in C$, $v \in V$, and the incentive constraint

$$(1-\delta)\, r(x,y) + \delta v \geq (1-\delta)\, r(x,\eta) + \delta \underline{v} \quad \text{for all } \eta \in Y. \tag{16.14}$$

Let $\tilde{v}$ be the continuation value that attains the right side of equation (16.13), and let $\sigma_{\tilde{v}}$ be the subgame perfect equilibrium that supports continuation value

$\tilde{v}$. Let $(\tilde{x}, \tilde{y})$ be the first-period outcome that attains the right side of equation (16.13). Since $\underline{v}$ is both the continuation value when first-period outcome $(x, y) \neq (\tilde{x}, \tilde{y})$ *and* the value associated with subgame perfect equilibrium $\sigma$, it follows that

$$\sigma_1 = (\tilde{x}, \tilde{y})$$
$$\sigma|_{(x,y)} = \begin{cases} \sigma & \text{if } (x,y) \neq (\tilde{x}, \tilde{y}) \\ \sigma_{\tilde{v}} & \text{if } (x,y) = (\tilde{x}, \tilde{y}). \end{cases} \tag{16.15}$$

Because of the double role played by $\underline{v}$ [i.e., $\underline{v}$ is both the value of equilibrium $\sigma$ and the "punishment" continuation value of the right side of the incentive constraint (16.14)], the equilibrium strategy $\sigma$ that supports $\underline{v}$ is self-enforcing.[8]

The preceding argument thus establishes this proposition:

PROPOSITION 2: The subgame perfect equilibrium $\sigma$ associated with $\underline{v} = \min V$ is self-enforcing.

Notice that the first subgame perfect equilibrium that we computed, whose outcome was infinite repetition of the one-period Nash equilibrium, is a self-enforcing equilibrium. However, in general, the infinite repetition of the one-period Nash equilibrium is not the *worst* subgame perfect equilibrium. This fact opens up the possibility to support repetition of the Ramsey outcome by threatening to revert to a worse equilibrium whenever it is not possible to support repetition of the Ramsey outcome by permanent reversion to the one-period Nash equilibrium.

## *Recursive strategies*

This section emphasizes similarities between credible government policies and the recursive contracts appearing in chapter 15. We will study situations where the strategy of the aggregate of households and of the government have a recursive representation. This approach substantially restricts the space of strategies because most history-dependent strategies cannot be represented recursively. Nevertheless, this class of strategies excludes no equilibrium payoffs $v \in V$. We use the following definitions:

[8] The structure of the programming problem, with the double role played by $\underline{v}$, makes it possible to compute the worst value directly.

DEFINITION 11: The aggregate of households and the government follow *recursive strategies* if there is a 3-tuple of functions $\phi = (z^h, z^g, \mathcal{V})$ and an initial condition $v_1$ with the following structure:

$$\begin{aligned} v_1 &\in I\!R \text{ is given} \\ x_t &= z^h(v_t) \\ y_t &= z^g(v_t) \\ v_{t+1} &= \mathcal{V}(v_t, x_t, y_t), \end{aligned} \tag{16.16}$$

where $v_t$ is a state variable designed to summarize the history of outcomes before $t$.

This recursive form of strategies operates much like an autoregression to let time-$t$ actions $(x_t, y_t)$ depend on the history $\{y_s, x_s\}_{s=1}^{t-1}$, as mediated through the state variable $v_t$. Representation (16.16) induces history-dependent government policies, and thereby allows for reputation. We shall soon see that beyond its role as descriptor of histories, $v_t$ also summarizes the future.[9]

A strategy description $(\phi, v)$ recursively generates an entire outcome path expressed as $(\vec{x}, \vec{y}) = (\vec{x}, \vec{y})(\phi, v)$. By substituting the outcome path into equation (16.4), we find that $(\phi, v)$ induces a value for the government, which we write as

$$\begin{aligned} V^g\big[(\vec{x}, \vec{y})(\phi, v)\big] =& (1-\delta)\, r\big[z^h(v), z^g(v)\big] \\ &+ \delta\, V^g\Big((x, y)\{\phi, \mathcal{V}[v, z^h(v), z^g(v)]\}\Big). \end{aligned} \tag{16.17}$$

So far, we have not interpreted the state variable $v$, except as a particular measure of the history of outcomes. The theory of credible policy ties past and future together by making the state variable $v$ a promised value, an outcome to be expressed

$$v = V^g\big[(\vec{x}, \vec{y})(\phi, v)\big]. \tag{16.18}$$

Equations (16.16), (16.17), and (16.18) assert a dual role for $v$. In equation (16.16), $v$ accounts for past outcomes. In equations (16.17) and (16.18), $v$ looks

[9] By iterating equations (16.16), we can construct a pair of sequences of functions indexed by $t \geq 1$ $\{Z_t^h(I_t), Z_t^g(I_t)\}$, mapping histories that are augmented by initial conditions $I_t = (\{x_s, y_s\}_{s=1}^{t-1}, v_1)$ into time-$t$ actions $(x_t, y_t) \in X \times Y$. Strategies for the repeated economy are a pair of sequences of such functions without the restriction that they have a recursive representation.

forward. The state $v_t$ is a discounted future value with which the government enters time $t$ based on past outcomes. Depending on the outcome $(x, y)$ and the entering promised value $v$, $\mathcal{V}$ updates the promised value with which the government leaves the period. We postpone struggling with the questions of which of two valid interpretations of the government's strategy should be emphasized: something chosen by the government, or a description of a system of public expectations to which the government conforms.

Evidently, we have the following:

DEFINITION 12: A recursive strategy description $(\phi, v)$ in equation (16.16) is a *subgame perfect equilibrium* (SPE) if and only if

(1) The outcome $x = z^h(v)$ is a competitive equilibrium, given that $y = z^g(v)$.
(2) For each $\eta \in Y$, $\mathcal{V}(v, z^h(v), \eta)$ is a value for a subgame perfect equilibrium.
(3) For each $\eta \in Y$,

$$\begin{aligned} v &= (1-\delta) r\big[z^h(v), z^g(v)\big] + \delta \mathcal{V}\big[v, z^h(v), z^g(v)\big] \\ &\geq (1-\delta) r\big[z^h(v), \eta\big] + \delta \mathcal{V}\big[v, z^h(v), \eta\big]. \end{aligned} \tag{16.19}$$

Condition (1) asserts that the first-period outcome pair $(x, y)$ is a competitive equilibrium. Each member of the private sector forms an expectation about the government's action according to $y_t = z^g(v_t)$, and the "market" responds with a competitive equilibrium $x_t$,

$$x_t = h(y_t) = h\big[z^g(v_t)\big] \equiv z^h(v_t). \tag{16.20}$$

This argument builds in rational expectations, because the private sector knows both the state variable $v_t$ and the government's decision rule $z^g$.

Besides the first-period outcome $(x, y)$, conditions (2) and (3) associate with a subgame perfect equilibrium three additional objects: a promised value $v$; a continuation value $v' = \mathcal{V}[v, z^h(v), z^g(v)]$ if the required first-period outcome is observed; and another continuation value $\tilde{v}(\eta) = \mathcal{V}[v, z^h(v), \eta]$ if the required first-period outcome is not observed but rather some pair $(x, \eta)$. All the continuation values must themselves be attained as subgame perfect equilibria. In terms of these objects, condition (3) is an incentive constraint inspiring the government to adhere to the equilibrium

$$\begin{aligned} v &= (1-\delta) r(x,y) + \delta v' \\ &\geq (1-\delta) r(y,\eta) + \delta \tilde{v}(\eta), \quad \forall \eta \in Y. \end{aligned}$$

This formula states that the government receives more if it adheres to an action called for by its strategy than if it departs. Part (2) of Definition 12 requires that the continuation values be values for subgame perfect equilibria to ensure that these values constitute "credible punishments." The definition is circular, because the same class of objects, namely equilibrium values $v$, occur on each side of expression (16.19). Circularity comes with recursivity.

One implication of the work of Abreu, Pearce, and Stacchetti (1986, 1990) is that recursive equilibria of form (16.16) can attain all subgame perfect equilibrium values. As we have seen, APS's innovation was to shift the focus away from the set of equilibrium strategies and toward the set of values $V$ attainable with subgame equilibrium strategies. They described a set $V$ such that for all $v \in V$, $v$ is the value associated with a subgame perfect equilibrium.

## *Examples of SPE with recursive strategies*

Our two earlier examples of subgame perfect equilibria were already of a recursive nature. But to highlight this property, we recast those SPE in the present notation for recursive strategies. Equilibria are constructed by using a guess-and-verify technique. First guess $(v_1, z^h, z^g, \mathcal{V})$ in equations (16.16), then verify parts (1), (2), and (3) of Definition 12.

The examples parallel the historical development of the theory. (1) The first example is infinite repetition of a one-period Nash outcome, which was Kydland and Prescott's (1977) time-consistent equilibrium. (2) Barro and Gordon (1983a, 1983b) and Stokey (1989) used the value from infinite repetition of the Nash outcome as a continuation value to deter deviation from the Ramsey outcome. For sufficiently high discount factors, the continuation value associated with repetition of the Nash outcome can deter the government from deviating from infinite repetition of the Ramsey outcome. This is not possible for low discount factors. (3) Abreu (1988) and Stokey (1991) showed that Abreu's "stick and carrot" strategy induces more severe punishments than repetition of the Nash outcome.

*Infinite repetition of Nash outcome*

It is easy to construct an equilibrium whose outcome path forever repeats the one-period Nash outcome. Let $v^N = r(x^N, y^N)$. The proposed equilibrium is

$$\begin{aligned} v_1 &= v^N, \\ z^h(v_t) &= x^N \ \forall\ v_t, \\ z^g(v_t) &= y^N \ \forall\ v_t, \text{ and} \\ \mathcal{V}(v_t, x_t, y_t) &= v^N, \ \forall\ (v_t, x_t, y_t). \end{aligned}$$

Here for each $t$, $v^N$ plays *all* the roles of $v, v'$, and $\tilde{v}$ in condition (3). Conditions (1) and (2) are satisfied by construction, and condition (3) collapses to

$$r(x^N, y^N) \geq r\left[x^N, H(x^N)\right],$$

which is satisfied at equality by the definition of a best response function. The equilibrium outcome forever repeats Kydland and Prescott's time-consistent equilibrium.

*Infinite repetition of a better than Nash outcome*

Let $v^b$ be a value associated with outcome $(x^b, y^b)$ such that $v^b = r(x^b, y^b) > v^N$, and assume that $(x^b, y^b)$ constitutes a competitive equilibrium. Suppose further that

$$r\left[x^b, H(x^b)\right] - r(x^b, y^b) \leq \frac{\delta}{1-\delta}(v^b - v^N). \tag{16.21}$$

The left side is the one-period return to the government from deviating from $y^b$; it is the gain from deviating. The right side is the difference in present values associated with conforming to the plan versus reverting forever to the Nash equilibrium; it is the cost of deviating. When the inequality is satisfied, the equilibrium presents the government with an incentive not to deviate from $y^b$. Then a SPE is

$$\begin{aligned} v_1 &= v^b \\ z^h(v) &= \begin{cases} x^b & \text{if } v = v^b; \\ x^N & \text{otherwise;} \end{cases} \\ z^g(v) &= \begin{cases} y^b & \text{if } v = v^b; \\ y^N & \text{otherwise;} \end{cases} \\ \mathcal{V}(v, x, y) &= \begin{cases} v^b & \text{if } (v, x, y) = (v^b, x^b, y^b); \\ v^N & \text{otherwise.} \end{cases} \end{aligned}$$

This strategy specifies outcome $(x^b, y^b)$ and continuation value $v^b$ as long as $v^b$ is the value promised at the beginning of the period. Any deviation from $y^b$ generates continuation value $v^N$. Inequality (16.21) validates condition (3) of Definition 12.

Barro and Gordon (1983a) considered a version of this equilibrium in which inequality (16.21) is satisfied with $(v^b, x^b, y^b) = (v^R, x^R, y^R)$. In this case, anticipated reversion to Nash forever supports Ramsey forever. When inequality (16.21) is *not* satisfied for $(v^b, x^b, y^b) = (v^R, x^R, y^R)$, we can solve for the best SPE value $v^b$ supportable by infinite reversion to Nash [with associated actions $(x^b, y^b)$] from

$$v^b = r(x^b, y^b) = (1-\delta) r\left[x^b, H(x^b)\right] + \delta v^N > v^N. \tag{16.22}$$

The payoff from following the strategy equals that from deviating and reverting to Nash. Any value lower than this can be supported, but none higher.

In a related context, Abreu (1988) searched for a way to support something better than $v^b$ when $v^b < v^R$. First, one must construct an equilibrium that yields a value worse than permanent repetition of the Nash outcome. The expectation of reverting to this equilibrium supports something better than $v^b$ in equation (16.22).

Somehow the government must be induced temporarily to take an action $y^{\#}$ that yields a worse period-by-period return than the Nash outcome, meaning that the government in general would be tempted to deviate. An equilibrium system of expectations has to be constructed that makes the government expect to do better in the future only by conforming to expectations that it temporarily adheres to the bad policy $y^{\#}$.

### *Something worse: a stick and carrot strategy*

Abreu (1988) proposed a "stick and carrot punishment." The "stick" part is an outcome $(x^{\#}, y^{\#}) \in C$, which is a bad competitive equilibrium from the government's viewpoint. The "carrot" part is the Ramsey outcome $(x^R, y^R)$, which the government attains forever after it has accepted the "stick" in the first period of its punishment.

We want a continuation value $v^*$ for deviating to support the first-period outcome $(x^{\#}, y^{\#})$ and attain the value[10]

$$\tilde{v} = (1-\delta) r(x^{\#}, y^{\#}) + \delta\, v^R \geq (1-\delta) r\left[x^{\#}, H(x^{\#})\right] + \delta\, v^*. \tag{16.23}$$

[10] This is a "one-period stick". Sometimes the worst SPE requires more than one period of a bad outcome.

Abreu proposed to set $v^* = \tilde{v}$. That is, the continuation value caused by deviating from the first-period action equals the original value. If the "stick" part is severe enough, this strategy attains a value worse than repetition of Nash. The strategy induces the government to accept the temporarily bad outcome by promising a high continuation value.

A SPE featuring "stick and carrot punishments" that attains $\tilde{v}$ is

$$\begin{aligned} v_1 &= \tilde{v} \\ z^h(v) &= \begin{cases} x^R & \text{if } v - v^R; \\ x^{\#} & \text{otherwise;} \end{cases} \\ z^g(v) &= \begin{cases} y^R & \text{if } v = v^R; \\ y^{\#} & \text{otherwise;} \end{cases} \\ \mathcal{V}(v,x,y) &= \begin{cases} v^R & \text{if } (x,y) = \left[z^h(v), z^g(v)\right]; \\ \tilde{v} & \text{otherwise.} \end{cases} \end{aligned} \tag{16.24}$$

The consequence of the government deviating from the bad prescribed first-period action $y^{\#}$ is to restart the equilibrium. In other words, the equilibrium is self-enforcing.

## *The best and the worst SPE*

We display a pair of simple programming problems to find the best and worst SPE values. APS (1990) showed how to find the entire set of equilibrium values $V$. In the current setting, their ideas imply the following:

1. The set of equilibrium values $V$ attainable by the government is a compact subset $[\underline{v}, \overline{v}]$ of $[\min_{(x,y)\in C} r(x,y), r(x^R, y^R)]$.

2. The worst equilibrium value $\underline{v}$ can be computed from a simple programming problem.

3. Given the worst equilibrium value $\underline{v}$, the best equilibrium value $\overline{v}$ can be computed from a programming problem.

4. Given a $v \in [\underline{v}, \overline{v}]$, it is easy to construct an equilibrium that attains it.

Recall from Proposition 2 that the worst equilibrium is self-enforcing, and here we repeat versions of equations (16.13) and (16.14),

$$\underline{v} = \min_{y \in Y,\, v_1 \in V} \left\{ (1-\delta)\, r\left[h(y), y\right] + \delta v_1 \right\} \tag{16.25}$$

where the minimization is subject to the incentive constraint

$$(1-\delta)\, r[h(y), y] + \delta v_1 \geq (1-\delta)\, r\{h(y), H[h(y)]\} + \delta \underline{v}. \tag{16.26}$$

In expression (16.26), we use the worst SPE as the continuation value in the event of a deviation. The minimum will be attained when the constraint is binding, which implies that $\underline{v} = r\{h(y), H[h(y)]\}$, for some government action $y$. Thus, the problem of finding the worst SPE reduces to solving

$$\underline{v} = \min_{y \in Y} r\{h(y), H[h(y)]\};$$

then computing $v_1$ from $(1-\delta) r[h(y^{\#}), y^{\#}] + \delta v_1 = \underline{v}$ where $y^{\#} = \arg\min r\{h(y), H[h(y)]\}$; and finally checking that $v_1$ is itself a value associated with a SPE. To check this condition, we need to know $\overline{v}$.

The computation of $\overline{v}$ utilizes the fact that the best SPE is self-rewarding; that is, the best SPE has continuation value $\overline{v}$ when the government follows the prescribed equilibrium strategy. Thus, after we have computed a candidate for the worst SPE value $\underline{v}$, we can compute a candidate for the *best* value $\overline{v}$ by solving the programming problem

$$\overline{v} = \max_{y \in Y} \; r\big[h(y), y\big]$$

$$\text{subject to} \quad r\big[h(y), y\big] \geq (1-\delta) r\big\{h(y), H\big[h(y)\big]\big\} + \delta \underline{v}.$$

Here we are using the fact that $\overline{v}$ is the maximizing continuation value available to reward adherence to the policy, so that $\overline{v} = (1-\delta) r[h(y), y] + \delta \overline{v}$. Let $y^b$ be the maximizing value of $y$. Once we have computed $\overline{v}$, we can check that the continuation value $v_1$ for supporting the worst value is within our candidate set $[\underline{v}, \overline{v}]$. If it is, we have succeeded in constructing $V$.

### *Attaining the worst, method 1*

We have seen that to evaluate the best sustainable value $\overline{v}$, we want to find the find the worst value $\underline{v}$. Many SPEs attain the worst value $\underline{v}$. To compute one such SPE strategy, we can use the following recursive procedure:

1. Set the first-period promised value $v_1 = \underline{v} = r\{h(y^{\#}), H[h(y^{\#})]\}$, where $y^{\#} = \arg\min r\{h(y), H[h(y)]\}$. The worst one-period value that is consistent with rational expectations is $r[h(y^{\#}), y^{\#}]$. Given expectations $x^{\#} = h(y^{\#})$, the government is tempted toward $H(x^{\#})$, which yields one-period utility to the government of $r\{h(y^{\#}), H[h(y^{\#})]\}$.

Then use $\underline{v}$ as continuation value in the event of a deviation, and construct an increasing sequence of continuation values to reward adherence, as follows:

2. Solve $\underline{v} = (1-\delta)r[h(y^\#), y^\#] + \delta v_2$ for continuation value $v_2$.
3. For $j = 2, 3, \cdots$, continue solving $v_j = (1-\delta)r[h(y^\#), y^\#] + \delta v_{j+1}$ for the continuation values $v_{j+1}$ as long as $v_{j+1} \leq \bar{v}$. If $v_{j+1}$ threatens to violate this constraint at step $j = \bar{j}$, then go to step 4.
4. Use $\bar{v}$ as the continuation value, and solve $v_j = (1-\delta)r[h(\tilde{y}), \tilde{y}] + \delta\bar{v}$ for the prescription $\tilde{y}$ to be followed if promised value $v_j$ is encountered.
5. Set $v_{j+s} = \bar{v}$ for $s \geq 1$.

*Attaining the worst, method 2.*

To construct another equilibrium supporting the worst SPE value, follow steps 1 and 2, and follow step 3 also, except that we continue solving $v_j = (1-\delta)r[h(y^\#), y^\#]+\delta v_{j+1}$ for the continuation values $v_{j+1}$ only so long as $v_{j+1} < v^N$. As soon as $v_{j+1} = v^{**} > v^N$, we use $v^{**}$ as both the promised value and the continuation value thereafter. In terms of our recursive strategy notation, whenever $v^{**} = r[h(y^{**}), y^{**}]$ is the promised value, $z^h(v^{**}) = h(y^{**})$, $z^g(v^{**}) = y^{**}$, and $v'[v^{**}, z^h(v^{**}), z^g(v^{**})] = v^{**}$.

*Attaining the worst, method 3.*

Here is another subgame perfect equilibrium that supports $\underline{v}$. Proceed as in step 1 to find continuation value $v_2$. Now set all the subsequent values and continuation values to $v_2$, with associated first-period outcome $\tilde{y}$ which solves $v_2 = r[h(\tilde{y}), \tilde{y}]$. It can be checked that the incentive constraint is satisfied with $\underline{v}$ the continuation value in the event of a deviation.

*Numerical example*

We now illustrate the concepts and arguments using the infinitely repeated version of the taxation example. To make the problem of finding $\underline{v}$ nontrivial, we impose an upper bound on admissible tax rates given by $\bar{\tau} = 1 - \alpha - \epsilon$ where $\epsilon \in (0, 0.5 - \alpha)$. Given $\tau \in Y \equiv [0, \bar{\tau}]$, the model exhibits a unique Nash equilibrium with $\tau = 0.5$. For a sufficiently small $\epsilon$, the worst one-period competitive equilibrium is $[\ell(\bar{\tau}), \bar{\tau}]$.

Set $[\alpha \quad \delta \quad \bar{\tau}] = [0.3 \quad 0.8 \quad 0.6]$. Compute $[\tau^R \quad \tau^N] = [0.3013 \quad 0.5000]$, $[v^R \quad v^N \quad \underline{v} \quad v_{\text{abreu}}] = [-0.6801 \quad -0.7863 \quad -0.9613 \quad -0.7370]$. In this numerical example, Abreu's "stick and carrot" strategy fails to attain a value lower

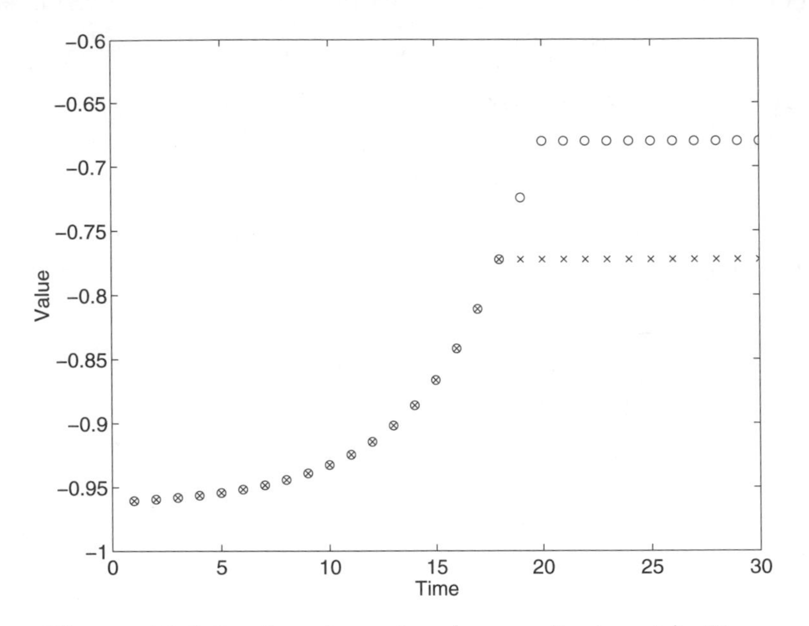


**Figure 16.6** Continuation values (on coordinate axis) of two SPE that attain $\underline{v}$.

than the repeated Nash outcome. The reason is that the upper bound on tax rates makes the least favorable one-period return (the "stick") not so bad.

Figure 16.6 describes two subgame perfect equilibria that attain the worst SPE value $\underline{v}$ with the depicted sequences of time-$t$ (promised value, tax rate) pairs. The circles represent the worst SPE attained with method 1, and the x-marks correspond to method 2. By construction, the continuation values of method 2 are less than or equal to the continuation values of method 1. Since both SPE attain the same promised value $\underline{v}$, it follows that method 2 must be associated with higher one-period returns in some periods. Figure 16.7 indicates that method 2 delivers those higher one-period returns around period 20 when the prescribed tax rates are closer to the Ramsey outcome $\tau^R = 0.3013$.

When varying the discount factor, we find that the cutoff value of $\delta$ below which reversion to Nash fails to support Ramsey forever is 0.2194.

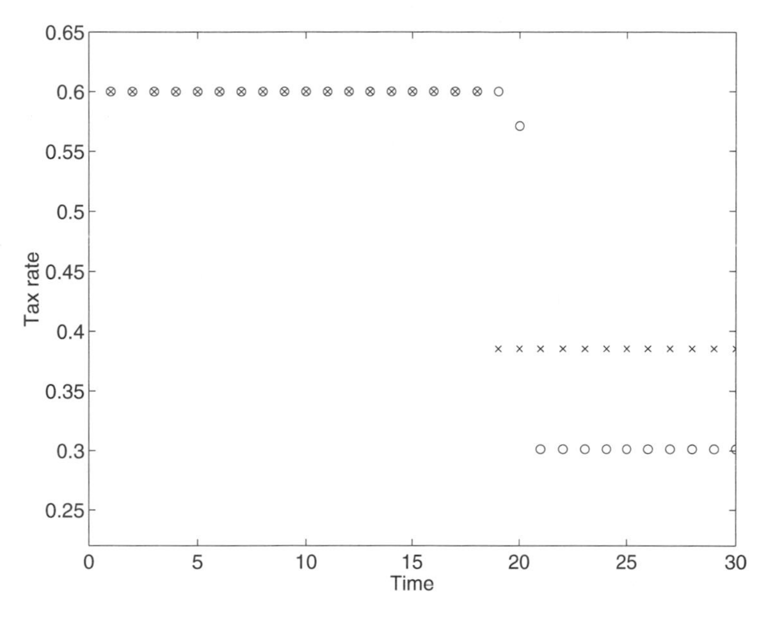


**Figure 16.7** Tax rates associated with the continuation values of Figure 16.6.

## Interpretations

The notion of credibility or sustainability emerges from a ruthless and complete application of two principles: rational expectations and self-interest. At each moment and for each possible history, individuals and the government act in their own best interests while expecting everyone else always to act in their best interests. A credible government policy is one that is in the interests of the government to implement at every moment and on every occasion.

The structures that we have studied have multiple equilibria, indexed by different systems of rational expectations. Multiple equilibria are essential to the construction, because what sustains a good equilibrium is a system of expectations that credibly raises the prospect of reverting to a bad equilibrium if the government deviates from the good equilibrium. For the expectations of reverting to the bad equilibrium to be credible, the bad equilibrium must itself be an equilibrium; that is, it must be in the self-interest of all agents to behave as they are expected to. Supporting a Ramsey outcome hinges on finding an equilibrium with outcomes bad enough to deter the government from surrendering to the temptation of a one-period deviation.[11]

Is the multiplicity of equilibria a strength or a weakness of such theories? In these structures, descriptions of preferences and technologies, supplemented by the restriction of rational expectations, don't pin down outcomes: an independent role is left for expectations. The theory is silent about which equilibrium will prevail; there is no sense in which the government chooses which equilibrium emerges. Depending on the purpose, the multiplicity of equilibria can be regarded either as a strength and as a weakness of these theories.

In inferior equilibria, the government is caught in an "expectations trap,"[12] an aspect of the theory that highlights how the government can be regarded not as choosing its element of the equilibrium strategy profile, $\sigma^g$, but as simply affirming the public's expectation of it. Within the theory, the government's strategy plays a dual role, as it does in any rational expectations model: one summarizing the government's choices, the other describing the public's rule for forecasting the government's behavior. In inferior equilibria, the government wishes that it could use a different strategy but nevertheless conforms to the public's expectations that it will adhere to an inferior rule.

## Exercises

*Exercise 16.1* Consider the following one-period economy. Let $(\xi, x, y)$ be the choice variables available to a representative agent, the market as a whole, and a benevolent government, respectively. In a rational expectations equilibrium or competitive equilibrium, $\xi = x = h(y)$, where $h(\cdot)$ is the "equilibrium response" correspondence that gives competitive equilibrium values of $x$ as a function of $y$; that is, $[h(y), y]$ is a competitive equilibrium. Let $C$ be the set of competitive equilibria.

Let $X = \{x_M, x_H\}, Y = \{y_M, y_H\}$. For the one-period economy, when $\xi_i = x_i$, the payoffs to the government and household are given by the values of $u(x_i, x_i, y_j)$ entered in the following table:

[11] This statement means that an equilibrium is supported by beliefs about behavior at prospective histories of the economy that might never be attained or observed. Part of the literature on learning in games and dynamic economies studies situations in which it is not reasonable to expect "adaptive" agents to learn so much. See Kreps and Fudenberg (1998), and Kreps (1990).

[12] See Chari, Christiano, and Eichenbaum (1998).

One-period payoffs to the government–household
[values of $u(x_i, x_i, y_j)$]

| | $x_M$ | $x_H$ |
|---|---|---|
| $y_M$ | 10* | 20 |
| $y_H$ | 4 | 15* |

*Denotes $(x, y) \in C$.

The values of $u(\zeta_k, x_i, y_j)$ not reported in the table are such that the competitive equilibria are the outcome pairs denoted by an asterisk (*).

**a.** Find the *Nash equilibrium* (in pure strategies) and *Ramsey outcome* for the one-period economy.

**b.** Suppose that this economy is repeated twice. Is it possible to support the Ramsey outcome in the first period by reverting to the Nash outcome in the second period in case of a deviation?

**c.** Suppose that this economy is repeated three times. Is it be possible to support the Ramsey outcome in the first period? In the second period?

Consider the following expanded version of the preceding economy. $Y = \{y_L, y_M, y_H\}$, $X = \{x_L, x_M, x_H\}$. When $\xi_i = x_i$, the payoffs are given by $u(x_i, x_i, y_j)$ entered here:

One-period payoffs to the government–household
[values of $u(x_i, x_i, y_j)$]

| | $x_L$ | $x_M$ | $x_H$ |
|---|---|---|---|
| $y_L$ | 3* | 7 | 9 |
| $y_M$ | 1 | 10* | 20 |
| $y_H$ | 0 | 4 | 15* |

*Denotes $(x, y) \in C$.

**d.** What are Nash equilibria in this one-period economy?

**e.** Suppose that this economy is repeated twice. Find a subgame perfect equilibrium that supports the Ramsey outcome in the first period. For what values of $\delta$ will this equilibrium work?

**f.** Suppose that this economy is repeated three times. Find a subgame perfect equilibrium that supports the Ramsey outcome in the first two periods (assume $\delta = 0.8$). Is it unique?

*Exercise 16.2* Consider a version of the setting studied by Stokey (1989). Let $(\xi, x, y)$ be the choice variables available to a representative agent, the market as a whole, and a benevolent government, respectively. In a rational expectations or competitive equilibrium, $\xi = x = h(y)$, where $h(\cdot)$ is the "equilibrium response" correspondence that gives competitive equilibrium values of $x$ as a function of $y$; that is, $[h(y), y]$ is a competitive equilibrium. Let $C$ be the set of competitive equilibria.

Consider the following special case. Let $X = \{x_L, x_H\}$ and $Y = \{y_L, y_H\}$. For the one-period economy, when $\xi_i = x_i$, the payoffs to the government are given by the values of $u(x_i, x_i, y_j)$ entered in the following table:

One-period payoffs to the government–household
[values of $u(x_i, x_i, y_j)$]

| | $x_L$ | $x_H$ |
|---|---|---|
| $y_L$ | 0* | 20 |
| $y_H$ | 1 | 10* |

* Denotes $(x, y) \in C$.

The values of $u(\xi_k, x_i, y_j)$ not reported in the table are such that the competitive equilibria are the outcome pairs denoted by an asterisk (*).

**a.** Define a *Ramsey plan* and a *Ramsey outcome* for the one-period economy. Find the Ramsey outcome.

**b.** Define a *Nash equilibrium* (in pure strategies) for the one-period economy.

**c.** Show that there exists no Nash equilibrium (in pure strategies) for the one-period economy.

**d.** Consider the infinitely repeated version of this economy, starting with $t = 1$ and continuing forever. Define a *subgame perfect equilibrium.*

**e.** Find the value to the government associated with the *worst* subgame perfect equilibrium.

**f.** Assume that the discount factor is $\delta = .8913 = (1/10)^{1/20} = .1^{.05}$. Determine whether infinite repetition of the Ramsey outcome is sustainable as a subgame perfect equilibrium. If it is, display the associated subgame perfect equilibrium.

**g.** Find the value to the government associated with the *best* subgame perfect equilibrium.

**h.** Find the outcome path associated with the *worst* subgame perfect equilibrium.

**i.** Find the one-period continuation value $v_1$ and the outcome path associated with the one-period continuation strategy $\sigma^1$ that induces adherence to the worst subgame perfect equilibrium.

**j.** Find the one-period continuation value $v_2$ and the outcome path associated with the one-period continuation strategy $\sigma^2$ that induces adherence to the first-period outcome of the $\sigma^1$ that you found in part i.

**k.** Proceeding recursively, define $v_j$ and $\sigma^j$, respectively, as the one-period continuation value and the continuation strategy that induces adherence to the first-period outcome of $\sigma^{j-1}$, where $(v_1, \sigma^1)$ were defined in part i. Find $v_j$ for $j = 1, 2, \ldots,$ and find the associated outcome paths.

**l.** Find the lowest value for the discount factor for which repetition of the Ramsey outcome is a subgame perfect equilibrium.

# 17
# *Fiscal-Monetary Theories of Inflation*

This chapter introduces some issues in monetary theory that mostly revolve around coordinating monetary and fiscal policies. We start from the observation that complete markets models have no role for inconvertible currency, and therefore assign zero value to it.[1] We describe one way to alter a complete markets economy so that a positive value is assigned to an inconvertible currency: we impose a transaction technology with shopping time and real money balances as inputs.[2] We use the model to illustrate ten doctrines in monetary economics. Most of these doctrines transcend many of the details of the model. The important thing about the transactions technology is that it makes demand for currency a decreasing function of the rate of return on currency. Our monetary doctrines mainly emerge from manipulating that demand function and the government's intertemporal budget constraint under alternative assumptions about government monetary and fiscal policy.[3]

---

[1] In complete markets models, money holdings would only serve as a store of value. The following transversality condition would hold in a nonstochastic economy:

$$\lim_{T\to\infty} \prod_{t=0}^{T-1} R_t^{-1} \frac{m_{T+1}}{p_T} = 0.$$

The real return on money, $p_t/p_{t+1}$, would have to equal the return $R_t$ on other assets, which substituted into the transversality condition yields

$$\lim_{T\to\infty} \prod_{t=0}^{T-1} \frac{p_{t+1}}{p_t} \frac{m_{T+1}}{p_T} = \lim_{T\to\infty} \frac{m_{T+1}}{p_0} = 0.$$

That is, an inconvertible money (i.e., one for which $\lim_{T\to\infty} m_{T+1} > 0$) must be valueless, $p_0 = \infty$.

[2] See Bennett McCallum (1983) for an early shopping time specification.

[3] Many of the doctrines were originally developed in setups differing in details from the one in this chapter.

After describing our ten doctrines, we use the model to analyze two important issues: the validity of Friedman's rule in the presence of distorting taxation, and its sustainability in the face of a time consistency problem. Here we use the methods for solving an optimal taxation problem with commitment in chapter 12, and for characterizing a credible government policy in chapter 16.

## A shopping time monetary economy

Consider an endowment economy with no uncertainty. A representative household has one unit of time. There is a single good of constant amount $y > 0$ each period $t \geq 0$. The good can be divided between private consumption $\{c_t\}_{t=0}^{\infty}$ and government purchases $\{g_t\}_{t=0}^{\infty}$, subject to

$$c_t + g_t = y. \tag{17.1}$$

The preferences of the household are ordered by

$$\sum_{t=0}^{\infty} \beta^t u(c_t, \ell_t), \tag{17.2}$$

where $\beta \in (0,1)$, $c_t \geq 0$ and $\ell_t \geq 0$ are consumption and leisure at time $t$, respectively, and $u_c$, $u_\ell > 0$, $u_{cc}$, $u_{\ell\ell} < 0$, and $u_{c\ell} \geq 0$. We use $u_c(t)$ and so on to denote the time-$t$ values of the indicated objects, evaluated at an allocation to be understood from the context.

The household must spend time shopping to acquire the consumption good. The amount of shopping time $s_t$ needed to purchase a particular level of consumption $c_t$ is negatively related to the household's holdings of real money balances $m_{t+1}/p_t$. Specifically, the shopping or transaction technology is

$$s_t = H\left(c_t, \frac{m_{t+1}}{p_t}\right), \tag{17.3}$$

where $H$, $H_c$, $H_{cc}$, $H_{m/p,m/p} \geq 0$, $H_{m/p}$, and $H_{c,m/p} \leq 0$. A parametric example of this transaction technology is

$$H\left(c_t, \frac{m_{t+1}}{p_t}\right) = \frac{c_t}{m_{t+1}/p_t}\epsilon, \tag{17.4}$$

which corresponds to a transaction cost that would arise in the frameworks of Baumol (1952) and Tobin (1956). That is, when a household spends money holdings for consumption purchases at a constant rate $c_t$ per unit of time, $c_t(m_{t+1}/p_t)^{-1}$ is the number of trips to the bank, and $\epsilon$ is the time cost per trip to the bank.

With one unit of time per period, the household's time constraint becomes

$$1 = \ell_t + s_t. \tag{17.5}$$

*Households*

The household maximizes expression (17.2) subject to the transaction technology (17.3) and the sequence of budget constraints

$$c_t + \frac{b_{t+1}}{R_t} + \frac{m_{t+1}}{p_t} = y - \tau_t + b_t + \frac{m_t}{p_t}. \tag{17.6}$$

Here $m_{t+1}$ is nominal balances held between times $t$ and $t+1$; $p_t$ is the price level; $b_t$ is the real value of one-period government bond holdings that mature at the beginning of period $t$, denominated in units of time-$t$ consumption; $\tau_t$ is a lump-sum tax at $t$; and $R_t$ is the real gross rate of return on one-period bonds held from $t$ to $t+1$. Maximization of expression (17.2) is subject to $m_{t+1} \geq 0$ for all $t \geq 0$,[4] no restriction on the sign of $b_{t+1}$ for all $t \geq 0$, and given initial stocks $m_0$, $b_0$.

After consolidating two consecutive budget constraints given by equation (17.6), we arrive at

$$\begin{aligned} c_t + \frac{c_{t+1}}{R_t} + \left(1 - \frac{p_t}{p_{t+1}}\frac{1}{R_t}\right)\frac{m_{t+1}}{p_t} + \frac{b_{t+2}}{R_t R_{t+1}} + \frac{m_{t+2}/p_{t+1}}{R_t} \\ = y - \tau_t + \frac{y - \tau_{t+1}}{R_t} + b_t + \frac{m_t}{p_t}. \end{aligned} \tag{17.7}$$

To ensure a bounded budget set, the expression in parentheses multiplying nonnegative holdings of real balances must be greater than or equal to zero. Thus, we have the arbitrage condition,

$$1 - \frac{p_t}{p_{t+1}}\frac{1}{R_t} = 1 - \frac{R_{mt}}{R_t} = \frac{i_{t+1}}{1 + i_{t+1}} \geq 0, \tag{17.8}$$

[4] Households cannot issue money.

where $R_{mt} \equiv p_t/p_{t+1}$ is the real gross return on money held from $t$ to $t+1$, that is, the inverse of the inflation rate, and $1 + i_{t+1} \equiv R_t/R_{mt}$ is the gross nominal interest rate. The real return on money $R_{mt}$ must be less than or equal to the return on bonds $R_t$, since agents would otherwise be able to make infinite profits by choosing arbitrarily large money holdings financed by issuing bonds. In other words, the net nominal interest rate $i_{t+1}$ cannot be negative.

The household's optimization problem in Lagrangian form is

$$\sum_{t=0}^{\infty} \beta^t \Bigg\{ u(c_t, \ell_t) + \lambda_t \Big( y - \tau_t + b_t + \frac{m_t}{p_t} - c_t - \frac{b_{t+1}}{R_t} - \frac{m_{t+1}}{p_t} \Big) + \mu_t \Big[ 1 - \ell_t - H\Big(c_t, \frac{m_{t+1}}{p_t}\Big) \Big] \Bigg\}.$$

At an interior solution, the first-order conditions with respect to $c_t$, $\ell_t$, $b_{t+1}$, and $m_{t+1}$ are

$$u_c(t) - \lambda_t - \mu_t H_c(t) = 0, \tag{17.9}$$

$$u_\ell(t) - \mu_t = 0, \tag{17.10}$$

$$-\lambda_t \frac{1}{R_t} + \beta \lambda_{t+1} = 0, \tag{17.11}$$

$$-\lambda_t \frac{1}{p_t} - \mu_t H_{m/p}(t) \frac{1}{p_t} + \beta \lambda_{t+1} \frac{1}{p_{t+1}} = 0. \tag{17.12}$$

From equations (17.9) and (17.10),

$$\lambda_t = u_c(t) - u_\ell(t) H_c(t). \tag{17.13}$$

The Lagrange multiplier on the budget constraint is equal to the marginal utility of consumption reduced by the marginal disutility of having to shop for that increment in consumption. By substituting equation (17.13) into equation (17.11), we obtain an expression for the real interest rate,

$$R_t = \frac{1}{\beta} \frac{u_c(t) - u_\ell(t) H_c(t)}{u_c(t+1) - u_\ell(t+1) H_c(t+1)}. \tag{17.14}$$

The combination of equations (17.11) and (17.12) yields

$$\frac{R_t - R_{mt}}{R_t} \lambda_t = -\mu_t H_{m/p}(t), \tag{17.15}$$

which sets the cost equal to the benefit of the marginal unit of real money balances held from $t$ to $t+1$, all expressed in time-$t$ utility. The cost of holding money balances instead of bonds is lost interest earnings $(R_t - R_{mt})$ discounted at the rate $R_t$ and expressed in time-$t$ utility when multiplied by the shadow price $\lambda_t$. The benefit of an additional unit of real money balances is the savings in shopping time $-H_{m/p}(t)$ evaluated at the shadow price $\mu_t$. By substituting equations (17.10) and (17.13) into equation (17.15), we get

$$\left(1 - \frac{R_{mt}}{R_t}\right)\left[\frac{u_c(t)}{u_\ell(t)} - H_c(t)\right] + H_{m/p}(t) = 0 \tag{17.16}$$

with $u_c(t)$ and $u_\ell(t)$ evaluated at $\ell_t = 1 - H(c_t, m_{t+1}/p_t)$. Equation (17.16) implicitly defines a money demand function

$$\frac{m_{t+1}}{p_t} = F(c_t, R_{mt}/R_t), \tag{17.17}$$

which is increasing in both of its arguments, as can be shown by applying the implicit function rule to expression (17.16).

*Government*

The government finances the purchase of the stream $\{g_t\}$ subject to the sequence of budget constraints

$$g_t = \tau_t + \frac{B_{t+1}}{R_t} - B_t + \frac{M_{t+1} - M_t}{p_t}, \tag{17.18}$$

where $B_0$ and $M_0$ are given. Here $B_t$ is government indebtedness to the private sector, denominated in time-$t$ goods, maturing at the beginning of period $t$, and $M_t$ is the stock of currency that the government has issued as of the beginning of period $t$.

*Equilibrium*

We use the following definitions:

DEFINITION: A *price system* is a pair of positive sequences $\{R_t\}_{t=0}^{\infty}, \{p_t\}_{t=0}^{\infty}$.

DEFINITION: We take as exogenous sequences $\{g_t, \tau_t\}_{t=0}^{\infty}$. We also take $B_0 = b_0$ and $M_0 = m_0 > 0$ as given. An *equilibrium* is a price system, a consumption sequence $\{c_t\}_{t=0}^{\infty}$, a sequence for government indebtedness $\{B_t\}_{t=1}^{\infty}$, and a positive sequence for the money supply $\{M_t\}_{t=1}^{\infty}$ for which the following statements are true: (a) given the price system and taxes, the household's optimum problem is solved with $b_t = B_t$ and $m_t = M_t$; (b) the government's budget constraint is satisfied for all $t \geq 0$; and (c) $c_t + g_t = y$.

*"Short run" versus "long run"*

We shall study government policies designed to ascribe a definite meaning to a distinction between outcomes in the "short run" (initial date) and the "long run" (stationary equilibrium). We assume

$$\begin{aligned} g_t &= g \ \forall t \geq 0 \\ \tau_t &= \tau \ \forall t \geq 1 \\ B_t &= B \ \forall t \geq 1. \end{aligned} \tag{17.19}$$

We permit $\tau_0 \neq \tau$ and $B_0 \neq B$.

These settings of policy variables are designed to let us study circumstances in which the economy is in a stationary equilibrium for $t \geq 1$, but starts from some other position at $t = 0$. We have enough free policy variables to discuss alternative meanings that the theoretical literature has attached to the phrase "open market operations".

*Stationary equilibrium*

We seek an equilibrium for which

$$\begin{aligned} p_t/p_{t+1} &= R_m \ \forall t \geq 0 \\ R_t &= R \ \forall t \geq 0 \\ c_t &= c \ \forall t \geq 0 \\ s_t &= s \ \forall t \geq 0. \end{aligned} \tag{17.20}$$

Substituting equations (17.20) into equations (17.14) and (17.17) yields

$$\begin{aligned} R &= \beta^{-1}, \\ \frac{m_{t+1}}{p_t} &= F(c, R_m/R) = f(R_m), \end{aligned} \tag{17.21}$$

where we have suppressed the constants $c$ and $R$ in the money demand function $f(R_m)$ for a stationary equilibrium.

Substituting equations (17.19), (17.20), and (17.21) into the government budget constraint (17.18), using the equilibrium condition $M_t = m_t$, and rearranging gives

$$g - \tau + B(R-1)/R = f(R_m)(1 - R_m). \tag{17.22}$$

Given the policy variables $(g, \tau, B)$, equation (17.22) is to be used to determine the stationary rate of return on currency $R_m$. In (17.22), $g - \tau$ is the net of interest deficit, sometimes called the operational deficit; $g - \tau + B(R-1)/R$ is the gross of interest government deficit; and $f(R_m)(1 - R_m)$ is the rate of seigniorage revenues from printing currency.

### *Initial date (time 0)*

Because $M_1/p_0 = f(R_m)$, the government budget constraint at $t = 0$ can be written

$$M_0/p_0 = f(R_m) - (g + B_0 - \tau_0) + B/R. \tag{17.23}$$

### *Equilibrium determination*

Given the policy parameters $(g, \tau, \tau_0, B)$, the initial stocks $B_0$ and $M_0$, and the equilibrium gross real interest rate $R = \beta^{-1}$, equations (17.22) and (17.23) determine $(R_m, p_0)$. The two equations are recursive: equation (17.22) determines $R_m$, then equation (17.23) determines $p_0$.

It is useful to illustrate the determination of an equilibrium with a parametric example. Let the utility function and the transaction technology be given by

$$\begin{aligned} u(c_t, l_t) &= \frac{c_t^{1-\delta}}{1-\delta} + \frac{l_t^{1-\alpha}}{1-\alpha}, \\ H(c_t, m_{t+1}/p_t) &= \frac{c_t}{1 + m_{t+1}/p_t}, \end{aligned}$$

where the latter is a modified version of equation (17.4), so that transactions can be carried out even in the absence of money.

For parameter values $(\beta, \delta, \alpha, c) = (0.96, 0.7, 0.5, 0.4)$, Figure 17.1 displays the function $f(R_m)(1 - R_m)$;[5] Figure 17.2 shows $M_0/p_0$. Stationary equilibrium is

[5] Figure 17.1 shows the stationary value of seigniorage per period,

$$\frac{M_{t+1} - M_t}{p_t} = \frac{M_{t+1}}{p_t} - \frac{M_t}{p_{t-1}} \frac{p_{t-1}}{p_t} = f(R_m)(1 - R_m).$$

determined as follows: Name a stationary gross of interest deficit $g - \tau + B(R - 1)/R$, then read an associated stationary value $R_m$ from Figure 17.1 that satisfies equation (17.22); for this value of $R_m$, compute $f(R_m) - (g + B_0 - \tau_0) + B/R$, then read the associated equilibrium price level $p_0$ from Figure 17.2 that satisfies equation (17.23).

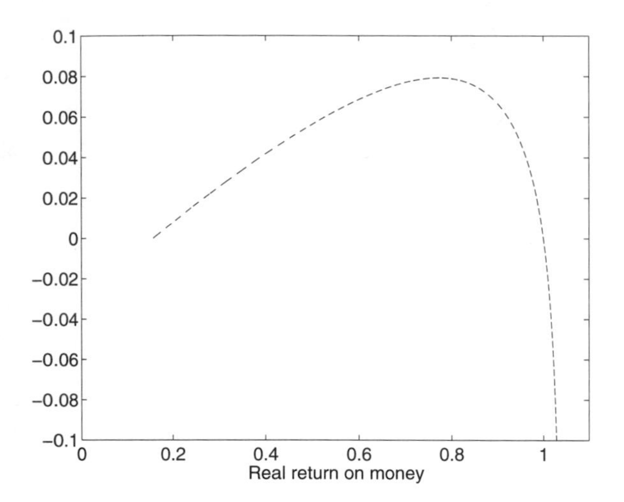


**Figure 17.1** Stationary seigniorage $f(R_m)(1 - R_m)$ as a function of the stationary rate of return on currency, $R_m$. An intersection of the stationary gross of interest deficit $g - \tau + B(R - 1)/R$ with $f(R_m)(1 - R_m)$ in this figure determines $R_m$.

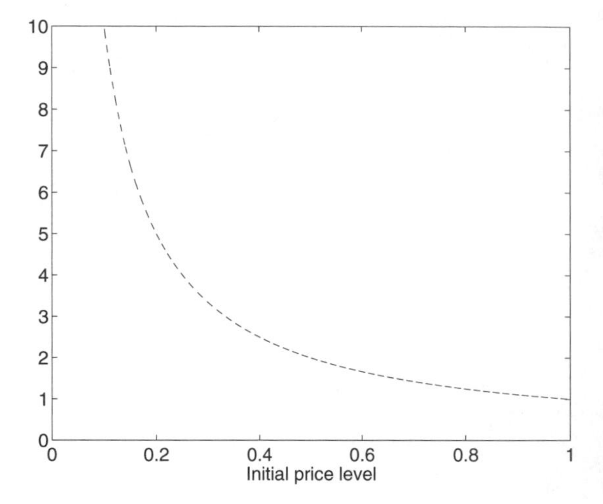


**Figure 17.2** Real value of initial money balances $M_0/p_0$ as a function of the price level $p_0$. Given $R_m$, an intersection of $f(R_m) - (g + B_0 - \tau_0) + B/R$ with $M_0/p_0$ in this figure determines $p_0$.

## *Ten monetary doctrines*

We now use equations (17.22) and (17.23) to explain some important doctrines about money and government finance.

---

For our parameterization, households choose to hold zero money balances for $R_m$ less than 0.15, so at these rates there is no seigniorage collected. Seigniorage turns negative for $R_m > 1$ because the government is then continuously withdrawing money from circulation to raise the real return on money above one.

### 1. *Sustained deficits cause inflation*

The parameterization in Figures 17.1 and 17.2 shows that there can be multiple values of $R_m$ that solve equation (17.22). As can be seen in Figure 17.1, some values of the gross-of-interest deficit $g - \tau + B(R-1)/R$ can be financed with either a low or high rate of return on money. The tax rate on real money balances is $(1 - R_m)$ in a stationary equilibrium, so the higher $R_m$ that solves equation (17.22) is on the good side of a "Laffer curve" in the inflation tax rate.

If there are multiple values of $R_m$ that solve equation (17.22), we shall always select the highest one for the purposes of doing our comparative dynamic exercises.[6] The stationary equilibrium with the higher rate of return on currency is associated with classical comparative dynamics: an increase in the stationary gross-of-interest government budget deficit causes a *decrease* in the rate of return on currency (i.e., an increase in the inflation rate). Notice how the stationary equilibrium associated with the lower rate of return on currency has "perverse" comparative dynamics, from the point of view of the classical doctrine that sustained government deficits cause inflation.

### 2. *Zero inflation policy*

Equation (17.22) implies a restriction on fiscal policy that is necessary and sufficient to sustain a zero inflation ($R_m = 1$) equilibrium:

$$g - \tau + B(R-1)/R = 0,$$

or

$$B/R = (\tau - g)/(R-1) = \sum_{t=1}^{\infty} R^{-t}(\tau - g).$$

This equation states that the real value of interest-bearing government indebtedness equals the present value of the net-of-interest government *surplus*, with zero revenues being contributed by an inflation tax.

[6] In chapter 8, we studied the perfect-foresight dynamics of a closely related system and saw that the stationary equilibrium selected here was *not* the limit point of those dynamics. Our selection of the higher rate of return equilibrium must be based on appeals to various forms of "adaptive" (nonrational) dynamics. See Bruno and Fischer (1990), Marcet and Sargent (1989), and Marimon and Sunder (1993). Also, see exercise 17.2.

### 3. *Unpleasant monetarist arithmetic*

Consider an open market sale of bonds at time $0$, defined as a *decrease* in $M_1$ accompanied by an *increase* in $B$, with all other government policy variables constant, including $(\tau_0, \tau)$. This policy can be analyzed by increasing $B$ in equations (17.22) and (17.23). The effect of the policy is to shift the permanent gross-of-interest deficit upward by $(R-1)/R$ times the increase in $B$, which decreases the real return on money $R_m$ in Figure 17.1. That is, the effect is unambiguously to *increase* the stationary inflation rate (the inverse of $R_m$). However, the effect on the initial price level $p_0$ can go either way, depending on the slope of the revenue curve $f(R_m)(1-R_m)$; the decrease in $R_m$ reduces the right-hand side of equation (17.23), $f(R_m)-(g+B_0-\tau_0)+B/R$, while the increase in $B$ raises the value. Thus, the new equilibrium can move us to the left or the right along the curve $M_0/p_0$ in Figure 17.2, that is, a decrease or an increase in the initial price level $p_0$.

The effect of a decrease in the money supply accomplished through such an open market operation is at best temporarily to drive the price level downward, at the cost of causing the inflation rate to be permanently higher. Sargent and Wallace (1981) called this "unpleasant monetarist arithmetic."

### 4. *Quantity theory of money*

The classic "quantity theory of money" experiment is to increase $M_0$ by some factor $\lambda > 1$ (a "helicopter drop" of money), leaving all of the other parameters of the model fixed [including the fiscal policy parameters $(\tau_0, \tau, g, B)$]. The effect is to multiply the initial equilibrium price and money supply sequences by $\lambda$ and to leave all other variables unaltered.

### 5. *An "open market" operation delivering neutrality*

We now alter the definition of open market operations used in the unpleasant monetarist arithmetic. We supplement the fiscal powers of the monetary authority in a way that lets open market operations have effects like those in the quantity theory experiment. Let there be an initial equilibrium with policy values denoted by bars over variables. Consider an open market sale or purchase defined as a decrease in $M_1$ and increases in $B$ and $\tau$ designed to satisfy

$$(1-1/R)(\hat{B}-\bar{B}) = \hat{\tau} - \bar{\tau}, \tag{17.24}$$

where variables with hats denote the new values of the corresponding variables. We assume that $\hat{\tau}_0 = \bar{\tau}_0$.

As long as the tax rate from time 1 on is adjusted according to equation (17.24), equation (17.22) will be satisfied at the initial value of $R_m$. Equation (17.24) imposes a requirement that the lump-sum tax $\tau$ be adjusted by just enough to service whatever additional interest payments are associated with the alteration in $B$ resulting from the exchange of $M_1$ for $B$.[7] Under this definition of an open market operation, increases in $M_1$ achieved by reductions in $B$ and the taxes needed to service $B$ cause proportionate increases in the paths of the money supply and the price level, and fulfill the pure quantity theory of money.

### 6. The "optimum quantity" of money

Friedman's (1969) ideas about the optimum quantity of money can be represented in Figures 17.1 and 17.2. Friedman noted that, given the stationary levels of $(g, B)$, the representative household prefers stationary equilibria with higher rates of return on currency. In particular, the higher the stationary level of real balances, the better the household likes it. By running a sufficiently large gross-of-interest surplus, that is, a negative value of $g - \tau + B(R-1)/R$, the government can attain any value of $R_m \in (1, \beta^{-1})$. Given $(g, B)$ and the target value of $R_m$ in this interval, a tax rate $\tau$ can be chosen to assure the required surplus. The proceeds of the tax are used to retire currency from circulation, thereby generating a deflation that makes the rate of return on currency equal to $R_m$. According to Friedman, the optimal policy is to satiate the system with real balances, insofar as it is possible to do so.

The social value of real money balances in our model is that they reduce households' shopping time. The optimum quantity of money is the one that minimizes the time allocated to shopping. For the sake of argument, suppose there is a satiation point in real balances $\psi(c)$ for any consumption level $c$, that is, $H_{m/p}(c, m_{t+1}/p_t) = 0$ for $m_{t+1}/p_t \geq \psi(c)$. According to condition (17.15), the government can attain this optimal allocation only by choosing $R_m = R$, since $\lambda_t, \mu_t > 0$. (Utility is assumed to be strictly increasing in both consumption and leisure.) Thus, welfare is at a maximum when the economy is satiated with real balances. For the transaction technology given by equation (17.4), the Friedman rule can only be approximately attained because money demand is insatiable.

[7] This definition of an "open market" operation imputes more power to a monetary authority than usual: central banks don't set tax rates.

### 7. Legal restrictions to boost demand for currency

If the government can somehow force households to increase their real money balances to $\tilde{f}(R_m) > f(R_m)$, it can finance a given stationary gross of interest deficit $g - \tau + B(R-1)/R$ at a higher stationary rate of return on currency $R_m$. The increased demand for money balances shifts the seigniorage curve in Figure 17.1 upward to $\tilde{f}(R_m)(1-R_m)$, thereby increasing the higher of the two intersections of the curve $\tilde{f}(R_m)(1-R_m)$ with the gross-of-interest deficit line in Figure 17.1. By increasing the base of the inflation tax, the rate $(1-R_m)$ of inflation taxation can be diminished.

Governments intent on raising revenues through the inflation tax have frequently resorted to legal restrictions and threats designed to promote the demand for its currency. In chapter 18, we shall study a version of Bryant and Wallace's (1984) theory of some of those restrictions. Sargent and Velde (1995) recount such restrictions in the Terror during the French Revolution, and the sharp tools used to enforce them.

To assess the welfare effects of policies forcing households to hold higher real balances, we must go beyond the reduced-form of the transaction process in equation (17.3). We need an explicit model of how money facilitates transactions and how the government interferes with markets to increase the demand for real balances. In such a model, there would be opposing effects on social welfare. On the one hand, our discussion of the optimum quantity of money says that a higher real return on money $R_m$ tends to improve welfare. On the other hand, the imposition of legal restrictions aimed at forcing households to hold higher real balances might elicit socially wasteful activities from the private economy trying to evade precisely those restrictions.

### 8. One big open market operation

Lucas (1988) and Wallace (1989) describe a policy where the government conducts a large open market purchase of private indebtedness at time 0. The purpose of the operation is to provide the government with a portfolio of interest-earning claims on the private sector, one that is sufficient to permit it to run a gross-of-interest surplus. The government uses the surplus to reduce the money supply each period, thereby engineering a deflation that raises the rate of return on money above one. That is, the government uses its own lending to arbitrage between its money and higher-yield bonds and, by doing so, drive the yield differential down. As we know from our discussion of the optimum quantity of money, the increase in the real return on money $R_m$ will lead to higher welfare.[8]

To highlight the effects of the described open market policy, we impose a nonnegative net-of-interest deficit, $g - \tau \geq 0$, which rules out deflation financed by direct taxation. The proposed operation is then to increase $M_1$ and decrease $B$, with $B < 0$ indicating private indebtedness to the government. We explore a candidate policy as follows: Given values of $(g, \tau)$, use equation (17.22) to pick a value of $B$ that solves equation (17.22) for a desired level of $R_m$, with $1 < R_m \leq \beta^{-1}$. Notice that a negative level of $B$ will be required, since $g - \tau \geq 0$. Substituting equation (17.23) into equation (17.22) [by eliminating $f(R_m)$] and rearranging gives

$$M_0/p_0 = \left(\frac{R - R_m}{1 - R_m}\right)\frac{B}{R} + \left(\frac{1}{1 - R_m}\right)(g - \tau) - (g + B_0 - \tau_0). \tag{17.25}$$

The first term on the right side is positive, while the remainder may be positive or negative. The candidate policy is only consistent with an equilibrium if $g, \tau, \tau_0$, and $B_0$ assume values for which the entire right side is positive. In this case, there exists a positive price level $p_0$ that solves equation (17.25).

As an example, assume that $g - \tau = 0$ and that $g + B_0 - \tau_0 = 0$, so that the government budget net of interest is balanced from time $t = 1$ onward. Then we know that the right-hand side of equation (17.25) is positive. In this case it is feasible to operate a scheme like this to support any return on currency $1 < R_m < 1/\beta$. However, it is instructive to notice that the policy cannot attain $R_m = 1/\beta$ (even if there is a point of satiation in money balances, as discussed earlier). The reason is once again that the scheme is based on the government doing arbitrage between its money and higher-yield bonds to finance the deflation. When there is no yield differential, $R_m = R$, the government earns no arbitrage income, so it cannot finance any deflation.

[8] Beatrix Paal (2000) describes how the stabilization of the second Hungarian hyperinflation had some features of "one big open market operation." After the stabilization the government lent the one-time seigniorage revenues gathered from remonetizing the economy. The severe hyperinflation (about $4 \times 10^{24}$ in the previous year) had reduced real balances of fiat currency virtually to zero. Paal argues that the fiscal aspects of the stabilization, dependent as they were on those one-time seigniorage revenues, were foreseen and shaped the dynamics of the preceding hyperinflation.

## 9. A fiscal theory of the price level

The preceding sections have illustrated what might be called a fiscal theory of *inflation.* This theory assumes a particular specification of exogenous variables that are chosen and committed to by the government. In particular, it is assumed that the government sets $g, \tau_0, \tau$, and $B$, that $B_0$ and $M_0$ are inherited from the past, and that the model then determines $R_m$ and $p_0$ via equations (17.22) and (17.23). In particular, the system is recursive: given $g, \tau$, and $B$, equation (17.22) determines the rate of return on currency $R_m$; then given $g, \tau, B$, and $R_m$, equation (17.23) determines $p_0$. After $p_0$ is determined, $M_1$ is determined from $\frac{M_1}{p_0} = f(R_m)$. In this setting, the government commits to a long-run gross-of-interest government deficit $g - \tau + B(R-1)/R$, and then the market determines $p_0, R_m$.

Woodford (1995) and Sims (1994) have converted a version of the same model into a fiscal theory of the *price level* by altering the assumptions about the variables that the government sets. Rather than assuming that the government sets $B$, and thereby the gross-of-interest government deficit, Woodford assumes that $B$ is endogenous and that instead the government sets in advance a present value of seigniorage $f(R_m)(1 - R_m)/(R - 1)$. This assumption is equivalent to saying that the government is able to commit to fix the nominal interest rate, or else the gross rate of inflation $R_m^{-1}$. Woodford emphasizes that in the present setting, such a nominal interest-rate-peg leaves the equilibrium-price-level process determinate.[9] To illustrate Woodford's argument in our setting, rearrange equation (17.22) to obtain

$$
\begin{aligned}
B/R &= \frac{1}{R-1}\left[(\tau - g) + f(R_m)(1 - R_m)\right] \\
&= \sum_{t=1}^{\infty} R^{-t}(\tau - g) + f(R_m)\frac{1 - R_m}{R - 1},
\end{aligned}
\tag{17.26}
$$

[9] Woodford (1995) interprets this finding against the background of a literature that occasionally asserted a different result, namely, that interest rate pegging led to price level indeterminacy because of the associated money supply endogeneity. That other literature focused on the homogeneity properties of conditions (17.14) and (17.16): the only ways in which the price level enters are as ratio to the money supply and in the form of the inflation rate. These properties suggested that a policy regime that leaves the money supply, as well as the price level, endogenous will not be able to determine the level of either.

which when substituted into equation (17.23) yields

$$\frac{M_0}{p_0} + B_0 = \sum_{t=0}^{\infty} R^{-t}(\tau_t - g_t) + f(R_m)\Big(1 + \frac{1 - R_m}{R - 1}\Big)$$
$$= \sum_{t=0}^{\infty} R^{-t}(\tau_t - g_t) + \sum_{t=1}^{\infty} R^{-t} f(R_m)(R - R_m). \tag{17.27}$$

In a stationary equilibrium, the real interest rate is equal to $1/\beta$, so by multiplying the nominal interest rate by $\beta$ we obtain the inverse of the corresponding value for $R_m$. Thus, pegging a nominal rate is equivalent to pegging the inflation rate and the steady-state flow of seigniorage $f(R_m)(1-R_m)$. Woodford uses such equations as follows: The government chooses $g, \tau, \tau_0$, and $R_m$ (or equivalently, $f(R_m)(1 - R_m)$. Then equation (17.26) determines $B$ as the present value of the government surplus from time 1 on, including seigniorage revenues. Equation (17.27) then determines $p_0$. Equation (17.27) says that the price level is set to equate the real value of *total* initial government indebtedness to the present value of the net-of-interest government surplus, including seigniorage revenues. Finally, the endogenous quantity of money is determined by the demand function for money (17.17),

$$M_1/p_0 = F(y - g,\ R_m/R). \tag{17.28}$$

Woodford uses this experiment to emphasize that without saying much more, the mere presence of a "quantity theory" equation of the form (17.28) does *not* imply the "monetarist" conclusion that it is necessary to make the money supply exogenous in order to determine the path of the price level.

Several commentators[10] have remarked that the Sims-Woodford use of these equations puts the government on a different setting than the private agents. Private agents' demand curves are constructed by requiring their budget constraints to hold for *all* hypothetical price processes, not just the equilibrium one. However, under Woodford's assumptions about what the government has already chosen *regardless* of the $(p_0, R_m)$ it faces, the only way an equilibrium can exist is if $p_0$ adjusts to make equation (17.27) satisfied.

By way of contrast, in the fiscal theory of *inflation* described by Sargent and Wallace (1981) and Sargent (1992), embodied in our description of unpleasant monetarist arithmetic, the focus is on how the one tax rate that is assumed to be free to adjust, the inflation tax, responds to fiscal conditions that the government inherits.

[10] See Ramon Marimon (1998).

## 10. Exchange rate (in)determinacy

Kareken and Wallace's (1981) exchange rate indeterminacy result provides a good laboratory for putting the fiscal theory of the price level to work. First, we will describe a version of the Kareken-Wallace result. Then we will show how it can be overturned by changing the assumptions about policy to ones like Woodford's.

To describe the theory of exchange rate indeterminacy, we change the preceding model so that two governments each issue currency. The government of country $i$ has $M_{it+1}$ units of its currency outstanding at the end of period $t$. The price level in terms of currency $i$ is $p_{it}$, and the exchange rate $e_t$ satisfies the purchasing power parity condition $p_{1t} = e_t p_{2t}$. The household is indifferent about which currency to use so long as both currencies bear the same rate of return, and will not hold one with an inferior rate of return. This fact implies that $\frac{p_{1t}}{p_{1t+1}} = \frac{p_{2t}}{p_{2t+1}}$, which in turn implies that $e_{t+1} = e_t = e$. Thus, the exchange rate is constant in a nonstochastic equilibrium with two currencies being valued. We let $M_{t+1} = M_{1t+1} + eM_{2t+1}$. The equilibrium condition in the money market is now

$$\frac{M_{t+1}}{p_{1t}} = f(R_m). \tag{17.29}$$

We let $g_i, \tau_i$, and $B_i$ denote constant steady-state values for government purchases, lump-sum taxes, and real government indebtedness for government $i = 1, 2$. We assume that $g_i$, and $\tau_i$ are the expenditure and tax rates for government $i$ for all $t \geq 0$. The budget constraint of government $i$ is

$$g_i = \tau_i + B_{it+1}/R - B_{it} + \frac{M_{it+1} - M_{it}}{p_{it}}. \tag{17.30}$$

We use variables without subscripts to denote worldwide totals. Then the sum of the government budget constraints for $t \geq 1$ can be written as

$$g - \tau + B(R-1)/R = f(R_m)(1 - R_m), \tag{17.31}$$

and

$$g = \tau + B/R - B_0 + f(R_m) - \frac{M_{10} + eM_{20}}{p_{10}}, \tag{17.32}$$

for $t = 0$.

Here is a version of Kareken and Wallace's exchange rate indeterminacy result: Assume that the governments of each country set $g_i, \tau_i$, and $B_i$, planning to adjust their rates of money creation to raise whatever revenues are needed to finance their budgets. Then equation (17.31) determines $R_m$, and equation (17.32) is left to determine two variables, $p_{10}$ and $e$. The system is underdetermined: if there is a solution of equation (17.32) with positive $(p_{10}, e)$, then for *any* positive $e$ there is another solution. Kareken and Wallace conclude that under such settings for government policy variables, something more is needed to set the exchange rate. With policy as specified here, the exchange rate is indeterminate.[11]

*10 (again). Determinacy of the exchange rate retrieved*

A version of Woodford's assumptions about the variables that governments choose can render the exchange rate determinate. Thus, suppose that rather than setting $B_i$, each government sets a constant rate of seigniorage $x_i = \frac{M_{it+1} - M_{it}}{p_{it}}$ for all $t \geq 0$. Then $B_i/R$ is determined by the following version of equation (17.26):

$$B_i/R = \sum_{\tau=1}^{\infty} R^{-t} \left[ (\tau_i - g_i) + x_i \right] \tag{17.33}$$

and

$$x_1 + x_2 = f(R_m)(1 - R_m).$$

The time-0 worldwide budget constraint becomes

$$\frac{M_0}{p_{10}} + B_0 = \tau - g + B/R + f(R_m). \tag{17.34}$$

The time-0 budget constraint for country 2 can be written

$$e\frac{M_{20}}{p_{10}} + B_{20} = \tau_2 - g_2 + B_2/R + \left[ f(R_m) - \frac{M_{11}}{p_{10}} \right],$$

or

$$e\frac{M_{20}}{p_{10}} + B_{20} = \tau_2 - g_2 + B_2/R + \left[ f(R_m) - \left( x_1 + \frac{M_{10}}{p_{10}} \right) \right], \tag{17.35}$$

Equations (17.34) and (17.35) are two independent equations in $e, p_{10}$ that can determine these two variables. Thus, with this Sims-Woodford structure of government commitments (i.e., setting of exogenous variables), the exchange rate is determinate.

[11] See Sargent and Velde (1990) for an application of this theory to events surrounding German monetary unification.

## Optimal inflation tax: The Friedman rule

Given lump-sum taxation, the sixth monetary doctrine (about the "optimum quantity of money") establishes the optimality of the Friedman rule. The optimal policy is to satiate the economy with real balances by generating a deflation that drives the net nominal interest rate to zero. In a stationary economy, there can be deflation only if the government retires currency with a government surplus. We now ask if such a costly scheme remains optimal when all government revenues must be raised through distortionary taxation. Or would the Ramsey plan then include an inflation tax on money holdings whose rate depends on the interest elasticity of money demand?

Following Correia and Teles (1996), we show that even with distortionary taxation the Friedman rule is the optimal policy under a transaction technology (17.3) that satisfies a homogeneity condition.

Earlier analyses of the optimal tax on money in models with transaction technologies include Kimbrough (1986), Faig (1988), and Guidotti and Vegh (1993). Chari, Christiano, and Kehoe (1996) also develop conditions for the optimality of the Friedman rule in models with cash and credit goods and money in the utility function.

### *Economic environment*

We convert our shopping-time monetary economy into a production economy with labor $n_t$ as the only input in a linear technology:

$$c_t + g_t = n_t. \tag{17.36}$$

The household's time constraint becomes

$$1 = \ell_t + s_t + n_t. \tag{17.37}$$

The shopping technology is now assumed to be homogeneous of degree $\nu \geq 0$ in consumption $c_t$ and real money balances $\hat{m}_{t+1} \equiv m_{t+1}/p_t$;

$$s_t = H(c_t, \hat{m}_{t+1}) = c_t^{\nu} H\left(1, \frac{\hat{m}_{t+1}}{c_t}\right), \qquad \text{for } c_t > 0. \tag{17.38}$$

By Euler's theorem we have

$$H_c(c, \hat{m})c + H_{\hat{m}}(c, \hat{m})\hat{m} = \nu H(c, \hat{m}). \tag{17.39}$$

For any consumption level $c$, we also assume a point of satiation in real money balances $\psi c$ such that

$$H_{\hat{m}}(c,\hat{m}) = H(c,\hat{m}) = 0, \qquad \text{for } \hat{m} \geq \psi c. \tag{17.40}$$

*Household's optimization problem*

After replacing net income $(y - \tau_t)$ in equation (17.7) by $(1-\tau_t)(1-\ell_t - s_t)$, consolidation of budget constraints yields the household's present-value budget constraint

$$\sum_{t=0}^{\infty} q_t^0 \left( c_t + \frac{i_{t+1}}{1+i_{t+1}} \hat{m}_{t+1} \right) = \sum_{t=0}^{\infty} q_t^0 (1-\tau_t)(1-\ell_t - s_t) + b_0 + \frac{m_0}{p_0}, \tag{17.41}$$

where we have used equation (17.8), and $q_t^0$ is the Arrow-Debreu price

$$q_t^0 = \prod_{i=1}^{t} R_i^{-1}$$

with the numeraire $q_0^0 = 1$. We have also imposed the transversality conditions,

$$\lim_{T\to\infty} q_{T-1}^0 \frac{b_{T+1}}{R_T} = 0, \tag{17.42a}$$

$$\lim_{T\to\infty} q_{T-1}^0 \hat{m}_{T+1} = 0. \tag{17.42b}$$

Given the satiation point in equation (17.40), real money balances held for transaction purposes are bounded from above by $\psi$. Real balances may also be held purely for savings purposes if money is not dominated in rate of return by bonds, but an agent would never find it optimal to accumulate balances that violate the transversality condition. Thus, for whatever reason money is being held, condition (17.42b) must hold in an equilibrium.

Substitute $s_t = H(c_t, \hat{m}_{t+1})$ into equation (17.41), and let $\lambda$ be the Lagrange multiplier on this present-value budget constraint. At an interior solution, the first-order conditions of the household's optimization problem become

$$c_t: \quad \beta^t u_c(t) - \lambda q_t^0 \big[(1-\tau_t) H_c(t) + 1\big] = 0, \tag{17.43a}$$

$$\ell_t: \quad \beta^t u_\ell(t) - \lambda q_t^0 (1-\tau_t) = 0, \tag{17.43b}$$

$$\hat{m}_{t+1}: \quad -\lambda q_t^0 \left[(1-\tau_t) H_{\hat{m}}(t) + \frac{i_{t+1}}{1+i_{t+1}}\right] = 0. \tag{17.43c}$$

From conditions (17.43a) and (17.43b), we obtain

$$\frac{u_\ell(t)}{1-\tau_t} = u_c(t) - u_\ell(t)H_c(t). \tag{17.44}$$

The left side of equation (17.44) is the utility of extra leisure obtained from giving up one unit of disposable labor income, which at the optimum should equal the marginal utility of consumption reduced by the disutility of shopping for the marginal unit of consumption, given by the right side of equation (17.44). Using condition (17.43b) and the corresponding expression for $t=0$ with the numeraire $q_0^0 = 1$, the Arrow-Debreu price $q_t^0$ can be expressed as

$$q_t^0 = \beta^t \frac{u_\ell(t)}{u_\ell(0)} \frac{1-\tau_0}{1-\tau_t}; \tag{17.45}$$

and by condition (17.43c),

$$\frac{i_{t+1}}{1+i_{t+1}} = -(1-\tau_t)H_{\hat{m}}(t). \tag{17.46}$$

This last condition equalizes the cost of holding one unit of real balances (the left side) with the opportunity value of the shopping time that is released by an additional unit of real balances, measured on the right side by the extra after-tax labor income that can be generated.

### *Ramsey plan*

Following the method for solving a Ramsey problem in chapter 12, we use the household's first-order conditions to eliminate prices and taxes from its present-value budget constraint. Specifically, we substitute equations (17.45) and (17.46) into equation (17.41), and then multiply by $u_\ell(0)/(1-\tau_0)$. After also using equation (17.44), the "adjusted budget constraint" becomes

$$\sum_{t=0}^{\infty} \beta^t \{ \left[u_c(t) - u_\ell(t)H_c(t)\right] c_t \; - \; u_\ell(t)H_{\hat{m}}(t)\hat{m}_{t+1} \; - \; u_\ell(t)(1-\ell_t - s_t)\} = 0,$$

where we have assumed zero initial assets, $b_0 = m_0 = 0$. Finally, we substitute $s_t = H(c_t, \hat{m}_{t+1})$ into this expression and invoke Euler's theorem (17.39), to arrive at

$$\sum_{t=0}^{\infty} \beta^t \left\{u_c(t)c_t \; - \; u_\ell(t)\left[1-\ell_t-(1-\nu)H(c_t,\, \hat{m}_{t+1})\right]\right\} = 0. \tag{17.47}$$

The Ramsey problem is to maximize expression (17.2) subject to equation (17.47) and a feasibility constraint that combines equations (17.36)–(17.38):

$$1 - \ell_t - H(c_t, \hat{m}_{t+1}) - c_t - g_t = 0. \tag{17.48}$$

Let $\Phi$ and $\{\theta_t\}_{t=0}^{\infty}$ be a Lagrange multiplier on equation (17.47) and a sequence of Lagrange multipliers on equation (17.48), respectively. First-order conditions for this problem are

$$\begin{aligned}
c_t\colon \quad & u_c(t) + \Phi\left\{u_{cc}(t)c_t + u_c(t)\right. \\
& \quad - u_{\ell c}(t)\left[1 - \ell_t - (1-\nu)H(c_t, \hat{m}_{t+1})\right] \\
& \quad \left. + (1-\nu)u_\ell(t)H_c(t)\right\} - \theta_t\left[H_c(t) + 1\right] = 0, && (17.49a) \\
\ell_t\colon \quad & u_\ell(t) + \Phi\left\{u_{c\ell}(t)c_t + u_\ell(t)\right. \\
& \quad \left. - u_{\ell\ell}(t)\left[1 - \ell_t - (1-\nu)H(c_t, \hat{m}_{t+1})\right]\right\} = -\theta_t, && (17.49b) \\
\hat{m}_{t+1}\colon \quad & H_{\hat{m}}(t)\left[\Phi(1-\nu)u_\ell(t) - \theta_t\right] = 0. && (17.49c)
\end{aligned}$$

The first-order condition for real money balances (17.49c) is satisfied when either $H_{\hat{m}}(t) = 0$ or

$$\theta_t = \Phi(1-\nu)u_\ell(t). \tag{17.50}$$

We now show that equation (17.50) cannot be a solution of the problem. Notice that when $\nu > 1$, equation (17.50) implies that the multipliers $\Phi$ and $\theta_t$ will either be zero or have opposite signs. Such a solution is excluded because $\Phi$ is nonnegative while the insatiable utility function implies that $\theta_t$ is strictly positive. When $\nu = 1$, a strictly positive $\theta_t$ also excludes equation (17.50) as a solution. To reject equation (17.50) for $\nu \in [0, 1)$, we substitute equation (17.50) into equation (17.49b),

$$u_\ell(t) + \Phi\left\{u_{c\ell}(t)c_t + \nu u_\ell(t) - u_{\ell\ell}(t)\left[1 - \ell_t - (1-\nu)H(c_t, \hat{m}_{t+1})\right]\right\} = 0,$$

which is a contradiction because the left-hand side is strictly positive, given our assumption that $u_{c\ell}(t) \geq 0$. We conclude that equation (17.50) cannot characterize the solution of the Ramsey problem when the transaction technology is homogeneous of degree $\nu \geq 0$, so the solution has to be $H_{\hat{m}}(t) = 0$. In other words, the social planner follows the Friedman rule and satiates the economy with real balances. According to condition (17.43c), this aim can be accomplished with a monetary policy that sustains a zero net nominal interest rate.

As an illustration of how the Ramsey plan is implemented, suppose that $g_t = g$ in all periods. Example 1 of chapter 12 presents the Ramsey plan for this case if there were no transaction technology and no money in the model. The optimal outcome is characterized by a constant allocation $(\hat{c}, \hat{n})$ and a constant tax rate $\hat{\tau}$ that supports a balanced government budget. We conjecture that the Ramsey solution to the present monetary economy shares that real allocation. But how can it do so in the present economy with its additional constraint in form of a transaction technology? First, notice that the preceding Ramsey solution calls for satiating the economy with real balances so there will be no time allocated to shopping in the Ramsey outcome. Second, the real balances needed to satiate the economy are constant over time and equal to

$$\frac{M_{t+1}}{p_t} = \psi \hat{c}, \qquad \forall t \geq 0, \tag{17.51}$$

and the real return on money is equal to the constant real interest rate,

$$\frac{p_t}{p_{t+1}} = R, \qquad \forall t \geq 0. \tag{17.52}$$

Third, the real balances in equation (17.51) also equal the real value of assets acquired by the government in period 0 from selling the money supply $M_1$ to the households. These government assets earn a net real return in each future period equal to

$$(R-1)\psi\hat{c} = R\,\frac{M_t}{p_{t-1}} - \frac{M_{t+1}}{p_t} = \frac{p_{t-1}}{p_t}\,\frac{M_t}{p_{t-1}} - \frac{M_{t+1}}{p_t} = \frac{M_t - M_{t+1}}{p_t},$$

where we have invoked equations (17.51) and (17.52) to show that the interest earnings just equal the funds for retiring currency from circulation in all future periods needed to sustain an equilibrium in the money market with a zero net nominal interest rate. It is straightforward to verify that households would be happy to incur the indebtedness of the initial period. They use the borrowed funds to acquire money balances and meet future interest payments by surrendering some of these money balances. Yet their real money balances are unchanged over time because of the falling price level. In this way, money holdings are costless to the households, and their optimal decisions with respect to consumption and labor are the same as in the nonmonetary version of this economy.

## Time consistency of monetary policy

The optimality of the Friedman rule was derived in the previous section under the assumption that the government can commit to a plan for its future actions. The Ramsey plan is not time consistent and requires that the government have a technology to bind itself to it. In each period along the Ramsey plan, the government is tempted to levy an unannounced inflation tax in order to reduce future distortionary labor taxes. Rather than examine this time consistency problem due to distortionary taxation, we now turn to another time consistency problem arising from a situation where surprise inflation can reduce unemployment.

Kydland and Prescott (1977) and Barro and Gordon (1983a, 1983b) study the time consistency problem and credible monetary policies in reduced-form models with a trade-off between surprise inflation and unemployment. In their spirit, Ireland (1997) proposes a model with microeconomic foundations that gives rise to such a trade-off because monopolistically competitive firms set nominal goods prices before the government sets monetary policy.[12] The government is here tempted to create surprise inflation that erodes firms' markups and stimulates employment above a suboptimally low level. But any anticipated inflation has negative welfare effects that arise as a result of a postulated cash-in-advance constraint. More specifically, anticipated inflation reduces the real value of nominal labor income that can be spent or invested first in the next period, thereby distorting incentives to work.

The following setup modifies Ireland's model and assumes that each household has some market power with respect to its labor supply while a single good is produced by perfectly competitive firms.

### *Model with monopolistically competitive wage setting*

There is a continuum of households indexed on the unit interval, $i \in [0,1]$. At time $t$, household $i$ consumes $c_{it}$ of a single consumption good and supplies labor $n_{it} \in [0,1]$. The preferences of the household are

$$\sum_{t=0}^{\infty} \beta^t \left( \frac{c_{it}^{\gamma}}{\gamma} - n_{it} \right), \tag{17.53}$$

where $\beta \in (0,1)$ and $\gamma \in (0,1)$. The parameter restriction on $\gamma$ ensures that the household's utility is well defined at zero consumption.

[12] Ireland's model takes most of its structure from those developed by Svensson (1986) and Rotemberg (1987). See Rotemberg and Woodford (1997) and King and Wolman (1999) for empirical implementations of related models.

The technology for producing the single consumption good is

$$y_t = \left( \int_0^1 n_{it}^{\frac{1-\alpha}{1+\alpha}} \mathrm{d}i \right)^{\frac{1+\alpha}{1-\alpha}}, \tag{17.54}$$

where $y_t$ is per capita output and $\alpha \in (0,1)$. The technology has constant returns to scale in labor inputs, and if all types of labor are supplied in the same quantity $n_t$, we have $y_t = n_t$. The marginal product of labor of type $i$ is

$$\frac{\partial y_t}{\partial n_{it}} = \left( \int_0^1 n_{it}^{\frac{1-\alpha}{1+\alpha}} \mathrm{d}i \right)^{\frac{2\alpha}{1-\alpha}} n_{it}^{\frac{-2\alpha}{1+\alpha}} = \left( \frac{y_t}{n_{it}} \right)^{\frac{2\alpha}{1+\alpha}} \equiv \hat{w}(y_t, n_{it}). \tag{17.55}$$

The single good is produced by a large number of competitive firms that are willing to pay a real wage to labor of type $i$ equal to the marginal product in equation (17.55).

The definition of the function $\hat{w}(y_t, n_{it})$ with its two arguments $y_t$ and $n_{it}$ is motivated by the first of the following two assumptions on households' labor-supply behavior.[13]

1. When maximizing the rent of its labor supply, household $i$ perceives that it can affect the marginal product $\hat{w}(y_t, n_{it})$ through the second argument while $y_t$ is taken as given.

2. The nominal wage for labor of type $i$ at time $t$ is chosen by household $i$ at the very beginning of period $t$. Given the nominal wage $w_{it}$, household $i$ is obliged to deliver any amount of labor $n_{it}$ that is demanded in the economy with feasibility as the sole constraint, $n_{it} \leq 1$.

The government's only task is to increase or decrease the money supply by making lump-sum transfers $(x_t - 1)M_t$ to the households, where $M_t$ is the per capita money supply at the beginning of period $t$ and $x_t$ is the gross growth rate of money in period $t$:

$$M_{t+1} = x_t M_t. \tag{17.56}$$

Following Ireland (1997), we assume that $x_t \in [\beta, \bar{x}]$. These bounds on money growth ensure the existence of a monetary equilibrium. The lower bound will

[13] Analogous assumptions are made implicitly by Ireland (1997), who takes the aggregate price index as given in the monopolistically competitive firms' profit maximization problem, and disregards firms' profitability when computing the output effect of a monetary policy deviation.

be shown to yield a zero net nominal interest rate in a stationary equilibrium, whereas the upper bound $\bar{x} < \infty$ guarantees that households never abandon the use of money altogether.

During each period $t$, events unfold as follows for household $i$: The household starts period $t$ with money $m_{it}$ and real private bonds $b_{it}$, and the household sets the nominal wage $w_{it}$ for its type of labor. After the wage is determined, the government chooses a nominal transfer $(x_t - 1)M_t$ to be handed over to the household. Thereafter, the household enters the asset market to settle maturing bonds $b_{it}$ and to pick a new portfolio composition with money and real bonds $b_{i,t+1}$. After the asset market has closed, the household splits into a shopper and a worker.[14] During period $t$, the shopper purchases $c_{it}$ units of the single good subject to the cash-in-advance constraint

$$\frac{m_{it}}{p_t} + \frac{(x_t - 1)M_t}{p_t} + b_{it} - \frac{b_{i,t+1}}{R_t} \geq c_{it}, \tag{17.57}$$

where $p_t$ and $R_t$ are the price level and the real interest rate, respectively. Given the household's predetermined nominal wage $w_{it}$, the worker supplies all the labor $n_{it} \in [0, 1]$ demanded by firms. At the end of period $t$ when the goods market has closed, the shopper and the worker reunite, and the household's money holdings $m_{i,t+1}$ now equal the worker's labor income $w_{it}n_{it}$ plus any unspent cash from the shopping round. Thus, the budget constraint of the household becomes[15]

$$\frac{m_{it}}{p_t} + \frac{(x_t - 1)M_t}{p_t} + b_{it} + \frac{w_{it}}{p_t} n_{it} = c_{it} + \frac{b_{i,t+1}}{R_t} + \frac{m_{i,t+1}}{p_t}. \tag{17.58}$$

### *Perfect foresight equilibrium*

We first study household $i$'s optimization problem under perfect foresight. Given initial assets $(m_{i0}, b_{i0})$ and sequences of prices $\{p_t\}$, real interest rates

---

[14] The interpretation that the household splits into a shopper and a worker follows Lucas's (1980b) cash-in-advance framework. It embodies the constraint on transactions recommended by Clower (1967).

[15] As long as the labor supply constraint $n_{it} \leq 1$ is not binding for any $i$, the assumptions of constant returns to scale and perfect competition in the goods market imply that profits of firms are zero. If any labor supply constraints were strictly binding, labor would have to be rationed among firms, and there would be strictly positive profits. But binding labor supply constraints cannot be part of a perfect foresight equilibrium because households would not be maximizing labor rents. When we later consider monetary surprises, we assume that monetary deviations are never so expansive that labor supply constraints become strictly binding.

$\{R_t\}$, output levels $\{y_t\}$, and nominal transfers $\{(x_t - 1)M_t\}$, the household maximizes expression (17.53) by choosing sequences of consumption $\{c_{it}\}$, labor supply $\{n_{it}\}$, money holdings $\{m_{i,t+1}\}$, real bond holdings $\{b_{i,t+1}\}$, and nominal wages $\{w_{it}\}$ that satisfy cash-in-advance constraints (17.57) and budget constraints (17.58), with the real wage equaling the marginal product of labor of type $i$ at each point in time, $w_{it}/p_t = \hat{w}(y_t, n_{it})$. The last constraint ensures that the household's choices of $n_{it}$ and $w_{it}$ are consistent with competitive firms' demand for labor of type $i$. Let us incorporate this constraint into budget constraint (17.58) by replacing the real wage $w_{it}/p_t$ by the marginal product $\hat{w}(y_t, n_{it})$. With $\beta^t \mu_{it}$ and $\beta^t \lambda_{it}$ as the Lagrange multipliers on the time-$t$ cash-in-advance constraint and budget constraint, respectively, the first-order conditions at an interior solution are

$$c_{it}: \quad c_{it}^{\gamma-1} - \mu_{it} - \lambda_{it} = 0, \tag{17.59a}$$

$$n_{it}: \quad -1 + \lambda_{it}\left[\frac{\partial\, \hat{w}(y_t, n_{it})}{\partial\, n_{it}} n_{it} + \hat{w}(y_t, n_{it})\right] = 0, \tag{17.59b}$$

$$m_{i,t+1}: \quad -\lambda_{it}\frac{1}{p_t} + \beta\left(\lambda_{i,t+1} + \mu_{i,t+1}\right)\frac{1}{p_{t+1}} = 0, \tag{17.59c}$$

$$b_{i,t+1}: \quad -\left(\lambda_{it} + \mu_{it}\right)\frac{1}{R_t} + \beta\left(\lambda_{i,t+1} + \mu_{i,t+1}\right) = 0. \tag{17.59d}$$

The first-order condition (17.59b) for the rent-maximizing labor supply $n_{it}$ can be rearranged to read

$$\hat{w}(y_t, n_{it}) = \frac{\lambda_{it}^{-1}}{1 + \epsilon_{it}^{-1}} = \frac{1+\alpha}{1-\alpha}\lambda_{it}^{-1}, \tag{17.60}$$

$$\text{where} \quad \epsilon_{it} = \left[\frac{\partial\, \hat{w}(y_t, n_{it})}{\partial\, n_{it}} \frac{n_{it}}{\hat{w}(y_t, n_{it})}\right]^{-1} = -\frac{1+\alpha}{2\alpha} < 0.$$

The Lagrange multiplier $\lambda_{it}$ is the shadow value of relaxing the budget constraint in period $t$ by one unit, measured in "utils" at time $t$. Since preferences (17.53) are linear in the disutility of labor, $\lambda_{it}^{-1}$ is the value of leisure in period $t$ in terms of the units of the budget constraint at time $t$. Equation (17.60) is then the familiar expression that the monopoly price $\hat{w}(y_t, n_{it})$ should be set as a markup above marginal cost $\lambda_{it}^{-1}$, and the markup is inversely related to the absolute value of the demand elasticity of labor type $i$, $|\epsilon_{it}|$.

First-order conditions (17.59c) and (17.59d) for asset decisions can be used to solve for rates of return,

$$\frac{p_t}{p_{t+1}} = \frac{\lambda_{it}}{\beta\left(\lambda_{i,t+1} + \mu_{i,t+1}\right)}, \tag{17.61a}$$

$$R_t = \frac{\lambda_{it} + \mu_{it}}{\beta\left(\lambda_{i,t+1} + \mu_{i,t+1}\right)}. \tag{17.61b}$$

Whenever the Lagrange multiplier $\mu_{it}$ on the cash-in-advance constraint is strictly positive, money has a lower rate of return than bonds, or equivalently, the net nominal interest rate is strictly positive as shown in equation (17.8).

Given initial conditions $m_{i0} = M_0$ and $b_{i0} = 0$, we now turn to characterizing an equilibrium under the additional assumption that the cash-in-advance constraint (17.57) holds with equality, even when it does not bind. Since all households are perfectly symmetric, they will make identical consumption and labor decisions, $c_{it} = c_t$ and $n_{it} = n_t$, so by goods market clearing and the constant-returns-to-scale technology (17.54), we have

$$c_t = y_t = n_t, \tag{17.62a}$$

and from the expression for the marginal product of labor in equation (17.55),

$$\hat{w}(y_t, n_t) = 1. \tag{17.62b}$$

Equilibrium asset holdings satisfy $m_{i,t+1} = M_{t+1}$ and $b_{i,t+1} = 0$. The substitution of equilibrium quantities into the cash-in-advance constraint (17.57) at equality yields

$$\frac{M_{t+1}}{p_t} = c_t, \tag{17.62c}$$

where a version of the "quantity theory of money" determines the price level, $p_t = M_{t+1}/c_t$. We now substitute this expression and conditions (17.59a) and (17.60) into equation (17.61a):

$$\frac{M_{t+1}/c_t}{M_{t+2}/c_{t+1}} = \frac{\left[\frac{1-\alpha}{1+\alpha}\,\hat{w}(y_t, n_t)\right]^{-1}}{\beta\, c_{t+1}^{\gamma-1}},$$

which can be rearranged to read

$$c_t = \frac{1-\alpha}{1+\alpha}\,\frac{\beta}{x_{t+1}}\,c_{t+1}^{\gamma},$$

where we have used equations (17.56) and (17.62b). After taking the logarithm of this expression, we get

$$\log(c_t) = \log\left(\frac{1-\alpha}{1+\alpha}\beta\right) + \gamma\log(c_{t+1}) - \log(x_{t+1}).$$

Since $0 < \gamma < 1$ and $x_{t+1}$ is bounded, this linear difference equation in $\log(c_t)$ can be solved forward to obtain

$$\log(c_t) = \frac{\log\left(\frac{1-\alpha}{1+\alpha}\beta\right)}{1-\gamma} - \sum_{j=0}^{\infty}\gamma^j\log(x_{t+1+j}), \tag{17.63}$$

where equilibrium considerations have prompted us to choose the particular solution that yields a bounded sequence.[16]

*Ramsey plan*

The Ramsey problem is to choose a sequence of monetary growth rates $\{x_t\}$ that supports the perfect foresight equilibrium with the highest possible welfare; that is, the optimal choice of $\{x_t\}$ maximizes the representative household's utility in expression (17.53) subject to expression (17.63) and $n_t = c_t$. From the expression (17.63) it is apparent that the constraints on money growth, $x_t \in [\beta, \bar{x}]$, translate into lower and upper bounds on consumption, $c_t \in [\underline{c}, \bar{c}]$, where

$$\underline{c} = \left(\frac{\beta}{\bar{x}}\frac{1-\alpha}{1+\alpha}\right)^{\frac{1}{1-\gamma}}, \qquad \text{and} \qquad \bar{c} = \left(\frac{1-\alpha}{1+\alpha}\right)^{\frac{1}{1-\gamma}} < 1. \tag{17.64}$$

The Ramsey plan then follows directly from inspecting the one-period return of the Ramsey optimization problem,

$$\frac{c_t^{\gamma}}{\gamma} - c_t, \tag{17.65}$$

which is strictly concave and reaches a maximum at $c = 1$. Thus, the Ramsey solution calls for $x_{t+1} = \beta$ for $t \geq 0$ in order to support $c_t = \bar{c}$ for $t \geq 0$. Notice that the Ramsey outcome can be supported by any initial money growth $x_0$. It is only future money growth rates that must be equal to $\beta$ in order to

[16] See the appendix to chapter 1 for the solution of scalar linear difference equations.

eliminate labor supply distortions that would otherwise arise from the cash-in-advance constraint if the return on money were to fall short of the return on bonds. The Ramsey outcome equalizes the returns on money and bonds; that is, it implements the Friedman rule with a zero net nominal interest rate.

It is instructive to highlight the inability of the Ramsey monetary policy to remove the distortions coming from monopolistic wage setting. Using the fact that the equilibrium real wage is unity, we solve for $\lambda_{it}$ from equation (17.60) and substitute into equation (17.59a),

$$c_{it}^{\gamma-1} = \mu_{it} + \frac{1+\alpha}{1-\alpha} > 1. \tag{17.66}$$

The left side of equation (17.66) is the marginal utility of consumption. Since technology (17.54) is linear in labor, the marginal utility of consumption should equal the marginal utility of leisure in a first-best allocation. But the right side of equation (17.66) exceeds unity, which is the marginal utility of leisure given preferences (17.53). While the Ramsey monetary policy succeeds in removing distortions from the cash-in-advance constraint by setting the Lagrange multiplier $\mu_{it}$ equal to zero, the policy cannot undo the distortion of monopolistic wage setting manifested in the "markup" $(1+\alpha)/(1-\alpha)$.[17] Notice that the Ramsey solution converges to the first-best allocation when the parameter $\alpha$ goes to zero, that is, when households' market power goes to zero.

To illustrate the time consistency problem, we now solve for the Ramsey plan when the initial nominal wages are taken as given, $w_{i0} = w_0 \in [\beta M_0, \bar{x} M_0]$. First, setting the initial period 0 aside, it is straightforward to show that the solution for $t \geq 1$ is the same as before. That is, the optimal policy calls for $x_{t+1} = \beta$ for $t \geq 1$ in order to support $c_t = \bar{c}$ for $t \geq 1$. Second, given $w_0$, the first-best outcome $c_0 = 1$ can be attained in the initial period by choosing $x_0 = w_0/M_0$. The resulting money supply $M_1 = w_0$ will then serve to transact $c_0 = 1$ at the equilibrium price $p_0 = w_0$. Specifically, firms are happy to hire any number of workers at the wage $w_0$ when the price of the good is $p_0 = w_0$. At the price $p_0 = w_0$, the goods market clears at full employment, since shoppers seek to spend their real balances $M_1/p_0 = 1$. The labor market also clears because workers are obliged to deliver the demanded $n_0 = 1$. Finally, money growth $x_1$ can be chosen freely and does not affect the real allocation of the Ramsey solution. The reason is that, because of the preset wage $w_0$, there cannot be any

[17] The government would need to use fiscal instruments, that is, subsidies and taxation, to correct the distortion from monopolistically competitive wage setting.

labor supply distortions at time 0 arising from a low return on money holdings between periods 0 and 1.

*Credibility of the Friedman rule*

Our comparison of the Ramsey equilibria with or without a preset initial wage $w_0$ hints at the government's temptation to create positive monetary surprises that will increase employment. We now ask if the Friedman rule is credible when the government lacks the commitment technology implicit in the Ramsey optimization problem. Can the Friedman rule be supported with a trigger strategy where a government deviation causes the economy to revert to the worst possible subgame perfect equilibrium?

Using the concepts and notation of chapter 16, we specify the objects of a strategy profile and state the definition of a subgame perfect equilibrium (SPE). Even though households possess market power with respect to their labor type, they remain atomistic vis-à-vis the government. We therefore stay within the framework of chapter 16 where the government behaves strategically, and the households' behavior can now be summarized as a "monopolistically competitive equilibrium" that responds nonstrategically to the government's choices. At every date $t$ for all possible histories, a strategy of the households $\sigma^h$ and a strategy of the government $\sigma^g$ specify actions $\tilde{w}_t \in \tilde{W}$ and $x_t \in X \equiv [\beta, \bar{x}]$, respectively, where

$$\tilde{w}_t = \frac{w_t}{M_t}, \qquad \text{and} \qquad x_t = \frac{M_{t+1}}{M_t}.$$

That is, the actions multiplied by the beginning-of-period money supply $M_t$ produce a nominal wage and a nominal money supply. (This scaling of nominal variables is used by Ireland, 1997, throughout his analysis, since the size of the nominal money supply at the beginning of a period has no significance *per se*.)

DEFINITION: A strategy profile $\sigma = (\sigma^h, \sigma^g)$ is a *subgame perfect equilibrium* if, for each $t \geq 0$ and each history $(\tilde{w}^{t-1}, x^{t-1}) \in \tilde{W}^t \times X^t$,

(1) Given the trajectory of money growth rates $\{x_{t-1+j} = x(\sigma|_{(\tilde{w}^{t-1}, x^{t-1})})_j\}_{j=1}^{\infty}$, the wage-setting outcome $\tilde{w}_t = \sigma_t^h(\tilde{w}^{t-1}, x^{t-1})$ constitutes a monopolistically competitive equilibrium.

(2) The government cannot strictly improve the households' welfare by deviating from $x_t = \sigma_t^g(\tilde{w}^{t-1}, x^{t-1})$, that is, by choosing some other money growth rate $\eta \in X$ with the implied continuation strategy profile $\sigma|_{(\tilde{w}^t; x^{t-1}, \eta)}$.

Besides changing to a "monopolistically competitive equilibrium," the main difference from Definition 6 of chapter 16 lies in requirement (1). The equilibrium in period $t$ can no longer be stated in terms of an isolated government action at time $t$ but requires the trajectory of the current and all future money growth rates, generated by the strategy profile $\sigma|_{(\tilde{w}^{t-1},x^{t-1})}$. The monopolistically competitive equilibrium in requirement (1) is understood to be the perfect foresight equilibrium described previously. When the government is contemplating a deviation in requirement (2), the equilibrium is constructed as follows: In period $t$ when the deviation takes place, equilibrium consumption $c_t$ is a function of $\eta$ and $\tilde{w}_t$ as implied by the cash-in-advance constraint at equality,

$$c_t = \frac{\eta M_t}{p_t} = \min\left\{\frac{\eta M_t}{w_t}, 1\right\} = \min\left\{\frac{\eta}{\tilde{w}_t}, 1\right\}, \tag{17.67}$$

where we use the equilibrium condition $p_t \geq w_t$ that holds with strict equality unless labor is rationed among firms at full employment.[18] Starting in period $t+1$, the deviation has triggered a switch to a new perfect foresight equilibrium with a trajectory of money growth rates given by $\{x_{t+j} = x(\sigma|_{(\tilde{w}^t;x^{t-1},\eta)})_j\}_{j=1}^{\infty}$.

We conjecture that the worst SPE has $c_t = \underline{c}$ for all periods, and the candidate strategy profile $\hat{\sigma}$ is

$$\hat{\sigma}_t^h = \frac{\bar{x}}{\underline{c}} \qquad \forall\, t\,, \quad \forall\, (\tilde{w}^{t-1}, x^{t-1});$$

$$\hat{\sigma}_t^g = \bar{x} \qquad \forall\, t\,, \quad \forall\, (\tilde{w}^{t-1}, x^{t-1}).$$

The strategy profile instructs the government to choose the highest permissible money growth rate $\bar{x}$ for all periods and for all histories. Similarly, the households are instructed to set the nominal wages that would constitute a perfect foresight equilibrium when money growth will always be at its maximum. Thus, requirement (1) of a SPE is clearly satisfied. It remains to show that the government has no incentive to deviate. Since the continuation strategy profile is $\hat{\sigma}$ regardless of the history, the government needs only to find the best response in terms of the one-period return (17.65). After substituting the household's action $\tilde{w}_t = \bar{x}/\underline{c}$ into equation (17.67), we get $c_t = \underline{c}\eta/\bar{x}$, so the best response of the government is to follow the proposed strategy $\bar{x}$. We conclude that the strategy

[18] Notice that all $\eta \geq \tilde{w}_t$ yield full employment. Under the assumption that firm profits are evenly distributed among households, it also follows that all $\eta \geq \tilde{w}_t$ share the same welfare implications. Without loss of generality, we can therefore restrict attention to choices of $\eta$ that are no larger than $\tilde{w}_t$, that is, the assumption referred to previously stating that monetary deviations are never so expansive that labor supply constraints become strictly binding.

profile $\hat{\sigma}$ is indeed a SPE, and it is the worst, since $\underline{c}$ is the lower bound on consumption in any perfect foresight equilibrium.

We are now ready to address the credibility of the Friedman rule. The best chance for the Friedman rule to be credible is if a deviation triggers a reversion to the worst possible subgame perfect equilibrium given by $\hat{\sigma}$. The condition for credibility becomes

$$\frac{\frac{\bar{c}^{\gamma}}{\gamma} - \bar{c}}{1-\beta} \geq \left(\frac{1}{\gamma} - 1\right) + \beta \frac{\frac{\underline{c}^{\gamma}}{\gamma} - \underline{c}}{1-\beta}. \tag{17.68}$$

By following the Friedman rule, the government removes the labor supply distortion coming from a binding cash-in-advance constraint and keeps output at $\bar{c}$. By deviating from the Friedman rule, the government creates a positive monetary surprise that increases output to its efficient level of unity, thereby eliminating the distortion caused by monopolistically competitive wage setting as well. However, this deviation destroys the government's reputation, and the economy reverts to an equilibrium that induces the government to inflate at the highest possible rate thereafter, and output falls to $\underline{c}$. Hence, the Friedman rule is credible if and only if equation (17.68) holds.

The Friedman rule is the more likely to be credible, the higher is the exogenous upper bound on money growth $\bar{x}$, since $\underline{c}$ depends negatively on $\bar{x}$. In other words, a higher $\bar{x}$ translates into a larger penalty for deviating, so the government becomes more willing to adhere to the Friedman rule to avoid this penalty. In the limit when $\bar{x}$ becomes arbitrarily large, $\underline{c}$ approaches zero and condition (17.68) reduces to

$$\left(\frac{1-\alpha}{1+\alpha}\right)^{\frac{\gamma}{1-\gamma}} \left(\frac{1}{\gamma} - \frac{1-\alpha}{1+\alpha}\right) \geq (1-\beta)\left(\frac{1}{\gamma} - 1\right),$$

where we have used the expression for $\bar{c}$ in equations (17.64). The Friedman rule can be sustained for a sufficiently large value of $\beta$. The government has less incentive to deviate when households are patient and put a high weight on future outcomes. Moreover, the Friedman rule is credible for a sufficiently small value of $\alpha$, which is equivalent to households having little market power. The associated small distortion from monopolistically competitive wage setting means that the potential welfare gain of a monetary surprise is also small, so the government is less tempted to deviate from the Friedman rule.

## *Concluding discussion*

Besides shedding light on a number of monetary doctrines, this chapter has brought out the special importance of the initial date $t = 0$ in the analysis. This point is especially pronounced in Woodford's (1995) model where the initial interest-bearing government debt $B_0$ is not indexed but rather denominated in nominal terms. So, although the construction of a perfect foresight equilibrium ensures that all future issues of nominal bonds will ex post yield the real rates of return that are needed to entice the households to hold these bonds, the realized real return on the initial nominal bonds can be anything depending on the price level $p_0$. Activities at the initial date were also important when we considered dynamic optimal taxation in chapter 12.

Monetary issues are also discussed in other chapters of the book. Chapters 8 and 14 study money in overlapping generation models and Bewley models, respectively. Chapters 18 and 19 present a couple of other explicit environments which give rise to a positive value of fiat money: Townsend's turnpike model and the Kiyotaki-Wright search model.

## Exercises

*Exercise 17.1* **Why deficits in Italy and Brazil were once extraordinary proportions of GDP**

The government's budget constraint can be written as

$$g_t - \tau_t + \frac{b_t}{R_{t-1}}(R_{t-1} - 1) = \frac{b_{t+1}}{R_t} - \frac{b_t}{R_{t-1}} + \frac{M_{t+1}}{p_t} - \frac{M_t}{p_t}. \tag{1}$$

The left side is the real gross-of-interest government deficit; the right side is change in the real value of government liabilities between $t-1$ and $t$.

Government budgets often report the *nominal* gross-of-interest government deficit, defined as

$$p_t(g_t - \tau_t) + p_t b_t \left(1 - \frac{1}{R_{t-1}p_t/p_{t-1}}\right),$$

and their ratio to nominal GNP, $p_t y_t$, namely,

$$\left[(g_t - \tau_t) + b_t\left(1 - \frac{1}{R_{t-1}p_t/p_{t-1}}\right)\right] / y_t.$$

For countries with a large $b_t$ (e.g., Italy) this number can be very big even with a moderate rate of inflation. For countries with a rapid inflation rate, like Brazil in 1993, this number sometimes comes in at 30 percent of GDP. Fortunately, this number overstates the magnitude of the government's "deficit problem," and there is a simple adjustment to the interest component of the deficit that renders a more accurate picture of the problem. In particular, notice that the real values of the interest component of the real and nominal deficits are related by

$$b_t\left(1-\frac{1}{R_{t-1}}\right)=\alpha_t b_t\left(1-\frac{1}{R_{t-1}p_t/p_{t-1}}\right),$$

where

$$\alpha_t=\frac{R_{t-1}-1}{R_{t-1}-p_{t-1}/p_t}.$$

Thus, we should multiply the real value of nominal interest payments $b_t(1-\frac{p_{t-1}}{R_{t-1}p_t})$ by $\alpha_t$ to get the real interest component of the debt that appears on the left side of equation (1).

**a.** Compute $\alpha_t$ for a country that has a $b_t/y$ ratio of .5, a gross real interest rate of 1.02, and a zero net inflation rate.

**b.** Compute $\alpha$ for a country that has a $b_t/y$ ratio of .5, a gross real interest rate of 1.02, and a 100 percent per year net inflation rate.

*Exercise 17.2* **A strange example of Brock (1974)**

Consider an economy consisting of a government and a representative household. There is one consumption good, which is not produced and not storable. The exogenous supply of the good at time $t \geq 0$ is $y_t = y > 0$. The household owns the good. At time $t$ the representative household's preferences are ordered by

$$\sum_{t=0}^{\infty}\beta^t\{\ln c_t+\gamma\ln(m_{t+1}/p_t)\}, \tag{17.69}$$

where $c_t$ is the household's consumption at $t$, $p_t$ is the price level at $t$, and $m_{t+1}/p_t$ is the real balances that the household carries over from time $t$ to $t+1$. Assume that $\beta \in (0,1)$ and $\gamma > 0$. The household maximizes equation (17.69) over choices of $\{c_t, m_{t+1}\}$ subject to the sequence of budget constraints

$$c_t+m_{t+1}/p_t=y_t-\tau_t+m_t/p_t, \quad t\geq 0, \tag{17.70}$$

where $\tau_t$ is a lump-sum tax due at $t$. The household faces the price sequence $\{p_t\}$ as a price taker and has given initial value of nominal balances $m_0$.

At time $t$ the government faces the budget constraint

$$g_t = \tau_t + (M_{t+1} - M_t)/p_t, \quad t \geq 0, \tag{17.71}$$

where $M_t$ is the amount of currency that the government has outstanding at the beginning of time $t$ and $g_t$ is government expenditures at time $t$. In equilibrium, we require that $M_t = m_t$ for all $t \geq 0$. The government chooses sequences of $\{g_t, \tau_t, M_{t+1}\}_{t=0}^{\infty}$ subject to the budget constraints (17.71) being satisfied for all $t \geq 0$ and subject to the given initial value $M_0 = m_0$.

**a.** Define a *competitive equilibrium.*

For the remainder of this problem assume that $g_t = g < y$ for all $t \geq 0$, and that $\tau_t = \tau$ for all $t \geq 0$. Define a *stationary equilibrium* as an equilibrium in which the rate of return on currency is constant for all $t \geq 0$.

**b.** Find conditions under which there exists a stationary equilibrium for which $p_t > 0$ for all $t \geq 0$. Derive formulas for real balances and the rate of return on currency in that equilibrium, given that it exists. Is the stationary equilibrium unique?

**c.** Find a first-order difference equation in the equilibrium level of real balances $h_t = M_{t+1}/p_t$ whose satisfaction assures equilibrium (possibly nonstationary).

**d.** Show that there is a fixed point of this difference equation with positive real balances, provided that the condition that you derived in part b is satisfied. Show that this fixed point agrees with the level of real balances that you computed in part b.

**e.** Under what conditions is the following statement true: If there exists a stationary equilibrium, then there also exist many other nonstationary equilibria. Describe these other equilibria. In particular, what is happening to real balances and the price level in these other equilibria? Among these other equilibria, within which one(s) are consumers better off?

**f.** Within which of the equilibria that you found in parts b and e is the following "old-time religion" true: "Larger sustained government deficits imply permanently larger inflation rates"?

*Exercise 17.3* **Optimal inflation tax in a cash-in-advance model**

Consider the version of Ireland's (1997) model described in the text but assume perfect competition (i.e., $\alpha = 0$) with flexible market-clearing wages. Suppose now that the government must finance a constant amount of purchases $g$ in each period by levying flat-rate labor taxes and raising seigniorage. Solve the optimal taxation problem under commitment.

*Exercise 17.4* **Deficits, inflation, and anticipated monetary shocks**, donated by Rodolfo Manuelli

Consider an economy populated by a large number of identical individuals. Preferences over consumption and leisure are given by,

$$\sum_{t=0}^{\infty} \beta^t c_t^{\alpha} \ell_t^{1-\alpha},$$

where $0 < \alpha < 1$. Assume that leisure is positively related - this is just a reduced form of a shopping-time model - to the stock of real money balances, and negatively related to a measure of transactions:

$$\ell_t = A(m_{t+1}/p_t)/c_t^{\eta}, \quad A > 0$$

and $\alpha - \eta(1-\alpha) > 0$. Each individual owns a tree that drops $y$ units of consumption per period (dividends). There is a government that issues one-period real bonds, money, and collects taxes (lump-sum) to finance spending. Per capita spending is equal to $g$. Thus, consumption equals $c = y - g$. The government's budget constraint is:

$$g_t + B_t = \tau_t + B_{t+1}/R_t + (M_{t+1} - M_t)/p_t$$

Let the rate of return on money be $R_{mt} = p_t/p_{t+1}$. Let the nominal interest rate at time $t$ be $1 + i_t = R_t p_{t+1}/p_t = R_t \pi_t$.

**a.** Derive the demand for money, and show that it decreases with the nominal interest rate.

**b.** Suppose that the government policy is such that $g_t = g$, $B_t = B$ and $\tau_t = \tau$. Prove that the real interest rate, $R$, is constant and equal to the inverse of the discount factor.

**c.** Define the deficit as $d$, where $d = g + (B/R)(R-1) - \tau$. What is the highest possible deficit that can be financed in this economy? An economist claims that — in this economy — increases in $d$, which leave $g$ unchanged, will result in increases in the inflation rate. Discuss this view.

**d.** Suppose that the economy is open to international capital flows and that the world interest rate is $R^* = \beta^{-1}$. Assume that $d = 0$, and that $M_t = M$. At $t = T$, the government increases the money supply to $M' = (1 + \mu)M$. This increase in the money supply is used to purchase (government) bonds. This, of course, results in a smaller deficit at $t > T$. (In this case, it will result in a surplus.) However, the government also announces its intention to cut taxes (starting at $T + 1$) to bring the deficit back to zero. Argue that this open market operation will have the effect of increasing prices at $t = T$ by $\mu$; $p' = (1 + \mu)p$, where $p$ is the price level from $t = 0$ to $t = T - 1$.

**e.** Consider the same setting as in d. Suppose now that the open market operation is announced at $t = 0$ (it still takes place at $t = T$). Argue that prices will increase at $t = 0$ and, in particular, that the rate of inflation between $T - 1$ and $T$ will be less than $1 + \mu$.

*Exercise 17.5* **Interest elasticity of the demand for money**, donated by Rodolfo Manuelli

Consider an economy in which the demand for money satisfies

$$m_{t+1}/p_t = F(c_t, R_{mt}/R_t),$$

where $R_{mt} = p_t/p_{t+1}$, and $R_t$ is the one-period interest rate. Consider the following open market operation: At $t = 0$, the government sells bonds and "destroys" the money it receives in exchange for those bonds. No other real variables — government spending or taxes — are changed. Find conditions on the income elasticity of the demand for money such that the decrease in money balances at $t = 0$ results in an increase in the price level at $t = 0$.

*Exercise 17.6* **Dollarization**, donated by Rodolfo Manuelli

In recent years, several countries — Argentina, and some of the countries hit by the Asian crisis, among others — have considered the possibility of giving up their currencies in favor of the U.S. dollar. Consider a country, say $A$, with deficit $d$ and inflation rate $\pi = 1/R_m$. Output and consumption are constant and, hence, the real interest rate is fixed with $R = \beta^{-1}$. The (gross of interest payments) deficit is $d$, with

$$d = g - \tau + (B/R)(R - 1).$$

Let the demand for money be $m_{t+1}/p_t = F(c_t, R_{mt}/R_t)$, and assume that $c_t = y - g$. Thus, the steady state government budget constraint is

$$d = F(y - g, \beta R_m)(1 - R_m) > 0.$$

Assume that the country is considering, at $t = 0$, the retirement of its money in exchange for dollars. The government promises to give to each person who brings a "peso" to the Central Bank $1/e$ dollars, where $e$ is the exchange rate (in pesos per dollar) between the country's currency and the U.S. dollar. Assume that the U.S. inflation rate (before and after the switch) is given and equal to $\pi^* = 1/R_m^* < \pi$, and that the country is in the "good" part of the Laffer curve.

**a.** If you are advising the government of $A$, how much would you say that it should demand from the U.S. government to make the switch? Why?

**b.** After the dollarization takes place, the government understands that it needs to raise taxes. Economist 1 argues that the increase in taxes (on a per period basis) will equal the loss of revenue from inflation — $F(y - g, \beta R_m)(1 - R_m)$ — while Economist 2 claims that this is an overestimate. More precisely, he/she claims that, if the government is a good negotiator vis-à-vis the U.S. government, taxes need only increase by $F(y - g, \beta R_m)(1 - R_m) - F(y - g, \beta R_m^*)(1 - R_m^*)$ per period. Discuss these two views.

*Exercise 17.7* **Currency boards**, donated by Rodolfo Manuelli

In the last few years several countries — Argentina (1991), Estonia (1992), Lithuania (1994), Bosnia (1997) and Bulgaria (1997) — have adopted the currency board model of monetary policy. In a nutshell, a currency board is a commitment on the part of the country to fully back its domestic currency with foreign denominated assets. For simplicity, assume that the foreign asset is the U.S. dollar.

The government's budget constraint is given by

$$g_t + B_t + B_{t+1}^* e/(Rp_t) = \tau_t + B_{t+1}/R + B_t^* e/p_t + (M_{t+1} - M_t)/p_t,$$

where $B_t^*$ is the stock of one period bonds – denominated in dollars – held by this country, $e$ is the exchange rate (pesos per dollar), and $1/R$ is the price of one-period bonds (both domestic and dollar denominated). Note that the budget constraint equates the real value of income and liabilities in units of consumption goods.

The currency board "contract" requires that the money supply be fully backed. One interpretation of this rule is that the domestic money supply is

$$M_t = eB_t^*.$$

Thus, the right side is the local currency value of foreign reserves (in bonds) held by the government, while the left side is the stock of money. Finally, let the law of one price hold: $p_t = ep_t^*$, where $p_t^*$ is the foreign (U.S.) price level.

**a.** Assume that $B_t = B$, and that foreign inflation is zero, $p_t^* = p^*$. Show that, even in this case, the properties of the demand for money — which you may take to be given by $F(y - g, \beta R_m)$ — are important in determining total revenue. In particular, explain how a permanent increase in $y$, income per capita, allows the government to lower taxes (permanently).

**b.** Assume that $B_t = B$. Let foreign inflation be positive, that is, $\pi^* > 1$. In this case, the price – in dollars – of a one-period dollar-denominated bond is $1/(R\pi^*)$. Go as far as you can describing the impact of foreign inflation on domestic inflation, and on per capita taxes, $\tau$.

**c.** Assume that $B_t = B$. Go as far as you can describing the effects of a once-and-for-all surprise devaluation – an unexpected and permanent increase in $e$ — on the level of per capita taxes.

*Exercise 17.8* **Growth and inflation**, donated by Rodolfo Manuelli

Consider an economy populated by identical individuals with instantaneous utility function given, by

$$u(c, \ell) = [c^{\varphi} \ell^{1-\varphi}]^{(1-\sigma)}/(1-\sigma).$$

Assume that shopping time is given by, $s_t = \psi c_t/(m_{t+1}/p_t)$. Assume that in this economy income grows exogenously at the rate $\gamma > 1$. Thus, at time $t$, $y_t = \gamma^t y$. Assume that government spending also grows at the same rate, $g_t = \gamma^t g$. Finally, $c_t = y_t - g_t$.

**a.** Show that for this specification, if the demand for money at $t$ is $x = m_{t+1}/p_t$, then the demand at $t+1$ is $\gamma x$. Thus, the demand for money grows at the same rate as the economy.

**b.** Show that the real rate of interest depends on the growth rate. (You may assume that $\ell$ is constant for this calculation.)

**c.** Argue that even for monetary policies that keep the price level constant, that is, $p_t = p$ for all $t$, the government raises positive amounts of revenue from printing money. Explain.

**d.** Use your finding in c to discuss why, following monetary reforms that generate big growth spurts, many countries manage to "monetize" their economies (this is just jargon for increases in the money supply) without generating inflation.

# 18
# Credit and Currency

## Credit and currency with long-lived agents

This chapter describes Townsend's (1980) turnpike model of money and puts it to work. The model uses a particular pattern of heterogeneity of endowments and locations to create a demand for currency. The model is more primitive than the shopping time model of chapter 17. As with the overlapping generations model, the turnpike model starts from a setting in which diverse intertemporal endowment patterns across agents prompt borrowing and lending. If something prevents loan markets from operating, it is possible that an unbacked currency can play a role in helping agents smooth their consumption over time. Following Townsend, we shall eventually appeal to locational heterogeneity as the force that causes loan markets to fail in this way.

The turnpike model can be viewed as a simplified version of the stochastic model proposed by Truman Bewley (1980). We use the model to study a number of interrelated issues and theories, including (1) a permanent income theory of consumption, (2) a Ricardian doctrine that government borrowing and taxes have equivalent economic effects, (3) some restrictions on the operation of private loan markets needed in order that unbacked currency be valued, (4) a theory of inflationary finance, (5) a theory of the optimal inflation rate and the optimal behavior of the currency stock over time, (6) a "legal restrictions" theory of inflationary finance, and (7) a theory of exchange rate indeterminacy.[1]

## Preferences and endowments

There is one consumption good. It cannot be produced or stored. The total amount of goods available each period is constant at $N$. There are $2N$ households, divided into equal numbers $N$ of two types, according to their endowment

[1] Some of the analysis in this chapter follows Manuelli and Sargent (1992). Also see Chatterjee and Corbae (1996) and Ireland (1994) for analyses of policies within a turnpike environment.

sequences. The two types of households, dubbed *odd* and *even*, have endowment sequences

$$\{y_t^o\}_{t=0}^{\infty} = \{1, 0, 1, 0, \ldots\},$$
$$\{y_t^e\}_{t=0}^{\infty} = \{0, 1, 0, 1, \ldots\}.$$

Households of both types order consumption sequences $\{c_t^h\}$ according to the common utility function

$$U = \sum_{t=0}^{\infty} \beta^t u(c_t^h),$$

where $\beta \in (0, 1)$, and $u(\cdot)$ is twice continuously differentiable, increasing and strictly concave, and satisfies

$$\lim_{c \downarrow 0} u'(c) = +\infty. \tag{18.1}$$

## Complete markets

As a benchmark, we study a version of the economy with complete markets. Later we shall more or less arbitrarily shut down many of the markets to make room for money.

### A Pareto problem

Consider the following Pareto problem: Let $\theta \in [0, 1]$ be a weight indexing how much a social planner likes odd agents. The problem is to choose consumption sequences $\{c_t^o, c_t^e\}_{t=0}^{\infty}$ to maximize

$$\theta \sum_{t=0}^{\infty} \beta^t u(c_t^o) + (1 - \theta) \sum_{t=0}^{\infty} \beta^t u(c_t^e), \tag{18.2}$$

subject to

$$c_t^e + c_t^o = 1, \quad t \geq 0. \tag{18.3}$$

The first-order conditions are

$$\theta u'(c_t^o) - (1 - \theta) u'(c_t^e) = 0.$$

Substituting the constraint (18.3) into this first-order condition and rearranging gives the condition

$$\frac{u'(c_t^o)}{u'(1-c_t^o)} = \frac{1-\theta}{\theta}. \tag{18.4}$$

Since the right side is independent of time, the left must be also, so that condition (18.4) determines the one-parameter family of optimal allocations

$$c_t^o = c^o(\theta), \quad c_t^e = 1 - c^o(\theta).$$

*A complete markets equilibrium*

A household takes the price sequence $\{q_t^0\}$ as given and chooses a consumption sequence to maximize $\sum_{t=0}^{\infty} \beta^t u(c_t)$ subject to the budget constraint

$$\sum_{t=0}^{\infty} q_t^0 c_t \leq \sum_{t=0}^{\infty} q_t^0 y_t.$$

The household's Lagrangian is

$$L = \sum_{t=0}^{\infty} \beta^t u(c_t) + \mu \sum_{t=0}^{\infty} q_t^0 (y_t - c_t),$$

where $\mu$ is a nonnegative Lagrange multiplier. The first-order conditions for the household's problem are

$$\beta^t u'(c_t) \leq \mu q_t^0, \quad = \text{ if } c_t > 0.$$

DEFINITION 1: A *competitive equilibrium* is a price sequence $\{q_t^o\}_{t=0}^{\infty}$ and an allocation $\{c_t^o, c_t^e\}_{t=0}^{\infty}$ that have the property that (a) given the price sequence, the allocation solves the optimum problem of households of both types, and (b) $c_t^o + c_t^e = 1$ for all $t \geq 0$.

To find an equilibrium, we have to produce an allocation and a price system for which we can verify that the first-order conditions of both households are satisfied. We start with a guess inspired by the constant-consumption property

of the Pareto optimal allocation. We guess that $c_t^o = c^o, c_t^e = c^e \ \forall t$, where $c^e + c^o = 1$. This guess and the first-order condition for the odd agents imply

$$q_t^0 = \frac{\beta^t u'(c^o)}{\mu^o},$$

or

$$q_t^0 = q_0^0 \beta^t, \tag{18.5}$$

where we are free to normalize by setting $q_0^0 = 1$. For odd agents, the right side of the budget constraint evaluated at the prices given in equation (18.5) is then

$$\frac{1}{1-\beta^2},$$

and for even households it is

$$\frac{\beta}{1-\beta^2}.$$

The left side of the budget constraint evaluated at these prices is

$$\frac{c^i}{1-\beta}, \qquad i = o, e.$$

For both of the budget constraints to be satisfied with equality we evidently require that

$$\begin{aligned} c^o &= \frac{1}{\beta+1} \\ c^e &= \frac{\beta}{\beta+1}. \end{aligned} \tag{18.6}$$

The price system given by equation (18.5) and the constant over time allocations given by equations (18.6) are a competitive equilibrium.

Notice that the competitive equilibrium allocation corresponds to a particular Pareto optimal allocation.

### *Ricardian proposition*

We temporarily add a government to the model. The government levies lump-sum taxes on agents of type $i = o, e$ at time $t$ of $\tau_t^i$. The government uses the

proceeds to finance a constant level of government purchases of $G \in (0,1)$ each period $t$. Consumer $i$'s budget constraint is

$$\sum_{t=0}^{\infty} q_t^0 c_t^i \leq \sum_{t=0}^{\infty} q_t^0 (y_t^i - \tau_t^i).$$

The government's budget constraint is

$$\sum_{t=0}^{\infty} q_t^0 G = \sum_{i=o,e} \sum_{t=0}^{\infty} q_t^0 \tau_t^i .$$

We modify Definition 1 as follows:

DEFINITION 2: A *competitive equilibrium* is a price sequence $\{q_t^0\}_{t=0}^{\infty}$, a tax system $\{\tau_t^o, \tau_t^e\}_{t=0}^{\infty}$, and an allocation $\{c_t^o, c_t^e, G_t\}_{t=0}^{\infty}$ such that given the price system and the tax system the following conditions hold: (a) the allocation solves each consumer's optimum problem, (b) the government budget constraint is satisfied for all $t \geq 0$, and (c) $N(c_t^o + c_t^e) + G_t = N$ for all $t \geq 0$.

Let the present value of the taxes imposed on consumer $i$ be $\tau^i \equiv \sum_{t=0}^{\infty} q_t^0 \tau_t^i$. Then it is straightforward to verify that the equilibrium price system is still equation (18.5) and that equilibrium allocations are

$$c^o = \frac{1}{\beta + 1} - \tau^o (1 - \beta)$$

$$c^e = \frac{\beta}{\beta + 1} - \tau^e (1 - \beta).$$

This equilibrium features a "Ricardian proposition":

RICARDIAN PROPOSITION:

The equilibrium is invariant to changes in the *timing* of tax collections that leave unaltered the present value of lump-sum taxes assigned to each agent.

*Loan market interpretation*

Define total time-$t$ tax collections as $\tau_t = \sum_{i=o,e} \tau_t^i$, and write the government's budget constraint as

$$(G_0 - \tau_0) = \sum_{t=1}^{\infty} \frac{q_t^0}{q_0^0} (\tau_t - G_t) \equiv B_1,$$

where $B_1$ can be interpreted as government debt issued at time 0 and due at time 1. Notice that $B_1$ equals the present value of the future (i.e., from time 1 onward) government *surpluses* $(\tau_t - G_t)$. The government's budget constraint can also be represented as

$$\frac{q_0^0}{q_1^0}(G_0 - \tau_0) + (G_1 - \tau_1) = \sum_{t=2}^{\infty} \frac{q_t^0}{q_1^0}(\tau_t - G_t) \equiv B_2,$$

or

$$R_1 B_1 + (G_1 - \tau_1) = B_2,$$

where $R_1 = \frac{q_0^0}{q_1^0}$ is the gross rate of return between time 0 and time 1, measured in time-1 consumption goods per unit of time-0 consumption good. More generally, we can represent the government's budget constraint by the sequence of budget constraints

$$R_t B_t + (G_t - \tau_t) = B_{t+1}, \quad t \geq 0,$$

subject to the boundary condition $B_0 = 0$. In the equilibrium computed here, $R_t = \beta^{-1}$ for all $t \geq 1$.

Similar manipulations of consumers' budget constraints can be used to express them in terms of sequences of one-period budget constraints. That no opportunities are lost to the government or the consumers by representing the budget sets in this way lies behind the following fact: the Arrow-Debreu allocation in this economy can be implemented with a sequence of one-period loan markets.

In the following section, we shut down *all* loan markets, and also set government expenditures $G = 0$.

## A monetary economy

We keep preferences and endowment patterns as they were in the preceding economy, but we rule out all intertemporal trades achieved through borrowing and lending or trading of future-dated consumptions. We replace complete markets with a fiat money mechanism. At time 0, the government endows each of the $N$ even agents with $M/N$ units of an unbacked or inconvertible currency. Odd agents are initially endowed with zero units of the currency. Let $p_t$ be the time-$t$ price level, denominated in dollars per time-$t$ consumption good. We seek an equilibrium in which currency is valued $(p_t < +\infty \; \forall t \geq 0)$ and in which each period agents not endowed with goods pass currency to agents who are endowed with goods. Contemporaneous exchanges of currency for goods are the only exchanges that we, the model builders, permit. (Later Townsend will give us a defense or reinterpretation of this high-handed shutting down of markets.)

Given the sequence of prices $\{p_t\}_{t=0}^{\infty}$, the household's problem is to choose nonnegative sequences $\{c_t, m_t\}_{t=0}^{\infty}$ to maximize $\sum_{t=0}^{\infty} \beta^t u(c_t)$ subject to

$$m_t + p_t c_t \leq p_t y_t + m_{t-1}, \quad t \geq 0, \tag{18.7}$$

where $m_t$ is currency held from $t$ to $t+1$. Form the household's Lagrangian

$$L = \sum_{t=0}^{\infty} \beta^t \{u(c_t) + \lambda_t (p_t y_t + m_{t-1} - m_t - p_t c_t)\},$$

where $\{\lambda_t\}$ is a sequence of nonnegative Lagrange multipliers. The household's first-order conditions for $c_t$ and $m_t$, respectively, are

$$u'(c_t) \leq \lambda_t p_t, \quad = \text{ if } c_t > 0,$$

$$-\lambda_t + \beta \lambda_{t+1} \leq 0, \quad = \text{ if } m_t > 0.$$

Substituting the first condition at equality into the second gives

$$\frac{\beta u'(c_{t+1})}{p_{t+1}} \leq \frac{u'(c_t)}{p_t}, \quad = \text{ if } m_t > 0. \tag{18.8}$$

DEFINITION 3: A *competitive equilibrium* is an allocation $\{c_t^o, c_t^e\}_{t=0}^{\infty}$, nonnegative money holdings $\{m_t^o, m_t^e\}_{t=-1}^{\infty}$, and a nonnegative price level sequence $\{p_t\}_{t=0}^{\infty}$ such that (a) given the price level sequence and $(m_{-1}^o, m_{-1}^e)$, the allocation solves the optimum problems of both types of households, and (b) $c_t^o + c_t^e = 1$, $m_{t-1}^o + m_{t-1}^e = M/N$, for all $t \geq 0$.

The periodic nature of the endowment sequences prompts us to guess the following two-parameter form of stationary equilibrium:

$$\begin{aligned} \{c_t^o\}_{t=0}^{\infty} &= \{c_0, 1-c_0, c_0, 1-c_0, \ldots\}, \\ \{c_t^e\}_{t=0}^{\infty} &= \{1-c_0, c_0, 1-c_0, c_0, \ldots\}, \end{aligned} \tag{18.9}$$

and $p_t = p$ for all $t \geq 0$. To determine the two undetermined parameters $(c_0, p)$, we use the first-order conditions and budget constraint of the odd agent at time

$0$. His endowment sequence for periods 0 and 1, $(y_0^o, y_1^o) = (1, 0)$, and the Inada condition (18.1), ensure that both of his first-order conditions at time 0 will hold with equality. That is, his desire to set $c_0^o > 0$ can be met by consuming some of the endowment $y_0^o$, and the only way for him to secure consumption in the following period 1 is to hold strictly positive money holdings $m_0^o > 0$. From his first-order conditions at equality, we obtain

$$\frac{\beta u'(1-c_0)}{p} = \frac{u'(c_0)}{p},$$

which implies that $c_0$ is to be determined as the root of

$$\beta - \frac{u'(c_0)}{u'(1-c_0)} = 0. \tag{18.10}$$

Because $\beta < 1$, it follows that $c_0 \in (1/2, 1)$. To determine the price level, we use the odd agent's budget constraint at $t = 0$, evaluated at $m_{-1}^o = 0$ and $m_0^o = M/N$, to get

$$pc_0 + M/N = p \cdot 1,$$

or

$$p = \frac{M}{N(1-c_0)}. \tag{18.11}$$

See Figure 18.1 for a graphical determination of $c_0$.

From equation (18.10), it follows that for $\beta < 1$, $c_0 > .5$ and $1 - c_0 < .5$. Thus, both types of agents experience fluctuations in their consumption sequences in this monetary equilibrium. Because Pareto optimal allocations have constant consumption sequences for each type of agent, this equilibrium allocation is not Pareto optimal.

## *Townsend's "Turnpike" Interpretation*

The preceding analysis of currency is artificial in the sense that it depends entirely on our having arbitrarily ruled out the existence of markets for private loans. The physical setup of the model itself provided no reason for those loan markets not to exist and indeed good reasons for them to exist. In addition, for

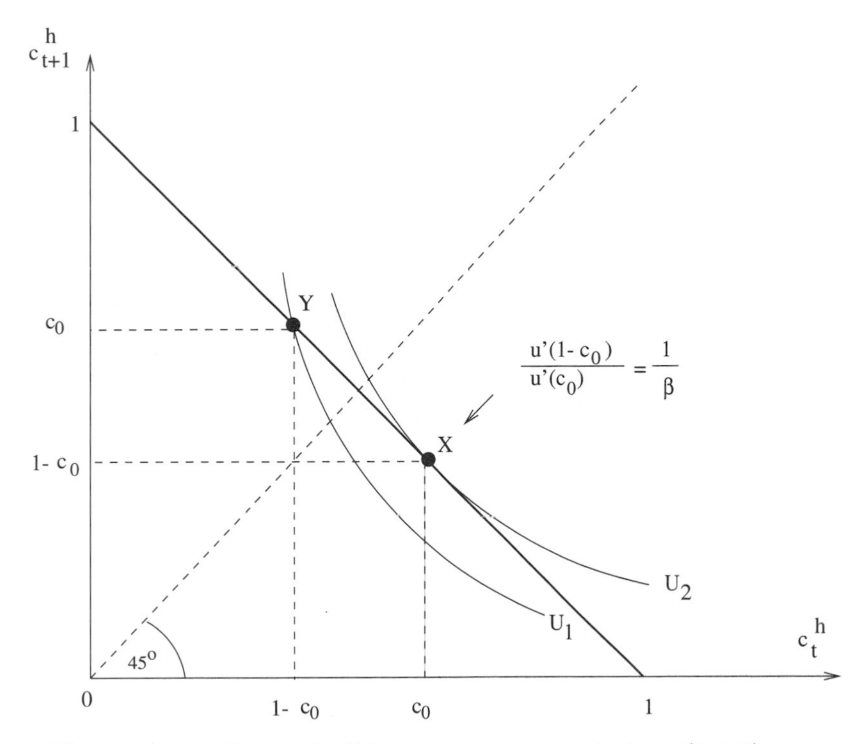


**Figure 18.1** The tradeoff between time-$t$ and time-$(t+1)$ consumption faced by agent $o(e)$ in equilibrium for $t$ even (odd). For $t$ even, $c_t^o = c_0$, $c_{t+1}^o = 1 - c_0$, $m_t^o = p(1-c_0)$, and $m_{t+1}^o = 0$. The slope of the indifference curve at $X$ is $-u'(c_t^h)/\beta u'(c_{t+1}^h) = -u'(c_0)/\beta u'(1-c_0) = -1$, and the slope of the indifference curve at $Y$ is $-u'(1-c_0)/\beta u'(c_0) = -1/\beta^2$.

many questions that we want to analyze, we want a model in which private loans and currency coexist, with currency being valued.[2]

Robert Townsend has proposed a model whose mathematical structure is identical with the preceding model, but in which a global market in private loans cannot emerge because agents are spatially separated. Townsend's setup can accommodate local markets for private loans, so that it meets the objections to the

[2] In the United States today, for example, $M_1$ consists of the sum of demand deposits (a part of which is backed by commercial loans and another, smaller part of which is backed by reserves or currency) and currency held by the public. Thus $M_1$ is not interpretable as the $m$ in our model.

model that we have expressed. But first, we will focus on a version of Townsend's model where local credit markets cannot emerge, which will be mathematically equivalent to our model above.

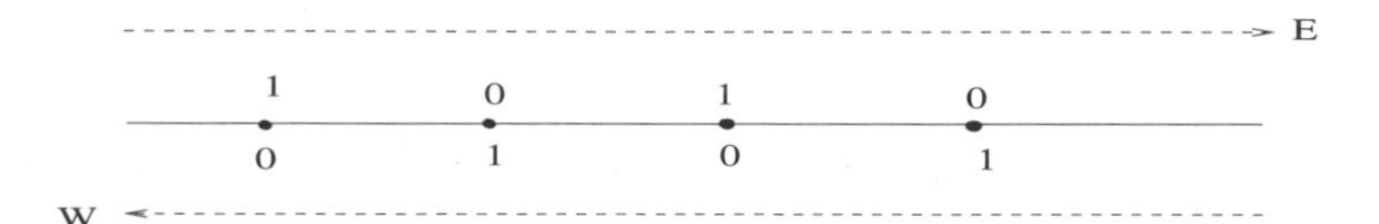


**Figure 18.2** Endowment pattern along a Townsend turnpike. The turnpike is of infinite extent in each direction, and has equidistant trading posts. Each trading post has equal numbers of east-heading and west-heading agents. At each trading post (the black dots) each period, for each east-heading agent there is a west-heading agent with whom he would like to borrow or lend. But itineraries rule out the possibility of repayment.

The economy starts at time $t = 0$, with $N$ east-heading migrants and $N$ west-heading migrants physically located at each of the integers along a "turnpike" of infinite length extending in both directions. Each of the integers $n = 0, \pm 1, \pm 2, \ldots$ is a trading post number. Agents can trade the one good only with agents at the trading post at which they find themselves at a given date. An east-heading agent at an even-numbered trading post is endowed with one unit of the consumption good, and an odd-numbered trading post has an endowment of zero units (see Figure 18.2). A west-heading agent is endowed with zero units at an even-numbered trading post and with one unit of the consumption good at an odd-numbered trading post. Finally, at the end of each period, each east-heading agent moves one trading post to the east, whereas each west-heading agent moves one trading post to the west. The turnpike along which the trading posts are located is of infinite length in each direction, implying that the east-heading and west-heading agents who are paired at time $t$ will never meet again. This feature means that there can be no private debt between agents moving in opposite directions. An IOU between agents moving in opposite directions can never be collected because a potential lender never meets the potential borrower again; nor does the lender meet anyone who ever meets the potential borrower, and so on, ad infinitum.

Let an agent who is endowed with one unit of the good $t = 0$ be called an agent of type $o$ and an agent who is endowed with zero units of the good at $t = 0$ be called an agent of type $e$. Agents of type $h$ have preferences summarized by

$\sum_{t=0}^{\infty} \beta^t u(c_t^h)$. Finally, start the economy at time 0 by having each agent of type $e$ endowed with $m_{-1}^e = m$ units of unbacked currency and each agent of type $o$ endowed with $m_{-1}^o = 0$ units of unbacked currency.

With the symbols thus reinterpreted, this model involves precisely the same mathematics as that which was analyzed earlier. Agents' spatial separation and their movements along the turnpike have been set up to produce a physical reason that a global market in private loans cannot exist. The various propositions about the equilibria of the model and their optimality that were already proved apply equally to the turnpike version.[3,4] Thus, in Townsend's version of the model, spatial separation is the "friction" that provides a potential social role for a valued unbacked currency. The spatial separation of agents and their endowment patterns give a setting in which private loan markets are limited by the need for people who trade IOUs to be linked together, if only indirectly, recurrently over time and space.

## *The Friedman rule*

Friedman's proposal to pay interest on currency by engineering a deflation can be used to solve for a Pareto optimal allocation in this economy. Friedman's proposal is to decrease the currency stock by means of lump-sum taxes at a properly chosen rate. Let the government's budget constraint be

$$M_t = (1 + \tau) M_{t-1}.$$

There are $N$ households of each type. At time $t$, the government transfers or taxes nominal balances in amount $\tau M_{t-1}/(2N)$ to each household of each type.

[3] A version of the model could be constructed in which local private markets for loans coexist with valued unbacked currency. To build such a model, one would assume some heterogeneity in the time patterns of the endowment of agents who are located at the same trading post and are headed in the same direction. If half of the east-headed agents located at trading post $i$ at time $t$ have present and future endowment pattern $y_t^h = (\alpha, \gamma, \alpha, \gamma \ldots)$, for example, whereas the other half of the east-headed agents have $(\gamma, \alpha, \gamma, \alpha, \ldots)$ with $\gamma \neq \alpha$, then there is room for local private loans among this cohort of east-headed agents. Whether or not there exists an equilibrium with valued currency depends on how nearly Pareto optimal the equilibrium with local loan markets is.

[4] Narayana Kocherlakota (1998) has analyzed the frictions in the Townsend turnpike and overlapping generations model. By permitting agents to use history-dependent decision rules, he has been able to support optimal allocations with the equilibrium of a gift-giving game. Those equilibria leave no room for valued fiat currency. Thus, Kocherlakota's view is that the frictions that give valued currency in the Townsend turnpike must include the restrictions on the strategy space that Townsend implicitly imposed.

The total transfer at time $t$ is thus $\tau M_{t-1}$, because there are $2N$ households receiving transfers.

The household's time-$t$ budget constraint becomes

$$p_t c_t + m_t \leq p_t y_t + \frac{\tau}{2} \frac{M_{t-1}}{N} + m_{t-1}.$$

We guess an equilibrium allocation of the same periodic pattern (18.9). For the price level, we make the "quantity theory" guess $M_t/p_t = k$, where $k$ is a constant. Substituting this guess into the government's budget constraint gives

$$\frac{M_t}{p_t} = (1+\tau) \frac{M_{t-1}}{p_{t-1}} \frac{p_{t-1}}{p_t}$$

or

$$k = (1+\tau) k \frac{p_{t-1}}{p_t},$$

or

$$p_t = (1+\tau) p_{t-1}, \tag{18.12}$$

which is our guess for the price level.

Substituting the price level guess and the allocation guess into the odd agent's first-order condition (18.8) at $t=0$ and rearranging shows that $c_0$ is now the root of

$$\frac{1}{(1+\tau)} - \frac{u'(c_0)}{\beta u'(1-c_0)} = 0. \tag{18.13}$$

The price level at time $t=0$ can be determined by evaluating the odd agent's time-0 budget constraint at $m_{-1}^o = 0$ and $m_0^o = M_0/N = (1+\tau) M_{-1}/N$, with the result that

$$(1-c_0) p_0 = \frac{M_{-1}}{N} \left(1 + \frac{\tau}{2}\right).$$

Finally, the allocation guess must also satisfy the even agent's first-order condition (18.8) at $t=0$ but not necessarily with equality since the stationary equilibrium has $m_0^e = 0$. After substituting $(c_0^e, c_1^e) = (1-c_0, c_0)$ and (18.12) into (18.8), we have

$$\frac{1}{1+\tau} \leq \frac{u'(1-c_0)}{\beta u'(c_0)}. \tag{18.14}$$

The substitution of (18.13) into (18.14) yields a restriction on the set of periodic allocations of type (18.9) that can be supported as one of our stationary monetary equilibria,

$$\left[\frac{u'(c_0)}{u'(1-c_0)}\right]^2 \le 1 \quad \Longrightarrow \quad c_0 \ge 0.5.$$

This restriction on $c_0$, together with (18.13), implies a corresponding restriction on the set of permissible monetary/fiscal policies, $1+\tau \ge \beta$.

*Welfare*

For allocations of the class (18.9), the utility functionals of odd and even agents, respectively, take values that are functions of the single parameter $c_0$, namely,

$$U^o(c_0) = \frac{u(c_0) + \beta u(1-c_0)}{1-\beta^2},$$

$$U^e(c_0) = \frac{u(1-c_0) + \beta u(c_0)}{1-\beta^2}.$$

Both expressions are strictly concave in $c_0$, with derivatives

$$U^{o\prime}(c_0) = \frac{u'(c_0) - \beta u'(1-c_0)}{1-\beta^2},$$

$$U^{e\prime}(c_0) = \frac{-u'(1-c_0) + \beta u'(c_0)}{1-\beta^2}.$$

The Inada condition (18.1) ensures strictly interior maxima with respect to $c_0$. For the odd agents, the preferred $c_0$ satisfies $U^{o\prime}(c_0) = 0$, or

$$\frac{u'(c_0)}{\beta u'(1-c_0)} = 1, \tag{18.15}$$

which by (18.13) is the zero-inflation equilibrium, $\tau = 0$. For the even agents, the preferred allocation given by $U^{e\prime}(c_0) = 0$ implies $c_0 < 0.5$, and can therefore not be implemented as a monetary equilibrium above. Hence, the even agents' preferred stationary monetary equilibrium is the one with the smallest permissible $c_0$, i.e., $c_0 = 0.5$. According to (18.13), this allocation can be supported by choosing money growth rate $1+\tau = \beta$ which is then also the equilibrium gross rate of deflation. Notice that all agents, both odd and even, are in agreement that they prefer no inflation to positive inflation, that is, they prefer $c_0$ determined by (18.15) to any higher value of $c_0$.

To abstract from the described conflict of interest between odd and even agents, suppose that the agents must pick their preferred monetary policy under a "veil of ignorance," before knowing their true identity. Since there are equal numbers of each type of agent, an individual faces a fifty-fifty chance of her identity being an odd or an even agent. Hence, prior to knowing one's identity, the expected lifetime utility of an agent is

$$\bar{U}(c_0) \equiv \frac{1}{2}U^o(c_0) + \frac{1}{2}U^e(c_0) = \frac{u(c_0) + u(1 - c_0)}{2(1-\beta)}.$$

The ex ante preferred allocation $c_0$ is determined by the first-order condition $\bar{U}'(c_0) = 0$, which has the solution $c_0 = 0.5$. Collecting equations (18.12), (18.13) and (18.14), this preferred policy is characterized by

$$\frac{p_t}{p_{t+1}} = \frac{1}{1+\tau} = \frac{u'(c_t^o)}{\beta u'(c_{t+1}^o)} = \frac{u'(c_t^e)}{\beta u'(c_{t+1}^e)} = \frac{1}{\beta}, \qquad \forall t \geq 0,$$

where $c_j^i = 0.5$ for all $j \geq 0$ and $i \in \{o, e\}$. Thus, the real return on money, $p_t/p_{t+1}$, equals a common marginal rate of intertemporal substitution, $\beta^{-1}$, and this return would therefore also constitute the real interest rate if there were a credit market. Moreover, since the gross real return on money is the inverse of the gross inflation rate, it follows that the gross real interest rate $\beta^{-1}$ multiplied by the gross rate of inflation is unity, or the net nominal interest rate is zero. In other words, all agents are ex ante in favor of Friedman's rule.

Figure 18.3 shows the "utility possibility frontier" associated with this economy. Except for the allocation associated with Friedman's rule, the allocations associated with stationary monetary equilibria lie inside the utility possibility frontier.

## *Inflationary finance*

The government prints new currency in total amount $M_t - M_{t-1}$ in period $t$ and uses it to purchase a constant amount $G$ of goods in period $t$. The government's time-$t$ budget constraint is

$$M_t - M_{t-1} = p_t G, \quad t \geq 0. \tag{18.16}$$

Preferences and endowment patterns of odd and even agents are as specified previously. We now use the following definition:

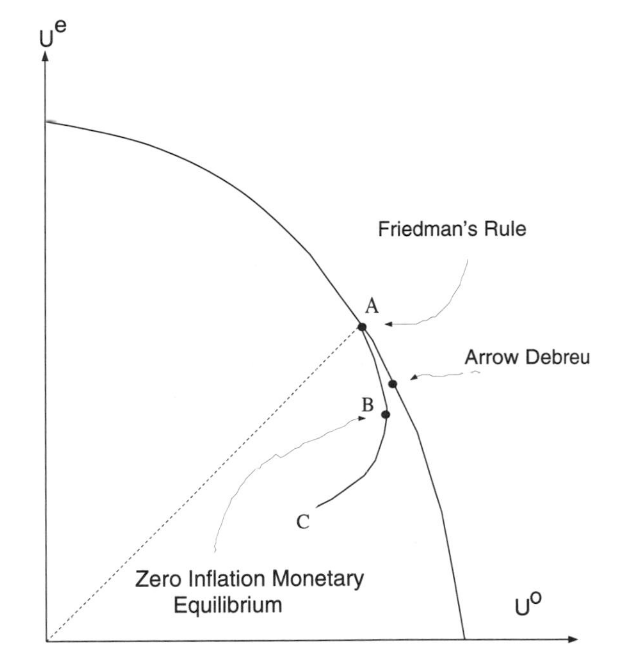


**Figure 18.3** Utility possibility frontier in Townsend turnpike. The locus of points $ABC$ denotes allocations attainable in stationary monetary equilibria. The point $B$ is the allocation associated with the zero-inflation monetary equilibrium. Point $A$ is associated with Friedman's rule, while points between $B$ and $C$ correspond to stationary monetary equilibria with inflation.

DEFINITION 4: A competitive equilibrium is a price level sequence $\{p_t\}_{t=0}^{\infty}$, a money supply process $\{M_t\}_{t=-1}^{\infty}$, an allocation $\{c_t^o, c_t^e, G_t\}_{t=0}^{\infty}$ and nonnegative money holdings $\{m_t^o, m_t^e\}_{t=-1}^{\infty}$ such that

(1) Given the price sequence and $(m_{-1}^o, m_{-1}^e)$, the allocation solves the optimum problems of households of both types.

(2) The government's budget constraint is satisfied for all $t \geq 0$.
(3) $N(c_t^o + c_t^e) + G_t = N$, for all $t \geq 0$; and $m_t^o + m_t^e = M_t/N$, for all $t \geq -1$.

For $t \geq 1$, write the government's budget constraint as

$$\frac{M_t}{Np_t} = \frac{p_{t-1}}{p_t} \frac{M_{t-1}}{Np_{t-1}} + \frac{G}{N},$$

or

$$\tilde{m}_t = R_{t-1}\tilde{m}_{t-1} + g, \tag{18.17}$$

where $g = G/N, \tilde{m}_t = M_t/(Np_t)$ is per-odd-person real balances, and $R_{t-1} = p_{t-1}/p_t$ is the rate of return on currency from $t-1$ to $t$.

To compute an equilibrium, we guess an allocation of the periodic form

$$\begin{aligned} \{c_t^o\}_{t=0}^{\infty} &= \{c_0, 1-c_0-g, c_0, 1-c_0-g, \ldots\}, \\ \{c_t^e\}_{t=0}^{\infty} &= \{1-c_0-g, c_0, 1-c_0-g, c_0, \ldots\}. \end{aligned} \tag{18.18}$$

We guess that $R_t = R$ for all $t \geq 0$, and again guess a "quantity theory" outcome

$$\tilde{m}_t = \tilde{m} \quad \forall t \geq 0.$$

Evaluating the odd household's time-0 first-order condition for currency at equality gives

$$\beta R = \frac{u'(c_0)}{u'(1-c_0-g)}. \tag{18.19}$$

With our guess, real balances held by each odd agent at the end of period 0, $m_0^o/p_0$, equal $1-c_0$, and time-1 consumption, which also is $R$ times the value of these real balances held from 0 to 1, is $1-c_0-g$. Thus, $(1-c_0)R = (1-c_0-g)$, or

$$R = \frac{1-c_0-g}{1-c_0}. \tag{18.20}$$

Equations (18.19) and (18.20) are two simultaneous equations that we want to solve for $(c_0, R)$.

Use equation (18.20) to eliminate $(1-c_0-g)$ from equation (18.19) to get

$$\beta R = \frac{u'(c_0)}{u'[R(1-c_0)]}.$$

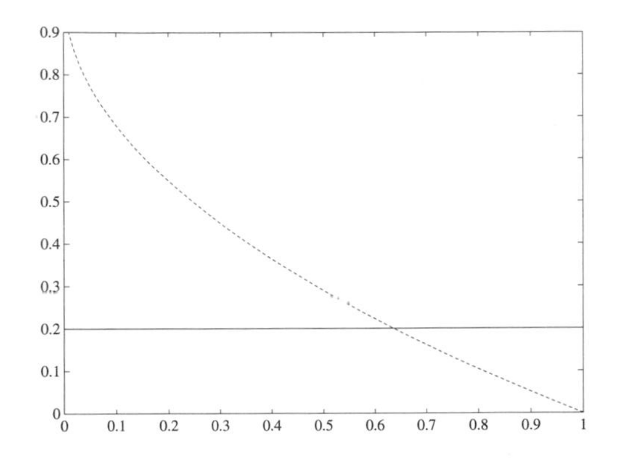


**Figure 18.4** Revenue from inflation tax $[m(R)(1-R)]$ and deficit for $\beta = .95, \delta = 2, g = .2$. The gross rate of return on currency is on the $x$-axis; the revenue from inflation and $g$ are on the $y$-axis.

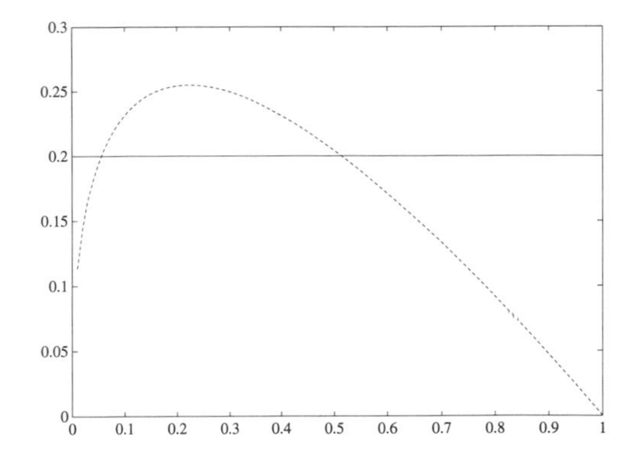


**Figure 18.5** Revenue from inflation tax $[m(R)(1-R)]$ and deficit for $\beta = .95, \delta = .7, g = .2$. The rate of return on currency is on the $x$-axis; the revenue from inflation and $g$ are on the $y$-axis. Here there is a Laffer curve.

Recalling that $(1 - c_0) = m_0$, this can be written

$$\beta R = \frac{u'(1-m_0)}{u'(Rm_0)}. \tag{18.21}$$

For the power utility function $u(c) = \frac{c^{1-\delta}}{1-\delta}$, this equation can be solved for $m_0$ to get the demand function for currency

$$m_0 = \tilde{m}(R) \equiv \frac{(\beta R^{1-\delta})^{1/\delta}}{1 + (\beta R^{1-\delta})^{1/\delta}}. \tag{18.22}$$

Substituting this into the government budget constraint (18.17) gives

$$\tilde{m}(R)(1-R) = g. \tag{18.23}$$

This equation equates the revenue from the inflation tax, namely, $\tilde{m}(R)(1-R)$ to the government deficit, $g$. The revenue from the inflation tax is the product of real balances and the inflation tax rate $1-R$. The equilibrium value of $R$ solves equation (18.23).

Figures 18.4 and 18.5 depict the determination of the stationary equilibrium value of $R$ for two sets of parameter values. For the case $\delta = 2$, shown in Figure 18.4, there is a unique equilibrium $R$; there is a unique equilibrium for every $\delta \geq 1$. For $\delta \geq 1$, the demand function for currency slopes upward as a function of $R$, as for the example in Figure 18.6. For $\delta < 1$, there can occur multiple stationary equilibria, as for the example in Figure 18.5. In such cases, there is a Laffer curve in the revenue from the inflation tax. Notice that the demand for real balances is downward sloping as a function of $R$ when $\delta < 1$.

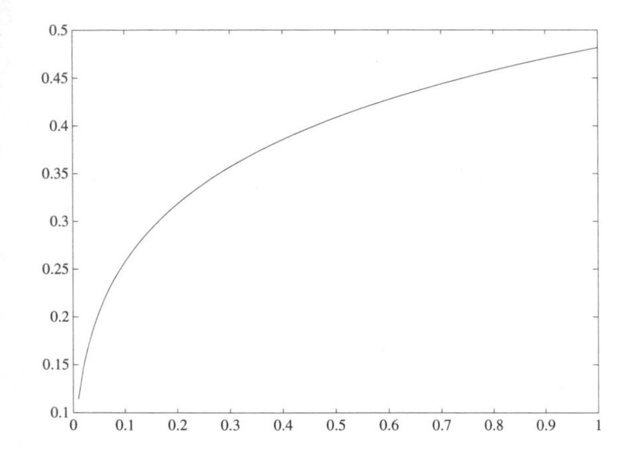


**Figure 18.6** Demand for real balances on the $y$-axis as function of the gross rate of return on currency on $x$-axis when $\beta = .95, \delta = 2$.

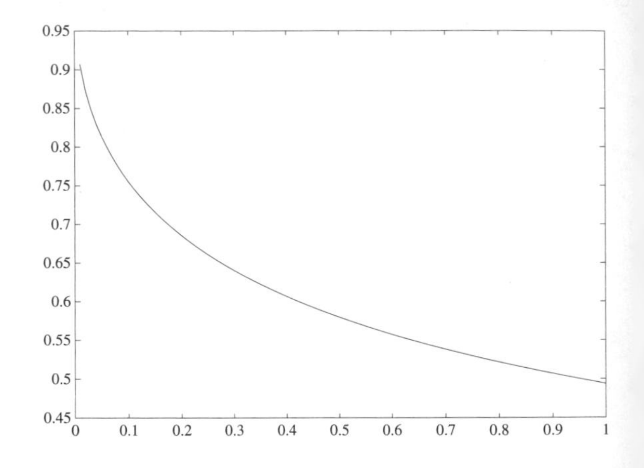


**Figure 18.7** Demand for real balances on the $y$-axis as function of the gross rate of return on currency on $x$-axis when $\beta = .95, \delta = .7$.

The initial price level is determined by the time-0 budget constraint of the government, evaluated at equilibrium time-0 real balances. In particular, the time-0 government budget constraint can be written

$$\frac{M_0}{Np_0} - \frac{M_{-1}}{Np_0} = g,$$

or

$$\tilde{m} - g = \frac{M_{-1}}{Np_0}.$$

Equating $\tilde{m}$ to its equilibrium value $1 - c_0$ and solving for $p_0$ gives

$$p_0 = \frac{M_{-1}}{N(1 - c_0 - g)}.$$

## Legal restrictions

This section adapts ideas of Bryant and Wallace (1984) to the turnpike environment. Bryant and Wallace and Villamil (1988) analyzed situations in which the government could make all savers better off by introducing a price discrimination scheme for marketing its debt. The analysis formalizes some ideas mentioned by John Maynard Keynes (1940).

Figure 18.8 depicts the terms on which an odd agent at $t = 0$ can transfer consumption between 0 and 1 in an equilibrium with inflationary finance. The agent is endowed at the point $(1, 0)$. The monetary mechanism allows him to transfer consumption between periods on the terms $c_1 = R(1 - c_0)$, depicted by the budget line connecting 1 on the $c_t$-axis with the point $B$ on the $c_{t+1}$-axis. The government insists on raising revenues in the amount $g$ for each pair of an odd and an even agent, which means that $R$ must be set so that the tangency between the agent's indifference curve and the budget line $c_1 = R(1 - c_0)$ occurs at the intersection of the budget line and the straight line connecting $1 - g$ on the $c_t$-axis with the point $1 - g$ on the $c_{t+1}$-axis. At this point, the marginal rate of substitution for odd agents is

$$\frac{u'(c_0)}{\beta u'(1 - c_0 - g)} = R,$$

(because currency holdings are positive). For even agents, the marginal rate of substitution is

$$\frac{u'(1 - c_0 - g)}{\beta u'(c_0)} = \frac{1}{\beta^2 R} > 1,$$

where the inequality follows from the fact that $R < 1$ under inflationary finance.

The fact that the odd agent's indifference curve intersects the solid line connecting $(1 - g)$ on the two axes indicates that the government could improve the welfare of the odd agent by offering him a higher rate of return subject to a minimal real balance constraint. The higher rate of return is used to send the line $c_1 = (1 - R)c_0$ into the "lens-shaped area" in Figure 18.8, onto a higher

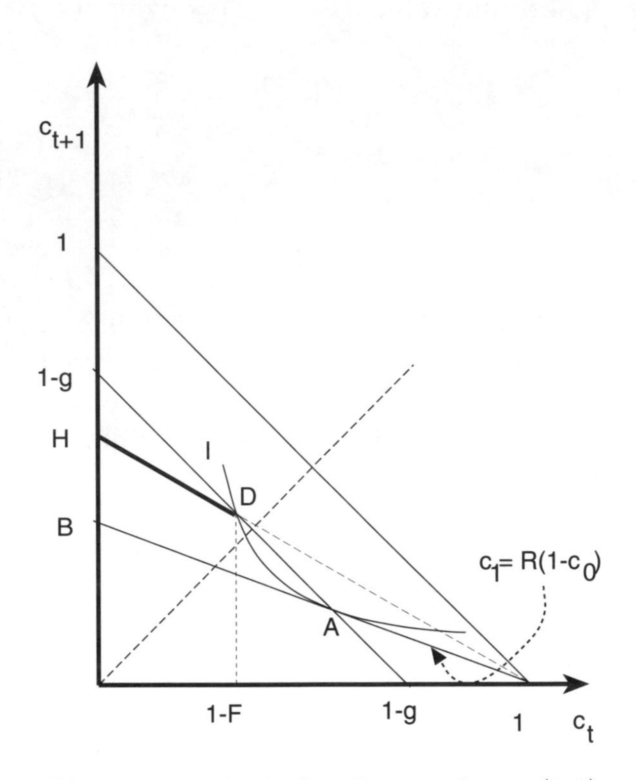


**Figure 18.8** The budget line starting at $(1, 0)$ and ending at the point B describes an odd agent's time-0 opportunities in an equilibrium with inflationary finance. Because this equilibrium has the "private consumption feasibility menu" intersecting the odd agent's indifference curve, a "forced saving" legal restriction can be used to put the odd agent onto a higher indifference curve than I, while leaving even agents better off and the government with revenue $g$. If the individual is confronted with a minimum denomination $F$ at the rate of return associated with the budget line ending at H, he would choose to consume $1 - F$.

indifference curve. The minimal real balance constraint is designed to force the agent onto the "post–government share" feasibility line connecting the points $1 - g$ on the two axes.

Thus, notice that in Figure 18.8, the government can raise the same revenue by offering odd agents the *higher* rate of return associated with the line connecting 1 on the $c_t$ axis with the point H on the $c_{t+1}$ axis, provided that the agent is required to save at least $F$, if he saves at all. This minimum saving requirement would make the household's budget set the point $(1,0)$ together with the heavy segment DH. With the setting of $F, R$ associated with the line DH in Figure 18.8, odd households have the same two-period utility as without this scheme. (Points $D$ and $A$ lie on the same indifference curve.) However, it is apparent that there is room to lower $F$ and lower $R$ a bit, and thereby move the odd household into the lens-shaped area. See Figure 18.9.

The marginal rates of substitution that we computed earlier indicate that this scheme makes both odd and even agents better off relative to the original equilibrium. The odd agents are better off because they move into the lens-shaped area in Figure 18.8. The even agents are better off because relative to the original equilibrium, they are being permitted to "borrow" at a gross rate of interest of one. Since their marginal rate of substitution at the original equilibrium is $1/(\beta^2 R) > 1$, this ability to borrow makes them better off.

## A two-money model

There are two types of currency being issued, in amounts $M_{it}, i = 1,2$ by each of two countries. The currencies are issued according to the rules

$$M_{it} - M_{it-1} = p_{it} G_{it}, \quad i = 1,2 \tag{18.24}$$

where $G_{it}$ is total purchases of time-$t$ goods by the government issuing currency $i$, and $p_{it}$ is the time-$t$ price level denominated in units of currency $i$. We assume that currencies of both types are initially equally distributed among the even agents at time $0$. Odd agents start out with no currency.

Household $h$'s optimum problem becomes to maximize $\sum_{t=0}^{\infty} \beta^t u(c_t^h)$ subject to the sequence of budget constraints

$$c_t^h + \frac{m_{1t}^h}{p_{1t}} + \frac{m_{2t}^h}{p_{2t}} \leq y_t^h + \frac{m_{1t-1}^h}{p_{1t}} + \frac{m_{2t-1}^h}{p_{2t}},$$

where $m_{jt-1}^h$ are nominal holdings of country $j$'s currency by household $h$. Currency holdings of each type must be nonnegative. The first-order conditions

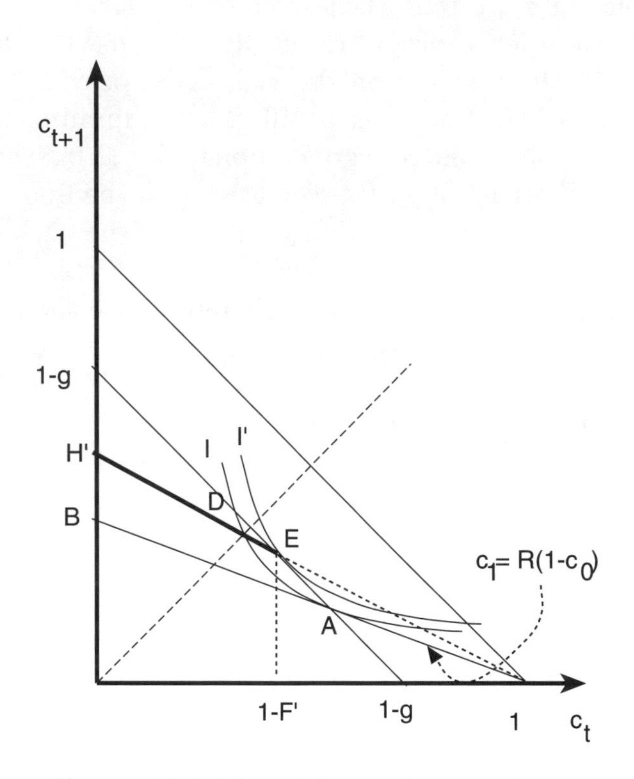


**Figure 18.9** The minimum denomination $F$ and the return on money can be lowered vis-à-vis their setting associated with line DH in Figure 18.8 to make the odd household better off, raise the same revenues for the government, and leave even households better off (as compared to no government intervention). The lower value of $F$ puts the odd household at $E$, which leaves him at the higher indifference curve I$'$.

for the household's problem with respect to $m^h_{jt}$ for $j = 1, 2$ are

$$\frac{\beta u'(c^h_{t+1})}{p_{1t+1}} \leq \frac{u'(c^h_t)}{p_{1t}}, \quad = \text{ if } m^h_{1t} > 0,$$

$$\frac{\beta u'(c^h_{t+1})}{p_{2t+1}} \leq \frac{u'(c^h_t)}{p_{2t}}, \quad = \text{ if } m^h_{2t} > 0.$$

If agent $h$ chooses to hold both currencies from $t$ to $t+1$, these first-order conditions imply that

$$\frac{p_{2t}}{p_{1t}} = \frac{p_{2t+1}}{p_{1t+1}},$$

or

$$p_{1t} = ep_{2t}, \quad \forall t \geq 0, \tag{18.25}$$

for some constant $e > 0$.[5] This equation states that if in each period there is some household that chooses to hold positive amounts of both types of currency, the rate of return from $t$ to $t+1$ must be equal for the two types of currencies, meaning that the exchange rate must be constant over time.[6]

We use the following definition:

DEFINITION 5: A *competitive equilibrium* with two valued fiat currencies is an allocation $\{c_t^o, c_t^e, G_{1t}, G_{2t}\}_{t=0}^{\infty}$, nonnegative money holdings $\{m_{1t}^o, m_{1t}^e, m_{2t}^o, m_{2t}^e\}_{t=-1}^{\infty}$, a pair of finite price level sequences $\{p_{1t}, p_{2t}\}_{t=0}^{\infty}$ and currency supply sequences $\{M_{1t}, M_{2t}\}_{t=-1}^{\infty}$ such that

(1) Given the price level sequences and $(m_{1,-1}^o, m_{1,-1}^e, m_{2,-1}^o, m_{2,-1}^e)$, the allocation solves the households' problems.
(2) The budget constraints of the governments are satisfied for all $t \geq 0$.
(3) $N(c_t^o + c_t^e) + G_{1t} + G_{2t} = N$, for all $t \geq 0$; and $m_{jt}^o + m_{jt}^e = M_{jt}/N$, for $j = 1, 2$ and all $t \geq -1$.

In the case of constant government expenditures $(G_{1t}, G_{2t}) = (Ng_1, Ng_2)$ for all $t \geq 0$, we guess an equilibrium allocation of the form (18.18), where we reinterpret $g$ to be $g = g_1 + g_2$. We also guess an equilibrium with a constant real value of the "world money supply," that is,

$$\tilde{m} = \frac{M_{1t}}{Np_{1t}} + \frac{M_{2t}}{Np_{2t}},$$

and a constant exchange rate, so that we impose condition (18.25). We let $R = p_{1t}/p_{1t+1} = p_{2t}/p_{2t+1}$ be the constant common value of the rate of return on the two currencies.

---

[5] Evaluate both of the first-order conditions at equality, then divide one by the other to obtain this result.

[6] As long as we restrict ourselves to nonstochastic equilibria.

With these guesses, the sum of the two countries' budget constraints for $t \geq 1$ and the conjectured form of the equilibrium allocation imply an equation of the form (18.23), where now

$$\tilde{m}(R) = \frac{M_{1t}}{p_{1t}N} + \frac{M_{2t}}{p_{2t}N}.$$

Equation (18.23) can be solved for $R$ in the fashion described earlier. Once $R$ has been determined, so has the constant real value of the world currency supply, $\tilde{m}$. To determine the time-$t$ price levels, we add the time-0 budget constraints of the two governments to get

$$\frac{M_{10}}{Np_{10}} + \frac{M_{20}}{Np_{20}} = \frac{M_{1,-1} + eM_{2,-1}}{Np_{10}} + (g_1 + g_2),$$

or

$$\tilde{m} - g = \frac{M_{1,-1} + eM_{2,-1}}{Np_{10}}.$$

In the conjectured allocation, $\tilde{m} = (1 - c_0)$, so this equation becomes

$$\frac{M_{1,-1} + eM_{2,-1}}{Np_{10}} = 1 - c_0 - g, \tag{18.26}$$

which, given any $e > 0$, has a positive solution for the initial country-1 price level. Given the solution $p_{10}$ and any $e \in (0, \infty)$, the price level sequences for the two countries are determined by the constant rate of return on currency $R$. To determine the values of the nominal currency stocks of the two countries, we use the government budget constraints (18.24).

Our findings are a special case of the following remarkable proposition:

PROPOSITION: (EXCHANGE RATE INDETERMINACY) Given the initial stocks of currencies $(M_{1,-1}, M_{2,-1})$ that are equally distributed among the even agents at time $0$, if there is an equilibrium for one constant exchange rate $e \in (0, \infty)$, then there exists an equilibrium for any $\hat{e} \in (0, \infty)$ with the same consumption allocation but different currency supply sequences.

PROOF: Let $p_{10}$ be the country 1 price level at time zero in the equilibrium that is assumed to exist with exchange rate $e$. For the conjectured equilibrium with exchange rate $\hat{e}$, we guess that the corresponding price level is

$$\hat{p}_{10} = p_{10} \frac{M_{1,-1} + \hat{e} M_{2,-1}}{M_{1,-1} + e M_{2,-1}}.$$

After substituting this expression into (18.26), we can verify that the real value at time 0 of the initial "world money supply" is the same across equilibria. Next, we guess that the conjectured equilibrium shares the same rate of return on currency, $R$, and constant end-of-period real value of the "world money supply", $\tilde{m}$, as the the original equilibrium. By construction from the original equilibrium, we know that this setting of the world money supply process guarantees that the consolidated budget constraint of the two governments is satisfied in each period. To determine the values of each country's prices and nominal money supplies, we proceed as above. That is, given $\hat{p}_{10}$ and $\hat{e}$, the price level sequences for the two countries are determined by the constant rate of return on currency $R$. The evolution of the nominal money stocks of the two countries is governed by government budget constraints (18.24). ▮

Versions of this proposition were stated by Kareken and Wallace (1980). See chapter 17 for a discussion of a possible way to alter assumptions to make the exchange rate determinate.

## A model of commodity money

Consider the following "small-country" model.[7] There are now two goods, the consumption good and a durable good, silver. Silver has a gross physical rate of return of one: storing one unit of silver this period yields one unit of silver next period. Silver is not valued domestically, but it can be exchanged abroad at a fixed price of $v$ units of the consumption good per unit of silver; $v$ is constant over time and is independent of the amount of silver imported or exported from this country. There are equal numbers $N$ of odd and even households, endowed with consumption good sequences

$$\{y_t^o\}_{t=0}^{\infty} = \{1, 0, 1, 0, \ldots\},$$

$$\{y_t^e\}_{t=0}^{\infty} = \{0, 1, 0, 1, \ldots\}.$$

[7] See Sargent and Wallace (1983), Sargent and Smith (1997), and Sargent and Velde (1999) for alternative models of commodity money.

Preferences continue to be ordered by $\sum_{t=0}^{\infty} \beta^t u(c_t^i)$ for each type of person, where $c_t$ is consumption of the consumption good.

Each *even* person is initially endowed with $S$ units of silver at time 0. Odd agents own no silver at $t = 0$.

Households are prohibited from borrowing or lending with each other, or with foreigners. However, they can exchange silver with each other and with foreigners. At time $t$, a household of type $i$ faces the budget constraint

$$c_t^i + m_t^i v \leq y_t^i + m_{t-1}^i v,$$

subject to $m_t^i \geq 0$, where $m_t^i$ is the amount of silver stored from time $t$ to time $t+1$ by agent $i$.

*Equilibrium*

DEFINITION 6: A *competitive equilibrium* is an allocation $\{c_t^o, c_t^e\}_{t=0}^{\infty}$ and nonnegative asset holdings $\{m_t^o, m_t^e\}_{t=-1}^{\infty}$ such that, given $(m_{-1}^o, m_{-1}^e)$, the allocation solves each agent's optimum problem.

Adding the budget constraints of the two types of agents with equality at time $t$ gives

$$c_t^o + c_t^e = 1 + v(S_{t-1} - S_t), \tag{18.27}$$

where $S_t = m_t^o + m_t^e$ is the total (per odd person) stock of silver in the country at time $t$. Equation (18.27) asserts that total domestic consumption at time $t$ is the sum of the country's endowment plus its imports of goods, where the latter equals its exports of silver, $v(S_{t-1} - S_t)$.

Given the opportunity to choose nonnegative asset holdings with a gross rate of return equal to one, the equilibrium allocation to the odd agent is $\{c_t^o\}_{t=0}^{\infty} = \{c_0, 1-c_0, c_0, 1-c_0, \ldots, \}$, where $c_0$ is the solution to equation (18.10). Thus, the odd agent holds $(1-c_0)$ units of silver from time 0 to time 1. He gets this silver either from even agents or from abroad.

Concerning the allocation to even agents, two types of equilibria are possible, depending on the value of $vS$ relative to the value $c_0$ that solves equation (18.10). If $u'(vS) \geq \beta u'(c_0)$, the equilibrium allocation to the even agent is $\{c_t^e\}_{t=0}^{\infty} = \{c_0^e, c_0, 1-c_0, c_0, 1-c_0, \ldots\}$, where $c_0^e = vS$. In this equilibrium, the even agent at time 0 sells all of his silver to support time-0 consumption. Net *exports* of silver for the country at time 0 are $S - (1-c_0)/v$, i.e., summing up the transactions of an even and an odd agent. For $t \geq 1$, the country's allocation and trade pattern is exactly as in the original model (with a stationary fiat money equilibrium).

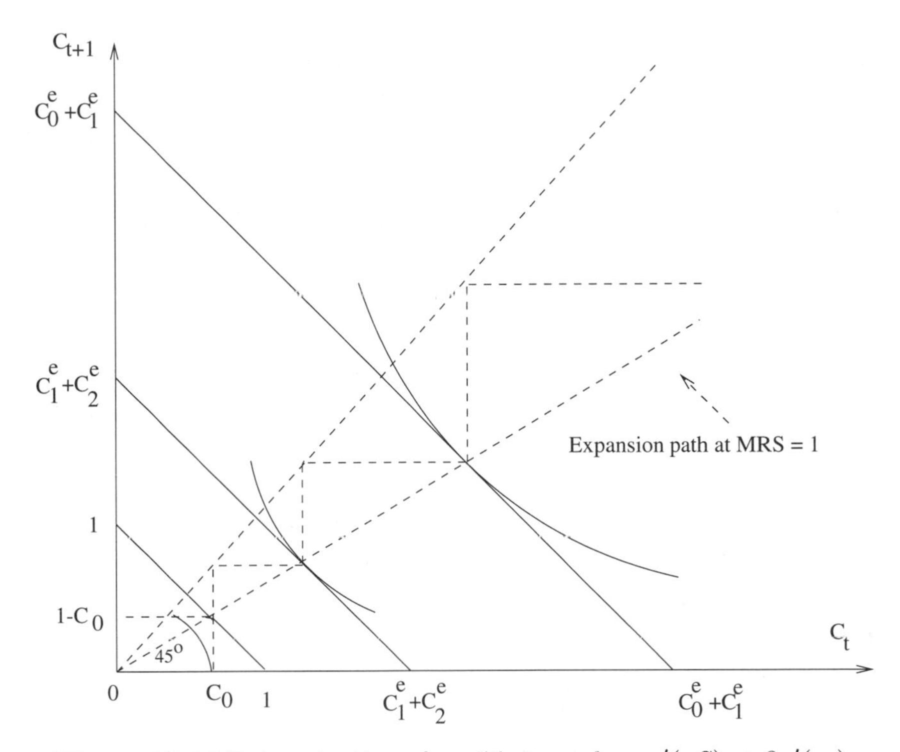


**Figure 18.10** Determination of equilibrium when $u'(vS) < \beta u'(c_0)$. For as long as it is feasible, the even agent sets $u'(c_{t+1}^e)/u'(c_t^e) = \beta$ by running down his silver holdings. This implies that $c_{t+1}^e < c_t^e$ during the run-down period. Eventually, the even agent runs out of silver, so that the "tail" of his allocation is $\{c_0, 1-c_0, c_0, 1-c_0, \ldots\}$, determined as before. The figure depicts how the spending of silver pushes the agent onto lower and lower two-period budget sets.

If the solution $c_0$ to equation (18.10) and $vS$ are such that $u'(vS) < \beta u'(c_0)$, the equilibrium allocation to the odd agents remains the same, but the allocation to the even agents is different. The situation is depicted in Figure 18.10. Even agents have so much silver at time 0 that they want to carry over positive amounts of silver into time 1 and maybe beyond. As long as they are carrying over positive amounts of silver from $t-1$ to $t$, the allocation to even agents has to satisfy

$$\frac{u'(c_{t-1}^e)}{\beta u'(c_t^e)} = 1, \tag{18.28}$$

which implies that $c_t^e < c_{t-1}^e$. Also, as long as they are carrying over positive amounts of silver, their first $T$ budget constraints can be used to deduce an intertemporal budget constraint

$$\sum_{t=0}^{T} c_t^e \leq \begin{cases} vS + (T+1)/2, & \text{if } T \text{ odd;} \\ vS + T/2, & \text{if } T \text{ even.} \end{cases} \tag{18.29}$$

The even agent finds the largest horizon $T$ over which he satisfies both (18.28) and (18.29) at equality with nonnegative carryover of silver for each period. This largest horizon $T$ will occur on an even date.[8] The equilibrium allocation to the even agents is determined by "gluing" this initial piece with declining consumption onto a "tail" of the allocation assigned to even agents in the original model, starting on an odd date, $\{c_t\}_{t=T+1}^{\infty} = \{c_0, 1-c_0, c_0, 1-c_0, \ldots\}$.[9]

*Virtue of fiat money*

This is a model with an exogenous price level and an endogenous stock of currency. The model can be used to express a version of Friedman's and Keynes's condemnation of commodity money systems: the equilibrium allocation can be Pareto dominated by the allocation in a fiat money equilibrium in which, in addition to the stock of silver at time 0, the even agents are endowed with $M$ units of an unbacked fiat currency. We can then show that there exists a monetary equilibrium with a constant price level $p$ satisfying (18.11),

$$p = \frac{M}{N(1-c_0)}.$$

In effect, the time-0 endowment of the even agents is increased by $1-c_0$ units of consumption good. Fiat money creates wealth by removing commodity money from circulation, which instead can be transformed into consumption.

[8] Suppose to the contrary that the largest horizon $T$ is an odd date. That is, up and until date $T$ both (18.28) and (18.29) are satisfied with nonnegative savings for each period. Now, let us examine what happens if we add one additional period and the horizon becomes $T+1$. Since that additional period is an even date, the right side of budget constraint (18.29) is unchanged. Therefore, condition (18.28) implies that the extra period induces the agent to reduce consumption in all periods $t \leq T$, in order to save for consumption in period $T+1$. Since the initial horizon $T$ satisfied (18.28) and (18.29) with nonnegative savings, it follows that so must also horizon $T+1$. Therefore, the largest horizon $T$ must occur on an even date.

[9] Is the equilibrium with $u'(vS) < \beta u'(c_0)$, a stylized model of Spain in the 16th century? At the beginning of the 16th century, Spain suddenly received a large claim on silver and gold from the New World. During the century, Spain exported gold and silver to the rest of Europe to finance government and private purchases.

## Concluding remarks

The model of this chapter is basically a "nonstochastic incomplete markets model," a special case of the stochastic incomplete markets models of chapter 14. The virtue of the model is that we can work out many things by hand. The limitation on markets in private loans leaves room for a consumption-smoothing role to be performed by a valued fiat currency. The reader might note how some of the monetary doctrines worked out precisely in this chapter have counterparts in the stochastic incomplete markets models of chapter 14.

## Exercises

### *Exercise 18.1* **Arrow-Debreu**

Consider an environment with equal numbers $N$ of two types of agents, odd and even, who have endowment sequences

$$\{y_t^o\}_{t=0}^{\infty} = \{1, 1, 0, 1, 1, 0, \ldots\}$$
$$\{y_t^e\}_{t=0}^{\infty} = \{0, 0, 1, 0, 0, 1, \ldots\}.$$

Households of each type $h$ order consumption sequences by $\sum_{t=0}^{\infty} \beta^t u(c_t^h)$. Compute the Arrow-Debreu equilibrium for this economy.

### *Exercise 18.2* **One-period consumption loans**

Consider an environment with equal numbers $N$ of two types of agents, odd and even, who have endowment sequences

$$\{y_t^o\}_{t=0}^{\infty} = \{1, 0, 1, 0, \ldots\}$$
$$\{y_t^e\}_{t=0}^{\infty} = \{0, 1, 0, 1, \ldots\}.$$

Households of each type $h$ order consumption sequences by $\sum_{t=0}^{\infty} \beta^t u(c_t^h)$. The only market that exists is for one-period loans. The budget constraints of household $h$ are

$$c_t^h + b_t^h \leq y_t^h + R_{t-1} b_{t-1}^h, \quad t \geq 0,$$

where $b_{-1}^h = 0, h = o, e$. Here $b_t^h$ is agent $h$'s lending (if positive) or borrowing (if negative) from $t$ to $t+1$, and $R_{t-1}$ is the gross real rate of interest on consumption loans from $t-1$ to $t$.

**a.** Define a competitive equilibrium with one-period consumption loans.

**b.** Compute a competitive equilibrium with one-period consumption loans.

**c.** Is the equilibrium allocation Pareto optimal? Compare the equilibrium allocation with that for the corresponding Arrow-Debreu equilibrium for an economy with identical endowment and preference structure.

*Exercise 18.3* **Stock market**

Consider a "stock market" version of an economy with endowment and preference structure identical to the one in the previous economy. Now odd and even agents begin life owning one of two types of "trees." Odd agents own the "odd" tree, which is a perpetual claim to a dividend sequence

$$\{y_t^o\}_{t=0}^{\infty} = \{1, 0, 1, 0, \ldots\},$$

while even agents initially own the "even" tree, which entitles them to a perpetual claim on dividend sequence

$$\{y_t^e\}_{t=0}^{\infty} = \{0, 1, 0, 1, \ldots\}.$$

Each period, there is a stock market in which people can trade the two types of trees. These are the only two markets open each period. The time-$t$ price of type $j$ trees is $a_t^j, j = o, e$. The time-$t$ budget constraint of agent $h$ is

$$c_t^h + a_t^o s_t^{ho} + a_t^e s_t^{he} \leq (a_t^o + y_t^o) s_{t-1}^{ho} + (a_t^e + y_t^e) s_{t-1}^{he},$$

where $s_t^{hj}$ is the number of shares of stock in tree $j$ held by agent $h$ from $t$ to $t+1$. We assume that $s_{-1}^{oo} = 1, s_{-1}^{ee} = 1, s_{-1}^{jk} = 0$ for $j \neq k$.

**a.** Define an equilibrium of the stock market economy.

**b.** Compute an equilibrium of the stock market economy.

**c.** Compare the allocation of the stock market economy with that of the corresponding Arrow-Debreu economy.

*Exercise 18.4* **Inflation**

Consider a Townsend turnpike model in which there are $N$ odd agents and $N$ even agents who have endowment sequences, respectively, of

$$\{y_t^o\}_{t=0}^{\infty} = \{1, 0, 1, 0, \ldots\}$$
$$\{y_t^e\}_{t=0}^{\infty} = \{0, 1, 0, 1, \ldots\}.$$

Households of each type order consumption sequences by $\sum_{t=0}^{\infty} \beta^t u(c_t)$. The government makes the stock of currency move according to

$$M_t = z M_{t-1}, \quad t \geq 0.$$

At the beginning of period $t$, the government hands out $(z-1)m_{t-1}^h$ to each type-$h$ agent who held $m_{t-1}^h$ units of currency from $t-1$ to $t$. Households of type $h = o, e$ have time-$t$ budget constraint of

$$p_t c_t^h + m_t^h \leq p_t y_t^h + m_{t-1}^h + (z-1) m_{t-1}^h.$$

**a.** Guess that an equilibrium endowment sequence of the periodic form (18.9) exists. Make a guess at an equilibrium price sequence $\{p_t\}$ and compute the equilibrium values of $(c_0, \{p_t\})$. *Hint:* Make a "quantity theory" guess for the price level.

**b.** How does the allocation vary with the rate of inflation? Is inflation "good" or "bad"? Describe odd and even agents' attitudes toward living in economies with different values of $z$.

*Exercise 18.5* **A Friedman-like scheme**

Consider Friedman's scheme to improve welfare by generating a deflation. Suppose that the government tries to boost the rate of return on currency above $\beta^{-1}$ by setting $\beta > (1+\tau)$. Show that there exists no equilibrium with an allocation of the class (18.9) and a price level path satisfying $p_t = (1+\tau)p_{t-1}$, with odd agents holding $m_0^o > 0$. [That is, the piece of the "restricted Pareto optimality frontier" does not extend above the allocation (.5,.5) in Figure 18.3.]

*Exercise 18.6* **Distribution of currency**

Consider an economy consisting of large and equal numbers of two types of infinitely lived agents. There is one kind of consumption good, which is nonstorable. "Odd" agents have period-2 endowment pattern $\{y_t^o\}_{t=0}^{\infty}$, while "even" agents have period-2 endowment pattern $\{y_t^e\}_{t=0}^{\infty}$. Agents of both types have preferences that are ordered by the utility functional

$$\sum_{t=0}^{\infty} \beta^t \ln(c_t^i), \quad i = o, e, \quad 0 < \beta < 1,$$

where $c_t^i$ is the time-$t$ consumption of the single good by an agent of type $i$.

Assume the following endowment pattern:

$$y_t^o = \{1, 0, 1, 0, 1, 0, \ldots\}$$

$$y_t^e = \{0, 1, 0, 1, 0, 1, \ldots\}.$$

Now assume that all borrowing and lending is prohibited, either ex cathedra through legal restrictions or by virtue of traveling and locational restrictions of the kind introduced by Robert Townsend. At time $t = 0$, all odd agents are endowed with $\alpha H$ units of an unbacked, inconvertible currency, and all even units are endowed with $(1 - \alpha)H$ units of currency, where $\alpha \in [0, 1]$. The currency is denominated in dollars and is perfectly durable. Currency is the only object that agents are permitted to carry over from one period to the next. Let $p_t$ be the price level at time $t$, denominated in units of dollars per time-$t$ consumption good.

**a.** Define an *equilibrium with valued fiat currency.*

**b.** Let an "eventually stationary" equilibrium with valued fiat currency be one in which there exists a $\bar{t}$ such that for $t \geq \bar{t}$, the equilibrium allocation to each type of agent is of period 2 (i.e., for each type of agent, the allocation is a periodic sequence that oscillates between two values). Show that for each value of $\alpha \in [0, 1]$, there exists such an equilibrium. Compute this equilibrium.

*Exercise 18.7* **Capital overaccumulation**

Consider an environment with equal numbers $N$ of two types of agents, odd and even, who have endowment sequences

$$\{y_t^o\}_{t=0}^{\infty} = \{1-\varepsilon, \varepsilon, 1-\varepsilon, \varepsilon, \ldots\}$$

$$\{y_t^e\}_{t=0}^{\infty} = \{\varepsilon, 1-\varepsilon, \varepsilon, 1-\varepsilon, \ldots\}.$$

Here $\varepsilon$ is a small positive number that is very close to zero. Households of each type $h$ order consumption sequences by $\sum_{t=0}^{\infty} \beta^t \ln(c_t^h)$ where $\beta \in (0,1)$. The one good in the model is storable. If a nonnegative amount $k_t$ of the good is stored at time $t$, the outcome is that $\delta k_t$ of the good is carried into period $t+1$, where $\delta \in (0,1)$. Households are free to store nonnegative amounts of the good.

**a.** Assume that there are no markets. Households are on their own. Find the autarkic consumption allocations and storage sequences for the two types of agents. What is the total per-period storage in this economy?

**b.** Now assume that there exists a fiat currency, available in fixed supply of $M$, all of which is initially equally distributed among the even agents. Define an equilibrium with valued fiat currency. Compute a stationary equilibrium with valued fiat currency. Show that the associated allocation Pareto dominates the one you computed in part a.

**c.** Suppose that in the storage technology $\delta = 1$ (no depreciation) and that there is a fixed supply of fiat currency, initially distributed as in part b. Define an "eventually stationary" equilibrium. Show that there is a continuum of eventually stationary equilibrium price levels and allocations.

*Exercise 18.8* **Altered endowments**

Consider a Bewley model identical to the one in the text, except that now the odd and even agents are endowed with the sequences

$$y_t^0 = \{1-F, F, 1-F, F, \ldots\}$$
$$y_t^e = \{F, 1-F, F, 1-F, \ldots\},$$

where $0 < F < (1-c^o)$, where $c^o$ is the solution of equation (18.10).

Compute the equilibrium allocation and price level. How do these objects vary across economies with different levels of $F$? For what values of $F$ does a stationary equilibrium with valued fiat currency exist?

*Exercise 18.9* **Inside money**

Consider an environment with equal numbers $N$ of two types of households, odd and even, who have endowment sequences

$$\{y_t^o\}_{t=0}^\infty = \{1, 0, 1, 0, \ldots\}$$
$$\{y_t^e\}_{t=0}^\infty = \{0, 1, 0, 1, \ldots\}.$$

Households of type $h$ order consumption sequences by $\sum_{t=0}^\infty \beta^t u(c_t^h)$. At the beginning of time 0, each even agent is endowed with $M$ units of an unbacked fiat currency and owes $F$ units of consumption goods; each odd agent is owed $F$ units of consumption goods and owns 0 units of currency. At time $t \geq 0$, a household of type $h$ chooses to carry over $m_t^h \geq 0$ of currency from time $t$ to $t+1$. (We start households out with these debts or assets at time 0 to support a stationary equilibrium.) Each period $t \geq 0$, households can issue indexed one-period debt in amount $b_t$, promising to pay off $b_t R_t$ at $t+1$, subject to the constraint that $b_t \geq -F/R_t$, where $F > 0$ is a parameter characterizing the borrowing constraint and $R_t$ is the rate of return on these loans between time $t$ and $t+1$. (When $F = 0$, we get the Bewley-Townsend model.) A household's period-$t$ budget constraint is

$$c_t + m_t/p_t + b_t = y_t + m_{t-1}/p_t + b_{t-1}R_{t-1},$$

where $R_{t-1}$ is the gross real rate of return on indexed debt between time $t-1$ and $t$. If $b_t < 0$, the household is borrowing at $t$, and if $b_t > 0$, the household is lending at $t$.

**a.** Define a competitive equilibrium in which valued fiat currency and private loans coexist.

**b.** Argue that, in the equilibrium defined in part a, the real rates of return on currency and indexed debt must be equal.

**c.** Assume that $0 < F < (1 - c^o)/2$, where $c^o$ is the solution of equation (18.10). Show that there exists a stationary equilibrium with a constant price level and that the allocation equals that associated with the stationary equilibrium of the $F = 0$ version of the model. How does $F$ affect the price level? Explain.

**d.** Suppose that $F = (1 - c^o)/2$. Show that there is a stationary equilibrium with private loans but that fiat currency is valueless in that equilibrium.

**e.** Suppose that $F = \frac{\beta}{1+\beta}$. For a stationary equilibrium, find an equilibrium allocation and interest rate.

**f.** Suppose that $F \in [(1 - c^o)/2, \frac{\beta}{1+\beta}]$. Argue that there is a stationary equilibrium (without valued currency) in which the real rate of return on debt is $R \in (1, \beta^{-1})$.

*Exercise 18.10* **Initial conditions and inside money**

Consider a version of the preceding model in which each odd person is initially endowed with no currency and no IOUs, and each even person is initially endowed with $M/N$ units of currency, but no IOUs. At every time $t \geq 0$, each agent can issue one-period IOUs promising to pay off $F/R_t$ units of consumption in period $t+1$, where $R_t$ is the gross real rate of return on currency or IOUs between periods $t$ and $t+1$. The parameter $F$ obeys the same restrictions imposed in exercise 18.9.

**a.** Find an equilibrium with valued fiat currency in which the "tail" of the allocation for $t \geq 1$ and the tail of the price level sequence, respectively, are identical with that found in exercise 18.9.

**b.** Find the price level, the allocation, and the rate of return on currency and consumption loans at period $0$.

*Exercise 18.11* **Real bills experiment**

Consider a version of the exercise 18.9. The initial conditions and restrictions on borrowing are as described in exercise 18.9. However, now the government augments the currency stock by an "open market operation" as follows: In period $0$, the government issues $\bar{M} - M$ units per each odd agent for the purpose of purchasing $\Delta$ units of IOUs issued at time 0 by the even agents. Assume that $0 < \Delta < F$. At each time $t \geq 1$, the government uses any net real interest payments from its stock IOUs from the private sector to decrease the outstanding stock of currency. Thus the government's budget constraint sequence is

$$\frac{\bar{M} - M}{p_0} = \Delta, \quad t = 0,$$

$$\frac{\bar{M}_t - \bar{M}_{t-1}}{p_t} = -(R_{t-1} - 1)\Delta \quad t \geq 1.$$

Here $R_{t-1}$ is the gross rate of return on consumption loans from $t-1$ to $t$, and $\bar{M}_t$ is the total stock of currency outstanding at the end of time $t$.

**a.** Verify that there exists a stationary equilibrium with valued fiat currency in which the allocation has the form (18.9) where $c_0$ solves equation (18.10).

**b.** Find a formula for the price level in this stationary equilibrium. Describe how the price level varies with the value of $\Delta$.

**c.** Does the "quantity theory of money" hold in this example?

# 19
# *Equilibrium Search and Matching*

## *Introduction*

This chapter presents various equilibrium models of search and matching. We describe (1) Lucas and Prescott's version of an island model, (2) some matching models in the style of Mortensen, Pissarides, and Diamond, and (3) a search model of money along the lines of Kiyotaki and Wright.

Chapter 5 studied the optimization problem of a single unemployed agent who searched for a job by drawing from an exogenous wage offer distribution. We now turn to a model with a continuum of agents who interact across a large number of spatially separated labor markets. Phelps (1970, introductory chapter) describes such an "island economy," and a formal framework is analyzed by Lucas and Prescott (1974). The agents on an island can choose to work at the market-clearing wage in their own labor market, or seek their fortune by moving to another island and its labor market. In an equilibrium, agents tend to move to islands that experience good productivity shocks, while an island with bad productivity may see some of its labor force departing. Frictional unemployment arises because moves between labor markets take time.

Another approach to model unemployment is the matching framework, as described by Diamond (1982), Mortensen (1982), and Pissarides (1990). These models postulate the existence of a matching function that maps measures of unemployment and vacancies into a measure of matches. A match pairs a worker and a firm who then have to bargain about how to share the "match surplus," that is, the value that will be lost if the two parties cannot agree and break the match. In contrast to the island model with price-taking behavior and no externalities, the decentralized outcome in the matching framework is in general not efficient. Unless parameter values satisfy a knife-edge restriction, there will either be too many or too few vacancies posted in an equilibrium. The efficiency problem is further exacerbated when assuming that heterogeneous jobs are created in the same labor market with a single matching function. This assumption creates a tension between an efficient mix of jobs and the efficient total supply of jobs.

As a reference point to models with search and matching frictions, we study a frictionless aggregate labor market but assume that labor is indivisible. For example, agents are constrained to work either full time or not at all. This kind of assumption has been used in the real business cycle literature to generate unemployment. If markets for contingent claims exist, Hansen (1985) and Rogerson (1988) show that employment lotteries can be welfare enhancing with the implication that only a fraction of agents will be employed in an equilibrium. Using this model and the other two frameworks that we have mentioned, we analyze how layoff taxes affect an economy's employment level. Different models are seen to yield diametrically different conclusions that shed further light on the economic forces at work in the various frameworks.

To illustrate another application of search and matching, we study Kiyotaki and Wright's (1993) search model of money. Agents who differ with respect to their taste for different goods meet pairwise and at random. In this model, fiat money can potentially ameliorate the problem of "double coincidence of wants."

## An island model

The model here is a simplified version of Lucas and Prescott's (1974) "island economy." There is a continuum of agents populating a large number of spatially separated labor markets. Each island is endowed with an aggregate production function $\theta f(n)$ where $n$ is the island's employment level and $\theta > 0$ is an idiosyncratic productivity shock. The production function satisfies

$$f' > 0, \qquad f'' < 0, \qquad \text{and} \qquad \lim_{n\to 0} f'(n) = \infty\,. \tag{19.1}$$

The productivity shock takes on $m$ possible values, $\theta_1 < \theta_2 < \ldots < \theta_m$, and the shock is governed by strictly positive transition probabilities, $\pi(\theta,\theta') > 0$. That is, an island with a current productivity shock of $\theta$ faces a probability $\pi(\theta,\theta')$ that its next period's shock is $\theta'$. The productivity shock is persistent in the sense that the cumulative distribution function, $\text{Prob}\,(\theta' \leq \theta_k|\theta) = \sum_{i=1}^{k} \pi(\theta,\theta_i)$, is a decreasing function of $\theta$.

At the beginning of a period, agents are distributed in some way over the islands. After observing the productivity shock, the agents decide whether or not to move to another island. A mover forgoes his labor earnings in the period of the move, while he can choose the destination with complete information about current conditions on all islands. An agent's decision to work or to move is taken so as to maximize the expected present value of his earnings stream. Wages are determined competitively, so that each island's labor market clears with a wage rate equal to the marginal product of labor.

In this environment, we will study stationary equilibria where an island's labor force at the beginning of next period is a function of its current productivity shock and labor force at the beginning of the current period.

*A single market (island)*

The state of a single market is given by its productivity level $\theta$ and its beginning-of-period labor force $x$. In an equilibrium, there will be functions mapping this state into an employment level, $n(\theta, x)$, and a wage rate, $w(\theta, x)$. These functions must satisfy the market-clearing condition

$$w(\theta, x) = \theta f'\left[n(\theta, x)\right]$$

and the labor supply constraint

$$n(\theta, x) \leq x .$$

Let $v(\theta, x)$ be the value of the optimization problem for an agent finding himself in market $(\theta, x)$ at the beginning of a period. The value associated with leaving the market is denoted $v_u$ (determined by conditions in the aggregate economy). The Bellman equation can then be written as

$$v(\theta, x) = \max\Big\{v_u \,,\; w(\theta, x) + \beta E\left[v(\theta', x')|\theta, x\right]\Big\} , \tag{19.2}$$

where the conditional expectation refers to the evolution of $\theta'$ and $x'$ if the agent remains in the same market.

The value function $v(\theta, x)$ is equal to $v_u$ whenever there are any agents leaving the market. It is instructive to examine the opposite situation when no one leaves the market. This means that the current employment level is $n(\theta, x) = x$ and the wage rate becomes $w(\theta, x) = \theta f'(x)$. Concerning the continuation value for next period, $\beta E\left[v(\theta', x')|\theta, x\right]$, there are two possibilities:

Case i. All agents remain, and some additional agents arrive next period. The arrival of new agents corresponds to a continuation value of $v_u$ in the market. Any value less than $v_u$ would not attract any new agents, and a value higher than $v_u$ would be driven down by a larger inflow of new agents. It follows that the current value function in equation (19.2) can under these circumstances be written as

$$v(\theta, x) = \theta f'(x) + v_u .$$

Case ii. All agents remain, and no additional agents arrive next period. In this case $x' = x$, and the lack of new arrivals implies that the market's continuation value is less than or equal to $v_u$. The current value function becomes

$$v(\theta, x) = \theta f'(x) + \beta E[v(\theta', x)|\theta] \le \theta f'(x) + v_u .$$

After putting both of these cases together, we can rewrite the value function in equation (19.2) as follows,

$$v(\theta, x) = \max\Big\{v_u , \theta f'(x) + \min\{v_u , \beta E[v(\theta', x)|\theta]\}\Big\}. \tag{19.3}$$

Given a value for $v_u$, this is a well-behaved functional equation with a unique solution $v(\theta, x)$. The value function is nondecreasing in $\theta$ and nonincreasing in $x$.

On the basis of agents' optimization behavior, we can study the evolution of the island's labor force. There are three possible cases:

Case 1. Some agents leave the market. An implication is that no additional workers will arrive next period when the beginning-of-period labor force will be equal to the current employment level, $x' = n$. The current employment level, equal to $x'$, can then be computed from the condition that agents remaining in the market receive the same utility as the movers, given by $v_u$,

$$\theta f'(x') + \beta E[v(\theta', x')|\theta] = v_u . \tag{19.4}$$

This equation implicitly defines $x^+(\theta)$ such that $x' = x^+(\theta)$ if $x \ge x^+(\theta)$.

Case 2. All agents remain in the market, and some additional workers arrive next period. The arriving workers must attain the value $v_u$, as discussed in case i. That is, next period's labor force $x'$ must be such that

$$\beta E[v(\theta', x')|\theta] = v_u . \tag{19.5}$$

This equation implicitly defines $x^{-}(\theta)$ such that $x' = x^{-}(\theta)$ if $x \le x^{-}(\theta)$. It can be seen that $x^{-}(\theta) < x^{+}(\theta)$.

Case 3. All agents remain in the market, and no additional workers arrive next period. This situation was discussed in case ii. It follows here that $x' = x$ if $x^{-}(\theta) < x < x^{+}(\theta)$.

### *The aggregate economy*

The previous section assumed an exogenous value to search, $v_u$. This assumption will be maintained in the first part of this section on the aggregate economy. The approach amounts to assuming a perfectly elastic outside labor supply with reservation utility $v_u$. We end the section by showing how to endogenize the value to search in the face of a given inelastic aggregate labor supply.

Define a set $X$ of possible labor forces in a market as follows.

$$X \equiv \begin{cases} \left\{ x \in \{x^{-}(\theta_i),\, x^{+}(\theta_i)\}_{i=1}^{m} \;:\; x^{+}(\theta_1) \le x \le x^{-}(\theta_m) \right\}, & \text{if } x^{+}(\theta_1) \le x^{-}(\theta_m); \\ \left\{ x \in [x^{-}(\theta_m),\, x^{+}(\theta_1)] \right\}, & \text{otherwise;} \end{cases}$$

The set $X$ is the ergodic set of labor forces in a stationary equilibrium. This can be seen by considering a single market with an initial labor force $x$. Suppose that $x > x^{+}(\theta_1)$; the market will then eventually experience the least advantageous productivity shock with a next period's labor force of $x^{+}(\theta_1)$. Thereafter, the island can at most attract a labor force $x^{-}(\theta_m)$ associated with the most advantageous productivity shock. Analogously, if the market's initial labor force is $x < x^{-}(\theta_m)$, it will eventually have a labor force of $x^{-}(\theta_m)$ after experiencing the most advantageous productivity shock. Its labor force will thereafter never fall below $x^{+}(\theta_1)$ which is the next period's labor force of a market experiencing the least advantageous shock [given a current labor force greater than or equal to $x^{+}(\theta_1)$]. Finally, in the case that $x^{+}(\theta_1) > x^{-}(\theta_m)$, any initial distribution of workers such that each island's labor force belongs to the closed interval $[x^{-}(\theta_m),\, x^{+}(\theta_1)]$ can constitute a stationary equilibrium. This would be a parameterization of the model where agents do not find it worthwhile to relocate in response to productivity shocks.

In a stationary equilibrium, a market's transition probabilities among states $(\theta, x)$ are given by

$$\begin{aligned}\Gamma(\theta', x'|\theta, x) \;=\; \pi(\theta, \theta') \;\cdot\; I(\, \big[x' = x^+(\theta) \text{ and } x \geq x^+(\theta)\big] \text{ or } \\ \big[x' = x^-(\theta) \text{ and } x \leq x^-(\theta)\big] \text{ or } \big[x' = x \text{ and } x^-(\theta) < x < x^+(\theta)\big]\,)\,, \\ \text{for } \; x, x' \in X \text{ and all } \; \theta,\, \theta'\,;\end{aligned}$$

where $I(\cdot)$ is the indicator function which takes on the value 1 if any of its arguments are true and 0 otherwise. These transition probabilities define an operator $P$ on distribution functions $\Psi_t(\theta, x; v_u)$ as follows: Suppose that at a point in time, the distribution of productivity shocks and labor forces across markets is given by $\Psi_t(\theta, x; v_u)$; then the next period's distribution is

$$\Psi_{t+1}(\theta', x'; v_u) \;=\; P\Psi_t(\theta', x'; v_u) \;=\; \sum_{x \in X} \sum_{\theta} \Gamma(\theta', x'|\theta, x)\, \Psi_t(\theta, x; v_u)\,.$$

Except for the case when the stationary equilibrium involves no reallocation of labor, the described process has a unique stationary distribution, $\Psi(\theta, x; v_u)$.

Using the stationary distribution $\Psi(\theta, x; v_u)$, we can compute the economy's average labor force per market,

$$\bar{x}(v_u) \;=\; \sum_{x \in X} \sum_{\theta} x\, \Psi(\theta, x; v_u)\,,$$

where the argument $v_u$ makes explicit that the construction of a stationary equilibrium rests on the maintained assumption that the value to search is exogenously given by $v_u$. The economy's equilibrium labor force $\bar{x}$ varies negatively with $v_u$. In a stationary equilibrium with labor movements, a higher value to search is only consistent with higher wage rates, which in turn require higher marginal products of labor, that is, a smaller labor force on the islands.

From an economy-wide viewpoint, it is the size of the labor force that is fixed, let's say $\hat{x}$, and the value to search that adjusts to clear the markets. To find a stationary equilibrium for a particular $\hat{x}$, we trace out the schedule $\bar{x}(v_u)$ for different values of $v_u$. The equilibrium pair $(\hat{x}, v_u)$ can then be read off at the intersection $\bar{x}(v_u) = \hat{x}$, as illustrated in Figure 19.1.

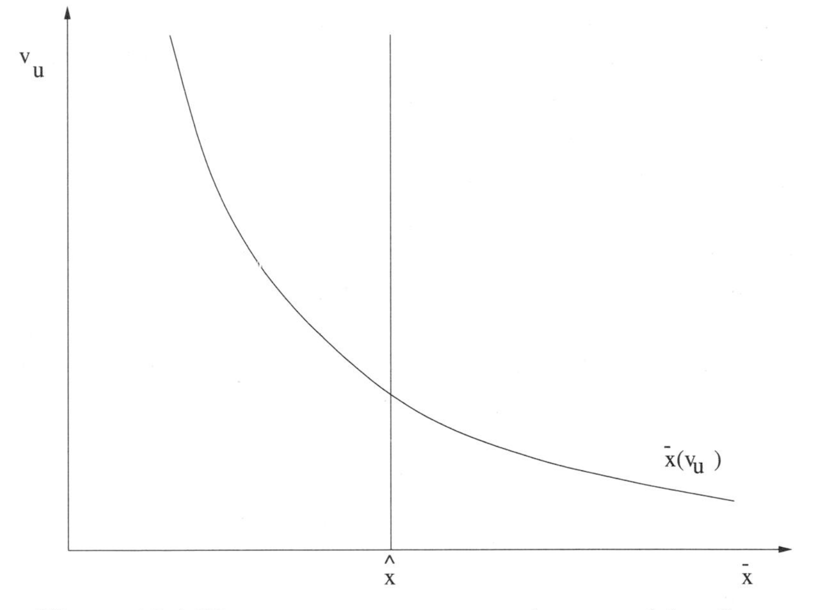


**Figure 19.1** The curve maps an economy's average labor force per market, $\bar{x}$, into the stationary-equilibrium value to search, $v_u$.

## A matching model

Another model of unemployment is the matching framework, as described by Diamond (1982), Mortensen (1982), and Pissarides (1990). The basic model is as follows: Let there be a continuum of identical workers with measure normalized to 1. The workers are infinitely lived and risk neutral. The objective of each worker is to maximize the expected discounted value of leisure and labor income. The leisure enjoyed by an unemployed worker is denoted $z$, while the current utility of an employed worker is given by the wage rate $w$. The workers' discount factor is $\beta = (1+r)^{-1}$.

The production technology is constant returns to scale with labor as the only input. Each employed worker produces $y$ units of output. Without loss of generality, suppose each firm employs at most one worker. A firm entering the economy incurs a vacancy cost $c$ in each period when looking for a worker, and in a subsequent match the firm's per-period earnings are $y - w$. All matches are exogenously destroyed with per-period probability $s$. Free entry implies that the expected discounted stream of a new firm's vacancy costs and earnings is equal to zero. The firms have the same discount factor as the workers (who would be the owners in a closed economy).

The measure of successful matches in a period is given by a matching function $M(u,v)$, where $u$ and $v$ are the aggregate measures of unemployed workers and vacancies. The matching function is increasing in both its arguments, concave, and homogeneous of degree 1. By the homogeneity assumption, we can write the probability of filling a vacancy as $q(v/u) \equiv M(u,v)/v$. The ratio between vacancies and unemployed workers, $\theta \equiv v/u$, is commonly labeled the *tightness* of the labor market. The probability that an unemployed worker will be matched in a period is $\theta q(\theta)$. We will assume that the matching function has the Cobb-Douglas form, which implies constant elasticities,

$$\begin{aligned} M(u,v) &= A u^{\alpha} v^{1-\alpha}, \\ \frac{\partial M(u,v)}{\partial u} \frac{u}{M(u,v)} &= -q'(\theta)\frac{\theta}{q(\theta)} = \alpha, \end{aligned}$$

where $A > 0$, $\alpha \in (0,1)$, and the last equality will be used repeatedly in our derivations that follow.

Finally, the wage rate is assumed to be determined in a Nash bargain between a matched firm and worker. Let $\phi \in [0,1)$ denote the worker's bargaining strength, or his weight in the Nash product, as described in the next subsection.

### *A steady state*

In a steady state, the measure of laid off workers in a period, $s(1-u)$, must be equal to the measure of unemployed workers gaining employment, $\theta q(\theta)u$. The steady-state unemployment rate can therefore be written as

$$u = \frac{s}{s + \theta q(\theta)}. \tag{19.6}$$

To determine the equilibrium value of $\theta$, we now turn to the situations faced by firms and workers, and we impose the no-profit condition for vacancies and the Nash-bargaining outcome on firms' and workers' payoffs.

A firm's value of a filled job $J$ and a vacancy $V$ are given by

$$J = y - w + \beta[sV + (1-s)J], \tag{19.7}$$

$$V = -c + \beta\{q(\theta)J + [1-q(\theta)]V\}. \tag{19.8}$$

That is, a filled job turns into a vacancy with probability $s$, and a vacancy turns into a filled job with probability $q(\theta)$. After invoking the condition that vacancies earn zero profits, $V = 0$, equation (19.8) becomes

$$J = \frac{c}{\beta q(\theta)}, \tag{19.9}$$

which we substitute into equation (19.7) to arrive at

$$w = y - \frac{r+s}{q(\theta)} c. \tag{19.10}$$

The wage rate in equation (19.10) ensures that firms with vacancies break even in an expected present-value sense. In other words, a firm's match surplus must be equal to $J$ in equation (19.9) in order for the firm to recoup its average discounted costs of filling a vacancy.

The worker's share of the match surplus is the difference between the value of an employed worker $E$ and the value of an unemployed worker $U$,

$$E = w + \beta\left[sU + (1-s)E\right], \tag{19.11}$$

$$U = z + \beta\left\{\theta q(\theta) E + [1 - \theta q(\theta)] U\right\}, \tag{19.12}$$

where an employed worker becomes unemployed with probability $s$ and an unemployed worker finds a job with probability $\theta q(\theta)$. The worker's share of the match surplus, $E - U$, has to be related to the firm's share of the match surplus, $J$, in a particular way to be consistent with Nash bargaining. Let the total match surplus be denoted $S = (E - U) + J$, which is shared according to the Nash product

$$\max_{(E-U),J} \; (E-U)^{\phi} J^{1-\phi} \tag{19.13}$$

$$\text{subject to} \quad S = E - U + J,$$

with solution

$$E - U = \phi S, \qquad \text{and} \quad J = (1-\phi) S. \tag{19.14}$$

After solving equations (19.7) and (19.11) for $J$ and $E$, respectively, and substituting them into equations (19.14), we get

$$w = \frac{r}{1+r} U + \phi \left( y - \frac{r}{1+r} U \right). \tag{19.15}$$

The expression is quite intuitive when seeing $r(1+r)^{-1}U$ as the annuity value of being unemployed. The wage rate is just equal to this outside option plus the worker's share $\phi$ of the one-period match surplus. The annuity value of being unemployed can be obtained by solving equation (19.12) for $E-U$ and substituting this expression and equation (19.9) into equations (19.14),

$$\frac{r}{1+r}U = z + \frac{\phi\,\theta\,c}{1-\phi}. \tag{19.16}$$

Substituting equation (19.16) into equation (19.15), we obtain still another expression for the wage rate,

$$w = z + \phi(y - z + \theta c). \tag{19.17}$$

That is, the Nash bargaining results in the worker receiving compensation for lost leisure $z$ and a fraction $\phi$ of both the firm's output in excess of $z$ and the economy's average vacancy cost per unemployed worker.

The two expressions for the wage rate in equations (19.10) and (19.17) determine jointly the equilibrium value for $\theta$,

$$y - z = \frac{r + s + \phi\,\theta\,q(\theta)}{(1-\phi)q(\theta)}\,c. \tag{19.18}$$

This implicit function for $\theta$ ensures that vacancies are associated with zero profits, and that firms' and workers' shares of the match surplus are the outcome of Nash bargaining.

## *Analysis of welfare*

A social planner would choose an allocation that maximizes the discounted value of output and leisure net of vacancy costs. The social optimization problem does not involve any uncertainty because the aggregate fractions of successful matches and destroyed matches are just equal to the probabilities of these events. The social planner's problem of choosing the measure of vacancies, $v_t$, and next period's employment level, $n_{t+1}$, can then be written as

$$\max_{\{v_t,n_{t+1}\}_t} \sum_{t=0}^{\infty} \beta^t \left[y n_t + z(1-n_t) - c v_t\right], \tag{19.19}$$

$$\text{subject to} \quad n_{t+1} = (1-s)n_t + q\left(\frac{v_t}{1-n_t}\right) v_t, \tag{19.20}$$

given $n_0$.

The first-order conditions with respect to $v_t$ and $n_{t+1}$, respectively, are

$$-\beta^t c + \lambda_t \left[q'(\theta_t)\,\theta_t + q(\theta_t)\right] = 0, \tag{19.21}$$

$$-\lambda_t + \beta^{t+1}(y-z) + \lambda_{t+1}\left[(1-s) + q'(\theta_{t+1})\,\theta_{t+1}^2\right] = 0, \tag{19.22}$$

where $\lambda_t$ is the Lagrangian multiplier on equation (19.20). Let us solve for $\lambda_t$ from equation (19.21), and substitute into equation (19.22) evaluated at a stationary solution,

$$y - z = \frac{r + s + \alpha\,\theta\,q(\theta)}{(1-\alpha)q(\theta)}\,c. \tag{19.23}$$

A comparison of this social optimum to the private outcome in equation (19.18) shows that the decentralized equilibrium is only efficient if $\phi = \alpha$. If the workers' bargaining strength $\phi$ exceeds (falls below) $\alpha$, the equilibrium job supply is too low (high). Recall that $\alpha$ is both the elasticity of the matching function with respect to the measure of unemployment, and the negative of the elasticity of the probability of filling a vacancy with respect to $\theta_t$. In its latter meaning, a high $\alpha$ means that an additional vacancy has a large negative impact on all firms' probability of filling a vacancy; the social planner would therefore like to curtail the number of vacancies by granting workers a relatively high bargaining power. Hosios (1990) shows how the efficiency condition $\phi = \alpha$ is a general one for the matching framework.

It is instructive to note that the social optimum is equivalent to choosing the worker's bargaining power $\phi$ such that the value of being unemployed is maximized in a decentralized equilibrium. To see this point, differentiate the value of being unemployed (19.16) to find the slope of the indifference in the space of $\phi$ and $\theta$,

$$\frac{\partial\theta}{\partial\phi} = -\frac{\theta}{\phi(1-\phi)},$$

and use the implicit function rule to find the corresponding slope of the equilibrium relationship (19.18),

$$\frac{\partial\theta}{\partial\phi} = -\frac{y - z + \theta c}{\left[\phi - (r+s)\,q'(\theta)\,q(\theta)^{-2}\right]c}.$$

We set the two slopes equal to each other because a maximum would be attained at a tangency point between the highest attainable indifference curve and equation (19.18) (both curves are negatively sloped and convex to the origin),

$$y - z = \frac{(r+s)\frac{\alpha}{\phi} + \phi\,\theta\,q(\theta)}{(1-\phi)q(\theta)}\,c\,. \tag{19.24}$$

When we also require that the point of tangency satisfies the equilibrium condition (19.18), it can be seen that $\phi = \alpha$ maximizes the value of being unemployed in a decentralized equilibrium. The solution is the same as the social optimum because the social planner and an unemployed worker share the same concern for an optimal investment in vacancies, which takes matching externalities into account.

*Size of the match surplus*

The size of the match surplus depends naturally on the output $y$ produced by the worker, which is lost if the match breaks up and the firm is left to look for another worker. In principle, this loss includes any returns to production factors used by the worker that cannot be adjusted immediately. It might then seem puzzling that a common assumption in the matching literature is to exclude payments to physical capital when determining the size of the match surplus (see, e.g., Pissarides, 1990). Unless capital can be moved without friction in the economy, this exclusion of payments to physical capital must rest on some implicit assumption of outside financing from a third party that is removed from the wage bargain between the firm and the worker. For example, suppose the firm's capital is financed by a financial intermediary that demands specific rental payments in order to not ask for the firm's bankruptcy. As long as the financial intermediary can credibly distance itself from the firm's and worker's bargaining, it would be rational for the two latter parties to subtract the rental payments from the firm's gross earnings and bargain over the remainder.

In our basic matching model, there is no physical capital, but there is investment in vacancies. Let us consider the possibility that a financial intermediary provides a single firm funding for this investment along the described lines. The simplest contract would be that the intermediary hands over funds $c$ to a firm with a vacancy in exchange for a promise that the firm pays $\epsilon$ in every future period of operation. If the firm cannot find a worker in the next period, it fails and the intermediary writes off the loan, and otherwise the intermediary receives the stipulated interest payment $\epsilon$ as long as a successful match stays in business. This agreement with a single firm will have a negligible effect on the economy-wide values of market tightness $\theta$ and the value of being unemployed $U$. Let us examine the consequences for the particular firm involved and the worker it meets.

Under the conjecture that a match will be acceptable to both the firm and the worker, we can compute the interest payment $\epsilon$ needed for the financial intermediary to break even in an expected present-value sense,

$$c = q(\theta)\,\beta \sum_{t=0}^{\infty} \beta^t (1-s)^t \epsilon \qquad \Longrightarrow \qquad \epsilon = \frac{r+s}{q(\theta)}\, c\,. \tag{19.25}$$

A successful match will then generate earnings net of the interest payment equal to $\tilde{y} = y - \epsilon$. To determine how the match surplus is split between the firm and the worker, we replace $y$, $w$, $J$, and $E$ in equations (19.7), (19.11), and (19.13) by $\tilde{y}$, $\tilde{w}$, $\tilde{J}$, and $\tilde{E}$. That is, $\tilde{J}$ and $\tilde{E}$ are the values to the firm and the worker, respectively, for this particular filled job. We treat $\theta$, $V$, and $U$ as constants, since they are determined in the rest of the economy. The Nash bargaining can then be seen to yield,

$$\tilde{w} = \frac{r}{1+r}U + \phi\left(\tilde{y} - \frac{r}{1+r}U\right) = \frac{r}{1+r}U + \phi\frac{\phi\,(r+s)}{(1-\phi)\,q(\theta)}\,c\,,$$

where the first equality corresponds to the previous equation (19.15). The second equality is obtained after invoking $\tilde{y} = y - \epsilon$ and equations (19.16), (19.18), and (19.25), and the resulting expression confirms the conjecture that the match is acceptable to the worker who receives a wage in excess of the annuity value of being unemployed. The firm will of course be satisfied for any positive $\tilde{y} - \tilde{w}$ because it has not incurred any costs whatsoever in order to form the match,

$$\tilde{y} - \tilde{w} = \frac{\phi\,(r+s)}{q(\theta)}\,c > 0\,,$$

where we once again have used $\tilde{y} = y - \epsilon$; equations (19.16), (19.18), and (19.25); and the preceding expression for $\tilde{w}$. Note that $\tilde{y} - \tilde{w} = \phi\epsilon$ with the following interpretation: If the interest payment on the firm's investment, $\epsilon$, was not subtracted from the firm's earnings prior to the Nash bargain, the worker would receive an increase in the wage equal to his share $\phi$ of the additional "match surplus." The present financial arrangement saves the firm this extra wage payment, and the saving becomes the firm's profit. Thus, a single firm with the described contract would have a strictly positive present value when entering the economy of the previous subsection. Since there cannot be such profits in an equilibrium with free entry, explain what would happen if the financing scheme became available to all firms? What would be the equilibrium outcome?

## Matching model with heterogeneous jobs

Acemoglu (1997), Bertola and Caballero (1994), and Davis (1995) explore matching models where heterogeneity on the job supply side must be negotiated through a single matching function, which gives rise to additional externalities. Here we will study an infinite horizon version of Davis's model, which assumes that heterogeneous jobs are created in the same labor market with only one matching function. We extend our basic matching framework as follows: Let there be $I$ types of jobs, and a filled job of type $i$ produces $y^i$. The cost in each period of creating a measure $v^i$ of vacancies of type $i$ is given by a strictly convex upward-sloping cost schedule, $C^i(v^i)$. In a decentralized equilibrium, we will assume that vacancies are competitively supplied at a price equal to the marginal cost of creating an additional vacancy, $C^{i'}(v^i)$, and we retain the assumption of firms employing at most one worker. Another implicit assumption is that $\{y^i, C^i(\cdot)\}$ are such that all types of jobs are created in both the decentralized steady state and the socially optimal steady state.

### A steady state

In a steady state, there will be a time-invariant distribution of employment and vacancies across types of jobs. Let $\eta^i$ be the fraction of type-$i$ jobs among all vacancies. With respect to a job of type $i$, the value of an employed worker, $E^i$, and a firm's values of a filled job, $J^i$, and a vacancy, $V^i$, are given by

$$J^i = y^i - w^i + \beta\big[sV^i + (1-s)J^i\big], \tag{19.26}$$

$$V^i = -C^{i'}(v^i) + \beta\big\{q(\theta)J^i + \big[1-q(\theta)\big]V^i\big\}, \tag{19.27}$$

$$E^i = w^i + \beta\left[sU + (1-s)E^i\right], \tag{19.28}$$

$$U = z + \beta\Big\{\theta q(\theta)\sum_j \eta^j E^j + \big[1-\theta q(\theta)\big]U\Big\}, \tag{19.29}$$

where the value of being unemployed, $U$, reflects that the probabilities of being matched with different types of jobs are equal to the fractions of these jobs among all vacancies.

After imposing a zero-profit condition on all types of vacancies, we arrive at the analogue to equation (19.10),

$$w^i = y^i - \frac{r+s}{q(\theta)} C^{i\prime}(v^i) . \tag{19.30}$$

As before, Nash bargaining can be shown to give rise to still another characterization of the wage,

$$w^i = z + \phi \Big[ y^i - z + \theta \sum_j \eta^j C^{j\prime}(v^j) \Big] , \tag{19.31}$$

which should be compared to equation (19.17). After setting the two wage expressions (19.30) and (19.31) equal to each other, we arrive at a set of equilibrium conditions for the steady-state distribution of vacancies and the labor market tightness,

$$y^i - z = \frac{r + s + \phi\,\theta\, q(\theta) \dfrac{\sum_j \eta^j C^{j\prime}(v^j)}{C^{i\prime}(v^i)}}{(1-\phi) q(\theta)} C^{i\prime}(v^i) . \tag{19.32}$$

When we next turn to the efficient allocation in the current setting, it will be useful to manipulate equation (19.32) in two ways. First, subtract from this equilibrium expression for job $i$ the corresponding expression for job $j$,

$$y^i - y^j = \frac{r+s}{(1-\phi) q(\theta)} \left[ C^{i\prime}(v^i) - C^{j\prime}(v^j) \right] . \tag{19.33}$$

Second, multiply equation (19.32) by $v^i$ and sum over all types of jobs,

$$\sum_i v^i (y^i - z) = \frac{r + s + \phi\,\theta\, q(\theta)}{(1-\phi) q(\theta)} \sum_i v^i C^{i\prime}(v^i) . \tag{19.34}$$

(This expression is reached after invoking $\eta^j \equiv v^j / \sum_h v^h$, and an interchange of summation signs.)

*Welfare analysis*

The social planner's optimization problem becomes

$$\max_{\{v_t^i, n_{t+1}^i\}_{t,i}} \sum_{t=0}^{\infty} \beta^t \Big[\sum_j y^j n_t^j + z\Big(1 - \sum_j n_t^j\Big) - \sum_j C^j(v_t^j)\Big], \tag{19.35a}$$

$$\text{subject to} \quad n_{t+1}^i = (1-s) n_t^i + q\left(\frac{\sum_j v_t^j}{1 - \sum_j n_t^j}\right) v_t^i, \quad \forall i,\ t \geq 0, \tag{19.35b}$$

$$\text{given} \quad \{n_0^i\}_i . \tag{19.35c}$$

The first-order conditions with respect to $v_t^i$ and $n_{t+1}^i$, respectively, are

$$-\beta^t C^{i\prime}(v_t^i) + \lambda_t^i q(\theta_t) + \frac{q'(\theta_t)}{1 - \sum_j n_t^j} \sum_j \lambda_t^j v_t^j = 0, \tag{19.36}$$

$$-\lambda_t^i + \beta^{t+1}(y^i - z) + \lambda_{t+1}^i (1-s) + \frac{q'(\theta_{t+1})\,\theta_{t+1}}{1 - \sum_j n_{t+1}^j} \sum_j \lambda_{t+1}^j v_{t+1}^j = 0. \tag{19.37}$$

To explore the efficient relative allocation of different types of jobs, we subtract from equation (19.36) the corresponding expression for job $j$,

$$\lambda_t^i - \lambda_t^j = \frac{\beta^t \left[C^{i\prime}(v_t^i) - C^{j\prime}(v_t^j)\right]}{q(\theta_t)}. \tag{19.38}$$

Next, we do the same computation for equation (19.37) and substitute equation (19.38) into the resulting expression evaluated at a stationary solution,

$$y^i - y^j = \frac{r+s}{q(\theta)} \left[C^{i\prime}(v^i) - C^{j\prime}(v^j)\right]. \tag{19.39}$$

A comparison of equation (19.39) to equation (19.33) suggests that there will be an efficient *relative* supply of different types of jobs in a decentralized equilibrium only if $\phi = 0$. For any strictly positive $\phi$, the difference in marginal costs of creating vacancies for two different jobs is smaller in the decentralized equilibrium as compared to the social optimum; that is, the decentralized equilibrium displays smaller differences in the distribution of vacancies across types of jobs. In other words, the decentralized equilibrium creates relatively too many "bad jobs" with low $y$'s or, equivalently, relatively too few "good jobs" with high $y$'s. The inefficiency in the mix of jobs disappears if the workers have no bargaining power so that the firms reap all the benefits of upgrading jobs.[1] But from before we know that workers' bargaining power is essential to correct an excess supply of the *total* number of vacancies.

To investigate the efficiency with respect to the total number of vacancies, multiply equation (19.36) by $v^i$ and sum over all types of jobs,

$$\sum_i \lambda_t^i v_t^i = \frac{\beta^t \sum_i v_t^i C^{i\prime}(v_t^i)}{q(\theta_t) + q'(\theta_t)\theta_t}. \tag{19.40}$$

Next, we do the same computation for equation (19.37) and substitute equation (19.40) into the resulting expression evaluated at a stationary solution,

$$\sum_i v^i(y^i - z) = \frac{r + s + \alpha\,\theta\, q(\theta)}{(1-\alpha)q(\theta)} \sum_i v^i C^{i\prime}(v^i). \tag{19.41}$$

A comparison of equations (19.41) and (19.34) suggests the earlier result from the basic matching model; that is, an efficient *total* supply of jobs in a decentralized equilibrium calls for $\phi = \alpha$.[2] Hence, Davis (1995) concludes that there is a

[1] The interpretation that $\phi = 0$, which is needed to attain an efficient relative supply of different types of jobs in a decentralized equilibrium, can be made precise in the following way: Let $v$ and $n$ denote any sustainable stationary values of the economy's measure of total vacancies and employment rate, that is, $sn = q\left(\frac{v}{1-n}\right) v$. Solve the social planner's optimization problem in equation (19.35) subject to the additional constraints

$$\sum_i v_t^i = v, \quad \sum_i n_{t+1}^i = n, \quad \forall t \geq 0;$$

given $\{n_0^i : \sum_i n_0^i = n\}$. After applying the steps in the main text to the first-order conditions of this problem, we arrive at the very same expression (19.39). Thus, if $\{v, n\}$ is taken to be the steady-state outcome of the decentralized economy, it follows that equilibrium condition (19.33) satisfies efficiency condition (19.39) when $\phi = 0$.

[2] The suggestion that $\phi = \alpha$, which is needed to attain an efficient total supply of jobs in a decentralized equilibrium, can be made precise in the following way. Suppose that the social planner is forever constrained to some arbitrary relative distribution, $\{\gamma^i\}$, of types of jobs

fundamental tension between the condition for an efficient mix of jobs ($\phi = 0$) and the standard condition for an efficient total supply of jobs ($\phi = \alpha$).

*The allocating role of wages I: Separate markets*

The last section clearly demonstrates Hosios's (1990) characterization of the matching framework: "Though wages in matching-bargaining models are completely flexible, these wages have nonetheless been denuded of any allocating or signaling function: this is because matching takes place before bargaining and so search effectively precedes wage-setting." In Davis's matching model, the problem of wages having no allocating role is compounded through the existence of heterogeneous jobs. But as discussed by Davis, this latter complication would be overcome if different types of jobs were ex ante sorted into separate markets. Equilibrium movements of workers across markets would then remove the tension between the optimal mix and the total supply of jobs. Different wages in different markets would serve an allocating role for the labor supply across markets, even though the equilibrium wage in each market would still be determined through bargaining after matching.

Let us study the outcome when there are such separate markets for different types of jobs and each worker can only participate in one market at a time. The modified model is described by equations (19.26), (19.27), and (19.28) where the market tightness variable is now also indexed by $i$ and $\theta^i$, and the new expression for the value of being unemployed is

$$U = z + \beta\{\theta^i q(\theta^i)E^i + [1 - \theta^i q(\theta^i)]U\}. \tag{19.42}$$

---

and vacancies, where $\gamma^i \geq 0$ and $\sum_i \gamma^i = 1$. The constrained social planner's problem is then given by equations (19.35) subject to the additional restrictions

$$v_t^i = \gamma^i v_t, \qquad n_t^i = \gamma^i n_t, \qquad \forall t \geq 0.$$

That is, the only choice variables are now total vacancies and employment, $\{v_t, n_{t+1}\}$. After consolidating the two first-order conditions with respect to $v_t$ and $n_{t+1}$, and evaluating at a stationary solution, we obtain

$$\sum_j y^j \gamma^j - z = \frac{r + s + \alpha\theta q(\theta)}{(1-\alpha)q(\theta)} \sum_j \gamma^j C^{j\prime}(\gamma^j v).$$

By multiplying both sides by $v$, we arrive at the very same expression (19.41). Thus, if the arbitrary distribution $\{\gamma^i\}$ is taken to be the steady-state outcome of the decentralized economy, it follows that equilibrium condition (19.34) satisfies efficiency condition (19.41) when $\phi = \alpha$.

In an equilibrium, an unemployed worker attains the value $U$ regardless of which labor market he participates in. The characterization of a steady state proceeds along the same lines as before. Let us here reproduce only three equations that will be helpful in our reasoning. The wage in market $i$ and the annuity value of an unemployed worker can be written as

$$w^i = \phi y^i + (1-\phi)\frac{r}{1+r}U\,, \tag{19.43}$$

$$\frac{r}{1+r}U = z + \frac{\phi\theta^i C^{i\prime}(v^i)}{1-\phi}\,, \tag{19.44}$$

and the equilibrium condition for market $i$ becomes

$$y^i - z = \frac{r + s + \phi\,\theta^i\, q(\theta^i)}{(1-\phi)q(\theta^i)}\, C^{i\prime}(v^i)\,. \tag{19.45}$$

The social planner's objective function is the same as expression (19.35a), but the earlier constraint (19.35b) is now replaced by

$$n^i_{t+1} = (1-s)n^i_t + q\left(\frac{v^i_t}{u^i_t}\right) v^i_t\,,$$

$$1 = \sum_j \left(u^j_t + n^j_t\right)\,,$$

where $u^i_t$ is the measure of unemployed workers in market $i$. At a stationary solution, the first-order conditions with respect to $v^i_t$, $u^i_t$, and $n^i_{t+1}$ can be combined to read

$$y^i - z = \frac{r + s + \alpha\,\theta^i\, q(\theta^i)}{(1-\alpha)q(\theta^i)}\, C^{i\prime}(v^i)\,. \tag{19.46}$$

Equations (19.45) and (19.46) confirm Davis's finding that the social optimum can be attained with $\phi = \alpha$ as long as different types of jobs are sorted into separate markets.

It is interesting to note that the socially optimal wages, that is, equation (19.43) with $\phi = \alpha$, imply wage differences for ex ante identical workers. Wage differences here are not a sign of any inefficiency but rather necessary to ensure an optimal supply and composition of jobs. Workers with higher pay are compensated for an unemployment spell in their job market that is on average longer.

*The allocating role of wages II: Wage announcements*

According to Moen (1997), we can reinterpret the socially optimal steady state in the last section as an economy with competitive wage announcements instead of wage bargaining with $\phi = \alpha$. Firms are assumed to freely choose a wage to announce, and then they join the market offering this wage without any bargaining. The socially optimal equilibrium is attained when workers as wage takers choose between labor markets so that the value of an unemployed worker is equalized in the economy.

To demonstrate that wage announcements are consistent with the socially optimal steady state, consider a firm with a vacancy of type $i$ which is free to choose any wage $\tilde{w}$ and then join a market with this wage. A labor market with wage $\tilde{w}$ has a market tightness $\tilde{\theta}$ such that the value of unemployment is equal to the economy-wide value $U$. After replacing $w$, $E$, and $\theta$ in equations (19.28) and (19.42) by $\tilde{w}$, $\tilde{E}$, and $\tilde{\theta}$, we can combine these two expressions to arrive at a relationship between $\tilde{w}$ and $\tilde{\theta}$,

$$\tilde{w} = \frac{r}{1+r}U + \frac{r+s}{\tilde{\theta}q(\tilde{\theta})}\Big(\frac{r}{1+r}U - z\Big). \tag{19.47}$$

The expected present value of posting a vacancy of type $i$ for one period in market $(\tilde{w}, \tilde{\theta})$ is

$$-C^{i\prime}(v^i) + q(\tilde{\theta})\beta \sum_{t=0}^{\infty} \beta^t (1-s)^t (y^i - \tilde{w}) = -C^{i\prime}(v^i) + q(\tilde{\theta})\frac{y^i - \tilde{w}}{r+s}.$$

After substituting equation (19.47) into this expression, we can compute the first-order condition with respect to $\tilde{\theta}$ as

$$q'(\tilde{\theta})\frac{y^i}{r+s} - \frac{z}{\tilde{\theta}^2} + \left[\frac{1}{\tilde{\theta}^2} - \frac{q'(\tilde{\theta})}{r+s}\right]\frac{r}{1+r}U = 0.$$

Since the socially optimal steady state is our conjectured equilibrium, we get the economy-wide value $U$ from equation (19.44) with $\phi$ replaced by $\alpha$. The substitution of this value for $U$ into the first-order condition yields

$$y^i - z = \frac{r + s + \alpha\,\tilde{\theta}\,q(\tilde{\theta})}{(1-\alpha)q(\tilde{\theta})}\,\frac{\theta^i}{\tilde{\theta}}\,C^{i\prime}(v^i). \tag{19.48}$$

The right-hand side is strictly decreasing in $\tilde{\theta}$, so by equation (19.46) the equality can only hold with $\tilde{\theta} = \theta^i$. We have therefore confirmed that the wages in an optimal steady state are such that firms would like to freely announce them and participate in the corresponding markets without any wage bargaining. The equal value of an unemployed worker across markets ensures also the participation of workers who now act as wage takers.

## Model of employment lotteries

Consider a labor market without search and matching frictions but where labor is indivisible. An individual can supply either one unit of labor or no labor at all, as assumed by Hansen (1985) and Rogerson (1988). In such a setting, employment lotteries can be welfare enhancing. The argument is best understood in Rogerson's static model, but with physical capital (and its implication of diminishing marginal product of labor) removed from the analysis. We assume that the good, $c$, can be produced with labor, $n$, as the sole input in a constant returns to scale technology,

$$c = \gamma n , \qquad \text{where} \quad \gamma > 0 .$$

Following Hansen and Rogerson, the preferences of an individual are assumed to be additively separable in consumption and labor,

$$u(c) - v(n) .$$

The standard assumptions are that both $u$ and $v$ are twice continuously differentiable and increasing, but while $u$ is strictly concave, $v$ is convex. However, as pointed out by Rogerson, the precise properties of the function $v$ are not essential because of the indivisibility of labor. The only values of $v(n)$ that matter are $v(0)$ and $v(1)$, let $v(0) = 0$ and $v(1) = A > 0$. An individual who can supply one unit of labor in exchange for $\gamma$ units of goods would then choose to do so if

$$u(\gamma) - A \geq u(0) ,$$

and otherwise, the individual would choose not to work.

The described allocation might be improved upon by introducing employment lotteries. That is, each individual chooses a probability of working, $\psi \in [0, 1]$, and he trades his stochastic labor earnings in contingency markets. We assume a continuum of agents so that the idiosyncratic risks associated with employment lotteries do not pose any aggregate risk and the contingency prices are then determined by the probabilities of events occurring. (See chapters 7 and 9.) Let $c_1$ and $c_2$ be the individual's choice of consumption when working and not working, respectively. The optimization problem becomes

$$\begin{aligned}
\max_{c_1,c_2,\psi} \quad & \psi\,[u(c_1) - A] + (1-\psi)\,u(c_2)\,, \\
\text{subject to} \quad & \psi c_1 + (1-\psi)c_2 \leq \psi\gamma\,, \\
& c_1, c_2 \geq 0\,, \quad \psi \in [0,1]\,.
\end{aligned}$$

At an interior solution for $\psi$, the first-order conditions for consumption imply that $c_1 = c_2$,

$$\begin{aligned}
\psi\, u'(c_1) &= \psi\,\lambda\,, \\
(1-\psi)\, u'(c_2) &= (1-\psi)\,\lambda\,,
\end{aligned}$$

where $\lambda$ is the multiplier on the budget constraint. Since there is no harm in also setting $c_1 = c_2$ when $\psi = 0$ or $\psi = 1$, the individual's maximization problem can be simplified to read

$$\begin{aligned}
\max_{c,\psi} \quad & u(c) - \psi\, A\,, \\
\text{subject to} \quad & c \leq \psi\gamma\,, \quad c \geq 0\,, \quad \psi \in [0,1]\,.
\end{aligned} \tag{19.49}$$

The welfare-enhancing potential of employment lotteries is implicit in the relaxation of the earlier constraint that $\psi$ could only take on two values, 0 or 1. With employment lotteries, the marginal rate of transformation between leisure and consumption is equal to $\gamma$.

The solution to expression (19.49) can be characterized by considering three possible cases:

Case 1. $A/u'(0) \geq \gamma$.
Case 2. $A/u'(0) < \gamma < A/u'(\gamma)$.
Case 3. $A/u'(\gamma) \leq \gamma$.

The introduction of employment lotteries will only affect individuals' behavior in the second case. In the first case, if $A/u'(0) \geq \gamma$, it will under all circumstances be optimal not to work ($\psi = 0$), since the marginal value of leisure in terms of consumption exceeds the marginal rate of transformation even at a zero consumption level. In the third case, if $A/u'(\gamma) \leq \gamma$, it will always be optimal

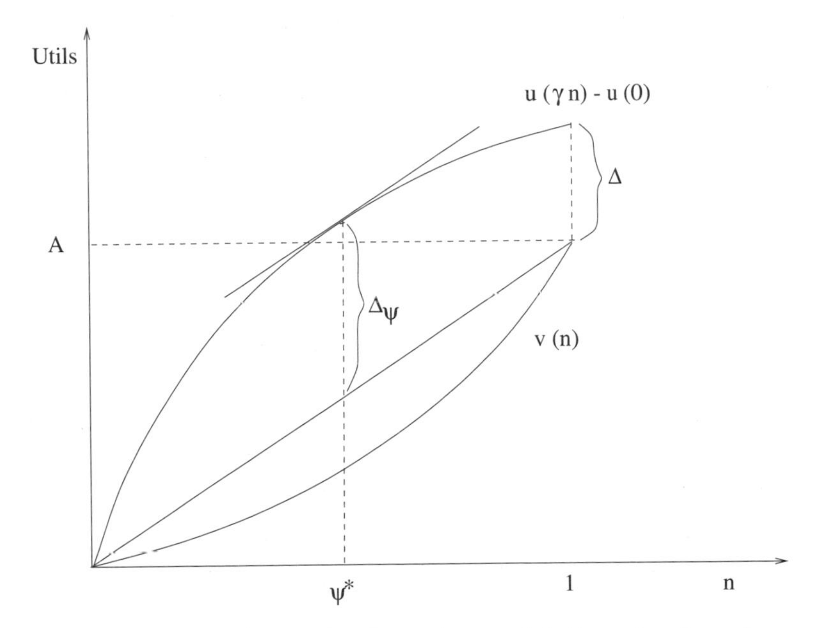


**Figure 19.2** The optimal employment lottery is given by probability $\psi^*$ of working, which increases expected welfare by $\triangle_\psi - \triangle$ as compared to working full-time $n = 1$.

to work ($\psi = 1$), since the marginal value of leisure falls short of the marginal rate of transformation when evaluated at the highest feasible consumption per worker. The second case implies that expression (19.49) has an interior solution with respect to $\psi$ and that employment lotteries are welfare enhancing. The optimal value, $\psi^*$, is then given by the first-order condition

$$\frac{A}{u'(\gamma\psi^*)} = \gamma \, .$$

An example of the second case is shown in Figure 19.2. The situation here is such that the individual would choose to work in the absence of employment lotteries, because the curve $u(\gamma n) - u(0)$ is above the curve $v(n)$ when evaluated at $n = 1$. After the introduction of employment lotteries, the individual chooses the probability $\psi^*$ of working, and his welfare increases by $\triangle_\psi - \triangle$.

## *Employment effects of layoff taxes*

The models of employment determination in this chapter can be used to address the question, How do layoff taxes affect an economy's employment? Hopenhayn and Rogerson (1993) apply the model of employment lotteries to this very question and conclude that a layoff tax would reduce the level of employment. Mortensen and Pissarides (1999b) reach the opposite conclusion in a matching model. We will here examine these results by scrutinizing the economic forces at work in different frameworks. The purpose is both to gain further insights into the workings of our theoretical models and to learn about possible effects of layoff taxes.[3]

Common features of many analyses of layoff taxes are as follows: The productivity of a job evolves according to a Markov process, and a sufficiently poor realization triggers a layoff. The government imposes a layoff tax $\tau$ on each layoff. The tax revenues are handed back as equal lump-sum transfers to all agents, denoted by $T$ per capita.

Here we assume the simplest possible Markov process for productivities. A new job has productivity $p_0$. In all future periods, with probability $\xi$, the worker keeps the productivity from last period, and with probability $1-\xi$, the worker draws a new productivity from a distribution $G(p)$.

In our numerical example, the model period is 2 weeks, and the assumption that $\beta = 0.9985$ then implies an annual real interest rate of 4 percent. The initial productivity of a new job is $p_0 = 0.5$, and $G(p)$ is taken to be a uniform distribution on the unit interval. An employed worker draws a new productivity on average once every two years when we set $\xi = 0.98$.

### *A model of employment lotteries with layoff taxes*

In a model of employment lotteries, there will be a market-clearing wage $w$ that will equate the demand and supply of labor. The constant returns to scale technology implies that this wage is determined from the supply side as follows: At the beginning of a period, let $V(p)$ be the firm's value of an employee with productivity $p$,

$$V(p) = \max\left\{p - w + \beta\Big[\xi V(p) + (1-\xi)\int V(p')\,dG(p')\Big],\ -\tau\right\}. \quad (19.50)$$

[3] The analysis is based on Ljungqvist's (1997) study of layoff taxes in different models of employment determination.

Given a value of $w$, the solution to this Bellman equation is a reservation productivity $\bar{p}$. If there exists an equilibrium with strictly positive employment, the equilibrium wage must be such that new hires exactly break even,

$$V(p_0) = p_0 - w + \beta\Big[\xi V(p_0) + (1-\xi)\int V(p')\,dG(p')\Big] = 0$$

$$\Rightarrow w = p_0 + \beta(1-\xi)\tilde{V}, \tag{19.51}$$

where

$$\tilde{V} \equiv \int V(p')\,dG(p').$$

In order to compute $\tilde{V}$, we first look at the value of $V(p)$ when $p \geq \bar{p}$,

$$\begin{aligned} V(p)\Big|_{p\geq\bar{p}} &= p - w + \beta\left[\xi V(p) + (1-\xi)\tilde{V}\right] \\ &= p - w + \beta\xi\Big\{p - w + \beta\big[\xi V(p) + (1-\xi)\tilde{V}\big]\Big\} + \beta(1-\xi)\tilde{V} \\ &= (1+\beta\xi)\left[p - w + \beta(1-\xi)\tilde{V}\right] + \beta^2\xi^2 V(p) \\ &= \frac{p - w + \beta(1-\xi)\tilde{V}}{1-\beta\xi} = \frac{p - p_0}{1-\beta\xi}, \end{aligned} \tag{19.52}$$

where the first equalities are obtained through successive substitutions of $V(p)$, and the last equality incorporates equation (19.51). We can then use equation (19.52) to find an expression for $\tilde{V}$,

$$\begin{aligned} \tilde{V} &= \int_{-\infty}^{\bar{p}} -\tau\,dG(p) + \int_{\bar{p}}^{\infty} V(p)\,dG(p) \\ &= -\tau\,G(\bar{p}) + \int_{\bar{p}}^{\infty} \frac{p-p_0}{1-\beta\xi}\,dG(p). \end{aligned} \tag{19.53}$$

From equation (19.50), the reservation productivity satisfies

$$\bar{p} - w + \beta\left[\xi V(\bar{p}) + (1-\xi)\tilde{V}\right] = -\tau,$$

and, after imposing equation (19.51) and $V(\bar{p}) = -\tau$,

$$\bar{p} = p_0 - (1-\beta\xi)\tau. \tag{19.54}$$

The equations (19.54), (19.53), and (19.51) can be used to solve for the equilibrium wage $w^*$.

Given the equilibrium wage $w^*$ and a gross interest rate $1/\beta$, the representative agent's optimization problem reduces to a static problem of the form,

$$\begin{aligned} &\max_{c,\psi} \; u(c) - \psi A, \\ &\text{subject to} \quad c \leq \psi w^* + \Pi + T, \quad c \geq 0, \quad \psi \in [0,1], \end{aligned} \tag{19.55}$$

where the profits from firms, $\Pi$, and the lump-sum transfer from the government, $T$, are taken as given by the agents. In a stationary equilibrium with $(w^*, \psi^*)$, we have

$$\Pi + T = \psi^* \int (p - w^*)\, dH(p),$$

where $H(p)$ is the equilibrium fraction of all jobs with a productivity less than or equal to $p$. Since all agents are identical including their asset holdings, the expected lifetime utility of an agent before seeing the outcome of any employment lottery is equal to

$$\sum_{t=0}^{\infty} \beta^t \left[ u\Big(\psi^* \int p\, dH(p)\Big) - \psi^* A \right].$$

Following Hopenhayn and Rogerson (1993), the preference specification is $u(c) = \log(c)$ and the disutility of work is calibrated to match an employment to population ratio equal to 0.6, which leads us to choose $A = 1.6$. Figures 19.3–19.7 show how equilibrium outcomes vary with the layoff tax. The curves labeled $L$ pertain to the model of employment lotteries. As derived in equation (19.54), the reservation productivity in Figure 19.3 falls when it becomes more costly to lay off workers. Figure 19.4 shows how the decreasing number of layoffs are outweighed by the higher tax per layoff, so total layoff taxes as a fraction of GNP are increasing for almost the whole range. Figure 19.5 reveals changing job prospects, where the probability of working falls with a higher layoff tax (which is equivalent to falling employment in a model of employment lotteries). The welfare loss associated with a layoff tax is depicted in Figure 19.6 as the amount of consumption that an agent would be willing to give up in order to rid the economy of the layoff tax, and the "willingness to pay" is expressed as a fraction of per capita consumption at a zero layoff tax.

Figure 19.7 reproduces Hopenhayn and Rogerson's (1993) result that employment falls with a higher layoff tax (except at the highest layoff taxes). Intuitively speaking, a higher layoff tax is synonymous from a private perspective with a deterioration in the production technology; the optimal change in the agents' employment lotteries will therefore depend on the strength of the substitution effect versus the income effect. The income effect is largely mitigated by the government's lump-sum transfer of the tax revenues back the to private economy. Thus, layoff taxes in models of employment lotteries have strong negative employment implications caused by the substitution away from work toward leisure. Formally, the logarithmic preference specification gives rise to an optimal choice of the probability of working, which is equal to the employment outcome, as given by

$$\psi^* = \frac{1}{A} - \frac{T + \Pi}{w^*} .$$

The precise employment effect here is driven by profit flows from firms gross of layoff taxes expressed in terms of the wage rate. Since these profits are to a large extent generated in order to pay for firms' future layoff taxes, a higher layoff tax tends to increase the accumulation of such funds with a corresponding negative effect on the optimal choice of employment.

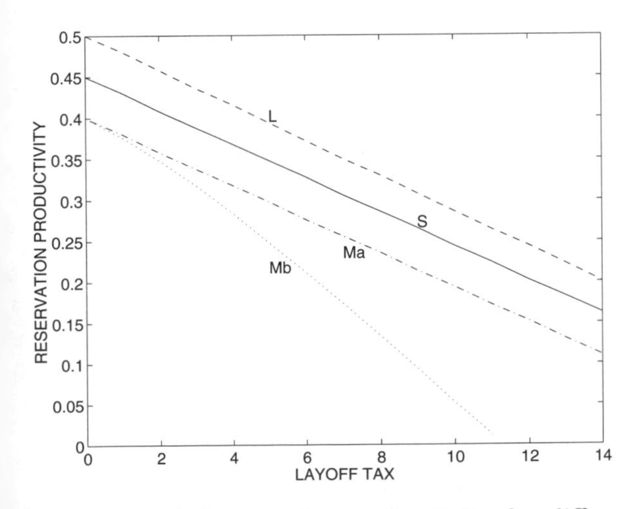


**Figure 19.3** Reservation productivity for different values of the layoff tax.

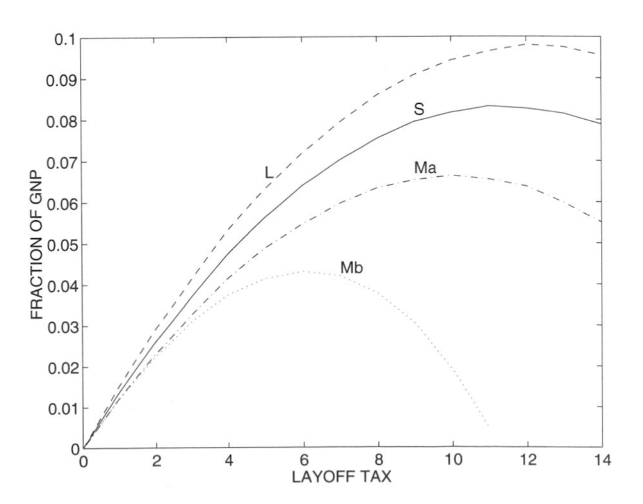


**Figure 19.4** Total layoff taxes as a fraction of GNP for different values of the layoff tax.

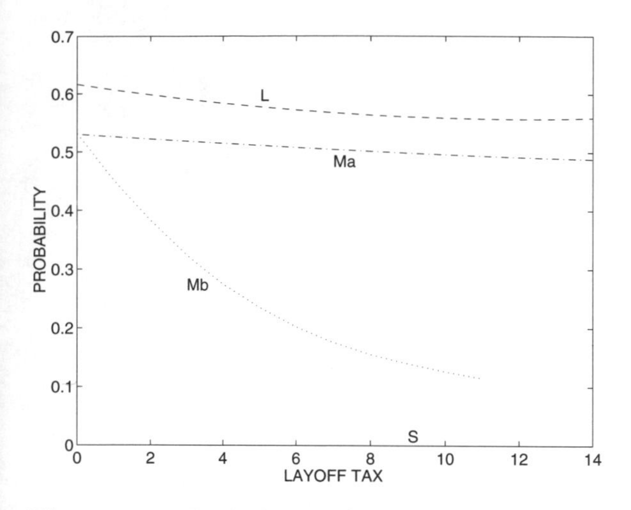


**Figure 19.5** Probability of working in the model with employment lotteries and probability of finding a job within 10 weeks in the other models, for different values of the layoff tax.

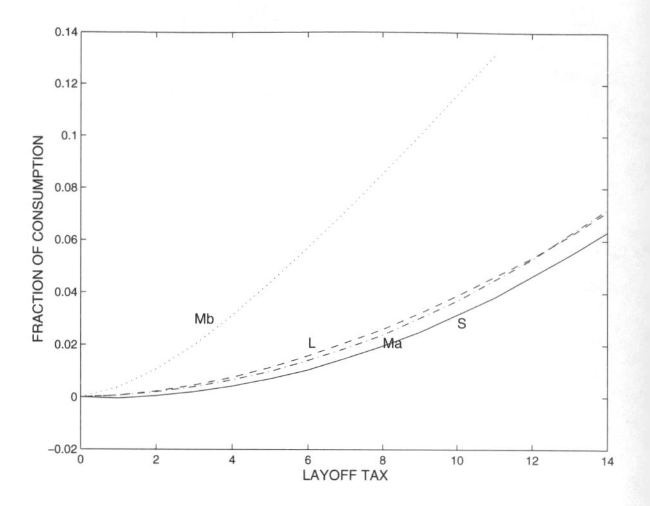


**Figure 19.6** A job finder's welfare loss due to the presence of a layoff tax, computed as a fraction of per capita consumption at a zero layoff tax.

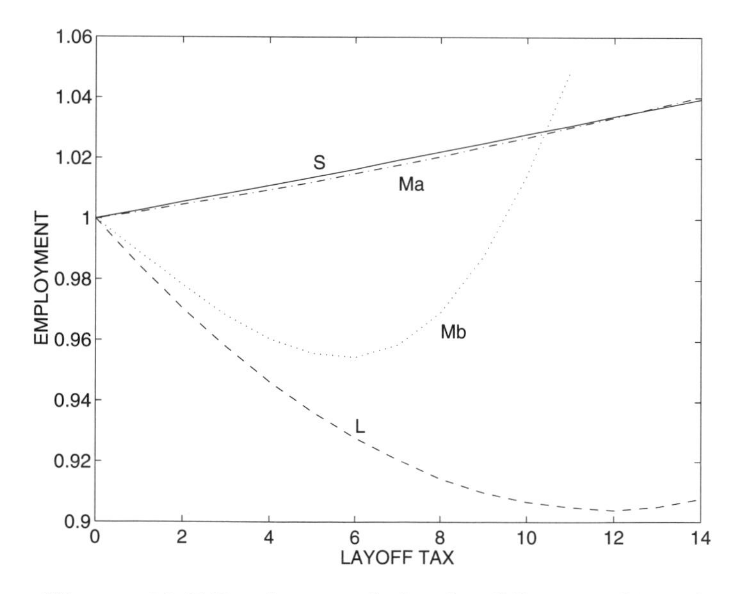


**Figure 19.7** Employment index for different values of the layoff tax. The index is equal to one at a zero layoff tax.

*An island model with layoff taxes*

To stay with the described technology in an island framework, let each job represent a separate island, and an agent moving to a new island experiences productivity $p_0$. We retain the feature that every agent bears the direct consequences of his decisions. He receives his marginal product $p$ when working and incurs the layoff tax $\tau$ if leaving his island. The Bellman equation can then be written as

$$V(p) = \max\left\{p - z + \beta\left[\xi V(p) + (1-\xi)\int V(p')\,dG(p')\right],\right.$$
$$\left. -\tau + \beta^T V(p_0)\right\}, \qquad (19.56)$$

where $z$ is the forgone utility of leisure when working and $T$ is the number of periods it takes to move to another island.[4] The solution to this equation is a reservation productivity $\bar{p}$.

If there exists an equilibrium with agents working, we must have

$$V(p_0) = p_0 - z + \beta\left[\xi V(p_0) + (1-\xi)\int V(p')\,dG(p')\right],$$
$$\Longrightarrow \quad \beta(1-\xi)\tilde{V} = (1-\beta\xi)V(p_0) + z - p_0\,, \qquad (19.57)$$

where

$$\tilde{V} \equiv \int V(p')\,dG(p')\,.$$

If the equilibrium entails agents moving between islands, the reservation productivity, by equation (19.56), satisfies

$$\bar{p} - z + \beta\left[\xi V(\bar{p}) + (1-\xi)\tilde{V}\right] = -\tau + \beta^T V(p_0)\,,$$

and, after imposing equation (19.57) and $V(\bar{p}) = -\tau + \beta^T V(p_0)$,

$$\bar{p} = p_0 - (1-\beta\xi)\left[\tau + (1-\beta^T)\,V(p_0)\right]. \qquad (19.58)$$

[4] Note that we have left out the lump-sum transfer from the government because it does not affect the optimization problem.

Note that if agents could move instantaneously between islands, $T = 0$, the reservation productivity would be the same as in the model of employment lotteries, given by equation (19.54).

A higher layoff tax does also reduce the reservation productivity in the island model; that is, an increase in $\tau$ outweighs the drop in the second term in square brackets in equation (19.58). For a formal proof, let us make explicit that the value function and the reservation productivity are functions of the layoff tax, $V(p;\tau)$ and $\bar{p}(\tau)$. Consider two layoff taxes, $\tau$ and $\tau'$, such that $\tau' > \tau \geq 0$ and denote the difference $\triangle\tau = \tau' - \tau$. We can then construct a lower bound for $V(p;\tau')$ in terms of $V(p;\tau)$. In response to the higher layoff tax $\tau'$, the agent can always keep his decision rule associated with $V(p;\tau)$ and an upper bound for his extra layoff tax payments would be that he paid $\triangle\tau$ in the current period and every $T$th period from there on,

$$V(p;\tau') > V(p,\tau) - \sum_{i=0}^{\infty} \beta^{iT} \triangle\tau \,, \tag{19.59}$$

where the strict inequality follows from the fact that it cannot be optimal to constantly move. In addition, the agent might be able to select a better decision rule than the one associated with $\tau$. In fact, the reservation productivity must fall in response to a higher layoff tax whenever there is an interior solution with respect to $\bar{p}$, as given by equation (19.58). By using equations (19.58) and (19.59), we have

$$\begin{aligned} \bar{p}(\tau') - \bar{p}(\tau) &= -(1-\beta\xi)\big\{\triangle\tau + (1-\beta^T)\big[V(p_0;\tau') - V(p_0;\tau)\big]\big\} \\ &< -(1-\beta\xi)\Big[\triangle\tau - (1-\beta^T)\sum_{i=0}^{\infty}\beta^{iT}\triangle\tau\Big] = 0 \,. \end{aligned}$$

The numerical illustration in Figures 19.3–19.7 is based on a value of leisure $z = 0.25$ and a length of transition between jobs $T = 7$; that is, unemployment spells last 14 weeks. The curves that pertain to the island model are labeled $S$. The effects of layoff taxes on the reservation productivity, the economy's total layoff taxes, and the welfare of a recent job finder are all similar to the outcomes in the model of employment lotteries. The sharp difference appears in Figure 19.7 depicting the effect on the economy's employment. In the island model where agents are left to fend for themselves, a lower reservation productivity is synonymous with both less labor reallocation and lower unemployment. Lower unemployment is thus attained at the cost of a less efficient labor allocation.

Mobility costs cause employment also to rise in the general version of the island model, as mentioned by Lucas and Prescott (1974, p. 205). For a given expected discounted value of arriving on a new island $v_u$, the value function in equation (19.3) is replaced by

$$v(\theta, x) = \max\Big\{v_u - \tau,\ \theta f'(x) + \min\big\{v_u,\ \beta E[v(\theta', x)|\theta]\big\}\Big\}, \tag{19.60}$$

which lies below equation (19.3) but with a drop of at most $\tau$. Similarly, equation (19.4) changes to

$$\theta f'(n) + \beta E\left[v(\theta', n)|\theta\right] = v_u - \tau. \tag{19.61}$$

An implication here is that $x^+(\theta)$ rises in response to a higher layoff tax. The unchanged expression (19.5) means that $x^-(\theta)$ falls as a result of the preceding drop in the value function. In other words, the range of an island's employment levels characterized by no labor movements is enlarged. This effect will shift the curve in Figure 19.1 downward and decrease the equilibrium value of $v_u$. Less labor reallocation maps directly into a lower unemployment rate.

### *A matching model with layoff taxes*

We now modify the matching model to incorporate a layoff tax, and the exogenous destruction of jobs is replaced by the described Markov process for a job's productivity. A job is now endogenously destroyed when the outside option, taking the layoff tax into account, is higher than the value of maintaining the match. The match surplus, $S_i(p)$, is a function of the job's current productivity $p$ and can be expressed as

$$S_i(p) + U_i = \max\Bigg\{p + \beta\left[\xi S_i(p) + (1-\xi)\int S_i(p')\,dG(p') + U_i\right],\ U_i - \tau\Bigg\}, \tag{19.62}$$

where $U_i$ is once again the agent's outside option, that is, the value of being unemployed. Both $S_i(p)$ and $U_i$ are indexed by $i$, since we will explore the implications of two alternative specifications of the Nash product, $i \in \{a, b\}$,

$$\left[E_a(p) - U_a\right]^{\phi} J_a(p)^{1-\phi}, \tag{19.63}$$

$$\left[E_b(p) - U_b\right]^{\phi} \left[J_b(p) + \tau\right]^{1-\phi}. \tag{19.64}$$

Specification (19.63) leads to the usual result that the worker receives a fraction $\phi$ of the match surplus, while the firm gets the remaining fraction $(1-\phi)$,

$$E_a(p) - U_a = \phi S_a(p) \qquad \text{and} \qquad J_a(p) = (1-\phi)S_a(p). \tag{19.65}$$

The alternative specification (19.64) adopts the assumption of Saint-Paul (1995) that the layoff cost changes the firm's threat point from 0 to $-\tau$, and thereby increases the worker's relative share of the match surplus. Solving for the corresponding surplus sharing rules, we get

$$E_b(p) - U_b = \phi\big(S_b(p) + \tau\big) \qquad \text{and} \qquad J_b(p) = (1-\phi)S_b(p) - \phi\tau. \tag{19.66}$$

The worker's continuation value outside of the match associated with Nash product (19.63) or (19.64), respectively, is

$$U_a = z + \beta\big[\theta q(\theta)\phi S_a(p_0) + U_a\big], \tag{19.67}$$
$$U_b = z + \beta\big\{\theta q(\theta)\phi\big[S_b(p_0) + \tau\big] + U_b\big\}. \tag{19.68}$$

The equilibrium conditions that firms post vacancies until the expected profits are driven down to zero become

$$(1-\phi)S_a(p_0) = \frac{c}{\beta q(\theta)}, \tag{19.69}$$
$$(1-\phi)S_b(p_0) - \phi\tau = \frac{c}{\beta q(\theta)}, \tag{19.70}$$

for Nash product (19.63) or (19.64), respectively.

In the calibration, we choose a matching function $M(u,v) = 0.01u^{0.5}v^{0.5}$, a worker's bargaining strength $\phi = 0.5$, and the same value of leisure as in the island model, $z = 0.25$. Qualitatively, the results in Figures 19.3–19.7 are the same across all the models considered here. The curve labeled $Ma$ pertains to the matching model where the workers' relative share of the match surplus is constant, while the curve $Mb$ refers to the model where the share is positively related to the layoff tax. However, matching model $Mb$ does stand out. Its reservation productivity plummets in response to the layoff tax in Figure 19.3, and is close to zero at $\tau = 11$. A zero reservation productivity means that labor reallocation comes to a halt and the economy's tax revenues fall to zero in Figure 19.4. The more dramatic outcomes under $Mb$ have to do with layoff taxes increasing workers' relative share of the match surplus. The equilibrium condition

(19.70) requiring that firms finance incurred vacancy costs with retained earnings from the matches becomes exceedingly difficult to satisfy when a higher layoff tax erodes the fraction of match surpluses going to firms. Firms can only break even if the expected time to fill a vacancy is cut dramatically; that is, there has to be a large number of unemployed workers for each posted vacancy. This equilibrium outcome is reflected in the sharply falling probability of a worker finding a job within 10 weeks in Figure 19.5. As a result, there are larger welfare costs in model $Mb$, as shown by the welfare loss of a job finder in Figure 19.6. The welfare loss of an unemployed agent is even larger in model $Mb$, whereas the differences between employed and unemployed agents in the three other model specifications are negligible (not shown in any figure).

In Figure 19.7, matching model $Ma$ looks very much as the island model with increasing employment, and matching model $Mb$ displays initially falling employment similar to the model of employment lotteries. The later sharp reversal of the employment effect in the $Mb$ model is driven by our choice of a Markov process with rather little persistence. (For a comparison, see Ljungqvist, 1997, who explores Markov formulations with more persistence.)

Mortensen and Pissarides (1999a) propose still another bargaining specification where expression (19.63) is the Nash product when a worker and a firm meet for the first time, while the Nash product in expression (19.64) characterizes all their consecutive negotiations. The idea is that the firm will not incur any layoff tax if the firm and worker do not agree upon a wage in the first encounter; that is, there is never an employment relationship. In contrast, the firm's threat point is weakened in future negotiations with an already employed worker because the firm would then have to pay a layoff tax if the match were broken up. We will here show that, except for the wage profile, this alternative specification is equivalent to just assuming Nash product (19.63) for all periods. The intuition is that the modified wage profile under the Mortensen and Pissarides assumption is equivalent to a new hire posting a bond equal to his share of the future layoff tax.

First, we compute the wage associated with expression (19.63), $w_a(p)$, from the expression for a firm's match surplus,

$$J_a(p) = p - w_a(p) + \beta\Big[\xi J_a(p) + (1-\xi)\int J_a(p')\, dG(p')\Big], \qquad (19.71)$$

which together with equation (19.65) implies

$$w_a(p) = p - (1-\phi)\, S_a(p) + \beta\Big[\xi(1-\phi)\, S_a(p)$$

$$+ (1-\xi)\int (1-\phi)\, S_a(p')\, dG(p')\Big]\,. \qquad (19.72)$$

Second, we verify that the present value of these wages is exactly equal to that of Mortensen and Pissarides' bargaining scheme for any completed job, under the maintained hypothesis that the two formulations have the same match surplus $S_a(p)$. Let $J_1(p)$ and $J_+(p)$ denote the firm's match surplus with Mortensen and Pissarides' specification in the first period and all future periods, respectively. The solutions to the maximization of their Nash products are

$$J_1(p) = (1-\phi)S_a(p) \qquad \text{and} \qquad J_+(p) = (1-\phi)S_a(p) - \phi\tau\,. \qquad (19.73)$$

The associated wage functions can be written as

$$\begin{aligned} w_1(p) &= p - J_1(p) + \beta\Big[\xi J_+(p) + (1-\xi)\int J_+(p')\, dG(p')\Big] \\ &= w_a(p) - \beta\,\phi\tau\,, \\ w_+(p) &= p - J_+(p) + \beta\Big[\xi J_+(p) + (1-\xi)\int J_+(p')\, dG(p')\Big] \\ &= w_a(p) + r\,\beta\,\phi\tau\,, \end{aligned}$$

where the second equalities follow from equations (19.72) and (19.73), and $r \equiv \beta^{-1} - 1$. It can be seen that the wage under the Mortensen and Pissarides' specification is reduced in the first period by the worker's share of any future layoff tax, and future wages are increased by an amount equal to the net interest on this posted "bond." In other words, the present value of a worker's total compensation for any completed job is identical for the two specifications. It follows that the present value of a firm's match surplus is also identical across specifications. We have thereby confirmed that the same equilibrium allocation is supported by Nash product (19.63) and Mortensen and Pissarides' alternative bargaining formulation.

## *Kiyotaki-Wright search model of money*

We now explore a discrete-time version of Kiyotaki and Wright's (1993) search model of money.[5] Let us first study their environment without money. The economy is populated by a continuum of infinitely lived agents, with total population

[5] Our only essential simplification is that the time to produce is deterministic rather than stochastic, and we alter the way money is introduced into the model.

normalized to unity. There is also a number of differentiated commodities, which are indivisible and come in units of size one. Agents have idiosyncratic tastes over these consumption goods as captured by a parameter $x \in (0,1)$. In particular, $x$ equals the proportion of commodities that can be consumed by any given agent, and $x$ also equals the proportion of agents that can consume any given commodity. If a commodity can be consumed by an agent, then we say that it is one of his consumption goods. An agent derives utility $U > 0$ from consuming one of his consumption goods, while the goods that he cannot consume yield zero utility.

Initially, let each agent be endowed with one good, and let these goods be randomly drawn from the set of all commodities. Goods are costlessly storable but each agent can store at most one good at a time. The only input in the production of goods is the agents' own prior consumption. After consuming one of his consumption goods, an agent produces next period a new good drawn randomly from the set of all commodities. We assume that agents can consume neither their own output nor their initial endowment, so for consumption and production to take place there must be exchange among agents.

Agents meet pairwise and at random. In each period, an agent meets another agent with probability $\theta \in (0,1]$ and he has no encounter with probability $1-\theta$. Two agents who meet will trade if there is a mutually agreeable transaction. Any transaction must be quid pro quo because private credit arrangements are ruled out by the assumptions of a random matching technology and a continuum of agents. We also assume that there is a transaction cost $\epsilon \in (0,U)$ in terms of disutility, which is incurred whenever accepting a commodity in trade. Thus, a trader who is indifferent between holding two goods will never trade one for the other.

Agents choose trading strategies in order to maximize their expected discounted utility from consumption net of transaction costs, taking as given the strategies of other traders. Following Kiyotaki and Wright (1993), we restrict our attention to symmetric Nash equilibria, where all agents follow the same strategies and all goods are treated the same, and to steady states, where strategies and aggregate variables are constant over time.

In a symmetric equilibrium, an agent will only trade if he is offered a commodity that belongs to his set of consumption goods and then consume it immediately. Accepting a commodity that is not one's consumption good would only give rise to a transaction cost $\epsilon$ without affecting expected future trading opportunities. This statement is true because no commodities are treated as special in a symmetric equilibrium, and therefore the probability of a commodity being accepted by the next agent one meets is independent of the type of commodity one has.[6] It follows that $x$ is the probability that a trader located at random is willing to accept any given commodity, and $x^2$ becomes the probability that two traders consummate a barter in a situation of "double coincidence of wants."

At the beginning of a period before the realization of the matching process, the value of an agent's optimization problem becomes

$$V_c^n = \theta x^2 (U - \epsilon) + \beta V_c^n ,$$

where $\beta \in (0,1)$ is the discount factor. The superscript and subscript of $V_c^n$ denote a nonmonetary equilibrium and a commodity trader, respectively, to set the stage for our next exploration of the role for money in this economy. How will fiat money affect welfare? Keep the benchmark of a barter economy in mind,

$$V_c^n = \frac{\theta x^2 (U - \epsilon)}{1 - \beta} . \tag{19.74}$$

*Monetary equilibria*

At the beginning of time, suppose a fraction $\bar{M} \in [0,1)$ of all agents are each offered one unit of fiat money. The money is indivisible, and an agent can store at most one unit of money or one commodity at a time. That is, fiat money will enter into circulation only if some agents accept money and discard their endowment of goods. These decisions must be based solely on agents' beliefs about other traders' willingness to accept money in future transactions because fiat money is by definition unbacked and intrinsically worthless. To determine whether or not fiat money will initially be accepted, we will therefore first have to characterize monetary equilibria.[7]

Fiat money adds two state variables in a symmetric steady state: the probability that a commodity trader accepts money, $\Pi \in [0,1]$, and the amount of money circulating, $M \in [0, \bar{M}]$, which is also the fraction of all agents carrying money. An equilibrium pair $(\Pi, M)$ must be such that an individual's choice of probability of accepting money when being a commodity trader, $\pi$, coincides with the

[6] Kiyotaki and Wright (1989) analyze commodity money in a related model—nonsymmetric equilibria where some goods become media of exchange.

[7] If money is valued in an equilibrium, the relative price of goods and money is trivially equal to one, since both objects are indivisible and each agent can carry at most one unit of the objects. Shi (1995) and Trejos and Wright (1995) endogenize the price level by relaxing the assumption that goods are indivisible.

economy-wide $\Pi$, and the amount of money $M$ is consistent with the decisions of those agents who are initially free to replace their commodity endowment with fiat money.

In a monetary equilibrium, agents can be divided into two types of traders. An agent brings either a commodity or a unit of fiat money to the trading process; that is, he is either a commodity trader or a money trader. At the beginning of a period, the values associated with being a commodity trader and a money trader are denoted $V_c$ and $V_m$, respectively. The Bellman equations can be written

$$V_c = \theta(1-M)x^2\big(U-\epsilon+\beta V_c\big) + \theta M x \max_{\pi}\big[\pi\beta V_m + (1-\pi)\beta V_c\big] + \big[1 - \theta(1-M)x^2 - \theta M x\big]\beta V_c\,, \tag{19.75}$$

$$V_m = \theta(1-M)x\Pi\big(U-\epsilon+\beta V_c\big) + \big[1 - \theta(1-M)x\Pi\big]\beta V_m\,. \tag{19.76}$$

The value of being a commodity trader in equation (19.75) equals the sum of three terms. The first term is the probability of the agent meeting other commodity traders, $\theta(1-M)$, times the probability that both want to trade, $x^2$, times the value of trading, consuming, and returning as a commodity trader next period, $U-\epsilon+\beta V_c$. The second term is the probability of the agent meeting money traders, $\theta M$, times the probability that a money trader wants to trade, $x$, times the value of accepting money with probability $\pi$, $\pi\beta V_m + (1-\pi)\beta V_c$, where $\pi$ is chosen optimally. The third term captures the complement to the two previous events when the agent stores his commodity to the next period with a continuation value of $\beta V_c$. According to equation (19.76), the value of being a money trader equals the sum of two terms. The first term is the probability of the agent meeting a commodity trader, $\theta(1-M)$, times the probability of both wanting to trade, $x\Pi$, times the value of trading, consuming, and becoming a commodity trader next period, $U-\epsilon+\beta V_c$. The second term is the probability of the described event not occurring times the value of keeping the unit of fiat money to the next period, $\beta V_m$.

The optimal choice of $\pi$ depends solely on $\Pi$. First, note that if $\Pi < x$ then equations (19.75) and (19.76) imply that $V_m < V_c$, so the individual's best response is $\pi = 0$. That is, if money is being accepted with a lower probability than a barter offer, then it is harder to trade using money than barter, so agents would never like to exchange a commodity for money. Second, if $\Pi > x$, then equations (19.75) and (19.76) imply that $V_m > V_c$, so the individual's best response is $\pi = 1$. If money is being accepted with a greater probability than a barter offer, then it is easier to trade using money than barter, and agents would

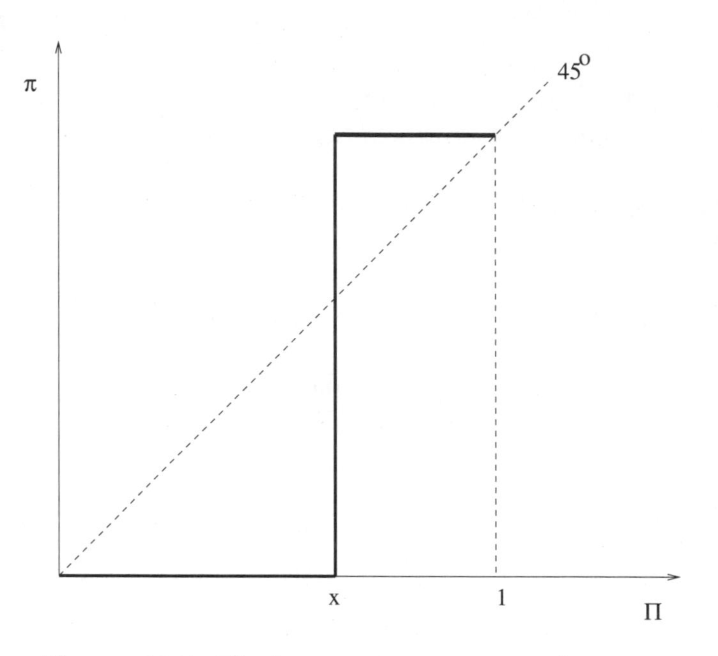


**Figure 19.8.** The best-response correspondence.

always like to exchange a commodity for money whenever possible. Finally, if $\Pi = x$, then equations (19.75) and (19.76) imply that $V_m = V_c$, so $\pi$ can be anything in $[0, 1]$. If monetary exchange and barter are equally easy then traders are indifferent between carrying commodities and fiat money, and they could accept money with any probability. Based on these results, the individual's best-response correspondence is as shown in Figure 19.8, and there are exactly three values consistent with $\Pi = \pi$: $\Pi = 0$, $\Pi = 1$, and $\Pi = x$.

We can now answer our very first question, How many of the agents who are initially free to exchange their commodity endowment for fiat money will choose to do so? The answer is already implicit in our discussion of the best-response correspondence. Thus, we have the following three types of symmetric equilibria:

a. A nonmonetary equilibrium with $\Pi = 0$ and $M = 0$, which is identical to the barter outcome in the previous section: Agents expect that money will be valueless, so they never accept it, and this expectation is self-fulfilling. All agents become commodity traders associated with a value of $V_c^n$, as given by equation (19.74).

b. A pure monetary equilibrium with $\Pi = 1$ and $M = \bar{M}$: Agents expect that money will be universally acceptable. From our previous discussion we know that agents will then prefer to bring money rather than commodities to the trading process. It is therefore a dominant strategy to accept money whenever possible; that is, expectation is self-fulfilling. Another implication is that the fraction $\bar{M}$ of agents who are initially free to exchange their commodity endowment for fiat money will also do so. Let $V_c^p$ and $V_m^p$ denote the values associated with being a commodity trader and a money trader, respectively, in a pure monetary equilibrium.

c. A mixed monetary equilibrium with $\Pi = x$ and $M \in [0, \bar{M}]$: Traders are indifferent between accepting and rejecting money as long as future trading partners take it with probability $\Pi = x$, so partial acceptability with agents setting $\pi = x$ can also be self-fulfilling. However, a mixed monetary equilibrium has no longer a unique mapping to the amount of circulating money $M$. Suppose the initial choices between commodity endowment and fiat money are separate from agents' decisions on trading strategies. It follows that any amount of money between $[0, \bar{M}]$ can constitute a mixed monetary equilibrium because of the indifference between a commodity endowment and a unit of fiat money. Of course, the allocation in a mixed-monetary equilibrium with $M = 0$ is identical to the one in a nonmonetary equilibrium. Let $V_c^i(M)$ and $V_m^i(M)$ denote the values associated with being a commodity trader and a money trader, respectively, in a mixed monetary equilibrium with an amount of money equal to $M \in [0, \bar{M}]$.

*Welfare*

To compare welfare across different equilibria, we set $\pi = \Pi$ in equations (19.75) and (19.76) and solve for the reduced-form expressions

$$V_c = \frac{\psi}{1-\beta}\{(1-\beta)x + \beta\theta x\Pi[M\Pi + (1-M)x]\}, \tag{19.77}$$

$$V_m = \frac{\psi}{1-\beta}\{(1-\beta)\Pi + \beta\theta x\Pi[M\Pi + (1-M)x]\}, \tag{19.78}$$

where $\psi = [\theta(1-M)x(U-\epsilon)]/[1-\beta(1-\theta x\Pi)] > 0$. The value $V_m$ is greater than or equal to $V_c$ in a monetary equilibrium, since a necessary condition is that monetary exchange is at least as easy as barter ($\Pi \geq x$),

$$V_m = V_c + \psi(\Pi - x).$$

After setting $\Pi = x$ in equations (19.77) and (19.78), we see that a mixed monetary equilibrium with $M > 0$ gives rise to a strictly lower welfare as compared to the barter outcome in equation (19.74),

$$V_c^i(M) \;=\; V_m^i(M) \;=\; (1-M)V_c^n \,.$$

Even though some agents are initially willing to switch their commodity endowment for fiat money, it is detrimental for the economy as a whole. Since money is accepted with the same probability as commodities, money does not ameliorate the problem of "double coincidence of wants" but only diverts real resources from the economy.[8] In fact, as noted by Kiyotaki and Wright (1990), the mixed monetary equilibrium is isomorphic to the nonmonetary equilibrium of another economy where the probability of meeting an agent is reduced from $\theta$ to $\theta(1-M)$.

In a pure monetary equilibrium ($\Pi = 1$), the value of being a money trader is strictly greater than the value of being a commodity trader. A natural welfare criterion is the ex ante expected utility before the quantity $\bar{M}$ of fiat money is randomly distributed,

$$\begin{aligned} W \;&=\; \bar{M}V_m^p + (1-\bar{M})V_c^p \\ &=\; \frac{\theta(1-\bar{M})x(U-\epsilon)}{1-\beta}\left[\bar{M} + (1-\bar{M})x\right]. \end{aligned} \tag{19.79}$$

The first and second derivatives of equation (19.79) are

$$\frac{\partial W}{\partial \bar{M}} \;=\; \frac{\theta x(U-\epsilon)}{1-\beta}\Big\{1 \;-\; 2\big[\bar{M} + (1-\bar{M})x\big]\Big\}, \tag{19.80}$$

$$\frac{\partial^2 W}{\partial \bar{M}^2} \;=\; -\,2\,\frac{\theta x(U-\epsilon)}{1-\beta}\,(1-x) \;<\; 0\,. \tag{19.81}$$

[8] This welfare result differs from that of Kiyotaki and Wright (1993), who assume that a fraction $\bar{M}$ of all agents are initially endowed with fiat money without any choice. It follows that those agents endowed with money are certainly better off in a mixed monetary equilibrium as compared to the barter outcome, while the other agents are indifferent. The latter agents are indifferent because the existence of the former agents has the same crowding-out effect on their consumption arrival rate in both types of equilibria. Our welfare results reported here are instead in line with Kiyotaki and Wright's (1990) original working paper based on a slightly different environment where agents can at any time dispose of their fiat money and engage in production.

Since the second derivative is negative, fiat money can only have a welfare-enhancing role if the first derivative is positive when evaluated at $\bar{M} = 0$. Thus, according to equation (19.80), money can (cannot) increase welfare if $x < 0.5$ ($x \geq 0.5$). Intuitively speaking, when $x \geq 0.5$, each agent is willing to consume (and therefore accept) at least half of all commodities, so barter is not very difficult. The introduction of money would here only reduce welfare by diverting real resources from the economy. When $x < 0.5$, barter is sufficiently difficult so that the introduction of some fiat money improves welfare. The optimum quantity of money is then found by setting equation (19.80) equal to zero, $\bar{M}^\star = (1 - 2x)/(2 - 2x)$. That is, $\bar{M}^\star$ varies negatively with $x$, and the optimum quantity of money increases when $x$ shrinks and the problem of "double coincidence of wants" becomes more difficult. In particular, $\bar{M}^\star$ converges to 0.5 when $x$ goes to zero.

## Concluding comments

The frameworks of search and matching present various ways of departing from the frictionless Arrow-Debreu economy where all agents meet in a complete set of markets. This chapter has mainly focused on labor markets as a central application of these theories. The presented models have the concept of frictions in common, but there are also differences. The island economy has frictional unemployment without any externalities. An unemployed worker does not inflict any injury on other job seekers other than what a seller of a good imposes on his competitors. The equilibrium value to search, $v_u$, serves the function of any other equilibrium price of signaling to suppliers the correct social return from an additional unit supplied. In contrast, the matching model with its matching function is associated with externalities. Workers and firms impose congestion effects when they enter as unemployed in the matching function or add another vacancy in the matching function. To arrive at an efficient allocation in the economy, it is necessary that the bilaterally bargained wage be exactly right. In a labor market with homogeneous firms and workers, efficiency prevails only if the workers' bargaining strength, $\phi$, is exactly equal to the elasticity of the matching function with respect to the measure of unemployment, $\alpha$. In the case of heterogeneous jobs in the same labor market with a single matching function, we established the impossibility of efficiency without government intervention.

The matching model offers unarguably a richer analysis through its extra interaction effects, but it comes at the cost of the model's microeconomic structure. In an explicit economic environment, feasible actions can be clearly envisioned for any population size, even if there is only one Robinson Crusoe. The island economy is an example of such a model with its microeconomic assumptions, such as the time it takes to move from one island to another. In contrast, the matching model with its matching function imposes relationships between aggregate outcomes. It is therefore not obvious how the matching function arises when gradually increasing the population from one Robinson Crusoe to an economy with more agents. Similarly, it is an open question what determines when heterogeneous firms and labor have to be matched through a common matching function and when they have access to separate matching functions.

Peters (1991) and Montgomery (1991) suggest some microeconomic underpinnings to labor market frictions, which are further pursued by Burdett, Shi and Wright (2000). Firms post vacancies with announced wages, and unemployed workers can only apply to one firm at a time. If the values of filled jobs differ across firms, firms with more valued jobs will have an incentive to post higher wages to attract job applicants. In an equilibrium, workers will be indifferent between applying to different jobs, and they are assumed to use identical mixed strategies in making their applications. In this way, vacancies may remain unfilled because some firms do not receive any applicants, and some workers may find themselves "second in line" for a job and therefore remain unemployed. When assuming a large number of firms that take market tightness as given for each posted wage, Montgomery finds that the decentralized equilibrium does maximize welfare for reasons similar to Moen's (1997) identical finding that was discussed earlier in this chapter.

Lagos (2000) derives a matching function from a model without any exogenous frictions at all. He studies a dynamic market for taxicab rides in which taxicabs seek potential passengers on a spatial grid and the fares are regulated exogenously. In each location, the shorter side determines the number of matches. It is shown that a matching function exists for this model, but this matching function is an equilibrium object that changes with policy experiments. Lagos sounds a warning that assuming an exogenous matching function when doing policy analysis might be misleading.

Throughout our discussion of search and matching models we have assumed risk-neutral agents. Gomes, Greenwood, and Rebelo (1997) and Acemoglu and Shimer (1999) analyze a search model and a matching model, respectively, where agents are risk averse and hold precautionary savings because of imperfect insurance against unemployment.

# Exercises

*Exercise 19.1* **An island economy** (Lucas and Prescott, 1974)

Let the island economy in this chapter have a productivity shock that takes on two possible values, $\{\theta_L, \theta_H\}$ with $0 < \theta_L < \theta_H$. An island's productivity remains constant from one period to another with probability $\pi \in (.5, 1)$, and its productivity changes to the other possible value with probability $1 - \pi$. These symmetric transition probabilities imply a stationary distribution where half of the islands experience a given $\theta$ at any point in time. Let $\hat{x}$ be the economy's labor supply (as an average per market).

**a.** If there exists a stationary equilibrium with labor movements, argue that an island's labor force has two possible values, $\{x_1, x_2\}$ with $0 < x_1 < x_2$.

**b.** In a stationary equilibrium with labor movements, construct a matrix $\Gamma$ with the transition probabilities between states $(\theta, x)$, and explain what the employment level is in different states.

**c.** In a stationary equilibrium with labor movements, we observe only four values of the value function $v(\theta, x)$ where $\theta \in \{\theta_L, \theta_H\}$ and $x \in \{x_1, x_2\}$. Argue that the value function takes on the same value for two of these four states.

**d.** Show that the condition for the existence of a stationary equilibrium with labor movements is

$$\beta(2\pi - 1)\theta_H > \theta_L , \tag{19.82}$$

and, if this condition is satisfied, an implicit expression for the equilibrium value of $x_2$ is

$$[\theta_L + \beta(1 - \pi)\theta_H] f'(2\hat{x} - x_2) = \beta\pi\theta_H f'(x_2) . \tag{19.83}$$

**e.** Verify that the allocation of labor in part d coincides with a social planner's solution when maximizing the present value of the economy's aggregate production. Starting from an initial equal distribution of workers across islands, condition (19.82) indicates when it is optimal for the social planner to increase the number of workers on high-productivity islands. The first-order condition for the social planner's choice of $x_2$ is then given by equation (19.83).
{*Hint*: Consider an employment plan $(x_1, x_2)$ such that the next period's labor force is $x_1$ ($x_2$) for an island currently experiencing productivity shock $\theta_L$ ($\theta_H$).

If $x_1 \leq x_2$, the present value of the economy's production (as an island average) becomes

$$0.5 \sum_{t=0}^{\infty} \beta^t \left[ \theta_L f(2\hat{x} - x_2) + (1-\pi)\theta_H f(2\hat{x} - x_2) + \pi\theta_H f(x_2) \right] .$$

Examine the effect of a once-and-for-all increase in the number of workers allocated to high-productivity islands.}

*Exercise 19.2* **Business cycles and search** (Gomes, Greenwood, and Rebelo, 1997)

**Part 1** *The worker's problem*

Think about an economy in which workers all confront the following common environment: Time is discrete. Let $t = 0, 1, 2, \ldots$ index time. At the beginning of each period, a previously employed worker can choose to work at her last period's wage or to draw a new wage. If she draws a new wage, the old wage is lost and she can start working at the new wage in the following period. New wages are independent and identically distributed from the cumulative distribution function $F$, where $F(0) = 0$, and $F(M) = 1$ for $M < \infty$. Unemployed workers face a similar problem. At the beginning of each period, a previously unemployed worker can choose to work at last period's wage offer or to draw a new wage from $F$. If she draws a new wage, the old wage offer is lost and she can start working at the new wage in the following period. Someone offered a wage is free to work at that wage for as long as she chooses (she cannot be fired). The income of an unemployed worker is $b$, which includes unemployment insurance and the value of home production. Each worker seeks to maximize $E_0 \sum_{t=0}^{\infty} (1-\mu)^t \beta^t I_t$, where $\mu$ is the probability that a worker dies at the end of a period, $\beta$ is the subjective discount factor, and $I_t$ is the worker's income in period $t$; that is, $I_t$ is equal to the wage $w_t$ when employed and the income $b$ when unemployed. Here $E_0$ is the mathematical expectation operator, conditioned on information known at time 0. Assume that $\beta \in (0,1)$ and $\mu \in (0,1)$.

**a.** Describe the worker's optimal decision rule. In particular, what should an employed worker do? What should an unemployed worker do?

**b.** How would an unemployed worker's behavior be affected by an increase in $\mu$?

**Part 2** *Equilibrium unemployment rate*

The economy is populated with a continuum of the workers just described. There is an exogenous rate of new workers entering the labor market equal to $\mu$, which equals the death rate. New entrants are unemployed and must draw a new wage.

**c.** Find an expression for the economy's unemployment rate in terms of exogenous parameters and the endogenous reservation wage. Discuss the determinants of the unemployment rate.

We now change the technology so that the economy fluctuates between booms $(B)$ and recessions $(R)$. In a boom, all employed workers are paid an extra $z > 0$. That is, the income of a worker with wage $w$ is $I_t = w + z$ in a boom and $I_t = w$ in a recession. Let whether the economy is in a boom or a recession define the *state* of the economy. Assume that the state of the economy is i.i.d. and that booms and recessions have the same probabilities of $0.5$. The state of the economy is publicly known at the beginning of a period before any decisions are made.

**d.** Describe the optimal behavior of employed and unemployed workers. When, if ever, might workers choose to quit?

**e.** Let $w_B$ and $w_R$ be the reservation wages in booms and recessions, respectively. Assume that $w_B < w_R$. Let $G_t$ be the fraction of workers employed at wages $w \in [w_B, w_R]$ in period $t$. Let $U_t$ be the fraction of workers unemployed in period $t$. Derive difference equations for $G_t$ and $U_t$ in terms of the parameters of the model and the reservation wages, $\{F, \mu, w_B, w_R\}$.

**f.** The following time series is a simulation from the solution of the model with booms and recessions. Interpret the time series in terms of the model.

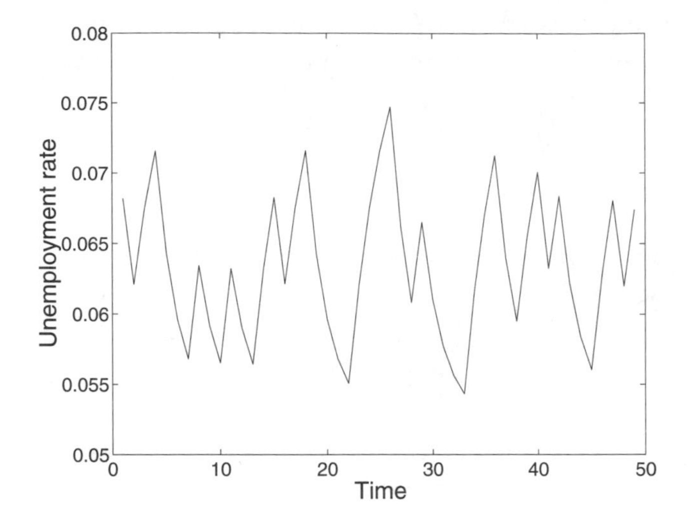


*Exercise 19.3* **Business cycles and search again**

The economy is either in a boom ($B$) or recession ($R$) with probability .5. The state of the economy ($R$ or $B$) is i.i.d. through time. At the beginning of each period, workers know the state of the economy for that period. At the beginning of each period, a previously employed worker can choose to work at her last period's wage or draw a new wage. If she draws a new wage, the old wage is lost, $b$ is received this period, and she can start working at the new wage in the following period. During recessions, new wages (for jobs to start next period) are i.i.d. draws from the c.d.f. $F$, where $F(0) = 0$ and $F(M) = 1$ for $M < \infty$. During booms, the worker can choose to quit and take *two* i.i.d. draws of a possible new wage (with the option of working at the higher wage, again for a job to start the next period) from the *same* c.d.f. $F$ that prevails during recessions. (This ability to choose is what "Jobs are more plentiful during booms" means to workers.) Workers who are unemployed at the beginning of a period receive $b$ this period and draw either one (in recessions) or two (in booms) wages offers from the c.d.f. $F$ to start work next period.

A worker seeks to maximize $E_0 \sum_{t=0}^{\infty} (1-\mu)^t \beta^t I_t$, where $\mu$ is the probability that a worker dies at the end of a period, $\beta$ is the subjective discount factor, and

$I_t$ is the worker's income in period $t$; that is, $I_t$ is equal to the wage $w_t$ when employed and the income $b$ when unemployed.

**a.** Write the Bellman equation(s) for a previously employed worker.

**b.** Characterize the worker's quitting policy. If possible, compare reservation wages in booms and recessions. Will employed workers ever quit? If so, who will quit and when?

*Exercises 19.4–19.6* **European unemployment**

The following three exercises are based on work by Ljungqvist and Sargent (1998), Marimon and Zilibotti (1999), and Mortensen and Pissarides (1999b), who calibrate versions of search and matching models to explain high European unemployment. Even though the specific mechanisms differ, they all attribute the rise in unemployment to generous benefits in times of more dispersed labor market outcomes for job seekers.

*Exercise 19.4* **Skill-biased technological change** (Mortensen and Pissarides, 1999b)

Consider a matching model in discrete time with infinitely lived and risk-neutral workers who are endowed with different skill levels. A worker of skill type $i$ produces $h_i$ goods in each period that she is matched to a firm, where $i \in \{1, 2, \ldots, N\}$ and $h_{i+1} > h_i$. Each skill type has its own but identical matching function $M(u_i, v_i) = A u_i^{\alpha} v_i^{1-\alpha}$, where $u_i$ and $v_i$ are the measures of unemployed workers and vacancies in skill market $i$. Firms incur a vacancy cost $ch_i$ in every period that a vacancy is posted in skill market $i$; that is, the vacancy cost is proportional to the worker's productivity. All matches are exogenously destroyed with probability $s \in (0, 1)$ at the beginning of a period. An unemployed worker receives unemployment compensation $b$. Wages are determined in Nash bargaining between matched firms and workers. Let $\phi \in [0, 1)$ denote the worker's bargaining weight in the Nash product, and we adopt the standard assumption that $\phi = \alpha$.

**a.** Show analytically how the unemployment rate in a skill market varies with the skill level $h_i$.

**b.** Assume an even distribution of workers across skill levels. For different benefit levels $b$, study numerically how the aggregate steady-state unemployment rate is affected by mean-preserving spreads in the distribution of skill levels.

**c.** Explain how the results would change if unemployment benefits are proportional to a worker's productivity.

*Exercise 19.5* **Dispersion of match values** (Marimon and Zilibotti, 1999)

We retain the matching framework of exercise 19.4 but assume that all workers have the same innate ability $h = \bar{h}$ and any earnings differentials are purely match specific. In particular, we assume that the meeting of a firm and a worker is associated with a random draw of a match-specific productivity $p$ from an exogenous distribution $G(p)$. If the worker and firm agree upon staying together, the output of the match is then $p \cdot h$ in every period as long as the match is not exogenously destroyed as in exercise 19.4. We also keep the assumptions of a constant unemployment compensation $b$ and Nash bargaining over wages.

**a.** Characterize the equilibrium of the model.

**b.** For different benefit levels $b$, study numerically how the steady-state unemployment rate is affected by mean-preserving spreads in the exogenous distribution $G(p)$.

*Exercise 19.6* **Idiosyncratic shocks to human capital** (Ljungqvist and Sargent, 1998)

We retain the assumption of exercise 19.5 that a worker's output is the product of his human capital $h$ and a job-specific component which we now denote $w$, but we replace the matching framework with a search model. In each period of unemployment, a worker draws a value $w$ from an exogenous wage offer distribution $G(w)$ and, if the worker accepts the wage $w$, he starts working in the following period. The wage $w$ remains constant throughout the employment spell that ends either because the worker quits or the job is exogenously destroyed with probability $s$ at the beginning of each period. Thus, in a given job with wage $w$, a worker's earnings $wh$ can only vary over time because of changes in human capital $h$. For simplicity, we assume that there are only two levels of human capital, $h_1$ and $h_2$ where $0 < h_1 < h_2 < \infty$. At the beginning of each period of employment, a worker's human capital is unchanged from last period with probability $\pi_e$ and is equal to $h_2$ with probability $1 - \pi_e$. Losses of human capital are only triggered by exogenous job destruction. In the period of an exogenous job loss, the laid off worker's human capital is unchanged from last period with probability $\pi_u$ and is equal to $h_1$ with probability $1 - \pi_u$. All unemployed workers receive unemployment compensation, and the benefits are equal to a replacement ratio $\gamma \in [0, 1)$ times a worker's last job earnings.

**a.** Characterize the equilibrium of the model.

**b.** For different replacement ratios $\gamma$, study numerically how the steady-state unemployment rate is affected by changes in $h_1$.

*Comparison of models*

**c.** Explain how the different models in exercises 19.4–19.6 address the observations that European welfare states have experienced less of an increase in earnings differentials as compared to the United States, but suffer more from long-term unemployment where the probability of gaining employment drops off sharply with the length of the unemployment spell.

**d.** Explain why the assumption of infinitely lived agents is innocuous for the models in exercises 19.4 and 19.5, but the alternative assumption of finitely lived agents can make a large difference for the model in exercise 19.6.

*Exercise 19.7* **Temporary jobs and layoff costs**

Consider a search model with temporary jobs. In each period, a previously employed worker loses her job with probability $\mu$, and she can keep her job and wage rate from last period with probability $1-\mu$. If she loses her job (or chooses to quit), she draws a new wage and can start working at the new wage in the following period with probability one. After a first period on the new job, she will again in each period face probability $\mu$ of losing her job. New wages are independent and identically distributed from the cumulative distribution function $F$, where $F(0) = 0$, and $F(M) = 1$ for $M < \infty$. The situation during unemployment is as follows. At the beginning of each period, a previously unemployed worker can choose to start working at last period's wage offer or to draw a new wage from $F$. If she draws a new wage, the old wage offer is lost and she can start working at the new wage in the following period. The income of an unemployed worker is $b$, which includes unemployment insurance and the value of home production. Each worker seeks to maximize $E_0 \sum_{t=0}^{\infty} \beta^t I_t$, where $\beta$ is the subjective discount factor, and $I_t$ is the worker's income in period $t$; that is, $I_t$ is equal to the wage $w_t$ when employed and the income $b$ when unemployed. Here $E_0$ is the mathematical expectation operator, conditioned on information known at time $0$. Assume that $\beta \in (0,1)$ and $\mu \in (0,1]$.

**a.** Describe the worker's optimal decision rule.

Suppose that there are two types of temporary jobs: short-lasting jobs with $\mu_s$ and long-lasting jobs with $\mu_l$, where $\mu_s > \mu_l$. When the worker draws a new wage from the distribution $F$, the job is now randomly designated as either short-lasting with probability $\pi_s$ or long-lasting with probability $\pi_l$, where $\pi_s + \pi_l = 1$. The worker observes the characteristics of a job offer, $(w, \mu)$.

**b.** Does the worker's reservation wage depend on whether a job is short-lasting or long-lasting? Provide intuition for your answer.

We now consider the effects of layoff costs. It is assumed that the government imposes a cost $\tau > 0$ on each worker that loses a job (or quits).

**c.** Conceptually, consider the following two reservation wages, for a given value of $\mu$: (i) a previously unemployed worker sets a reservation wage for accepting last period's wage offer; (ii) a previously employed worker sets a reservation wage for continuing working at last period's wage. For a given value of $\mu$, compare these two reservation wages.

**d.** Show that an unemployed worker's reservation wage for a short-lasting job exceeds her reservation wage for a long-lasting job.

**e.** Let $\bar{w}_s$ and $\bar{w}_l$ be the reservation wages for short-lasting jobs and long-lasting jobs, respectively. In period $t$, let $N_{st}$ and and $N_{lt}$ be the fractions of workers employed in short-lasting jobs and long-lasting jobs, respectively. Let $U_t$ be the fraction of workers unemployed in period $t$. Derive difference equations for $E_{st}$, $E_{lt}$ and $U_t$ in terms of the parameters of the model and the reservation wages, $\{F, \mu_s, \mu_l, \pi_s, \pi_l, \bar{w}_s, \bar{w}_l\}$.

*Exercise 19.8* **Productivity shocks, job creation, and job destruction**, donated by Rodolfo Manuelli

Consider an economy populated by a large number of identical individuals. The utility function of each individual is

$$\sum_{t=0}^{\infty} \beta^t x_t,$$

where $0 < \beta < 1$, $\beta = 1/(1+r)$, and $x_t$ is income at time $t$. All individuals are endowed with one unit of labor that is supplied inelastically: if the individual is working in the market, its productivity is $y_t$, while if he/she works at home productivity is $z$. Assume that $z < y_t$. Individuals who are producing at home can also — at no cost — search for a market job. Individuals who are searching and jobs that are vacant get randomly matched. Assume that the number of matches per period is given by

$$M(u_t, x_t),$$

where $M$ is concave, increasing in each argument, and homogeneous of degree one. In this setting, $u_t$ is interpreted as the total number of unemployed workers, and $v_t$ is the total number of vacancies. Let $\theta \equiv v/u$, and let $q(\theta) = M(u,v)/v$ be the probability that a vacant job (or firm) will meet a worker. Similarly, let $\theta q(\theta) = M(u,v)/u$ be the probability that an unemployed worker is matched with a vacant job. Jobs are exogenously destroyed with probability $s$. In order to create a vacancy a firm must pay a cost $c > 0$ per period in which the vacancy is "posted" (i.e., unfilled). There is a large number of potential firms (or jobs) and this guarantees that the expected value of a vacant job, $V$, is zero. Finally, assume that, when a worker and a vacant job meet, they bargain according to the Nash Bargaining solution, with the workers' share equal to $\varphi$. Assume that $y_t = y$ for all $t$.

**a.** Show that the zero profit condition implies that,

$$w = y - (r+s)c/q(\theta).$$

**b.** Show that if workers and firms negotiate wages according to the Nash Bargaining solution (with worker's share equal to $\varphi$), wages must also satisfy

$$w = z + \varphi(y - z + \theta c).$$

**c.** Describe the determination of the equilibrium level of market tightness, $\theta$.

**d.** Suppose that at $t = 0$, the economy is at its steady state. At this point, there is a once and for all permanent increase in productivity. The new value of $y$ is $y' > y$. Show how the new steady state value of $\theta$, $\theta'$, compares with the previous value. Argue that the economy "jumps" to the new value right away. Explain why there are no "transitional dynamics" for the level of market tightness, $\theta$.

**e.** Let $u_t$ be the unemployment rate at time $t$. Assume that at time 0 the economy is at the steady state unemployment rate corresponding to $\theta$ — the "old" market tightness — and display this rate. Denote this rate as $u_0$. Let

$\theta_0 = \theta'$. Note that that change in unemployment rate is equal to the difference between Job Destruction at $t$, $JD_t$ and Job Creation at $t$, $JC_t$. It follows that

$$\begin{aligned} JD_t &= (1-u_t)s, \\ JC_t &= \theta_t q(\theta_t) u_t, \\ u_{t+1} - u_t &= JD_t - JC_t. \end{aligned}$$

Go as far as you can characterizing job creation and job destruction at $t = 0$ (after the shock). In addition, go as far as you can describing the behavior of both $JC_t$ and $JD_t$ during the transition to the new steady state (the one corresponding to $\theta'$).

interpreted as the total number of unemployed workers, and $\sum_{j=1}^I v^j$ is the total number of vacancies. Let $\theta \sum_{j=1}^I v^j/u$, and let $q(\theta) = M(u, \sum_{j=1}^I v^j)/(\sum_{j=1}^I v^j)$.

*Exercise 19.9* **Workweek restrictions, unemployment, and welfare**, donated by Rodolfo Manuelli

Recently, France has moved to a shorter workweek of about 35 hours per week. In this exercise you are asked to evaluate the consequences of such a move. To this end, consider an economy populated by risk-neutral, income-maximizing workers with preferences given by

$$U = E_t \sum_{j=0}^{\infty} \beta^j y_{t+j}, \quad 0 < \beta < 1, \quad 1 + r = \beta^{-1}.$$

Assume that workers produce $z$ at home if they are unemployed, and that they are endowed with one unit of labor. If a worker is employed, he/she can spend $x$ units of time at the job, and $(1-x)$ at home, with $0 \leq x \leq 1$. Productivity on the job is $yx$, and $x$ is perfectly observed by both workers and firms.

Assume that if a worker works $x$ hours, his/her wage is $wx$.

Assume that all jobs have productivity $y > z$, and that to create a vacancy firms have to pay a cost of $c > 0$ units of output per period. Jobs are destroyed with probability $s$. Let the number of matches per period be given by

$$M(u, v),$$

where $M$ is concave, increasing in each argument, and homogeneous of degree one. In this setting, $u$ is interpreted as the total number of unemployed workers, and $v$ is the total number of vacancies. Let $\theta \equiv v/u$, and let $q(\theta) = M(u,v)/v$.

Assume that workers and firms bargain over wages, and that the outcome is described by a Nash Bargaining outcome with the workers' bargaining power equal to $\varphi$.

**a.** Go as far as you can describing the unconstrained (no restrictions on $x$ other than it be a number between zero and one) market equilibrium.

**b.** Assume that $q(\theta) = A\theta^{-\alpha}$, for some $0 < \alpha < 1$. Does the solution of the planner's problem coincide with the market equilibrium?

**c.** Assume now that the workweek is restricted to be less than or equal to $x^* < 1$. Describe the equilibrium.

**d.** For the economy in part c go as far as you can (if necessary make additional assumptions) describing the impact of this workweek restriction on wages, unemployment rates, and the total number of jobs. Is the equilibrium optimal?

*Exercise 19.10* **Costs of creating a vacancy and optimality**, donated by Rodolfo Manuelli

Consider an economy populated by risk-neutral, income-maximizing workers with preferences given by

$$U = E_t \sum_{j=0}^{\infty} \beta^j y_{t+j}, \quad 0 < \beta < 1, \quad 1 + r = \beta^{-1}.$$

Assume that workers produce $z$ at home if they are unemployed. Assume that all jobs have productivity $y > z$, and that to create a vacancy firms have to pay $p_A$, with $p_A = C'(v)$, per period when they have an open vacancy, with $v$ being the total number of vacancies. Assume that the function $C(v)$ is strictly convex, twice differentiable and increasing. Jobs are destroyed with probability $s$.

Let the number of matches per period be given by

$$M(u, v),$$

where $M$ is concave, increasing in each argument, and homogeneous of degree one. In this setting, $u$ is interpreted as the total number of unemployed workers, and $v$ is the total number of vacancies. Let $\theta \equiv v/u$, and let $q(\theta) = M(u,v)/v$.

Assume that workers and firms bargain over wages, and that the outcome is described by a Nash Bargaining outcome with the workers' bargaining power equal to $\varphi$.

**a.** Go as far as you can describing the market equilibrium. In particular, discuss how changes in the exogenous variables, $z$, $y$ and the function $C(v)$, affect the equilibrium outcomes.

**b.** Assume that $q(\theta) = A\theta^{-\alpha}$, for some $0 < \alpha < 1$. Does the solution of the planner's problem coincide with the market equilibrium? Describe instances, if any, in which this is the case.

*Exercise 19.11* **Financial wealth, heterogeneity, and unemployment**, donated by Rodolfo Manuelli

Consider the behavior of a risk-neutral worker that seeks to maximize the expected present discounted value of wage income. Assume that the discount factor is fixed and equal to $\beta$, with $0 < \beta < 1$. The interest rate is also constant and satisfies $1 + r = \beta^{-1}$. In this economy, jobs last forever. Once the worker has accepted a job, he/she never quits and the job is never destroyed. Even though preferences are linear, a worker needs to consume a minimum of $a$ units of consumption per period. Wages are drawn from a distribution with support on $[a, b]$. Thus, any employed individual can have a feasible consumption level. There is no unemployment compensation.

Individuals of type $i$ are born with wealth $a^i$, $i = 0, 1, 2$, where $a^0 = 0$, $a^1 = a$, $a^2 = a(1 + \beta)$. Moreover, in the period that they are born, all individuals are unemployed. Population, $N_t$, grows at the constant rate $1 + n$. Thus, $N_{t+1} = (1+n)N_t$. It follows that, at the beginning of period $t$, at least $nN_{t-1}$ individuals — those born in that period — will be unemployed. Of the $nN_{t-1}$ individuals born at time $t$, $\varphi^0$ are of type $0$, $\varphi^1$ of type $1$, and the rest, $1 - \varphi^0 - \varphi^1$, are of type $2$. Assume that the mean of the offer distribution (the mean offered, not necessarily accepted, wage) is greater than $a/\beta$.

**a.** Consider the situation of an unemployed worker who has $a^0 = 0$. Argue that this worker will have a reservation wage $w^*(0) = a$. Explain.

**b.** Let $w^*(i)$ be the reservation wage of an individual with wealth $i$. Argue that $w^*(2) > w^*(1) > w^*(0)$. What does this say about the cross sectional relationship between financial wealth and employment probability? Discuss the economic reasons underlying this result.

**c.** Let the unemployment rate be the number of unemployed individuals at $t, U_t$, relative to the population at $t, N_t$. Thus, $u_t = U_t/N_t$. Argue that in this economy the unemployment rate is constant.

**d.** Consider a policy that redistributes wealth in the form of changes in the fraction of the population that is born with wealth $a^i$. Describe as completely as you can the effect upon the unemployment rate of changes in $\varphi^i$. Explain your results.

*Extra Credit:* Go as far as you can describing the distribution of the random variable "number of periods unemployed" for an individual of type 2.

# 20
# Appendix on Functional Analysis

This appendix provides an introduction to the analysis of functional equations (functional analysis). It describes the contraction mapping theorem, a workhorse for studying dynamic programs.

## Metric spaces and operators

We begin with the definition of a metric space, which is a pair of objects, a set $X$, and a function $d$.[1]

DEFINITION: *A metric space is a set $X$ and a function $d$ called a metric, $d: X \times X \rightarrow R$. The metric $d(x,y)$ satisfies the following four properties:*

M1. *Positivity: $d(x,y) \geq 0$ for all $x, y \in X$.*

M2. *Strict positivity: $d(x,y) = 0$ if and only if $x = y$.*

M3. *Symmetry: $d(x,y) = d(y,x)$ for all $x, y \in X$.*

M4. *Triangle inequality: $d(x,y) \leq d(x,z) + d(z,y)$ for all $x, y$, and $z \in X$.*

We give some examples of the metric spaces with which we will be working:

**Example 20.1.** $l_p[0,\infty)$. We say that $X = l_p[0,\infty)$ is the set of all sequences of complex numbers $\{x_t\}_{t=0}^{\infty}$ for which $\sum_{t=0}^{\infty} |x_t|^p$ converges, where $1 \leq p < \infty$. The function $d_p(x,y) = (\sum_{t=0}^{\infty} |x_t - y_t|^p)^{1/p}$ is a metric. Often we will say that $p = 2$ and will work in $l_2[0,\infty)$.

---

[1] General references on the mathematics described in this appendix are Luenberger (1969) and Naylor and Sell (1982).

**Example 20.2.** $l_\infty[0,\infty)$. The set $X = l_\infty[0,\infty)$ is the set of bounded sequences $\{x_t\}_{t=0}^\infty$ of real or complex numbers. The metric is $d_\infty(x,y) = \sup_t |x_t - y_t|$.

**Example 20.3.** $l_p(-\infty,\infty)$ is the set of "two-sided" sequences $\{x_t\}_{t=-\infty}^\infty$ such that $\sum_{t=-\infty}^\infty |x_t|^p < +\infty$, where $1 \le p < \infty$. The associated metric is $d_p(x,y) = (\sum_{t=-\infty}^\infty |x_t - y_t|^p)^{1/p}$.

**Example 20.4.** $l_\infty(-\infty,\infty)$ is the set of bounded sequences $\{x_t\}_{t=-\infty}^\infty$ with metric $d_\infty(x,y) = \sup|x_t - y_t|$.

**Example 20.5.** Let $X = C[0,T]$ be the set of all continuous functions mapping the interval $[0,T]$ into $R$. We consider the metric

$$d_p(x,y) = \left[ \int_0^T |x(t) - y(t)|^p dt \right]^{1/p},$$

where the integration is in the Riemann sense.

**Example 20.6.** Let $X = C[0,T]$ be the set of all continuous functions mapping the interval $[0,T]$ into $R$. We consider the metric

$$d_\infty(x,y) = \sup_{0 \le t \le T} |x(t) - y(t)|.$$

We now have the following important definition:

DEFINITION: *A sequence $\{x_n\}$ in a metric space $(X,d)$ is said to be a Cauchy sequence if for each $\epsilon > 0$ there exists an $N(\epsilon)$ such that $d(x_n, x_m) < \epsilon$ for any $n, m \ge N(\epsilon)$. Thus a sequence $\{x_n\}$ is said to be Cauchy if $\lim_{n,m\to\infty} d(x_n, x_m) = 0$.*

We also have the following definition of convergence:

DEFINITION: *A sequence $\{x_n\}$ in a metric space $(X,d)$ is said to converge to a limit $x_0 \in X$ if for every $\epsilon > 0$ there exists an $N(\epsilon)$ such that $d(x_n, x_0) < \epsilon$ for $n \ge N(\epsilon)$.*

The following lemma asserts that every convergent sequence in $(X, d)$ is a Cauchy sequence:

LEMMA: *Let* $\{x_n\}$ *be a convergent sequence in a metric space* $(X, d)$. *Then* $\{x_n\}$ *is a Cauchy sequence.*

PROOF: Fix any $\epsilon > 0$. Let $x_0 \in X$ be the limit of $\{x_n\}$. Then for all $m, n$ one has

$$d(x_n, x_m) \leq d(x_n, x_0) + d(x_m, x_0)$$

by virtue of the triangle inequality. Because $x_0$ is the limit of $\{x_n\}$, there exists an $N$ such that $d(x_n, x_0) < \epsilon/2$ for $n \geq N$. Together with the preceding inequality, this statement implies that $d(x_n, x_m) < \epsilon$ for $n, m \geq N$. Therefore, $\{x_n\}$ is a Cauchy sequence. ▮

We now consider two examples of sequences in metric spaces. The examples are designed to illustrate aspects of the concept of a Cauchy sequence. We first consider the metric space $\{C[0,1], d_2(x, y)\}$. We let $\{x_n\}$ be the sequence of continuous functions $x_n(t) = t^n$. Evidently this sequence converges pointwise to the function

$$x_0(t) = \begin{cases} 0, & 0 \leq t < 1 \\ 1, & t = 1. \end{cases}$$

Now, in $\{C[0,1], d_2(x, y)\}$, the sequence $x_n(t)$ is a Cauchy sequence. To verify this point, calculate

$$d_2(t^m, t^n)^2 = \int_0^1 (t^n - t^m)^2 dt = \frac{1}{2n+1} + \frac{1}{2m+1} - \frac{2}{m+n+1}.$$

Clearly, for any $\epsilon > 0$, it is possible to choose an $N(\epsilon)$ that makes the square root of the right side less than $\epsilon$ whenever $m$ and $n$ both exceed $N$. Thus $x_n(t)$ is a Cauchy sequence. Notice, however, that the limit point $x_0(t)$ does *not* belong to $\{C[0,T], d_2(x, y)\}$ because it is not a continuous function.

As our second example, we consider the space $\{C[0,T], d_\infty(x, y)\}$. We consider the sequence $x_n(t) = t^n$. In $(C[0,1], d_\infty)$, the sequence $x_n(t)$ is *not* a Cauchy sequence. To verify this point, it is sufficient to establish that, for any fixed $m > 0$, there is a $\delta > 0$ such that

$$\sup_{n>0} \sup_{0 \leq t \leq 1} |t^n - t^m| > \delta.$$

Direct calculations show that, for fixed $m$,

$$\sup_{n} \sup_{0 \le t \le 1} |t^n - t^m| = 1.$$

Parenthetically we may note that

$$\begin{aligned}\sup_{n>0} \sup_{0\le t\le 1} |t^n - t^m| &= \sup_{0\le t\le 1} \sup_{n>0} |t^n - t^m| = \sup_{0\le t\le 1} \lim_{n\to\infty} |t^n - t^m| \\ &= \sup_{0\le t\le 1} \lim_{n\to\infty} t^m |t^{n-m} - 1| = \sup_{0\le t\le 1} t^m = 1.\end{aligned}$$

Therefore, $\{t^n\}$ is not a Cauchy sequence in $(C[0,1], d_\infty)$.

These examples illustrate the fact that whether a given sequence is Cauchy depends on the metric space within which it is embedded, in particular on the metric that is being used. The sequence $\{t^n\}$ is Cauchy in $(C[0,1], d_2)$, and more generally in $(C[0,1], d_p)$ for $1 \le p < \infty$. The sequence $\{t^n\}$, however, is *not* Cauchy in the metric space $(C[0,1], d_\infty)$. The first example also illustrates the fact that a Cauchy sequence in $(X, d)$ need *not* converge to a limit point $x_0$ belonging to the metric space. The property that Cauchy sequences converge to points lying in the metric space is desirable in many applications. We give this property a name.

DEFINITION: *A metric space $(X, d)$ is said to be complete if each Cauchy sequence in $(X, d)$ is a convergent sequence in $(X, d)$. That is, in a complete metric space, each Cauchy sequence converges to a point belonging to the metric space.*

The following metric spaces are complete:

$$\begin{aligned}&(l_p[0,\infty), d_p), \qquad 1 \le p < \infty \\ &(l_\infty[0,\infty), d_\infty) \\ &(C[0,T], d_\infty)\end{aligned}$$

The following metric spaces are not complete:

$$(C[0,T], d_p), \qquad 1 \le p < \infty.$$

Proofs that $(l_p[0,\infty), d_p)$ for $1 \le p \le \infty$ and $(C[0,T], d_\infty)$ are complete are contained in Naylor and Sell (1982, chap. 3). In effect, we have already shown by counterexample that the space $(C[0,1], d_2)$ is not complete, because we displayed a Cauchy sequence that did not converge to a point in the metric space. A definition may now be stated:

DEFINITION: *A function $f$ mapping a metric space $(X, d)$ into itself is called an operator.*

We need a notion of continuity of an operator.

DEFINITION: *Let $f : X \to X$ be an operator on a metric space $(X, d)$. The operator $f$ is said to be continuous at a point $x_0 \in X$ if for every $\epsilon > 0$ there exists a $\delta > 0$ such that $d[f(x), f(x_0)] < \epsilon$ whenever $d(x, x_0) < \delta$. The operator $f$ is said to be continuous if it is continuous at each point $x \in X$.*

We shall be studying an operator with a particular property, the application of which to any two distinct points $x, y \in X$ brings them closer together.

DEFINITION: *Let $(X, d)$ be a metric space and let $f : X \to X$. We say that $f$ is a contraction or contraction mapping if there is a real number $k, 0 \leq k < 1$, such that*

$$d[f(x), f(y)] \leq kd(x, y) \qquad \text{for all} \quad x, y \in X.$$

It follows directly from the definition that a contraction mapping is a continuous operator.

We now state the following theorem:

THEOREM 20.1: *Contraction Mapping.*

*Let $(X, d)$ be a complete metric space and let $f : X \to X$ be a contraction. Then there is a unique point $x_0 \in X$ such that $f(x_0) = x_0$. Furthermore, if $x$ is any point in $X$ and $\{x_n\}$ is defined inductively according to $x_1 = f(x), x_2 = f(x_1), \ldots, x_{n+1} = f(x_n)$, then $\{x_n\}$ converges to $x_0$.*

PROOF: Let $x$ be any point in $X$. Define $x_1 = f(x), x_2 = f(x_1), \ldots$. Express this as $x_n = f^n(x)$. To show that the sequence $x_n$ is Cauchy, first assume that $n > m$. Then

$$\begin{aligned} d(x_n, x_m) &= d[f^n(x), f^m(x)] = d[f^m(x_{n-m}), f^m(x)] \\ &\leq kd[f^{m-1}(x_{n-m}), f^{m-1}(x)] \end{aligned}$$

By induction, we get

$$(*) \qquad d(x_n, x_m) \leq k^m d(x_{n-m}, x).$$

When we repeatedly use the triangle inequality, the preceding inequality implies that

$$d(x_n, x_m) \leq k^m [d(x_{n-m}, x_{n-m-1}) + \ldots + d(x_2, x_1) + d(x_1, x)].$$

Applying $(*)$ gives

$$d(x_n, x_m) \leq k^m (k^{n-m-1} + \ldots + k + 1) d(x_1, x).$$

Because $0 \leq k < 1$, we have

$$(\dagger) \qquad d(x_n, x_m) \leq k^m \sum_{i=0}^{\infty} k^i d(x_t, x) = \frac{k^m}{1-k} d(x_1, x).$$

The right side of $(\dagger)$ can be made arbitrarily small by choosing $m$ sufficiently large. Therefore, $d(x_n, x_m) \to 0$ as $n, m \to \infty$. Thus $\{x_n\}$ is a Cauchy sequence. Because $(X, d)$ is complete, $\{x_n\}$ converges to an element of $(X, d)$.

The limit point $x_0$ of $\{x_n\} = \{f^n(x)\}$ is a fixed point of $f$. Because $f$ is continuous, $\lim_{n\to\infty} f(x_n) = f(\lim_{n\to\infty} x_n)$. Now $f(\lim_{n\to\infty} x_n) = f(x_0)$ and $\lim_{n\to\infty} f(x_n) = \lim_{n\to\infty} x_{n+1} = x_0$. Therefore $x_0 = f(x_0)$.

To show that the fixed point $x_0$ is unique, assume the contrary. Assume that $x_0$ and $y_0$, $x_0 \neq y_0$, are two fixed points of $f$. But then

$$0 < d(x_0, y_0) = d[f(x_0), f(y_0)] \leq k d(x_0, y_0) < d(x_0, y_0),$$

which is a contradiction. Therefore $f$ has a unique fixed point. ∎

We now restrict ourselves to sets $X$ whose elements are functions. The spaces $C[0, T]$ and $l_p[0, \infty)$ for $1 \leq p \leq \infty$ are examples of spaces of functions. Let us define the notion of inequality of two functions.

DEFINITION: *Let $X$ be a space of functions, and let $x, y \in X$. Then $x \geq y$ if and only if $x(t) \geq y(t)$ for every $t$ in the domain of the functions.*

Let $X$ be a space of functions. We use the $d_\infty$ metric, defined as $d_\infty(x, y) = \sup_t |x(t) - y(t)|$, where the supremum is over the domain of definition of the function.

A pair of conditions that are sufficient for an operator $T : (X, d_\infty) \to (X, d_\infty)$ to be a contraction appear in the following theorem.[2]

THEOREM 20.2: *Blackwell's Sufficient Conditions for T to be a Contraction.*

*Let $T$ be an operator on a metric space $(X, d_\infty)$, where $X$ is a space of functions. Assume that $T$ has the following two properties:*
(a) *Monotonicity: For any $x, y \in X$, $x \geq y$ implies $T(x) \geq T(y)$.*
(b) *Discounting: Let $c$ denote a function that is constant at the real value $c$ for all points in the domain of definition of the functions in $X$. For any positive real $c$ and every $x \in X$, $T(x + c) \leq T(x) + \beta c$ for some $\beta$ satisfying $0 \leq \beta < 1$.*
*Then $T$ is a contraction mapping with modulus $\beta$.*

PROOF: Suppose that $x \geq y$ implies that $T(x) \geq T(y)$. We have $x = y + x - y$. It follows, because $x - y \leq d_\infty(x, y)$ for all $x, y \in X$, that $x \leq y + d_\infty(x, y)$. Then by monotonicity property (a), $T(x) \leq T[y + d_\infty(x, y)]$. Applying the discounting property (b) to this inequality gives $T(x) \leq T(y) + \beta d_\infty(x, y)$ or $T(x) - T(y) \leq \beta d_\infty(x, y)$ for all $x, y \in X$. It follows that $d_\infty[T(x), T(y)] \leq \beta d_\infty(x, y)$. Therefore $T$ is a contraction. ▮

We can also state the following alternative sufficient conditions for $T$ to be a contraction:

THEOREM 20.3: *Let $T$ be an operator on a metric space $(X, d_\infty)$, where $X$ is a space of functions. Assume that $T$ has the following two properties:*
(a) *Monotonicity: For any $x, y \in X, x \geq y$ implies $T(x) \leq T(y)$.*
(b) *Discounting: Let $c$ denote a function that is constant at the real value $c$ for all points in the domain of definition of the functions in $X$. For any positive real $c$ and every $x \in X$, $T(x - c) \leq T(x) + \beta c$ for some $0 \leq \beta < 1$.*
*Then $T$ is a contraction mapping with modulus $\beta$.*

PROOF: Suppose $x \geq y$ implies that $T(x) \leq T(y)$. Note that $y \geq x - d_\infty(x, y)$. Monotonicity property (a) and discounting property (b) of $T$ imply that $T(y) \leq T[x - d_\infty(x, y)] \leq T(x) + \beta d_\infty(x, y)$ for $0 \leq \beta < 1$. Therefore, $0 \leq T(y) - T(x) \leq \beta d_\infty(x, y)$. Therefore, $d_\infty[T(x), T(y)] \leq \beta d_\infty(x, y)$. Therefore, $T$ is a contraction. ▮

[2] See Blackwell's (1965) Theorem 5. This theorem is used extensively by Stokey and Lucas with Prescott (1989).

## Discounted dynamic programming

We study the functional equation associated with a discounted dynamic programming problem:

$$v(x) = \max_{u \in R^k} \{r(x,u) + \beta v(x')\}, \qquad x' \leq g(x,u), \qquad 0 < \beta < 1. \tag{20.1}$$

We assume that $r(x,u)$ is real valued, continuous, concave, and bounded and that the set $[x', x, u : x' \leq g(x,u), u \in R^k]$ is convex and compact.

We define the operator

$$Tv = \max_{u \in R^k} \{r(x,u) + \beta v(x')\}, \qquad x' \leq g(x,u), \quad x \in X.$$

We work with the space of continuous bounded functions mapping $X$ into the real line. We use the $d_\infty$ metric,

$$d_\infty(v,w) = \sup_{x \in X} |v(x) - w(x)|.$$

This metric space is complete.

The operator $T$ maps a continuous bounded function $v$ into a continuous bounded function $Tv$. (For a proof, see Stokey and Lucas with Prescott, 1989.)[3]

We now establish that $T$ is a contraction by verifying Blackwell's pair of sufficient conditions. First, suppose that $v(x) \geq w(x)$ for all $x \in X$. Then

$$\begin{aligned} Tv &= \max_{u \in R^k} \{r(x,u) + \beta v(x')\}, \qquad x' \leq g(x,u) \\ &\geq \max_{u \in R^k} \{r(x,u) + \beta w(x')\}, \qquad x' \leq g(x,u) \\ &= Tw. \end{aligned}$$

Thus $T$ is monotone. Next, notice that for any positive constant $c$,

$$\begin{aligned} T(v+c) &= \max_{u \in R^k} \{r(x,u) + \beta[v(x') + c]\}, \qquad x' \leq g(x,u) \\ &= \max_{u \in R^k} \{r(x,u) + \beta v(x') + \beta c\}, \qquad x' \leq g(x,u) \\ &= Tv + \beta c. \end{aligned}$$

[3] The assertions in the preceding two paragraphs are the most difficult pieces of the argument to prove.

Thus $T$ discounts. Therefore $T$ satisfies both of Blackwell's conditions. It follows that $T$ is a contraction on a complete metric space. Therefore the functional equation (20.1), which can be expressed as $v = Tv$, has a unique fixed point in the space of bounded continuous functions. This fixed point is approached in the limit in the $d_\infty$ metric by iterations $v^n = T^n(v^0)$ starting from any bounded and continuous $v^0$. Convergence in the $d_\infty$ metric implies uniform convergence of the functions $v^n$.

Stokey and Lucas with Prescott (1989) show that $T$ maps concave functions into concave functions. It follows that the solution of $v = Tv$ is a concave function.

*Policy improvement algorithm*

For ease of exposition, in this section we shall assume that the constraint $x' \leq g(x,u)$ holds with equality. For the purposes of describing an alternative way to solve dynamic programming problems, we introduce a new operator. We use one step of iterating on the Bellman equation to define the new operator $T_\mu$ as follows:

$$T_\mu(v) = T(v)$$

or

$$T_\mu(v) = r[x,\, \mu(x)] + \beta v\{g[x,\, \mu(x)]\}\ ,$$

where $\mu(x)$ is the policy function that attains $T(v)(x)$. For a fixed $\mu(x)$, $T_\mu$ is an operator that maps bounded continuous functions into bounded continuous functions. Denote by $C$ the space of bounded continuous functions mapping $X$ into $X$.

For any admissible policy function $\mu(x)$, the operator $T_\mu$ is a contraction mapping. This fact can be established by verifying Blackwell's pair of sufficient conditions:

1. $T_\mu$ is monotone.

Suppose that $v(x) \geq w(x)$. Then

$$\begin{aligned} T_\mu v &= r[x,\, \mu(x)] + \beta v\{g[x,\, \mu(x)]\} \\ &\geq r[x,\, \mu(x)] + \beta w\{g[x,\, \mu(x)]\} = T_\mu w\ . \end{aligned}$$

2. $T_\mu$ discounts.

For any positive constant $c$

$$\begin{aligned} T_\mu(v+c) &= r(x,\mu) + \beta\left(v\{g[x,\mu(x)]+c\}\right) \\ &= T_\mu v + \beta c\ . \end{aligned}$$

Because $T_\mu$ is a contraction operator, the functional equation

$$v_\mu(x) = T_\mu[v_\mu(x)]$$

has a unique solution in the space of bounded continuous functions. This solution can be computed as a limit of iterations on $T_\mu$ starting from any bounded continuous function $v_0(x) \in C$,

$$v_\mu(x) = \lim_{k \to \infty} T_\mu^k(v_0)(x) .$$

The function $v_\mu(x)$ is the value of the objective function that would be attained by using the stationary policy $\mu(x)$ each period.

The following proposition describes the *policy iteration* or *Howard improvement* algorithm.

PROPOSITION: Let $v_\mu(x) = T_\mu[v_\mu(x)]$. Define a new policy $\bar{\mu}$ and an associated operator $T_{\bar{\mu}}$ by

$$T_{\bar{\mu}}[v_\mu(x)] = T[v_\mu(x)] ;$$

that is, $\bar{\mu}$ is the policy that solves a one-period problem with $v_\mu(x)$ as the terminal value function. Compute the fixed point

$$v_{\bar{\mu}}(x) = T_{\bar{\mu}}[v_{\bar{\mu}}(x)] .$$

Then $v_{\bar{\mu}}(x) \geq v_\mu(x)$. If $\mu(x)$ is not optimal, then $v_{\bar{\mu}}(x) > v_\mu(x)$ for at least one $x \in X$.

PROOF: From the definition of $\bar{\mu}$ and $T_{\bar{\mu}}$, we have

$$\begin{aligned} T_{\bar{\mu}}[v_\mu(x)] = r[x,\, \bar{\mu}(x)] + \beta v_\mu\{g[x,\, \bar{\mu}(x)]\} = \\ T(v_\mu)(x) \geq r[x,\, \mu(x)] + \beta v_\mu\{g[x,\, \mu(x)]\} \\ = T_\mu[v_\mu(x)] = v_\mu(x) \end{aligned}$$

or

$$T_{\bar{\mu}}[v_\mu(x)] \geq v_\mu(x) .$$

Apply $T_{\bar{\mu}}$ repeatedly to this inequality and use the monotonicity of $T_{\bar{\mu}}$ to conclude

$$v_{\bar{\mu}}(x) = \lim_{n \to \infty} T_{\bar{\mu}}^n\,[v_\mu(x)] \geq v_\mu(x) .$$

This establishes the asserted inequality $v_{\bar{\mu}}(x) \geq v_\mu(x)$. If $v_{\bar{\mu}}(x) = v_\mu(x)$ for all $x \in X$, then

$$\begin{aligned} v_\mu(x) &= T_{\bar{\mu}}[v_\mu(x)] \\ &= T[v_\mu(x)] \, , \end{aligned}$$

where the first equality follows because $T_{\bar{\mu}}[v_{\bar{\mu}}(x)] = v_{\bar{\mu}}(x)$, and the second equality follows from the definitions of $T_{\bar{\mu}}$ and $\bar{\mu}$. Because $v_\mu(x) = T[v_\mu(x)]$, the Bellman equation is satisfied by $v_\mu(x)$. ∎

The *policy improvement* algorithm starts from an arbitrary feasible policy and iterates to convergence on the two following steps:[4]

Step 1. For a feasible policy $\mu(x)$, compute $v_\mu = T_\mu(v_\mu)$.

Step 2. Find $\bar{\mu}$ by computing $T(v_\mu)$. Use $\bar{\mu}$ as the policy in step 1.

In many applications, this algorithm proves to be much faster than iterating on the Bellman equation.

### *A search problem*

We now study the functional equation associated with a search problem of chapter 5. The functional equation is

$$v(w) = \max \left\{ \frac{w}{1-\beta}, \beta \int v(w') dF(w') \right\}, \qquad 0 < \beta < 1. \tag{20.2}$$

Here the wage offer drawn at $t$ is $w_t$. Successive offers $w_t$ are independently and identically distributed random variables. We assume that $w_t$ has cumulative distribution function $\text{prob}\{w_t \leq w\} = F(w)$, where $F(0) = 0$ and $F(\bar{w}) = 1$ for some $\bar{w} < \infty$. In equation (20.2), $v(w)$ is the optimal value function for a currently unemployed worker who has offer $w$ in hand. We seek a solution of the functional equation (20.2).

We work in the space of bounded continuous functions $C[0, \bar{w}]$ and use the $d_\infty$ metric

$$d_\infty(x, y) = \sup_{0 \leq w \leq \bar{w}} |x(w) - y(w)|.$$

[4] A policy $\mu(x)$ is said to be *unimprovable* if it is optimal to follow it for the first period, given a terminal value function $v(x)$. In effect, the policy improvement algorithm starts with an arbitrary value function, then by solving a one-period problem, it generates an improved policy and an improved value function. The proposition states that optimality is characterized by the features, first, that there is no incentive to deviate from the policy during the first period, and second, that the terminal value function is the one associated with continuing the policy.

The metric space $(C[0,\bar{w}], d_\infty)$ is complete.

We consider the operator

$$T(z) = \max\Big\{\frac{w}{1-\beta}, \beta \int z(w')dF(w')\Big\}. \tag{20.3}$$

Evidently the operator $T$ maps functions $z$ in $C[0,\bar{w}]$ into functions $T(z)$ in $C[0,\bar{w}]$. We now assert that the operator $T$ defined by equation (20.3) is a contraction. To prove this assertion, we verify Blackwell's sufficient conditions. First, assume that $f(w) \geq g(w)$ for all $w \in [0,\bar{w}]$. Then note that

$$\begin{aligned} Tg &= \max\left\{\frac{w}{1-\beta}, \beta \int g(w')dF(w')\right\} \\ &\leq \max\left\{\frac{w}{1-\beta}, \beta \int f(w')dF(w')\right\} \\ &= Tf. \end{aligned}$$

Thus $T$ is monotone. Next, note that for any positive constant $c$,

$$\begin{aligned} T(f+c) &= \max\left\{\frac{w}{1-\beta}, \beta \int [f(w')+c]dF(w')\right\} \\ &= \max\left\{\frac{w}{1-\beta}, \beta \int f(w')dF(w') + \beta c\right\} \\ &\leq \max\left\{\frac{w}{1-\beta}, \beta \int f(w')dF(w')\right\} + \beta c \\ &= Tf + \beta c. \end{aligned}$$

Thus $T$ satisfies the discounting property and is therefore a contraction.

Application of the contraction mapping theorem, then, establishes that the functional equation $Tv = v$ has a unique solution in $C[0,\bar{w}]$, which is approached in the limit as $n \to \infty$ by $T^n(v^0) = v^n$, where $v^0$ is any point in $C[0,\bar{w}]$. Because the convergence in the space $C[0,\bar{w}]$ is in terms of the metric $d_\infty$, the convergence is uniform.

# 21
# Appendix on Control and Filtering

## Introduction

By recursive techniques we mean the application of dynamic programming to control problems, and of Kalman filtering to the filtering problems. We describe classes of problems in which the dynamic programming and the Kalman filtering algorithms are formally equivalent, being tied together by *duality*. By exploiting their equivalence, we reap double dividends from any results that apply to one or the other problem.[1]

The next-to-last section of this appendix contains statements of a few facts about linear least squares projections. The final section briefly describes filtering problems where the state evolves according to a finite-state Markov process.

## The optimal linear regulator control problem

We briefly recapitulate the *optimal linear regulator* problem. Consider a system with a $(n \times 1)$ *state* vector $x_t$ and a $(k \times 1)$ *control* vector $u_t$. The system is assumed to evolve according to the law of motion

$$x_{t+1} = A_t x_t + B_t u_t \qquad t = t_0, t_0 + 1, \ldots, t_1 - 1, \tag{21.1}$$

where $A_t$ is an $(n \times n)$ matrix and $B_t$ is an $(n \times k)$ matrix. Both $A_t$ and $B_t$ are known sequences of matrices. We define the *return function* at time $t$, $r_t(x_t, u_t)$, as the quadratic form

$$r_t(x_t, u_t) = [x_t' \, u_t'] \begin{bmatrix} R_t & W_t \\ W_t' & Q_t \end{bmatrix} \begin{bmatrix} x_t \\ u_t \end{bmatrix} \qquad t = t_0, \ldots, t_1 - 1$$

[1] The concepts of controllability and reconstructibility are used to establish conditions for the convergence and other important properties of the recursive algorithms.

where $R_t$ is $(n \times n)$, $Q_t$ is $(k \times k)$, and $W_t$ is $(n \times k)$. We shall initially assume that the matrices $\begin{bmatrix} R_t & W_t \\ W_t' & Q_t \end{bmatrix}$ are negative semidefinite, though subsequently we shall see that the problem can still be well posed even if this assumption is weakened. We are also given an $(n \times n)$ negative semidefinite matrix $P_t$, which is used to assign a terminal value of the state $x_{t_1}$.

The *optimal linear regulator* problem is to maximize

$$\sum_{t=t_0}^{t_1-1} [x_t' \, u_t'] \begin{bmatrix} R_t & W_t \\ W_t' & Q_t \end{bmatrix} \begin{bmatrix} x_t \\ u_t \end{bmatrix} + x_{t_1}' P_{t_1} x_{t_1}$$

$$\text{subject to} \qquad x_{t+1} = A_t x_t + B_t u_t, \qquad x_{t_0} \text{ given.} \tag{21.2}$$

The maximization is carried out over the sequence of controls $(u_{t_0}, u_{t_0+1}, \ldots, u_{t_1-1})$. This is a recursive or serial problem, which is appropriate to solve using the method of dynamic programming. In this case, the *value functions* are defined as the quadratic forms, $s = t_0, t_0+1, \ldots, t_1-1$,

$$x_s' P_s x_s = \max \left\{ \sum_{t=s}^{t_1-1} [x_t' \, t_{1-1} \, u_t'] \begin{bmatrix} R_t & W_t \\ W_t' & Q_t \end{bmatrix} \begin{bmatrix} x_t \\ u_t \end{bmatrix} + x_{t_1}' P_{t_1} x_{t_1} \right\} \tag{21.3}$$

$$\text{subject to} \quad x_{t+1} = A_t x_t + B_t u_t,$$

$x_s$ given, $s = t_0, t_0+1, \ldots, t_1-1$. The *Bellman equation* becomes the following backward recursion in the quadratic forms $x_t' P_t x_t$:

$$\begin{aligned} x_t' P_t x_t = \max_{u_t} \Big\{ & x_t' R_t x_t + u_t' Q_t u_t + 2 x_t' W_t u_t + (A_t x_t + B_t u_t)' \\ & P_{t+1}(A_t x_t + B_t u_t) \Big\}, \\ & t = t_1 - 1, t_1 - 2, \ldots, t_0 \\ & P_{t_1} \text{ given .} \end{aligned} \tag{21.4}$$

Using the rules for differentiating quadratic forms, the first-order necessary condition for the problem on the right side of equation (21.4) is found by differentiating with respect to the vector $u_t$:

$$\{Q_t + B_t' P_{t+1} B_t\} u_t = -(B_t' P_{t+1} A_t + W_t') x_t.$$

Solving for $u_t$ we obtain

$$u_t = -(Q_t + B_t' P_{t+1} B_t)^{-1} (B_t' P_{t+1} A_t + W_t') x_t. \tag{21.5}$$

The inverse $(Q_t + B_t' P_{t+1} B_t)^{-1}$ is assumed to exist. Otherwise, it could be interpreted as a generalized inverse, and most of our results would go through.

Equation (21.5) gives the optimal control in terms of a *feedback rule* upon the state vector $x_t$, of the form

$$u_t = -F_t x_t \tag{21.6}$$

where

$$F_t = (Q_t + B_t' P_{t+1} B_t)^{-1} (B_t' P_{t+1} A_t + W_t'). \tag{21.7}$$

Substituting equation (21.5) for $u_t$ into equation (21.4) and rearranging gives the following recursion for $P_t$:

$$\begin{aligned} P_t = R_t + A_t' P_{t+1} A_t - (A_t P_{t+1} B_t + W_t) \; (Q_t + B_t' P_{t+1} B_t)^{-1} \\ (B_t' P_{t+1} A_t + W_t'). \end{aligned} \tag{21.8}$$

Equation (21.8) is a version of the *matrix Riccati difference equation.*

Equations (21.8) and (21.5) provide a recursive algorithm for computing the optimal controls in feedback form. Starting at time $(t_1 - 1)$, and given $P_{t_1}$, equation (21.5) is used to compute $u_{t_1 - 1} = -F_{t_1 - 1} x_{t_1 - 1}$. Then equation (21.8) is used to compute $P_{t_1 - 1}$. Then equation (21.5) is used to compute $u_{t_1 - 2} = F_{t_1 - 2} x_{t_1 - 2}$, and so on.

By substituting the optimal control $u_t = -F_t x_t$ into the state equation (21.1), we obtain the optimal *closed loop system* equations

$$x_{t+1} = (A_t - B_t F_t) x_t.$$

Eventually, we shall be concerned extensively with the properties of the optimal closed loop system, and how they are related to the properties of $A_t$, $B_t$, $Q_t$, $R_t$, and $W_t$.

## *Converting a problem with cross-products in states and controls to one with no such cross-products*

For our future work it is useful to introduce a problem that is equivalent with equations (21.2) and (21.3), and has a form in which no cross-products between states and controls appear in the objective function. This is useful because our theorems about the properties of the solutions (21.5) and (21.8) will be in terms of the special case in which $W_t = 0 \quad \forall t$. The equivalence between the problems (21.2) and (21.3) and the following problem implies that no generality is lost by restricting ourselves to the case in which $W_t = 0 \quad \forall t$.

The equivalent problem

$$\max_{\{u_t^*\}} \sum_{t=t_0}^{t_1-1} \left\{ x_t'(R_t - W_t Q_t^{-1} W_t')x_t + u_t^{*\prime} Q_t u_t^* \right\} + x_{t_1}' P_{t_1} x_{t_1} \tag{21.9}$$

subject to

$$x_{t+1} = (A_t - B_t Q_t^{-1} W_t')x_t + B_t u_t^*, \tag{21.10}$$

and $x_{t_0}$, $P_{t_0}$ are given. The new control variable $u_t^*$ is related to the original control $u_t$ by

$$u_t^* = Q_t^{-1} W_t' x_t + u_t. \tag{21.11}$$

We can state the problem (21.9)–(21.10) in a more compact notation as being to maximize

$$\sum_{t=t_0}^{t_1-1} \left\{ x_t' \bar{R}_t x_t + u_t^{*\prime} Q_t u_t^* \right\} + x_i, P_t, x_t, \tag{21.12}$$

subject to

$$x_{t+1} = \bar{A}_t x_t + B_t u_t^* \tag{21.13}$$

where

$$\bar{R}_t = R_t - W_t Q_t^{-1} W_t' \tag{21.14}$$

and

$$\bar{A}_t = A_t - B_t Q_t^{-1} W_t'. \tag{21.15}$$

With these specifications, the solution of the problem can be computed using the following versions of equations (21.5) and (21.8)

$$u_t^* = -\bar{F}_t x_t \equiv -(Q_t + B_t' P_{t+1} B_t)^{-1} B_t P_{t+1} \bar{A}_t \tag{21.16}$$

$$P_t = \bar{R}_t + \bar{A}_t' P_{t+1} \bar{A}_t - \bar{A}_t' P_{t+1} B_t (Q_t + B_t' P_{t+1} B_t)^{-1} B_t' P_{t+1} \bar{A}_t \tag{21.17}$$

We ask the reader to verify the following facts:

a. Problems (21.2)–(21.3) and (21.9)–(21.10) are equivalent.

b. The feedback laws $\bar{F}_t$ and $F_t$ for $u_t^*$ and $u_t$, respectively, are related by

$$F_t = \bar{F}_t + Q_t^{-1} W_t'.$$

c. The Riccati equations (21.8) and (21.17) are equivalent.

d. The "closed loop" transition matrices are related by

$$A_t - B_t F_t = \bar{A}_t - B_t \bar{F}_t.$$

## An example

We now give an example of a problem for which the preceding transformation is useful. A consumer wants to maximize

$$\sum_{t=t_0}^{\infty} \beta^t \left\{ u_1 c_t - \frac{u_2}{2} c_t^2 \right\} \quad 0 < \beta < 1 \quad , \ u_1 > 0, u_2 > 0 \tag{21.18}$$

subject to the intertemporal budget constraint

$$k_{t+1} = (1 + r)\ (k_t + y_t - c_t), \tag{21.19}$$

the law of motion for labor income

$$y_{t+1} = \lambda_0 + \lambda_1 y_t, \tag{21.20}$$

and a given level of initial assets, $k_{t_0}$. Here $\beta$ is a discount factor, $u_1$ and $u_2$ are constants, $c_t$ is consumption, $k_t$ is "nonhuman" assets at the beginning of time $t$, $r > -1$ is the interest rate on nonhuman assets, and $y_t$ is income from labor at time $t$.

We define the transformed variables

$$\tilde{k}_t = \beta^{t/2} k_t$$
$$\tilde{y}_t = \beta^{t/2} y_t$$
$$\tilde{c}_t = \beta^{t/2} c_t.$$

In terms of these transformed variables, the problem can be rewritten as follows: maximize

$$\sum_{t=t_0}^{\infty} \left\{ u_1 \beta^{t/2} \cdot \tilde{c}_t - \frac{u_2}{2} \tilde{c}_t^2 \right\} \tag{21.21}$$

subject to

$$\tilde{k}_{t+1} = (1+r)\beta^{1/2}\left(\tilde{k}_t + \tilde{y}_t - \tilde{c}_t\right) \quad \text{and}$$
$$\tilde{y}_{t+1} = \lambda_0 \beta^{\frac{t+1}{2}} + \lambda_1 \beta^{1/2} \tilde{y}_t \tag{21.22}$$

and $k_{t_0}$ given. We write this problem in the state-space form:

$$\max_{\{\tilde{u}_t\}} \sum_{t=t_0}^{\infty} \left\{ \tilde{x}_t' R \tilde{x}_t + 2\tilde{x}_t' W \tilde{u}_t + \tilde{u}_t' Q \tilde{u}_t \right\}$$
$$\text{subject to } \tilde{x}_{t+1} = A\tilde{x}_t + B\tilde{u}_t.$$

We take

$$\tilde{x}_t = \begin{bmatrix} \tilde{k}_t \\ \tilde{y}_t \\ \beta^{t/2} \end{bmatrix}, \ \tilde{u}_t = \tilde{c}_t,$$

$$R = \begin{bmatrix} 0 & 0 & 0 \\ 0 & 0 & 0 \\ 0 & 0 & 0 \end{bmatrix}, \ W' = \begin{bmatrix} 0 & 0 & \frac{u_1}{2} \end{bmatrix},$$

$$Q = -\frac{u_2}{2}, \ A = \begin{bmatrix} (1+r) & (1+r) & 0 \\ 0 & \lambda_1 & \lambda_0 \\ 0 & 0 & 1 \end{bmatrix} \beta^{1/2}, \quad B = \begin{bmatrix} -(1+r) \\ 0 \\ 0 \end{bmatrix} \beta^{1/2}.$$

To obtain the equivalent transformed problem in which there are no cross-product terms between states and controls in the return function, we take

$$\bar{A} = A - BQ^{-1}W' = \begin{bmatrix} (1+r) & (1+r) & -\frac{u_1(1+r)}{u_2} \\ 0 & \lambda_1 & \lambda_0 \\ 0 & 0 & 1 \end{bmatrix} \beta^{1/2}$$

$$\bar{R} = R - WQ^{-1}W' = \begin{bmatrix} 0 & 0 & 0 \\ 0 & 0 & 0 \\ 0 & 0 & \frac{u_1^2}{2u_2} \end{bmatrix} \tag{21.23}$$
$$u_t^* = \tilde{u}_t + Q^{-1}W'\tilde{x}_t$$
$$c_t^* = \tilde{c}_t - \frac{u_1}{u_2}\beta^{t/2}.$$

Thus, our original problem can be expressed as

$$\max_{\{u_t^*\}} \sum_{t=t_0}^{\infty} \left\{ \tilde{x}_t' \bar{R} \tilde{x}_t + u_t^{*\prime} Q u_t^* \right\}$$
$$\text{subject to} \quad \tilde{x}_{t+1} = \bar{A}\tilde{x}_t + B u_t^*. \tag{21.24}$$

## The Kalman filter

Consider the linear system

$$x_{t+1} = A_t x_t + B_t u_t + G_t w_{1t+1} \tag{21.25}$$

$$y_t = C_t x_t + H_t u_t + w_{2t} \tag{21.26}$$

where $[w_{1t+1}', w_{2t}']$ is a vector white noise with contemporaneous covariance matrix

$$E\begin{bmatrix} w_{1t+1} \\ w_{2t} \end{bmatrix}\begin{bmatrix} w_{1t+1} \\ w_{2t} \end{bmatrix}' = \begin{bmatrix} V_{1t} & V_{3t} \\ V_{3t}' & V_{2t} \end{bmatrix} \geq 0.$$

The $[w_{1t+1}', w_{2t}']$ vector for $t \geq t_0$ is assumed orthogonal to the initial condition $x_{t_0}$, which represents the initial state. Here, $A_t$ is $(n \times n)$, $B_t$ is $(n \times k)$, $G_t$ is $(n \times N)$, $C_t$ is $(\ell \times n)$, $H_t$ is $(\ell \times k)$, $w_{1t+1}$ is $(N \times 1)$, $w_{2t+1}$ is $(\ell \times 1)$, $x_t$ is an $(n \times 1)$ vector of *state* variables, $u_t$ is a $(k \times 1)$ vector of *controls*, and $y_t$ is an $(\ell \times 1)$ vector of *output* or observed variables. The matrices $A_t, B_t, G_t, C_t$, and $H_t$ are known, though possibly time varying. The noise vector $w_{1t+1}$ is the state disturbance, while $w_{2t}$ is the measurement error.

The analyst does not directly observe the $x_t$ process. So from his point of view, $x_t$ is a "hidden state vector." The system is assumed to start up at time $t_0$, at which time the state vector $x_{t_0}$ is regarded as a random variable with mean $Ex_{t_0} = \hat{x}_{t_0}$, and given covariance matrix $\sum_{t_0} = \sum_0$. The pair $(\hat{x}_{t_0}, \sum_0)$ can be regarded as the mean and covariance of the analyst's Bayesian prior distribution on $x_{t_0}$.

It is assumed that for $s \geq 0$, the vector of random variables $\left[\begin{smallmatrix} w_{1t_0+s+1} \\ w_{2t_0+s} \end{smallmatrix}\right]$ is orthogonal to the random variable $x_{t_0}$ and to the random variables $\left[\begin{smallmatrix} w_{1t_0+r+1} \\ w_{2t_0+r} \end{smallmatrix}\right]$ for $r \neq s$. It is also assumed that $E\left[\begin{smallmatrix} w_{1t_0+s+1} \\ w_{2t+s} \end{smallmatrix}\right] = 0$ for $s \geq 0$. Thus, $\left[\begin{smallmatrix} w_{1t} \\ w_{2t} \end{smallmatrix}\right]$ is a serially uncorrelated or white noise process. Further, from equations (21.25) and (21.26) and the orthogonality properties posited for $\left[\begin{smallmatrix} w_{1t+1} \\ w_{2t} \end{smallmatrix}\right]$ and $x_{t_0}$, it follows that $\begin{bmatrix} w_{1t+1} \\ w_{2t} \end{bmatrix}$ is orthogonal to $\{x_s, y_{s-1}\}$ for $s \leq t$. This conclusion follows because $y_t$ and $x_{t+1}$ are in the space spanned by current and lagged $u_t, w_{1t+1}, w_{2t}$, and $x_{t_0}$.

The analyst is assumed to observe at time $t\,\{y_s,\, u_s \,:\, s = t_0, t_0+1, \ldots t\}$, for $t = t_0, t_0+1, \ldots t_1$. The object is then to compute the linear least squares projection of the state $x_{t+1}$ on this information, which we denote $\widehat{E}_t x_{t+1}$. We write this projection as

$$\widehat{E}_t x_{t+1} \equiv \widehat{E}[x_{t+1} \mid y_t, y_{t-1}, \ldots, y_{t_0}, \hat{x}_{t_0}] \tag{21.27}$$

where $\hat{x}_{t_0}$ is the initial estimate of the state. It is convenient to let $Y_t$ denote the information on $y_t$ collected through time $t$:

$$Y_t = \{y_t, y_{t-1}, \ldots, y_{t_0}\}.$$

The linear least squares projection of $y_{t+1}$ on $Y_t$, and $\hat{x}_{t_0}$ is, from equations (21.26) and (21.27), given by

$$\begin{aligned} \widehat{E}_t y_{t+1} &\equiv \widehat{E}[y_{t+1} \mid Y_t, \hat{x}_0] \\ &= C_{t+1} \widehat{E}_t x_{t+1} + E_t\, u_{t+1}, \end{aligned} \tag{21.28}$$

since $w_{2t+1}$ is orthogonal to $\{w_{1s+1},\, w_{2s}\}$, $s \leq t$, and $\hat{x}_{t_0}$ and is therefore orthogonal to $\{Y_t,\, \hat{x}_{t_0}\}$.

In the interests of conveniently constructing the projections $\widehat{E}_t x_{t+1}$ and $\widehat{E}_t y_{t+1}$, we now apply a Gram-Schmidt orthogonalization procedure to the set of random variables $\{\hat{x}_{t_0}, y_{t_0}, y_{t_0+1}, \cdots y_{t_1}\}$. An orthogonal basis for this set of random variables is formed by the set $\{\hat{x}_{t_0}, \tilde{y}_{t_0} \tilde{y}_{t_0+1}, \ldots, \tilde{y}_{t_1}\}$ where

$$\tilde{y}_t = y_t - \widehat{E}[y_t \mid \tilde{y}_{t-1}, \tilde{y}_{t-2}, \ldots \tilde{y}_{t_0}, \hat{x}_{t_0}]. \tag{21.29}$$

For convenience, let us write $\widetilde{Y}_t = \{\tilde{y}_{t_0}, \tilde{y}_{t_0+1}, \ldots, \tilde{y}_t\}$. We note that the linear spaces spanned by $(\hat{x}_{t_0}, Y_t)$ equal the linear spaces spanned by $(\hat{x}_{t_0}, \tilde{Y}_t)$. This follows because (a) $\tilde{y}_t$ is formed as indicated previously as a linear function of $Y_t$ and $\hat{x}_{t_0}$, and (b) $y_t$ can be recovered from $\tilde{Y}_t$ and $\hat{x}_{t_0}$ by noting that $y_t = \widehat{E}[y_t \mid \hat{x}_{t_0}, \tilde{Y}_{t-1}] + \tilde{y}_t$. It follows that $\widehat{E}[y_t \mid \hat{x}_{t_0}, Y_{t-1}] = \widehat{E}[y_t \mid \hat{x}_{t_0}, \tilde{Y}_{t-1}] = E_{t-1} y_t$. In equation (21.29), we use equation (21.26) to write

$$\widehat{E}[y_{t_0} \mid \hat{x}_{t_0}] = C_{t_0}\hat{x}_{t_0} + E_{t_0}u_{t_0}.$$

Here we are implying $\hat{x}_{t_0} = Ex_0$. To summarize developments up to this point, we have defined the *innovations process*

$$\begin{aligned} \tilde{y}_t &= y_t - \widehat{E}[y_t \mid \hat{x}_{t_0},\ Y_{t-1}] \\ &= y_t - \widehat{E}[y_t \mid \hat{x}_{t_0}, \tilde{Y}_{t-1}],\ t \geq t_0 + 1 \\ \tilde{y}_{t_0} &= y_{t_0} - \widehat{E}[y_{t_0} \mid \hat{x}_{t_0}]. \end{aligned}$$

The innovations process is *serially uncorrelated* ($\tilde{y}_t$ is orthogonal to $\tilde{y}_s$ for $t \neq s$) and spans the same linear space as the original $Y$ process.

We now use the innovations process to get a recursive procedure for evaluating $\widehat{E}_t x_{t+1}$. Using Theorem 21.4 about projections on orthogonal bases gives

$$\begin{aligned} &\widehat{E}\left[x_{t+1} \mid \hat{x}_{t_0}, \tilde{y}_{t_0}, \tilde{y}_{t_0+1}, \ldots, \tilde{y}_t\right] \\ &\quad = \widehat{E}[x_{t+1} \mid \tilde{y}_t] + \widehat{E}[x_{t+1} \mid \hat{x}_{t_0}, \tilde{y}_{t_0}, \tilde{y}_{t_0+1}, \ldots, \tilde{y}_{t-1}] - Ex_{t+1} \end{aligned} \tag{21.30}$$

We have to evaluate the first two terms on the right side of equation (21.30).

From Theorem 21.1, we have the following:[2]

$$\widehat{E}[x_{t+1} \mid \tilde{y}_t] = Ex_{t+1} + \text{cov}\ (x_{t+1}, \tilde{y}_t)\ \left[\text{cov}\ (\tilde{y}_t, \tilde{y}_t)\right]^{-1} \tilde{y}_t. \tag{21.31}$$

To evaluate the covariances that appear in equation (21.31), we shall use the covariance matrix of one-step-ahead errors, $\tilde{x}_t = x_t - \widehat{E}_{t-1}x_t$, in estimating $x_t$. We define this covariance matrix as $\Sigma_t = E\tilde{x}_t\tilde{x}_t'$. It follows from equations

[2] Here we are using $E\tilde{y}_t = 0$.

(21.25) and (21.26) that

$$
\begin{aligned}
\operatorname{cov}(x_{t+1},\tilde{y}_t) &= \operatorname{cov}(A_t x_t + B_t u_t - G_t w_{1t+1}, y_t - \widehat{E}_{t-1} y_t) \\
&= \operatorname{cov}(A_t x_t + B_t u_t + G_t w_{1t+1}, C_t x_t + w_{2t} - C_t \widehat{E}_{t-1} x_t) \\
&= \operatorname{cov}(A_t x_t + B_t u_t + G_t w_{1t+1}, C_t \tilde{x}_t + w_{2t}) \\
&= E\{[A_t x_t + B_t u_t + G_t w_{1t+1} - E(A_t x_t + B_t u_t + G_t w_{1t+1})] \\
&\qquad [C_t \tilde{x}_t + w_{2t} - E(C_t \tilde{x}_t + w_{2t})']\} \\
&= E[(A_t x_t + G_t w_{1t+1} - A_t E x_t)(\tilde{x}_t' C_t' + w_{2t}')] \qquad (21.32) \\
&= E(A_t x_t \tilde{x}_t' C_t') + G_t E(w_{1t+1} \tilde{x}_t' C_t') - A_t E x_t E \tilde{x}_t' C_t' \\
&\qquad + A_t E(x_t w_{2t}') + G_t E(w_{1t+1} w_{2t}') - A_t E x_t E w_{2t}' \\
&= E(A_t x_t \tilde{x}_t' C_t') + G_t E(w_{1t+1} w_{2t}') \\
&= E[A_t(\tilde{x}_t + \widehat{E}_{t-1} x_t) \tilde{x}_t' C_t'] + G_t E(w_{1t+1} w_{2t}') \\
&= A_t E \tilde{x}_t \tilde{x}_t' C_t' + G_t E(w_{1t+1}\, w_{2t}') = A_t \Sigma_t C_t' + G_t V_{3t}.
\end{aligned}
$$

The second equality uses the fact that $\widehat{E}_{t-1} w_{2t} = 0$, since $w_{2t}$ is orthogonal to $\{x_s, y_{s-1}\}$, $s \leq t$. To get the fifth equality, we use the fact that $E\tilde{x}_t = E(x_t - \widehat{E}_{t-1} x_t) = 0$ by the unbiased property of linear projections when one of the regressors is a constant. We also use the facts that $u_t$ is known and that $w_{1t+1}$ and $w_{2t}$ have zero means. The seventh equality follows from the orthogonality of $w_{1t+1}$ and $w_{2t}$ to variables dated $t$ and earlier and the means of $w_{2t}'$ and $\tilde{x}_t'$ being zero. Finally, the ninth equation relies on the fact that $\tilde{x}_t$ is orthogonal to the subspace generated by $y_{t-1}, y_{t-2}, \ldots, \hat{x}_{t_0}$ and $\widehat{E}_{t-1} x_t$ is a function of these vectors.

Next we evaluate

$$
\begin{aligned}
\operatorname{cov}(\tilde{y}_t, \tilde{y}_t) &= E(C_t \tilde{x}_t + w_{2t})(C_t \tilde{x}_t + w_{2t})' \\
&= C_t \Sigma_t C_t' + V_{2t},
\end{aligned}
$$

since $E\tilde{y}_t = 0$ and $E\tilde{x}_t w_{2t}' = 0$. Therefore, equation (21.32) becomes

$$
\widehat{E}(x_{t+1} \mid \tilde{y}_t) = E(x_{t+1}) + (A_t \Sigma_t C_t' + G_t V_{3t})(C_t \Sigma_t C_t' + V_{2t})^{-1} \tilde{y}_t. \qquad (21.33)
$$

Using equation (21.25), we evaluate the second term on the right side of equation (21.30),

$$
\widehat{E}(x_{t+1} \mid \tilde{Y}_{t-1}, \hat{x}_{t_0}) = A_t \widehat{E}(x_t \mid \tilde{Y}_{t-1}, \hat{x}_{t_0}) + B_t u_t
$$

or

$$\widehat{E}_{t-1}x_{t+1} = A_t\widehat{E}_{t-1}x_t + B_tu_t. \tag{21.34}$$

Using equations (21.33) and (21.34) in equation (21.30) gives

$$\widehat{E}_tx_{t+1} = A_t\widehat{E}_{t-1}x_t + B_tu_t + K_t(y_t - \widehat{E}_{t-1}y_t) \tag{21.35}$$

where

$$K_t = \left(A_t\Sigma_tC_t' + G_tV_{3t}\right)\left(C_t\Sigma_tC_t' + V_{2t}\right)^{-1}. \tag{21.36}$$

Using $\widehat{E}_{t-1}y_t = C_t\widehat{E}_{t-1}x_t + E_tu_t$, equation (21.35) can also be written

$$\widehat{E}_tx_{t+1} = (A_t - K_tC_t)\widehat{E}_{t-1}x_t + (B_t - K_tE_t)u_t + K_ty_t. \tag{21.37a}$$

We now aim to derive a recursive formula for the covariance matrix $\Sigma_t$. From equation (21.26) we know that $\widehat{E}_{t-1}y_t = C_t\widehat{E}_{t-1}x_t + E_tu_t$. Subtracting this expression from $y_t$ in equation (21.26) gives

$$y_t - \widehat{E}_{t-1}y_t = C_t(x_t - \widehat{E}_{t-1}x_t) + w_{2t}. \tag{21.37b}$$

Substituting this expression in equation (21.35) and subtracting the result from equation (21.26) gives

$$\begin{aligned}(x_{t+1} - \widehat{E}_tx_{t+1}) = (A_t - K_tC_t)&(x_t - \widehat{E}_{t-1}x_t)\\ &+ G_tw_{1t+1} - K_tw_{2t}\end{aligned}$$

or

$$\tilde{x}_{t+1} = (A_t - K_tC_t)\tilde{x}_t + G_tw_{1t+1} - K_tw_{2t}. \tag{21.38}$$

From equation (21.38) and our specification of the covariance matrix

$$E\begin{bmatrix} w_{1t+1} \\ w_{2t} \end{bmatrix}\begin{bmatrix} w_{1t+1} \\ w_{2t} \end{bmatrix}' = \begin{bmatrix} V_{1t} & V_{3t} \\ V_{3t}' & V_{2t} \end{bmatrix}$$

we have

$$\begin{aligned}E\tilde{x}_{t+1}\tilde{x}_{t+1}' = &\left(A_t - K_tC_t\right)E\tilde{x}_t\tilde{x}_t'\left(A_t - K_tC_t\right)'\\ &+ G_tV_{1t}G_t' + K_tV_{2t}K_t'\\ &- G_tV_{3t}K_t' - K_tV_{3t}'G_t'\end{aligned}$$

We have defined the covariance matrix of $\tilde{x}_t$ as $\Sigma_t = E\tilde{x}_t\tilde{x}_t' = E(x_t - \widehat{E}_{t-1}x_t)$ $(x_t - \widehat{E}_{t-1}x_t)'$. So we can express the preceding equation as

$$\begin{aligned}\Sigma_{t+1} = &\Big(A_t - K_tC_t\Big)\Sigma_t\Big(A_t - K_tC_t\Big)' \\ &+ G_tV_{1t}G_t' + K_tV_{2t}K_t' - G_tV_{3t}K_t' \\ &- K_tV_{3t}'G_t'.\end{aligned} \tag{21.39}$$

Equation (21.39) can be rearranged to the equivalent form

$$\begin{aligned}\Sigma_{t+1} &= A_t\Sigma_tA_t' + G_tV_{1t}G_t' \\ &- \Big(A_t\Sigma_tC_t' + G_tV_{3t}\Big)\Big(C_t\Sigma_tC_t' + V_{2t}\Big)^{-1}\Big(A_t\Sigma_tC_t + G_tV_{3t}\Big)'\end{aligned} \tag{21.40}$$

Starting from the given initial condition for $\Sigma_{t_0} = E(x_{t_0} - Ex_{t_0})(x_{t_0} - Ex_{t_0})'$, equations (21.39) and (21.36) give a recursive procedure for generating the "Kalman gain" $K_t$, which is the crucial unknown ingredient of the recursive algorithm (21.35) for generating $\widehat{E}_tx_{t+1}$. The Kalman filter is used as follows: Starting from time $t_0$ with $\Sigma_{t_0} = \Sigma_0$ and $\hat{x}_{t_0} = Ex_0$ given, equation (21.36) is used to form $K_{t_0}$, and equation (21.35) is used to obtain $\widehat{E}_{t_0}x_{t_0+1}$ with $\widehat{E}_{t_0-1}x_{t_0} = \hat{x}_0$. Then equation (21.39) is used to form $\Sigma_{t_0+1}$, equation (21.36) is used to form $K_{t_0+1}$, equation (21.35) is used to obtain $\widehat{E}_{t_0+1}x_{t_0+2}$, and so on.

Define $\hat{x}_t = \widehat{E}_{t-1}x_t$ and $\hat{y}_t = \widehat{E}_{t-1}y_t$. Set

$$a_t = w_{2t} + C_t(x_t - \hat{x}_t) \tag{21.41}$$

From equation (21.37b), we have

$$y_t - \hat{y}_t = C_t(x_t - \hat{x}_t) + w_{2t}$$

or

$$y_t - \hat{y}_t = a_t. \tag{21.42}$$

We know that $Ea_ta_t' = C_t\Sigma_tC_t' + V_{2t}$. The random process $a_t$ is the "innovation" in $y_t$, that is, the part of $y_t$ that cannot be predicted linearly from past $y$'s.

From equations (21.25) and (21.42) we get $y_t = C_t\hat{x}_{t+1}H_tu_t + a_t$. Substituting this expression into equation (21.37a) produces the following system:

$$\begin{aligned}\hat{x}_{t+1} &= A_t\hat{x}_t + B_tu_t + K_ta_t \\ y_t &= C_t\hat{x}_t + H_tu_t + a_t\end{aligned} \tag{21.43}$$

System (21.43) is called an *innovations representation.*

Another representation of the system that is useful is obtained from equation (21.37a):

$$\begin{aligned} \hat{x}_{t+1} &= (A_t - K_t C_t)\hat{x}_t + (B_t - K_t H_t)\, u_t + K_t y_t \\ a_t &= y_t - C_t \hat{x}_t - H_t u_t \end{aligned} \tag{21.44}$$

This is called a *whitening filter.* Starting from a given $\hat{x}_{t_0}$, this system accepts as an "input" a history of $y_t$ and gives as an output the sequence of innovations $a_t$, which by construction are serially uncorrelated.

We shall often study situations in which the system is time invariant, that is, $A_t = A$, $B_t = B$, $G_t = G$, $H_t = H$, $C_t = C$, and $V_{jt} = V_j$ for all $t$. We shall later describe regulatory conditions on $A, C, V_1, V_2$, and $V_3$ which imply that (1) $K_t \to K$ as $t \to \infty$ and $\Sigma_t \to \Sigma$ as $t \to \infty$; and (2) $\mid \lambda_i(A - KC) \mid < 1$ for all $i$, where $\lambda_i$ is the $i$th eigenvalue of $(A - KC)$. When these conditions are met, the limiting representation for equation (21.44) is time invariant and is an (infinite dimensional) innovations representation. Using the lag operator $L$ where $L\hat{x}_t = \hat{x}_{t-1}$, imposing time invariance in equation (21.43), and rearranging gives the representation

$$y_t = [I + C(L^{-1}I - A)^{-1}K]a_t + \left[H + C(L^{-1}I - A)\, B\right] u_t \tag{21.45}$$

which expresses $y_t$ as a function of $[a_t, a_{t-1}, \ldots]$. In order that $[y_t, y_{t-1}, \ldots]$ span the same linear space as $[a_t, a_{t-1}, \ldots]$, it is necessary that the following condition be met:

$$\det\,[I + C(zI - A)^{-1}K] = 0 \;\Rightarrow \mid z \mid < 1.$$

Now by a theorem from linear algebra we know that[3]

$$\det[I + C(zI - A)^{-1}K] = \frac{\det[zI - (A - KC)]}{\det(zI - A)}.$$

The formula shows that the zeros of $\det[I + C(zI - A)^{-1}K]$ are zeros of $\det[zI - (A - KC)]$, which are eigenvalues of $A - KC$. Thus, if the eigenvalues of $(A - KC)$ are all less than unity in modulus, then the spaces $[a_t, a_{t-1}, \ldots]$ and $[y_t, y_{t-1}, \ldots]$ in representation (21.45) are equal.

[3] See Noble and Daniel (1977, exercises 6.49 and 6.50, p. 210).

## *Duality*

For purposes of highlighting their relationship, we now repeat the Kalman filtering formulas for $K_t$ and $\Sigma_t$ and the optimal linear regulator formulas for $F_t$ and $P_t$

$$K_t = \left(A_t \Sigma_t C_t' + G_t V_{3t}\right)\left(C_t \Sigma_t C_t' + V_{2t}\right)^{-1}. \tag{21.46}$$

$$\begin{aligned} \Sigma_{t+1} = & A_t \Sigma_t A_t' + G_t V_{1t} G_t' \\ & - \left(A_t \Sigma_t C_t' + G_t V_{3t}\right)\left(C_t \Sigma_t C_t' + V_{2t}\right)^{-1} \\ & \times \left(A_t \Sigma_t C_t' + G_t V_{3t}\right)' \end{aligned} \tag{21.47}$$

$$F_t = (Q_t + B_t' P_{t+1} B_t)^{-1} (B_t' P_{t+1} A_t + W_t'). \tag{21.48}$$

$$\begin{aligned} P_t = & R_t + A_t' P_{t+1} A_t \\ & - (A_t' P_{t+1} B_t + W_t)\,(Q_t + B_t' P_{t+1} B_t)^{-1} \\ & \times \left(B_t' P_{t+1} A_t + W_t'\right) \end{aligned} \tag{21.49}$$

for $t = t_0, t_0 + 1, \ldots, t_1$. Equations (21.46) and (21.47) are solved forward from $t_0$ with $\Sigma_{t_0}$ given, while equations (21.48) and (21.49), are solved backward from $t_1 - 1$ with $P_{t_1}$ given.

The equations for $K_t$ and $F_t$ are intimately related, as are the equations for $P_t$ and $\Sigma_t$. In fact, upon properly reinterpreting the various matrices in equations (21.46), (21.47), (21.48), and (21.49), the equations for the Kalman filter and the optimal linear regulator can be seen to be identical. Thus, where $A$ appears in the Kalman filter, $A'$ appears in the corresponding regulator equation; where $C$ appears in the Kalman filter, $B'$ appears in the corresponding regulator equation; and so on. The correspondences are listed in detail in Table 21.1. By taking account of these correspondences, a single set of computer programs can be used to solve either an optimal linear regulator problem or a Kalman filtering problem.

The concept of *duality* helps to clarify the relationship between the optimal regulator and the Kalman filtering problem.

**Table 21.1**

| Object in Optimal Linear Regulator Problem | Object in Kalman Filter |
|---|---|
| $A_{t_0+s}, s = 0, \ldots, t_1 - t_0 - 1$ | $A'_{t_1-1-s}, s = 0, \ldots, t_1 - t_0 - 1$ |
| $B_{t_0+s}$ | $C'_{t_1-1-s}$ |
| $R_{t_0+s}$ | $-G_{t_1-1-s} V_{1t_1-1-s} G'_{t_1-1-s}$ |
| $Q_{t_0+s}$ | $-V_{2t_1-1-s}$ |
| $W_{t_0+s}$ | $-G_{t_1-1-s} V_{3t_1-1-s}$ |
| $P_{t_0+s}$ | $-\Sigma_{t_1-s}$ |
| $F_{t_0+s}$ | $K'_{t_1-1-s}$ |
| $P_{t_1}$ | $-\Sigma_{t_0}$ |
| $A_{t_0+s} - B_{t_0+s} F_{t_0+s}$ | $A'_{t_1-1-s} - C'_{t_1-1-s} K'_{t_1-1-s}$ |

DEFINITION 21.1: Consider the time-varying linear system.

$$\begin{aligned} x_{t+1} &= A_t x_t + B_t u_t \\ y_t &= C_t x_t, \ t = t_0, \ldots, t_1 - 1 \end{aligned} \tag{21.50}$$

The *dual* of system (21.50) (sometimes called the "dual with respect to $t_1 - 1$") is the system

$$\begin{aligned} x^*_{t+1} &= A'_{t_1-1-t} x^*_t + C'_{t_1-1-t} u^*_t \\ y^*_t &= B'_{t_1-1-t} x^*_t \end{aligned}$$

with $t = t_0, t_0 + 1, \ldots, t_1 - 1$.

With this definition, the correspondence exhibited in Table 21.1 can be summarized succinctly in the following proposition:

PROPOSITION 21.1: Let the solution of the optimal linear regulator problem defined by the given matrices $\{A_t, B_t, R_t, Q_t, W_t; t = t_0, \ldots, t_1 - 1;\ P_{t_1}\}$ be given by $\{P_t, F_t,\ t = t_0, \ldots, t_1 - 1\}$. Then the solution of the Kalman filtering problem defined by the matrices $\{A'_{t_1-1-t}, C'_{t_1-1-t}, -G_{t_1-1-t} V_{1t_1-1-t} G'_{t_1-1-t}, -V_{2t_1-1-t}, -G_{t_1-1-t} V_{3t_1-1-t};\ t = t_0, \ldots, t_1 - 1;\ \Sigma_{t_0}\}$ is given by $\{K'_{t_1-t-1} = F_t, -\Sigma_{t_1-t} = P_t;\ t = t_0,\ t_0 + 1, \ldots, t_1 - 1\}$.

This proposition describes the sense in which the Kalman filtering problem and the optimal linear regulator problems are "dual" to one another. As is also true of so-called classical control and filtering methods, the same equations arise in solving both the filtering problem and the control problem. This fact implies that almost everything that we learn about the control problem applies to the filtering problem, and vice versa.

As an example of the use of duality, recall the transformations (21.13) and (21.14) that we used to convert the optimal linear regulator problem with cross-products between *states* and *controls* into an equivalent problem with no such cross-products. The preceding discussion of duality and Table 21.1 suggest that the same transformation will convert the original dual filtering problem, which has nonzero covariance matrix $V_3$ between *state noise* and *measurement noise*, into an equivalent problem with covariances zero. This hunch is correct. The transformations, which can be obtained by duality directly from equations (21.13) and (21.14), are for $t = t_0, \ldots, t_1 - 1$

$$\begin{aligned}\bar{A}'_{t_1-1-t} &= A'_{t_1-1-t} - C'_{t_1-1-t}V^{-1}_{2t_1-1-t}V'_{3t_1-1-t}G'_{t_1-1-t}\\ -\bar{V}_{1t_1-1-t} &= -V_{1t_1-1-t} + V_{3t_1-1-t}V^{-1}_{2t_1-1-t}V'_{3t_1-1-t}\end{aligned}$$

The Kalman filtering problem defined by the matrices $\{\bar{A}_t, C_t, -G_t\bar{V}_{1t}G'_t - V_{2t}, 0;$ $t = t_0, \ldots, t_1 - 1;\ \Sigma_0\}$ is equivalent to the original problem in the sense that

$$A_t - K_tC_t = \bar{A}_t - \bar{K}_tC_t$$

where $\bar{K}_t$ is the solution of the transformed problem. We also have, by the results for the regulator problem and duality, the following:

$$\bar{K}_t = K_t - G_tV_{3t}V_{2t}^{-1}.$$

## Examples of Kalman filtering

This section contains several examples that have been widely used by economists and that fit into the Kalman filtering setting. After the reader has worked through our examples, no doubt many other examples will occur.

*a. Vector autoregression*: We consider an $(n \times 1)$ stochastic process $y_t$ that obeys the linear stochastic difference equation

$$y_t = A_1y_{t-1} + \ldots + A_my_{t-m} + \varepsilon_t$$

where $\varepsilon_t$ is an $(n\times 1)$ vector white noise, with mean zero and $E\varepsilon_t\varepsilon_t' = V_{1t}$, $E\varepsilon_t y_s' = 0,\ t > s$. We define the state vector $x_t$ and shock vector $w_t$ as

$$x_t = \begin{bmatrix} y_{t-1} \\ y_{t-2} \\ \vdots \\ y_{t-m} \end{bmatrix}, \quad \begin{bmatrix} w_{1t+1} \\ w_{2t} \end{bmatrix} = \begin{pmatrix} \varepsilon_t \\ \varepsilon_t \end{pmatrix}.$$

The law of motion of the system then becomes

$$\begin{bmatrix} y_t \\ y_{t-1} \\ y_{t-2} \\ \vdots \\ y_{t-m+1} \end{bmatrix} = \begin{bmatrix} A_1 & A_2 & \dots & A_m \\ I & 0 & \dots & 0 \\ 0 & I & \dots & 0 \\ \vdots & \vdots & \ddots & \vdots \\ 0 & \dots & I & 0 \end{bmatrix} \begin{pmatrix} y_{t-1} \\ y_{t-2} \\ y_{t-3} \\ \vdots \\ y_{t-m} \end{pmatrix} + \begin{bmatrix} I \\ 0 \\ 0 \\ \vdots \\ 0 \end{bmatrix} \varepsilon_t.$$

The measurement equation is

$$y_t = [A_1\ A_2 \dots A_m]\, x_t + \varepsilon_t.$$

For the filtering equations, we have

$$A_t = \begin{bmatrix} A_1 & A_2 & \dots & A_m \\ I & 0 & \dots & 0 \\ 0 & I & \dots & 0 \\ \vdots & \vdots & \ddots & \vdots \\ 0 & \dots & I & 0 \end{bmatrix}, \quad G_t = G = \begin{bmatrix} I \\ 0 \\ 0 \\ \vdots \\ 0 \end{bmatrix}$$

$$C_t = [A_1, \dots, A_n]$$

$$V_{1t} = V_{2t} = V_{3t}.$$

Starting from $\Sigma_{t_0} = 0$, which means that the system is imagined to start up with $m$ lagged values of $y$ having been observed, equation (21.35) implies

$$K_{t_0} = G,$$

while equation (21.39) implies that $\Sigma_{t_0+1} = 0$. It follows recursively that $K_t = G$ for all $t \geq t_0$ and that $\Sigma_t = 0$ for all $t \geq t_0$. Computing $(A - KC)$, we find that

$$\widehat{E}_t x_{t+1} = \begin{bmatrix} 0 & 0 & \dots & 0 \\ I & 0 & \dots & 0 \\ 0 & I & \dots & 0 \\ \vdots & & & \\ 0 & \dots & I & 0 \end{bmatrix} \widehat{E}_{t-1} x_t + \begin{bmatrix} I \\ 0 \\ \vdots \\ 0 \end{bmatrix} y_t,$$

which is equivalent with

$$\hat{E}_t x_{t+1} = \begin{bmatrix} y_t \\ y_{t-1} \\ \vdots \\ y_{t-m} \end{bmatrix}.$$

The equation $\hat{E}_t y_{t+1} = C\hat{E}_t x_{t+1}$ becomes

$$\hat{E}_t y_{t+1} = A_1 y_t + A_2 y_{t-1} + \ldots + A_m y_{t-m+1}.$$

Evidently, the preceding equation for forecasting a vector autoregressive process can be obtained in a much less roundabout manner, with no need to use the Kalman filter.

*b. Univariate moving average*: We consider the model

$$y_t = w_t + c_1 w_{t-1} + \ldots + c_n w_{t-n}$$

where $w_t$ is a univariate white noise with mean zero and variance $V_{1t}$. We write the model in the state-space form

$$x_{t+1} = \begin{bmatrix} w_t \\ w_{t-1} \\ \vdots \\ w_{t-n+1} \end{bmatrix} = \begin{bmatrix} 0 & 0 & \ldots & 0 \\ 1 & 0 & \ldots & 0 \\ \vdots & \vdots & \ddots & \vdots \\ 0 & \ldots & 1 & 0 \end{bmatrix} \begin{bmatrix} w_{t-1} \\ w_{t-2} \\ \vdots \\ w_{t-n} \end{bmatrix} + \begin{bmatrix} 1 \\ 0 \\ \vdots \\ 0 \end{bmatrix} w_t$$
$$y_t = [c_1 \; c_2 \ldots c_n] x_t + w_t.$$

We assume that $\Sigma_{t_0} = 0$, so that the initial state is known. In this setup, we have $A, G$, and $C$ as indicated previously, and $w_{1t+1} = w_t, w_{2t} = w_t$, and $V_1 = V_2 = V_3$. Iterating on the Kalman filtering equations (21.39) and (21.35) with $\Sigma_{t_0} = 0$, we obtain $\Sigma_t = 0,\ t \geq t_0,\ K_t = G,\ t \geq t_0$, and

$$(A - KC) = \begin{pmatrix} -c_1 & -c_2 & \ldots & -c_{n-1} & -c_n \\ 1 & 0 & \ldots & 0 & 0 \\ 0 & 1 & \ldots & 0 & 0 \\ \vdots & \vdots & \ddots & \vdots & \vdots \\ 0 & 0 & \ldots & 1 & 0 \end{pmatrix}.$$

It follows that

$$\widehat{E}_t x_{t+1} = \widehat{E}_t \begin{pmatrix} w_t \\ w_{t-1} \\ \vdots \\ w_{t-n+1} \end{pmatrix} = \begin{pmatrix} -c_1 & -c_2 & \dots & -c_{n-1} & -c_n \\ 1 & 0 & \dots & 0 & 0 \\ 0 & 1 & \dots & 0 & 0 \\ \vdots & \vdots & \ddots & \vdots & \vdots \\ 0 & 0 & \dots & 1 & 0 \end{pmatrix}$$

$$\widehat{E}_{t-1} \begin{pmatrix} w_{t-1} \\ w_{t-2} \\ \vdots \\ w_{t-n} \end{pmatrix} + \begin{pmatrix} 1 \\ 0 \\ \vdots \\ 0 \end{pmatrix} y_t .$$

With $\Sigma_{t_0} = 0$, this equation implies

$$\widehat{E}_t w_t = y_t - c_1 w_{t-1} - \dots - c_n w_{t-n}.$$

Thus the innovation $w_t$ is recoverable from knowledge of $y_t$ and $n$ past innovations.

c. *Mixed moving average–autoregression*: We consider the univariate, mixed second-order autoregression, first-order moving average process

$$y_t = A_1 y_{t-1} + A_2 y_{t-2} + v_t + B_1 v_{t-1}$$

where $v_t$ is a white noise with mean zero, $E v_t^2 = V_1$ and $E v_t y(s) = 0$ for $s < t$. The trick in getting this system into the state-space form is to define the state variables $x_{1t} = y_t - v_t$, and $x_{2t} = A_2 y_{t-1}$. With these definitions the system and measurement equations become

$$x_{t+1} = \begin{pmatrix} A_1 & 1 \\ A_2 & 0 \end{pmatrix} x_t + \begin{pmatrix} B_1 + A_1 \\ A_2 \end{pmatrix} v_t \tag{21.51}$$

$$y_t = [1\ 0] x_t + v_t. \tag{21.52}$$

Notice that using equation (21.51) and (21.52) repeatedly, we have

$$\begin{aligned} y_t = x_{1t} + v_t &= A_1 x_{1t-1} + x_{2t-1} + (B_1 + A_1) v_{t-1} + v_t \\ &= A_1 (x_{1t-1} + v_{t-1}) + v_t + B_1 v_{t-1} + A_2 (x_{1t-2} + v_{t-2}) \\ &= A_1 y_{t-1} + A_2 y_{t-2} + v_t + B_1 v_{t-1} \end{aligned}$$

as desired. With the state and measurement equations (21.51) and (21.52), we have $V_1 = V_2 = V_3$,

$$A = \begin{pmatrix} A_1 & 1 \\ A_2 & 0 \end{pmatrix}, G = \begin{pmatrix} B_1 + A_1 \\ A_2 \end{pmatrix}, \ C = [1\ 0].$$

We start the system off with $\Sigma_{t_0} = 0$, so that the initial state is imagined to be known. With $\Sigma_{t_0} = 0$, recursions on equations (21.35) and (21.39) imply that $\Sigma_t = 0$ for $t \geq t_0$ and $K_t = G$ for $t \geq t_0$. Computing $A - KC$ we find

$$(A - KC) = \begin{pmatrix} -B_1 & 1 \\ 0 & 0 \end{pmatrix}$$

and we have

$$\widehat{E}_t x_{t+1} = \begin{bmatrix} -B_1 & 1 \\ 0 & 0 \end{bmatrix} \hat{\imath}_{t-1} x_t + \begin{bmatrix} B_1 + A_1 \\ A_2 \end{bmatrix} y_t.$$

Therefore the recursive prediction equations become

$$\widehat{E}_t y_{t+1} = [1 \quad 0]\, \widehat{E}_{t+1} x_{t+1} = \widehat{E}_t x_{1t+1}.$$

Recalling that $x_{2t} = A_2 y_{t-1}$, the preceding two equations imply that

$$\widehat{E}_t y_{t+1} = -B_1 \widehat{E}_{t-1} y_t + A_2 y_{t-1} + (B_1 + A_1) y_t. \tag{21.53}$$

Consider the special case in which $A_2 = 0$, so that the $y_t$ obeys a first-order moving average, first-order autoregressive process. In this case equation (21.53) can be expressed

$$\widehat{E}_t y_{t+1} = B_1 (y_t - \widehat{E}_{t-1} y_t) + A_1 y_t,$$

which is a version of the Cagan-Friedman "error-learning" model. The solution of the preceding difference equation for $\widehat{E}_t y_{t+1}$ is given by the geometric distributed lag

$$\begin{aligned} \widehat{E}_t y_{t+1} = (B_1 + A_1) \sum_{j=0}^{m} (-B_1)^j y_{t-j} \\ + (-B_1)^{m+1} \widehat{E}_{t-m-1} y_{t-m}. \end{aligned}$$

For the more general case depicted in equation (21.53) with $A_2 \neq 0$, $\widehat{E}_t y_{t+1}$ can be expressed as a convolution of two geometric lag distributions in current and past $y_t$'s.

*d. Linear regressions*: Consider the standard linear regression model

$$y_t = z_t \beta + \varepsilon_t, \quad t = 1, 2, \ldots, T$$

where $z_t$ is a $1 \times n$ vector of independent variables, $\beta$ is an $n \times 1$ vector of parameters, and $\varepsilon_t$ is a serially uncorrelated random term with mean zero and variance $E\varepsilon_t^2 = \sigma^2$, and satisfying $E\varepsilon_t z_s = 0$ for $t \geq s$. The least squares estimator of $\beta$ based on $t$ observations, denoted $\hat{\beta}_{t+1}$, is obtained as follows. Define the stacked matrices

$$Z_t = \begin{bmatrix} z_1 \\ z_2 \\ \vdots \\ z_t \end{bmatrix}, \quad Y_t = \begin{bmatrix} y_1 \\ y_2 \\ \vdots \\ y_t \end{bmatrix}.$$

Then the least squares estimator based on data through time $t$ is given by

$$\hat{\beta}_{t+1} = (Z_t' Z_t)^{-1} Z_t' Y_t \tag{21.54}$$

with covariance matrix

$$E(\hat{\beta}_{t+1} - E\hat{\beta}_{t+1})(\hat{\beta}_{t+1} - E\hat{\beta}_{t+1})' = \sigma^2 (Z_t' Z_t)^{-1}. \tag{21.55}$$

For reference, we note that

$$\begin{aligned} \hat{\beta}_t &= (Z_{t-1}' Z_{t-1})^{-1} Z_{t-1}' Y_{t-1} \\ E(\hat{\beta}_t - E\hat{\beta}_t)(\hat{\beta}_t - E\hat{\beta}_t)' &= \sigma^2 (Z_{t-1}' Z_{t-1})^{-1}. \end{aligned} \tag{21.55e}$$

If $\hat{\beta}_t$ has been computed by equation (21.55e), it is computationally inefficient to compute $\hat{\beta}_{t+1}$ by equation (21.54) when new data $(y_t, z_t)$ arrive at time $t$. In particular, we can avoid inverting the matrix $(Z_t' Z_t)$ directly, by employing a recursive procedure for inverting it. This approach can be viewed as an application of the Kalman filter. We explore this connection briefly.

We begin by noting how least squares estimators can be computed recursively by means of the Kalman filter. We let $y_t$ in the Kalman filter be $y_t$ in the regression model. We then set $x_t = \beta$ for all $t$, $V_{1t} = 0$, $V_{3t} = 0$, $V_{2t} = \sigma^2$, $w_{1t+1} = 0$, $w_{2t} = \varepsilon_t$, $A = I$, and $C_t = z_t$. Let

$$\hat{\beta}_{t+1} = E\left[\beta \mid y_t, y_{t-1}, \ldots y_1, z_t, z_{t-1}, \ldots, z_1, \hat{\beta}_0\right],$$

where $\hat{\beta}_0$ is $\hat{x}_0$. Also, let $\Sigma_t = E(\hat{\beta}_t - E\hat{\beta}_t)(\hat{\beta}_t - E\hat{\beta}_t)'$. We start things off with a "prior" covariance matrix $\Sigma_0$. With these definitions, the recursive formulas (21.35) and (21.39) become

$$\begin{aligned} K_t &= \Sigma_t z_t'(\sigma^2 + z_t \Sigma_t z_t')^{-1} \\ \Sigma_{t+1} &= \Sigma_t - \Sigma_t z_t'(\sigma^2 + z_t \Sigma_t z_t')^{-1} z_t \Sigma_t \end{aligned} \tag{21.56}$$

Applying the formula $\hat{x}_{t+1} = (A - K_t C_t)\hat{x}_t + K_t y_t$ to the present problem with the preceding formula for $K_t$ we have

$$\hat{\beta}_{t+1} = (I - K_t z_t)\hat{\beta}_t + K_t y_t. \tag{21.57}$$

We now show how equations (21.56) and (21.57) can be derived directly from equations (21.54) and (21.55). From a matrix inversion formula (see Noble and Daniel, 1977, p. 194), we have

$$\begin{aligned} (Z_t'Z_t)^{-1} &= (Z_{t-1}'Z_{t-1})^{-1} \\ &\quad - (Z_{t-1}'Z_{t-1})^{-1} z_t'[1 + z_t(Z_{t-1}'Z_{t-1}^1)^{-1} z_t']^{-1} z_t (Z_{t-1}'Z_{t-1})^{-1} \end{aligned} \tag{21.58}$$

Multiplying both sides of equation (21.58) by $\sigma^2$ immediately gives equation (21.56). Use the right side of equation (21.58) to substitute for $(Z_t'Z_t)^{-1}$ in equation (21.54) and write

$$Z_t'Y_t = Z_{t-1}'Y_{t-1} + z_t' y_t$$

to obtain

$$\begin{aligned} \hat{\beta}_{t+1} &= \frac{1}{\sigma^2}\{\Sigma_t - \Sigma_t z_t'(\sigma^2 + z_t \Sigma_t z_t')^{-1} z_t \Sigma_t\} \\ &\qquad \cdot \{Z_{t-1}'Y_{t-1} + z_t' y_t\} \\ &= \underbrace{\frac{1}{\sigma^2}\Sigma_t Z_{t-1}'Y_{t-1}}_{\hat{\beta}_t} - \underbrace{\Sigma_t z_t'(\sigma^2 + z_t \Sigma_t z_t')^{-1}}_{K_t} \underbrace{z_t}_{C_t} \underbrace{\frac{1}{\sigma^2}\Sigma_t Z_{t-1}'Y_{t-1}}_{\beta_t} \\ &\qquad + \underbrace{\Sigma_t Z_t'(\sigma^2 + z_t \Sigma_t Z_t')^{-1}}_{K_t} y_t \end{aligned}$$

$$\hat{\beta}_{t+1} = (A - K_t C_t)\hat{\beta}_t + K_t y_t.$$

These formulas are evidently equivalent with those asserted earlier.

## Linear projections

For reference we state the following theorems about linear least squares projections. We let $Y$ be an $(n \times 1)$ vector of random variables and $X$ be an $(h \times 1)$ vector of random variables. We assume that the following first and second moments exist:

$$EY = \mu_Y, \; EX = \mu_X,$$
$$EXX' = S_{XX}, \; EYY' = S_{YY}, \; EYX' = S_{YX}.$$

Letting $x = X - EX$ and $y = Y - EY$, we define the following covariance matrices

$$Exx' = \Sigma_{xx}, \; E'_{yy} = \Sigma_{yy}, \; Eyx' = \Sigma_{yx}.$$

We are concerned with estimating $Y$ as a linear function of $X$. The estimator of $Y$ that is a linear function of $X$ and that minimizes the mean squared error between each component $Y$ and its estimate is called the *linear projection of Y on X*.

DEFINITION 21.2: The *linear projection* of $Y$ on $X$ is the affine function $\hat{Y} = AX + a_0$ that minimizes $E$ trace $\{(Y - \hat{Y})(Y - \hat{Y})'\}$ over all affine functions $a_0 + AX$ of $X$. We denote this linear projection as $\widehat{E}[Y \mid X]$, or sometimes as $\widehat{E}[Y \mid x, 1]$ to emphasize that a constant is included in the "information set."

The linear projection of $Y$ on $X$, $\widehat{E}[Y \mid X]$ is also sometimes called the *wide sense expectation of Y conditional on X*. We have the following theorems:

THEOREM 21.1:

$$\widehat{E}[Y \mid X] = \mu_y + \Sigma_{yx}\Sigma_{xx}^{-1}(X - \mu_x). \tag{21.59}$$

PROOF: The theorem follows immediately by writing out $E\,\text{trace}\,(Y-\hat{Y})(Y-\hat{Y})'$ and completing the square, or else by writing out $E\,\text{trace}(Y - \hat{Y})(Y - \hat{Y})'$ and obtaining first-order necessary conditions ("normal equations") and solving them. ∎

THEOREM 21.2:

$$\widehat{E}\left[\left(Y - \widehat{E}[Y \mid x]\right) \mid X'\right] = 0$$

This equation states that the errors from the projection are orthogonal to each variable included in $X$.

PROOF: Immediate from the normal equations. ∎

THEOREM 21.3: Orthogonality principle:

$$E\Big[[Y - \widehat{E}\,(Y \mid x)]\,x'\Big] = 0.$$

PROOF: Follows from Theorem 21.3. ∎

THEOREM 21.4: Orthogonal regressions:

Suppose that $X' = (X_1, X_2, \ldots, X_h)', EX' = \mu' = (\mu_{x1}, \ldots, \mu_{xh})'$, and $E(X_i - \mu_{xi})\,(X_j - \mu_{xj}) = 0$ for $i \neq j$. Then

$$\widehat{E}\,[Y \mid x_1, \ldots, x_n, 1] = \widehat{E}\,[Y \mid x_1] + \widehat{E}\,[Y \mid x_2] + \ldots + \widehat{E}\,[Y \mid x_n] - (n-1)\mu_y \quad (21.60)$$

PROOF: Note that from the hypothesis of orthogonal regressors, the matrix $\Sigma_{xx}$ is diagonal. Applying equation (21.59) then gives equation (21.60). ∎

## *Hidden Markov chains*

This section gives a brief introduction to hidden Markov chains, a tool that is useful to study a variety of nonlinear filtering problems in finance and economics. We display a solution to a nonlinear filtering problem that a reader might want to compare to the linear filtering problem described earlier.

Consider an $N$-state Markov chain. We can represent the state space in terms of the unit vectors $S_x = \{e_1, \ldots, e_N\}$, where $e_i$ is the $i$th $N$-dimensional unit vector. Let the $N \times N$ transition matrix be $P$, with $(i, j)$ element

$$P_{ij} = \text{Prob}(x_{t+1} = e_j \mid x_t = e_i).$$

With these definitions, we have

$$E x_{t+1} \mid x_t = P' x_t.$$

Define the "residual"

$$v_{t+1} = x_{t+1} - P'x_t,$$

which implies the linear "state-space" representation

$$x_{t+1} = P'x_t + v_{t+1}.$$

Notice how it follows that $E\ v_{t+1} \mid x_t = 0$, which qualifies $v_{t+1}$ as a "martingale process adapted to $x_t$."

We want to append a "measurement equation." Suppose that $x_t$ is not observed, but that $y_t$, a noisy function of $x_t$, is observed. Assume that $y_t$ lives in the $M$-dimensional space $S_y$, which we represent in terms of $M$ unit vectors: $S_y = \{f_1, \ldots, f_M\}$, where $f_i$ is the $i$th $M$-dimensional unit vector. To specify a linear measurement equation $y_t = C(x_t, u_t)$, where $u_t$ is a measurement noise, we begin by defining the $N \times M$ matrix $Q$ with

$$\text{Prob}\ (y_t = f_j \mid x_t = e_i) = Q_{ij}.$$

It follows that

$$E\ (y_t \mid x_t) = Q'x_t.$$

Define the residual

$$u_t \equiv y_t - E\ y_t \mid x_t,$$

which suggests the "observer equation"

$$y_t = Q'x_t + u_t.$$

It follows from the definition of $u_t$ that $E\ u_t \mid x_t = 0$. Thus, we have the linear state-space system

$$\begin{aligned} x_{t+1} &= P'x_t + v_{t+1} \\ y_t &= Q'x_t + u_t. \end{aligned}$$

Using the definitions, it is straightforward to calculate the conditional second moments of the error processes $v_{t+1}, u_t$.[4]

---

[4] Notice that

$$\begin{aligned} x_{t+1}x'_{t+1} = P'x_t(P'x_t)' + P'x_t v'_{t+1} \\ + v_{t+1}(P'x_t)' + v_{t+1}v'_{t+1} \end{aligned}$$

*Optimal filtering*

We seek a recursive formula for computing the conditional distribution of the hidden state:

$$\rho_i(t) = \text{Prob}\{x_t = i \mid y_1 = \eta_1, \ldots, y_t = \eta_t\}.$$

Denote the history of observed $y_t$'s up to $t$ as $\eta^t = \text{col}\,(\eta_1, \ldots, \eta_t)$. Define the conditional probabilities

$$p(\xi_t, \eta_1, \ldots, \eta_t) = \text{Prob}\,(x_t = \xi_t, y_1 = \eta_1, \ldots, y_t = \eta_t),$$

and assume $p(\eta_1, \ldots, \eta_t) \neq 0$. Then apply the calculus of conditional expectations to compute[5]

$$\begin{aligned} p(\xi_t \mid \eta^t) &= \frac{p(\xi_t, \eta_t \mid \eta^{t-1})}{p(\eta_t \mid \eta^{t-1})} \\ &= \frac{\sum_{\xi_{t-1}} p(\eta_t \mid \xi_t)\, p(\xi_t \mid \xi_{t-1}) p(\xi_{t-1} \mid \eta^{t-1})}{\sum_{\xi_t} \sum_{\xi_{t-1}} p(\eta_t \mid \xi_t) p(\xi_t \mid \xi_{t-1}) p(\xi_{t-1} \mid \eta^{t-1})} \end{aligned}$$

---

Substituting into this equation the facts that $x_{t+1} x_{t+1}' = \text{diag}\, x_{t+1} = \text{diag}\,(P' x_t) + \text{diag}\, v_{t+1}$ gives

$$\begin{aligned} v_{t+1} v_{t+1}' = \text{diag}\,(P' x_t) + \text{diag}\,(v_{t+1}) - P' \text{diag}\, x_t P \\ - P' x_t v_{t+1}' (P' x_t)'. \end{aligned}$$

It follows that

$$E\,[v_{t+1} v_{t+1}' \mid x_t] = \text{diag}\,(P' x_t) - P' \text{diag}\, x_t P.$$

Similarly,

$$E\,[u_t\, u_t' \mid x_t] = \text{diag}\,(Q' x_t) - Q' \text{diag}\, x_t Q.$$

[5] Notice that

$$\begin{aligned} p(\xi_t, \eta_t \mid \eta^{t-1}) &= \sum_{\xi_{t-1}} p(\xi_t, \eta_t, \xi_{t-1} \mid \eta^{t-1}) \\ &= \sum_{\xi_{t-1}} p(\xi_t, \eta_t \mid \xi_{t-1}, \eta^{t-1}) p(\xi_{t-1} \mid \eta^{t-1}) \end{aligned}$$

$$\begin{aligned} p(\xi_t, \eta_t \mid \xi_{t-1}, \eta^{t-1}) &= p(\xi_t \mid \xi_{t-1}, \eta^{t-1}) p(\eta_t \mid \xi_t,\ \xi_{t-1}, \eta^{t-1}) \\ &= p(\xi_t \mid \xi_{t-1}) p(\eta_t \mid \xi_t) \end{aligned}$$

Combining these results gives the formula in the text.

This result can be written

$$\rho_i(t+1) = \frac{\sum_s Q_{ij} P_{si} \rho_s(t)}{\sum_s \sum_i Q_{ij} P_{si} \rho_s(t)}$$

where $\eta_{t+1} = j$ is the value of $y$ at $t+1$ We can represent this recursively as

$$\begin{aligned} \tilde{\rho}(t+1) &= \text{ diag } (Q_j) P' \rho(t) \\ \rho(t+1) &= \frac{\tilde{\rho}(t+1)}{< \tilde{\rho}(t+1), \underline{1} >} \end{aligned}$$

where $Q_j$ is the $j$th column of $Q$, and diag$(Q_j)$ is a diagonal matrix with $Q_{ij}$ as the $i$th diagonal element; here $< \cdot, \cdot >$ denotes the inner product of two vectors, and $\underline{1}$ is the unit vector.

# References

Abel, Andrew B., N. Gregory Mankiw, Lawrence H. Summers, and Richard J. Zeckhauser. 1989. "Assessing Dynamic Efficiency: Theory and Evidence." *Review of Economic Studies*, Vol. 56, pp. 1–20.

Abreu, Dilip. 1988. "On the Theory of Infinitely Repeated Games with Discounting." *Econometrica*, Vol. 56, pp. 383–396.

Abreu, Dilip, David Pearce, and Ennio Stacchetti. 1986. "Optimal Cartel Equilibria with Imperfect Monitoring." *Journal of Economic Theory*, Vol. 39, pp. 251–269.

Abreu, Dilip, David Pearce, and Ennio Stacchetti. 1990. "Toward a Theory of Discounted Repeated Games with Imperfect Monitoring." *Econometrica*, Vol. 58(5), pp. 1041–1063.

Acemoglu, Daron. 1997. "Good Jobs versus Bad Jobs: Theory and Some Evidence." Mimeo. CEPR Discussion Paper No. 1588.

Acemoglu, Daron, and Robert Shimer. 1999. "Efficient Unemployment Insurance." *Journal of Political Economy*, Vol. 107, pp. 893–928.

Aghion, Philippe, and Peter Howitt. 1992. "A Model of Growth through Creative Destruction." *Econometrica*, Vol. 60, pp. 323–351.

Aghion, Philippe, and Peter Howitt. 1998. *Endogenous Growth Theory*. Cambridge, MA. MIT Press.

Aiyagari, S. Rao. 1985. "Observational Equivalence of the Overlapping Generations and the Discounted Dynamic Programming Frameworks for One-Sector Growth." *Journal of Economic Theory*, Vol. 35(2), pp. 201–221.

Aiyagari, S. Rao. 1987. "Optimality and Monetary Equilibria in Stationary Overlapping Generations Models with Long Lived Agents." *Journal of Economic Theory*, Vol. 43, pp. 292–313.

Aiyagari, S. Rao. 1993. "Explaining Financial Market Facts: The Importance of Incomplete Markets and Transaction Costs." *Quarterly Review*, Federal Reserve Bank of Minneapolis, Vol. 17(1), pp. 17–31.

Aiyagari, S. Rao. 1994. "Uninsured Idiosyncratic Risk and Aggregate Saving." *Quarterly Journal of Economics*, Vol. 109(3), pp. 659–684.

Aiyagari, S. Rao. 1995. "Optimal Capital Income Taxation with Incomplete Markets and Borrowing Constraints." *Journal of Political Economy*, Vol. 103(6), pp. 1158–1175.

Aiyagari, S. Rao, and Mark Gertler. 1991. "Asset Returns with Transactions Costs and Uninsured Individual Risk." *Journal of Monetary Economics*, Vol. 27, pp. 311–331.

Aiyagari, S. Rao, and Ellen R. McGrattan. 1998. "The Optimum Quantity of Debt." *Journal of Monetary Economics*, Vol. 42(3), pp. 447–469.

Aiyagari, S. Rao, and Neil Wallace. 1991. "Existence of Steady States with Positive Consumption in the Kiyotaki-Wright Model." *Review of Economic Studies*, Vol. 58(5), pp. 901–916.

Albrecht, James, and Bo Axell. 1984. "An Equilibrium Model of Search Unemployment." *Journal of Political Economy*, Vol. 92(5), pp. 824–840.

Altug, Sumru. 1989. "Time-to-Build and Aggregate Fluctuations: Some New Evidence." *International Economic Review*, Vol. 30(4), pp. 889–920.

Altug, Sumru, and Pamela Labadie. 1994. *Dynamic Choice and Asset Markets.* San Diego: Academic Press.

Alvarez, Fernando, and Urban J. Jermann. 1999. "Measuring the Cost of Business Cycles." Mimeo. University of Chicago and Wharton School, University of Pennsylvania.

Anderson, Evan W., Lars P. Hansen, Ellen R. McGrattan, and Thomas J. Sargent. 1996. "Mechanics of Forming and Estimating Dynamic Linear Economies." In Hans M. Amman, David A. Kendrick, and John Rust (eds.), *Handbook of Computational Economics Vol. 1, Handbooks in Economics, Vol. 13.* Amsterdam: Elsevier Science, North-Holland, pp. 171–252.

Apostol, Tom M. 1975. *Mathematical Analysis.* 2nd ed. Reading, MA: Addison-Wesley.

Arrow, Kenneth J. 1962. "The Economic Implications of Learning by Doing." *Review of Economic Studies*, Vol. 29, pp. 155–173.

Arrow, Kenneth J. 1964. "The Role of Securities in the Optimal Allocation of Risk-Bearing." *Review of Economic Studies*, Vol. 31, pp. 91–96.

Åström, K. J. 1965. "Optimal Control of Markov Processes with Incomplete State Information." *Journal of Mathematical Analysis and Applications*, Vol. 10, pp. 174–205.

Atkeson, Andrew G. 1988. "Essays in Dynamic International Economics." Ph.D. dissertation, Stanford University.

Atkeson, Andrew G. 1991. "International Lending with Moral Hazard and Risk of Repudiation." *Econometrica*, Vol. 59(4), pp. 1069–1089.

Atkeson, Andrew, and Robert E. Lucas, Jr. 1992. "On Efficient Distribution with Private Information." *Review of Economic Studies*, Vol. 59(3), pp. 427–453.

Atkeson, Andrew, and Robert E. Lucas, Jr. 1995. "Efficiency and Equality in a Simple Model of Efficient Unemployment Insurance." *Journal of Economic Theory*, Vol. 66(1), pp. 64–88.

Atkeson, Andrew, and Christopher Phelan. 1994. "Reconsidering the Costs of Business Cycles with Incomplete Markets." In Julio J. (ed.), *Fischer, Stanley; Rotemberg, NBER Macroeconomics Annual.* Cambridge, MA: MIT Press, pp. 187–207.

Attanasio, Orazio P. 2000. "Consumption." In John Taylor and Michael Woodford (eds.), *Handbook of Macroeconomics*. Amsterdam: North-Holland.

Attanasio, Orazio P., and Steven J. Davis. 1996. "Relative Wage Movements and the Distribution of Consumption." *Journal of Political Economy*, Vol. 104(6), pp. 1227–1262.

Attanasio, Orazio P., and Guglielmo Weber. 1993. "Consumption Growth, the Interest Rate and Aggregation." *Review of Economic Studies*, Vol. 60(3), pp. 631–649.

Auerbach, Alan J., and Laurence J. Kotlikoff. 1987. *Dynamic Fiscal Policy*. New York: Cambridge University Press.

Auernheimer, Leonardo. 1974. "The Honest Government's Guide to the Revenue from the Creation of Money." *Journal of Political Economy*, Vol. 82, pp. 598–606.

Azariadis, Costas. 1993. *Intertemporal Macroeconomics*. Cambridge, MA: Blackwell Press.

Balasko, Y., and Karl Shell. 1980. "The Overlapping-Generations Model I: The Case of Pure Exchange without Money." *Journal of Economic Theory*, Vol. 23, pp. 281–306.

Barro, Robert J. 1974. "Are Government Bonds Net Wealth?" *Journal of Political Economy*, Vol. 82(6), pp. 1095–1117.

Barro, Robert J. 1979. "On the Determination of Public Debt." *Journal of Political Economy*, Vol. 87, pp. 940–971.

Barro, Robert J., and David B. Gordon. 1983a. "A Positive Theory of Monetary Policy in a Natural Rate Model." *Journal of Political Economy*, Vol. 91, pp. 589–610.

Barro, Robert J., and David B. Gordon. 1983b. "Rules, Discretion, and Reputation in a Model of Monetary Policy." *Journal of Monetary Economics*, Vol. 12, pp. 101–121.

Barro, Robert J., and Xavier Sala-i-Martin. 1995. *Economic Growth*. New York: McGraw-Hill.

Barsky, Robert B., Gregory N. Mankiw, and Stephen P. Zeldes. 1986. "Ricardian Consumers with Keynesian Propensities." *American Economic Review*, Vol. 76(4), pp. 676–691.

Basar, Tamer, and Geert Jan Olsder. 1982. *Dynamic Noncooperative Game Theory*. New York: Academic Press.

Baumol, William J. 1952. "The Transactions Demand for Cash: An Inventory Theoretic Approach." *Quarterly Journal of Economics*, Vol. 66, pp. 545–556.

Bellman, Richard. 1957. *Dynamic Programming*. Princeton, NJ: Princeton University Press.

Bellman, Richard, and Stuart E. Dreyfus. 1962. *Applied Dynamic Programming*. Princeton, NJ: Princeton University Press.

Benassy, Jean-Pascal. 1998. "Is There Always Too Little Research in Endogenous Growth with Expanding Product Variety?" *European Economic Review*, Vol. 42, pp. 61–69.

Benoit, Jean-Pierre, and Vijay Krishna. 1985. "Finitely Repeated Games." *Econometrica*, Vol. 53, pp. 905–922.

Benveniste, Lawrence, and Jose Scheinkman. 1979. "On the Differentiability of the Value Function in Dynamic Models of Economics." *Econometrica*, Vol. 47(3), pp. 727–732.

Benveniste, Lawrence, and Jose Scheinkman. 1982. "Duality Theory for Dynamic Optimization Models of Economics: The Continuous Time Case." *Journal of Economic Theory*, Vol. 27, pp. 1–19.

Bernheim, B. Douglas, and Kyle Bagwell. 1988. "Is Everything Neutral?" *Journal of Political Economy*, Vol. 96(2), pp. 308–338.

Bertola, Giuseppe, and Ricardo J. Caballero. 1994. "Cross-Sectional Efficiency and Labour Hoarding in a Matching Model of Unemployment." *Review of Economic Studies*, Vol. 61, pp. 435–456.

Bertsekas, Dimitri P. 1976. *Dynamic Programming and Stochastic Control.* New York: Academic Press (esp. chaps. 2, 6.).

Bertsekas, Dimitri P. 1987. *Dynamic Programming: Deterministic and Stochastic Models.* Englewood Cliffs, NJ: Prentice-Hall.

Bertsekas, Dimitri P., and Steven E. Shreve. 1978. *Stochastic Optimal Control: The Discrete Time Case.* New York: Academic Press.

Bewley, Truman F. 1977. "The Permanent Income Hypothesis: A Theoretical Formulation." *Journal of Economic Theory*, Vol. 16(2), pp. 252–292.

Bewley, Truman F. 1980. "The Optimum Quantity of Money." In J. H. Kareken and N. Wallace (eds.), *Models of Monetary Economies.* Minneapolis: Federal Reserve Bank of Minneapolis, pp. 169–210.

Bewley, Truman F. 1983. "A Difficulty with the Optimum Quantity of Money." *Econometrica*, Vol. 51, pp. 1485–1504.

Bewley, Truman F. 1986. "Stationary Monetary Equilibrium with a Continuum of Independently Fluctuating Consumers." In Werner Hildenbrand and Andreu Mas-Colell (eds.), *Contributions to Mathematical Economics in Honor of Gerard Debreu.* Amsterdam: North-Holland, pp. 79–102.

Black, Fisher, and Myron Scholes. 1973. "The Pricing of Options and Corporate Liabilities." *Journal of Political Economy*, Vol. 81, pp. 637–654.

Blackwell, David. 1965. "Discounted Dynamic Programming." *Annals of Mathematical Statistics*, Vol. 36(1), pp. 226–235.

Blanchard, Olivier J.. 1985. "Debt, Deficits, and Finite Horizons." *Journal of Political Economy*, Vol. 93(2), pp. 223–247.

Blanchard, Olivier Jean and Stanley Fischer. 1989. *Lectures on Macroeconomics.* Cambridge: MIT Press.

Blanchard, Olivier Jean, and Charles M. Kahn. 1980. "The Solution of Linear Difference Models under Rational Expectations." *Econometrica*, Vol. 48(5), pp. 1305–1311.

Bohn, Henning. 1995. "The Sustainability of Budget Deficits in a Stochastic Economy." *Journal of Money, Credit, and Banking*, Vol. 27(1), pp. 257–271.

Bond, Eric W., and Jee-Hyeong Park. 1998. "Gradualism in Trade Agreements with Asymmetric Countries." Mimeo. Pennsylvania State University, October.

Breeden, Douglas T. 1979. "An Intertemporal Asset Pricing Model with Stochastic Consumption and Investment Opportunities." *Journal of Financial Economics*, Vol. 7(3), pp. 265–296.

Brock, William A. 1972. "On Models of Expectations Generated by Maximizing Behavior of Economic Agents Over Time." *Journal of Economic Theory*, Vol. 5, pp. 479–513.

Brock, William A. 1974. "Money and Growth: The Case of Long Run Perfect Foresight." *International Economic Review*, Vol. 15, pp. 750–777.

Brock, William A. 1982. "Asset Prices in a Production Economy." In J. J. McCall (ed.), *The Economics of Information and Uncertainty.* Chicago: University of Chicago Press, pp. 1–43.

Brock, William A. 1990. "Overlapping Generations Models with Money and Transactions Costs." In B. M. Friedman and F. H. Hahn (eds.), *Handbook of Monetary Economics, Vol. 1.* Amsterdam: North-Holland, pp. 263-295.

Brock, William A., and Leonard Mirman. 1972. "Optimal Economic Growth and Uncertainty: The Discounted Case." *Journal of Economic Theory*, Vol. 4(3), pp. 479–513.

Browning, Martin, Lars P. Hansen, and James J. Heckman. 2000. "Micro Data and General Equilibrium Models." In John Taylor and Michael Woodford (eds.), *Handbook of Macroeconomics.* Amsterdam: North-Holland.

Bruno, Michael, and Stanley Fischer. 1990. "Seigniorage, Operating Rules, and the High Inflation Trap." *Quarterly Journal of Economics*, Vol. 105, pp. 353–374.

Bryant, John, and Neil Wallace. 1984. "A Price Discrimination Analysis of Monetary Policy." *Review of Economic Studies*, Vol. 51(2), pp. 279–288.

Burdett, Kenneth, Shouyong Shi, and Randall Wright. 2000. "Pricing and Matching with Frictions." Mimeo. University of Essex, Queen's University, and University of Pennsylvania.

Bulow, Jeremy, and Kenneth Rogoff. 1989. "Sovereign Debt: Is to Forgive to Forget?" *American Economic Review*, Vol. 79, pp. 43–50.

Burnside, C., M. Eichenbaum, and S. Rebelo. 1993. "Labor Hoarding and the Business Cycle." *Journal of Political Economy*, Vol. 101(2), pp. 245–273.

Burnside, C., and M. Eichenbaum. 1996a. "Factor Hoarding and the Propagation of Business Cycle Shocks." *American Economic Review*, Vol. 86(5), pp. 1154–74.

Burnside, C., and M. Eichenbaum. 1996b. "Small Sample Properties of GMM Based Wald Tests." *Journal of Business and Economic Statistics*, Vol. 14(3), pp. 294–308.

Caballero, Ricardo J. 1990. "Consumption Puzzles and Precautionary Saving" *Journal of Monetary Economics*, Vol. 25, No. 1, pp. 113-136.

Cagan, Phillip. 1956. "The Monetary Dynamics of Hyperinflation." In Milton Friedman (ed.), *Studies in the Quantity Theory of Money.* Chicago: University of Chicago Press, pp. 25–117.

Calvo, Guillermo A. 1978. "On the Time Consistency of Optimal Policy in a Monetary Economy." *Econometrica*, Vol. 46(6), pp. 1411–1428.

Campbell, John Y., Andrew W. Lo, and A. Craig MacKinlay. 1997. *The Econometrics of Financial Markets.* Princeton: Princeton University Press.

Campbell, John Y., and John H. Cochrane. 1999. "By Force of Habit: A Consumption-Based Explanation of Aggregate Stock Market Behavior." *Journal of Political Economy*, Vol. 107(2), pp. 205–251.

Carroll, Christopher D., and Miles S. Kimball. 1996. "On the Concavity of the Consumption Function." *Econometrica*, Vol. 64(4), pp. 981–992.

Casella, Alessandra, and Jonathan S. Feinstein. 1990. "Economic Exchange during Hyperinflation." *Journal of Political Economy*, Vol. 98(1), pp. 1–27.

Cass, David. 1965. "Optimum Growth in an Aggregative Model of Capital Accumulation." *Review of Economic Studies*, Vol. 32(3), pp. 233–240.

Cass, David, and M. E. Yaari. 1966. "A Re-examination of the Pure Consumption Loans Model." *Journal of Political Economy*, Vol. 74, pp. 353–367.

Chamberlain, Gary, and Charles Wilson. 1984. "Optimal Intertemporal Consumption Under Uncertainty." Mimeo. Social Systems Research Institute Working Paper 8422, University of Wisconsin.

Chamley, Christophe. 1986. "Optimal Taxation of Capital Income in General Equilibrium with Infinite Lives." *Econometrica*, Vol. 54(3), pp. 607–622.

Chamley, Christophe, and Heraklis Polemarchakis. 1984. "Assets, General Equilibrium, and the Neutrality of Money." *Review of Economic Studies*, Vol. 51, pp. 129–138.

Champ, Bruce, and Scott Freeman. 1994. *Modeling Monetary Economies.* New York: Wiley.

Chang, Roberto. 1998. "Credible Monetary Policy in an Infinite Horizon Model: Recursive Approaches." *Journal of Economic Theory*, Vol. 81(2), pp. 431–461.

Chari, V. V., Lawrence J. Christiano, and Martin Eichenbaum. 1998. "Expectations Traps." *Journal of Economic Theory*, Vol. 81(2), pp. 462–492.

Chari, V. V., Lawrence J. Christiano, and Patrick J. Kehoe. 1994. "Optimal Fiscal Policy in a Business Cycle Model." *Journal of Political Economy*, Vol. 102(4), pp. 617–652.

Chari, V. V., Lawrence J. Christiano, and Patrick J. Kehoe. 1996. "Optimality of the Friedman Rule in Economies with Distorting Taxes." *Journal of Monetary Economics*, Vol. 37(2), pp. 203–223.

Chari, V. V., and Patrick J. Kehoe. 1990. "Sustainable Plans." *Journal of Political Economy*, Vol. 98, pp. 783–802.

Chari, V. V., and Patrick J. Kehoe. 1993a. "Sustainable Plans and Mutual Default." *Review of Economic Studies*, Vol. 60, pp. 175–195.

Chari, V. V., and Patrick J. Kehoe. 1993b. "Sustainable Plans and Debt." *Journal of Economic Theory*, Vol. 61, pp. 230–261.

Chari, V. V., Patrick J. Kehoe, and Edward C. Prescott. 1989. "Time Consistency and Policy." In Robert Barro (ed.), *Modern Business Cycle Theory.* Cambridge, MA: Harvard University Press, pp. 265–305.

Chatterjee, Satyajit, and Dean Corbae. 1996. "Money and Finance with Costly Commitment." *Journal of Monetary Economics*, Vol. 37(2), pp. 225–248.

Chen, Ren-Raw, and Louis Scott. 1993. "Maximum Likelihood Estimation for a Multifactor Equilibrium Model of the Term Structure of Interest Rates." *Journal of Fixed Income*, No. 95-09.

Cho, In-Koo, and Akihiko Matsui. 1995. "Induction and the Ramsey Policy." *Journal of Economic Dynamics and Control*, Vol. 19(5-7), pp. 1113–1140.

Chow, Gregory. 1981. *Econometric Analysis by Control Methods.* New York: Wiley.

Chow, Gregory. 1997. *Dynamic Economics: Optimization by the Lagrange Method.* New York: Oxford University Press.

Christiano, Lawrence J. 1990. "Linear-Quadratic Approximation and Value-Function Iteration: A Comparison." *Journal of Business and Economic Statistics*, Vol. 8(1), pp. 99–113.

Christiano, Lawrence J., and M. Eichenbaum. 1992. "Current Real Business Cycle Theories and Aggregate Labor Market Fluctuations." *American Economic Review*, Vol. 82(3).

Clower, Robert. 1967. "A Reconsideration of the Microfoundations of Monetary Theory." *Western Economic Journal*, Vol. 6, pp. 1–9.

Cochrane, John H. 1991. "A Simple Test of Consumption Insurance." *Journal of Political Economy*, Vol. 99(5), pp. 957–976.

Cochrane, John H. 1997. "Where Is the Market Going? Uncertain Facts and Novel Theories." *Economic Perspectives*, Vol. 21(6), pp. 3–37.

Cochrane, John H., and Lars Peter Hansen. 1992. "Asset Pricing Explorations for Macroeconomics." In Olivier Jean Blanchard and Stanley Fischer (eds.), *NBER Macroeconomics Annual.* Cambridge, MA: MIT Press, pp. 115–165.

Cogley, Timothy. 1999. "Idiosyncratic Risk and the Equity Premium: Evidence from the Consumer Expenditure Survey." Mimeo. Arizona State University.

Cole, Harold L., and Narayana Kocherlakota. 1998a. "Efficient Allocations with Hidden Income and Hidden Storage." Mimeo. Federal Reserve Bank of Minneapolis Staff Report: 238, 36. May.

Cole, Harold L., and Narayana Kocherlakota. 1998b. "Dynamic Games with Hidden Actions and Hidden States." Mimeo. Federal Reserve Bank of Minneapolis Staff Report: 254, 13. September.

Constantinides, George M., and Darrell Duffie. 1996. "Asset Pricing with Heterogeneous Consumers." *Journal of Political Economy*, Vol. 104 (2), pp. 219–240.

Cooley, Thomas F. 1995. *Frontiers of Business Cycle Research.* Princeton, NJ: Princeton University Press.

Cooper, Russell W. 1999. *Coordination Games: Complementarities and Macroeconomics.* New York: Cambridge University Press.

Correia, Isabel H. 1996. "Should Capital Income Be Taxed in the Steady State?" *Journal of Public Economics*, Vol. 60(1), pp. 147–151.

Correia, Isabel, and Pedro Teles. 1996. "Is the Friedman Rule Optimal When Money Is an Intermediate Good?" *Journal of Monetary Economics*, Vol. 38, pp. 223–244.

Cox, John C., Jonathan E. Ingersoll, Jr., and Stephen A. Ross. 1985a. "An Intertemporal General Equilibrium Model of Asset Prices." *Econometrica*, Vol. 53(2), pp. 363–384.

Cox, John C., Jonathan E. Ingersoll, Jr., and Stephen A. Ross. 1985b. "A Theory of the Term Structure of Interest Rates." *Econometrica*, Vol. 53(2), pp. 385–408.

Dai, Qiang, and Kenneth J. Singleton. Forthcoming. "Specification Analysis of Affine Term Structure Models." *Journal of Finance*, In press.

Davis, Steven J. 1995. "The Quality Distribution of Jobs and the Structure of Wages in Search Equilibrium." Mimeo. University of Chicago.

Deaton, Angus. 1992. *Understanding Consumption.* New York: Oxford University Press.

Debreu, Gerard. 1954. "Valuation Equilibrium and Pareto Optimum." *Proceedings of the National Academy of Sciences*, Vol. 40, pp. 588–592.

Debreu, Gerard. 1959. *Theory of Value.* New York: Wiley.

Den Haan, Wouter J., and Albert Marcet. 1990. "Solving the Stochastic Growth Model by Parametering Expectations." *Journal of Business and Economic Statistics*, Vol. 8(1), pp. 31–34.

Diamond, Peter A. 1965. "National Debt in a Neoclassical Growth Model." *American Economic Review*, Vol. 55, pp. 1126–1150.

Diamond, Peter A. 1981. "Mobility Costs, Frictional Unemployment, and Efficiency." *Journal of Political Economy*, Vol. 89(4), pp. 798–812.

Diamond, Peter A. 1982. "Wage Determination and Efficiency in Search Equilibrium." *Review of Economic Studies*, Vol. 49, pp. 217–227.

Diamond, Peter A. 1984. "Money in Search Equilibrium." *Econometrica*, Vol. 52, pp. 1–20.

Diamond, Peter A., and Joseph Stiglitz. 1974. "Increases in Risk and in Risk Aversion." *Journal of Economic Theory*, Vol. 8(3), pp. 337–360.

Diaz-Giménez, J., Edward C. Prescott, T. Fitzgerald, and Fernando Alvarez. 1992. "Banking in Computable General Equilibrium Economies." *Journal of Economic Dynamics and Control*, Vol. 16, pp. 533–560.

Dixit, Avinash, Gene Grossman, and Faruk Gul. 1998. "A Theory of Political Compromise." Mimeo. Princeton University, May.

Dixit, Avinash K. and Joseph E. Stiglitz. 1977. "Monopolistic Competition and Optimum Product Diversity." *American Economic Review*, Vol. 67, pp. 297–308.

Domeij, David, and Jonathan Heathcote. 2000. "Capital versus Labor Income Taxation with Heterogeneous Agents." Mimeo. Stockholm School of Economics.

Doob, Joseph L. 1953. *Stochastic Processes*. New York: Wiley.

Dornbusch, Rudiger. 1976. "Expectations and Exchange Rate Dynamics." *Journal of Political Economy*, Vol. 84, pp. 1161–1176.

Dow, James R., Jr., and Lars J. Olson. 1992. "Irreversibility and the Behavior of Aggregate Stochastic Growth Models." *Journal of Economic Dynamics and Control*, Vol. 16, pp. 207–233.

Duffie, Darrell. 1996. *Dynamic Asset Pricing Theory*. Princeton, NJ: Princeton University Press, Princeton, pp. xvii, 395.

Duffie, Darrell, J. Geanakoplos, A. Mas-Colell, and A. McLennan. 1994. "Stationary Markov Equilibria." *Econometrica*, Vol. 62, No. 4, pp. 745–781.

Duffie, Darrell, and Rui Kan. 1996. "A Yield-Factor Model of Interest Rates." *Mathematical Finance*, Vol. 6(4), pp. 379–406.

Eichenbaum, Martin. 1991. "Real Business-Cycle Theory: Wisdom or Whimsy?" *Journal of Economic Dynamics and Control*, Vol. 15, No. 4, pp. 607-626.

Eichenbaum, Martin, and Lars P. Hansen. 1990. "Estimating Models with Intertemporal Substitution Using Aggregate Time Series Data." *Journal of Business and Economic Statistics*, Vol. 8, pp. 53–69.

Eichenbaum, Martin, Lars P. Hansen, and S.F. Richard. 1984. "The Dynamic Equilibrium Pricing of Durable Consumption Goods." Mimeo. Carnegie-Mellon University, Pittsburgh.

Elliott, Robert J., Lakhdar Aggoun, and John B. Moore. 1995. *Hidden Markov Models: Estimation and Control*. New York: Springer-Verlag..

Epstein, Larry G., and Stanley E. Zin. 1989. "Substitution, Risk Aversion, and the Temporal Behavior of Consumption and Asset Returns: A Theoretical Framework." *Econometrica*, Vol. 57(4), pp. 937–969.

Epstein, Larry G., and Stanley E. Zin. 1991. "Substitution, Risk Aversion, and the Temporal Behavior of Consumption and Asset Returns: An Empirical Analysis." *Journal of Political Economy*, Vol. 99(2), pp. 263–286.

Ethier, Wilfred J. 1982. "National and International Returns to Scale in the Modern Theory of International Trade." *American Economic Review*, Vol. 72, pp. 389–405.

Faig, Miquel. 1988. "Characterization of the Optimal Tax on Money When It Functions as a Medium of Exchange." *Journal of Monetary Economics*, Vol. 22(1), pp. 137–148.

Fama, Eugene F. 1976a. *Foundations of Finance: Portfolio Decisions and Securities Prices*. New York: Basic Books.

Fama, Eugene F. 1976b. "Inflation Uncertainty and Expected Returns on Treasury Bills." *Journal of Political Economy*, Vol. 84(3), pp. 427–448.

Farmer, Roger E. A. 1993. *The Macroeconomics of Self-fulfilling Prophecies.* Cambridge, MA: MIT Press.

Fischer, Stanley. 1983. "A Framework for Monetary and Banking Analysis." *Economic Journal*, Vol. 93, Supplement, pp. 1–16.

Fisher, Irving. 1913. *The Purchasing Power of Money: Its Determination and Relation to Credit, Interest and Crises.* New York: Macmillan.

Fisher, Irving. [1907] 1930. *The Theory of Interest.* London: Macmillan.

Frankel, Marvin. 1962. "The Production Function in Allocation and Growth: A Synthesis." *American Economic Review*, Vol. 52, pp. 995–1022.

Friedman, Milton. 1956. *A Theory of the Consumption Function.* Princeton, NJ: Princeton University Press.

Friedman, Milton. 1967. "The Role of Monetary Policy." *American Economic Review*, Vol. 58, 1968, pp. 1–15. Presidential Address delivered at the 80th Annual Meeting of the American Economic Association, Washington, DC, December 29.

Friedman, Milton. 1969. "The Optimum Quantity of Money." In Milton Friedman (ed.), *The Optimum Quantity of Money and Other Essays.* Chicago: Aldine, pp. 1–50.

Friedman, Milton, and Anna J. Schwartz. 1963. *A Monetary History of the United States, 1867–1960.* Princeton, NJ: Princeton University Press and N.B.E.R.

Fudenberg, Drew, Bengt Holmstrom, and Paul Milgrom. 1990. "Short-Term Contracts and Long-Term Agency Relationships." *Journal of Economic Theory*, Vol. 51(1).

Gabel, R. A., and R. A. Roberts. 1973. *Signals and Linear Systems.* New York: Wiley.

Gale, David. 1973. "Pure Exchange Equilibrium of Dynamic Economic Models." *Journal of Economic Theory*, Vol. 6, pp. 12-36.

Gali, Jordi. 1991. "Budget Constraints and Time-Series Evidence on Consumption." *American Economic Review*, Vol 81(5), pp. 1238–1253.

Gallant, R., L. P. Hansen, and G. Tauchen. 1990. "Using Conditional Moments of Asset Payoffs to Infer the Volatility of Intertemporal Marginal Rates of Substitution." *Journal of Econometrics*, Vol. 45, pp. 145–179.

Gittins, J.C. 1989. *Multi-armed Bandit and Allocation Indices.* New York: Wiley.

Gomes, Joao, Jeremy Greenwood, and Sergio Rebelo. 1997. "Equilibrium Unemployment." Mimeo. NBER Working Paper No. 5922.

Gong, Frank F., and Eli M. Remolona. 1997. "A Three Factor Econometric Model of the U.S. Term Structure." Mimeo. Federal Reserve Bank of New York, Staff Report 19.

Gourinchas, Pierre-Olivier, and Jonathan A. Parker. 1999. "Consumption over the Life Cycle." Mimeo. NBER Working Paper No. 7271.

Granger, C. W. J. 1966. "The Typical Spectral Shape of an Economic Variable." *Econometrica*, Vol. 34(1), pp. 150–161.

Granger, C. W. J. 1969. "Investigating Causal Relations by Econometric Models and Cross-Spectral Methods." *Econometrica*, Vol. 37(3), pp. 424–438.

Green, Edward J. 1987. "Lending and the Smoothing of Uninsurable Income." In Edward C. Prescott and Neil Wallace (eds.), *Contractual Arrangements for Intertemporal Trade, Minnesota Studies in Macroeconomics series, Vol. 1.* Minneapolis: University of Minnesota Press, pp. 3–25.

Green, Edward J., and Robert H. Porter. 1984. "Non-Cooperative Collusion under Imperfect Price Information." *Econometrica*, Vol. 52, pp. 975–993.

Grossman, Gene M., and Elhanan Helpman. 1991. "Quality Ladders in the Theory of Growth." *Review of Economic Studies*, Vol. 58, pp. 43–61.

Grossman, Sanford J., and Robert J. Shiller. 1981. "The Determinants of the Variability of Stock Market Prices." *American Economic Review*, Vol. 71(2), pp. 222–227.

Gul, Faruk and Wolfgang Pesendorfer. 2000. "Self-Control and the Theory of Consumption." Mimeo. Princeton University.

Guidotti, Pablo E., and Carlos A. Vegh. 1993. "The Optimal Inflation Tax When Money Reduces Transactions Costs: A Reconsideration." *Journal of Monetary Economics*, Vol. 31(2), pp. 189–205.

Hall, Robert E. 1971. "The Dynamic Effects of Fiscal Policy in an Economy with Foresight." *Review of Economic Studies*, Vol. 38, pp. 229–244.

Hall, Robert E. 1978. "Stochastic Implications of the Life Cycle-Permanent Income Hypothesis: Theory and Evidence." *Journal of Political Economy*, Vol. 86(6), pp. 971–988. (Reprinted in *Rational Expectations and Econometric Practice*, ed. Thomas J. Sargent and Robert E. Lucas, Jr., Minneapolis: University of Minnesota Press, 1981, pp. 501–520.).

Hamilton, James D. 1994. *Time Series Analysis.* Princeton, NJ: Princeton University Press.

Hamilton, James D., and Marjorie A. Flavin. 1986. "On the Limitations of Government Borrowing: A Framework for Empirical Testing." *American Economic Review*, Vol. 76(4), pp. 808–819.

Hansen, Gary D. 1985. "Indivisible Labor and the Business Cycle." *Journal of Monetary Economics*, Vol. 16, pp. 309–327.

Hansen, Gary D., and Ayşe İmrohoroğlu. 1992. "The Role of Unemployment Insurance in an Economy with Liquidity Constraints and Moral Hazard." *Journal of Political Economy*, Vol. 100 (1), pp. 118–142.

Hansen, Lars P. 1982a. "Consumption, Asset Markets, and Macroeconomic Fluctuations: A Comment." *Carnegie-Rochester Conference Series on Public Policy*, Vol. 17, pp. 239–250.

Hansen, Lars P. 1982b. "Large Sample Properties of Generalized Method of Moments Estimators." *Econometrica*, Vol. 50, pp. 1029–1060.

Hansen, Lars P., Dennis Epple, and Will Roberds. 1985. "Linear-Quadratic Duopoly Models of Resource Depletion." In Thomas J. Sargent (ed.), *Energy, Foresight, and Strategy.* Washington, DC: Resources for the Future, pp. 101–142.

Hansen, Lars P., and Ravi Jagannathan. 1991. "Implications of Security Market Data for Models of Dynamic Economies." *Journal of Political Economy*, Vol. 99, pp. 225–262.

Hansen, Lars P., and Ravi Jagannathan. 1997. "Assessing Specification Errors in Stochastic Discount Factor Models." *Journal of Finance*, Vol. 52(2), pp. 557–590.

Hansen, Lars P., William T. Roberds, and Thomas J. Sargent. 1991. "Time Series Implications of Present Value Budget Balance and of Martingale Models of Consumption and Taxes." In L. P. Hansen and T. J. Sargent (eds.), *Rational Expectations and Econometric Practice*. Boulder, CO: Westview Press, pp. 121–161.

Hansen, Lars P., and Thomas J. Sargent. 1981. "Linear Rational Expectations Models for Dynamically Interrelated Variables." In R. E. Lucas, Jr. and T. J. Sargent (eds.), *Rational Expectations and Econometric Practice*. Minneapolis: University of Minnesota Press, pp. 127–156.

Hansen, Lars P., and Thomas J. Sargent. 1982. "Instrumental Variables Procedures for Estimating Linear Rational Expectations Models." *Journal of Monetary Economics*, Vol. 9(3), pp. 263–296.

Hansen, Lars P., and Thomas J. Sargent. 1980. "Formulating and Estimating Dynamic Linear Rational Expectations Models." *Journal of Economic Dynamics and Control*, Vol. 2(1), pp. 7–46.

Hansen, Lars P., and Thomas J. Sargent. 1995. "Discounted Linear Exponential Quadratic Gaussian Control." *IEEE Transactions on Automatic Control*, Vol. 40, pp. 968–971.

Hansen, Lars P., Thomas J. Sargent, and Thomas D. Tallarini, Jr. 1999. "Robust Permanent Income and Pricing." *Review of Economic Studies*, Vol. 66(4), pp. 873–907.

Hansen, Lars P., and Kenneth J. Singleton. 1983. "Stochastic Consumption, Risk Aversion, and the Temporal Behavior of Asset Returns." *Journal of Political Economy*, Vol. 91(2), pp. 249–265.

Hansen, Lars P., and Kenneth J. Singleton. 1986. "Generalized Instrumental Variables Estimation of Nonlinear Rational Expectations Models." *Econometrica*, Vol. 50(5), pp. 1269–1286.

Hansen, Lars P., and Thomas J. Sargent. 2000. "Recursive Models of Dynamic Linear Economies" Mimeo. University of Chicago and Stanford University.

Harrison, Michael, and David Kreps. 1979. "Martingales and Arbitrage in Multiperiod Security Markets." *Journal of Economic Theory*, Vol. 20, pp. 381–408.

Heaton, John, and Deborah J. Lucas. 1996. "Evaluating the Effects of Incomplete Markets on Risk Sharing and Asset Pricing." *Journal of Political Economy*, Vol. 104(3), pp. 443–487.

Helpman, Elhanan. 1981. "An Exploration in the Theory of Exchange-Rate Regimes." *Journal of Political Economy*, Vol. 89(5), pp. 865–890.

Hirshleifer, Jack. 1966. "Investment Decision under Uncertainty: Applications of the State Preference Approach." *Quarterly Journal of Economics*, Vol. 80(2), pp. 252–277.

Hopenhayn, Hugo A., and Juan Pablo Nicolini. 1997. "Optimal Unemployment Insurance." *Journal of Political Economy*, Vol. 105(2), pp. 412–438.

Hopenhayn, Hugo A., and Edward C. Prescott. 1992. "Stochastic Monotonicity and Stationary Distributions for Dynamic Economies." *Econometrica*, Vol. 60 (6), pp. 1387–1406.

Hopenhayn, Hugo, and Richard Rogerson. 1993. "Job Turnover and Policy Evaluation: A General Equilibrium Analysis." *Journal of Political Economy*, Vol. 101, pp. 915–938.

Hosios, Arthur, J. 1990. "On the Efficiency of Matching and Related Models of Search and Unemployment." *Review of Economic Studies*, Vol. 57, pp. 279–298.

Hubbard, R. Glenn, Jonathan Skinner, and Stephen P. Zeldes. 1995. "Precautionary Saving and Social Insurance." *Journal of Political Economy*, Vol. 103(2), pp. 360–399.

Huffman, Gregory. 1986. "The Representative Agent, Overlapping Generations, and Asset Pricing." *Canadian Journal of Economics*, Vol. 19(3), pp. 511–521.

Huggett, Mark. 1993. "The Risk Free Rate in Heterogeneous-Agent, Incomplete-Insurance Economies." *Journal of Economic Dynamics and Control*, Vol. 17(5-6), pp. 953–969.

Huggett, Mark, and Sandra Ospina. 2000. "Aggregate Precautionary Savings: When is the Third Derivative Irrelevant?" Mimeo. Georgetown University.

İmrohoroğlu, Ayşe. 1992. "The Welfare Cost of Inflation Under Imperfect Insurance." *Journal of Economic Dynamics and Control*, Vol. 16(1), pp. 79–92.

İmrohoroğlu, Ayşe, Selahattin İmrohoroğlu, and Douglas Joines. 1995. "A Life Cycle Analysis of Social Security." *Economic Theory*, Vol. 6 (1), pp. 83–114.

Ireland, Peter N. 1994. "Inflationary Policy and Welfare with Limited Credit Markets." *Journal of Financial Intermediation*, Vol. 3(3), pp. 245–271.

Ireland, Peter N. 1997. "Sustainable Monetary Policies." *Journal of Economic Dynamics and Control*, Vol. 22, pp. 87–108.

Jacobson, David H. 1973. "Optimal Stochastic Linear Systems with Exponential Performance Criteria and Their Relation to Deterministic Differential Games." *IEEE Transactions on Automatic Control*, Vol. 18(2), pp. 124–131.

Johnson, Norman, and Samuel Kotz. 1971. *Continuous Univariate Distributions*. New York: Wiley.

Jones, Charles I. 1995. "R&D-Based Models of Economic Growth." *Journal of Political Economy*, Vol. 103, pp. 759–784.

Jones, Larry E., and Rodolfo Manuelli. 1990. "A Convex Model of Equilibrium Growth: Theory and Policy Implications." *Journal of Political Economy*, Vol. 98, pp. 1008–1038.

Jones, Larry E., and Rodolfo E. Manuelli. 1992. "Finite Lifetimes and Growth." *Journal of Economic Theory*, Vol. 58, pp. 171–197.

Jones, Larry E., Rodolfo E. Manuelli, and Peter E. Rossi. 1993. "Optimal Taxation in Models of Endogenous Growth." *Journal of Political Economy*, Vol. 101, pp. 485–517.

Jones, Larry E., Rodolfo E. Manuelli, and Peter E. Rossi. 1997. "On the Optimal Taxation of Capital Income." *Journal of Economic Theory*, Vol. 73(1), pp. 93–117.

Jovanovic, Boyan. 1979a. "Job Matching and the Theory of Turnover." *Journal of Political Economy*, Vol. 87(5), pp. 972–990.

Jovanovic, Boyan. 1979b. "Firm-Specific Capital and Turnover." *Journal of Political Economy*, Vol. 87(6), pp. 1246–1260.

Jovanovic, Boyan and Yaw Nyarko. 1996. "Learning by Doing and the Choice of Technology." *Econometrica*, Vol. 64, No. 6, pp. 1299–1310.

Judd, Kenneth L. 1985a. "On the Performance of Patents." *Econometrica*, Vol. 53, pp. 567–585.

Judd, Kenneth L. 1985b. "Redistributive Taxation in a Simple Perfect Foresight Model." *Journal of Public Economics*, Vol. 28, pp. 59–83.

Judd, Kenneth L. 1990. "Cournot versus Bertrand: A Dynamic Resolution." Mimeo. Hoover Institution, Stanford University; available at http://bucky.stanford.edu.

Judd, Kenneth L. 1996. "Approximation, Perturbation, and Projection Methods in Economic Analysis." In Hans Amman, David Kendrick, and John Rust (eds.), *Handbook of Computational Economics, Vol. 1.* Amsterdam: North-Holland.

Judd, Kenneth L. 1998. *Numerical Methods in Economics.* Cambridge, MA: MIT Press.

Judd, Kenneth L. and Andrew Solnick. 1994. "Numerical dynamic programming with shape preserving splines." Mimeo. Hoover Institution.

Kahn, Charles, and William Roberds. 1998. "Real-Time Gross Settlement and the Costs of Immediacy." Mimeo. Federal Reserve Bank of Atlanta, Working Paper 98-21, December.

Kalman, R. E. 1960. "Contributions to the Theory of Optimal Control." *Bol. Soc. Mat. Mexicana*, Vol. 5, pp. 102–119.

Kalman, R. E., and R. S. Bucy. 1961. "New Results in Linear Filtering and Prediction Theory." *J. Basic Eng., Trans. ASME, Ser. D*, Vol. 83, pp. 95–108.

Kandori, Michihiro. 1992. "Repeated Games Played by Overlapping Generations of Players." *Review of Economic Studies*, Vol. 59 (1), pp. 81–92.

Kareken, John, T. Muench, and N. Wallace. 1973. "Optimal Open Market Strategy: The Use of Information Variables." *merican Economic Review*, Vol. 63(1), pp. 156–172.

Kareken, John, and Neil Wallace. 1980. *Models of Monetary Economies.* Minneapolis: Federal Reserve Bank of Minneapolis, pp. 169–210.

Kareken, John, and Neil Wallace. 1981. "On the Indeterminacy of Equilibrium Exchange Rates." *Quarterly Journal of Economics*, Vol. 96, pp. 207–222.

Kehoe, Patrick, and Fabrizio Perri. 1998. "International Business Cycles with Endogenous Incomplete Markets." Mimeo. University of Pennsylvania, February.

Kehoe, Timothy J., and David K. Levine. 1984. "Intertemporal Separability in Overlapping-Generations Models." *Journal of Economic Theory*, Vol. 34, pp. 216–226.

Kehoe, Timothy J., and David K. Levine. 1985. "Comparative Statics and Perfect Foresight in Infinite Horizon Economies." *Econometrica*, Vol. 53, pp. 433–453.

Kehoe, Timothy J., and David K. Levine. 1993. "Debt-Constrained Asset Markets." *Review of Economic Studies*, Vol. 60(4), pp. 865–888.

Keynes, John Maynard. 1940. *How to Pay for the War: A Radical Plan for the Chancellor of the Exchequer*. London: Macmillan.

Kihlstrom, Richard E., and Leonard J. Mirman. 1974. "Risk Aversion with Many Commodities." *Journal of Economic Theory*, Vol. 8, pp. 361–388.

Kim, Chang-Jin, and Charles R. Nelson. 1999. *State Space Models with Regime Switching*. Cambridge, MA: MIT Press.

Kimball, M. S. 1990. "Precautionary Saving in the Small and in the Large." *Econometrica*, Vol. 58, pp. 53–73.

Kimball, M. S. 1993. "Standard Risk Aversion." *Econometrica*, Vol. 63(3), pp. 589–611.

Kimbrough, Kent P. 1986. "The Optimum Quantity of Money Rule in the Theory of Public Finance." *Journal of Monetary Economics*, Vol. 18, pp. 277–284.

King, Robert G., and Charles I. Plosser. 1988. "Real Business Cycles: Introduction." *Journal of Monetary Economics*, Vol. 21, pp. 191–193.

King, Robert G., Charles I. Plosser, and Sergio T. Rebelo. 1988. "Production, Growth and Business Cycles: I. The Basic Neoclassical Model." *Journal of Monetary Economics*, Vol. 21, pp. 195–232.

King, Robert G. and Alexander L. Wolman. 1999. "What Should the Monetary Authority Do When Prices are Sticky." In John B. Taylor (ed.), *Monetary Policy Rules*. University of Chicago Press, pp. 349-398.

Kiyotaki, Nobuhiro, and Randall Wright. 1989. "On Money as a Medium of Exchange." *Journal of Political Economy*, Vol. 97(4), pp. 927–954.

Kiyotaki, Nobuhiro, and Randall Wright. 1990. "Search for a Theory of Money." Mimeo. National Bureau of Economic Research, Working Paper No. 3482.

Kiyotaki, Nobuhiro, and Randall Wright. 1993. "A Search-Theoretic Approach to Monetary Economics." *American Economic Review*, Vol. 83(1), pp. 63–77.

Kocherlakota, Narayana R. 1996a. "The Equity Premium: It's Still a Puzzle." *Journal of Economic Literature*, Vol. 34(1), pp. 42–71.

Kocherlakota, Narayana R. 1996b. "Implications of Efficient Risk Sharing without Commitment." *Review of Economic Studies*, Vol. 63(4), pp. 595–609.

Kocherlakota, Narayana R. 1998. "Money Is Memory." *Journal of Economic Theory*, Vol. 81 (2), pp. 232–251.

Kocherlakota, Narayana, and Neil Wallace. 1998. "Incomplete Record-Keeping and Optimal Payment Arrangements." *Journal of Economic Theory*, Vol. 81(2), pp. 272–289.

Koopmans, Tjalling C. 1965. *On the Concept of Optimal Growth.* The Econometric Approach to Development Planning.Chicago: Rand McNally

Kreps, David M. 1979. "Three Essays on Capital Markets." Mimeo. Technical Report 298. Institute for Mathematical Studies in the Social Sciences, Stanford University.

Kreps, David M. 1988. *Notes on the Theory of Choice.* Boulder, CO: Westview Press.

Kreps, David M. 1990. *Game Theory and Economic Analysis.* New York: Oxford University Press.

Krueger, Dirk. 1999. "Risk Sharing in Economies with Incomplete Markets." Mimeo. Stanford University.

Krusell, Per, and Anthony Smith. 1998. "Income and Wealth Heterogeneity in the Macroeconomy." *Journal of Political Economy*, Vol. 106(5), pp. 867–896.

Kwakernaak, Huibert, and Raphael Sivan. 1972. *Linear Optimal Control Systems.* New York: Wiley.

Kydland, Finn E., and Edward C. Prescott. 1977. "Rules Rather than Discretion: The Inconsistency of Optimal Plans." *Journal of Political Economy*, Vol. 85(3), pp. 473–491.

Kydland, Finn E., and Edward C. Prescott. 1980. "Dynamic Optimal Taxation, Rational Expectations and Optimal Control." *Journal of Economic Dynamics and Control*, Vol. 2(1), pp. 79–91.

Kydland, Finn E., and Edward C. Prescott. 1982. "Time to Build and Aggregate Fluctuations." *Econometrica*, Vol. 50(6), pp. 1345–1371.

Labadie, Pamela. 1986. "Comparative Dynamics and Risk Premia in an Overlapping Generations Model." *Review of Economic Studies*, Vol. 53(1), pp. 139–152.

Lagos, Ricardo. 2000. "An Alternative Approach to Search Frictions." *Journal of Political Economy*, In press.

Laibson, David I. 1994. "Hyperbolic Discounting and Consumption." Mimeo. Massachusetts Institute of Technology.

Leland, Hayne E. 1968. "Saving and Uncertainty: The Precautionary Demand for Saving." *Quarterly Journal of Economics*, Vol. 82, No. 3, pp. 465–473.

LeRoy, Stephen F. 1971. "The Determination of Stock Prices." Ph.D. dissertation, unpublished, University of Pennsylvania.

LeRoy, Stephen F. 1973. "Risk Aversion and the Martingale Property of Stock Prices." *International Economic Review*, Vol. 14(2), pp. 436–446.

LeRoy, Stephen F. 1982. "Risk Aversion and the Term Structure of Interest Rates." *Economics Letters*, Vol. 10(3–4), pp. 355–361. (Correction in *Economics Letters* [1983] 12(3–4): 339–340.).

LeRoy, Stephen F. 1984a. "Nominal Prices and Interest Rates in General Equilibrium: Money Shocks." *Journal of Business*, Vol. 57(2), pp. 177–195.

LeRoy, Stephen F. 1984b. "Nominal Prices and Interest Rates in General Equilibrium: Endowment Shocks." *Journal of Business*, Vol. 57(2), pp. 197–213.

LeRoy, Stephen F., and Richard D. Porter. 1981. "The Present-Value Relation: Tests Based on Implied Variance Bounds." *Econometrica*, Vol. 49(3), pp. 555–574.

Levhari, David, and Leonard J. Mirman. 1980. "The Great Fish War: An Example Using a Dynamic Cournot-Nash Solution." *Bell Journal of Economics*, Vol. 11(1).

Levhari, David, and T. N. Srinivasan. 1969. "Optimal Savings under Uncertainty." *Review of Economic Studies*, Vol. 36(2), pp. 153–163.

Levine, David K., and Drew Fudenberg. 1998. *The Theory of Learning in Games.* Cambridge, MA: MIT Press.

Levine, David K., and William R. Zame. 1999. "Does Market Incompleteness Matter?" Mimeo. Department of Economics, University of California at Los Angeles.

Lippman, Steven A., and John J. McCall. 1976. "The Economics of Job Search: A Survey." *Economic Inquiry*, Vol. 14(3), pp. 347–368.

Ljungqvist, Lars. 1997. "How Do Layoff Costs Affect Employment?" Mimeo. Stockholm School of Economics.

Ljungqvist, Lars, and Thomas J. Sargent. 1998. "The European Unemployment Dilemma." *Journal of Political Economy*, Vol. 106, pp. 514–550.

Lucas, Robert E., Jr. 1972. "Expectations and the Neutrality of Money." *Journal of Economic Theory*, Vol. 4, pp. 103–124.

Lucas, Robert E., Jr. 1973. "Some International Evidence on Output-Inflation Trade-Offs." *American Economic Review*, Vol. 63, pp. 326–334.

Lucas, Robert E., Jr. 1976. "Econometric Policy Evaluation: A Critique." In K. Brunner and A. H. Meltzer (eds.), *The Phillips Curve and Labor Markets.* Amsterdam: North-Holland, pp. 19–46.

Lucas, Robert E., Jr. 1978. "Asset Prices in an Exchange Economy." *Econometrica*, Vol. 46(6), pp. 1426–1445.

Lucas, Robert E., Jr. 1980a. "Equilibrium in a Pure Currency Economy." In J. H. Kareken and N. Wallace (eds.), *Economic Inquiry.* Vol. 18(2), pp. 203-220. (Reprinted in *Models of Monetary Economies*, Federal Reserve Bank of Minneapolis, 1980, pp. 131–145.).

Lucas, Robert E., Jr. 1980b. "Two Illustrations of the Quantity Theory of Money." *American Economic Review*, Vol. 70, pp. 1005–1014.

Lucas, Robert E., Jr. 1981. "Econometric Testing of the Natural Rate Hypothesis." In Robert E. Lucas, Jr. (ed.), *Studies of Business-Cycle Theory.* Cambridge, MA: MIT Press, pp. 90–103. Reprinted from *The Econometrics of Price Determination Conference*, ed. Otto Eckstein. Washington, DC: Board of Governors of the Federal Reserve System, 1972, pp. 50–59.

Lucas, Robert E., Jr. 1982. "Interest Rates and Currency Prices in a Two-Country World." *Journal of Monetary Economics*, Vol. 10(3), pp. 335–360.

Lucas, Robert E., Jr. 1987. *Models of Business Cycles.* Yrjo Jahnsson Lectures Series. London: Blackwell.

Lucas, Robert E., Jr. 1988. "On the Mechanics of Economic Development." *Journal of Monetary Economics*, Vol. 22, pp. 3–42.

Lucas, Robert E., Jr. 1992. "On Efficiency and Distribution." *Economic Journal*, Vol. 102, No. 4, pp. 233–247.

Lucas, Robert E., Jr., and Edward C. Prescott. 1971. "Investment under Uncertainty." *Econometrica*, Vol. 39(5), pp. 659–681.

Lucas, Robert E., Jr., and Edward C. Prescott. 1974. "Equilibrium Search and Unemployment." *Journal of Economic Theory*, Vol. 7(2), pp. 188–209.

Lucas, Robert E., Jr., and Nancy Stokey. 1983. "Optimal Monetary and Fiscal Policy in an Economy without Capital." *Journal of Monetary Economics*, Vol. 12(1), pp. 55–94.

Luenberger, David G. 1969. *Optimization by Vector Space Methods.* New York: John Wiley and Sons, Inc..

Lustig, Hanno. 2000. "Secured Lending and Asset Prices." Mimeo. Department of Economics, Stanford University.

Mankiw, Gregory N. 1986. "The Equity Premium and the Concentration of Aggregate Shocks." *Journal of Financial Economics*, Vol. 17(1), pp. 211–219.

Manuelli, Rodolfo, and Thomas J. Sargent. 1988. "Models of Business Cycles: A Review Essay." *Journal of Monetary Economics*, Vol. 22 (3), pp. 523–542.

Manuelli, Rodolfo and Thomas J. Sargent. 1992. "Alternative Monetary Policies in a Turnpike Economy." Mimeo. Stanford University and Hoover Institution.

Marcet, Albert, and Ramon Marimon. 1992. "Communication, Commitment, and Growth." *Journal of Economic Theory*, Vol. 58(2), pp. 219–249.

Marcet, Albert, and Ramon Marimon. 1999. "Recursive Contracts." Mimeo. Universitat Pompeu Fabra, Barcelona.

Marcet, Albert, and Juan Pablo Nicolini. 1999. "Recurrent Hyperinflations and Learning." Mimeo. Universitat Pampeu Fabra, Barcelona.

Marcet, Albert, and Thomas J. Sargent. 1989. "Least Squares Learning and the Dynamics of Hyperinflation." In William Barnett, John Geweke, and Karl Shell (eds.), *Economic Complexity: Chaos, Sunspots, and Nonlinearity.* Cambridge University Press.

Marcet, Albert, Thomas J. Sargent, and Juha Seppälä. 1996. "Optimal Taxation without State-Contingent Debt." Mimeo. Universitat Pompeu Fabra and Stanford University.

Marcet, Albert and Kenneth J. Singleton. 1999. "Equilibrium Asset Prices and Savings of Heterogeneous Agents in the Presence of Incomplete Markets and Portfolio Constraints." *Macroeconomic Dynamics*, Vol. 3, No. 2, pp. 243–277.

Marimon, Ramon. Forthcoming. "The Fiscal Theory of Money as an Unorthodox Financial Theory of the Firm." In Axel Leijonhufvud (ed.), *Monetary Theory as a Basis for Monetary Policy.* International Economic Association (IEA).

Marimon, Ramon, and Shyam Sunder. 1993. "Indeterminacy of Equilibria in a Hyperinflationary World: Experimental Evidence." *Econometrica*, Vol. 61(5), pp. 1073–1107.

Marimon, Ramon, and Fabrizio Zilibotti. 1999. "Unemployment vs. Mismatch of Talents: Reconsidering Unemployment Benefits." *Economic Journal*, Vol. 109, pp. 266–291.

Mas-Colell, Andrew, Michael D. Whinston, and Jerry R. Green. 1995. *Microeconomic Theory.* New York: Oxford University Press.

Matsuyama, Kiminori, Nobuhiro Kiyotaki, and Akihiko Matsui. 1993. "Toward a Theory of International Currency." *Review of Economic Studies*, Vol. 60(2), pp. 283–307.

McCall, John J. 1970. "Economics of Information and Job Search." *Quarterly Journal of Economics*, Vol. 84(1), pp. 113–126.

McCallum, Bennett T. 1983. "The Role of Overlapping-Generations Models in Monetary Economics." *Carnegie-Rochester Conference Series on Public Policy*, Vol. 18(0), pp. 9–44.

McCandless, George T., and Neil Wallace. 1992. *Introduction to Dynamic Macroeconomic Theory: An Overlapping Generations Approach.* Cambridge, MA: Harvard University Press.

McGrattan, Ellen R. 1994. "A Note on Computing Competitive Equilibria in Linear Models." *Journal of Economic Dynamics and Control*, Vol. 18(1), pp. 149–160.

McGrattan, Ellen R. 1996. "Solving the Stochastic Growth Model with a Finite Element Method." *Journal of Economic Dynamics and Control*, Vol. 20(1-3), pp. 19–42.

Mehra, Rajnish, and Edward C. Prescott. 1985. "The Equity Premium: A Puzzle." *Journal of Monetary Economics*, Vol. 15(2), pp. 145–162.

Miller, Bruce L. 1974. "Optimal Consumption with a Stochastic Income Stream." *Econometrica*, Vol. 42(2), pp. 253–266.

Miller, Robert A. 1984. "Job Matching and Occupational Choice." *Journal of Political Economy*, Vol. 92(6), pp. 1086–1120.

Modigliani, Franco, and Richard Brumberg. 1954. "Utility Analysis and the Consumption Function: An Interpretation of Cross-Section Data." In K. K. Kurihara (ed.), *Post-Keynesian Economics.* New Brunswick, NJ: Rutgers University Press.

Modigliani, F., and M. H. Miller. 1958. "The Cost of Capital, Corporation Finance, and the Theory of Investment." *American Economic Review*, Vol. 48(3), pp. 261–297.

Moen, Espen R. 1997. "Competitive Search Equilibrium." *Journal of Political Economy*, Vol. 105(2), pp. 385–411.

Montgomery, James D. 1991. "Equilibrium Wage Dispersion and Involuntary Unemployment." *Quarterly Journal of Economics*, Vol. 106, pp. 163–179.

Mortensen, Dale T. 1982. "The Matching Process as a Noncooperative Bargaining Game." In John J. McCall (ed.), *The Economics of Information and Uncertainty.* Chicago: University of Chicago Press for the National Bureau of Economic Research, pp. 233–258.

Mortensen, Dale T. 1994. "The Cyclical Behavior of Job and Worker Flows." *Journal of Economic Dynamics and Control*, Vol. 18, pp. 1121–1142.

Mortensen, Dale T., and Christopher A. Pissarides. 1994. "Job Creation and Job Destruction in the Theory of Unemployment." *Review of Economic Studies*, Vol. 61, pp. 397–415.

Mortensen, Dale T., and Christopher A. Pissarides. 1999a. "New Developments in Models of Search in the Labor Market." In Orley Ashenfelter and David Card (eds.), *Handbook of Labor Economics, Vol. 3B.* Amsterdam: Elsevier/North-Holland.

Mortensen, Dale T., and Christopher A. Pissarides. 1999b. "Unemployment Responses to "Skill-Biased" Technology Shocks: The Role of Labour Market Policy." *Economic Journal*, Vol. 109, pp. 242–265.

Muth, J. F. [1960] 1981. "Estimation of Economic Relationships Containing Latent Expectations Variables." In R. E. Lucas, Jr. and T. J. Sargent (eds.), *Rational Expectations and Econometric Practice.* University of Minnesota, pp. 321–328.

Muth, John F. 1960. "Optimal Properties of Exponentially Weighted Forecasts." *Journal of the American Statistical Association*, Vol. 55, pp. 299-306.

Muth, John F. 1961. "Rational Expectations and the Theory of Price Movements." *Econometrica*, Vol. 29, pp. 315–335.

Naylor, Arch, and George Sell. 1982. *Linear Operator Theory in Engineering and Science.* New York: Springer.

Neal, Derek. 1999. "The Complexity of Job Mobility among Young Men." *Journal of Labor Economics*, Vol. 17(2), pp. 237–261.

Nerlov, Marc. 1967. "Distributed Lags and Unobserved Components in Economic Time Series." In William Fellner et al. (eds.), *Ten Economic Studies in the Tradition of Irving Fisher.* New York: Wiley.

Noble, Ben, and James W. Daniel. 1977. *Applied Linear Algebra.* Englewood Cliffs, NJ: Prenctice Hall.

O'Connell, Stephen A., and Stephen P. Zeldes. 1988. "Rational Ponzi Games." *International Economic Review*, Vol. 29(3), pp. 431–450.

Paal, Beatrix. 2000. "Destabilizing Effects of a Successful Stabilization: A Forward-Looking Explanation of the Second Hungarian Hyperinflation." *Journal of Economic Theory*, In press.

Peled, Dan. 1984. "Stationary Pareto Optimality of Stochastic Asset Equilibria with Overlapping Generations." *Journal of Economic Theory*, Vol. 34, pp. 396–403.

Persson, Mats, Torsten Persson, and Lars E. O. Svensson. 1988. "Time Consistency of Fiscal and Monetary Policy." *Econometrica*, Vol. 55, pp. 1419–1432.

Peters, Michael. 1991. "Ex Ante Price Offers in Matching Games Non-Steady States." *Econometrica*, Vol. 59(5), pp. 1425–1454.

Phelan, Christopher. 1994. "Incentives and Aggregate Shocks." *Review of Economic Studies*, Vol. 61(4), pp. 681–700.

Phelan, Christopher, and Robert M. Townsend. 1991. "Computing Multi-period, Information-Constrained Optima." *Review of Economic Studies*, Vol. 58(5), pp. 853–881.

Phelan, Chrisopher, and Ennio Stacchetti. 1999. "Sequential Equilibria in a Ramsey Tax Model." Mimeo. Federal Reserve Bank of Minneapolis, Staff Report 258.

Phelps, Edmund S. 1970. *Introduction to Microeconomic Foundations of Employment and Inflation Theory*. New York: Norton.

Phelps, Edmund S and Robert A. Pollak. 1968. "On Second-Best National Saving and Game-Equilibrium Growth." *Review of Economic Studies*, Vol. 35, No. 2, pp. 185–199.

Piazzesi, Monika. 2000. "An Econometric Model of the Yield Curve with Macroeconomic Jump Effects." Mimeo. Stanford University, Department of Economics.

Pissarides, Christopher A. 1983. "Efficiency Aspects of the Financing of Unemployment Insurance and other Government Expenditures." *Review of Economic Studies*, Vol. 50(1), pp. 57–69.

Pissarides, Christopher A. 1990. *Equilibrium Unemployment Theory*. Cambridge, MA: Basil Blackwell.

Pratt, John W. 1964. "Risk Aversion in the Small and in the Large." *Econometrica*, Vol. 32(1-2), pp. 122–136.

Prescott, Edward C., and Rajnish Mehra. 1980. "Recursive Competitive Equilibrium: The Case of Homogeneous Households." *Econometrica*, Vol. 48(6), pp. 1365–1379.

Prescott, Edward C., and Robert M. Townsend. 1980. "Equilibrium under Uncertainty: Multiagent Statistical Decision Theory." In Arnold Zellner (ed.), *Bayesian Analysis in Econometrics and Statistics*. Amsterdam: North-Holland, pp. 169–194.

Prescott, Edward C., and Robert M. Townsend. 1984a. "General Competitive Analysis in an Economy with Private Information." *International Economic Review*, Vol. 25, pp. 1–20.

Prescott, Edward C., and Robert M. Townsend. 1984b. "Pareto Optima and Competitive Equilibria with Adverse Selection and Moral Hazard." *Econometrica*, Vol. 52, pp. 21–45.

Putterman, Martin L., and Shelby Brumelle. 1979. "On the Convergence of Policy Iteration on Stationary Dynamic Programming." *Mathematics of Operations Research*, Vol. 4(1), pp. 60–67.

Putterman, Martin L., and M. C. Shin. 1978. "Modified Policy Iteration Algorithms for Discounted Markov Decision Problems." *Management Science*, Vol. 24(11), pp. 1127–1137.

Quah, Danny. 1990. "Permanent and Transitory Movements in Labor Income: An Explanation for "Excess Smoothness" in Consumption" *Journal of Political Economy*, Vol. 98(3), pp. 449–475.

Razin, Assaf, and Efraim Sadka. 1995. "The Status of Capital Income Taxation in the Open Economy." *FinanzArchiv*, Vol. 52(1), pp. 21–32.

Rebelo, Sergio. 1991. "Long-Run Policy Analysis and Long-Run Growth." *Journal of Political Economy*, Vol. 99, pp. 500–521.

Reinganum, Jennifer F. 1979. "A Simple Equilibrium Model of Price Dispersion." *Journal of Political Economy*, Vol. 87(4), pp. 851–858.

Ríos-Rull, Víctor José. 1994a. "Life-Cycle Economies and Aggregate Fluctuations." Mimeo. University of Pennsylvania.

Ríos-Rull, Víctor José. 1994b. "Population Changes and Capital Accumulation: The Aging of the Baby Boom." Mimeo. University of Pennsylvania.

Ríos-Rull, Víctor José. 1994c. "On the Quantitative Importance of Market Completeness." *Journal of Monetary Economics*, Vol. 34(3), pp. 463–496.

Ríos-Rull, Víctor José. 1995. "Models with Heterogeneous Agents." In Thomas F. Cooley (ed.), *Frontiers of Business Cycle Research.* Princeton, NJ: Princeton University Press, pp. 98–125.

Ríos-Rull, Víctor José. 1996. "Life-Cycle Economies and Aggregate Fluctuations." *Review of Economic Studies*, Vol. 63(3), pp. 465–489.

Roberds, William T. 1996. "Budget Constraints and Time-Series Evidence on Consumption: Comment." *American Economic Review*, Vol. 86(1), pp. 296–297.

Rogerson, Richard. 1988. "Indivisible Labor, Lotteries, and Equilibrium." *Journal of Monetary Economics*, Vol. 21, pp. 3–16.

Rogoff, Kenneth. 1989. "Reputation, Coordination, and Monetary Policy." In Robert J. Barro (ed.), *Modern Business Cycle Theory.* Cambridge, MA: Harvard University Press, pp.236–264.

Roll, Richard. 1970. *The Behavior of Interest Rates: An Application of the Efficient Market Model to U.S. Treasury Bills.* New York: Basic Books.

Romer, David. 1996. *Advanced Macroeconomics.* New York: McGraw Hill.

Romer, Paul M. 1986. "Increasing Returns and Long-Run Growth." *Journal of Political Economy*, Vol. 94, pp. 1002–1037.

Romer, Paul M. 1987. "Growth Based on Increasing Returns Due to Specialization." *American Economic Review Paper and Proceedings*, Vol. 77, pp. 56–62.

Romer, Paul M. 1990. "Endogenous Technological Change." *Journal of Political Economy*, Vol. 98, pp. S71–S102.

Rosen, Sherwin, and Robert H. Topel. 1988. "Housing Investment in the United States." *Journal of Political Economy*, Vol. 96, No. 4, pp. 718–740.

Rosen, Sherwin, Kevin M. Murphy, and Jose A. Scheinkman. 1994. "Cattle Cycles." *Journal of Political Economy*, Vol. 102(3), pp. 468-492.

Ross, Stephen A. 1976. "The Arbitrage Theory of Capital Asset Pricing." *Journal of Economic Theory*, Vol. 13(3), pp. 341–360.

Rotemberg, Julio J. 1987. "The New Keynesian Microfoundations." In Stanley Fischer (ed.), *NBER Macroeconomics Annual 1987.* Cambridge, MA: MIT Press, pp. 69–104.

Rotember, Julio J. and Michael Woodford. 1997. "An Optimization-Based Econometric Framework for the Evaluation of Monetary Policy." In Olivier Blanchard and Stanley Fischer (eds.), *NBER Macroeconomic Annual, 1987.* Cambridge, Mass.: MIT Press, pp. 297-345.

Rothschild, Michael, and Joseph Stiglitz. 1970. "Increasing Risk I: A Definition." *Journal of Economic Theory*, Vol. 2(3), pp. 225–243.

Rothschild, Michael, and Joseph Stiglitz. 1971. "Increasing Risk II: Its Economic Consequences." *Journal of Economic Theory*, Vol. 3(1), pp. 66–84.

Rubinstein, Mark. 1974. "An Aggregation Theorem for Security Markets." *Journal of Financial Economics*, Vol. 1, No.3, pp. 225-244.

Saint-Paul, Gilles. 1995. "The High Unemployment Trap." *Quarterly Journal of Economics*, Vol. 110, pp. 527–550.

Samuelson, Paul A. 1958. "An Exact Consumption-Loan Model of Interest with or without the Social Contrivance of Money." *Journal of Political Economy*, Vol. 66, pp. 467–482.

Samuelson, Paul A. 1965. "Proof that Properly Anticipated Prices Fluctuate Randomly." *Industrial Management Review*, Vol. 6(1), pp. 41–49.

Sandmo, Agnar. 1970. "The Effect of Uncertainty on Saving Decisions." *Review of Economic Studies*, Vol. 37, pp. 353–360.

Sargent, Thomas J. 1980. "Lecture notes on Filtering, Control, and Rational Expectations." Mimeo. University of Minnesota, Minneapolis.

Sargent, Thomas J. 1987a. *Macroeconomic Theory,.* 2nd ed. New York: Academic Press.

Sargent, Thomas J. 1987b. *Dynamic Macroeconomic Theory.* Cambridge, Mass.: Harvard University Press.

Sargent, Thomas J. 1980. "Tobin's $q$ and the Rate of Investment in General Equilibrium." In K. Brunner and A. Meltzer (eds.), *On the State of Macroeconomics.* Carnegie-Rochester Conference Series 12, pp. 107–154. Amsterdam: North-Holland.

Sargent, Thomas J. 1991. "Equilibrium with Signal Extraction from Endogeneous Variables." *Journal of Economic Dynamics and Control*, Vol. 15, pp. 245–273.

Sargent, Thomas J. 1992. *Rational Expectations and Inflation.* 2nd ed. Harper and Row.

Sargent, Thomas J., and Bruce Smith. 1997. "Coinage, Debasements, and Gresham's Laws." *Economic Theory*, Vol. 10, pp. 197–226.

Sargent, Thomas J., and François R. Velde. 1990. "The Analytics of German Monetary Reform." *Quarterly Review*, Federal Reserve Bank of San Francisco, Vol. 0, n4, pp. 33-50.

Sargent, Thomas J., and Francois R. Velde. 1995. "Macroeconomic Features of the French Revolution." *Journal of Political Economy*, Vol. 103(3), pp. 474–518.

Sargent, Thomas J., and François R. Velde. 1999. "The Big Problem of Small Change." *Journal of Money, Credit, and Banking*, Vol. 31(2), pp. 137–161.

Sargent, Thomas J., and Neil Wallace. 1973. "Rational Expectations and the Dynamics of Hyperinflation." *International Economic Review*, Vol. 14, pp. 328–350.

Sargent, Thomas J., and Neil Wallace. 1981. "Some Unpleasant Monetarist Arithmetic." *Quarterly Review*, Federal Reserve Bank of Minneapolis, Vol. 5(3), pp. 1–17.

Sargent, Thomas J., and Neil Wallace. 1982. "The Real Bills Doctrine vs. the Quantity Theory: A Reconsideration." *Journal of Political Economy*, Vol. 90(6), pp. 1212–1236.

Sargent, Thomas J., and Neil Wallace. 1983. "A Model of Commodity Money." *Journal of Monetary Economics*, Vol. 12(1), pp. 163–187.

Seater, John J.. 1993. "Ricardian Equivalence." *Journal of Economic Literature*, Vol. 31(1), pp. 142–190.

Segerstrom, Paul S. 1998. "Endogenous Growth without Scale Effects." *American Economic Review*, Vol. 88, pp. 1290–1310.

Segerstrom, Paul S., T. C. A. Anant, and Elias Dinopoulos. 1990. "A Schumpeterian Model of the Product Life Cycle." *American Economic Review*, Vol. 80, pp. 1077–1091.

Shavell, Steven, and Laurence Weiss. 1979. "The Optimal Payment of Unemployment Insurance Benefits Over Time." *Journal of Political Economy*, Vol. 87, pp. 1347–1362.

Shi, Shouyong. 1995. "Money and Prices: A Model of Search and Bargaining." *Journal of Economic Theory*, Vol. 67, pp. 467–496.

Shiller, Robert J. 1972. "Rational Expectations and the Structure of Interest Rates." Ph.D. dissertation, Massachusetts Institute of Technology.

Shiller, Robert J. 1981. "Do Stock Prices Move Too Much to be Justified by Subsequent Changes in Dividends?" *American Economic Review*, Vol. 71(3), pp. 421–436.

Sibley, David S.. 1975. "Permanent and Transitory Income Effects in a Model of Optimal Consumption with Wage Income Uncertainty." *Journal of Economic Theory*, Vol. 11, pp. 68–82.

Sidrauski, Miguel. 1967. "Rational Choice and Patterns of Growth in a Monetary Economy." *American Economic Review*, Vol. 57(2), pp. 534–544.

Sims, Christopher A. 1972. "Money, Income, and Causality." *American Economic Review*, Vol. 62(4), pp. 540–552.

Sims, Christopher A. 1989. "Solving Nonlinear Stochastic Optimization and Equilibrium Problems Backwards." Mimeo. Institute for Empirical Macroeconomics, Federal Reserve Bank of Minneapolis, 15.

Sims, Christopher A. 1994. "A Simple Model for the Determination of the Price Level and the Interaction of Monetary and Fiscal Policy." *Economic Theory*, Vol. 4, pp. 381–399.

Smith, Bruce. 1988. "Legal Restrictions, "Sunspots," and Peel's Bank Act: The Real Bills Doctrine versus the Quantity Theory of Reconsidered" *Journal of Political Economy*, Vol. 96, No. 1, pp. 3-19.

Smith, Lones. 1992. "Folk Theorems in Overlapping Generations Games." *Games and Economic Behavior*, Vol. 4 (3), pp. 426–449.

Solow, Robert M. 1956. "A Contribution to the Theory of Economic Growth." *Quarterly Journal of Economics*, Vol. 70, pp. 65–94.

Sotomayor, Marlida A. de Oliveira. 1984. "On Income Fluctuations and Capital Gains." *Journal of Economic Theory*, Vol. 32, No. 1, pp. 14–35.

Spear, Stephen E., and Sanjay Srivastava. 1987. "On Repeated Moral Hazard with Discounting." *Review of Economic Studies*, Vol. 54(4), pp. 599–617.

Stacchetti, Ennio. 1991. "Notes on Reputational Models in Macroeconomics." Mimeo. Stanford University, September.

Stigler, George. 1961. "The Economics of Information." *Journal of Political Economy*, Vol. 69(3), pp. 213–225.

Stiglitz, Joseph E. 1969. "A Reexamination of the Modigliani-Miller Theorem." *American Economic Review*, Vol. 59(5), pp. 784–793.

Stiglitz, Joseph E. 1987. "Pareto Efficient and Optimal Taxation and the New New Welfare Economics." In Alan J. Auerbach, and Martin Feldstein (eds.), *Handbook of Public Economics, Vol. 2.* Amsterdam: Elsevier/North-Holland.

Stokey, Nancy L. 1989. "Reputation and Time Consistency." *American Economic Review*, Vol. 79, pp. 134–139.

Stokey, Nancy L. 1991. "Credible Public Policy." *Journal of Economic Dynamics and Control*, Vol. 15(4), pp. 627–656.

Stokey, Nancy, and Robert E. Lucas, Jr. (with Edward C. Prescott). 1989. *Recursive Methods in Economic Dynamics.* Cambridge, MA: Harvard University Press.

Storesletten, Kjetil, Chris Telmer, and Amir Yaron. 1998. "Persistent Idiosyncratic Shocks and Incomplete Markets." Mimeo. Carnegie Mellon University and Wharton School, University of Pennsylvania.

Svensson, Lars E. O. 1986. "Sticky Goods Prices, Flexible Asset Prices, Monopolistic Competition, and Monetary Policy." *Review of Economic Studies*, Vol. 53, pp. 385–405.

Tallarini, Thomas D., Jr. 1996. "Risk-Sensitive Real Business Cycles." Ph.D. dissertation, University of Chicago.

Tallarini, Thomas D., Jr. 2000. "Risk-Sensitive Real Business Cycles." *Journal of Monetary Economics*, Vol. 45, No. 3, pp. 507–532.

Tauchen, George. 1986. "Finite State Markov Chain Approximations to Univariate and Vector Autoregressions." *Economic Letters*, Vol. 20, pp. 177–181.

Taylor, John B. 1977. "Conditions for Unique Solutions in Stochastic Macroeconomic Models with Rational Expectations." *Econometrica*, Vol. 45, pp. 1377–1185.

Taylor, John B. 1980. "Output and Price Stability: An International Comparison." *Journal of Economic Dynamics and Control*, Vol. 2, pp. 109–132.

Thomas, Jonathan, and Tim Worrall. 1988. "Self-Enforcing Wage Contracts." *Review of Economic Studies*, Vol. 55, pp. 541–554.

Thomas, Jonathan, and Tim Worrall. 1990. "Income Fluctuation and Asymmetric Information: An Example of a Repeated Principal-Agent Problem." *Journal of Economic Theory*, Vol. 51(2), pp. 367–390.

Tirole, Jean. 1982. "On the Possibility of Speculation under Rational Expectations." *Econometrica*, Vol. 50, pp. 1163–1181.

Tirole, Jean. 1985. "Asset Bubbles and Overlapping Generations." *Econometrica*, Vol. 53, pp. 1499–1528.

Tobin, James. 1956. "The Interest Elasticity of the Transactions Demand for Cash." *Review of Economics and Statistics*, Vol. 38, pp. 241–247.

Tobin, James. 1961. "Money, Capital, and Other Stores of Value." *American Economic Review*, Vol. 51(2), pp. 26–37.

Tobin, James. 1963. "An Essay on the Principles of Debt Management." In William Fellner et al. (eds.), *Fiscal and Debt Management Policies.*. Englewood Cliffs, NJ: Prentice-Hall, pp. 141–215. (Reprinted in James Tobin, *Essays in Economics*, 2 vols., Vol. 1. Amsterdam: North-Holland, 1971, pp. 378–455.).

Topel, Robert H., and Sherwin Rosen. 1988. "Housing Investment in the United States." *Journal of Political Economy*, Vol. 96(4), pp. 718–740.

Townsend, Robert M. 1980. "Models of Money with Spatially Separated Agents." In J. H. Kareken and N. Wallace (eds.), *Models of Monetary Economies.* Minneapolis: Federal Reserve Bank of Minneapolis, pp. 265–303.

Townsend, Robert M. 1983. "Forecasting the Forecasts of Others." *Journal of Political Economy*, Vol. 91, pp. 546–588.

Trejos, Alberto, and Randall Wright. 1995. "Search, Bargaining, Money and Prices." *Journal of Political Economy*, Vol. 103, pp. 118–139.

Turnovsky, Stephen J., and William A. Brock. 1980. "Time Consistency and Optimal Government Policies in Perfect Foresight Equilibrium." *Journal of Public Economics*, Vol. 13, pp. 183–212.

Uzawa, Hirofumi. 1965. "Optimum Technical Change in an Aggregative Model of Economic Growth." *International Economic Review*, Vol. 6, pp. 18–31.

Villamil, Anne P. 1988. "Price Discriminating Monetary Policy: A Nonuniform Pricing Approach." *Journal of Public Economics*, Vol. 35(3), pp. 385–392.

Wallace, Neil. 1980. "The Overlapping Generations Model of Fiat Money." In J. H. Kareken and N. Wallace (eds.), *Models of Monetary Economies.* Minneapolis: Federal Reserve Bank of Minneapolis, pp. 49–82.

Wallace, Neil. 1981. "A Modigliani-Miller Theorem for Open-Market Operations." *American Economic Review*, Vol. 71, pp. 267–274.

Wallace, Neil. 1983. "A Legal Restrictions Theory of the Demand for 'Money' and the Role of Monetary Policy." *Quarterly Review*, Federal Reserve Bank of Minneapolis, Vol. 7(1), pp. 1–7.

Wallace, Neil. 1989. "Some Alternative Monetary Models and Their Implications for the Role of Open-Market Policy." In Robert J. Barro (ed.), *Modern Business Cycle Theory.* Cambridge, MA: Harvard University Press, pp. 306–328.

Walsh, Carl E. 1998. *Monetary Theory and Policy*. Cambridge: MIT Press.

Wang, Cheng, and Stephen D. Williamson. 1996. "Unemployment Insurance with Moral Hazard in a Dynamic Economy." *Carnegie-Rochester Conference Series on Public Policy*, Vol. 44(0), pp. 1–41.

Watanabe, Shinichi. 1984. "Search Unemployment, the Business Cycle, and Stochastic Growth." Mimeo. Ph.D. dissertation, University of Minnesota.

Weil, Philippe. 1989. "The Equity Premium Puzzle and the Risk-Free Rate Puzzle." *Journal of Monetary Economics*, Vol. 24(2), pp. 401–421.

Weil, Philippe. 1990. "Nonexpected Utility in Macroeconomics." *Quarterly Journal of Economics*, Vol. 105, pp. 29–42.

Weil, Philippe. 1993. "Precautionary Savings and the Permanent Income Hypothesis." *Review of Economic Studies*, Vol. 60(2), pp. 367–383.

Whiteman, Charles H. 1983. *Linear Rational Expectations Models: A Users Guide.* Minneapolis: University of Minnesota Press.

Whittle, Peter. 1963. *Prediction and Regulation by Linear Least-Square Methods.* Princeton, NJ: Van Nostrand-Reinhold.

Whittle, Peter. 1990. *Risk-Sensitive Optimal Control.* New York: Wiley.

Wilcox, David W. 1989. "The Sustainability of Government Deficits: Implications of the Present-Value Borrowing Constraint." *Journal of Money, Credit, and Banking*, Vol. 21(3), pp. 291–306.

Woodford, Michael. 1994. "Monetary Policy and Price Level Determinacy in a Cash-in-Advance Economy." *Economic Theory*, Vol. 4, pp. 345–380.

Woodford, Michael. 1995. "Price-Level Determinacy without Control of a Monetary Aggregate." *Carnegie-Rochester Conference Series on Public Policy*, Vol. 43(0), pp. 1–46.

Woodford, Michael. 1999. "Optimal Monetary Policy Inertia." Mimeo. Princeton University, June.

Woodford, Michael. 2000. "Interest and Prices." Mimeo. Princeton University.

Wright, Randall. 1986. "Job Search and Cyclical Unemployment." *Journal of Political Economy*, Vol. 94(1), pp. 38–55.

Young, Alwyn. 1998. "Growth without Scale Effects." *Journal of Political Economy*, Vol. 106, pp. 41–63.

Zeira, Joseph. 1999. "Informational Overshooting, Booms, and Crashes." *Journal of Monetary Economics*, Vol. 43, No. 1, pp. 237–257.

Zeldes, Stephen P. 1989. "Optimal Consumption with Stochastic Income: Deviations from Certainty Equivalence." *Quarterly Journal of Economics*, Vol. 104(2), pp. 275–298.

Zhao, Rui. 1999. "The Optimal Unemployment Insurance Contract: Why a Replacement Ratio?" Mimeo. University of Chicago, November.

Zhu, Xiaodong. 1992. "Optimal Fiscal Policy in a Stochastic Growth Model." *Journal of Economic Theory*, Vol. 58, pp. 250–289.

# *Author Index*

# Index

# *Matlab Programs Index*